Objective
Life Science

MCQs for Life Science Examination

— 4th Edition —

The Authors

Dr. Kailash Choudhary is the Director, Life Science, at Institute for Advanced Studies (IFAS), Pune, Maharashtra. He is actively involved in teaching and research for the last 18 years on different aspects of Lifesciences. Dr. Choudhary was born in Jodhpur and completed his school from Jodhpur followed by Bachelor in Science and then Masters in Botany, in 2000 from Jai Narian Vyas University securing Gold Medal. In June 2006 he completed his Ph.D degree on "Somatic Cell genetics and cloning in Acacia and Vigna Species". In 2004 he joined Lachoo Memorial College, Jodhpur where he served as Assistant Professor in Department of Botany and Biotechnology upto 2016. Dr Choudhary has been the recipient of several honours and awards. He secured All India rank 1 in prestigious CSIR NET JRF exam of Country for 10 consecutive time. He also cleared GATE lifesciences with top rank. He has cleared ARS-NET, ICMR-JRF and Rajasthan SET exams. He is imparting guidance to NET and GATE Lifesciences/Biotechnology aspirants from last 18 years at IFAS Jodhpur, Pune, Hyderabad, Kolkata, online and over 2500 students have qualified various national level exams in stream of life science. He has published around 23 articles, reviews, book and conference chapters, including nearly 20 original research articles in National and International peer reviewed Journals. His areas of interest include Conservation Biology, Plant Physiology and Biotechnology especially Plant Tissue Culture and Science and Society Related Programmes in the context of the Indian Thar Desert Region.

Dr. R.P. Saran is presently working as Assistant Professor in Zoology at Jai Narain Vyas University, Jodhpur (Rajasthan). He did his M.Sc. in Zoology from Jai Narayan Vyas University, Jodhpur (Rajasthan) in the year 2005 with specialization in cell biology. He was throughout first class in studies up to M.Sc. He has cleared many national level exams like UGC-CSIR NET, GATE lifesciences, SET lifesciences. He is engaged in researches on threatened bird species of India with special emphasis on vultures, cytotoxicity of veterinary drugs on avian fauna and trophic transfer of nanoparticles and organic pollutants. At present he is involved in teaching - diversity among animal forms, animal physiology, Immunology, Developmental biology, Animal behaviour and biological evolution.

Dr. Ramesh Raliya is a Research Scientist at Department of Energy, Environmental and Chemical Engineering, Washington University in St. Louis, USA. Prior to this, he was Research Associate at National Agriculture Innovation Project of Nanotechnology at Central Arid Zone Research Institute, Jodhpur (An Institute of Indian Council of Agriculture Research, Government of India).

Raliya's research expertise is in the product and technology development with the aim of reducing fertilizer runoff while improving sustainable agriculture. A. Develop a fertilizer and its delivery techniques that has nearly 100% use efficiency so zero runoff; B. Climate-Smart Agriculture: Nanoscale composite formulation showing spatial and temporal release profile of nutrient, pesticides, and water. C. Energy efficient recovery, recycle and reuse of nutrients and water from waste-water; D. Integration of data science and artificial intelligence for precision and sustainable agriculture.

Dr. Raliya developed methods for synthesizing novel inorganic and organic nanomaterials with independently controlled properties. These nanomaterials have been used to study their impact on plant nutrition, plant protection, biomedical uses including drug delivery, therapeutics, and imaging, water chemistry, energy storage materials, sensors, and environmental interactions.

At Washington University in St. Louis, Dr. Raliya associated with Nano Research Facility, and Aerosol and Air Quality Research Laboratory of School of Engineering & Applied Science, and Center for Multiple Myeloma Nanotherapy of Washington University School of Medicine.

Dr. Raliya is a Co-founder of Smart Aerosol Technologies (SmArT) LLC, USA; Academy of Translational Research, India and RASTA – USA.

Objective
Life Science

MCQs for Life Science Examination

> CSIR > DBT > ICAR > ICMR > ASRB > IARI > SET > **NET**

— 4th Edition —

Kailash Choudhary
R.P. Saran
Ramesh Raliya

Published by:

SCIENTIFIC PUBLISHERS (INDIA)
5A, New Pali Road, P.O. Box 91
Jodhpur 342 001, India
E-mail: info@scientificpub.com
Website: www.scientificpub.com

ISBN: 978-93-89832-86-0
eISBN: 978-93-89832-87-7

Printed in India

PREFACE

The idea of the book entitled "Objective Life Science: MCQs for Life Science Examination" was born because of the lack of any comprehensive book covering all the aspects of various entry level life science competitive examinations in particular conducted by CSIR, DBT, ICAR, ICMR, ASRB, IARI, State and National Eligibility Test, but not limited to.

This book, covers all the subjects of life science under 13 section namely, 1. Molecules and their interaction relevant to biology; 2. Cellular organization; 3. Fundamental processes; 4. Cell communication and cell signaling; 5. Developmental biology; 6. System physiology – Plant; 7. System physiology – Animal; 8. Inheritance biology; 9. Diversity of life forms; 10. Ecological principles; 11. Evolution and behavior; 12. Applied biology and 13. Methods in biology. Each Section has been further divided into two parts with 200 short tricky questions and 100 applied conceptual questions.

The ultimate purpose of this book is to equip the reader with brainstorming challenges and solution for life science and applied aspect examinations. It contains predigested information on all the academic subject of life science for good understanding, assimilation, self-evaluation, and reproducibility. Although we have made every effort to make the book error free, we are under no illusion. We welcome comments, criticism and suggestions from the readers to evolve the contents.

Kailash Choudhary
R.P. Saran
Ramesh Raliya

LIFE SCIENCES
EXAM SCHEME

TIME: 3 HOURS **MAXIMUM MARKS: 200**

From June, 2011 CSIR-UGC (NET) Exam for Award of Junior Research Fellowship and Eligibility for Lecturership shall be a Single Paper Test having Multiple Choice Questions (MCQs). The question paper is divided in three parts

Part 'A' : This part shall carry 20 questions pertaining to General Science, Quantitative Reasoning & Analysis and Research Aptitude. The candidates shall be required to answer any 15 questions. Each question shall be of two marks. The total marks allocated to this section shall be 30 out of 200.

Part 'B': This part shall contain 50 Multiple Choice Questions (MCQs) generally covering the topics given in the syllabus. A candidate shall be required to answer any 35 questions. Each question shall be of two marks. The total marks allocated to this section shall be 70 out of 200.

Part 'C': This part shall contain 75 questions that are designed to test a candidate's knowledge of scientific concepts and/or application of the scientific concepts. The questions shall be of analytical nature where a candidate is expected to apply the scientific knowledge to arrive at the solution to the given scientific problem. A candidate shall be required to answer any 25 questions. Each question shall be of four marks. The total marks allocated to this section shall be 100 out of 200.

- There will be negative marking @25% for each wrong answer.
- To enable the candidates to go through the questions, the question paper booklet shall be distributed 15 minutes before the scheduled time of the exam. The Answer sheet shall be distributed at the scheduled time of the exam.

SYLLABUS

1. MOLECULES AND THEIR INTERACTION RELAVENT TO BIOLOGY

A. **Structure of atoms, molecules and chemical bonds.**

B. **Composition, structure and function of biomolecules:** Carbohydrates, lipids, proteins, nucleic acids and vitamins.

C. **Stablizing interactions:** Van der Waals, electrostatic, hydrogen bonding, hydrophobic interaction, etc.

D. **Principles of biophysical chemistry:** pH, buffer, reaction kinetics, thermodynamics, colligative properties.

E. **Bioenergetics:** Glycolysis, oxidative phosphorylation, coupled reaction, group transfer, biological energy transducers.

F. **Principles of catalysis:** Enzymes and enzyme kinetics, enzyme regulation, mechanism of enzyme catalysis, isozymes

G. **Conformation of proteins:** Ramachandran plot, secondary structure, domains, motif and folds.

H. **Conformation of nucleic acids:** helix (A, B, Z), t-RNA, micro-RNA.

I. **Stability of proteins and nucleic acids.**

J. **Metabolism of carbohydrates**, lipids, amino acids nucleotides and vitamins.

2. CELLULAR ORGANIZATION

A. **Membrane structure and function:** Structure of model membrane, lipid bilayer and membrane protein diffusion, osmosis, ion channels, active transport, membrane pumps, mechanism of sorting and regulation of intracellular transport, electrical properties of membranes.

B. **Structural organization and function of intracellular organelles:** Cell wall, nucleus, mitochondria, Golgi bodies, lysosomes, endoplasmic reticulum, peroxisomes, plastids, vacuoles, chloroplast, structure & function of cytoskeleton and its role in motility.

C. **Organization of genes and chromosomes:** Operon, unique and repetitive DNA, interrupted genes, gene families, structure of chromatin and chromosomes, heterochromatin, euchromatin, transposons.

D. **Cell division and cell cycle:** Mitosis and meiosis, their regulation, steps in cell cycle, regulation and control of cell cycle.

E. **Microbial Physiology:** Growth yield and characteristics, strategies of cell division, stress response.

3. FUNDAMENTAL PROCESSES

A. **DNA replication, repair and recombination:** Unit of replication, enzymes involved, replica-tion origin and replication fork, fidelity of replication, extrachromosomal replicons, DNA damage and repair mechanisms, homologous and site-specific recombination.

B. **RNA synthesis and processing:** Transcription factors and machinery, formation of initiation complex, transcription activator and repressor, RNA polymerases, capping, elongation, and termination, RNA processing, RNA editing, splicing, and polyadenylation, structure and function of different types of RNA, RNA transport.

C. **Protein synthesis and processing:** Ribosome, formation of initiation complex, initiation factors and their regulation, elongation and elongation factors, termination, genetic code, aminoacylation of tRNA, tRNA-identity, aminoacyl tRNA synthetase, and translational proof-reading, translational inhibitors, Post- translational modification of proteins.

D. **Control of gene expression at transcription and translation level:** Regulating the expression of phages, viruses, prokaryotic and eukaryotic genes, role of chromatin in gene expression and gene silencing.

4. CELL COMMUNICATION AND CELL SIGNALING

A. **Host parasite interaction**: Recognition and entry processes of different pathogens like bacteria, viruses into animal and plant host cells, alteration of host cell behavior by pathogens, virus-induced cell transformation, pathogen-induced diseases in animals and plants, cell-cell fusion in both normal and abnormal cells.

B. **Cell signaling:** Hormones and their receptors, cell surface receptor, signaling through G-protein coupled receptors, signal transduction pathways, second messengers, regulation of signaling pathways, bacterial and plant twocomponent systems, light signaling in plants, bacterial chemotaxis and quorum sensing.

C. **Cellular communication**: Regulation of hematopoiesis, general principles of cell communication, cell adhesion and roles of different adhesion molecules, gap junctions, extracellular matrix, integrins, neurotransmission and its regulation.

D. Cancer: Genetic rearrangements in progenitor cells, oncogenes, tumor suppressor genes, cancer and the cell cycle, virus-induced cancer, metastasis, interaction of cancer cells with normal cells, apoptosis, therapeutic interventions of uncontrolled cell growth.

E. Innate and adaptive immune system: Cells and molecules involved in innate and adaptive immunity, antigens, antigenicity and immunogenicity. B and T cell epitopes, structure and function of antibody molecules. generation of antibody diversity, monoclonal antibodies, antibody engineering, antigen-antibody interactions, MHC molecules, antigen processing and presentation, activation and differentiation of B and T cells, B and T cell receptors, humoral and cell mediated immune responses, primary and secondary immune modulation, the complement system, Toll-like receptors, cell-mediated effector functions, inflammation, hypersensitivity and autoimmunity, immune response during bacterial (tuberculosis), parasitic (malaria) and viral (HIV) infections, congenital and acquired immunodeficiencies, vaccines.

5. DEVELOPMENTAL BIOLOGY

A. Basic concepts of development : Potency, commitment, specification, induction, competence, determination and differentiation; morphogenetic gradients; cell fate and cell lineages; stem cells; genomic equivalence and the cytoplasmic determinants; imprinting; mutants and transgenics in analysis of development

B. Gametogenesis, fertilization and early develo-pment: Production of gametes, cell surface molecules in sperm-egg recognition in animals; embryo sac development and double fertilization in plants; zygote formation, cleavage, blastula formation, embryonic fields, gastrulation and formation of germ layers in animals; embryogenesis, establishment of symmetry in plants; seed formation and germination.

C. Morphogenesis and organogenesis in animals: Cell aggregation and differentiation in Dictyostelium; axes and pattern formation in Drosophila, amphibia and chick; organogenesis–vulva formation in Caenorhabditis elegans, eye lens induction, limb development and regeneration in vertebrates; differentiation of neurons, post embryonic development-larval formation, metamorphosis; environmental regulation of normal development; sex determination.

D. Morphogenesis and organogenesis in plants: Organization of shoot and root apical meristem; shoot and root development; leaf development and phyllotaxy; transition to flowering, floral meristems and floral development in Arabidopsis and Antirrhinum

E Programmed cell death, aging and senescence

6. SYSTEM PHYSIOLOGY – PLANT

A. Photosynthesis: Light harvesting complexes; mechanisms of electron transport; photoprotective mechanisms; CO_2 fixation-C3, C4 and CAM pathways.

B. Respiration and photorespiration: Citric acid cycle; plant mitochondrial electron transport and ATP synthesis; alternate oxidase; photorespiratory pathway.

C. Nitrogen metabolism: Nitrate and ammonium assimilation; amino acid biosynthesis.

D. Plant hormones: Biosynthesis, storage, break-down and transport; physiological effects and mechanisms of action.

E. Sensory photobiology: Structure, function and mechanisms of action of phytochromes, cryptochromes and phototropins; stomatal movement; photoperiodism and biological clocks.

F. Solute transport and photoassimilate translocation – uptake, transport and translocation of water, ions, solutes and macromolecules from soil, through cells, across membranes, through xylem and phloem; transpiration; mechanisms of loading and unloading of photoassimilates.

G. **Secondary metabolites**: Biosynthesis of terpenes, phenols and nitrogenous compounds and their roles.

H. **Stress physiology**: Responses of plants to biotic (pathogen and insects) and abiotic (water, temperature and salt) stresses.

7. SYSTEM PHYSIOLOGY - ANIMAL

A. **Blood and circulation**: Blood corpuscles, haemopoiesis and formed elements, plasma function, blood volume, blood volume regulation, blood groups, haemoglobin, immunity, haemostasis.

B. **Cardiovascular System:** Comparative anatomy of heart structure, myogenic heart, specialized tissue, ECG – its principle and significance, cardiac cycle, heart as a pump, blood pressure, neural and chemical regulation of all above.

C. **Respiratory system**: Comparison of respiration in different species, anatomical considerations, transport of gases, exchange of gases, waste elimination, neural and chemical regulation of respiration.

D. **Nervous system**: Neurons, action potential, gross neuroanatomy of the brain and spinal cord, central and peripheral nervous system, neural control of muscle tone and posture.

E. **Sense organs**: Vision, hearing and tactile response.

F. **Excretory system**: Comparative physiology of excretion, kidney, urine formation, urine concentration, waste elimination, micturition, regulation of water balance, blood volume, blood pressure, electrolyte balance, acid-base balance.

G. **Thermoregulation**: Comfort zone, body temperature – physical, chemical, neural regulation, acclimatization.

H. **Stress and adaptation**

I. **Digestive system**: Digestion, absorption, energy balance, BMR.

J. **Endocrinology and reproduction**: Endocrine glands, basic mechanism of hormone action, hormones and diseases; reproductive processes, gametogenesis, ovulation, neuroendocrine regulation

8. INHERITANCE BIOLOGY

A. **Mendelian principles:** Dominance, segregation, independent assortment.

B. **Concept of gene:** Allele, multiple alleles, pseudoallele, complementation tests

C. **Extensions of Mendelian principles**: Codominance, incomplete dominance, gene interactions, pleiotropy, genomic imprinting, penetrance and expressivity, phenocopy, linkage and crossing over, sex linkage, sex limited and sex influenced characters.

D. **Gene mapping methods:** Linkage maps, tetrad analysis, mapping with molecular markers, mapping by using somatic cell hybrids, development of mapping population in plants.

E. **Extra chromosomal inheritance**: Inheritance of Mitochondrial and chloroplast genes, maternal inheritance.

F. **Microbial genetics**: Methods of genetic transfers – transformation, conjugation, transduction and sex-duction, mapping genes by interrupted mating, fine structure analysis of genes.

G. **Human genetics**: Pedigree analysis, lod score for linkage testing, karyotypes, genetic disorders.

H. **Quantitative genetics**: Polygenic inheritance, heritability and its measurements, QTL mapping.

I. **Mutation**: Types, causes and detection, mutant types – lethal, conditional, biochemical, loss of function, gain of function, germinal verses somatic mutants, insertional mutagenesis.

J. **Structural and numerical alterations of chromosomes**: Deletion, duplication, inversion, translocation, ploidy and their genetic implications.

K. **Recombination:** Homologous and non-homologous recombination including transposition.

9. DIVERSITY OF LIFE FORMS

A. **Principles & methods of taxonomy:** Concepts of species and hierarchical taxa, biological nomenclature, classical & quantititative methods of taxonomy of plants, animals and microorganisms.

B. **Levels of structural organization:** Unicellular, colonial and multicellular forms. Levels of organization of tissues, organs & systems. Comparative anatomy, adaptive radiation, adaptive modifications.

C. **Outline classification of plants, animals & microorganisms:** Important criteria used for classification in each taxon. Classification of plants, animals and microorganisms. Evolutionary relationships among taxa.

D. **Natural history of Indian subcontinent**: Major habitat types of the subcontinent, geographic origins and migrations of species. Common Indian mammals, birds. Seasonality and phenology of the subcontinent.

E. **Organisms of health & agricultural importance:** Common parasites and pathogens of humans, domestic animals and crops.

F. **Organisms of conservation concern:** Rare, endangered species. Conservation strategies.

10. ECOLOGICAL PRINCIPLES

A. **The Environment:** Physical environment; biotic environment; biotic and abiotic interactions.

B. **Habitat and Niche:** Concept of habitat and niche; niche width and overlap; fundamental and realized niche; resource partitioning; character displacement.

C. **Population Ecology:** Characteristics of a population; population growth curves; population regulation; life history strategies (r and K selection); concept of metapopulation – demes and dispersal, interdemic extinctions, age structured populations.

D. **Species Interactions:** Types of interactions, interspecific competition, herbivory, carnivory, pollination, symbiosis.

E. **Community Ecology:** Nature of communities; community structure and attributes; levels of species diversity and its measurement; edges and ecotones.

F. **Ecological Succession:** Types; mechanisms; changes involved in succession; concept of climax.

G. **Ecosystem Ecology**: Ecosystem structure; ecosystem function; energy flow and mineral cycling (C, N, P); primary production and decomposition; structure and function of some Indian ecosystems: terrestrial (forest, grassland) and aquatic (fresh water, marine, eustarine).

H. **Biogeography:** Major terrestrial biomes; theory of island biogeography; biogeographical zones of India.

I. **Applied Ecology:** Environmental pollution; global environmental change; biodiversity: status, monitoring and documentation; major drivers of biodiversity change; biodiversity management approaches.

J. **Conservation Biology:** Principles of conservation, major approaches to management, Indian case studies on conservation/ management strategy (Project Tiger, Biosphere reserves).

11. EVOLUTION AND BEHAVIOUR

A. **Emergence of evolutionary thoughts** Lamarck; Darwin–concepts of variation, adaptation, struggle, fitness and natural selection; Mendelism; Spontaneity of mutations; The evolutionary synthesis.

B. **Origin of cells and unicellular evolution:** Origin of basic biological molecules; Abiotic synthesis of organic monomers and polymers; Concept of Oparin and Haldane; Experiement of Miller (1953); The first cell; Evolution of prokaryotes; Origin of eukaryotic cells; Evolution of unicellular eukaryotes; Anaerobic metabolism, photosynthesis and aerobic metabolism.

C. **Paleontology and Evolutionary History:** The evolutionary time scale; Eras, periods and epoch; Major events in the evolutionary time scale; Origins of unicellular and multi cellular organisms; Major groups of plants and animals; Stages in primate evolution including Homo.

D. **Molecular Evolution:** Concepts of neutral evolution, molecular divergence and molecular clocks; Molecular tools in phylogeny, classification and identification; Protein and nucleotide sequence analysis; origin of new genes and proteins; Gene duplication and divergence.

E. **The Mechanisms:** Population genetics – Populations, Gene pool, Gene frequency; Hardy-Weinberg Law; concepts and rate of change in gene frequency through natural selection, migration and random genetic drift; Adaptive radiation; Isolating mechanisms; Speciation; Allopatricity and Sympatricity; Convergent evolution; Sexual selection; Co-evolution.

F. **Brain, Behavior and Evolution:** Approaches and methods in study of behavior; Proximate and ultimate causation; Altruism and evolution-Group selection, Kin selection, Reciprocal altruism; Neural basis of learning, memory, cognition, sleep and arousal; Biological clocks; Development of behavior; Social communication; Social dominance; Use of space and territoriality; Mating systems, Parental investment and Reproductive success; Parental care; Aggressive behavior; Habitat selection and optimality in foraging; Migration, orientation and navigation; Domestication and behavioral changes.

12. APPLIED BIOLOGY

A. Microbial fermentation and production of small and macro molecules.

B. Application of immunological principles, vaccines, diagnostics. Tissue and cell culture methods for plants and animals.

C. Transgenic animals and plants, molecular approaches to diagnosis and strain identification.

D. Genomics and its application to health and agriculture, including gene therapy.

E. Bioresource and uses of biodiversity.

F. Breeding in plants and animals, including marker – assisted selection

G. Bioremediation and phytoremediation

H. Biosensors

13. METHODS IN BIOLOGY

A. **Molecular Biology and Recombinant DNA methods:** Isolation and purification of RNA, DNA (genomic and plasmid) and proteins, different separation methods. Analysis of RNA, DNA and proteins by one and two dimensional gel electrophoresis, Isoelectric focusing gels. Molecular cloning of DNA or RNA fragments in bacterial and eukaryotic systems. Expression of recombinant proteins using bacterial, animal and plant vectors. Isolation of specific nucleic acid sequences Generation of genomic and cDNA libraries in plasmid, phage, cosmid, BAC and YAC vectors. In vitro mutagenesis and deletion techniques, gene knock out in bacterial and eukaryotic organisms. Protein sequencing methods, detection of post translation modification of proteins. DNA sequencing methods, strategies for genome sequencing. Methods for analysis of gene expression at RNA and protein level, large scale expression, such as micro array based techniques Isolation, separation and analysis of carbohydrate and lipid molecules RFLP, RAPD and AFLP techniques

B. **Histochemical and Immunotechniques:** Antibody generation, Detection of molecules using ELISA, RIA, western blot, immunoprecipitation, fluocytometry and immunofluorescence microscopy, detection of molecules in living cells, in situ localization by techniques such as FISH and GISH.

C **Biophysical Method:** Molecular analysis using UV/visible, fluorescence, circular dichroism, NMR and ESR spectroscopy Molecular structure determination using X-ray diffraction and NMR, Molecular analysis using light scattering, different types of mass spectrometry and surface plasma resonance methods.

D **Statisitcal Methods:** Measures of central tendency and dispersal; probability distributions (Binomial, Poisson and normal); Sampling distribution; Difference between parametric and non-parametric statistics; Confidence Interval; Errors; Levels of significance; Regression and Correlation; t-test; Analysis of variance; X2 test;; Basic introduction to Muetrovariate statistics

E. **Radiolabeling techniques:** Detection and measurement of different types of radioisotopes normally used in biology, incorporation of radioisotopes in biological tissues and cells, molecular imaging of radioactive material, safety guidelines.

F. **Microscopic techniques**: Visulization of cells and subcellular components by light microscopy, resolving powers of different microscopes, microscopy of living cells, scanning and transmission microscopes, different fixation and staining techniques for EM, freeze-etch and freeze fracture methods for EM, image processing methods in microscopy.

G. **Electrophysiological methods:** Single neuron recording, patch-clamp recording, ECG, Brain activity recording, lesion and stimulation of brain, pharmacological testing, PET, MRI, fMRI, CAT.

H. **Methods in field biology:** Methods of estimating population density of animals and plants, ranging patterns through direct, indirect and remote observations, sampling methods in the study of behavior, habitat characterization: ground and remote sensing methods.

CONTENTS

Part - A

Part - B

UNIT-1

BIOCHEMISTRY

1. Collagen consists of 3 helical chains containing Glycine and proline amino acids in each chain. The overall structure of each polypeptide in the collagen molecule is a
(1) polyproline I (2) polyproline II
(3) α–helix (4) Polyglycine I

2. Which of the following conditions would not promote denaturation of double-stranded DNA?
(1) Heating to 100 degrees Celsius
(2) Adding high concentrations of sodium citrate
(3) Decreasing the ionic strength of the solution
(4) Adding a protein that binds to single stranded, but not to double-stranded DNA

3. Biological washing powders remove stains by enzyme action. Which of the following combinations would be most effective in removing an egg stain?
(1) Amylase and protease
(2) Catalase and lipase
(3) Lipase and maltase
(4) Lipase and protease

4. The peptide bond is rigid because it is a
(1) single bond (2) partial double bond
(3) double bond (4) triple bond

5. The main difference between active transport and facilitated diffusion is that
(1) in active transport, the molecules move from areas of high concentration to areas of low concentration
(2) carrier protein is involved only in case of active transport
(3) in active transport, energy is consumed to move molecules against a concentration gradient
(4) in active transport, only water molecules are transported

6. What is the correct order of molecular weights?
(1) Human antibody>albumin>insulin> glutathione
(2) Albumin>insulin>antibody> glutathione
(3) Glutathione>insulin>albumin> antibody
(4) Insulin>antibody>glutathione> albumin

7. Arrange the three residues, glycine, phenylalanine and proline, in the decreasing order of backbone flexibility
(1) Gly > Phe > Pro
(2) Pro > Gly > Phe
(3) Phe > Pro > Gly
(4) It is not possible to comment

8. In competitive inhibition
(1) K_m increases, V_{max} constant
(2) K_m decreases, V_{max} constant
(3) K_m constant, V_{max} increases
(4) K_m decreases, V_{max} increases

9. For a reaction to be spontaneous,
(1) ΔG is negative
(2) ΔG is positive
(3) $\Delta G = 0$
(4) ΔH increases

10. For the folding of a linear polypeptide into a compact tertiary structure, globular in nature, the change in entropy is known to be negative. In order for the folding process to be thermodynamically feasible, the overall change is enthalpy based in intermolecular interaction should be

(1) Positive (2) Negative
(3) Zero (4) Endothermic

11. If the equilibrium constant for a chemical reaction at 20ºC is 20, the standard free energy change associated with the reaction will be

(1) -1.74 kcals (2) 1.74 kcals
(3) 0.76 kcals (4) 0.12 kcals

12. A particular enzyme loses its activity if just stored in normal saline. However, if normal saline contains 10 mM β-mercaptoethanol, a reducing agent the enzyme retains its activity. What can you conclude about the enzyme from above?

(1) It has methionine residues that are necessary for activity
(2) It has sulphydryl groups that are necessary for activity
(3) It has disulphide bonds that are necessary for activity
(4) It has histidine residues that are necessary for activity

13. The major contribution to the stability of Watson-Crick structure of DNA in aqueous solution comes from

(1) hydrogen bonds between Watson –Crick base pairs
(2) stacking interaction of bases
(3) counter-ion condensation on phosphates
(4) entropic contribution

14. In an enzyme catalyzed reaction, Km = 4×10^{-5} µmol/l, and the rate of reaction (V) at substrate concentration [S] = 1.2×10^{-2} M is 80 µmol/l-min. Assume no inhibitor is present. Vmax is practically equal to

(1) 40 µmol/l-min
(2) 80 µmol/l-min
(3) 120 µmol/l-min
(4) 4.8×10^{2} µmol/l-min

15. The repeating galactose units of all carrageenans is joined by

(1) α 1-3 glycosisdic linkages
(2) β 1-4 glycosisdic linkages
(3) both (1) and (2)
(4) none

16. Fish can survive inside a frozen lake because

(1) Fish hibernate in ice
(2) Fish are warm blooded animals
(3) Ice is a good counductor of heat
(4) Water near the bottom does not freeze

17. A left handed alpha helix falls in the Ramachandran plot under

(1) allowed region
(2) partially allowed region
(3) disallowed region
(4) line joining allowed and partially allowed region

18. The Greek key motif is composed of

(1) Four alpha helices
(2) Three alpha helices and one beta strand
(3) Two alpha helices and two beta strands
(4) Four beta strands

19. Which of the following statements is true regarding a secondary amide?

(1) It can only participate in hydrogen bonding as a hydrogen bond donor
(2) It can only participate in hydrogen bonding as a hydrogen bond acceptor
(3) It can participate in hydrogen bonding both as a hydrogen bond donor and a hydrogen bond acceptor
(4) It cannot participate in hydrogen bonding at all

0. If systematic conformational search is performed for a molecule with six rotatable bonds and step size is 30 degree then number of conformers will be

(1) 1, 895, 672 (2) 2, 985, 984
(3) 2, 008, 672 (4) 1, 895, 760

1. Which of the following amino acids are more likely to occur in alpha helices?

(1) A,E,L,M (2) P,G,Y,S
(3) A,G,Y,W (4) A,C,G,S

2. The alpha helix can be called a 3.6^{13} helix . The numbers refer to

(1) the number of residues and the pitch of the helix
(2) the number of residues and number of atoms in the helix
(3) the number of residues in a turn of the helix and the number of atoms in the hydrogen bond ring
(4) the number of turns and diameter of the helix

23. If a DNA sequence predominantly contains alternating pyrimidines and purines, which of the following DNA structures is highly favored ?

(1) A-DNA (2) B-DNA
(3) Z-DNA (4) A-B-DNA

24. The number of different isomers of a aldehyde pentose sugar (saccharide) are

(1) 4 (2) 8
(3) 16 (4) 32

25. Active site of all serine proteases consists of

(1) Ser- Glu –Asp (2) Ser- Glu - Met
(3) Ser-His-Asp (4) Ala-Glu-Met

26. Conversion of glucose to glucose-6- phosphate requires energy. However, critically ill patients are treated with intravenous infusion of glucose rather than glucose -6-phosphate because

(1) glucose-6-phosphate is unable to enter into cells
(2) glucose-6-phosphate is degraded very fast
(3) exogenous glucose-6-phosphate is toxic to the cells
(4) exogenous glucose-6-phosphate will competitively inhibit endogenous enzymes

27. Analysis of a nucleotide sequence reveals the proportion of A : T : C : G :: 0.40 : 0.85 : 1.56 : 1. Type of DNA concluded from this study is a

(1) purine rich DNA
(2) cruciform DNA
(3) double stranded DNA
(4) single stranded DNA

28. A dNTP master mix is prepared by combining 50μl each of 10mM dNTP stock. Two micro liters from this dNTP mix are added to the PCR master mix of 25μl reaction volume. What is the total dNTP concentration in the PCR reaction?

(1) 200μM
(2) 400 μM
(3) 800 μM
(4) 250 μM

29. Which of the following statements is correct for a reaction A + B ←→AB?

(1) Larger the value of the equilibrium constant, weaker is the binding between A and B
(2) Lower the value of the equilibrium constant, stronger is the binding between A and B
(3) Larger the value of the equilibrium constant, stronger is the binding between A and B
(4) This is a third order reaction

30. The amino acids with Phi and Psi values (-60, -40); (-59,-47) and (-80, 120) will be adopting which of the following conformation?

(1) Helix-helix-extended
(2) Helix-coil-extended
(3) Extended-extended-loop
(4) Loop-loop-coil

31. Large RNA molecules show greater backbone conformational variation than DNA double helices because of
(1) presence of unusual modified bases
(2) the occurrence of single stranded regions
(3) the presence of ribose sugar
(4) presence of triplex regions

32. In which of the following cases, the enzyme substrate complex is irreversible in nature?
(1) Competitive inhibition
(2) Non-competitive inhibition
(3) Un-competitive inhibition
(4) Covalent/permanent inhibition

33. Independently folded functional unit of a protein is called a
(1) motif (2) fold
(3) domain (4) module

34. In Ramachandran plot, the values of the dihedral angle ψ (psi) is based on rotation around
(1) N-Cα bond (2) Cα-C' bond
(3) C'-N bond (4) N-H bond

35. Which of the following does *not* participate in the formation of antigen-antibody/ligand-receptor complexes?
(1) Hydrophobic bonds
(2) Covalent bonds
(3) Electrostatic interactions
(4) Hydrogen bonds

36. Deoxy position of deoxyribose in DNA is at
(1) 1st Carbon (2) 3rd Carbon
(3) 2nd Carbon (4) 5th Carbon

37. For a new chemical entity, to be a good enzyme inhibitor, it should have a
(1) higher dissociation constant [Ki] for enzyme–inhibitor complex
(2) lower dissociation constant [Ki] for enzyme–inhibitor complex
(3) competitive type of inhibition
(4) uncompetitive type of inhibition

38. The first, second, third and fourth number in EC stands for
(1) Class name, subclass, hydroxyl group acceptor, phosphoryl group acceptor
(2) Class name, subclass, phoshorylgroup acceptor, acetyl group acceptor
(3) Class name, subclass, phoshoryl group acceptor, hydroxyl group acceptor
(4) Class name, subclass, acetyl group acceptor, hydroxyl group acceptor

39. A prochiral ketone can be reduced by oxidoreductase up to a maximum of
(1) 25% reduction
(2) 50% reduction
(3) 75% reduction
(4) 100% reduction

40. Hydrogen bonds in α-helices are
(1) more numerous than vander waals interactions
(2) not present at phenylalanine residues
(3) analogous to the steps in a spiral staircase
(4) roughly parallel to the helix axis

41. Z-DNA helix
(1) is the primary form in any living organism
(2) is favoured by an alternating GC sequence
(3) tends to be formed at 3' end of genes
(4) formation is inhibited by methylation of bases

42. In H-DNA bases of the third strand pairs with
(1) Pyrimidines of the double helix
(2) Purines of the double helix
(3) Both Purines and Pyrimidines of the duplex
(4) Remains unpaired

43. The Ramachandran map of a protein representation allows you to identify
(1) The most stable structure
(2) The tertiary allowed structure
(3) The sterically disallowed conformations
(4) the secondary structure elements

44. Methotrexate, an analogue of dihydro-folate, is an inhibitor of dihydrofolate reductase. Methotrexate would be expected to
(1) Decrease the Vm of the enzyme
(2) Increase the Km of the enzyme for the substrate
(3) Decrease both the Km and Vm of the enzyme
(4) Increase the affinity of the enzyme for the substrate

45. Silk fibroin displays
(1) alpha helix
(2) loop structure
(3) Antiparallel- β pleated sheets
(4) parallel β sheet

46. Enzymes catalyze reactions by
(1) binding regulatory proteins
(2) covalently modifying active-site residues
(3) binding substrates with great affinity
(4) selectively binding the transition state of a reaction with high affinity

47. The angle between the two long helical arms of tRNA is about
(1) 180° (2) 0°
(3) 45° (4) 90°

48. Which one of the following amino acids interrupts α-helices, and also disrupts β-sheets?
(1) Phe (2) Cys
(3) His (4) Pro

49. How many milliliters of 0.05 N HCl are required to neutralize eight grams of NaOH?
(1) 5000 (2) 4000
(3) 4500 (4) 5050

50. What is the pH of 10^{-8} M solution of HCl?
(1) 8.99 (2) 6.99
(3) 7.99 (4) 7.00

51. How many microliters of 0.1 M solution of sodium chloride will make 10 ml of 5 mM sodium chloride?
(1) 200 (2) 100
(3) 500 (4) 10

52. Cα- C' distance plot might be useful in
(1) Identifying secondary structures in proteins
(2) Identifying globular domains in a protein
(3) Identifying active sites in enzymes
(4) For docking of inhibitors on protein's surface

53. The parts of proteins having the highest flexibility are
(1) α-helices (2) β-sheets
(3) peptide bonds (4) surface side chains

54. What are the concentrations of HOAc and OAc^- in 0.2 M "acetate buffer", pH 5.00? The Ka for HOAc is 1.70×10^{-5}
(1) OAc- = 0.126 M, HOAc = 0.074 M
(2) OAc- = 0.150 M, HOAc = 0.084 M
(3) OAc- = 0.200 M, HOAc = 0.184 M
(4) OAc- = 0.250 M, HOAc = 0.284 M

55. Calculate the instantaneous buffering capacity in both directions of 0.05 M Tricine buffer, pH 7.5. Tricine is an N-tris (hydroxymethyl)-methyl glycine. pKa = 8.15 (Ka = 7.08×10^{-9})
(1) 0.017 M (2) 0.217 M
(3) 0.257 M (4) 0.175 M

56. Hydrolysis of ATP over ADP and AMP generates highest energy because
(1) ATP on hydrolysis generates thermodynamically stable structure
(2) ATP is highly unstable
(3) Hydrolysis of ATP is pH dependent
(4) Hydrolysis of ADP and AMP do not generate thermodynamically stable structure

57. The equilibrium constant ((Keq) for the reaction S↔ P is 5
Suppose we have a mixture of [S] = 2 x 10^{-4} M and [P] = 3 x 10^{-4} M. In which direction will the reaction proceed on addition of appropriate enzyme?
(1) Proceeds in a forward direction
(2) Proceeds in a reverse direction
(3) Proceeds in both the directions
(4) Proceeds sometimes in forward and sometimes in reverse direction

58. Polyproline II structure found in some proteins and having helical orientation is known to have
(1) 3 residues per turn
(2) 3.3 residues per turn
(3) 3.6 residues per turn
(4) 4 residues per turn

59. If the pKa of the ε-amino group of lysine is 10.5, the side chain at pH 7 is likely to be
(1) neutral
(2) half dissociated
(3) negatively charged
(4) positively charged

60. The dihedral angle ψ for a peptide used in the Ramachandran plot is given a value based on rotation around
(1) N-Cα bond (2) Cα-C' bond
(3) C'-N bond (4) Cα-Cβ bond

61. Water has a high dielectric constant value of 80. Because of its presence in biological systems compared with low dielectric solvents like hydrocarbons, it should
(1) strengthen electrostatic interaction but weaken hydrophobic interactions
(2) weaken electrostatic interaction but strengthen hydrophobic interactions
(3) strengthen both electrostatic as well as hydrophobic interactions
(4) weaken both electrostatic as well as hydrophobic interactions

62. In which of the following processes eukaryotic mitochondria is known to be least involved?
(1) ATP production
(2) Apoptosis
(3) Tricarboxylic acid cycle
(4) Fatty acid biosynthesis

63. The pI calculated from the sequence may differ from the experimentally determined value because
(1) pI is not accurately determined experimentally
(2) pKa of the amino acid side chains depend on the micro structural environment
(3) pI calculation from the sequence does not take the N and C termini into account
(4) pH is not known theoretically

64. Which of these sets of amino acids are not capable of forming hydrogen bonds through their side chains
(1) Val, Ile, Phe (2) Trp, Tyr, His
(3) Ser, Thr, Asn (4) Arg, Lys, Asp

65. Hydrogen bonds in anti parallel β-sheets
(1) occur in more number than van der Waals interactions
(2) are not present at Phe residues
(3) occur roughly perpendicular to the polypeptide chain direction
(4) are about five Angstroms in length

66. The hydrophobic moment is
(1) zero for amphiphilic secondary structures
(2) negative for amphiphilic secondary structures
(3) positive for amphiphilic secondary structures
(4) indeterminate for amphiphilic secondary structures

67. Generally ---- puckered sugar residues are found in A-DNA structures
(1) C3'-Exo (2) C3'-Endo
(3) C4'-Exo (4) C4'-Endo

68. Major gluconeogenesis occurs in
(1) Liver and kidney
(2) Liver and heart
(3) Liver and skeletal muscle
(4) Liver and adrenal gland

69. What is the force among the following that is primarily responsible for stabilizing the tertiary structure of globular proteins?
(1) Disulfide bonding
(2) The hydrophobic effect
(3) Hydrogen bonding
(4) Ionic interactions

70. Isoelectric point of lysozyme is 9.2. When the enzyme solution at this pH in water was titrated with HCl to give a pH of 5, it was observed that six ionized glutamic acid side chains got protonated. The net charge on the enzyme at pH 5 would therefore, be
(1) +5 (2) +6
(3) -5 (4) -6

71. 6M Guanidium hydrochloride is known to denature a number of proteins. Such a high concentration is able to break down the noncovalent forces sustaining the structure of proteins by affecting
(1) electrostatic interactions only
(2) electrostatic and hydrophobic interactions
(3) intrapeptide hydrogen bonding only
(4) electrostatic, hydrophobic, and hydrogen bonding interactions

72. Which of the following amino acids is critical in the folding of proteins due to the slow isomerization of the peptide bond preceding this amino acid depending upon the solvent environment?
(1) Tryptophan (2) Leucine
(3) Proline (4) Histidine

73. In peptides the values of dihedral angle phi is based on rotation around
(1) N-Cα bond (2) Cα –C' bond
(3) C'-N bond (4) N-H bond

74. The double-helical Watson-Crick structure of DNA was first obtained form
(1) Fiber diffraction only
(2) Fiber diffraction and molecular modeling
(3) X-ray diffraction from single crystals
(4) Diffraction from single crystals and molecular modeling

75. The major and minor grooves of B-form DNA correspond to the following feature
(1) minor and major grooves
(2) major and minor grooves
(3) deep and shallow grooves
(4) wide and shallow grooves

76. Which is the amino acid among the following that can occupy positions in the Ramachandran map that are disallowed for other 19 amino acids, but allowed for D-amino acids?
(1) Ala (2) Gly
(3) Pro (4) Cys

77. If side chains of amino acids interact with each other, which of the following would be termed as a salt bridge?
(1) Tyr- Phe
(2) Cys- Cys
(3) Lys- Glu
(4) Ala- Val

78. All of the following enzymes are linked to reduction of NADH except
(1) Isocitrate dehydrogenase
(2) Lactate dehydrogenase
(3) Succinate dehydrogenase
(4) Pyruvate dehydrogenase

79. Which of the following are components of a phospholipid?
(1) Cholesterol, glycerol, fatty acids
(2) Fatty acids, phosphate group, glycerol
(3) Glycerol, amino acids, phosphate group
(4) Phosphate group, cholesterol, monosaccharides

80. Amino acid analysis of 1.0 mg of pure enzyme yielded 58.1 µg of leucine (MW = 131.2) What is the minimum MW of the enzyme based on leucine content?

(1) 2000
(2) 2300
(3) 2350
(4) 2258

81. The first metabolic intermediate that is common to the aerobic metabolism of glucose and fatty acids is

(1) acetyl CoA (2) aceto-acetyl CoA
(3) pyruvate (4) citrate

82. Prosthetic groups such as iron-sulfur clusters and heme function to

(1) donate electrons to NADH
(2) allow proteins to diffuse within the mitochondrial inner membrane
(3) both accept and donate electrons during electron transport
(4) transport protons across the mitochondrial inner membrane

83. Glucose labeled with ^{14}C at C-6 is added to a solution containing the enzymes and cofactors of the oxidative phase of the pentose phosphate pathway. The radioactive label will be observed at

(1) C5 of ribulose 5-phosphate
(2) C3 of ribulose 5-phosphate
(3) C1 of ribulose 5-phosphate
(4) C4 of ribulose 5-phosphate

84. The molarity of 15 % NaCl is

(1) 2.56 (2) 0.256
(3) 25.6 (4) 0.025

85. Competitive inhibition of an enzyme by a competitive inhibitor can be overcome by simply

(1) decreasing the concentration of substrate
(2) increasing the concentration of substrate
(3) decreasing the temperature of reaction
(4) increasing the temperature of reaction

86. In which of the following reactions, is the unit of rate constant and rate of reaction the same?

(1) 1st order reaction
(2) 2nd order reaction
(3) 3rd order reaction
(4) Zero order reaction

87. In an enzymatic reaction, if the enzyme concentration is increased from 1mg to 2 mg, which of the following statements is correct?

(1) K_m and V_{max} will remain constant
(2) K_m will change while V_{max} will remain constant
(3) K_m will remain constant but V_{max} will be more
(4) K_m and V_{max} will increase

88. Among the different amino acid side chains in proteins, which of the following pairs might form side chain-side chain hydrogen bonding interaction with each other?

(1) Valine- Glutamic acid
(2) Alanine- Glycine
(3) Asparagine- Serine
(4) Proline- Lysine

89. If a protein is known to bind Ca^{2+} ions, which of the following side chains is likely to be involved in Ca2+ binding?

(1) Aspartic acid (2) Lysine
(3) Proline (4) Methionine

90. Warburg effect is characterised by

(1) Increased glycolysis
(2) Decreased glycolysis
(3) Absence of glycolysis
(4) Malfunctional glycolysis

91. Phenylmethylsulfonyl fluoride (PMSF) is a/an

(1) competive inhibitor of serine proteases
(2) non competive inhibitor of serine proteases
(3) reversible inhibitor of serine proteases
(4) irreversible inhibitor of serine proteases

92. Deficiency of HGPRT and Glucose-6-phosphatase leads to
(1) Gaucher's disease
(2) Phenylketoneuria
(3) Gout
(4) Alkaptonuria

93. Which one of the following amino acids has pKa value near physiological pH?
(1) Histidine (2) Arginine
(3) Lysine (4) Serine

94. The transition state of a catalyzed reaction (ES^*) is a highly-populated intermediate on the reaction pathway which is
(1) higher in energy than that of an uncatalyzed reaction
(2) lower in energy than that of an uncatalyzed reaction
(3) lower in energy than the reaction substrate
(4) bound very weakly to the catalyst

95. ETC has a potential to produce highly reactive free radicals that can damage the cells. Which one of the following is useful to prevent oxidative damage in cells?
(1) Superoxide molecule
(2) Glutathione peroxidase
(3) Antimycin A
(4) Rotenone

96. Gangliosides derived from glucosylceramide contain one or more molecules of
(1) sialic acid (2) glycerol
(3) diacylglycerol (4) hyaluronic acid

97. Esterification of cholesterol in plasma is catalyzed by
(1) Lecithin: Cholesterol acyl transferase
(2) Acyl Co A: Cholesterol acyl transferase
(3) Succinyl CoA: Cholesterol acyl transferase
(4) Malonyl CoA : Cholesterol acyl transferase

98. Poly-L-leucine solution in dioxane is
(1) α-helical (2) random coil
(3) β-sheet (4) α-helix

99. Which one of the following is best suited for hydrolyzing peptide bonds on the carboxyl side of aromatic residues?
(1) Chymotrypsin (2) Trypsin
(3) CNBr (4) Performic acid

100. Regulation of blood glucose level is by
(1) opposing effects of insulin and glucagon
(2) additive effect of insulin and glucagon
(3) independent of glucagon
(4) only by insulin

101. Treatment of the inside-out vesicles with an uncoupler like DNP that allows protons to freely move across the vesicle membrane would be expected to
(1) increase ATP synthesis
(2) inhibit ATP synthesis
(3) increase O_2 consumption
(4) inhibit electron transport

102. In the active site of Transaldolase, which amino acid forms the Schiff base?
(1) Arginine (2) Methionine
(3) Lysine (4) Glutamine

103. An enzyme has an activation energy of 10,700 cal/mole. How many times faster (approximately) will the reaction proceed at 37°C compared to 15°C?
(1) 5 times (2) 10 times
(3) 4 times (4) 8 times

104. A TLC run of rat liver phospholipids is sprayed with ninhydrin, and the colour is allowed to develop. Which phospho-lipids can be detected in this way?
(1) DP PC (2) DS PC
(3) PS (4) Lecithin

105. N-ethylmaleimide (NEM) is known to selectively inhibit bacterial lactose transporters. Which amino acid of the transporter is responsible for this inhibition?
(1) Cysteine (2) Cystine
(3) Methionine (4) Arginine

106. Identify the small protein (167 amino acids) that is produced in adipocytes and acts on receptors in the hypothalamus to curtail appetite
(1) LDL
(2) Lutenizing hormone
(3) HDL (4) Leptin

107. Highest turnover number of an enzymatic reaction so far known is exhibited by
(1) Aspartate transcarbamylase
(2) ATPase
(3) Lysozyme
(4) Carbonic anhydrase

108. The chemical nature of covalent linkage in a disaccharide is known as
(1) ester (2) ether
(3) amide (4) diester

109. Glutathione is
(1) L-Glu-Cys-Gly
(2) L--Glu-Met-Gly
(3) D-Glu-Cys-Ala
(4) D-Glu-Cys-Ala

110. Which one of the following amino acids is optically inactive?
(1) Glycine (2) Methionine
(3) Phenylalanine (4) Glutamine

111. Protein responsible for transport of O_2 from alveoli to the tissue is
(1) Leghaemoglobin
(2) Oxyhaemoglobin
(3) Myoglobin
(4) Carbaminohaemoglobin

112. Process of formation of ATP from ADP during photosynthesis is referred to as
(1) photophosphorylation
(2) photorespiration
(3) phosphorylation
(4) oxidative phosphorylation

113. Glycogen and cellulose are
(1) helical and β-sheet structure, respectively
(2) helical structures but with different degree of helicity
(3) β-sheet structures
(4) helical but glycogen is extensively branched molecule

114. One side of a 30-residue α-helix faces the hydrophobic interior of a protein. If the helical wheel of this helix were to be drawn, which of the following residues are likely to be hydrophobic?
(1) 10th, 20th and 30th
(2) 1st, 6th, 11th, 16th, 21st and 26th
(3) 1st, 4th, 7th, 10th, 13th, 16th, 19th, 22nd, 25th and 28th
(4) 1st, 5th, 8th, 12th, 15th, 19th, 22nd, 26th and 30th

115. The dielectric constant of protein interiors is likely to be
(1) similar to that of water's, i.e. close to 80
(2) much smaller than that of water
(3) much more than that of water
(4) at some places in the interior more and at some other places less than that of water

116. In a helix-turn-helix motif in proteins, which one of the following is true for binding to nucleic acids?
(1) Both the helices bind
(2) Neither of the two helices bind
(3) Only the 1st helix is important in binding
(4) The 2nd helix is important in binding

117. Which of the following does not apply to triplex DNA?
(1) It is triple stranded in nature
(2) It requires only Hoogstein hydrogen bonding
(3) It requires Watson-Crick hydrogen bonding
(4) It forms at neutral or acidic pH

118. The hydrogen-bonding pattern in alpha helices is
(1) n to n+4 (2) n to n+3
(3) n to n+5 (4) n-1 to n

119. The rate of renaturation of DNA is governed by the equation
(1) $dc/dt = -kC^2$ (2) $dt/dc = -kC^2$
(3) $dt/dc = kC^2$ (4) $dc/dt = 2kC$

120. A peptide bond
(1) has a partial double bond character
(2) is stable in strong acids
(3) occurs most commonly in *cis* configuration
(4) is cleaved by agents that denature proteins, such as organic solvents and high concentrations of urea

121. The complete denaturation of a protein leads to a loss of the following structure(s)
(1) primary
(2) primary and tertiary
(3) primary and secondary
(4) secondary and tertiary

122. Which of the following ionizes at physiological pH?
(1) Glycine (2) Alanine
(3) Histidine (4) Purine

123. Catalytic efficiency allows a comparison of different enzymes. It is the ratio of
(1) Kcat/Km (2) Km/Kcat
(3) Vmax/Kcat (4) Kcat/Vmax

124. Random single displacement enzyme reactions resemble
(1) competitive inhibition
(2) uncompetitive inhibition
(3) irreversible inhibition
(4) mixed inhibition

125. In an enzyme catalyzed reaction, K_m = 4×10^{-5} µmol/l, and the rate of reaction (V) at substrate concentration [S] = 1.2×10^{-2} M is 80 µmol/l-min. Assuming no inhibitor is present, V_{max} is practically equal to
(1) 40 µmol/l-min
(2) 80 µmol/l-min
(3) 120 µmol/l-min
(4) 4.8×10^2 µmol/l-min

126. Regulation of fatty acid biosynthesis occurs at the enzymatic step catalyzed by
(1) carnitine acyltransferase I
(2) acetyl CoA carboxylase
(3) pyruvate carboxylase
(4) citrate synthase

127. The non-oxidative branch of the pentose phosphate pathway does not include which of the following reactions?
(1) Ribulose 5-P →ribose 5-P
(2) Xylulose5-P + ribose 5-P → sedoheptulose 7-P +glyceraldehydes 3-P
(3) Ribulose 5-P + glyceraldehydes 3-P→ sedoheptulose 7-P
(4) Sedoheptulose 7-P + glyceraldehyde 3- P →fructose 6-P + erythrose 4-P

128. Gluconeogenesis is not capable of making glucose from
(1) adenine
(2) lactate
(3) acetyl CoA
(4) malony CoA

129. The rate of reaction increases by increase in temperature because
(1) collision frequency increases
(2) energy of products decreases
(3) fraction of molecules possessing energy ≥ ET (Threshold energy) increases
(4) mechanism of reaction is changed

130. Non-superimposable mirror images having similar molecular formula are called
(1) *cis-trans* isomers
(2) geometric isomers
(3) anomers
(4) enantiomers

131. Which of the following is the best way to detoxify the methanol toxicity if a person drinks methanol?
(1) Make the patient drink glucose water
(2) Intravenous injection with steroid
(3) Make the patient drink ethanol
(4) Make the patient drink lemon juice

132. During vigorous exercise lactic acid gets accumulated in skeletal muscle due to
(1) lack of NADH
(2) lack of NAD^+
(3) excess supply of CO_2
(4) excess supply of O_2

133. phi, psi angles of a peptide segment adopting alpha helical conformation would be around
(1) –78, +59 (2) +49,+26
(3) –57,-78 (4) –60,-40

134. The allowed region in the Ramachandran Plot for three residues (alanine, glycine and proline) decreases in the order
(1) Pro > Gly > Ala
(2) Gly > Ala > Pro
(3) Ala > Pro > Gly
(4) Gly > Pro = Ala

135. The potential energy for the interaction of two atoms is given by $U = A/r^{12} - B/r^6$. The bottom of the potential well corresponds to
(1) the sum of van der Waals radii of the atoms
(2) the existence of the maximum electrostatic interaction
(3) the situation when the first term vanishes
(4) the situation when the atoms get bonded covalently

136. A protein with mostly hydrophobic residues in the surface is likely to be a
(1) fibrous protein
(2) globular protein
(3) membrane protein
(4) glycosylated protein

137. Quaternary structure of a protein consists of
(1) arrangement of one protein chain in a protein with a single subunit
(2) arrangement of separate protein chains in a protein molecule with more than one subunit
(3) arrangement of only parallel and ant-iparallel β- sheets in a protein chain
(4) occurrence of an alpha-helix bundle in a protein chain

138. Which of the following statements about hydrogen bonds are true?
(1) They are weak covalent bonds that are easily disrupted by heat
(2) They are weak bonds formed between hydrocarbons in water
(3) They are weak bonds only formed in the presence of water
(4) They are weak bonds involved in maintaining the 3-D structure of macromolecules

139. Based on what you know about the properties of water, which of the following statements about methanol (CH_3OH) are true?
(1) Methanol molecules form more hydrogen bonds than water molecules do
(2) The boiling point of methanol is higher than that of water
(3) Salts such as NaCl are less soluble in methanol than in water
(4) Methanol is a more cohesive liquid than water

140. On the molar scale which of the following interaction in a nonpolar environment provides the highest contribution to the bio-molecule?
(1) Van der Walls interaction
(2) Hydrogen bonding
(3) Salt bridge
(4) Hydrophobic interaction

141. Osmosis is movement of a
(1) charged solute molecule (ion) across a membrane
(2) gas molecule across a membrane
(3) nonpolar solute molecule across a membrane
(4) water molecule across a membrane

142. Which of the following statements about cystine is correct?
(1) Cystine forms when the —CH_2—SH R group is oxidized to form a —CH_2—S—S—CH_2— disulfide bridge between two cysteines
(2) Cystine is an example of a nonstandard amino acid, derived by linking two standard amino acids
(3) Cystine is formed by the oxidation of the carboxylic acid group on cysteine
(4) Two cystines are released when a —CH_2—S—S—CH_2— disulfide bridge is reduced to —CH_2—SH

143. For amino acids with neutral R groups, at any pH below the pI of the amino acid, the population of amino acids in solution will have
(1) a net negative charge
(2) a net positive charge
(3) no charged groups
(4) positive and negative charges in equal concentration

144. Which of the following correctly describes linoleic or linolenic acid?
(1) They can be produced from palmitate via desaturases
(2) Are readily converted to arachidonic acid
(3) Linoleic acid is an example of an omega-3 fatty acid
(4) Can be produced from leukotrienes via malonyl CoA transferase

145. Which one of the following amino-acids has highest fluorescence quantum yield (φ) in aqueous solution?
(1) Tyrosine (2) Tryptophan
(3) Phenylalanine (4) Histidine

146. Among the following which amino acid do not absorb UV light in range of 250-300 nm?
(1) Phenyl alanine
(2) Tyrosine
(3) Tryptophan
(4) Cysteine

147. Which one of the following statements concerning glutamine is correct?
(1) Contains three titrable groups
(2) Contains an amide group in side chain
(3) Has E as its one-letter symbol
(4) Migrates to the cathode (negative electrode) during electrophoresis at 7.0

148. Which of the following molecules orient themselves into sandwich like membranes because of hydrophobic components within the molecule?
(1) Amylose (2) Glycogen
(3) Cellulose (4) Phospholipid

149. Each of the following molecules is a polymer except
(1) protein (2) lipids
(3) cellulose (4) starch

150. If you were to take 1.0 ml of a solution of HCl with a pH of 4.0 and add it to 9.0 ml of distilled water, what would be the pH of the final solution?
(1) The pH would remain unchanged
(2) The pH would rise to 5.5
(3) The pH would rise to 5.0
(4) The pH would rise to 7.0

151. What is the pH of an aqueous solution that has a $[H^+]$ of 6.65×10^{-10} M?
(1) 9.18 (2) 4.82
(3) 5.47 (4) 6.99

152. What is the pI of Cys? (pKa α-NH: 10.7; pKa α-COOH: 1.9; pKa R: 8.4)
(1) 8.14 (2) 7.0
(3) 6.3 (4) 5.15

153. Amphiphilic or amphipathic substances
(1) are hydrocarbons
(2) do not aggregate
(3) have one end that will dissolve in water and one end that will dissolve in a nonpolar environment
(4) behaves as zwitter ion

154. Hydrogen bonds
(1) are carbon-hydrogen bonds
(2) are nonpolar
(3) do not involve electronegative atoms
(4) occur when a hydrogen atom covalently bonded to one electronegative atom interacts with another electronegative atom

155. Which of the following amino acid plays important role in stress among plants and microbes?
(1) Glutamate (2) Glutamine
(3) Aspartate (4) Proline

156. Which of the following is an important functional group in fatty acids?
(1) Hydroxyl (2) Phosphate
(3) Carboxyl (4) carbonyl

157. Consider the two molecules below. Which label describes them accurately?

$H_3C-C(OH)=C(OH)-CH_3$ $H_3C-C(OH)=C(OH)-CH_3$

(1) Two structural isomers
(2) Two geometric isomers
(3) Two stereoisomers
(4) Two enantiomers

158. Homeostasis refers to a constancy which is the result of
(1) no input or output
(2) imbalance of input and output
(3) controlled balance of input and output
(4) constant surrounding environment

159. If the pH of a solution goes from 8 to 7, this means that the hydrogen ion concentration has
(1) decreased by 1 mole/liter
(2) increased by 1 mole/liter
(3) decreased 10 fold
(4) increased 10 fold

160. A molecule which releases hydrogen ion when added to a solution (thus raising the hydrogen ion concentration of the solution and lowering the pH) is called
(1) a base (2) an acid
(3) an antacid (4) a buffer

161. Which functional group is found in both aldehydes and ketones?
(1) hydroxyl (2) carbonyl
(3) carboxyl (4) amino

162. How does aspirin reduce inflammation?
(1) It oxygenates arachidonate
(2) It acetylates prostaglandins
(3) It inhibits the formation of prostaglandins
(4) It inhibits the formation of arachidonate

163. Which of the following amino acid residues are likely to be found on the inside of a watersoluble protein?
(1) Val (2) Arg
(3) Asn (4) Asp

164. Which one of the following statements about prostaglandins is incorrect?
(1) Prostaglandins have a very short half-life
(2) Prostaglandins are synthesized only in the liver and the adrenal cortex
(3) Prostaglandins generally act locally on or near the tissue that produced them
(4) The common precursor of the prostaglandins is arachidonic acid

165. In the reaction drawn below, what has happened to molecule 1 to convert it to molecule 2?

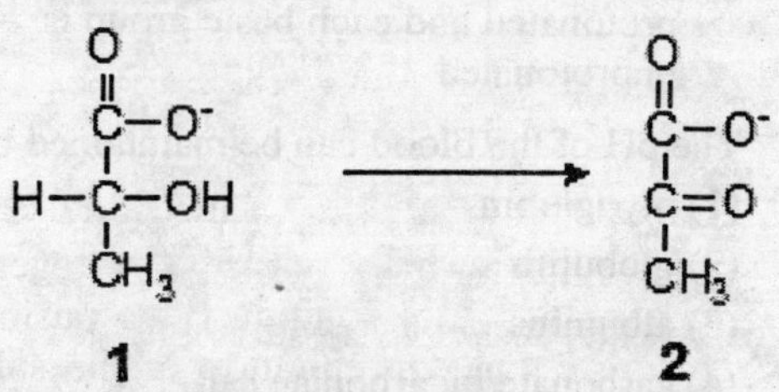

(1) Oxidation (2) Isomerization
(3) Dehydration (4) Reduction

166. Phospholipids are involved in all except one of the following
(1) cell-cell recognition
(2) signal transduction
(3) surfactant function in the lungs
(4) mediator of hypersensitivity

167. Which pair of amino acids will have the highest absorbance at 280 nm? (Assume equimolar concentrations)
(1) Thr and His
(2) Trp and Tyr
(3) Phe and Pro
(4) Phe and His

168. Structural features that are common to all prostaglandins include
(1) 20-carbon atoms
(2) a peroxide group at C-15
(3) two double bonds
(4) a ketone group

169. Amino acids considered nonessential for humans are
(1) those not incorporated into protein
(2) not necessary in the diet if sufficient amounts of precursors are present
(3) the same for adults as for children
(4) the ones made in specific proteins by posttranslational modifications

170. Which of the following amino acid if added to a acidic solution will have buffer action on solution at the physiological pH
(1) alanine (2) glycine
(3) histidine (4) arginine

171. Linoleate and α-linolenate are "essential fatty acids" because
(1) mammals can use them to make other fatty acids
(2) mammals cannot synthesize fatty acid chains with more than 16 carbons
(3) mammals cannot introduce more than one double bond in fatty acid chains
(4) mammals cannot introduce double bonds into fatty acid chains

172. In a highly acidic solution, pH = 1.3, the dominant form of glycine is
(1) NH_2-CH_2-COOH
(2) NH_3^+-CH_2-COOH
(3) NH_2-CH_2-CO^-
(4) NH_3^+-CH_2-COO^-

173. Which of the following fatty acids has the lowest melting point?
(1) Fatty acids with sites of unsaturation with cis double bonds
(2) Fatty acids with sites of unsaturation with trans double bonds
(3) Fatty acids with no sites of unsaturation
(4) Fatty acids with longer hydrophobic tails

174. Which of the following statements is true about the cyclic (ring) form of D–glucose?
(1) The cyclic form of D–glucose is a furanose
(2) The cyclic form of D–glucose involves the formation of a covalent bond between the carbonyl group and the oxygen of a hydroxyl group along the chain
(3) The cyclic form of D–glucose is the result of a bond forming between the C–1 and C–6 carbon atoms
(4) In aqueous solution, the cyclic form of D–glucose is rare

175. Lipids are called macromolecule because they
(1) form liposomes
(2) are major component of membranes
(3) form aggregates in water
(4) are hydrolysed further into smaller components

176. You have decided to take your doctor's advice and increase your dietary intake of fish. What effect would you hope this dietary change would accomplish?
(1) You would increase your intake of omega-6 fatty acids
(2) You would increase your production of arachidonate
(3) You would inhibit prostaglandin and leukotriene synthesis
(4) You would increase your intake of linolenic acid, an omega-3 fatty acid, which is the preferred substrate for arachidonate synthesis

177. The glycolipids are
(1) membrane lipids containing a phosphate-alcohol headgroup
(2) lipids that are synthesized using phosphatidate as the key intermediate of the pathway
(3) degraded to form diacylglycerol as a second messenger
(4) lipids in which the backbone is initially synthesized from palmitoyl CoA and serine and seen on extrinsic phase of plasma membrane

178. Which of the following amino acids has a major role in the transport of amino groups from muscle tissue to the liver?
(1) Methionine (2) Glutamate
(3) Arginine (4) Alanine

179. The isoelectric point of a protein is defined as
(1) the pH at which the net charge on the molecule is zero
(2) the pH at which all groups are protonated
(3) the pH at which all groups are unprotonated
(4) the pH at which each acidic group is protonated and each basic group is unprotonated

180. The pH of the blood can be maintained by
(1) myoglobin
(2) globulins
(3) albumins
(4) carbonate/bicarbonate salt

181. Which of the following describes a universal property of buffers?
(1) Buffers are usually composed of a mixture of strong acids and strong bases
(2) Buffers work best at the pH at which they are completely dissociated
(3) Buffers work best at the pH at which they are 50% dissociated
(4) Buffers work best at one pH unit lower than the pKa value

182. Which one of the following amino-containing buffering molecules will be the most able to resist change in pH when in a pH 7.0 solution at a total (ionized and non-ionic) concentration of 100mM?
(1) Tris (pKa = 8.1)
(2) HEPES (pKa = 7.7)
(3) MOPS (pKa = 7.3)
(4) PIPES (pKa = 7.1)

183. A feature of amino acids not found in carbohydrates is the presence of
(1) phosphorous
(2) nitrogen
(3) hydrogen
(4) oxygen

184. Identify the relationship between cysteine and cystine
(1) carboxylic acid and amide
(2) amide and carboxylic acid
(3) thiol and disulfide
(4) disulfide and thiol

185. Amphipathic molecules may spontaneously assemble to form micelles or bilayers in water solution through the formation of
(1) hydrophobic and hydrophilic interactions
(2) covalent bonds
(3) interactions with membrane proteins
(4) ionic bonds

186. The dominant interaction that drives a water-soluble protein to fold is
(1) H-bonding
(2) formation of strong, covalent bonds
(3) the hydrophobic interaction
(4) charged and polar residues on the outside, in contact with the water

187. Many biosystems require hydrogen bonds. In other words
(1) the hydrogen gives up an electron to another molecule
(2) hydrogen shares an electron with another molecule
(3) hydrogen shifts between molecules
(4) the electronegative pole of one molecule attracts the electropositive pole of another

188. Among the following which is weakest interaction?
(1) Vanderwall interaction
(2) Ionic interaction in non-polar environment
(3) Hydrogen bonding
(4) Hydrophobic interaction

189. A decrease in entropy in a reaction (+ ΔS_{system}) can be overcome by
(1) a decrease in entropy in the surroundings ($-\Delta S_{surrounding}$)
(2) if the temperature of the reaction is lower than the ratio of $\Delta H/\Delta S$
(3) if the temperature of the reaction is greater than the ratio of $\Delta H/\Delta S$
(4) Increase in free energy

190. If a polypeptide has 400 amino acid residues, what is its approximate mass?
(1) 11,000 daltons
(2) 44,000 daltons
(3) 22,000 daltons
(4) 88,000 daltons

191. Which of the following statements is correct in case of glycoproteins?
(1) May contain 95% or more of carbohydrate
(2) Have the carbohydrate linked to the protein by either N- or O- glycosidic bonds
(3) Found only on cell membranes
(4) Always contain a serially repeating carbohydrate unit

192. A solution that has a pH of 5
(1) has twice the hydrogen ion concentration of a solution with a pH of 10
(2) has half the hydrogen ion concentration of a solution with a pH of 10
(3) is acidic
(4) is basic

193. Water is a polar molecule. This means that
(1) the opposite ends of the molecule have opposite charges
(2) water molecules are linear, like a pole
(3) water is one of the many hydrophobic molecules
(4) the atoms in water have same electronegativity

194. Buffers
(1) release a hydrogen ion when pH drops
(2) release a hydrogen ion when pH rises
(3) absorb a OH^- ion when pH drops
(4) absorb a OH ion when pH rises

195. L and D forms of a molecule
(1) are distereoisomers
(2) rotate light in different directions
(3) have atoms arranged differently around all carbons
(4) are identical

196. Carbon dioxide (CO_2) is readily soluble in water, according to the equation CO_2 + $H_2O \leftrightarrow H_2CO_3$. Carbonic acid ($H_2CO_3$) is a weak acid. Respiring cells release CO_2 into the bloodstream. What will be the effect on pH of blood as that blood first comes in contact with respiring cells?
(1) Blood pH will decrease slightly
(2) Blood pH will increase slightly
(3) Blood pH will remain unchanged
(4) Blood pH will first increase, then decrease as CO_2 combines with hemoglobin

197. Measurements show that the pH of a particular lake is 4.0. What is the hydrogen ion concentration of the lake?
(1) 4.0 M (2) 10^{-10} M
(3) 10^{-4} M (4) 10^{4} M

198. A slice of pizza has 500 kcal. If we could burn the pizza and use all the heat to warm a 50-L container of cold water, what would be the approximate increase in the temperature of the water? (*Note*: A liter of cold water weighs about 1 kg)
(1) 50°C (2) 5°C
(3) 1°C (4) 10°C

199. How many grams of acetic acid ($C_2H_4O_2$) would you use to make 10 L of a 0.1 *M* aqueous solution of acetic acid? (*Note*: The atomic masses, in daltons, are approximately 12 for carbon, 1 for hydrogen, and 16 for oxygen)
(1) 10 g (2) 0.1 g
(3) 6.0 g (4) 60 g

200. Which is the major transport form of sugar in phloem of plants?
(1) Maltose (2) Glucose
(3) Sucrose (4) Lactose

ANSWER SHEET

Unit 1: Biochemistry

1. 2	**2.** 2	**3.** 4	**4.** 2	**5.** 3	**6.** 1	**7.** 1	**8.** 1	**9.** 1	**10.** 2
11. 1	**12.** 2	**13.** 2	**14.** 2	**15.** 3	**16.** 4	**17.** 1	**18.** 4	**19.** 3	**20.** 2
21. 1	**22.** 2	**23.** 3	**24.** 2	**25.** 3	**26.** 1	**27.** 4	**28.** 3	**29.** 3	**30.** 1
31. 3	**32.** 3	**33.** 3	**34.** 2	**35.** 2	**36.** 2	**37.** 2	**38.** 1	**39.** 4	**40.** 4
41. 2	**42.** 2	**43.** 4	**44.** 2	**45.** 3	**46.** 4	**47.** 4	**48.** 4	**49.** 2	**50.** 2
51. 3	**52.** 1	**53.** 4	**54.** 1	**55.** 1	**56.** 1	**57.** 1	**58.** 1	**59.** 4	**60.** 2
61. 2	**62.** 4	**63.** 2	**64.** 1	**65.** 2	**66.** 1	**67.** 2	**68.** 1	**69.** 2	**70.** 2
71. 4	**72.** 3	**73.** 1	**74.** 2	**75.** 3	**76.** 2	**77.** 3	**78.** 3	**79.** 2	**80.** 4
81. 1	**82.** 3	**83.** 1	**84.** 1	**85.** 2	**86.** 4	**87.** 3	**88.** 3	**89.** 1	**90.** 1
91. 4	**92.** 3	**93.** 1	**94.** 2	**95.** 2	**96.** 1	**97.** 1	**98.** 1	**99.** 1	**100.** 1
101. 3	**102.** 3	**103.** 3	**104.** 3	**105.** 1	**106.** 4	**107.** 4	**108.** 2	**109.** 1	**110.** 1
111. 2	**112.** 1	**113.** 1	**114.** 4	**115.** 2	**116.** 4	**117.** 2	**118.** 1	**119.** 1	**120.** 1

121. 4	**122.** 3	**123.** 1	**124.** 4	**125.** 2	**126.** 2	**127.** 2	**128.** 3	**129.** 3	**130.** 4
131. 3	**132.** 2	**133.** 4	**134.** 2	**135.** 1	**136.** 1	**137.** 2	**138.** 4	**139.** 3	**140.** 3
141. 4	**142.** 1	**143.** 2	**144.** 4	**145.** 2	**146.** 4	**147.** 2	**148.** 4	**149.** 2	**150.** 3
151. 1	**152.** 4	**153.** 3	**154.** 4	**155.** 4	**156.** 3	**157.** 2	**158.** 3	**159.** 4	**160.** 2
161. 1	**162.** 3	**163.** 1	**164.** 2	**165.** 1	**166.** 1	**167.** 2	**168.** 1	**169.** 2	**170.** 3
171. 3	**172.** 2	**173.** 1	**174.** 2	**175.** 3	**176.** 4	**177.** 4	**178.** 4	**179.** 1	**180.** 4
181. 3	**182.** 4	**183.** 2	**184.** 3	**185.** 1	**186.** 3	**187.** 4	**188.** 1	**189.** 2	**190.** 2
191. 2	**192.** 3	**193.** 1	**194.** 2	**195.** 2	**196.** 1	**197.** 3	**198.** 1	**199.** 4	**200.** 3

UNIT - 2

CELL BIOLOGY

1. What is the activated reactant in the biosynthesis of phosphatidylinositol from inositol?
   (1) CDP-ethanolamine
   (2) CDP-diacylglycerol
   (3) Geranyl-pyrophosphate
   (4) UDP-inositol

2. Best method in tracking the synthesis and maturation of a protein in a living cell is
   (1) Making a fusion protein with an appropriate epitope
   (2) Making a specific and high affinity monoclonal antibody
   (3) Making a specific and high affinity polyclonal antibody
   (4) Making a fusion protein with GFP

3. A signal sequence KDEL is removed from a ER resident protein. Assuming that there is no change in tertiary structure of protein and on other signal sequences present in protein, the changed protein will now have following fate
   (1) It will remain in ER and be degraded
   (2) It will be targeted to Golgi apparatus
   (3) It will be secreted outside the cell
   (4) It will be targeted to lysosome for degradation

4. Co-transport of nutrients across the intestinal cell membranes is an active process that can move glucose against a concentration gradient. The energy requiring step for co-transport involves
   (1) The Na^+K^+ ATPase that pumps Na+ from the cell into the lumen of the intestine
   (2) The permease that allows glucose and Na+ into the cell requires ATP
   (3) The permease that pumps glucose from the cell into the blood requires ATP
   (4) The Na^+K^+ ATPase that pumps Na+ from the cell into the blood, maintaining low Na^+ levels in the cell

5. A researcher made an interesting observation about a protein made by the rough ER and eventually used to build a cell's plasma membrane. The protein in the membrane was slightly different from the protein made in the ER. The protein was probably changed in the
   (1) Golgi apparatus
   (2) Smooth ER
   (3) Mitochondrion
   (4) Nucleus

6. Driving force for a membrane process is only the concentration difference in the case of
   (1) Reverse osmosis
   (2) Ultrafiltration
   (3) Microfiltration
   (4) Dialysis

7. Which of the following organelles is surrounded by a single membrane?
   (1) Chloroplast (2) Mitochondria
   (3) Peroxisomes (4) Nucleus

8. A C-terminal KDEL motif will most often ensure
(1) the protein to be folded by hsc70
(2) the protein to be degraded by the ubiquitin proteasome pathway
(3) secretion of the protein
(4) ER-retention of the protein

9. Mitochondria are involved in the following except
(1) ATP production
(2) Glycosylation
(3) Fatty acid catabolism
(4) TCA cycle

10. Mycoplasmas are bacterial cells that
(1) fail to reproduce in artificial media
(2) have a rigid cell wall
(3) are resistant to penicillin
(4) stain well with Gram's stain

11. Vinblastine, a chemotherapeutic agent, inhibits
(1) microtubule polymerization
(2) microtubule depolymerization
(3) spindle formation
(4) actin polarisation

12. Porins
(1) are cytoskeletal proteins
(2) form channels which allow passage of hydrophilic molecules
(3) are fatty acids
(4) are pores in the stem of a plant

13. Fluorescence recovery after photobleaching (FRAP) is a powerful technique for calculating
(1) the diffusion coefficient of membrane lipids and proteins
(2) the rate of synthesis of membrane proteins
(3) the distance between a lipid and a membrane protein
(4) the extent of signal transduction in membrane upon ligand-receptor interactions

14. An inhibitor of sodium dependent glucose transport across the plasma membrane is
(1) ouabain (2) sodium azide
(3) dicumarol (4) phlorhizin

15. Which of the following marine sources acts as a Na^+ channel blocker?
(1) Tetradotoxin (2) Conotoxin
(3) Carageneen (4) Acetyl choline

16. Sperm DNA is covered by
(1) lipids
(2) protamines
(3) carbohydrates
(4) histones

17. SRP selectively recognizes ER signal sequences on newly synthesized proteins. This is an outcome of
(1) hydrogen bonding
(2) hydrophobic interactions
(3) ionic interactions
(4) formation of covalent intermediate

18. Resting membrane potential of a biological membrane is close to the theoretical Nernst potential for the ions that are
(1) least abundant
(2) most abundant
(3) impermeable
(4) permeable

19. Cholesterol contributes to which of the following properties of biological membranes?
(1) Membrane rigidity
(2) Membrane fluidity
(3) Membrane permeability
(4) Membrane osmolarity

20. Which of the following properties is common to all cytoskeletal motor proteins like kinesins, dyneins and myosins?
(1) GTPase activity
(2) ATPase activity
(3) Actin binding domain
(4) DNA binding domain

21. Gaucher's disease is caused due to deficiency of
(1) glucocerebrosidase
(2) streptokinase
(3) uricase
(4) serratiopeptidase

22. A lectin can bind to plasma membrane vesicles if it is
(1) right side out
(2) inside out
(3) both right side out and inside out
(4) digested with protease

23. Membrane asymmetry can be regulated by
(1) ABC transporters
(2) Flippase
(3) Glycosylation of integral membrane proteins
(4) F1-ATPase

24. Which one of the following molecules has the highest diffusion coefficient in plasma membrane?
(1) F_0-F_1 ATPase
(2) Glycophorin
(3) ABC transporters
(4) Insulin receptor

25. Which one of the following is the correct statement for an active sodium-potassium ATPase?
(1) It pumps out 3 Na-ions and pumps in 2 K-ions
(2) It pumps out 3 Na-ions and pumps in 3 K-ions
(3) It pumps out 3 Ca-ions and pumps in 2 K-ions
(4) It pumps out 3 Na-ions and pumps in 2 Ca-ions

26. Information for which of the following functions is not encoded in the genome of virus?
(1) Replication of the genome
(2) Modulation of host defense
(3) Envelope membrane biosynthesis
(4) Assembly and packaging of the genome

27. Order glucose, urea, H_2O, CO_2 and steroid hormones according to their ability to diffuse through the lipid bilayer, beginning with the one that crosses the bilayer most readily
(1) CO_2, steroid hormone, H_2O, urea, glucose
(2) CO_2, H_2O, urea, glucose, steroid hormone
(3) H_2O, CO_2, glucose, urea, steroid hormone
(4) steroid hormone, CO_2, H_2O, urea, glucose

28. The typical length of an integral membrane protein domain (glycophorin fold) is
(1) 20-25 amino acids
(2) 10-15 amino acids
(3) 30-40 amino acids
(4) 5-10 amino acids

29. Fluid mosaic nature of a biological membrane can be proved by
(1) Patch clamp technology
(2) FRAP technique
(3) Electron spin resonance technique
(4) Cell-cell fusion technique

30. Complete fusion of two cells takes place only when
(1) two outer bilayers of the plasma membrane are fused
(2) two inner bilayers of the plasma membrane are fused
(3) both outer and inner bilayers of the plasma membrane are fused
(4) both outer and inner bilayers of the plasma membrane are fused and aqueous connection is established

31. Which one of the following is the ideal molecular marker of a mature lysosome?

(1) Glucose -6-P receptor
(2) Endoprotease
(3) Mannose-6-P receptor
(4) Endoglycosidase

32. Which of the following has the highest affinity of ligand-receptor interaction?

(1) Insulin-insulin receptor
(2) Lectin-sugar receptor
(3) Biotin-avidin
(4) Cytokine-cytokine receptor

33. Lateral diffusion rate of an individual membrane protein in a biological membrane generally depends on

(1) the extent of glycosylation
(2) the extent of phosphorylation
(3) the number of integral membrane domains
(4) the extent of its association with cholesterol

34. Maturation of endosome in a living cell is strictly dependent on

(1) appearance of specific glycolipids on its membrane
(2) appearance of specific phospholipids on its membrane
(3) acidification of its aqueous compartment
(4) kinetics of ligand receptor interaction on cell surface

35. Asymmetry of biological membranes is regulated by

(1) Phospholipase A
(2) Flippase
(3) Phospholipase C
(4) N-glycosidase

36. Which of the following statements concerning biological membrane is correct?

(1) Lipid rafts are fixed in position in membrane
(2) Lipid composition of two layers of the membrane equilibrate
(3) The membrane is most fluid at the surfaces
(4) Lipid transporters catalyze unidirectional movement of specific lipids from one layer to other

37. Signal recognition particle consists of

(1) a single large RNA
(2) multiprotein complex
(3) a complex of single RNA and multiple proteins
(4) a complex of multiple RNA and multiple proteins

38. Which of the following amino acids of a membrane protein are most likely to be buried in the interior of protein structure?

(1) Aspartic acid and phenylalanine
(2) Isoleucine and glutamic acid
(3) Aspartic acid and glutamic acid
(4) Methionine and Tyrosine

39. Twin membrane proteins vSNARE and tSNARE act as anchors when two vesicles fuse into one during molecular transport in the cell. These snare proteins are found in membranes of all except

(1) mitochondria
(2) golgi complex
(3) early endosome
(4) endoplasmic reticulum

40. Which of the following is not a function of molecular chaperon in protein folding?

(1) Molecular chaperones assist in protein folding into their correct structure
(2) Molecular chaperones specify the tertiary structure of a protein
(3) Molecular chaperones can stabilize partially folded proteins and prevent them aggregating with other proteins
(4) Molecular chaperones can shield an protect exposed hydrophobic regions of proteins

41. The true source of tetradotoxin production reported in puffer fish, porcupine fish and polyclad flat worm is attributed to
(1) Small cone snails attached to puffer fish
(2) Presence of *Microbulbifer elongates* in digestive tract
(3) Presence of *Vibrio alginolyticus*
(4) *Stachybotrys chartarum* a known pathogen infecting above organisms

42. The neurotoxin that is produced by marine dianoflagellates and is responsible for paralytic shell fish poisoning in humans by acting on voltage-gated sodium channels of nerve cell is
(1) Shigatoxin (2) Saxitoxin
(3) Amatoxin (4) Gliotoxin

43. A signal sequence KDEL is added to cytosolic protein. Assuming that there is no change in tertiary structure of such plant protein and no other signal sequences present in protein, the changed protein will now have following fate
(1) It will remain in ER and be degraded
(2) It will be targeted to Golgi apparatus
(3) It will remain in cytosol
(4) It will be targeted to lysosome for degradation

44. Glycosylation of proteins occurs in the
(1) peroxisome
(2) Mitochondrion
(3) lysosome
(4) endoplasmic reticulum

45. Which of the following is not found inside the eukaryotic nucleus?
(1) Nucleolus
(2) Cajal bodies
(3) PML bodies
(4) Centrosomes

46. Cystic Fibrosis Transmembrane Regulator (CFTR) Protein is a
(1) sodium transporter
(2) calcium antiporter
(3) chloride transporter
(4) potassium transporter

47. The observation that plasma membrane proteins mix after cell fusion provides evidence for
(1) rotational movement of plasma membrane proteins
(2) the bilayer structure of biomolecules
(3) the fluid mosaic model
(4) interactions of plasma membrane proteins of two different cell types

48. SNARE proteins are found in the membranes of all of the following compartments except
(1) mitochondria
(2) golgi complex
(3) early endosome
(4) endoplasmic reticulum

49. Satellite DNAs are not typically found within which of the following parts of a chromosome?
(1) Heterochromatin
(2) Euchromatin
(3) Telomeres
(4) Centromeres

50. Which one of the following is an incorrect association?
(1) Lysosome : synthesizes molecules for extracellular protein degradation
(2) Mitochondria : cellular respiration
(3) Endoplasmic reticulum: synthesizes proteins and sends them into Golgi-apparatus
(4) Polysomes : make large quantities of a particular protein

51. The function of molecular chaperones is
(1) To escort a protein to different organelles in the cell such as the nucleus
(2) To assist in protein folding
(3) To phosphorylate proteins with ATP
(4) To participate in the targeting of ubiquitin

52. Which of the following statements about hsp-70 is false?

(1) hsp-70 is an enzyme that hydrolizes ATP

(2) hsp-70 binds to nascent proteins

(3) hsp-70 delivers proteins to the Proteasome

(4) hsp-70 binds to patches of hydrophobic amino acids on denatured proteins

53. A major difference between hsp-60 and HSP-70 is that

(1) hps-60 is monomeric while hsp-70 is multimeric

(2) hsp-60 is a protease while hsp-70 is a chaperone

(3) The primary structure of hsp-60 is larger than that of hsp-70

(4) hsp-60 has two folding chambers while hsp-70 does not

54. Which cell have have routinely used to understand structure of plasma membrane?

(1) Neurons

(2) Mylein cells

(3) RBC

(4) Stem cells

55. Which of the following statements about Prokaryotes is true?

(1) Prokaryotes lack a nucleus and therefore lack DNA

(2) Prokaryotes form proteins from 22 different amino acids

(3) Prokaryotes include bacteria, yeast, and protozoa

(4) Prokaryotes lack a Golgi apparatus

56. A protein can be unfolded into an extended conformation by

(1) the addition of ubiquitin onto cysteine amino acids on the protein

(2) the addition of ubiquitin onto arginine amino acids on the protein

(3) the binding of HSP60 onto the protein

(4) the binding of HSP70 onto the protein

57. A bottle containing yeast cells that are fermenting maltose are taken out of a warm room and put in the cold refrigerator. What adjustments might the yeast cells make to their plasma membrane to maintain the same level of fluidity?

(1) Decrease the proportion of glycolipids in their membrane

(2) Decrease the proportion of cholesterol in their membrane

(3) Increase the length of hydrocarbon tails in its membrane phospholipids

(4) Increase the proportion of unsaturated hydrocarbon tails in their membrane phospholipids

58. Bilayers in membranes form primarily because of

(1) Double bonds found in fatty acid tails of phospholipids

(2) Ionic interactions between the polar head groups of phospholipids

(3) Hydrophobic interactions between fatty acid tails of phospholipids

(4) The amphipathic properties of transmembrane proteins

59. You are studying a Protein X on the plasma membrane. You find that this protein can be extracted only with buffer containing SDS from the membrane, and not with a buffer containing high salt. You identify a sequence on the carboxyl-terminus of the protein, -K-K-N-K-K-A-A-C, that is essential for the attachment of Protein X to the membrane. What is the most likely way that this sequence atttaches Protein X to the membrane?

(1) The cysteine in this peptide is covalently attached to a membrane lipid

(2) The peptide is part of an alpha-helix that spans the membrane

(3) The positively charged lysines interact with negatively charged integral membrane proteins

(4) The peptide is part of a peripheral membrane protein that interacts with the membrane by ionic interactions

60. Which of the following statements about peripheral membrane proteins (PMP's) is false

(1) PMP's can be extracted from membranes with buffers containing detergent

(2) PMP's can be extracted from membranes with high salt buffers

(3) PMP's can bind membranes via noncovalent interactions with integral membrane proteins

(4) PMP's can bind membranes via covalent bonds with a fatty acid tail

61. Membrane rafts are enriched in

(1) cholesterol and sphingolipids

(2) cholesterol and integral membrane proteins

(3) glycolipids and phospholipids

(4) sphingolipids and phospholipids

62. Considering the radius of potassium (K^+) and sodium (Na^+) ions in their hydrated and dehydrated states, which is the correct order for their relative size?

(1) Hydrated Na^+ > Hydrated K^+ > dehydrated Na^+ > dehydrated K^+

(2) Hydrated K^+ > dehydrated K^+ > Hydrated Na^+ > dehydrated Na^+

(3) Hydrated K^+ > Hydrated Na^+ > dehydrated K^+ > dehydrated Na^+

(4) Hydrated Na^+ > dehydrated Na^+ > Hydrated K^+ > dehydrated K^+

63. Which of the following statements about ion channels is false?

(1) they are integral membrane proteins

(2) they use ATP to transport ions across membranes

(3) they can open and close by mechanical stimulation

(4) they can open and close by changes in membrane potential

64. Which of the following statements about the N^+/K^+ pump is false?

(1) the pump binds 3 Na^+ ions in the inside of the cell and releases them outside of the cell

(2) the pump binds 2 K^+ ions from the outside of the cell and releases them inside of the cell

(3) the pump uses the energy of ATP to move ions against their gradients

(4) the pump hydrolyses ATP and retains ADP while releasing P_i

65. Which of the following statements about Secondary Transport is false?

(1) secondary transport is accomplished by symport carriers

(2) secondary transport is accomplished by antiport carriers

(3) secondary transport involves the movement of ions against their gradient using the energy of ATP

(4) secondary transport involves the movement of a molecule against its own concentration gradient using the energy of a previously established gradient of a second molecule

66. Which of the following statements about Nucleoporins is false?

(1) nucleoporins are just one of many subunits of the nuclear pore complex

(2) nucleoporins contain multiple copies of the amino acids Phenylalanine and Glycine

(3) nucleoporins form a hydrogel that blocks the passage of large cytoplasmic proteins into the nucleus

(4) nucleoporins give shape to the nucleus by forming a structure called the nuclear lamina

67. Which of the following statements in regard to the nuclear localization signal (NLS) is false?

(1) the NLS can be found at the beginning, middle and end of a protein sequence

(2) the NLS is covalently attached to a protein by the import receptor

(3) the NLS is recognized by the import receptor

(4) ranGTP does not bind the NLS

68. Which of the following does not occur during mitosis?
(1) Condensation of the chromosomes
(2) Replication of the DNA
(3) Separation of sister chromatids
(4) Spindle formation

69. A particular cell has half as much DNA as some other cells in a mitotically active tissue. The cell in question is most likely in
(1) G_1.   (2) G_2
(3) prophase.   (4) anaphase

70. Most common type of phospholipid in the cell membrane of nerve cells is
(1) phosphatidylcholine
(2) phosphatidylinositol
(3) phoshphatidylserine
(4) Sphingomylein

71. Which of the following statement is correct for archeabacteria adapted to higher temperature condition?
(1) Membrane have high proportion of unsaturated fatty acids
(2) Cell wall is composed of peptidoglycans
(3) Genome is GC rich
(4) Proteins have high methione and low cysteine content

72. Cholesterol contributes rigidity in biological membranes of ?
(1) Yeast   (2) *Mycoplasma*
(3) *E. coli*   (4) Human

73. Which of the following is not found inside the eukaryotic nucleus?
(1) Nucleolus   (2) Chromatin
(3) Solenoid   (4) Chromatids

74. The number of nucleosomes present in a 30 nm solenoid structure of a chromatin is
(1) 2   (2) 4
(3) 6   (4) 8

75. Sister chromatids are separated during which stage of meiosis?
(1) Metaphase I   (2) Metaphase II
(3) Anaphase I   (4) Anaphase II

76. Polyribosomes (polysomes) present in the cytosol
(1) can make plasma membrane integral membrane proteins
(2) can synthesize Golgi apparatus integral membrane proteins
(3) can synthesize proteins that are secreted in a regulated manner
(4) can synthesize mitochondrial proteins

77. Ten bacteria were inoculated into a rich medium. If at the end of ten hours the total number of cells is 10^4, then the number of elapsed generations and the generation time respectively is
(1) 10, 120 minutes
(2) 10, 60 minutes
(3) 20, 30 minutes
(4) 40, 15 minutes

78. Which of the following statement is not true?
(1) Mer B is a bacterial protein similar to eukaryotic actin
(2) FtsZ protein found in most of the bacteria is a tubulin homologue
(3) AmiC hydrolyzes peptidoglycan to separate daughter cells
(4) FtsZ is an inhibitor of Z ring assembly

79. Which one of the following is common to both chloroplast and mitochondria?
(1) Oxidative phosphorylation
(2) Electron donor – acceptor pair
(3) Photophosphorylation
(4) Independent genomes

80. The decline of MPF activity at the end of mitosis is due to
(1) the destruction of the protein kinase Cdk
(2) decreased synthesis of Cdk
(3) the degradation of cyclin
(4) the accumulation of cyclin

81. In the cells of some organisms, mitosis occurs without cytokinesis. This will result in
(1) cells with more than one nucleus
(2) cells that are unusually small
(3) cells lacking nuclei
(4) destruction of chromosomes

82. Which is not a correct statement about chloroplasts?
(1) Grana and stroma are membrane components of the thylakoids
(2) The membranes that are staked in the thylakoid are called stroma
(2) Photo-harvesting apparatus are embedded in the thylakoid membrane
(4) Like mitochondrial membrane, chloroplast membranes are also capable of generating ATP using proton motif force

83. Some studies indicated that diffusion directly across the lipid bilayer was not sufficient to account for observed. Rates of water movement across membranes, but the evidence in support of microscopic pores was not compelling. This uncertainty was put to rest with the recent discovery of
(1) P-type pumps
(2) ABC transporter
(3) Aquaporins
(4) Reverse osmosis

84. Which one of the following events does not take place in the lumen of the endoplasmic reticulum?
(1) N-linked glycosylation
(2) Folding of the polypeptide chain
(3) Disulphide bond formation and assembly of subunits
(4) Phospholipid synthesis

85. The ability of very small amounts of protease to release protein fragments from preparations of rough ER was critical in the discovery of which of the following?
(1) Mannose 6-P receptor
(2) Signal peptide
(3) mRNA
(4) SRP receptor

86. Permanent insertion of integral proteins in membranes can not be mediated by
(1) a stop-transfer sequence
(2) a single membrane-spanning α-helical domain
(3) a GPI membrane anchor
(4) the Asn-X-Ser/Thr sequence

87. During protein modification in the secretory pathway, proteins move through organelles in the following order
(1) mitochondria → peroxisome → lysosome
(2) chloroplast → peroxisome → lysosome
(3) mitochondria → chloroplast → nucleus
(4) ER → Golgi complex → plasma membrane

88. The protein ______ is found in the nucleus and it causes the ____________
(1) RanGEF, activation of Ran
(2) RanGEF, inactivation of Ran
(3) RanGAP, inactivation of Ran
(4) RanGAP, activation of Ran

89. An important similarity between nuclear import and nuclear export is that
(1) RanGTP binds tightly to both the import receptor and the export receptor
(2) They are both used to transport mRNA molecules across the nuclear envelope
(3) They both use RanGAP to dissociate their transport complexes
(4) They both use RanGEF to dissociate their transport complexes

90. Which of the following statements about Ran is correct?
(1) Ran can bind to GTP but not to GDP
(2) Ran can convert GDP to GTP by phosphorylation of GDP
(3) Ran can bind to the nuclear export signal on a protein
(4) Ran can diffuse across the nuclear pore

91. Which of the following statements do you think is the most likely direct consequence of removing the protein Porin from a cell?
(1) the pH of the endoplasmic reticulum would increase
(2) the pH of the lysosome would decrease
(3) the pH of the mitochondrial matrix would increase
(4) the pH of the mitochondrial intermembrane space would decrease

92. Which of the following statements about the Electron Transport Chain (ETC) is false?
(1) the proteins of the ETC are integral membrane proteins
(2) the ETC uses the energy of the proton motive force to produce ATP
(3) the proteins of the ETC are found in the inner mitochondrial membrane
(4) the ETC consumes molecular oxygen to produce water

93. A difference between the chaperones found in the cytosol and those found in the mitochondria is that
(1) the cytosol has both HSP-70 & HSP-60 but the mitochondria has only HSP-70
(2) mitochondrial HSP-70 is synthesized in the matrix while cytosolic HSP-70 is synthesized in the cytosol
(3) the mRNA for mitochondrial HSP-70 encodes a mitochondrial targeting signal but the mRNA for cytosolic HSP-70 does not
(4) cytosolic HSP-70 has the ability to fold proteins but the mitochondrial HSP-70 does not and Cytosolic HSP-70 uses the energy of ATP while mitochondrial HSP-70 uses the energy of the proton motive force

94. Which of the following statements about the TIM-23 complex is correct?
(1) the TIM-23 complex is used to translocate proteins across the mitochondrial outer membrane
(2) the TIM-23 complex uses the energy of ATP to translocate proteins into the mitochondrial matrix
(3) the TIM-23 complex has membrane spanning segments in both inner and outer mitochondrial membranes
(4) the TIM-23 complex is used to insert multi-pass transmembrane proteins into the inner mitochondrial membrane

95. The mitochondrial targeting signal (MTS) is used to
(1) transport the amino terminal portion of a protein into the mitochondrial matrix
(2) bind to the Signal Recognition Particle (SRP) and stop translation by the ribosome
(3) bind to the TOM receptor and stop translation by the ribosome
(4) bind to the TIM-23 receptor and transport the protein across the outer mitochondrial membrane

96. ATP Synthase can bind to all of the following ions or molecules except
(1) Protons (2) $NADH^+$
(3) ADP (4) ATP

97. After isolating the rough endoplasmic reticulum from a cell, you purify the RNA molecules that are attached to it. You could expect these RNAs to encode all of the following proteins except
(1) mitochondrial membrane proteins
(2) ER membrane proteins
(3) extracellular proteins
(4) proteins that contain the signal K-D-E-L at their carboxyl-terminus

98. Which of the following statements is true about the Signal Recognition Particle
(1) SRP is found in the ER lumen
(2) SRP is required for the translation of cytosolic proteins
(3) SRP is a complex of polypeptides and a small RNA molecule
(4) SRP binds to Stop-transfer sequences

99. If you looked at the sequence of mRNAs encoding the following proteins, which of the following would not begin with an ER signal sequence?
(1) Immunoglobulin proteins that are secreted by white blood cells
(2) Glycosyltransferases responsible for O-linked glycosylation
(3) PDI (protein disulfide isomerase)
(4) Actin

100. A protein has a Start-transfer sequence, and no other membrane-trafficking signals. What will be the final destination for this protein?
(1) The extracellular matrix
(2) The plasma membrane
(3) The lysosome
(4) The cytosol

101. Consider the sequence of events that direct a secretory protein into the Endoplasmic Reticulum. In the following list, which of the following is the last event in this sequence of reactions?
(1) SRP binds the SRP receptor
(2) The translocon binds the ER signal sequence
(3) The translocon binds the SRP receptor
(4) The ribosome binds the 5' end of the mRNA encoding the secretory protein

102. Which of the following statements about clathrin coated vesicles is false?
(1) Clathrin is a protein that is found in the cytosol
(2) Clathrin coats assemble on the TGN
(3) Clathrin coats assemble on the plasma membrane
(4) The disassembly of the clathrin coat requires NSF and clathrin coated vesicles transport proteins from the ER to the cis-Golgi

103. Which of the following statements about a vesicle that came from the ER and fused with the cis-Golgi membrane is true?
(1) The protein NSF is needed at the step where the vesicle docks to the target membrane
(2) Fusion of the vesicle requires the correct Rab protein bound to GDP
(3) The vesicle must have the right T-snare and the cis-Golgi must have the right V-snare
(4) The COPII coat must come off the vesicle before the vesicle can fuse

104. A transmembrane protein is glycosylated i the ER. Which of the following statement about this glycoprotein is false?
(1) During the ER glycosylation, a complex carbohydrate was added to the amino acid Asparagine
(2) More sugars were added and removed as the protein passed through the Golgi stacks
(3) When the protein reached the plasma membrane, the carbohydrate was on the cytoplasmic tail of the protein
(4) During the ER glycosylation reaction, a complex carbohydrate was transferred from dolicol to the protein

105. What is O-linked glycosylation?
(1) The modification of carbohydrates that are attached to Asparagines in the Golgi
(2) The phosphorylation of mannoses in the Golgi
(3) The addition of carbohydrates to serines or threonines in the Golgi
(4) The phosphorylation of carbohydrates in the ER

106. Which of the following statements about th proteins AP-1 and AP-2 is false?
(1) AP-1 associates with membranes on the TGN
(2) AP-2 must hydrolyze GTP to regulate the assembly of clathrin triskelions into coat
(3) AP-1 and AP-2 stimulate the assembly of clathrin triskelions into a lattice
(4) AP-2 can bind to a signal on the cytoplasmic tail of LDL receptors

107. Which of the following vesicles or organelles would not contain Man-6-P receptors?

(1) COPI coated vesicles

(2) COPII coated vesicles

(3) Clathrin coated vesicles

(4) Golgi Apparatus

108. Which of the following statements about telomeres is false?

(1) Telomeres are synthesized using an RNA template

(2) Telomeres consist of repeated sequences at the ends of eukaryotic chromosomes

(3) Telomerase uses an RNA primer to synthesize the telomeres

(4) On a telomere, one DNA strand is longer than the other

109. The nature of membrane lipids is

(1) Zwitterion (2) Hydrophobic

(3) Amphipathic (4) Uncharged

110. Cells that produce large amounts of secretary proteins are abundant in

(1) mitochondria

(2) lysosomes

(3) smooth endoplasmic reticulum

(4) rough endoplasmic reticulum

111. Which of the following contain enzymes that detoxify xenobiotics?

(1) rough endoplasmic reticulum

(2) endosomes

(3) lysosomes (4) peroxisomes

112. An unknown membrane protein X can be extracted from disrupted RBC membranes into a concentrated salt solution. Isolated X can be cleaved into fragments by proteolytic enzymes. But treatment of RBC, first with proteolytic enzymes, followed by disruption and extraction of membrane component, yields intact X. In contrast, treatment of RBC ghosts (which consist of only membranes) with proteolytic enzymes, followed by disruption and extraction, yields extensively fragmented X. On the basis of this information, these properties resemble those of

(1) ankyrin (2) glycophorin

(3) band 3 (4) haemoglobin

113. Which of the statements regarding Cdc6/Cdt1 and origin recognition complex is true?

(1) In late G1 phase Cdc6 and Cdt1 bind to origin recognition complex

(2) In early mitotic phase Cdt1 and Cdc6 bind to origin recognition complex

(3) In early S phase Cdt 1 and Cdc6 binds to origin recognition complex

(4) In early G1 phase Cdt1 and Cdc6 bind to origin recognition complex

114. Mitochondria were isolated and permeabilized with a detergent. They were incubated with succinate, ADP and Pi in the presence of O_2. ATP synthesis and O_2 consumption were measured. Which one of the following statements is true?

(1) ATP synthesis will occur but O_2 consumption will not occur

(2) O_2 consumption will occur but ATP synthesis will not occur

(3) Both O_2 consumption and ATP synthesis will occur

(4) Both O_2 synthesis and ATP synthesis will not occur

115. Which one of the following is not com-mon to both chloroplast and mitochondria?

(1) Electron transport

(2) Electron donor – acceptor pair

(3) Proton pump

(4) Independent genome

116. Which of the following statements about KDEL receptors is false?

(1) KDEL receptors can return proteins from the cytosol to the ER lumen

(2) KDEL receptors can be found in the membrane of COPII coated vesicles

(3) KDEL receptors can be found in the membrane of COPI coated vesicles

(4) KDEL receptors can bind the hsp70 chaperone, BiP

117. Which of the following statements about the active ingredient of BOTOX, the botulinum toxin is true

(1) The botulinum toxin blocks the formation of secretory vesicles
(2) The botulinum toxin requires GTP to be active
(3) The botulinum toxin inhibits the formation of clathrin coated vesicles and it stimulates secretion along the constitutive secretory pathway
(4) The botulinum toxin is a protease that breaks peptide bonds within t-snares

118. Which of the following trafficking steps does not require a signal on the cargo and/or receptor?

(1) Packaging plasma membrane receptors into clathrin coated vesicles
(2) Packaging acid hydrolases into clathrin coated vesicles
(3) Packaging secretory proteins into COPII coated vesicles
(4) Packaging BIP and PDI into COPI coated vesicles and packaging Man-6-P receptors into clathrin coated vesicles

119. Which of the following is a true statement regarding p53 tumor suppressor gene?

(1) inherited mutations are the commonest cause of loss of function
(2) p53 does not have a role in the regulation of the expression of genes involved in the control of apoptosis
(3) p53 is activated when DNA damage occurs
(4) p53 phosphorylation results in loss of its ability in blocking entry of the cell into S-phase

120. You add a drug that changes the pH compartments in the secretory pathway so that all are now neutral. Which of the following events will happen?

(1) KDEL receptors will be unable to release PDI or BiP into the ER lumen
(2) Man-6-P receptors will be unable to fold properly in the TGN
(3) Lysosomal enzymes will be activated in the lysosome
(4) Lysosomal enzymes will be secreted from the cell

121. You mutated an acid hydrolase so that you added an extra sequence that encoded a nuclear localization signal at the c-terminus. What would be the final destination for the mutant acid hydrolase?

(1) In the cytosol
(2) In the extracellular space
(3) In the ER
(4) In the lysosome

122. The JD mutation of the LDL receptor is associated with high levels of LDL outside the cell. This mutation changes a single tyrosine amino acid in the LDL receptor tail. This particular mutation results in

(1) Failure of the mutant LDL receptor to bind LDL on the surface of the cell
(2) Failure of the mutant LDL receptor to traffic from the trans-Golgi to the plasma membrane
(3) Failure of the mutant LDL receptor to enter clathrin coated vesicles on the plasma membrane
(4) Failure of the mutant LDL receptor to traffic from the ER to the cis-Golgi network

123. Regarding COP I and COP II coated vesicles, which statement is false

(1) Both of these vesicles include small GTPase proteins in their coats
(2) Both of these vesicles require signal sequences for their cargo proteins
(3) Both of these vesicles transport KDEL receptors
(4) Both vesicles are surrounded by peripheral membrane proteins

124. Regarding rab proteins, which of the following statements is false

(1) Rab proteins catalyze this reaction: GTP → GDP + Pi

(2) Cells contain many different kinds of rab proteins

(3) Rab proteins are important for fusing the correct vesicle with its appropriate target membrane and are associated with both the endocytic & with secretory pathways

(4) Rab proteins are important for the assembly of "coat proteins" onto a vesicle

125. You have isolated a mutant cell that completely lacks AP-2. Which of the following phenotypes would you see in this mutant cell?

(1) The cell would not be able to transport Man-6-P receptors from the TGN

(2) The cell would not be able to endocytose LDL receptors from the plasma membrane

(3) The cell would not have any clathrin coated vesicles

(4)The cell would not have any COPII and COP I coated vesicles

126. Which of the following is not a property of COPII coat proteins?

(1) COPII proteins function to bend and curve vesicles from the ER

(2) COPII proteins are required for fusion of the vesicle with the cis-Golgi network

(3) Receptors do not require a signal to enter COPII coated vesicles

(4) The COPII coat also contains Sar1p and can be found in the cytosol AND bound to membranes

127. Which of the following vesicles or compartments is LEAST likely to contain glycoproteins?

(1) Cytosol

(2) Golgi Apparatus

(3) Integral Membrane Proteins on the Plasma Membrane

(4) Proteins in the Extracellular Space and Endoplasmic Reticulum

128. Which of the following statements is true in regard to properties of the cytoskeleton

(1) Cytoskeletal filaments can have one or more protofilaments

(2) The three major cytoskeletal filaments are each made from identical subunits

(3) The three major cytoskeletal filaments can coexist in the same cell

(4) The three major cytoskeletal systems consist of filaments with polarity

129. Which of the following statements describes what happens to a microtubule after loosing its GTP cap?

(1) Tubulin dimers dissociate from the microtubule and then alpha-tubulin dissociates from beta-tubulin

(2) Protofilaments dissociate from the microtubule end, pulling away from each other like a pealing banana

(3) Tubulin dimers dissociate from the microtubule and then alpha-tubulin exchanges GDP for GTP

(4) EB1 dissociates from the microtu-bule, exposing the microtubule's seam

130. Which of the following statements about centrosomes is false?

(1) Centrosomes duplicate to form the poles from which the mitotic spindle emerges

(2) The centrosome is also known as the microtubule organizing center

(3) Centrosomes contain centrioles which are surrounded by pericentriolar material

(4) Microtubules grow from the centrio-les contained within the centrosome

131. The γ-tubulin ring complex is

(1) a protein complex made of alpha, beta and gamma tubulins

(2) responsible for the formation of centrioles

(3) a protein complex that caps the plus ends of microtubules

(4) responsible for the nucleation of new microtubules

132. Which of the following statements about dyneins is false?

(1) Cytoplasmic dyneins are large proteins with two globular, force-generating heads
(2) Dyneins are associated with cilia and flagella but are also found in the cytoplasm
(3) Dyneins are used to move the ER membranes to the cell periphery
(4) Dyneins are involved in chromosomal movement during mitosis and are used to move vesicles in the axon toward the cell body

133. You observe the behavior of two microtubules sitting side by side on a glass slide. One microtubule is growing at its plus end and the other is shrinking from the plus end. The best explanation for this observation is that

(1) The growing microtubule has a GDP cap and the shrinking microtubule has lost its GDP cap
(2) The growing microtubule is surrounded by a higher concentration of tubulin dimers than the shrinking microtubule
(3) The growing microtubule has a lower critical concentration for assembly than the shrinking microtubule
(4) Some dimers on the growing microtubule have not hydrolyzed their GTP while all dimers on the shrinking microtubule have hydrolyzed their GTP

134. All of the following are properties of actin filaments except

(1) Actin filaments are flexible and have polarity
(2) Actin filaments act as tracts for different motors
(3) Actin filaments can be bundled together
(4) Actin filaments display dynamic instability

135. Which of the following statements about the organization of the sarcomere is false?

(1) The actin filaments are capped at both ends
(2) The protein Nebulin serves as a ruler for the actin filaments
(3) The actin filaments are interdigitated with myosin II filaments
(4) The myosin II filaments are anchored in place by the protein Tropomodulin

136. Muscle contraction is induced by high levels of calcium, these ions bind to__________ causing _________ to shift closer to the center of the actin filament groove

(1) tropomyosin, tropomodulin
(2) troponin, tropomyosin
(3) troponin, tropomodulin
(4) myosin, actin microfilaments

137. Which of the following statements best describes what happens during muscle contraction?

(1) The distance between Z-disks increases
(2) The number of sarcomeres in a muscle cell is reduced
(3) The number of sarcomeres in a muscle cell increases
(4) The size of the I band is reduced

138. Kinesin I moves vesicles

(1) from the microtubule (+) end to the (-) end
(2) from the microtubule (-) end to the (+) end
(3) bidirectionally along the same microtubule
(4) along both microtubules and microfilaments

139. Treatment of quiescent cells in G0 with a drug that blocks phosphorylation of lamins, and hence keeps them in a dephosphorylated state, would

(1) have no effect on the structure of the nucleus
(2) would prevent reformation of the nucleus
(3) would prevent nuclear pores from associating with the nuclear membrane
(4) would disrupt nucleolar structure

140. Which of the cell junctions listed below involves adhesion to laminin?
(1) tight junction
(2) adherens junction
(3) hemi-desmosome
(4) desmosome

141. Which of the following pairs is INCORRECT?
(1) cristae - inner mitochondrial membrane
(2) nucleotide triphosphates - intermembrane space
(3) porin-inner mitochondrial membrane
(4) ATP synthetase - inner mitochondrial membrane

142. Which of the following statements about the glycosylation of a protein is true?
(1) requires the presence of mannose-6-phosphate
(2) occurs in the condensing vesicles of the late endosome
(3) occurs on tyrosine residues
(4) begins in the rough endoplasmic reticulum

143. Which of the following is not true for archebacteria?
(1) Membrane lipids have branch chain fatty acids
(2) Membrane lipids have ether linked fatty acids
(3) Cell wall is composed of peptidoglycan
(4) They are more similar to eukaryotes and originated from eubacteria

144. At which points during the cell cycle is the nuclear envelope broken down and then reassembled?
(1) broken down during prophase, reassembled in telophase
(2) broken down during prometaphase, reassembled in anaphase
(3) broken down during G2, reassembled in G1
(4) broken down during prophase, reassembled in metaphase

145. Colchicine treatment of cells is used for karyotype analyses. Its mechanism of action is to depolymerize spindle microtubules and thus
(1) arrest mitosis during the relatively long period of prophase
(2) prevent inactivation of MPF, and hence block mitosis in metaphase
(3) arrest mitosis during cytokinesis
(4) activate MPF, and hence allow all the cells to leave interphase and enter mitosis

146. If an organism has a diploid number of 8, how many CHROMATIDS are present during prophase (beginning) of MITOSIS?
(1) 4
(2) 6
(3) 8
(4) 16

147. Which of the following-proteins is not found on shuttle vesicles that transport proteins from the cis to the medial Golgi apparatus?
(1) sv-SNARE
(2) ARF
(3) clathrin
(4) Rab

148. The signal recognition partical (SRP)
(1) binds to ribophorin I and II
(2) directs lysosomal enzymes to endosomes
(3) binds to a receptor that is an integral protein in the rough endoplasmic reticulum
(4) directs nuclear proteins into the nucleus

149. Which of the following molecules binds in a heterophilic manner?
(1) cadherin
(2) N-CAM
(3) selectin
(4) desmoglein

150. Which of the following is not a function performed by the Golgi apparatus?
(1) cleavage of sugars from the N-linked precursor oligosaccharide
(2) addition of sugars to the N-linked precursor oligosaccharide
(3) addition of sialic acid to N-linked oligosaccharides in the cis Golgi
(4) addition of sulfate and phosphate modifications to proteins

151. Which of the following statements about Condensins is CORRECT?
(1) Condensins are ubiquitinated by SCF
(2) Condensins are ubiquitinated by the APC
(3) Condensins are phosphorylated by M-Cdk
(4) Condensins are phosphorylated by S-Cdk

152. Which of the following statements about MPF is CORRECT?
(1) MPF is activated at the end of G1
(2) MPF is degraded at the end of mitosis
(3) Activation of MPF leads to fragmentation of the nuclear envelope
(4) Activation of MPF leads to the assembly of the nuclear lamina

153. Centriole duplication is
(1) A process that occurs in the presence of S-Cdk
(2) A process that occurs in G1 phase
(3) Initiated by the activation of MPF
(4) Caused by the inactivation of MPF

154. Which of the following statements about Cohesins is false?
(1) Cohesins are ubiquitinated by APC-Cdc20
(2) Cohesins are bound to DNA during Prophase
(3) Cohesins are bound to DNA during S and G2
(4) Cohesins are bound to DNA during Anaphase

155. Chromokinesins are
(1) motors found on the kinetochores to move chromosomes toward the poles
(2) motors found on the kinetochores to move chromosomes toward the middle of the spindle
(3) motors found on the middle of the spindle to push the poles apart
(4) motors found on chromosome arms to move them toward the middle of the spindle

156. Which of the following is not a property of the kinetochores?
(1) Kinetochores cap the kinetochore microtubules
(2) Kinetochores can move toward the minus ends of microtubules
(3) Kinetochores can move toward the plus ends of microtubules
(4) Signals emanating from the kinetochores can stop the cell cycle

157. Which of the following events do not occur during cytokinesis?
(1) Actin and myosin II form a contractile ring
(2) Myosin II filaments are assembled at the equator of the cell
(3) Cytokinesis occurs at places where microtubules of opposite polarity interact with each other
(4) Microtubules pull on the membrane at the equator to form the cleavage furrow

158. If you inhibited the activity of S-Cdk in a cell near the end of G1 you would expect to observe the following
(1) DNA replication would start prematurely
(2) The origin recognition complex (ORC) would not bind to DNA
(3) MCM proteins would not bind to the Pre-Replication complex
(4) The protein Cdc6 would not get degraded

159. What would happen if you were to stop the activity of Cdc25 at the end of G2?

(1) The chromosomes would begin to decondense

(2) MPF would remain inactive

(3) Weel protein would become inactivated

(4) Synthesis of M-cyclin would stop and APC-cdc20 complex would get activated

160. Which of the following steps is involved in the regulation of S-cyclin synthesis?

(1) The protein Rb binds to the protein E2F to stimulate S-cyclin synthesis

(2) The protein Rb gets phosphorylated by M-Cdk

(3) The protein E2F gets phosphorylated by G1-Cdk

(4) The protein E2F binds to the promoter of the G1/S cyclin gene

161. What do you think would happen if you made a mutant form of M-cyclin that cannot be ubiquitinated?

(1) Nuclear Lamins would remain phosphorylated

(2) The cell would enter Anaphase prematurely

(3) The cell would not be able to enter into Mitosis

(4) The spindle would start to dismantle and the sister chromatids would not be able to separate

162. What would the most likely consequence of inhibiting the APC/cdc20 complex before Metaphase?

(1) The chromosomes would start decondensing

(2) The sister chromatids would not be able to separate

(3) The cell would enter Anaphase prematurely

(4) The spindle would start to elongate and cohesins would get degraded

163. You treat a growing cell with the drug Caffeine. What will happen if the cell does not have any DNA damage?

(1) The cell will enter mitosis normally

(2) The cell will arrest at metaphase

(3) The cell will stop DNA synthesis

(4) The cell will arrest at the G2/M transition

164. Levels of cyclin-dependent kinase (Cdk) activity change during the cell cycle as a result of

(1) the Cdks phosphorylating each other

(2) the Cdks activating the cyclins

(3) the level of production of Cdks changing throughout the cell cycle

(4) the Cdks binding different cyclins to become active

165. M-phase promoting factor

(1) is a kinase

(2) contains a kinase

(3) was purified from extracts of frog oocytes in G2 phase

(4) is found only in Xenopus and acts to initiate both S phase and M phase

166. The concentration of mitotic cyclin

(1) rises dramatically during M phase

(2) changes as Cdk concentration changes

(3) falls during M phase due to ubiquitination

(4) is highest in G1 phase

167. Cells in the G_0 state

(1) cannot reenter the cell cycle

(2) can remain in that state for a lifetime

(3) have entered from G1 or G2 checkpoints

(4) do not divide even when conditions are favourable

168. In proliferating cells

(1) Rb protein is active

(2) nutrients are limiting

(3) growth factors secreted by other cells are bound to receptors

(4) cyclin-Cdk complexes are inactive

169. If cells isolated from an animal become senescent after a limited number of divisions in culture, how are the cell lines used in experimental cell biology able to proliferate indefinitely in culture? Explain your answer

(1) They are continually given fresh supplies of growth factor

(2) They have undergone mutations that inactivate the genes involved in programmed cell death

(3) They have undergone mutations that abolish cell senescence

(4) They have become differentiated

170. Heyflick limit after 50-60 cell generation time. However, in a rare situation or employing certain agents, cells can be induced to immortality. Which of the following genes/ proteins have a major relationship with this?

(1) Cdc2

(2) Cyclin

(3) P53

(4) Proteasomes

171. SNARE proteins are found in the membranes of all of the following compartments except

(1) Mitochondria

(2) Golgi complex

(3) Early endosome

(4) Endoplasmic reticulum

172. Starting with a fertilized egg (zygote), a series of five cell divisions would produce an early embryo with how many cells?

(1) 4 (2) 8

(3) 16 (4) 32

173. For a newly evolving protist, what would be the advantage of using eukaryote-like cell division rather than binary fission?

(1) Binary fission would not allow for the formation of new organisms

(2) Cell division would allow for the orderly and efficient segregation of multiple linear chromosomes

(3) Cell division would be faster than binary fission

(4) Cell division allows for lower rates of error per chromosome replication

174. If cells in the process of dividing are subjected to colchicine, a drug that interferes with the formation of the spindle apparatus, at which stage will mitosis be arrested?

(1) Anaphase

(2) Prophase

(3) Telophase

(4) Metaphase

175. If there are 20 centromeres in a cell at anaphase, how many chromosomes are there in each daughter cell following cytokinesis?

(1) 10 (2) 20

(3) 30 (4) 40

176. Taxol is an anticancer drug extracted from the Pacific yew tree. In animal cells, Taxol disrupts microtubule formation by binding to microtubules and accelerating their assembly from the protein precursor, tubulin. Surprisingly, this stops mitosis. Specifically, Taxol must affect

(1) the formation of the mitotic spindle

(2) anaphase

(3) formation of the centrioles

(4) chromatid assembly

177. Which of the following are primarily responsible for cytokinesis in plant cells but not in animal cells?

(1) Kinetochores

(2) Golgi-derived vesicles

(3) Actin and myosin

(4) Centrioles and centromeres

178. In which group of eukaryotic organisms does the nuclear envelope remain intact during mitosis?

(1) Seedless plants

(2) Dinoflagellates

(3) Dinoflagellates and diatoms only

(4) Seedless plants, dinoflagellates, and diatoms

179. Movement of the chromosomes during anaphase would be most affected by a drug that

(1) reduces cyclin concentrations
(2) increases cyclin concentrations
(3) prevents elongation of microtubules
(4) prevents shortening of microtubules

180. Measurements of the amount of DNA per nucleus were taken on a large number of cells from a growing fungus. The measured DNA levels ranged from 3 to 6 picograms per nucleus. In which stage of the cell cycle did the nucleus contain 6 picograms of DNA?

(1) G_0
(2) G1
(3) S
(4) G2

181. A group of cells is assayed for DNA content immediately following mitosis and is found to have an average of 8 picograms of DNA per nucleus. How many picograms would be found at the end of S and the end of G2?

(1) 8; 8
(2) 8; 16
(3) 16; 8
(4) 16; 16

182. For anaphase to begin, which of the following must occur?

(1) Chromatids must lose their kinetochores
(2) Cohesin must attach the sister chromatids to each other
(3) Cohesin must be cleaved enzymatically
(4) Kinetochores must attach to the metaphase plate

183. Which of the following best describes how chromosomes move toward the poles of the spindle during mitosis?

(1) The chromosomes are "reeled in" by the contraction of spindle microtubules
(2) Motor proteins of the kinetochores move the chromosomes along the spindle microtubules
(3) Nonkinetochore spindle fibers serve to push chromosomes in the direction of the poles
(4) The chromosomes are "reeled in" by the contraction of spindle microtubules, and motor proteins of the kinetochores move the chromosomes along the spindle microtubules

184. Which of the following is a function of those spindle microtubules that do not attach to kinetochores?

(1) Maintaining an appropriate spacing among the moving chromosomes
(2) Producing a cleavage furrow when telophase is complete
(3) Providing the ATP needed by the fibers attached to kinetochores
(4) Maintaining the region of overlap of microtubules in the cell's center

185. Which of the following is (are) required for motor proteins to function in the movement of chromosomes toward the poles of the mitotic spindle?

(1) intact centromeres
(2) an MTOC (microtubule organizing center)
(3) a kinetochore attached to the metaphase plate
(4) ATP as an energy source

186. Which of the following proteins are involved in binary fission as well as eukaryotic mitotic division?

(1) Cyclins (2) Cdks
(3) MPF (4) Actin and tubulin

187. Using which of the following techniques would enable your lab group to distinguish between a cell in G_2 and a cell from the same organism in G_1?

(1) Fluorescence microscopy
(2) Electron microscopy
(3) Spectrophotometry
(4) Radioactive-labeled nucleotides

188. You have the technology necessary to measure each of the following in a sample of animal cells: chlorophylls, organelle density, picograms of DNA, cell wall components, and enzymatic activity. Which would you expect to increase significantly from M to G_1?

(1) Organelle density and enzymatic activity
(2) Cell wall components and DNA
(3) Chlorophyll and cell walls
(4) Organelle density and cell walls

189. A plant-derived protein known as colchicine can be used to poison cells by blocking the formation of the spindle. Which of the following would result if colchicine is added to a sample of cells in G_2?

(1) The cells would immediately die
(2) The cells would be unable to begin M and stay in G_2
(3) The chromosomes would coil & shorten but have no spindle to which to attach
(4) The chromosomes would segregate but in a disorderly pattern

190. A student is looking through his light microscope (~450 X) at a squashed and stained onion root tip. Some, but not all, of the cells have clearly visible chromosome strands. When the cell has just completed telophase, which of the following does he see?

(1) Chromosomes clustered at the poles
(2) Individual chromatids separating
(3) Formation of vesicles at the midline
(4) Two small cells with chromatin

191. Through a microscope, you can see a cell plate beginning to develop across the middle of a cell and nuclei forming on either side of the cell plate. This cell is most likely

(1) an animal cell in the process of cytokinesis
(2) a plant cell in the process of cytokinesis
(3) an animal cell in the S phase of the cell cycle
(4) a bacterial cell dividing

192. Vinblastine is a standard chemotherapeutic drug used to treat cancer. Because it interferes with the assembly of microtubules, its effectiveness must be related to

(1) disruption of mitotic spindle formation
(2) inhibition of regulatory protein phosphorylation
(3) suppression of cyclin production
(4) myosin denaturation and inhibition of cleavage furrow formation

193. The decline of MPF activity at the end of mitosis is due to

(1) the destruction of the protein kinase Cdk
(2) decreased synthesis of Cdk
(3) the degradation of cyclin
(4) the accumulation of cyclin

194. In the cells of some organisms, mitosis occurs without cytokinesis. This will result in

(1) cells with more than one nucleus
(2) cells that are unusually small
(3) cells lacking nuclei
(4) destruction of chromosomes

195. Which of the following does not occur during mitosis?

(1) Condensation of the chromosomes
(2) Replication of the DNA
(3) Separation of sister chromatids
(4) Spindle formation

196. A particular cell has half as much DNA as some other cells in a mitotically active tissue. The cell in question is most likely in

(1) G_1.
(2) G_2
(3) prophase.
(4) anaphase

197. The drug cytochalasin B blocks the function of actin. Which of the following aspects of the cell cycle would be most disrupted by cytochalasin B?

(1) Spindle formation
(2) Spindle attachment to kinetochores
(3) Cell elongation during anaphase
(4) Cleavage furrow formation and cytokinesis

198. The binding of acting and myosin is prevented by

(1) Calmodulin (2) Troponin
(3) Tropomyosin (4) Colagen

199. Cell organalle responsible for Detoxification of Xenobiotics and Oxidative reaction is:

(1) Golgi (2) Mitochondria
(3) Plastid (4) Peroxisome

200. Helicase and Licensing factor bind to origin recognization complex during

(1) G1 stage (2) G2 stage
(3) M stage (3) S stage

ANSWER SHEET

Unit 2 : Cell Biology

1. 2	**2.** 4	**3.** 3	**4.** 4	**5.** 1	**6.** 4	**7.** 3	**8.** 4	**9.** 2	**10.** 3
11. 1	**12.** 2	**13.** 1	**14.** 1	**15.** 1	**16.** 2	**17.** 2	**18.** 4	**19.** 2	**20.** 2
21. 1	**22.** 1	**23.** 2	**24.** 2	**25.** 1	**26.** 3	**27.** 1	**28.** 1	**29.** 2	**30.** 3
31. 3	**32.** 3	**33.** 3	**34.** 2	**35.** 2	**36.** 3	**37.** 3	**38.** 4	**39.** 1	**40.** 2
41. 1	**42.** 2	**43.** 3	**44.** 4	**45.** 4	**46.** 3	**47.** 3	**48.** 1	**49.** 2	**50.** 1
51. 2	**52.** 3	**53.** 4	**54.** 3	**55.** 4	**56.** 4	**57.** 4	**58.** 3	**59.** 1	**60.** 4
61. 1	**62.** 3	**63.** 2	**64.** 4	**65.** 3	**66.** 4	**67.** 2	**68.** 2	**69.** 1	**70.** 4
71. 3	**72.** 4	**73.** 4	**74.** 3	**75.** 4	**76.** 4	**77.** 2	**78.** 4	**79.** 4	**80.** 2
81. 1	**82.** 2	**83.** 2	**84.** 2	**85.** 2	**86.** 4	**87.** 4	**88.** 1	**89.** 1	**90.** 3
91. 4	**92.** 2	**93.** 2	**94.** 3	**95.** 1	**96.** 2	**97.** 1	**98.** 3	**99.** 3	**100.** 2
101. 2	**102.** 4	**103.** 4	**104.** 3	**105.** 3	**106.** 2	**107.** 1	**108.** 3	**109.** 3	**110.** 4
111. 4	**112.** 1	**113.** 4	**114.** 2	**115.** 2	**116.** 1	**117.** 4	**118.** 4	**119.** 3	**120.** 4
121. 4	**122.** 3	**123.** 2	**124.** 4	**125.** 2	**126.** 2	**127.** 1	**128.** 3	**129.** 2	**130.** 4
131. 4	**132.** 3	**133.** 4	**134.** 4	**135.** 4	**136.** 2	**137.** 4	**138.** 2	**139.** 1	**140.** 3
141. 3	**142.** 4	**143.** 3	**144.** 1	**145.** 1	**146.** 4	**147.** 3	**148.** 3	**149.** 3	**150.** 4
151. 3	**152.** 3	**153.** 1	**154.** 4	**155.** 4	**156.** 1	**157.** 4	**158.** 4	**159.** 2	**160.** 4
161. 1	**162.** 2	**163.** 1	**164.** 4	**165.** 2	**166.** 3	**167.** 2	**168.** 3	**169.** 3	**170.** 3
171. 1	**172.** 4	**173.** 2	**174.** 4	**175.** 1	**176.** 2	**177.** 2	**178.** 3	**179.** 4	**180.** 4
181. 4	**182.** 3	**183.** 4	**184.** 4	**185.** 4	**186.** 4	**187.** 4	**188.** 4	**189.** 3	**190.** 1
191. 2	**192.** 1	**193.** 3	**194.** 1	**195.** 2	**196.** 1	**197.** 4	**198.** 3	**199.** 4	**200.** 1

UNIT - 3

MOLECULAR BIOLOGY

1. Which subunit of *E. coli* RNA polymerase is responsible for gene selection?
   (1) alpha (2) beta
   (3) omega (4) sigma

2. Which of the statements about tRNAs is false?
   (1) All organisms have more than 20 tRNA genes
   (2) The three-dimensional structure of tRNAs looks like a cloverleaf
   (3) tRNAs contain modified bases
   (4) The sequence of the last 3 nt at the 3' end of all tRNAs is the same

3. Cyclic adenosine monophosphate (cAMP) regulates the lactose (*lac*) operon by
   (1) binding to the operator to turn on transcription
   (2) binding to the *lac* repressor to prevent transcription
   (3) combining with the catabolite activator protein (CAP) to form a complex that enhances transcription upon binding to the promoter
   (4) combining with the CAP to remove CAP's inhibition of transcription

4. Which of the following is not found in an *E. coli* replication fork?
   (1) DnaA
   (2) Primase
   (3) PCNA
   (4) Single-stranded DNA binding protein

5. One strand of double-stranded DNA is mutated, changing all cytosines to uracils. After one round of replication of the mutated DNA strand, the melting temperature of the resulting DNA will
   (1) be higher
   (2) be lower
   (3) remain the same
   (4) be double

6. Which of the statements about translation is false?
   (1) During translocation in the "hybrid sites" model, the tRNA attached to the nascent polypeptide chain is in the P site of the small subunit and the A site of the large subunit
   (2) Fusidic acid prevents the release of EFG-GDP from the ribosome
   (3) Puromycin leads to premature release of the polypeptide chain
   (4) IF-3 preferentially binds to 30S ribosomes

7. Which of the following elements is not a characteristic of factor independent terminators in *E. coli*?
   (1) a C-rich sequence
   (2) an RNA sequence that can form a stem-loop
   (3) a run of single-stranded U residues
   (4) a GC rich sequence

8. Okazaki fragments are
(1) RNA primers for DNA synthesis
(2) Short DNA fragments after nuclease digestion
(3) Newly synthesized DNA fragments
(4) Short stretches of DNA attached to RNA primers on lagging strand

9. When synthetic mRNA consisting of alternating A and C residues (ACACAC....) was translated in a bacterial extract, only one kind of polypeptide consisting of alternating threonine and histidine residues was made. When the base sequence was ... AACAACAAC.... three different polypeptides were formed : polyasparagine, polythreonine and polyglutamine. Hence we may conclude that one codon for histidine is
(1) ACA (2) CAC
(3) AAC (4) CAA

10. During RNA polymerase II transcriptional initiation, phosphorylation of the following factor is essential to commence transcription
(1) Polymerase II CTD
(2) TFIID
(3) TFIIH
(4) TFIIE

11. Typical nucleosomal organization of a gene is not found in
(1) human liver nuclei
(2) malarial parasite
(3) human sperm
(4) Neuron

12. In a temperature sensitive mutant bacteria, at non-permissive temperature, there are huge accumulation of Okazaki fragments. The bacteria are mutant for
(1) DNA ploymerase I
(2) DNA topoisomerase I
(3) DNA ligase
(4) DNA gyrase

13. What is the sugar–sugar linkage among the following in the 5' cap on mRNA molecules?
(1) 5'-3' (2) 3'-5'
(3) 3'-3' (4) 5'-5'

14. Introduction of a non-sense mutation in a eukaryotic protein coding gene often leads to
(1) retention of mRNA in a nucleus so it is not translated
(2) increased degradation of the mRNA in the nucleus
(3) increased degradation of the mRNA in the cytoplasm
(4) decreased ribosome binding in the cytoplasm in mRNA due to its shorter size

15. Alu elements are
(1) SINEs (2) LINEs
(3) retroposon (4) DNA transposon

16. Telomeres consist of simple sequence repeats of
(1) AA rich strands that interact with protein
(2) GG rich strands that interact with protein
(3) CC rich strands that interact with protein
(4) TG rich strands that interact with protein

17. True activators of transcription are transcription factors that bind to
(1) other proteins to enhance transcription
(2) promoters
(3) enhancers
(4) promoters and enhancers

18. The U2 snRNA basepairs with
(1) a sequence spanning the exon intron splicing site
(2) the 3' splice site of the intron
(3) a sequence spanning the intron exon splicing site
(4) the branched sequence in the intron

19. If the sequence of the DNA sense strand is 5'GATCCTATGCTAC 3', then the transcribed mRNA sequence will be
(1) 5' GAUCCUAUGCUAC 3'
(2) 5' CUAGGUAUCGAUC 3'
(3) 5' CAUCGUAUACCUAG3'
(4) 5' GUAGCAUAGGAUC 3'

20. DNA polymerase processivity
(1) is a measure of the number of nucleotides joined before the polymerase dissociates
(2) is determined by the ability of the enzyme to also have nuclease activity
(3) is a measure of thermal stability of the enzyme
(4) is a measure of rate of elongation of newly synthesized strands

21. In human genome, approximately--------of the DNA codes for proteins
(1) 10% (2) 2%
(3) 50% (4) 20%

22. A linear fragment of DNA will be unstable if it carries
(1) Two origins of replication
(2) Two centromeres
(3) Two telomeres
(4) Two selection markers

23. The genomic DNA fraction which has highest value of cot ½ on Cot curve represents
(1) Highly repetitive DNA
(2) Moderately repetitive DNA
(3) Minisatellite DNA
(4) Unique DNA

24. The antibiotic that resembles the 3' end of the charged tRNA molecule is
(1) Tetracycline (2) Puromycin
(3) Kanamycin (4) Steptomycin

25. Which of the following amino acids is coded by maximum number of codons ?
(1) Leucine (2) Tryptophan
(3) Valine (4) Alanine

26. Integration of phage lambda genome into *E. Coli* chromosome is by
(1) COS sites
(2) random integration by the function of e-14 element in the chromosome
(3) site specific recombination
(4) red gene mediated recombination

27. Full expression of the lac operon requires
(1) lactose and cAMP
(2) allolactose and cAMP
(3) cAMP
(4) lactose

28. An enzyme that induces double strand breaks in DNA and rejoins them is called
(1) Restriction endonuclease
(2) DNA gyrase / Topoisomerase
(3) DNA ligase
(4) DNA polymerase

29. RNA is analyzed for the location of hairpin folds. Which of the following sequences could form a minihairpin?
(1) AGGUUUCCU
(2) AAAAAAAAA
(3) AGGUUUGGA
(4) AGGUUUAGG

30. Increasing the concentration of which of the following would most effectively antagonize the inhibition of protein synthesis by puromycin?
(1) ATP (2) eIF2.GTP
(3) aminoacyl-tRNAs (4) peptidyl-tRNAs

31. Induction of β-galactosidase activity by IPTG is due to
(1) Stimulation of lac repressor function
(2) IPTG binding to lac operon and inducing transcription
(3) IPTG binding to lac I gene product and inhibiting its activity
(4) Inhibition of β-galactosidase degradation

32. Which of the following enzymes doesn't require a primer?
(1) RNA dependent DNA polymerase
(2) DNA dependent DNA polymerase
(3) DNA dependent RNA polymerase
(4) Taq DNA polymerase

33. Which one of the following antibiotics attaches to 50S ribosome and inhibits its peptidyl-transferase activity?
(1) Penicillin (2) Chloramphenicol
(3) Trimethoprim (4) Amphotericin

34. Operon having positive and negative regulation by single regulatory protein is
(1) lac operon (2) trp operon
(3) ara operon (4) his operon

35. Uvr ABC endonuclease is present in which repair system ?
(1) Mismatch repair
(2) Nucleotide excision repair
(3) Base excision repair
(4) SOS repair

36. What would be the likely explanation for the existence of pseudogenes?
(1) gene duplication
(2) gene duplication and mutation events
(3) mutation events
(4) unequal crossing over

37. Processing of rRNA involves
(1) Sn-RNA (2) Mi-RNA
(3) Sno-RNA (4) g-RNA

38. Ultraviolet radiation causes DNA damage by formation of
(1) cytidine dimer (2) thymidine dimer
(3) adenine dimer (4) guanine dimer

39. Which of the following antibiotics/toxicants acts by interfering with the phosphodiester bond formation?
(1) Rifampian (2) Actinomycin
(3) α – amanitin (4) Penicillin

40. Diphtheria toxin blocks protein synthesis by
(1) ADP ribosylation of EF-2
(2) Phosphorylating EF-1
(3) Inhibiting peptidyl transferase activity
(4) ADP ribosylation of EF-G

41. Blocking of 'A' site on the ribosome and thereby inhibiting protein synthesis is the mechanism of action of
(1) Streptomycin (2) Tetracycline
(3) Chloramphenicol (4) Erythromycin

42. Which of the following repair mechanisms is absent in a cell arrested in cell cycle?
(1) Transcriptional coupled repair mechanism
(2) Excision repair mechanism
(3) DNA synthesis annealing repair mechanism
(4) Recombination repair mechanism

43. Bacteria utilize glucose preferentially over other sugars through a mechanism called
(1) Operon repression
(2) Enzyme repression
(3) Catabolite repression
(4) Catabolite induction

44. Which of the following compounds mimics aminoacyl t-RNA and blocks protein synthesis ?
(1) puromycin (2) kirromycin
(3) streptomycin (4) neomycin

45. A mammalian cell has an outstretched double stranded DNA of 1.2 meter which duplicates in 4hrs. If it duplicates at the rate of 20μ meter/min, how many origins of replication are there in the DNA?
(1) 2500 (2) 250
(3) 25 (4) 1

46. The ATP required for ligation during base excision repair is generated from
(1) NAD (2) Poly (ADP-ribose)
(3) DNA (4) AMP

47. Which of the following features is not found in heterogeneous nuclear RNAs (hnRNAs)?
(1) intron
(2) polycistronic coding
(3) polyadenylation at 3'-end
(4) 5-' cap structure

48. The average length attained by a chromosome varies from
(1) 30 to 1000 nm (2) 0.5 to 30 μm
(3) 30 μm to 1 mm (4) 1 mm to 10 mm

49. *E. coli* with mutation in operator region of lac operon and containing suppressors will
(1) produce β-galactosidase even when lactose is absent
(2) produce β-galactosidase only in the presence of lactose
(3) will not produce β-galactosidase even in the presence of lactose
(4) will produce β-galactosidase even in the presence of glucose

50. Which one of the following statements about prion proteins is incorrect?
(1) Prion proteins form cross-beta filaments
(2) Prion proteins are heat resistant
(3) Prion proteins are protease sensitive
(4) Prion proteins can convert the normally folded prion protein to pathological form

51. Which of the following directly reverses DNA damage?
(1) AP endonuclease (2) UVr-ABC
(3) MutS and MutL (4) Methyltransferase

52. When DNA molecules from a complex genome are denatured and then returned to conditions that favor duplex formation, the strands reanneal. Which of the following statements about the renaturation is incorrect?
(1) strands with the same overall A+T composition will anneal in the fastest category
(2) the slowly annealing fraction contains most of the genes
(3) only strands with complementary base sequences will anneal stably
(4) strands derived from highly repeated sequences anneal rapidly because the rate of the reaction is concentration dependent

53. Which one of the following codons is used for selenocysteine during protein biosynthesis?
(1) UGA (2) UAG
(3) UGC (4) UCG

54. Rifampicin is a specific inhibitor of
(1) Bacterial RNA polymerase
(2) RNA polymerase II
(3) RNA polymerase I
(4) DNA polymerase II

55. In the prokaryotic ribosome structure, the peptidyl transferase activity of the ribosomes is found
(1) Entirely in the ribosomal RNA
(2) Entirely in the ribosomal proteins
(3) Partially embedded in the RNA and partially in the proteins
(4) Ribosomal RNA binds the m-RNA but proteins catalyze peptidyl transferase activity

56. *Amanita phalloides*, which is responsible for the majority of fatal mushroom poisonings contains alpha-amanitin that specifically inhibits
(1) DNA Polymerase
(2) RNA Polymerase
(3) Telomerase
(4) Topoisomerase

57. Which of the following antibiotics inhibits bacterial protein synthesis by interfering with peptidyl transferase activity?
(1) Streptomycin
(2) Erythromycin
(3) Puromycin
(4) Cycloheximide

58. Disulphide bonds in proteins could be formed by cysteines using
(1) Iodoacetamide
(2) oxidized glutathione
(3) reduced glutathione
(4) β-mercaptoethanol

59. During DNA replication in *E. coli*, RNA primers are synthesized by
(1) dna B (2) dna G
(3) dna C (4) dna A

60. Which of the following statements about the trp operon is not correct?
(1) It is normally turned off if tryptophan is present
(2) Tryptophan acts as the corepressor
(3) The regulator gene product is inactive by itself
(4) Tryptophan binds to the repressor protein and inactivates it

61. In a lac operon, a mutation is created so that lactose cannot bind. Now, if lactose is provided, what would happen?
(1) the β-galactosidase will be over-expressed
(2) lac I repressor will remain inactivated
(3) expression of lac operon will remain unaffected
(4) lac I repressor would remain bound to operon preventing expression

62. For transcription to occur in the lactose operon, an inducer must be present so that
(1) the repressor can bind to the operator
(2) the repressor does not bind to the operator
(3) the inducer can bind to the operator
(4) the inducer does not bind to the operator

63. Which of the following is not a feature of mutagenic action of 5-Bromo-deoxyuridine?
(1) It acts on growing cells
(2) It forms base pair with A in its rare form
(3) It induces transitions
(4) It affects only one strand of DNA

64. The 20 different amino acids found in proteins are normally coded by
(1) 59 codons (2) 60 codons
(3) 61 codons (4) 63 codons

65. The following are important differences between Topoisomerases I and II except
(1) Topo I is made of one peptide while Topo II is made of two peptides
(2) Topo I makes a single strand cut while Topo II makes two strand cuts
(3) Topo I binds to one DNA double helix while Topo II binds to two double helices
(4) Topo I does not make a covalent bond with DNA while Topo II makes a covalent bond with DNA

66. Which of the following proteins is not a component of the Replication Fork Complex?
(1) Primase
(2) Helicase
(3) Ligase
(4) DNA polymerase

67. What would be the most likely consequence of inhibiting the activity of Primase on a replication fork that is already advancing?
(1) The lagging strand would stop being synthesized
(2) The leading strand would stop being synthesized
(3) The DNA helicase would stop advancing
(4) The Okasaki fragments would get longer

68. Mitochondrial genes are different from nuclear genes in that, mitochondrial genes
(1) have fewer stop codons
(2) do not have introns
(3) have a different start codon
(4) are shorter-than nuclear genes

69. Intracellular transport in mammalian cells through vesicular fusion is regulated by which among the following GTPases ?
(1) Rho (2) Ran
(3) Rab (4) Ras

70. hsp70 chaperons are not present in which among the following organelles?

(1) Endoplasmic reticulum
(2) Golgi bodies
(3) Nucleus
(4) Mitochondria

71. The covalent modification of histones that is not known to play a role in regulation of gene expression is

(1) ADP-ribosylation
(2) Acetylation
(3) Methylation
(4) glycosylation

72. The bacterial ftsZ gene is required for

(1) septum formation
(2) periseptal annulus formation and localization
(3) DNA replication
(4) transport of DNA

73. The common type of duplication generates a second copy of the gene

(1) in close proximity of the first copy
(2) in a distant location on a same chromosome
(3) on a different chromosome
(4) in any chromosome at random probability

74. Which of the following statement is true for eukaryotic telomeres?

(1) Telomere are located at centre of chromosome
(2) It have GC rich strands that interact with protein
(3) Telomerase replicate telomeric DNA in somatic cells
(4) It have TG/AC rich strands that interact with protein

75. 5-Bromouracil induces mutations because it

(1) replaces a T and binds to G rather than A
(2) replaces a G and binds to A rather than C
(3) changes the binding affinity of G
(4) changes the binding affinity of T

76. Which of the following statements concerning the Edman degradation method is incorrect?

(1) Phenyl isothiocyanate is coupled to the amino-terminal residue
(2) Under mildly acidic conditions, the modified peptide is cleaved into a cyclic derivative of the terminal amino acids and a shortened peptide (minus the first amino acid)
(3) Once the PTH amino acid is separated from the original peptide, a new cycle of sequential degradation can begin
(4) Phenyl isothiocyanate is coupled to the carboxy-terminal residue

77. In the ds-break recombination model, which of the following steps is the first one that happens after formation of the Holliday junction?

(1) DNA cleavage catalyzed by RuvC
(2) Coating ssDNA with RecA
(3) Branch migration catalyzed by RuvAB
(4) Dissociation of RecD from RecBCD at a chi site

78. The following statements regarding spacer DNA are correct except

(1) Spacer DNA includes repetitive DNA, which accounts for greater than 50% of the genome
(2) SINES and LINES are distinguished by the size of the repeat units
(3) Spacer DNA includes the protein-coding regions
(4) Alpha-satellite DNA is located near centromeres

79. At a DNA replication fork, both parental DNA strands act simultaneously as templates for synthesis of complementary daughter strands, but only one daughter strand is synthesized continuously. The other daughter strand is synthesized discontinuously because

(1) DNA polymerase III can only add a dNTP to the 3′ end of a growing strand

(2) the lagging strand is synthesized by DNA polymerase I, which is slower than DNA polymerase III

(3) DNA polymerase III lacks a 5′ to 3′ exonuclease activity

(4) DNA polymerase III requires a pre-existing primer

80. Telomerase, which is required for preventing the shortening of eukaryotic chromosomes during replication, may serve as a mitotic clock that determines the replicative lifespan of cells. The RNA component of telomerase

(1) acts as the template for telomere DNA synthesis

(2) proofreads the newly incorporated DNA nucleotides

(3) binds to the 5' end of the parent DNA strand for extension

(4) catalyzes the formation of phosphodiester bonds on the newly synthesized DNA

81. A 35 yr old male with hypercholesterolemia is found to have an LDL receptor protein that is 865 amino acids long, rather than the 840 amino acids found in the normal protein. Which of the following point mutations is consistent with this abnormality?

(1) UAG → UAC (2) UAG → UAA

(3) GGA → UGA (4) AUC → AUA

82. An anemic patient produces β-globin protein containing a mutation resulting in protein misfolding. The patient's cells modify this abnormal protein by attaching ubiquitin molecules to it. What is the fate of the modified β-globin protein?

(1) It is secreted from the cell

(2) It is degraded by the proteosome

(3) It is degraded by the lysosome

(4) It is placed into storage vesicles

83. A 28 yr old male presents with anemia and low levels of measured erythropoietin due to a high rate of turnover. He is found to have a mutation in the erythropoietin gene resulting in an Asn- Asp amino acid substitution. Of the following options, what is the most likely cause of his condition?

(1) mislocalization due to a change in targeting signal

(2) lack of O-glycosylation

(3) improper protein hydroxylation

(4) lack of N-glycosylation

84. cis-trans isomerization of the peptide bond preceding an amino acid X is known to be critical in the folding of proteins by slowing down the folding reaction. The amino acid X is

(1) isoleucine (2) tryptophan

(3) proline (4) histidine

85. Holiday junction is observed during

(1) Mitosis (2) Interphase

(3) Recombination (4) DNA Repair

86. The lac repressor binds its operator with a $K_d = 10^{-10}$ M. In a strain of *E. coli*, the intracellular repressor concentration is 10^{-12} M. Induction of the lac operon in this strain

(1) does not require lactose

(2) requires lactose

(3) requires glucose

(4) requires both glucose and lactose

87. RNA interference is induced by

(1) double strand DNA

(2) double strand RNA

(3) both double strand DNA and RNA

(4) specific proteins

88. Which of the following represents a commonly found order of arrangement of core promoter elements in eukaryotes?
(1) TATA-BRE-Inr
(2) TATA-Inr-DPE
(3) BRE-Inr-DPE
(4) BRE-DPE-Inr

89. Reverse transcriptase has both ribonuclease and polymerase activities. Ribonuclease activity is required for
(1) the synthesis of new RNA strand
(2) the degradation RNA strand
(3) the synthesis of new DNA strand
(4) the degradation of DNA strand

90. The germ cells of animals express a distinct class of small RNAs, that suppress the movement of transposable elements in the germ line are termed as
(1) Small interfering RNA (siRNA)
(2) Micron RNA (miRNA)
(3) Piwi-interacting RNA (piRNA)
(4) Transfer-messenger RNA (tmRNA)

91. Of the following are true about transpositions except
(1) transposons move from one location to a different one within a chromosome
(2) both the donor and target sites must be homologous
(3) transposons have insertion sequences that are recognized by transposases
(4) the transposon may either be excised and moved or be replicated with the replicated piece moving

92. Targeting a protein to be degraded within proteasomes usually requires ubiquitin. In the functions of ubiquitin all of the following are true except
(1) ATP is required for activation of ubiquitin
(2) a peptide bond forms between the carboxyl terminal of ubiquitin and an E-amino group of a lysine
(3) linkage of a protein to ubiquitin does not always mark it for degradation
(4) ATP is required by the enzyme that transfers the ubiquitin to the protein to be degraded

93. What is the main damaging effect of UV radiation on DNA?
(1) Depurination
(2) formation of thymine dimers
(3) Single strand break
(4) Double strand break

94. Why do scientists think RNA, rather than DNA, may have been the original genetic material?
(1) Most organisms on the earth use RNA as their genetic materials
(2) The simplest life forms, viruses, have RNA as their genetic materials
(3) RNA is more stable than DNA
(4) RNA has both informational and catalytic capabilities

95. The yeast genome is 0.004 times the size of the human genome and yet it contains approximately 0.2 times fewer genes. The explanation for this is
(1) the genes in yeast contain far fewer codons compared to human genes
(2) yeast chromosomes contain much smaller centromere and telomere
(3) the yeast genome consists much less intergenic DNA and fewer introns
(4) the yeast genome consists of many overlapping genes

96. When bacterium lambda infects a sensitive bacterium, one of the first messenger RNA species synthesized is very short, beginning at a site P_L and extending just through an adjacent gene N. After the appearance of the gene N protein, messages become much longer, still beginning at P_L but extending far beyond gene N. The gene N encodes

(1) an anti-terminator acting just beyond gene N

(2) a new sigma factor acting on a. promoter beyond gene N

(3) an activator for a promoter beyond gene N

(4) a protein that stabilizes the longer message

97. A 4 year old female presents with an insatiable appetite and morbid obesity. Genetic analysis shows that the patient's gene for the appetite control hormone, leptin, contains a mutation in its promoter region that decreases TFII-D binding. What is the most likely effect of this mutation?

(1) Capping of the leptin mRNA will be defective

(2) Splicing of the leptin mRNA will be defective

(3) Initiation of leptin transcription will be defective

(4) Termination of leptin transcription will be defective

98. Which of the following statements regarding prokaryotic DNA replication is incorrect?

(1) The pre-priming stage is the regulated step

(2) DNA is synthesized in the 5' to 3' direction

(3) Gyrase unwinds the DNA in front of the fork

(4) DNA polymerase III synthesizes only the lagging strand

99. Which of the following correctly describes the enzymatic activities of DNA polymerase I?

(1) The polymerase synthesizes DNA in the 3' to 5' direction

(2) The RNA primer nucleotides are removed by the 5' to 3' exonuclease

(3) The 3' to 5' exonuclease extends the 3' end of the parent strand

(4) The 3' to 5' exonuclease replaces the RNA primer with deoxynucleotides

100. Which of the following is not required for eukaryotic DNA replication?

(1) Telomerase

(2) DNA polymerase α (alpha)

(3) RNA polymerase III

(4) DNA polymerase δ (delta)

101. A 10-year-old Caucasian girl is brought in by her parents for evaluation of a skin disorder. The child has many freckles on her face, arms, and legs. The parents were told by previous physicians that she suffers from Xeroderma pigmentosum and that they should limit her exposure to sun light. What is the most likely etiology of this disorder?

(1) deficient DNA replication

(2) impaired DNA repair by nucleotide excision

(3) defective RNA transcription from TATA-less promoter

(4) impaired ribosomal protein translation

102. The addition or removal of acetyl groups from histones can affect the rate at which genes are transcribed. What is the effect on transcription if histone acetyl transferases (HAT) are removed in a cell through mutations?

(1) Transcription levels increase due to the tightening of histone-DNA interactions

(2) There is no effect on transcription levels because histones are only used to package DNA

(3) Transcription of genes will be blocked due to the tightening of histone-DNA interactions

(4) Transcription levels will increase due to a loosening of histone-DNA interactions

103. Which one of the following modification leads to protein degradation?

(1) Methylation (2) Acetylation

(3) Phosphorylation (4) Ubiquitination

104. A protozoan hnRNA splicing out its own intron to produce a mature mRNA

(1) is a classic example of enzymatic catalysis

(2) is an energy independent process

(3) is dependent on the presence of one or more proteins

(4) is termed a ribozyme

105. Aminoacyl-tRNAval synthetase mistakenly attaches the structurally similar amino acid, isoleucine, to the tRNAval containing a valine anticodon. Assuming this error is uncorrected by the aminoacyl-tRNAval synthetase, what will be the fate of this isoleucine residue during protein synthesis?

(1) it will remain attached to the tRNA and not be used

(2) it will be corrected in the A-site of the ribosome

(3) it will be incorporated into a protein in response to an isoleucine codon

(4) it will be incorporated into a protein in response to a valine codon

106. In the process of prokaryotic DNA replication, which of the following enzymes acts first?

(1) helicase (2) primase

(3) DNA polymerase III (4) DNA ligase

107. An important strategy for developing new antibiotics relies on differences in the enzymes used by prokaryotes and eukaryotes to replicate DNA. For example, a small molecule inhibitor of bacterial DNA ligases has been shown to have broad spectrum antibacterial activity. Analysis of DNA in bacterial cells exposed to this compound would show

(1) nicks in the phosphodiester backbone of the lagging strand

(2) an increased frequency of transition mutations

(3) RNA primers remaining in the Okazaki fragments

(4) inability to initiate replication at the origin

108. Mutations causing xeroderma pigmentosum (XP) reside in any of several proteins required for

(1) mismatch repair

(2) nucleotide excision repair

(3) base excision repair

(4) recombination repair

109. A 5-year-old girl presents with basal and squamous cell carcinomas, predominantly on the face, head, and neck. Clinical tests confirm the suspected diagnosis of xeroderma pigmentosum (XP) Cells from individuals with XP have elevated mutation rates due to a deficiency in the ability to correct UV-induced damage. Which segment of DNA is most likely to be mutated in this patient?

(1) GCATATGC (2) ATTCGCAT

(3) GTACGTAC (4) GCGCGCGC

110. The diagram below schematically illustrates a eukaryotic transcription bubble, in which RNA polymerase is transcribing a single-stranded RNA on the basis of the sequence of nucleotides on the template strand of the double-stranded DNA

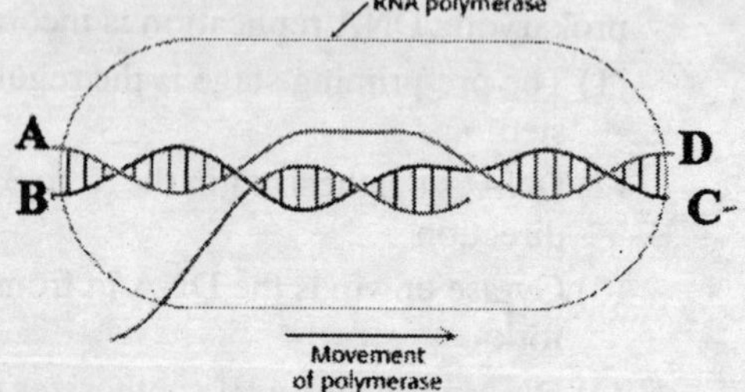

Which letter (A - D) marks the 5'-end of the template strand?

(1) A (2) B

(3) C (4) D

111. A strain of *M. tuberculosis* that is resistant to the drug rifampicin most likely produces an altered form of
(1) DNA polymerase
(2) RNA polymerase
(3) TFII-D
(4) 30 S ribosomal subunit

112. Each year, more than 100 people worldwide die after eating poisonous mushrooms. The death cap mushroom, *Amanita*, produces a toxin which strongly binds to and inhibits RNA polymerase II, thereby blocking
(1) proofreading during transcription
(2) recognition of the TATA box promoter element
(3) production of tRNA
(4) production of mRNA

113. Ciprofloxin antibiotics inhibit DNA replication by acting on
(1) DNA polymerase
(2) DNA promase
(3) DNA helicase
(4) DNA Topoisomerase

114. An 18 month old female with chronic anemia is found to have β-thalassemia. Genetic analysis shows that one of her β-globin genes has a mutation that creates a new splice donor site 25 nucleotides upstream of the normal splice donor site for the 3rd intron. Which of the following best describes the new mRNA molecule that can be produced from this mutant gene?
(1) Exon 3 will be too short
(2) Exon 3 will be too long
(3) Exon 3 will be missing
(4) Exon 4 will be too short

115. An hnRNA consisting of 8 exons and 7 introns is normally processed in skeletal muscle to omit exon 5. The same hnRNA is processed in heart muscle to include all exons. This is an example of
(1) Abberant splicing
(2) Ordered splicing
(3) Alternative splicing
(4) Self splicing

116. Bacterial mRNAs contain
(1) Only one open reading frame
(2) Formyl-methionine
(3) 7-methyl-guanosine
(4) Sequence specific ribosomal binding sites

117. Of the following options, the inability for all possible tRNA anticodons to exist in nature best exemplifies what principle of the genetic code?
(1) the code is degenerate
(2) tRNAs are isoaccepting
(3) the code is unambiguous
(4) the code allows for "wobble"

118. Which one of the following antibiotics attaches to 50S ribosome and inhibits peptidyl-transferase activity?
(1) Penicillin
(2) Chloramphenicol
(3) Trimethoprim
(4) Amphotericin

119. Which of the following is absent in heterogeneous nuclear RNAs (hnRNAs)?
(1) Intron
(2) Polycistronic coding
(3) Polyadenylation at 3'-end
(4) 5-' cap structure

120. Telomeric DNA does not contain
(1) G-rich sequences
(2) Quadruplex
(3) T and D loops
(4) AT rich sequences

121. Which of the following is also called the "suicide enzyme"?
(1) DNA glycosylase
(2) DNA photolyase
(3) O^6 alkyl guanine transferase (ada)
(4) Exonuclease

122. Two plasmids are of the same compatibility group if they
(1) can co-exist in the same bacterial cell
(2) cannot co-exist in the same bacterial cell
(3) carry the same antibiotic gene
(4) carry the same toxin gene

123. Which of the following enzymatic activities is not a property of the 10^9-kDa polypeptide *E. coli* DNA polymerase I ?
(1) 5' - 3' polymerase
(2) 5' - 3' exonuclease
(3) 3' - 5' exonuclease
(4) 3' - 5' polymerase

124. RNAs that catalyze biological reactions such as self-splicing introns, are known as
(1) micro RNAs (2) ribozymes
(3) spliceozymes
(4) small nucleolar RNAs

125. Addition of tryptophan to an *E. coli* culture caused rapid transcriptional repression of genes encoding the enzymes of the TRP operon. Which of the following is the correct regulatory mechanism?
(1) Interaction of tryptophan with RNA polymerase causes inactivation of the enzyme
(2) Binding of the TRP repressor to the operator is stimulated upon tryptophan binding to the repressor and consequent eviction of the polymerase
(3) Upon tryptophan binding to TRP repressor, the repressor is tightly bound on RNA polymerase
(4) TRP repressor binding to RNA polymerase is stimulated by tryptophan

126. Blocking of 'A' site on the ribosome and thereby inhibiting protein synthesis is the mechanism of action of
(1) Streptomycin
(2) Tetracycline
(3) Chloramphenicol
(4) Erythromycin

127. Which of the following statements regarding mRNA processing is incorrect?
(1) Pre-mRNA processing is coupled to transcriptional elongation
(2) co-transcriptional splicing is tightly coupled to transcriptional elongation
(3) Mature mRNAs are capped in the cytosol
(4) mRNAs with premature stop codons are degraded after primary round of translation

128. Which of the following is not true of the C-terminal domain (CTD) of RNA polymerase II?
(1) CTD phosphorylation leads to promoter escape and transcript elongation
(2) Mediator complex binds to the unphosphorylated CTD
(3) Mediator complex remains associated with the CTD at all times
(4) Complete CTD deletion leads to lethality of yeast cells

129. Select the technique most appropriate to demonstrate that lactose induces the synthesis of β-galactosidase enzyme
(1) Northern Blot
(2) Western Blot
(3) Quantitative PCR
(4) Southern Blot

130. Entry of λ-phage lysogen to lytic phase is triggered by
(1) mutation in the λ-genome
(2) loss of co-operativity in binding of λ-repressor
(3) increase in the λ-repressor concentration
(4) decrease in recA function

131. An *E. coli* mutant constitutive for the lac operon was mated with a wild type strain. The merodiploid thus obtained was inducible by lactose. This observation indicates that the original mutation is
(1) dominant (2) trans-dominant
(3) recessive (4) cis-dominant

132. Small RNAs with internally complementary sequences that form hairpin-like structure, synthesized as percursor RNAs and cleaved by endonucleases to form short duplexes are called
(1) snRNA (2) mRNA
(3) tRNA (4) miRNA

133. Chromatin loops are formed by periodic attachment of the following onto the nuclear skeleton
(1) Histones (2) MARs
(3) Promoters (4) Introns

134. FOS, JUN and MYC are
(1) proteins expressed on the surface of cancerous cells
(2) protein kinases that phosphorylate transcription factors in cancerous cells
(3) proteins involved in regulation of expression of genes involved in growth promotion
(4) proteins involved in ion transport in cancerous cells

135. While replicating DNA, the rate of misincorporation by DNA polymerase is 1 in 10^5 nucleotides. However, the actual error rate in the replicated DNA is 1 in 10^9 nucleotides incorporated. This is achieved mainly due to
(1) spontaneous excision of misincorporated nucleotides
(2) 3'→ 5' proofreading activity of DNA polymerase
(3) termination of DNA polymerase at misincorporated sites
(4) 5' → 3' proof reading activity

136. Ignoring proofreading steps, what is the total energy cost or ATP equivalents, i.e. GTP = ATP energetically), for the synthesis of a 150 residue polypeptide in *E. coli*?
(1) 150 (2) 300
(3) 450 (4) 600

137. All of the following are true about nucleotide excision repair except
(1) removal of the damaged bases occurs on only one strand of the DNA
(2) it removes thymine dimers generated by UV light
(3) it involves the activity of an excision nuclease, which is an endonuclease
(4) only the damaged nucleotides are removed

138. During the elongation stage of eukaryotic protein synthesis
(1) the incoming aminoacyl-tRNA binds to the P site
(2) a new peptide bond is synthesized by peptidyl transferase site of the large ribosomal subunit in a GTP-requiring reaction
(3) the peptide, still bound to a tRNA molecule, is translocated to a different site on the ribosome
(4) peptide bond formation occurs by the attack of the carboxyl group of the incoming aminoacyl-tRNA on the amino group of the growing peptide chain

139. Termination of protein synthesis
(1) requires a stop codon to be located at the P site of the large ribosomal subunit
(2) occurs when a non-ribosomal protein release factor binds to the ribosome
(3) requires the action of a non-ribosomal hydrolase to release the peptide
(4) does not require energy

140. Diphtheria toxin
(1) acts catalytically
(2) activates translocase
(3) releases incomplete polypeptide chains from the ribosome
(4) attacks the RNA of the large subunit

141. The excision repair of UV-induced DNA damage is defective in individuals suffering from

(1) hereditary nonpolyposis colon cancer
(2) Crohn's disease
(3) classical xeroderma pigmentosum
(4) Cocyne syndrome

142. Which one of the following statements about transcription are correct?

(1) RNA synthesis occurs in the 3' to 5' direction
(2) the transcribed RNA is complementary to the template strand
(3) the RNA polymerase adds ribonucleotides to the 5' end of the growing RNA chain
(4) the RNA polymerase adds deoxyribonucleotides to the 3' end of the growing RNA chain

143. Which one of the following statements about *E. coli* RNA polymerase is false?

(1) the holoenzyme includes the sigma factor
(2) the core enzyme includes the sigma factor
(3) it requires Mg^{2+} for its activity
(4) it requires Zn^{2+} for its activity

144. Which one of the following statements is incorrect?

(1) there are two a α-subunits in the *E. coli* RNA polymerase
(2) *E. coli* has one sigma factor
(3) the β subunit of *E. coli* RNA polymerase is inhibited by rifampicin
(4) the streptolydigins inhibit transcription elongation

145. Which one of the following statements about transcription in *E. coli* is true?

(1) the -10 sequence is always exactly 10 bp upstream from the transcription start site
(2) the initiating nucleotide is always a G
(3) the intervening sequence between the -35 and -10 sequences is conserved
(4) the distance between the -35 and -10 sequences is critical for transcription efficiency

146. Which one of the following statement about transcription in *E. coli* is not true?

(1) loose binding of the RNA polymerase core enzyme to DNA is non-specific and unstable
(2) sigma factor dramatically increases the relative affinity of the enzyme for correct promoter sites
(3) almost all RNA start sites consist of a purine residue, with A being more common than G
(4) all promoters are inhibited by negative supercoiling

147. Which of the following statements is correct?

(1) the double stranded DNA sequence that has the upper strand sequence 5'-GGATCCTAGG-3' palindrome
(2) the Lac repressor inhibits binding of the polymerase to the lac promoter
(3) the lac operon is directly induced by lactose
(4) binding of Lac repressor to allolactose reduces its affinity for the lac operator

148. All of the following arc potential control mechanisms for regulation of gene expression in eukaryotic organisms except

(1) Gene amplification
(2) The degradation of mRNA
(3) The lactose operon
(4) Transcription

149. It is theoretically possible for a gene from any organism to function in any other organism. Why is this possible?

(1) All organisms have similar nuclei
(2) All organisms have the same genetic code
(3) All organisms are made up of cells
(4) All organisms have transfer RNA

150. The sequence given below is a short segment of a coding strand of a DNA molecule
ATG TCC CTG
The complementary mRNA strand that this DNA sequence could produce would be
(1) UAC AGG GAC
(2) TAC TGG GTC
(3) UAG ACC GAC
(4) AUG UCC CUG

151. The correct sequence of mRNA codons that would produce valine-valine-alanine-tyrosine is
(1) CAA CAA CGA AUA
(2) CGA CAA UCU AUA
(3) AUA CGA CAA UCU
(4) GUU GUU GCU UAU

152. Treatment that inhibits the complete expression of immediate early genes of T4 bacteriophage with double-stranded DNA genome during infection of an *E. coli* host cell is
(1) nalidixic acid that inhibits DNA replication
(2) rifampicin that inhibits host RNA polymerase
(3) mutation that inactivates phage DNA polymerase
(4) chloramphenicol that inhibits protein synthesis

153. The features that distinguish the fMet-tRNA initiator from Met-tRNA and other tRNAs used during elongation are located
(1) at the 5' end of the tRNA
(2) at the3' end of the tRNA
(3) in the anticodon loop
(4) in the anticodon stem

154. Small nucleolar RNA (snoRNA) is involved in
(1) inhibition of translation from mRNA
(2) processing of mRNA
(3) processing of pre-rRNA to mature rRNA species
(4) site-specific base modifications in rRNA

155. Topoisomerases help in unwinding and rewinding of the DNA. Which or the following statements about these enzymes is correct?
(1) Will not affect transcription in anyways as they are active only when DNA is, replicating and positive supercoiling occurs
(2) Will interfere with transcription by rewinding the DNA which has melted due to the binding of RNA polymerase
(3) Will facilitate transcription because rewinding occurs only after unwinding. These enzymes help in unwinding of the DNA and thus facilitate initiation of transcription
(4) The effect will, vary depending upon the replication status of the DNA

156. Mutation or deletion of which of the following will result in permanent lysogenic state of the bacteriophage lambda?
(1) N (2) CII
(3) CI (4) Cro

157. An inducer is a protein that
(1) combines with a repressor and prevents it from binding the promoter
(2) combines with a repressor and prevents it from binding the operator
(3) binds to the promoter and prevents the repressor from binding to the operator
(4) binds to the operator and prevents the repressor from binding at this site

158. All of the following statements are correct with regard to chromatin, except
(1) core nucleosomes consist of 8 histone protein molecules and about 146 base pairs of DNAs
(2) histones have been highly conserved during evolution
(3) almost all of the DNAs in a eukaryotic cell are assembled into nucleosomes
(4) assembly of DNAs into nucleosomes compacts it about l00-fold in length

159. Which of the following proteins is functionally similar to the translation factor EF - Ts?
(1) G-protein-coupled receptor
(2) Phospholipase C
(3) Tyrosine kinase
(4) Heterotrimeric G-protein

160. Aminoacyl tRNA synthetases are enzymes that charge the amino acids to their cognate tRNA. They have proofreading activity and remove the wrongly attached amino acid. For this activity, the editing domain of the amino acyl tRNA synthetases recognizes which of the following and uses as a substrate?
(1) Nascent polypeptide with incorrect amino acid
(2) Incorrect aminoacyl-ATP
(3) Incorrect aminoacyl-ADP
(4) Incorrect aminoacyl-AMP

161. The proteins synthesized early during infection from early in mRNA of a lytic phage in a susceptible bacterial cell are required for
(1) attachment of the phage to its host cell
(2) entry of the phage nucleic acid into the host cell
(3) replication of the phage nucleic acid
(4) lysis of the host cell

162. Which of the following proteins serves as a repressor only in the presence of the small molecule indicated together?
(1) Tet repressor/ tetracycline
(2) AraC protein/ arabinose
(3) Trp repressor / tryptophan
(4) Lac repressor / IPTG

163. During the second step in RNA splicing
(1) the U1 snRNP binds to the branch point
(2) the branch point 2'-OH attacks the 5' intron-exon junction
(3) the upstream exon attacks the 3' exon-intron junction
(4) formation of the GpppG cap takes place

164. Which of the following is a correct statement?
(1) Not all lytic phages bring about transduction
(2) Mutation in gene encoding RNAI would, result in uncontrolled replication and unlimited copy number
(3) As the name indicates, Kil-Kor system only helps in getting rid of plasmidless cells
(4) Cytidine deaminase encoded by T 4is an essential enzyme

165. In gene regulation, Open Reading frame (ORF) implies
(1) interviening nucleotide sequence in between two genes
(2) a series of triplet codons not interrupted by a stop codon
(3) a series of triplet codons that begins with a start codon and ends with a stop codon
(4) the exonic sequence of a gene that corresponds to 5' UTR of the mRNA and thus does not code for the protein

166. Amino acid selenocysteine (Sec) is incorporated into polypeptide chain during translation by
(1) charging of Sec into tRNAser followed by incorporation through serine codon
(2) charging of serine into tRNAser followed by modification of serine into selenocysteine and then incorporation through serine codon
(3) charging of Sec into tRNAsec and then incorporation through selenocysteine codon
(4) charging of serine into tRNAser followed by modification of serine into selcnocysteine and then incorporation through a specially placed stop codon

167. α-amanitin inhibits
(1) only RNA polymerase I
(2) only RNA polymerase II
(3) only RNA polymerase III
(4) all RNA polymerases

168. How many different transfer RNA molecules are possible in a hypothetical organism where the ribosome detect a quadruplet instead of a triplet codon including the start and stop codon?
(1) 64 (2) 128
(3) 20 (4) 256

169. Which of the following does not contribute to protein diversity (variants of the same protein)?
(1) RNA editing
(2) RNA splicing
(3) RNA interference
(4) alternative initiation of translation

170. If a protein coding gene has an 'amber mutation' and if you want to overcome the effect of this mutation on the protein, which of the second mutations will you introduce into another gene?
(1) Reading frame mutation
(2) Suppressor tRNA
(3) Stop codon mutation
(4) Start codon mutation

171. Which of the following best describes a source or function of short interfering RNA?
(1) Processed from transcripts of endogenous genes
(2) Defends against viral infections
(3) Knock out expression of specific genes
(4) Processed from a self-complimentary stem-loop RNA structure

172. The linear and circular forms of the same DNA molecule can be distinguished using
(1) Absorbance at 260 nm
(2) Endonuclease digestion
(3) Viscosity of the solution
(4) Exonuclease digestion

173. Studying the presence of mRNAs in a sample is useful because
(1) it contains introns, which are the protein coding sequences
(2) it lacks exons, which do not code for proteins
(3) it reflects the genes that are being actively transcribed
(4) it reflects the total number of genes in the genome

174. Which one of the following codons is used for selenocysteine during protein biosynthesis?
(1) UGA
(2) UAG
(3) UGC
(4) UCG

175. Protein-protein interaction can be evaluated by all of the following except
(1) Fat-Western blotting
(2) Chromatin immunoprecipitation
(3) Yeast-two hybrid system
(4) Co-immunoprecipitation

176. RNAi methodology uses double-stranded pieces of RNA to trigger a breakdown or blocking of mRNA. This is often used
(1) To increase the rate of production of an enzyme of pharmacological significance
(2) To decrease the production from a harmful gain-of-function of mutated gene
(3) To mutate an unwanted allele in a homozygous individual
(4) To form a knockout organism that will not pass the deleted sequence to its progeny

177. Which of the following statements about telomeres is false?
(1) Telomeres are synthesized using an RNA template
(2) Telomeres consist of repeated sequences at the ends of eukaryotic chromosomes
(3) Telomerase uses an RNA primer to synthesize the telomeres
(4) A normal adult brain cell has shorter telomeres than a brain cancer cell

178. Which enzymes remove supercoiling in replicating DNA ahead of the replication fork?

(1) Helicases (2) DNA polymerases
(3) Primases (4) Topoisomerases

179. Which reaction in DNA replication is catalysed by DNA ligase?

(1) Addition of new nucleotides to the lagging strand
(2) Addition of new nucleotides to the leading strand
(3) Base pairing of the template and the newly formed DNA strand
(4) Formation of a phosphodiester bond between the 3'-OH of one Okazaki fragment and the 5'-phosphate of the next on the lagging strand

180. In which of the following would you find telomeres?

(1) Human mitochondrial DNA
(2) Human chromosomes
(3) Bacterial chromosomes
(4) The influenza virus genome

181. Which of the following reactions is required for proofreading (i.e. correcting replication errors) during DNA replication by DNA polymerase III?

(1) 3' - 5' exonuclease activity
(2) 5' - 3' exonuclease activity
(3) 3' - 5' endonuclease activity
(4) 5' - 3' endonuclease activity

182. How does the mismatch repair system distinguish between the parental (i.e. correct) DNA strand and the newly synthesised strand containing the mismatched base?

(1) Thymine in the parental strand of the helix is methylated at GATC
(2) Thymine in the new strand of the helix is methylated at GATC
(3) Adenine in the parental strand of the helix is methylated at GATC
(4) Adenine in the new strand of the helix is methylated at GATC

183. What is the name of the DNA repair system in *E. coli* in which dual incisions are made in the damaged part of the double helix, and a 12-13 base segment is removed and replaced with new DNA?

(1) Mismatch repair
(2) Base excision repair
(3) Nucleotide excision repair
(4) AP site repair

184. Which of the following is the name of the human genetic disorder resulting from defects in nucleotide excision repair?

(1) Hereditary nonpolyposis colorectal cancer (HNPC 3)
(2) Xeroderma pigmentosum (XP)
(3) Lynch syndrome
(4) Diabetes

185. During which phase of the cell cycle is DNA replicated?

(1) G1 phase (2) S phase
(3) G2 phase (4) M phase

186. Which of the following statements about the *E. coli* chromosome is correct?

(1) The *E. coli* chromosome is a multiple replicon
(2) Replication begins at oriC
(3) Replication can start at any point in the chromosome
(4) A single replication fork moves around the molecule until the chromosome is completely replicated

187. Which of the following statements is correct?

(1) Telomeres become shorter as cells age
(2) Somatic cells lack telomerase gene
(3) Immortal cancer cell lines always have high levels of telomerase
(4) Telomeres become longer as cells age

188. Which of the following is true of RNA synthesis (transcription)?
(1) RNA synthesis is always in the 5' - 3' direction
(2) RNA polymerase needs a primer to initiate transcription
(3) In transcription, U is inserted opposite T
(4) New nucleotides are added on to the 2' OH of the ribose sugar

189. In bacterial promoters, which of the following describes the 'Pribnow box'?
(1) The 5' untranslated region
(2) The -10 box
(3) The -35 box
(4) The termination sequence

190. The role of the sigma factor in bacterial RNA polymerase is
(1) to catalyse RNA synthesis
(2) to position RNA polymerase correctly on the template DNA
(3) to terminate RNA synthesis
(4) to unwind the DNA template

191. Which of the following statements regarding termination of transcription in prokaryotes is correct?
(1) In Rho dependent termination the Rho factor moves along the DNA template ahead of the RNA polymerase
(2) Rho factor has topoisomerase activity for relieving supercoiling
(3) Termination often involves a stem-loop structure forming in the RNA transcript
(4) Termination often involves a stem-loop structure forming in the DNA template

192. What is the role of eukaryotic RNA polymerase I?
(1) Transcription of mRNA only
(2) Transcription of mRNA, rRNA & tRNA
(3) Transcription of 'small' RNAs including tRNAs, 5S RNAs and snRNAs
(4) Transcription of the major rRNA transcript

193. Which of the following does the abbreviation TBP stand for?
(1) TATA-box binding protein
(2) Transcription associated factor
(3) Transcription factor binding protein
(4) TATA box polymerase

194. Which of the following best describes the 'cap' modification of eukaryotic mRNA?
(1) A modified guanine nucleotide added to the 3' end of the transcript
(2) A modified guanine nucleotide added to the 5' end of the transcript
(3) A string of adenine nucleotides added to the 3' end of the transcript
(4) A string of adenine nucleotides added to the 5' end of the transcript

195. Which of the following statements regarding splicing of eukaryotic mRNA transcripts is correct?
(1) Exons are spliced out and introns are retained in the mature mRNA transcript
(2) Several reactions in the splicing process involve hydrolysis of ATP
(3) Small nuclear RNAs are retained in the mature mRNA transcript
(4) Splicing takes place in the cytosol

196. Which of the following statements about mRNA splicing is true?
(1) The existence of split genes has no advantage
(2) Self-splicing introns do not require the help of any protein for splicing to occur accurately
(3) β-thalassaemia results from a genetic defect in the spliceosome
(4) Splicing occurs in the cytosol

197. Which of the following occur in both eukaryotic and bacterial transcription?
(1) 5' cap
(2) polyA tail
(3) Promoter
(4) RNA-dependent RNA polymerase

198. In terms of lac operon regulation, what happens when *E. coli* is grown in medium containing both glucose and lactose?

(1) Both CAP and the lac repressor are bound to the DNA

(2) CAP is bound to the DNA but the lac repressor is not

(3) Lac repressor is bound to the DNA but CAP is not

(4) Neither CAP nor the lac repressor are bound to the DNA

199. Which of the following can be described as 'a sequence that can be several thousand base pairs upstream or downstream of a eukaryotic promoter and which increases gene expression as much as 200-fold?

(1) CAAT box (2) Enhancer

(3) Insulator (4) TATA box

200. Nuclear receptors belong to which class of transcription factor?

(1) Helix-loop-helix proteins

(2) Helix-turn-helix proteins

(3) Leucine zipper proteins

(4) Zinc finger proteins

ANSWER SHEET
Unit 3: Molecular Biology

1. 4	**2.** 2	**3.** 3	**4.** 3	**5.** 2	**6.** 1	**7.** 1	**8.** 4	**9.** 2	**10.** 1
11. 3	**12.** 3	**13.** 4	**14.** 3	**15.** 1	**16.** 4	**17.** 3	**18.** 4	**19.** 1	**20.** 1
21. 2	**22.** 2	**23.** 4	**24.** 2	**25.** 1	**26.** 3	**27.** 2	**28.** 2	**29.** 1	**30.** 3
31. 3	**32.** 3	**33.** 2	**34.** 3	**35.** 2	**36.** 2	**37.** 3	**38.** 2	**39.** 2	**40.** 1
41. 2	**42.** 3	**43.** 3	**44.** 1	**45.** 2	**46.** 2	**47.** 2	**48.** 2	**49.** 1	**50.** 2
51. 4	**52.** 1	**53.** 1	**54.** 1	**55.** 1	**56.** 2	**57.** 4	**58.** 2	**59.** 2	**60.** 3
61. 4	**62.** 2	**63.** 2	**64.** 3	**65.** 4	**66.** 3	**67.** 1	**68.** 4	**69.** 3	**70.** 2
71. 4	**72.** 1	**73.** 1	**74.** 4	**75.** 1	**76.** 4	**77.** 3	**78.** 3	**79.** 1	**80.** 1
81. 1	**82.** 2	**83.** 4	**84.** 3	**85.** 3	**86.** 1	**87.** 2	**88.** 2	**89.** 2	**90.** 3
91. 2	**92.** 4	**93.** 2	**94.** 4	**95.** 3	**96.** 1	**97.** 3	**98.** 4	**99.** 2	**100.** 3
101. 2	**102.** 3	**103.** 4	**104.** 4	**105.** 4	**106.** 1	**107.** 1	**108.** 2	**109.** 2	**110.** 3
111. 2	**112.** 4	**113.** 4	**114.** 1	**115.** 3	**116.** 4	**117.** 1	**118.** 2	**119.** 1	**120.** 4
121. 3	**122.** 1	**123.** 4	**124.** 2	**125.** 2	**126.** 2	**127.** 3	**128.** 3	**129.** 2	**130.** 2
131. 3	**132.** 4	**133.** 2	**134.** 3	**135.** 2	**136.** 2	**137.** 4	**138.** 3	**139.** 2	**140.** 1
141. 3	**142.** 2	**143.** 2	**144.** 2	**145.** 4	**146.** 4	**147.** 4	**148.** 3	**149.** 2	**150.** 1
151. 4	**152.** 2	**153.** 3	**154.** 3	**155.** 3	**156.** 4	**157.** 2	**158.** 4	**159.** 4	**160.** 4
161. 3	**162.** 3	**163.** 3	**164.** 1	**165.** 3	**166.** 4	**167.** 2	**168.** 4	**169.** 3	**170.** 2
171. 3	**172.** 4	**173.** 3	**174.** 1	**175.** 2	**176.** 2	**177.** 3	**178.** 4	**179.** 4	**180.** 2
181. 1	**182.** 4	**183.** 3	**184.** 2	**185.** 2	**186.** 2	**187.** 1	**188.** 1	**189.** 2	**190.** 2
191. 3	**192.** 4	**193.** 1	**194.** 2	**195.** 2	**196.** 2	**197.** 3	**198.** 4	**199.** 2	**200.** 4

UNIT-4

SIGNALLING, IMMUNOLOGY AND CANCER

1. Specificity and versatility of G-protein-linked receptors is (are) due to
   (1) the receptor
   (2) the G-protein composition, especially the alpha (α) subunit
   (3) the final effector system, e.g. adenylyl cyclase, ion channel, phospholipase C
   (4) all of the above are correct

2. Which of the following kinases evolved later in the time scale of biological evolution?
   (1) Tyrosine kinase
   (2) Serine-threonine kinase
   (3) Receptor tyrosine kinase
   (4) Histidine kinase

3. Immunoglobulin G molecules can be characterized by which of the following statements?
   (1) They are maintained at a constant level in the serum
   (2) They contain mostly carbohydrate
   (3) They can be separated into subunits with a reducing agent and urea
   (4) They can be separated into subunits with a proteolytic enzyme and urea

4. Which of the following statements correctly describes immunoglobulin G?
   (1) Polypeptide chains composing immunoglobulins are held together by hydrogen bonds
   (2) Each immunoglobulin has two antigen-binding sites per molecule
   (3) Heavy immunoglobulin chains have constant N-terminal regions and light immunoglobulin chains have variable C-terminal regions
   (4) Antigen-binding sites of all different antibodies are determined after encountering specific antigens

5. Patient comes with symptoms of autonomic hemolysis. The diagnostic tests reveal that he has auto-antibodies to red blood cells (RBCs) Which one of the following mechanisms is the cause of this condition?
   (1) Neutrophils release granzymes which lyse RBCs
   (2) Complement is activated and membrane attack complex lyse RBCs
   (3) Cytotoxic T-cells lyse RBCs
   (4) Interleukin-2 binds to the receptor on RBCs

6. Immunoglobulin G was fractioned by size exclusion chromatography and it was observed to have a molecular mass of 150 kd. SDS PAGE analysis under reducing condition revealed two band of size 50 kd and 25 kd. The oligomeric status of the protein consist of
   (1) 3 polypeptide chains of molecular mass 50 kd
   (2) 6 polypeptide chains of molecular mass 25 kd
   (3) one polypeptide chain of 50 kd and 4 polypeptide of 25 kd
   (4) two polypeptides of 50 kd and two of 25 kd

7. Among the following live vaccine is-
(1) Poliomycetes (2) Small pox
(3) Diptheria (4) Tetanus

8. *Neisseria meningitides* is a
(1) Gram positive bacillus
(2) Gram positive coccus
(3) Gram negative bacillus
(4) Gram negative coccus

9. The major protease involved in apoptosis has at its active site
(1) serine (2) aspartate
(3) cysteine (4) histidine

10. Diacylglycerol is known to activate
(1) phospholipase A
(2) phospholipase C
(3) protein kinase A
(4) protein kinase C

11. Hormones that act on cells near the point of their synthesis and not transported through blood circulation are
(1) estradiol and cortisol
(2) prostaglandins and thromboxane
(3) thyroxine and glucagon
(4) prednisolone and prednisone

12. Which of the following is correct about dominant-negative mutation?
(1) Loss-of-function mutation that completely abolishes the activity of a gene
(2) Either reduces or abolishes the activity of the gene
(3) Blocks gene activity, causing a loss-of-function even in presence of normal gene product
(4) Increases the activity of the gene or makes it active in inappropriate circumstances

13. Immuno-suppression is mediated by T-cells having
(1) $CD4^{+}CD25^{-}$ (2) $CD8^{+}CD25^{-}$
(3) $CD8^{+}CD25^{+}$ (4) $CD4^{+}CD25^{+}$

14. The target cells for ADA gene therapy are
(1) bone marrow cells
(2) B-lymphocytes
(3) Liver cells
(4) Spleen cells

15. All of the following are angiogenic factors except
(1) VEGF (2) Ang-1
(3) Endostatin (4) Cox-2

16. Pasteurization of milk is done by
(1) boiling the milk for 20 minutes
(2) heating the milk at 72^0 C for 30 min
(3) heating the milk at 72^0 C for 20 min
(4) heating the milk at 62^0 C for 20 min

17. Long terminal repeats are found in
(1) proviral DNA
(2) retroviral RNA
(3) reoviral genome
(4) influenza virus

18. Penicillin is not effective in treating *mycoplasma* infection, because mycoplasma
(1) produce penicillinase
(2) are gram-positive
(3) are gram-negative
(4) do not have cell wall

19. Quorum sensing refers to the mechanism by which
(1) aquatic microbes can determine the depth of water
(2) certain microbes sense the day-length
(3) certain microbes sense their population size
(4) migratory birds determine their path of flight

20. What are the tiny molecules of naked circular RNA that infect plants?
(1) Prions
(2) Viroids
(3) RNA viruses
(4) Tobacco mosaic virus

21. Which of the following kinases is involved in signalling in growth promotion pathway?
(1) Tyrosine kinase
(2) Serine-threonine kinase
(3) Receptor tyrosine kinase
(4) Histidine kinase

22. Bacteria after phagocytosis by macrophages are confined to phagosomal compartments only under such conditions
(1) only CD_4^+ T cells can be activated
(2) only CD_8^+ T cells can be activated
(3) no T cells can be activated
(4) both CD_4^+ and CD_8^+ T cells can be activated

23. The avidity of antigen-antibody reaction will differ between antibody in serum and on the cell surface
(1) only in the case of IgM antibodies
(2) only in the case of IgG antibodies
(3) only in the case of IgE antibodies
(4) only in the case of IgD antibodies

24. Which one of the following viruses is not transmitted sexually
(1) Human immunodeficiency virus
(2) Hepatitis B virus
(3) Adenovirus
(4) Rubella virus

25. Which of the following cell surface markers is normally associated with both T and B cells?
(1) CD19 antigens
(2) Receptors for tumor necrosis factor (TNF)
(3) CD3 antigens
(4) MHC class I (MHC-I) gene products

26. Which of the following is not required for an immediate hypersensitivity reaction?
(1) IgE
(2) Mast cells
(3) IgE receptors
(4) IgD

27. If you attempt to isolate DNA from matured human erythrocyte, you will get
(1) no DNA as there are no nucleus and mitochondria
(2) only mitochondrial DNA
(3) nuclear DNA as chromosomes are present in the cytoplasm like prokaryotes
(4) both mitochondrial DNA and nuclear chromosomal DNA

28. Which of the following is not true for neutrophils?
(1) Their nucleus can have multiple lobes
(2) They are first to reach the site of inflammation
(3) Their cytoplasmic granules are strongly stained with basic dyes
(4) They constitute the most predominant class of leukocytes in blood

29. Which of the following is an example of a G-protein coupled receptor signaling?
(1) Rhodopsin
(2) Platelet derived growth factor receptor
(3) Insulin receptor
(4) Nerve growth factor receptor

30. A number of viruses have RNA genomes. In positive-strand viruses,
(1) the RNA genome codes directly for viral protein
(2) a complementary copy of the genome is translated to form viral proteins
(3) both the strands can be utilized for protein synthesis
(4) none of the strands have information for proteins

31. Lyme disease, caused by *Borrelia burgdorferi*, can be passed directly from deer to humans when ticks simply transfer infected blood from one to the other. *Borrelia* may be said, therefore, to have
(1) alternative hosts and a vector
(2) alternative and alternate hosts
(3) alternate hosts and a vector
(4) neither alternate nor alternative hosts

32. Which of the following is least likely to be true in relation to SCID (Severe Combined Immuno Deficiency), caused due to complete lack of adenosine deaminase?

(1) Loss of the enzyme causes increased levels of dATP because there is less turn over of adenosine nucleosides in general

(2) Increased dATP decreases the concentration of all rNTPs, blocking RNA synthesis

(3) Increased dATP inhibits ribonucleotide reductase, such that de novo production of all dNDPs is inhibited

(4) Adenosine deaminase loss causes SCID, because T cells are particularly sensitive to DNA replication inhibition

33. A typical receptor senses extracellular stimuli by virtue of its localization on plasma membrane. The receptor to which of the following ligands is an exception to this rule?

(1) γ-amino butyric acid

(2) Acetylcholine

(3) Estrogen

(4) Luteinizing hormone

34. Which one of the following mutants is used to carry out genetic analysis to determine the function of an essential gene?

(1) Knock out mutant

(2) Deletion mutant

(3) Insertion mutant

(4) Temperature sensitive mutant

35. In eukaryotes which of the following is not a secondary messenger?

(1) DAG

(2) IP_3

(3) Ca^{2+}

(4) Cyclin-A

36. Choose the incorrect statement?

(1) Phospholipase C generates DAG and IP3 from PIP2

(2) Phospholipase C-γ is not a membrane bound enzyme

(3) DAG activates protein kinase C under increased Ca^{2+} concentration

(4) IP3 induces the opening of Ca^{2+} channel present on ER

37. Which of the following is involved in cell signaling process

(1) cytochrome P450

(2) protein kinase C

(3) zinc finger

(4) helix turn helix

38. All of the following statement apply to G proteins except

(1) G proteins transmit a signal from the cell surface to the interior of cell

(2) G protein do not use second messangers but transmits the signal directly into the nucleus

(3) all G proteins have similar structure

(4) G proteins act to amplify the signal creating a cascade response in the cell

39. Which one of the following bacterial toxin does not have ADP-ribosylation activity

(1) *Cholera* toxin (2) *Diptheria* toxin

(3) Exotoxin A (4) *S. aureus* toxin

40. The receptor for which of the following hormone acts as transcription factor?

(1) Insulin (2) Glucagon

(3) Cortisol (4) Adrenalin

41. Which of the following statement is false about nitric oxide gas?

(1) An intracellular and inter cellular signaling molecule

(2) Deamination of histidine results in NO production

(3) Stimulates guanylyl cyclase to produce cGMP

(4) Can be produced by activated neutrophils

42. Calmodulin activates protein kinase in response to a transient increase in

(1) IP_3 (2) DAG

(3) Ca^{2+} (4) NO

43. Which of the following subunit of trimeric G-protein acts as GTPase switch protein subunit?

(1) α (2) β
(3) γ (4) βγ

44. Synaptic signaling is example of
(1) endocrine signaling
(2) paracrine signaling
(3) aurocrine signaling
(4) neuro signalling

45. The receptor of NO is
(1) intercellular (2) intracellular
(3) extracellular (4) transcellular

46. What is the main damaging effect of X radiation on DNA?
(1) Depurination
(2) Formation of thymine dimers
(3) Substitution of bases
(4) Double strand break

47. Messenger molecules released by virus infected cells that bind to the surfaces of healthy cells and stimulate them to synthesize proteins that prevent viral reproduction are called
(1) Interferons (2) Antibiotics
(3) Lymphokines (4) Cytotoxins

48. Which of the following statements about cAMP and its functioning in hormone action are correct?
(1) Most effects of cAMP in eukaryotic cells are exerted through the activation of protein kinase A (PKA)
(2) Cyclic AMP binds the catalytic subunits of PKA and activates the enzyme allosterically
(3) Cyclic AMP binds the regulatory subunits of PKA and inactivates the enzyme by releasing the catalytic subunits
(4) Cyclic AMP is bound by the activated hormone receptor and PKA simultaneously to convey the hormonal signal in order to activate the kinase

49. P53 protein is associated with all of the following except
(1) tumor suppression
(2) programmed cell death
(3) transcription
(4) post-transcriptional modifications

50. Proto-oncogenes are
(1) oncogenes found in transforming retro viruses
(2) oncogenes present in humans
(3) genes encoding oncogene related proteins in extinct organism
(4) cellular genes encoding proteins related to viral oncogenes

51. Mitochondria is involved in all of the following except
(1) ATP production
(2) Apoptosis
(3) Tri carboxylic acid cycle
(4) Fatty acid biosynthesis

52. Comparative genomic hybridization is a technique to look for certain types of genetic changes in cancer cell by comparing the DNA from cancer cells with that of corresponding normal cells. Which types of genetic changes are detected here?
(1) Base mismatch
(2) Gene deletion and amplification
(3) Loss of single base
(4) Promoter methylation

53. The normal cellular counterpart of oncogenes are important for the following function except
(1) inhibition of apoptosis
(2) promotion of cell cycle progression
(3) promotion of DNA replication
(4) promotion of repair and apoptosis

54. Which of the following biochemical reactions is most commonly utilized by living cells to propagate intracellular signals?
(1) Acylation (2) Phosphorylation
(3) Methylation (4) Decarboxylation

55. Which of the following hormone passes through the cell membrane, binds to its intracellular receptors and activates it?
(1) Insulin (2) Thyroxine
(3) Prostaglandins (4) Acetyl choline

56. The amino acid sequence of a novel membrane protein contains four immunoglobulin-like domains and six fibronectin-like repeats. This protein is most likely a
(1) hormone responsive ion channel
(2) cell adhesion molecule
(3) G-protein
(4) transcription factor

57. Which of the following receptors is not a signalling receptor?
(1) Cytokine receptor
(2) Chemokine receptor
(3) T-cell receptor
(4) Mannose receptor

58. Which of the following is a lipid with a signal transducing activity?
(1) Phosphatidyl serine
(2) Phosphatidyl ethanolamine 5
(3) Phosphatidyl inositol 4,5- bisphosphate
(4) Phospholipase A2

59. An extracellular ligand will
(1) elicit the same response in various cells that have a receptor for the ligand
(2) elicit the same response but to varying degrees in various cells that have a receptor for the ligand
(3) may elicit different responses in various cells that have a receptor for the ligand
(4) elicit the same response in all types of cells because receptors have to be identical to bind to the same ligand

60. Elevation of intracellular inositol triphosphate (IP3) results in the release of Ca^{2+} from which of the following organelles?
(1) Mitochondria
(2) Smooth endoplasmic reticulum
(3) Peroxisome
(4) Golgi-complex

61. Receptor tyrosine kinases
P. are seven-transmembrane-helix receptors
Q. are integral membrane enzyme
R. activate their targets via the G protein cascade
S. are often activated by ligand-induced dimerization
(1) P and Q (2) R and S
(3) P and R (4) Q and S

62. Which of the following are not correct pairs of characteristics that differentiate benign and malignant tumors?
(1) Polyclonal / monoclonal
(2) Rarely may result / in death frequently results in death
(3) Cannot invade / can invade
(4) Cannot metastasize / may metastasize

63. Which of the following is a true statement regarding p53 tumor suppressor gene?
(1) Inherited mutations are the commonest cause of loss of function
(2) p53 does not have a role in the regulation of the expression of genes involved in the control of apoptosis
(3) p53 is activated when DNA damage occurs
(4) p53 phosphorylation results in loss of its ability in blocking entry of the cell into S-phase

64. Angiogenesis contributes to cancer development by
(1) providing nutrients and oxygen to tumors, and by facilitating metastasis
(2) initiating a mutation in a gene that causes uncontrolled cell growth
(3) inhibiting apoptosis (programmed cell death)
(4) preventing the normal inhibition of growth that occurs between cells upon contact

65. Bcl-2 and Bcl-XL proteins involved in apoptosis are
(1) proapoptic and anti-apoptic
(2) both proapoptic
(3) both anti-apoptic
(4) anti-apoptic and Pro-apoptic

66. If you wish to produce a subunit vaccine for a positive sense RNA that will stimulate the production of neutralizing antibody in the recipient, which of the following proteins would be a logical candidate for such a vaccine?
(1) Viral capsomere protein
(2) Viral protein that binds to its RNA genome
(3) Viral RNA polymerase
(4) Viral matrix protein

67. Influenza viruses require the presence of the nucleus in their host cells, because
(1) they use reverse transcriptase to make a cDNA which is integrated into the host genome
(2) they scavenge poly [A] tails from host mRNAs in the nucleus
(3) they use the host RNA polymerase II to transcribe viral mRNAs
(4) they use the host RNA splicing enzymes to process some viral mRNAs

68. Which immunoglobulin is produced first by neonate, present on membrane of mature B cells and effective in activating classical complement pathway?
(1) IgG (2) IgA
(3) IgM (4) IgE

69. Although all members of a species inherit the same set of isotype genes, multiple alleles exist for some genes. These alleles encode subtle amino acid differences in structure of antibody. Such antibodies are termed as
(1) Isotype
(2) Allotype
(3) Idiotype
(4) Epitope

70. During serum sickness, kidney damage occurs as immune complexes form. Why?
(1) The immune complexes are filtered by the kidneys and damage results from concomitant complement activation and neutrophil activity
(2) Antigen presenting cells rapidly bind all of the complexes in their MHC-encoded receptors
(3) Free antigen activates PAMP receptors in the kidneys which rapidly activate adaptive immunity
(4) The immune complexes bind to Mast cells and are destroyed

71. Contact dermatitis generally occurs against substances that are too small to induce an immune response. How do these substance induce an immune response?
(1) These substances form depots and are then slowly released into the blood
(2) These low molecular weight substances react with liver enzymes and are difficult to eliminate
(3) These substances bind to tissues & cells, resulting in a larger total antigenic size
(4) The substances trigger the complement cascade and cause neutrophils to accumulate and to serve as antigen presenting cells

72. Innate host defense mechanisms are critical to the protection of the body because
(1) they utilize pre-committed antigen presenting cells that have already been induced by other immune responses
(2) the antibodies derived from the innate response are critical to neutralize bacterial toxins
(3) they are highly specific for the invading pathogens that avoid PAMP receptor recognition
(4) they provide immediate, continuous protection in the absence of a specific immune response

73. Plutonian immunoglobulin molecules follow the same rules of proportions that are found in human immunoglobulins. If you were told that Plutonian light chains had a molecular weight of 4 kDa (each), what would you expect the molecular weight of their IgG molecules to be?
(1) 150,000 kDa
(2) 900,000 kDa
(3) 12 kDa
(4) 24 kDa

74. Antibody diversity is generated by
(1) Somatic hypermutation
(2) DNA rearrangement
(3) Nucleotide insertion
(4) All of the above

75. Gell and Coombs classified hypersensitivity into 4 types. Immunocomplex mediated hypersensitivity belongs to
(1) Type I (2) Type II
(3) Type III (4) Type IV

76. Which of the following is correct for two component signaling pathway
(1) it is present in prokaryotes only
(2) receptor has histidine kinase activity which always becomes active on binding with compatible ligand
(3) phosporylation of histidine of sensory kinase is transferred to conserved aspartic acid on respose regulators
(4) it is generally used by bacteria, plants, animals and viruses for non-polar signals

77. Which of the following statement is false about apoptosis?
(1) bcl-2 and bcl-xl are anti-apoptic
(2) cytochrome C activates intrinsic pathway of apoptosis
(3) The executioner of apoptosis are caspaces
(4) It is reversible process

78. Which of the following antibiotics do not affect the cell wall of synthesis of bacteria?
(1) Penicillin (2) Ampicillin
(3) Vancomycin (4) Isoniazid

79. The cell wall of microorganisms is coated with certain plasma protein promoting the attachment of microbe to phagocytes, only then they can be phagocytosised. The coat protein are called as
(1) globins (2) opsonins
(3) ovulbumins (4) phagosonins

80. The membrane of mature B cells have
(1) both IgG & IgM
(2) both IgG & IgD
(3) both IgM & IgE
(4) both IgM & IgD

81. Antibody coated pathogens are recognized by effector cells through
(1) CD4 receptor (2) FC receptor
(3) CD8 receptor (4) IFN gamma receptor

82. Idiotypic determinants of an antibody are associated with the
(1) constant region of the heavy chains
(2) constant region of the light chains
(3) variable region
(4) constant regions of light and heavy chains

83. Identification of blood groups involves
(1) precipitation (2) neutralization
(3) opsonization (4) agglutination

84. B-lymphocytes originate from the bone marrow whereas T-lymphocytes originate from
(1) thymus (2) bone marrow
(3) spleen (4) liver

85. Choose the correct signal transduction pathway
(1) Hormone → 7 TM receptor → G protein → cAMP → PKA
(2) Hormone → G protein → 7 TM receptor → cAMP → PKA
(3) Hormone → 7 TM receptor → G protein → PKA → cAMP
(4) Hormone → 7 TM receptor → cAMP → G protein → PKA

86. Graft rejection does not involve
(1) erythrocyte
(2) T cells
(3) macrophages
(4) polymorphonuclear leukocytes

87. Which of the following statement is false regarding antibody diversity?
(1) Vk gene segement sometimes joins Cλ gene segments
(2) With exception of a switch to IgD, immunoglobulin class switching is mediated by DNA rearrangement
(3) Separate exon encode the transmembrane portion of each membrane immunoglobulin
(4) Although each B cell caries two alleles encoding the immunoglobulin heavy and light chain, only one allele is expressed

88. B-cells are distinguished from T-cells by the presence of
(1) CD3 (2) CD4
(3) Class I MHC (4) Surface Igm

89. Ig heavy chains are
(1) are decided by a constant exon of gene
(2) not glycosylated
(3) not important to binding of antigen
(4) expressed by T-cells

90. Which of the following statements regarding monoclonal antibodies is false?
(1) Monoclonal antibodies are capable of binding to only a single antigenic determinant
(2) Monoclonal antibodies produced in rats can be used clinically in the treatment of disease in humans
(3) Monoclonal antibodies can be used in immunoassays
(4) Monoclonal antibodies can be used for immunotherapy

91. Which of the following statements regarding immunoglobulins is false?
(1) The constant regions determine the 'class' of the antibody
(2) Immunoglobulins consist of four polypeptide chains
(3) Polypeptide chains in immunoglobulins are held together by disulfide bonds
(4) The antigen-binding sites on an immunoglobulin are different and allow binding to two different antigens

92. Which technique can be used for derivation of monoclonal antibodies without immunization?
(1) Knock out technology
(2) Site directed mutagenesis
(3) Phage display libraries
(4) Hybridoma technology

93. A mouse has been immunized with HSA, and separately with ARS-KLH. Which of the following would elicit a least anti-ARS IgG response if the mouse is challenged four weeks later? (HSA=human serum albumin; ARS=arsonate hapten; KLH=the protein keyhole limpet hemocyanin; the two proteins are unrelated)
(1) ARS-KLH alone
(2) ARS-HSA alone
(3) ARS-HSA together with KLH
(4) KLH alone

94. Many cells in the body divide only rarely, if at all; neurons, red blood cells, and keratinocytes are extreme examples. In which portion of the cell cycle would such cells be considered to be?
(1) M phase (2) G1 phase
(3) G0 phase (4) Senescence phase

95. Th2 response is generated and maintained mainly by which of the following pair of cytokines?
(1) IL-4 and IL-10
(2) IL-12 and IFN-γ
(3) IFN-γ and TNF-α
(4) IL-2 and IL-12

96. Allotypes are
(1) antigenic determinants which segregate within a species
(2) critical to the function of the antibody combining site
(3) involved in specificity
(4) involved in memory

97. Which of the following is not a member of the Immunoglobulin supergene family?
(1) Antibodies
(2) lymphokines
(3) TCR
(4) Fc receptor on leukocytes

98. HAT selection is based on
(1) TK and HPRT genes
(2) APRT and ATK genes
(3) HK and AP genes
(4) HAT gene

99. The endogenous GTPase activity of G-proteins serves to
(1) stimulate the activity of enzymes by producing energy
(2) synthesize cGMP as a second messenger
(3) synthesize GTP as an energy source
(4) hydrolyze GTP returning the G protein to a pre-stimulated level of activity

100. Which of the following statements about tumor suppressors is true?
(1) Tumor suppressors are mutant tRNAs that recognize stop codons
(2) Tumor suppressors are mutated viral versions of cellular proteins involved in signal ransduction
(3) Recessive mutations that inactivate the Rb tumor suppressor are found in families with high incidence of retinoblastomas
(4) Viral oncogenes can act by increasing the activity of cellular tumor suppressor p53

101. The insulin receptor functions as a
(1) receptor with 7 transmembrane spanning regions
(2) nuclear protein that acts as a transcription factor
(3) receptor guanylate cyclase
(4) tyrosine kinase

102. The gene that increases susceptibility to breast cancer is
(1) p53 (2) BRCA-1
(3) Rb-1 (4) H-Ras

103. In the urine of Burkitt's lymphoma patient abnormal quantities of the following is detected
(1) Bence-Jones Proteins
(2) Human Chorionic Gonadotrophin (hCG)
(3) Carcinoembryonic antigen (CEA)
(4) Alpha-fetoprotein (AFP)

104. What would be the functional consequence for the immune system in a knock out mice lacking β2 microglobulin?
(1) Loss of TCR expression
(2) Loss of phagocytic ability
(3) Loss of structural integrity of MHC II
(4) Loss of structural integrity of MHC I

105. Frequent development of primary tumours of reticulo-endothelial system is due to
(1) Acquired haemolytic anemia
(2) Hypergammaglobulinemia
(3) Corticosteroid abuse
(4) Impairment of CMI

106. Negative selection of T-cells depends on
(1) high affinity to self antigen
(2) high affinity to thymosin
(3) intermediate affinity to self antigen
(4) low affinity to self MHC

107. Junctional diversity affects primarily the amino acid sequence in
(1) all CDR equally (2) CDR1
(3) CDR2 (4) CDR3

108. A child stung by a bee experiences respiratory distress within minutes and lapses into unconsciousness. This reaction is probably mediated by
(1) IgE antibody
(2) IgG antibody
(3) sensitized T cells
(4) complement

109. In the immune response to a hapten-protein conjugate, in order to get anti-hapten antibodies it is essential that
(1) the hapten be recognized by helper T cells
(2) the protein be recognized by helper T cells
(3) the protein be recognized by B cells
(4) the hapten be recognized by suppressor T-cells

110. The infiltration of cells at site of the adjuvant infection often results in formation of dense, macrophage rich mass of cell called
(1) Sarcoma (2) Lymphoma
(3) Granuloma (4) Myeloma

111. Which of the following is a characteristic of antigen recognized by T-cells
(1) interaction with antigen results in formation of binary complex between membrane Igm and Ag
(2) it can bind to soluble antigen which chemically may be protein, polysaccharide or lipid
(3) epitope is linear peptide produced by processing of antigen and bound to MHC molecules
(4) antigens must be B-dependent to be recognized by T cells

112. Immunoglobulins belongs to
(1) alpha globulin fraction
(2) gamma globulin fraction
(3) albumin fraction
(4) all of the above

113. What substance might you add to a culture dish to block the migration of neural crest cells?
(1) Trypsin, which digests proteins; a peptide containing RGD
(2) Neuraminidase, which removes sialic acid from an oligosaccharide;
(3) Collagenase, which digests collagen or hyaluronidase, which digests hyaluronic acid
(4) EDTA, which binds Ca^{2+} ions in the medium

114. The p21 and p15 proteins are examples of
(1) Cdk inhibitors (2) cyclins
(3) oncogenes (4) growth factors

115. Which of the following statement regarding hapten and carrier is true?
(1) Haptens are large protein molecules such as BSA
(2) Carrier are needed only if one wants to elicit cell mediated response
(3) It is necessary to immunize with hapten carrier complex in order to obtain antibodies directed against the hapten
(4) Carriers include small molecules such as dinitrophenol, penicillenic acid and amino benzoic acid

116. Toxic shock is caused by
(1) toxins produced by some bacteria
(2) excessive stimulation of a large proportion of T cells by bacterial superantigens
(3) abnormal cytokine production by B cells
(4) excessive production of immunoglobulins

117. The complementary determining region of antibody lies in
(1) entire Fab segment
(2) variable and Constant portion of light chain
(3) variable and constant portion of heavy chain
(4) variable portion of light and heavy chain

118. The process by which small cluster of cancerous cells dislodge from a tumor, invade the blood or lymphatic vessels, and are carried to other tissues where they continue to proliferate is termed as

(1) neoplasm (2) angiogenesis
(3) metastasis (4) diapedesis

119. The cancers which are most common world wide is

(1) lung Cancer (2) breast Cancer
(3) colon Cancer (4) prostrate Cancer

120. The most common viral-borne cancer world wide is

(1) breast (2) liver
(3) prostrate (4) cervical

121. Which of the following virus is capable of uncontrolled growth in human cell

(1) SV40 (2) Polioma Virus
(3) Adenovirus (4) Papilloma Virus

122. Which of the following is not an oncogenes

(1) Growth factor receptors (fus, erbB, neu, erbA)
(2) Signal tranducers (src, abl, ras)
(3) Cell Cycle arresting proteins (P53, P21, Rb, ATM, ATR)
(4) Transription factors (jun, fos, myc)

123. Conversion of proto-oncogene into oncogene involves

(1) Mutation in Promoter
(2) Gene fragmentation
(3) Gene amplification
(4) Mutation in intron

124. What do the detection of odorants and sweet tastants have in common?

(1) Detection is communicated through hyperpolarization of the plasma membrane
(2) The second messenger Ca^{2+} is generated
(3) Receptors for these stimuli are found on neurons
(4) Receptor on depolarization releases the neurotransmitters

125. The major cell of immune system involved in tumor recognition are

(1) NK cells and macrophage
(2) NK cells and CTL cells
(3) Macrophage and CTL cells
(4) B-lymphocytes and T helper cells

126. Which statement is false regarding tumor evasion of the immune system?

(1) Antitumor monoclonal antibody are not effective in repress tumor growth
(2) Antibodies can modulate tumor antigens
(3) Tumor cells frequently express low levels of MHC-I molecules
(4) Tumor cells may provide poor co-stimulatory signals

127. Which of the following is mismatch between monoclonal antibody and cancer against which it is used?

(1) Herceptin (Trastuzumab) → Colorectal cancer
(2) Avastin (Bevacizumab)→Colorectal cancer
(3) Rituxan (Rituximab)→Non-hodgkin's lymphoma
(4) Erbitux (Cetuximab) → Head and neck cancers

128. Antibodies raised against idiotypic determinants on cancer cells can be used for treatment of B-cell lymphoma. They are termed as

(1) Anti-idiotype antibody
(2) Anti-Isotype antibody
(3) Polyclonal antibody
(4) Idiotypic antibody

129. Which of the following can be used as anti-HIV drugs for clinical use

(1) Reverse transrcipatase inhibitors (ddI, AZT, ddC)
(2) Protease inhibitors (Crisivan, Viracept, Norvir)
(3) Fusion inhibitors (Fuzeon)
(4) All of the above

130. Which of the following statement is false?
(1) Hereditary retinoblastomas results from loss of a cellular Rb anti-oncogene
(2) Multiple copies of cellular oncogenes are sometimes observed in cancer cells
(3) All oncogenic reteroviruses carry viral oncogenes
(4) Viral integration into the cellular genome may convert a protooncogene into transforming oncogene

131. Loss-of-function mutations in which of the following genes typifies colorectal carcinomas?
(1) myc (2) K-ras
(3) Rb (4) APC

132. Innate immunity includes molecular and cellular mechanisms predeployed before an infection and poised to prevent or eliminate it. Among the following which is not a component of innate response
(1) Natural Killer cells
(2) Macrophages
(3) Complement proteins
(4) Immunoglobulins

133. Which of the following is not an antigen presenting cell
(1) T-lymphocyte (2) B-lymphocyte
(3) Dendritic Cell (4) Macrophage

134. Which property is not shared by malignant tumor cells and transformed cells in culture?
(1) Reduced growth factor requirement
(2) Attachment-dependent growth
(3) Loss of actin microfilaments
(4) Altered morphology

135. Transcription factor essential for differentiation of lymphoid lineage is
(1) GATA 1 and GATA 2
(2) Ikaros
(3) Oct-2
(4) Bmi-1

136. The gene which is over-expressed in B-cell lymphoma due to translocation into immunoglobulin heavy chain locus is
(1) ras
(2) myc
(3) bcl-2
(4) bax

137. Which of the following can lead to loss-of-heterozygosity in a tumor-suppressor gene?
(1) Deletion of the normal copy
(2) Nondysjunction during mitosis
(3) Mitotic recombination between a wild-type and mutant allele
(4) All of the above

138. With regard to p53 tumour suppressor gene, which of the following statement is false?
(1) It is located on chromosome 17
(2) It encodes 53 kDa protein
(3) It holds the cell cycle in the G1 phase
(4) Both copies of gene must be mutated for loss of function

139. The gene for which of the following would be considered a protooncogene?
(1) p53
(2) Rb protein
(3) p21 (Cdk inhibitor protein)
(4) Ras

140. Which one of the following donot act as secondary messanger during signaling process?
(1) DAG (2) PIP_3
(3) PIP_2 (4) IP_3

141. Which of the following is most commonly stylized by living cells to propagate intracellular signals
(1) Acylation
(2) Phosphorylation
(3) Protelytic cleavage
(4) Methylation

142. Which of the following is a mis-match
(1) Diacyl glycerol synthesis – Phospholipase A
(2) CREB phosphorylation – Protein Kinase A
(3) GTP hydrolysis – Ras
(4) Increase in cytosolic Ca^{2+} – IP_3

143. Which of the following will have membrane receptors?
(1) Progestrone
(2) Thyroxine
(3) Prostaglandin
(4) Cortosol

144. Two component signaling pathway is present in
(1) Bacteria only
(2) Bacteria and plants
(3) Bacteria, plants and animals
(4) Bacteria, plants, animals and viruses

145. Protooncogenes are
(1) Oncogenes found in transforming viruses
(2) Oncogenes present in bacteria
(3) Genes encoding oncogene related proteins in extinct organism
(4) Genes encoding protein related to growth promotion

146. Choose the mis-match
(1) Rho and Ras –GTP binding proteins
(2) Raf – Serine/threonine kinase
(3) p53 – tumor suppressor gene
(4) myc- protooncogene

147. Which statement is false about elastin fibre protein?
(1) Elastin is rich in aminoacids glycine and proline
(2) In contrast to collagen the proline is not extensively hydroxylated
(3) Hydoxy lysine is absent in elastin
(4) Elastin molecules are crosslinked to each other by covalent bonds involving amino acid hydroxy proline

148. Cell-cell recognition and adhesion is mainly due to
(1) Collagen
(2) Fibronectin
(3) Cadherins
(4) Glycosaminglycans

149. Which of the following statements about familial hypercholesterolemia is false?
(1) Familial hypercholesteremia is caused by a defect in membrane transport
(2) Familial hypercholesteremia can cause early heart attack and stroke
(3) Sufferers from this disease have high cholesterol levels in their liver cells
(4) Familial hypercholesterolemia can be due to a single point mutation

150. Cell junctions which acts as permeability barrier especially in organs like bladder where seepage of urine back to body tissue is prevented and also keeps the plasma membrane of epithelial cell polarized is
(1) Gap junction
(2) Adherens junction
(3) Tight junction
(4) Desmosomes

151. Which of the following is not a cell adhesion molecule (CAM)
(1) Firbronectin
(2) Cadherin
(3) Integrin
(4) Selectin

152. Which is not a cell to cell interaction?
(1) Plasmodesmata
(2) Desmosomes
(3) Hemidesmosomes
(4) Gap junctions

153. Which of the following is not a signaling molecule?
(1) CO (2) NO
(3) O_2 (4) Ethylene

154. Which of the following statements about the actions or targets of the second messengers of the phosphoinositide cascade is not correct?
(1) Diacylglycerol (DAG) activates protein kinase C (PKC)
(2) Most of the effects of IP3 and DAG are antagonistic
(3) DAG increases the affinity of PKC for Ca^{2+}
(4) PKC requires Ca^{2+} for its activity

155. Which of the following statements about the phosphoinositide cascade is not correct?
(1) The phosphoinositide cascade depends on the hydrolysis of a phospholipid component of the plasma membrane
(2) In some cases, a G protein system acts to transduce the stimulus from the receptor to the phosphoinositidase
(3) At least four kinds of phospholipase C play a crucial role in the phosphoinositide cascade
(4) The phosphoinositide cascade directly produces a single unique second messenger molecule

156. Apoptosis refers to
(1) limb regeneration in vertebrates
(2) metamorphosis in *C. elegans*
(3) genetically programmed cell death
(4) formation of imaginal disc in Drosophila

157. Protein phosphorylation is part of most pathways of intracellular signal transduction. All of the following statements about phosphorylation are correct except
(1) phosphorylation can be on a serine or threonine
(2) phosphorylation changes the structure and function of the protein
(3) protein kinase cascades result in amplification of the signal
(4) tyrosine kinases are always part of the receptor protein

158. The receptors for cytokines and some growth factors mainly belongs to
(1) RTK Family
(2) Non-RTK family
(3) GPCR family
(4) Receptor STK family

159. Which of the following bacterial toxin do not have catalytic ability for ADP-ribosylation?
(1) Diphtheria toxin
(2) Cholera toxin
(3) Pertusis toxin
(4) Anthrax toxin

160. Which of the following antibiotic do not inhibit cell wall synthesis in bacteria?
(1)Penicillin and Ampicillin
(2) Carbenicillin and Methicillin
(3) Cyclosporin, Vancomycin and Bacitracin
(4) Ciprofloxin, Rifampicin & polymxyin B

161. Which of the following used as receptor for entry of Rabies virus in target cell?
(1) Sialic acid containing glycoprotein
(2) C_3d complement receptor
(3) Immunoglobulin superfamily
(4) Acetyl choline receptor

162. Which of the following is not a property of attenuated vaccine?
(1) Avirulent forms of pathogen derived from virulent pathogens by culturing them under stress conditions
(2) Generally requires a single booster
(3) They are highly stable vaccines cannot revert back
(4) They induces both humoral and cell mediated immunity

163. Among the organs of the body that are rich in mature T cells and B cells are the
(1) Liver and gall bladder
(2) Brain and spinal cord
(3) Small and large intestine
(4) Spleen and lymph nodes

164. Which of the following statement is true?
(1) Infection has influence on rate of haematopoesis
(2) Follicular dendritic cells can process and present antigen to T lymphocytes
(3) All lymphoid cells have antigen specific receptors on their membrane
(4) All vertebrates produce both B or T lymphocytes and most produce both

165. Which of following is not a primary lymphoid organ which functions as sites for B-cell and T-cell maturation?
(1) Bone marrow
(2) Spleen
(3) Bursa of Fabricius
(4) Thymus

166. At what age does the thymus reach its maximal size?
(1) During the first year of life
(2) Teenage years (puberty)
(3) Between 40 and 50 years of age
(4) After 70 years of age

167. The antimicrobial material secreted from human skin which kills *E. coli* is
(1) Psoriasin (2) Lysozyme
(3) IgA (4) Complement protein

168. Which statement is false about macrophages?
(1) They are phagocytotic cells originated from monocytes and are deployed at different tissues
(2) They are involved in antigen presentation
(3) They produces reactive oxygen and nitrogen species
(4) They secrete antimicrobial peptides

169. Which statement is correct regarding dendritic cells?
(1) They are major phagocytotic cells
(2) They always express antigen on MHC-II and present it to TH cells
(3) They are responsible for lysis of virus infected cells
(4) They secrete interferon and cytokines

170. Which component of immunity is absent in plants?
(1) Innate immunity
(2) Adaptive immunity
(3) Invasion induced protective enzymes and enzyme cascades
(4) Pattern recognition receptors

171. The properties of immunogen which contributes immunogenicity includes
(1) foreignness
(2) molecular size greater than 5kd
(3) chemical composition and heterogeneity
(4) all of the above

172. The adjuvant increases the immunogenicity of antigen due to following effect
(1) antigen persistence is prolonged
(2) co-stimulatory signal are reduced
(3) local inflammation is reduced
(4) the non-specific proliferation of lymphocytes is inhibited

173. Which of the following component is absent in Freund's incomplete adjuvant?
(1) Antigen in aquous solution, mineral oil
(2) An emulsifing agent such as mannide monooleate
(3) Heat killed mycobacterium (muramyl dipeptide)
(4) All are present

174. Which of the following statements is true for autoimmune response?
(1) Antibodies are generated against mutant proteins
(2) Antibodies are generated against self non-altered proteins
(3) No Antibodies are generated against self antigen
(4) Mainly viruses and bacteria are responsible for this response

75. A tetanus booster shot results in the increased production of
(1) tetanus-specific NK cells
(2) T cells that recognize tetanus toxoid but not tetanus toxin
(3) antibodies which neutralize tetanus toxin
(4) T-cells which kill Clostridium tetani

176. Which of the following virus replicate in cytoplasm?
(1) SV40
(2) Adenovirus
(3) Vaccinia virus
(4) Herpes simplex virus

177. The genome of CaMV is
(1) Positive strand RNA
(2) Negative strand RNA
(3) Single stranded DNA
(4) Double stranded DNA

178. Hydroxy methyl cytosine is found in DNA of
(1) Simain virus 40 (2) Phage T_4
(3) Phage Q_B (4) Adenovirus

179. Which of the following statement about HIV is incorrect?
(1) HIV is an enveloped reterovirus
(2) HIV infects T-cell that carry the CD_4 antigen on their surface
(3) HIV contains two copies of negative strand RNA genome
(4) HIV-I need coreceptors CCXR4/CCR5 for entry into host cell

180. Live vaccine is –
(1) low dose of infectious bacteria administered as prophylactic
(2) a dose of bacterium strain in a modified form which retains immunogenecity but is not pathogenic
(3) a low dose of toxin produced by bacterium
(4) drug against specific pathogen

181. Which of the following would you expect to be coded by genome of an enveloped animal virus-
(1) protein that regulate the virus life cycle
(2) enzyme that control enzyme metabolism
(3) ribosomal RNA
(4) enzymes required for protein biosynthesis

182. Something that is not characteristic of cancer
(1) evading apoptosis
(2) limitless replication potential
(3) metastasis
(4) loss of both copies of a proto oncogene

183. SDS page analysis of a monoclonal pentameric IgM molecule will produce
(1) one protein band
(2) three protein bands
(3) five protein bands
(4) twenty-one protein bands

184. Which of the following can act as opsonins?
(1) Interleukin-2
(2) Interferons-α
(3) Mixture of potassium and magnesium ions
(4) Immunoglobulin-G

185. Antigen receptors on cytotoxic T cells recognize
(1) viral peptides in association with class I MHC molecules
(2) tumor specific peptides in association with class II MHC molecules
(3) bacterial antigens in association with class III MHC antigens
(4) free carbohydrate antigens released by bacterial pathogens

186. Which of the following is not an autoimmune disorder
(1) AIDS
(2) Systemic Lupus Erythematosus
(3) Rheumatoid Arthritis
(4) Myesthenia Gravis

187. The formation of antigen-antibody complex helps in disposing antigen through the following pathways except
(1) neutralizing the antigen by blocking its activity
(2) by directly hydrolyzing the antigen
(3) by promoting the precipitation of antigen
(4) by activating cell lysis pathway

188. Retroviruses, like the influenza virus, escape the detection of pre-existing antibodies in the host by generating surface antigen variants. They do so
(1) by editing the surface antigen post-translationally
(2) because RNA polymerase of these viruses display high mutation rate during RNA synthesis
(3) because the surface antigens are attacked by the proteases present in the host cell
(4) because DNA polymerase of the host mutates the viral genome in the infected cell

189. In addition to the leukocytes found in the blood, immune cells tend to be found concentrated in all of the following areas except
(1) the appendix
(2) the kidneys
(3) the lymph nodes
(4) the skin

190. The major histocompatibility complex (MHC) serves as a way for cells to
(1) present possible antigens to immune cells in a way that identifies the presenting cell as "self."
(2) kill pathogens living inside the cell
(3) kill pathogens living outside the cell
(4) initiate a cascade of events that results in the insertion of a pore into the membrane of bacteria

191. A body cell infected with a virus is a likely target for attack by a
(1) plasma cell. (2) cytotoxic T cell
(3) helper T cell. (4) macrophage

192. Which of the following statements about the activation of B cells is correct?
(1) B cells have "I" shaped antigen receptors on their surface
(2) B cells do not have a memory cell form, so their response to a second infection by a pathogen is no different than the first
(3) To start proliferating, B cells usually need to be activated by both an antigen and the cytokine interleukin-2
(4) Activated B cells kill bacteria by phagocytosis

193. Receptor tyrosine kinases (RTKs) are found at high levels on various cancer cells. A protein, Herceptin, has been found to bind to an RTK known as HER2. This information can now be utilized in breast cancer treatment if which of the following is true?
(1) If Herceptin is found in the breast lymph nodes of the patient
(2) If HER2, administered by injection, is in sufficient concentration
(3) If the patient's cancer cells have detectable HER2
(4) If the patient's genome codes for the HER2 receptor

194. The receptors for a group of signaling molecules known as growth factors are often
(1) ligand-gated ion channels
(2) G protein-coupled receptors
(3) cyclic AMP
(4) receptor tyrosine kinases

195. In general, a signal transmitted via phosphorylation of a series of proteins
(1) brings a conformational change to each protein
(2) requires binding of a hormone to a cytosol receptor
(3) cannot occur in yeasts because they lack protein phosphatases
(4) requires phosphorylase activity

196. Sutherland discovered that epinephrine signals
(1) a decrease in levels of cAMP as a result of bypassing the plasma membrane
(2) lower blood glucose by binding to liver cells
(3) interactions with insulin inside muscle cells
(4) elevation of cytosolic concentrations of cyclic AMP

197. Which of the following is the best explanation for the inability of a specific animal cell to reduce the Ca^{2+} concentration in its cytosol compared with the extracellular fluid?
(1) Blockage of the synaptic signal
(2) Loss of transcription factors
(3) Insufficient ATP levels in the cytoplasm
(4) Low oxygen concentration around the cell

198. Which of the following statements about signaling is false?
(1) Trimeric G-proteins are activated by cell surface receptors that dimerize after the ligand binds
(2) Activated trimeric G-proteins can activate phospholipase C and adenylate cyclase
(3) cAMP is a second messenger that activates protein kinase A
(4) Activated protein kinase A can activate enzymes and gene regulatory proteins

199. Which of the following is the most active CaM-Kinase?
(1) Dephosphorylated CaM-Kinase bound to Calcium-loaded Calmodulin
(2) Phosphorylated CaM-Kinase bound to Calcium-loaded Calmodulin
(3) Phosphorylated CaM-Kinase without Calcium-loaded Calmodulin
(4) Dephosphorylated CaM-Kinase without Calcium-loaded Calmodulin

200. In unstimulated cells, calcium levels are highest in
(1) The cytoplasm
(2) The nucleus
(3) The golgi apparatus
(4) The endoplasmic reticulum

ANSWER SHEET

Unit 4: Signalling, Immunology and Cancer

---

1. 4	**2.** 3	**3.** 3	**4.** 2	**5.** 2	**6.** 4	**7.** 1	**8.** 4	**9.** 3	**10.** 4
11. 2	**12.** 3	**13.** 4	**14.** 1	**15.** 4	**16.** 4	**17.** 1	**18.** 4	**19.** 3	**20.** 2
21. 1	**22.** 1	**23.** 1	**24.** 4	**25.** 4	**26.** 4	**27.** 1	**28.** 3	**29.** 1	**30.** 1
31. 1	**32.** 4	**33.** 3	**34.** 4	**35.** 4	**36.** 2	**37.** 2	**38.** 2	**39.** 4	**40.** 3
41. 2	**42.** 3	**43.** 1	**44.** 2	**45.** 2	**46.** 4	**47.** 1	**48.** 1	**49.** 4	**50.** 4
51. 4	**52.** 2	**53.** 4	**54.** 2	**55.** 2	**56.** 2	**57.** 4	**58.** 3	**59.** 3	**60.** 2
61. 4	**62.** 1	**63.** 3	**64.** 1	**65.** 3	**66.** 1	**67.** 3	**68.** 3	**69.** 2	**70.** 1
71. 3	**72.** 4	**73.** 4	**74.** 4	**75.** 3	**76.** 3	**77.** 4	**78.** 4	**79.** 2	**80.** 4
81. 2	**82.** 3	**83.** 4	**84.** 2	**85.** 1	**86.** 1	**87.** 1	**88.** 4	**89.** 1	**90.** 2
91. 4	**92.** 3	**93.** 4	**94.** 3	**95.** 1	**96.** 1	**97.** 2	**98.** 1	**99.** 4	**100.** 3

101. 4	**102.** 2	**103.** 1	**104.** 4	**105.** 4	**106.** 1	**107.** 4	**108.** 1	**109.** 1	**110.** 3
111. 3	**112.** 2	**113.** 1	**114.** 1	**115.** 3	**116.** 2	**117.** 4	**118.** 3	**119.** 4	**120.** 4
121. 4	**122.** 3	**123.** 3	**124.** 4	**125.** 2	**126.** 1	**127.** 1	**128.** 1	**129.** 4	**130.** 3
131. 4	**132.** 4	**133.** 1	**134.** 2	**135.** 2	**136.** 3	**137.** 4	**138.** 4	**139.** 4	**140.** 3
141. 2	**142.** 1	**143.** 3	**144.** 2	**145.** 4	**146.** 4	**147.** 4	**148.** 3	**149.** 3	**150.** 3
151. 1	**152.** 3	**153.** 3	**154.** 2	**155.** 4	**156.** 3	**157.** 4	**158.** 2	**159.** 4	**160.** 4
161. 4	**162.** 3	**163.** 4	**164.** 1	**165.** 2	**166.** 2	**167.** 1	**168.** 4	**169.** 4	**170.** 2
171. 4	**172.** 1	**173.** 3	**174.** 2	**175.** 3	**176.** 3	**177.** 4	**178.** 2	**179.** 3	**180.** 2
181. 1	**182.** 4	**183.** 1	**184.** 4	**185.** 1	**186.** 1	**187.** 2	**188.** 2	**189.** 2	**190.** 1
191. 2	**192.** 3	**193.** 3	**194.** 4	**195.** 1	**196.** 1	**197.** 3	**198.** 1	**199.** 2	**200.** 4

UNIT - 5

DEVELOPMENTAL BIOLOGY

1. Zygotic genes required for the formation of a group of adjacent segment in the developing *Drosophila* embryo is called

(1) maternal gene (2) pair rule gene

(3) homeotic gene (4) gap gene

2. The blastopore region of amphibian embryo that secretes BMP inhibitors and dorsalizes the surrounding tissue is known as

(1) Brachet's cleft

(2) Nieuwkoop center

(3) Spemann's organizer

(4) Hensen's node

3. A growth in which an animal that, increases its weight (volume) two fold will increase its length only 1.26 times is

(1) isometric (2) allometric

(3) determinate (4) indeterminate

4. Among the following which invertebrate phylum is a deutrostome

(1) arthropod (2) mollusc

(3) nematodes (4) echinodermate

5. Which of the following protein is not involved in intrinsic pathway of apoptosis

(1) Tumor necrosis factor α

(2) Cytochrome C

(3) Apaf-1

(4) Caspase 3

6. In regulative development, the prospective potency of cells

(1) equal to prospective fate

(2) more than prospective fate

(3) lesser than prospective fate

(4) not determined

7. The major function of cortical granules in cytoplasm of egg is to

(1) early block to polyspermy

(2) late block to polyspermy

(3) allowing meiosis to complete

(4) organize sperm

8. Many cells in the body divide only rarely, if at all; neurons, red blood cells, and keratinocytes are extreme examples. In which portion of the cell cycle would such cells be considered to be?

(1) M phase (2) G1 phase

(3) G_0 phase (4) Senescence phase

9. Meiosis in Pteridophytes results in-

(1) spores are formed

(2) prothallium is formed

(3) gametes are formed

(4) sex organs are formed

10. In elauata (elu) mutant of Arabidopsis, shoot apical meristem (SAM) size as well as expression of wuschel (wus) are increased whereas SAM size is reduced in wus mutant plants. With this information, choose the correct statement which best describes the function of wus and elu genes

(1) elu positively regulates wus expression and negatively regulates SAM size

(2) elu negatively regulates wus expression and wus positively regulates SAM size

(3) wus negatively regulates SAM size and elu expression

(4) wus and elu independently regulate. SAM size

11. The plant species that is extensively used as a model system and considered as *Drosophila* of the plant world is
(1) *Antirrhinum majus*
(2) *Arabidopsis thaliana*
(3) *Zea mays*
(4) *Avena sativa*

12. The major function of alteration of membring potential after interactions of sperm is
(1) early block to polyspermy
(2) late block to polyspermy
(3) allowing meiosis to complete
(4) organize sperm

13. Which of the following statements concerning cytoplasmic determinants is false?
(1) They can be distributed unequally in egg cytoplasm
(2) They play a role in directing the embryonic development of many organisms
(3) They were revealed during experiments with developmental polarity
(4) They are responsible for forming an animal's polarity

14. Which of the statements about quiescent centre is not true?
(1) This occupies the central position in root meristem
(2) This is present in apical and as well as lateral root meristem
(3) This connects to all initial cells of root meristem
(4) This is the most actively dividing cells in root meristem

15. The process of epithelial cell formation in leaf starting from a lateral meristem may be referred as
(1) transformation
(2) regeneration
(3) differentiation
(4) dedifferentiation

16. In vp1 mutant of maize, the seeds germinate while they are still maturing in mother plant. Most likely, vp1 mutation leads to
(1) increased ABA signaling
(2) increased auxin signaling
(3) reduced ABA signaling
(4) reduced GA signaling

17. What prevents the cells adjacent to the progenitor vulva cells in *C. elegans* from differentiating into vulva cells?
(1) These cells are too distant from the anchor cell to receive the LIN-3 signal
(2) These cells lack the ability to bind and respond to the LIN-3 signal
(3) These cells receive a signaling compound from a hypodermal cell that deactivates the LIN-3 signal
(4) These cells have condensed their chromatin in the regions responsive to the LIN-3 signal

18. Which of the following is the last to occur after the binding of a sea urchin sperm to an egg?
(1) Initiation of mRNA synthesis
(2) Increase in cytosolic pH
(3) Increase in intracellular calcium concentration
(4) Exocytosis of cortical granules

19. Monopteros mutant of *Arabidopsis* has a defective cell polarity in embryo. In most likelihood Monopteros gene is involved in which hormone signaling?
(1) Cytokinin (2) Auxin
(3) Brassinosteroid (4) Gibbrellins

20. A mutation in a representative dicot plant resulted into having altered floral organs as sepal, sepal, carpel, carpel. According to ABC model of flowering which class of mutation is there in the plant?
(1) A class gene mutant
(2) B class gene mutant
(3) A and B double mutant
(4) C class gene mutant

21. Hydrophobic interaction is likely to be the highest contributory factor in stabilizing the structure of the
(1) mitochondrial matrix
(2) chromosome
(3) plasma membrane
(4) cytoskeletal structure

22. Which of the following proteins represents an Extracellular Matrix (ECM) associated signal transduction pathway?
(1) Actin (2) Myosin
(3) Integrin (4) Tubulin

23. Nerve cord in invertebrates is
(1) dorsally present
(2) ventrally present
(3) laterally present
(4) absent

24. Antiserum raised in rabbits against a monoclonal mouse IgG will not react with
(1) mouse IgM (2) mouse light chains
(3) rabbit IgG (4) rat IgG

25. Which of the following is not a function of the immune system?
(1) Immune memory
(2) Elimination of pathogens
(3) Production of red blood cells
(4) Inflammatory response

26. The grafting of a second polarizing region into the anterior of a limb bud results in
(1) formation of a second bud at the site of the graft
(2) formation of extra copies of the anterior-most digit, digit 2, is the only possible result
(3) formation of extra copies of the posterior-most digit, digit 4, is the only possible result
(4) a new axis can be formed if the second polarizing region is large, and a second full set of digits can form with mirror-image orientation to the normal set of digits

27. Which of the following proteins is produced by anchor cell during the vulval development of *C. elegans*?
(1) Lin 3 (2) Let 23
(3) Par 2 (4) Lin 12

28. If bicoid mRNA is injected into the posterior pole of a wild type *Drosophila* embryo, the expected phenotype is
(1) head structures at bo.th the posterior and anterior poles
(2) head structures at the posterior pole only
(3) head structures at the anterior pole only
(4) no head structures at all

29. Tissue boundaries are largely determined by
(1) Notch signaling
(2) TGF-β signaling
(3) Hedgehog signaling
(4) Wingless signaling

30. When the experimentally isolated parts of an embryo develop according to their fate map, it is referred to as
(1) mosaic development
(2) regulative development
(3) clonal development
(4) syncytial development

31. The Spemann organizer of the amphibians are
(1) mesodermal in origin
(2) endodermal in origin
(3) ectodermal in origin
(4) epidermal in origin

32. The mode of development is which fate of embryo is fixed at very early stage and prospective potency is equal to prospective fate is teremed as
(1) mosaic development
(2) morphylaxsis
(3) regulative development
(4) epimorphosis

33. How many founder cells are there based on the early cell lineage pattern in *C. elegans*?
(1) 4 (2) 5
(3) 6 (4) 7

34. The limb bud of a tetrapod is specified by
(1) Hox genes only
(2) retinoic acid only
(3) both Hox genes and retinoic acid
(4) fibroblast growth factor

35. Which amongst the following statements is incorrect in the context of transcription factor NFkB?
(1) It mediates antiviral response
(2) It resides in the nucleus
(3) It is a member of the Rel family of proteins
(4) Its activity is blocked by lkB

36. In plant, haploid number of chromosomes are found in
(1) antipodal cell and endosperm
(2) egg cell and antipodal cell
(3) egg cell and megaspore-mother cell
(4) megaspore mother cell and synergids

37. Double fertilization is a characteristic of
(1) thallophytes (2) pteridophytes
(3) angiosperm (4) gymnosperm

38. Which of the following genes of Drosophila determine the identification of the segments of the developing embryo?
(1) The gap genes
(2) The pair rule genes
(3) The segment polarity genes
(4) The homeotic selector genes

39. Quiescent Centre is
(1) relatively slow dividing central area of root meristem
(2) peripherals of root apical meristem
(3) centre of shoot apical meristem
(4) origin points of lateral meristem

40. Regarding plants in general
(1) the sporophyte is haploid and the gametophyte is diploid
(2) both sporophyte and gametophyte are diploid
(3) the sporophyte is diploid and the gametophyte is haploid
(4) both sporophyte and gametophyte are haploid

41. Which statement best explains why invertebrates regenerate lost tissue more readily than most vertebrates do?
(1) Invertebrates contain specialized cells that produce the hormones necessary for this process
(2) Invertebrate cells exhibit a higher degree of uncontrolled cell division than vertebrate cells do
(3) Invertebrate animals reproduce asexually, but vertebrate animals reproduce sexually
(4) Invertebrate animals have more undifferentiated cells than vertebrate animals

42. Which of the following maternal effect gene products regulate production of anterior structures in Drosophila embryo?
(1) Bicoid and Nanos
(2) Bicoid and Hunchback
(3) Bicoid and Caudal
(4) Nanos and Caudal

43. The first organogenic event observed in embryological development is formation of
(1) skin (2) nervous System
(3) heart (4) kidney

44. In *Dictyostelium discoides* the factor necessary for stalk cell differentiation is
(1) Catenin
(2) c-AMP
(3) c-GMP
(4) Differentiation inducing factor (DIF)

45. *Dictyostelium* has been a wonderful experimental organism for developmental biologists because
(1) it shows social behavior
(2) its gives idea about multicellularity
(3) it behaves like stem cells
(4) it is true multicellular organism

46. The adaptive evolution that allowed vertebrates to roam on land, far from existing ponds was
(1) Amniotic egg (2) Limbs
(3) Cutaneous skin (4) Uricotellism

47. The part of the embryo from which the ectoderm, mesoderm and endoderm are formed in chick is known as
(1) primitive streak (2) hypoblast
(3) epiblast. (4) cytotrophoblast

48. Which of the following is mismatch regarding differentiated cell and marker product
(1) Keratinocyte (epidermal cell): Keratin
(2) Lens cell: Crystallins
(3) Leydig cell: Progestrone
(4) Hepatocyte: Serum albumin

49. Cell death in the inter digital necrotic zone of the autopod is an example of
(1) phagocytosis (2) apoptosis
(3) necrosis (4) involution

50. The lateral separation of amphibian embryo into two parts at four celled stage will result in
(1) identical twins
(2) two embryos joined at belly region
(3) single embryo
(4) two embryos missing various organs

51. Which protein secreted by the amphibian organizer induces neural tissue formation by inhibiting Bone Morphogenetic Protein?
(1) β-catenin (2) Noggin
(3) Dickkopf (4) Dishevelled

52. An embryo lacking bicoid protein is injected with bicoid protein in middle portion. It will result into
(1) two heads and no tarsons
(2) head in middle and tarsons at both end
(3) no head and Tarson at both ends
(4) normal phenotype

53. Which of the following is a species specific protein in sea urchin which plays important role in recognization during acrosomal reaction
(1) bindin (2) avidin
(3) fertilin (4) cortical granule

54. Which of the following is not an extra-cellular matrix protein
(1) cadherin (2) elastin
(3) collagen (4) fibronectin

55. In regulative development, the prospective potency of cells
(1) equal to prospective fate
(2) more than prospective fate
(3) lesser than prospective fate
(4) not determined

56. The homologue of β-catenin in *Drosophila* is
(1) fushi tarazu (2) engrailed
(3) armadillo (4) cubitus interruptus

57. The grafting of the dorsal lip of the blastopore from an early Xenopus gastrula onto the ventral side of an early embryo will result in two complete embryos. Thus dorsal lip can be designated as
(1) primary organizer
(2) cytoplasmic determinant
(3) morphogen (4) commitment

58. Which of the following is not a step in therapeutic cloning?
(1) Eggs are taken from the recipient
(2) A cell is removed from the recipient
(3) Embryonic stem cells form
(4) Stem cells differentiate into the desired cell types

59. The determining factor in Zone of polarization activity in limb bud in tetrapods is
(1) β-catenin
(2) GSK-3
(3) sonic hedgehog
(4) bone Morphogenic protein

60. Which statement is correct for capacitation
(1) is the maturation of mammalian spermatozoa after entering into oviduct of female
(2) meiotic division in egg after penetration of sperm
(3) maturation of egg in oviduct after fertilization
(4) release of egg after maturation

61. Mutations in homeotic genes can lead to what type of developmental defect?
(1) The anterior portion of the embryo does not develop
(2) Several adjacent segments will be missing from embryo
(3) The embryo will develop with every other segment failing to form
(4) The development of segments will be changed wholesale from their normal identity to that of a different segment

62. Cytoplasmic determinants coding for anterior structure of *Drosophila embryo* if injected elsewhere in the recipient embryo, would lead to
(1) normal development
(2) formation of additional ectopic head
(3) degeneration
(4) a phenotype with two heads and two tails

63. In amphibian oocyte, the germplasm which gets segregated during cleavage to give rise to primordial germ cells (PGC's) is normally
(1) distributed evenly throughout the oocyte
(2) localized at animal pole
(3) localized at vegetal pole
(4) aggregated in central part of oocyte

64. If you treat insulin-responsive mammalian cells in culture and immune precipitate with the antibody against insulin receptor and detect the proteins bound to the receptor by Western blotting using an antibody against phosphorylated tyrosine, which of the following proteins you are expected to detect?
(1) Insulin receptor and IRS-l
(2) Insulin receptor and Pl3-kinase
(3) Insulin receptor and Ras
(4) Insulin receptor and Grb2

65. In *Drosophila*, one of the most important maternal effect genes is bicoid that is responsible for the anterior-posterior axis generation. A bicoid mutant does not form head and thorax region. When bicoid mRNA was injected in the middle of the deficient embryo, which of the following phenotypes would be seen?
(1) Acron-head-thorax-abdomen-telson
(2) Telson-thorax -head -thorax-abdomen - telson
(3) Acron-thorax-head-thorax-abdomen telson
(4) Head-thorax-abdomen

66. Two experiments were carried out in frog embryos. In the first experiment with early gastrula, prospective neural ectoderm from the donor embryo was transplanted onto the prospective epidermis region of the acceptor embryo. The neural ectoderm developed into epidermis. In a similar experiment carried out in late gastrula, the transplanted prospective neural ectoderm developed into neural ectoderm itself. What conclusion would you draw?
(1) The early gastrula shows conditional specification
(2) The early gastrula shows autonomous specification
(3) The late gastrula shows syncytial specification
(4) The late gastrula shows conditional specification

67. A model plant that follows ABC rule of flowering has mutation in both B and C class genes. What will be floral phenotype from outside to inside?

(1) (sepal-petal-petal-petal)

(2) (sepal-petal-sepal-petal)

(3) $(sepal\text{-}sepal\text{-}sepal\text{-}sepal)_n$

(4) $(petal\text{-}petal\text{-}petal\text{-}petal)_n$

68. Which of the following second messengers can stimulate a mammalian cell to become a cancer cell?

(1) Ca^{+2} (2) cAMP

(3) Oncogene (4) PIP3

69. The Apical Ectodermal Ridge (AER) plays a very important role in the development of limb of any tetrapod organism. In a given experiment during chick development, an extra AER was grafted onto the existing limb bud. Which of the following could have been possible outcome?

(1) Formation of supernumerary structure usually towards the distal end of the limb

(2) Formation of a complete extra limb

(3) The addition of the new AER would regress the original AER thereby ceasing the limb development

(4) Development of the distal limb skeletal elements ceases

70. In the sperm, the function of the acrosomal vesicle is

(1) to form the centrosome of the zygote

(2) to provide enzymes that digest the outer coat of the egg

(3) to contain mitochondria that power sperm movement

(4) to provide proteins necessary for sperm movement

71. Which of the following are not necessary for movement of the sperm flagellum

(1) dynein (2) microtubules

(3) mitochondria (4) filopodia

72. The main purpose of gametogenesis is to

(1) make a haploid cell

(2) specialize the gametes

(3) organize the 3 germ layers

(4) 1 and 2

73. In the egg, ________ contain(s) proteolytic enzymes that are released by exocytosis after fertilization and are important for preventing polyspermy

(1) cortical granules

(2) vitelline envelope

(3) acrosome

(4) jelly layer

74. Which of the following is an example of a factor that helps to achieve species specificity in sea urchin fertilization?

(1) Actin (2) Microfilaments

(3) Bindin (4) Cortical granules

75. Which is not a property of resact?

(1) Influences flagellar movement

(2) Confers species specificity

(3) Degrades jelly layer

(4) Acts as a chemokinetic

76. Cells in a specific region of late gastrula stage of embryo always give rise to the blood cells in frog. These cells in the gastrula embryo are best described as

(1) Differentiated

(2) Specified

(3) Determined

(4) De-differentiated

77. A specific chemical can prevent the degradation of the mitosis promoting factor (MPF) What results if this chemical were added to early cleavage stage frog embryos?

(1) The cells cannot exit mitosis and enter S phase

(2) The cells cannot exit S phase and enter mitosis

(3) The cell enters mid blastula transition

(4) The cell begins zygotic transcription

78. The first step in sea urchin gastrulation is
(1) the ingression of the primary mesenchyme into the blastocoel
(2) the invagination of the primary mesenchyme into the blastocoel
(3) the convergent extension of the primary mesenchyme into the blastocoel
(4) the involution of the ectoderm into the blastocoel

79. During sea urchin gastrulation,
(1) the cells that will form the ectoderm move inside the embryo
(2) the cells that will form the endoderm and mesoderm move inside the embryo
(3) the cells that will form the endoderm and mesoderm move outside the embryo
(4) the cells that will form the ectoderm and mesoderm move inside the embryo

80. You have a female fly that has mutations in both copies of the gurken gene. In her ovary, her cells cannot produce functional gurken protein. What phenotype might you reasonably expect to obtain in her progeny?
(1) The embryos will develop normally
(2) The embryos will be dorsalized
(3) The embryos will be ventralized
(4) Half the embryos will be normal, and half will be dorsalized

81. The following are important for setting up the anterior-posterior body axis in *Drosphila*
(1) bicoid and pipe
(2) bicoid and nanos
(3) cactus and nanos
(4) cactus and pipe

82. Frog embryos are injected with a morpholino against fibronectin. Which of the following developmental processes will be affected
(1) neural tube closure
(2) migration of cells during gastrulation
(3) the first cell division will be blocked
(4) the first nuclear division will be blocked

83. Dorsal endoderm with is cultured with ectoderm. The ectoderm is converted to
(1) ventral ectoderm
(2) ventral endoderm
(3) organizer
(4) nieuwkoop Center

84. These 3 signaling pathways are important for induction and formation of the organizer
(1) TGF-β, Wnt, Nodal
(2) Notch, Wnt, Nodal
(3) BMP, Fibronectin, Nodal
(4) TGF-β, FGF, Nodal

85. In a mammalian embryo, the trophectoderm
(1) gives rise to the inner cell mass
(2) gives rise to the embryo
(3) gives rise to extraembryonic tissues
(4) is pluripotent

86. The neural tube becomes separated from the epidermis by
(1) expression of N-cadherin in the epidermis
(2) differential cadherin expression in epidermis versus neural plate
(3) induction of medial hinge point cells by the notochord
(4) pushing forces exerted by the migrating neural crest

87. Which pathway(s) relies on activated smads to transduce its signals?
(1) Notch signaling
(2) TGF-β signaling
(3) Wnt signaling
(4) Hox gene signaling

88. Which of the following is not derived from the ectoderm
(1) neural tube
(2) spinal cord
(3) epidermis
(4) muscle

89. Determination of dorsal ectoderm to become neural rather than epidermal requires
(1) BMP signals from the organizer
(2) BMP antagonist signals from the organizer
(3) BMP antagonist signals from the ventral mesoderm
(4) BMP signals from the ventral mesoderm

90. Injection of Noggin mRNA in cells that will become the future ventral side of a frog embryo mimics the effect of an organizer graft to the ventral side. This experiment demonstrates
(1) that Noggin is a transcription factor
(2) that Noggin is committed to organizer fates
(3) that Noggin is required to induce a secondary axis
(4) that Noggin is sufficient to induce a secondary axis

91. The neural crest is a population of cells that
(1) form at the border between the epidermis and neural tube
(2) undergo an epithelial to mesenchymal transition
(3) migrate to distant parts in the body
(4) all of the above

92. After removing the epidermis, the neural tube
(1) lacks ventral neural fates
(2) lacks dorsal neural fates
(3) an ectopic roof plate is induced
(4) more sensory relay neurons are formed

93. Removing the epidermis from a developing embryo and observing the effect on the neural tube, is a
(1) loss-of-function experiment
(2) gain-of-function experiment
(3) fate map experiment
(4) knock-down experiment

94. Fate mapping in a 16-cell stage sea urchin embryo shows that the mesomere cells, which are in the animal pole, always give rise to the ectoderm. You then perform a series of experiments. In the first experiment, you remove the mesomeres, culture them alone in a dish, and see that they still give rise to ectoderm. In the next experiment, you remove the mesomeres and transplant them adjacent to the vegetal pole of another 16-cell stage embryo (in a region that will become endoderm). This time, your transplanted cells develop into the endoderm. From these experiments, you can conclude that the mesomere cells at the 16-cell stage are
(1) specified, but not determined to become ectoderm
(2) specified but not determined to become endoderm
(3) determined but not specified to become ectoderm
(4) both specified and determined to become ectoderm

95. Which technique provides information about mRNA in a tissue sample?
(1) Southern blot
(2) Fate mapping
(3) Microarray
(4) Immunohistochemistry

96. Fate mapping in a frog species shows that cells in a specific region of a blastula stage which kept on original position or neutral medium always develop into skin but if its position is shifted and placed near the mesodermal cells it get converts in muscles. Based on this information, these cells in the blastula embryo are best described as
(1) specified
(2) committed
(3) determined
(4) differentiated

97. The process of commitment proceeds in which of the following ways
(1) differentiation —> specification —> determination
(2) specification —> determination —> differentiation
(3) determination —> specification —> differentiation
(4) determination —> differentiation —> specification

98. The concept of epigenesis meant that
(1) embryos were pre-formed in the germ cells
(2) embryos were pre-formed in certain genders' gametes
(3) embryos were born from eggs
(4) embryos were formed from new materials

99. Which of the following does not require you to know the gene you want to study
(1) gene targeting
(2) forward genetic screen
(3) RNAi
(4) mRNA overexpression using microinjection

100. Inductive interactions between two cells can involve
(1) secretion of a morphogen from one cell
(2) direct signaling between membrane-bound proteins in neighboring cells
(3) a receptor of one cell binding a secreted molecule
(4) all of the above are possible

101. Specialization, or the process by which cells become structurally and functionally distinct, is known as
(1) specification
(2) determination
(3) differentiation
(4) both 1 and 3

102. The reverse transcriptase-polymerase chain reaction (RT-PCR) is useful because
(1) it can be used to measure protein levels in a cell population
(2) it can be used to study chromatin
(3) it can be used to detect the presence of specific transcripts in a tissue
(4) it can be used to visualize distribution of transcripts in a tissue section

103. You wish to visualize muscle formation in live zebrafish embryos so you decide to make a transgenic fish. To do this, you use a piece of DNA 5' to the start of a muscle-specific gene, containing the ______ to express __________ in muscle cells
(1) Control region, GFP
(2) Control region, LacZ
(3) Reporter, LacZ
(4) Reporter, myosin

104. Transgenic animals are best described as the result of the ______of a piece of DNA that is ______the next generation
(1) removal, removed from
(2) addition, removed from
(3) addition, transmitted to
(4) removal, transmitted to

105. Cancers of epithelial origin are called
(1) sarcoma (2) leukemia
(3) carcinoma (4) lymphoma

106. The most permanent way to fate map a group of cells is
(1) performing in situ hybridization to detect transcripts
(2) making chick-quail chimeras
(3) labeling cells with vital dye
(4) labeling cells with fluorescent dye

107. The result of transferring AER from late limb blastema to early limb bud will lead to
(1) inhibition of limb growth
(2) only distil structure would be formed
(3) limb will elongate with proximal values leading to doubling of zeugopod
(4) normal development of limb

108. The sex of the echiuroid worm, *Bonellia viridis* depended on
(1) chromosomes
(2) temperature
(3) ploidy level
(4) hormonal environment

109. If hydra is fragmented into various parts, separate group of cells repattern themselves into various small hydra. Such an mode of development is termed as
(1) regeneration
(2) morphlaxis
(3) epimorphogenesis
(4) all of the above

110. Morphylaxis can be defined as
(1) Production of lost organ by division in remaining cell
(2) Reinitiation of cell division in existing cells, followed by repatterning of those cells
(3) Production of complete organism by single cell
(4) Movement of organism toward stimulus

111. Which of the following represents the gametophyte generation in plants?
(1) Microspore (2) Megaspore
(3) Embryo sac (4) Pollen

112. Sex determination in Ants, Bees and Wasps depends on
(1) XX-XY System
(2) XX-XO System
(3) AA-AO System
(4) ZZ-ZW System

113. Which of the following phenotypes is obtained upon deletion of the ultrabithorax gene in Drosophila?
(1) Fly with four wings
(2) Fly with six wings
(3) Fly with no wings
(4) Fly with two wings

114. Specification of the limb field is by
(1) hox genes
(2) retinoic acid
(3) hox genes and retinoic acid
(4) FGF

115. When bicoid mRNA is injected at the posterior pole of a wild type Drosophila embryo, which one of the following phenotypes is expected?
(1) Head structures at both the poles
(2) Head structures at the posterior pole only
(3) Head structures at the anterior pole only
(4) No head structures at all

116. During embryogenesis, discoidal cleavage occurs in
(1) Chicken (2) *Drosophila*
(3) Mouse (4) *C. elegans*

117. Nuclei from frog kidney, upon transplantation into newt oocytes, support the synthesis of
(1) frog kidney-specific proteins
(2) frog oocyte-specific proteins
(3) newt kidney-specific proteins
(4) newt oocyte-specific proteins

118. True avian embryo forms almost entirely from the
(1) epiblast (2) hypoblast
(3) meroblast (4) marginal cells

119. Mouse embryonic stem cells are used in gene inactivation experiments because they
(1) can be cloned to give rise to a stable cell line
(2) are chimeric and will produce cells heterozygous for the gene
(3) are the only mouse cells that can be genetically engineered to inactivate genes
(4) are totipotent and can give rise to all types of differentiated cells

120. What is the best method to identify the cellular location of a protein *in vivo*?

(1) Place a reporter gene next to the promoter of the gene encoding the protein and identify the cellular location of the reporter protein

(2) Use a labelled antibody to identify the cellular location of the protein

(3) Separate the cellular compartments by centrifugation and screen the different compartments with an antibody

(4) Tag the protein with fluorescent amino acids and identify the cellular location by fluorescent microscopy

121. Which of the following techniques is useful in determining the movement of proteins within the nucleus?

(1) Electron microscopy

(2) Fluorescence recovery after photobleaching

(3) Fluorescent in situ hybridization

(4) Confocal light microscopy

122. According to 'ABC-model' of flowering, if a flower is mutated for both A and 'B' classes of genes, the mutant flowers will have

(1) sepal, petal, stamen, carpel

(2) sepal, sepal, sepal, sepal

(3) leaf, stamen/carpel, stamen/carpel, leaf

(4) carpel, carpel, carpel, carpel

123. Which of the following genes of *Drosophila* are expressed in every alternate segments of the developing embryo?

(1) The gap genes

(2) The pair rule genes

(3) The segment polarity genes

(4) The homeotic selector genes

124. Testicular feminization syndrome is a genetic condition wherein an individual with a XY genotype will have an external female-like phenotype. This is caused by

(1) functional loss of androgen receptor

(2) increased production of estrogen and its receptor

(3) functional loss of Mullerian inhibiting hormone

(4) functional loss of androgen receptor and Mullerian inhibiting hormone

125. Which one of the following defects do you expect to see if you were able to specifically block apoptosis in the developing limb bud of a frog embryo?

(1) The digits will remain connected through a web-like extension

(2) The bones will not form, and the limb would look like a paddle

(3) The limb would look normal but would be larger in size

(4) The anterior-posterior polarity of the limb will be lost

126. In *Drosophila*, mutations in homeotic genes result in which one of the following developmental defects?

(1) The anterior portion of the embryo will not develop

(2) Several segments in the anterior region of the embryo will be lost

(3) Segmentation will be lost, and the embryo will have only one segment

(4) Segment-specific identities will be lost

127. What molecular marker of head identity is found in the apical tip of a *Hydra* during head regeneration?

(1) Hox genes from the 3' (anterior) end of the *Hydra* Hox complex

(2) BMPs

(3) Nodal

(4) Wnt

128. Insects such as *Drosophila* undergo three molts before becoming a pupa and undergoing metamorphosis. Molting, which is also called "ecdysis", is controlled by what hormone?

(1) Ecdysone. (2) Juvenile hormone

(3) Cytokinin (4) Growth hormone

129. Metamorphosis of amphibians is triggered by environmental cues that act on the
(1) thyroid (2) pituitary
(3) hypothalamus (4) embryo

130. In plants gene 'sepallata' is not involved in formation of organ
(1) Sepals (2) Petals
(3) Stamens (4) Carpels

131. The lateral separation of amphibian embryo at two celled stage will result in
(1) identical twins
(2) two embryos joined at belly region
(3) single embryo
(4) two embryos missing various organs

132. An embryo lacking biocoid is injected with biocoid m-RNA at middle portion. It will result into
(1) two heads and no tarsons
(2) head in middle and tarsons at both end
(3) no head and Tarson at both ends
(4) normal phenotype

133. Zygotic genes required for the formation of a group of adjacent segment in the developing *Drosophila* embryo is called
(1) maternal gene
(2) pair rule gene
(3) homeotic gene
(4) gap gene

134. The blastopore region of amphibian embryo that secretes GSK3 inhibitors and dorsalizes the surrounding tissue into dorsal lip is known as
(1) Brachet's cleft
(2) Nieuwkoop center
(3) Spemann's organizer
(4) Hensen's node

135. A growth in which an animal that, increases its weight (volume) two fold will increase its length only 1.26 times is
(1) isometric (2) allometric
(3) determinate (4) indeterminate

136. Among the following which invertebrate phylum is a deutrostome
(1) arthropod
(2) mollusk
(3) nematodes
(4) echinodermate

137. Capacitation of sperm takes place in the reproductive tract of cows due to presence of
(1) amino acids
(2) proteins like osteopontin
(3) galactosamine
(4) glycosaminoglycan

138. Changes in the shape of a cell usually involve a reorganization of the
(1) nucleus
(2) cytoskeleton
(3) extracellular matrix
(4) transport proteins

139. Fertilized single cell cattle egg is what type of stem cell?
(1) Totipotent stem cell
(2) Pluripotent stem cell
(3) Multipotent stem cell
(4) None of the above

140. Which of the following genes in *Arabidopsis* mediates interactions between floral meristem and floral organ identity genes?
(1) SRE (2) MADS box
(3) UFO (4) AP2

141. The structure of the *Drosophila* gene called Tinman is similar to a gene in humans that also
(1) promotes ear development
(2) specifies the location of the heart
(3) determines structures in the eyes
(4) specifies limb elongation points

142. As an embryo develops, new cells are produced as the result of
(1) differentiation (2) preformation
(3) cell division (4) morphogenesis

143. Fertilization of an egg without activation is most like
(1) placing the key in the ignition of a car but not starting the engine
(2) resting during halftime of a basketball game
(3) preparing a pie from scratch and baking it in the oven
(4) walking to the cafeteria and eating lunch

144. Contact of a sperm with signal molecules in the coat of an egg causes the sperm to undergo
(1) mitosis
(2) depolarization
(3) apoptosis
(4) the acrosomal reaction

145. Even in the absence of sperm, metabolic activity in an egg can be artificially activated by
(1) abnormally high levels of carbonic acid in the cytosol
(2) abnormally low levels of extracellular oxygen
(3) injection of calcium ions into the cytosol
(4) exposure to the low pH of the uterus

146. The formation of the fertilization membrane requires an increase in the availability of
(1) bicarbonate ions. (2) calcium ions
(3) hydrogen ions. (4) potassium ions

147. A sea urchin zygote undergoes its first cell division
(1) 5 seconds after fertilization
(2) 30 minutes after fertilization
(3) 90 minutes after fertilization
(4) 4 hours after fertilization

148. A human zygote undergoes its first cell division
(1) 5 seconds after fertilization
(2) 30 minutes after fertilization
(3) 90 minutes after fertilization
(4) 24 hours after fertilization

149. A reproductive difference between sea urchins and humans is
(1) the sea urchin egg completes meiosis prior to fertilization, but meiosis in humans is completed after fertilization
(2) sea urchin eggs are produced by meiosis, but human eggs are produced by mitosis
(3) sea urchin eggs and sperm are of equal size, but human eggs are much bigger than human sperm
(4) sea urchins, but not humans, have a need to block polyspermy because only in sea urchins can there be more than one source of sperm to fertilize the eggs

150. Contact of a sea urchin egg with signal molecules on sperm causes the egg to undergo a brief
(1) mitosis
(2) membrane depolarization
(3) apoptosis
(4) vitellogenesis

151. During fertilization, the acrosomal contents
(1) block polyspermy
(2) help propel more sperm toward the egg
(3) digest the protective jelly coat on the surface of the egg
(4) nourish the mitochondria of the sperm

152. The vitelline layer of the sea urchin egg
(1) is outside of the fertilization membrane
(2) releases calcium, which initiates the cortical reaction
(3) has receptor molecules that are specific for binding acrosomal proteins
(4) is first visible only when organogenesis is nearly completed

153. In a newly fertilized egg, the vitelline layer
(1) opens the egg's nuclear membrane to allow haploid sperm DNA to enter
(2) hardens to form a protective cover
(3) secretes hormones that enhance steroidogenesis by the ovary
(4) reduces the loss of water from the egg and prevents desiccation

154. From earliest to latest, the overall sequence of early development proceeds in which of the following sequences?

(1) first cell division → synthesis of embryo's DNA begins → acrosomal reaction → cortical reaction

(2) cortical reaction → synthesis of embryo's DNA begins → acrosomal reaction → first cell division

(3) cortical reaction → acrosomal reaction → first cell division → synthesis of embryo's DNA begins

(4) acrosomal reaction → cortical reaction → synthesis of embryo's DNA begins → first cell division

155. The cortical reaction functions directly in the

(1) formation of a fertilization envelope

(2) production of a fast block to polyspermy

(3) release of hydrolytic enzymes from the sperm cell

(4) generation of a nerve-like impulse by the egg cell

156. In sea urchins, the "fast block" and the longer lasting "slow block" to polyspermy, respectively, are

(1) the acrosomal reaction and the formation of egg white

(2) the cortical reaction and the formation of yolk protein

(3) the jelly coat of the egg and the vitelline membrane

(4) membrane depolarization and the cortical reaction

157. In an egg cell treated with EDTA, a chemical that binds calcium and magnesium ions, the

(1) acrosomal reaction would be blocked

(2) fusion of sperm and egg nuclei would be blocked

(3) fast block to polyspermy would not occur

(4) fertilization envelope would not be formed

158. In mammals, the nuclei resulting from the union of the sperm and the egg are first truly diploid at the end of the

(1) acrosomal reaction

(2) completion of spermatogenesis

(3) initial cleavage

(4) activation of the egg

159. Fertilization normally

(1) reinstates diploidy

(2) follows gastrulation

(3) is required for parthenogenesis

(4) merges two diploid cells into one haploid cell

160. In mammalian eggs, the receptors for sperm are found in the

(1) fertilization membrane

(2) zona pellucida

(3) cytosol of the egg

(4) nucleus of the egg

161. Compared to sea urchin eggs, those of mammals

(1) complete the fertilization process more rapidly

(2) have not already completed meiosis at the time of ovulation

(3) have a more distinct animal pole

(4) have a more distinct vegetal pole

162. A human blastomere is

(1) an embryonic cell that is much smaller than the ovum

(2) an embryonic structure that includes a fluid-filled cavity

(3) that part of the acrosome that opens the egg's membrane

(4) a component of the zona pellucida

163. At the moment of sperm penetration, human eggs
(1) have used flagellar propulsion to move from the ovary to the oviduct
(2) accept as many sperm as possible in order to select the one with the highest fertility
(3) are still surrounded by follicular cells
(4) have a paper-thin cell of calcium carbonate that prevents desiccation

164. In a developing frog embryo, most of the yolk is
(1) located near the animal pole
(2) located near the vegetal pole
(3) found within the cleavage furrow
(4) found within the blastocoels

165. Among these choices, the largest cell involved in frog reproduction is
(1) the spermatozoon
(2) the ovum
(3) a blastomere in the vegetal pole
(4) a blastomere in the animal pole

166. The pattern of embryonic development in which only the cells lacking yolk subsequently undergo cleavage is called
(1) arcadian development, which is typical of insects
(2) holoblastic development, which is typical of marsupial mammals
(3) meroblastic development, which is typical of humans
(4) meroblastic development, which is typical of birds

167. As cleavage continues during frog development, the size of the blastomeres
(1) increases as the number of the blastomeres decreases
(2) increases as the number of the blastomeres increases
(3) decreases as the number of the blastomeres increases
(4) decreases as the number of the blastomeres decreases

168. During the early part of the cleavage stage in frog development, the rapidly developing cells
(1) skip the mitosis phase of the cell cycle
(2) skip the S phase of the cell cycle
(3) skip the G_1 and G_2 phases of the cell cycle
(4) rapidly increase the volume and mass of the embryo

169. The vegetal pole of the zygote differs from the animal pole in that
(1) the vegetal pole has a higher concentration of yolk
(2) the blastomeres originate only in the vegetal pole
(3) the posterior end of the embryo forms at the vegetal pole
(4) the vegetal pole cells undergo mitosis but not cytokinesis

170. The small portion of the embryo that will become its dorsal side develops from the
(1) morula. (2) primitive streak
(3) archenteron. (4) gray crescent

171. The yolk of the frog egg
(1) prevents gastrulation
(2) is concentrated at the animal pole
(3) is homogeneously arranged in the egg
(4) supports the higher rate of cleavage at the animal pole compared to the vegetal pole

172. An embryo with meroblastic cleavage, extraembryonic membranes, and a primitive streak must be that of
(1) an insect. (2) a fish
(3) an amphibian. (4) a bird

173. Meroblastic cleavage occurs in
(1) sea urchins, but not humans or birds
(2) humans, but not sea urchins or birds
(3) birds, but not sea urchins or humans
(4) both sea urchins and birds, but not humans

174. Which of the following correctly displays the sequence of developmental milestones?
(1) cleavage → blastula → gastrula → morula
(2) cleavage → gastrula → morula → blastula
(3) cleavage → morula → blastula → gastrula
(4) gastrula → morula → blastula → cleavage

175. Cells move to new positions as an embryo establishes its three germ tissue layers during
(1) determination. (2) cleavage
(3) fertilization. (4) gastrulation

176. The outer-to-inner sequence of tissue layers in a post-gastrulation vertebrate embryo is
(1) endoderm → ectoderm → mesoderm
(2) mesoderm → endoderm → ectoderm
(3) ectoderm → mesoderm → endoderm
(4) ectoderm → endoderm → mesoderm

177. If gastrulation was blocked by an environmental toxin, then
(1) cleavage would not occur in the zygote
(2) embryonic germ layers would not form
(3) fertilization would be blocked
(4) the blastula would not be formed

178. The archenteron of the developing frog eventually develops into the
(1) reproductive organs
(2) blastocoel
(3) heart and lungs
(4) digestive tract

179. The vertebrate ectoderm is the origin of the
(1) nervous system (2) liver
(3) pancreas. (4) heart

180. In frog embryos, the blastopore becomes the
(1) anus. (2) ears
(3) eyes. (4) nose

181. In a frog embryo, gastrulation
(1) produces a blastocoel displaced into the animal hemisphere
(2) occurs along the primitive streak in the animal hemisphere
(3) is impossible because of the large amount of yolk in the ovum
(4) proceeds by involution as cells roll over the lip of the blastopore

182. The earliest developmental stage among these choices is
(1) germ layers. (2) morula
(3) blastopore. (4) gastrulation

183. Which of the following is a correct description of an anatomical relationship?
(1) The mesoderm gives rise to the notochord
(2) The endoderm gives rise to the hair follicles
(3) The ectoderm gives rise to the liver
(4) The mesoderm gives rise to the lungs

184. An open space within the gastrula is the
(1) ectoderm. (2) mesoderm
(3) archenteron (4) endoderm

185. Although it contributes no cells to the embryo, the formation of the primitive streak is guided by the structure known as the
(1) endoderm. (2) mesoderm
(3) ectoderm. (4) hypoblast

186. The primitive streak in a bird is the functional equivalent of
(1) the lip of the blastopore in the frog
(2) the archenteron in a frog
(3) polar bodies in a sea urchin
(4) the notochord in a mammal

187. In all vertebrate animals, development requires
(1) a large supply of yolk
(2) an aqueous environment
(3) extraembryonic membranes
(4) an amnion

188. The least amount of yolk would be found in the egg of a
(1) bird
(2) fish
(3) frog
(4) eutherian (placental) mammal

189. A primitive streak forms during the early embryonic development of
(1) birds, but not frogs or humans
(2) frogs, but not birds or humans
(3) humans, but not birds or frogs
(4) humans and birds, but not frogs

190. Extraembryonic membranes develop in
(1) mammals, but not birds or lizards
(2) birds, but not mammals or lizards
(3) lizards, but not mammals or birds
(4) mammals, birds, and lizards

191. At the time of implantation, the human embryo is called a
(1) blastocyst
(2) gastrula
(3) fetus
(4) somite

192. Uterine implantation due to enzymatic digestion of the endometrium is initiated by the
(1) inner cell mass
(2) endoderm
(3) chorion
(4) trophoblast

193. In placental mammals, the yolk sac
(1) transfers nutrients from the yolk to the embryo
(2) differentiates into the placenta
(3) becomes a fluid-filled sac that surrounds and protects the embryo
(4) produces blood cells that then migrate into the embryo

194. Gases are exchanged in a mammalian embryo in the
(1) amnion. (2) hypoblast
(3) chorion. (4) trophoblast

195. Thalidomide, now banned for use as a sedative in pregnancy, was used in the early 1960s by many women in their first trimester of pregnancy. Some of these women gave birth to children with arm and leg deformities, suggesting that the drug most likely influenced
(1) early cleavage divisions
(2) determination of the polarity of the zygote
(3) differentiation of bone tissue
(4) morphogenesis

196. The migratory neural crest cells
(1) form most of the central nervous system
(2) serve as precursor cells for the notochord
(3) form the spinal cord in the frog
(4) form neural and non-neural structures in the periphery

197. From earliest to latest, the overall sequence of early development proceeds in which of the following sequences?
(1) gastrulation → organogenesis → cleavage
(2) ovulation → gastrulation → fertilization
(3) cleavage → gastrulation → organogenesis
(4) gastrulation → blastulation → neurulation

198. Changes in both cell shape and cell position occur extensively during
(1) gastrulation, but not organogenesis or cleavage
(2) organogenesis, but not gastrulation or cleavage
(3) cleavage, but not gastrulation or organogenesis
(4) gastrulation and organogenesis and cleavage

199. Animal development compares to plant development in that

(1) plant cells, but not animal cells, migrate during morphogenesis

(2) animal cells, but not plant cells, migrate during morphogenesis

(3) plant cells and animal cells migrate extensively during morphogenesis

(4) neither plant cells nor animal cells migrate during morphogenesis

200. The triggering of the intrinsic pathway of apoptosis involves a balance between pro-apoptotic and anti-apoptotic proteins. Which of the following is anti-apoptotic?

(1) Bax (2) Bad

(3) Bcl-2 (4) Cytochrome c

ANSWER SHEET

Unit 5: Developmental Biology

---

1. 4	**2.** 2	**3.** 1	**4.** 4	**5.** 1	**6.** 2	**7.** 2	**8.** 3	**9.** 1	**10.** 2
11. 2	**12.** 1	**13.** 4	**14.** 4	**15.** 3	**16.** 3	**17.** 2	**18.** 1	**19.** 2	**20.** 2
21. 3	**22.** 3	**23.** 2	**24.** 3	**25.** 3	**26.** 4	**27.** 1	**28.** 1	**29.** 3	**30.** 1
31. 1	**32.** 1	**33.** 2	**34.** 3	**35.** 2	**36.** 2	**37.** 3	**38.** 4	**39.** 1	**40.** 3
41. 4	**42.** 2	**43.** 2	**44.** 4	**45.** 2	**46.** 1	**47.** 3	**48.** 3	**49.** 2	**50.** 1
51. 2	**52.** 2	**53.** 1	**54.** 1	**55.** 2	**56.** 3	**57.** 1	**58.** 1	**59.** 3	**60.** 1
61. 4	**62.** 2	**63.** 3	**64.** 1	**65.** 2	**66.** 1	**67.** 3	**68.** 4	**69.** 1	**70.** 2
71. 4	**72.** 4	**73.** 1	**74.** 3	**75.** 3	**76.** 3	**77.** 1	**78.** 1	**79.** 2	**80.** 2
81. 2	**82.** 2	**83.** 3	**84.** 1	**85.** 3	**86.** 2	**87.** 2	**88.** 4	**89.** 2	**90.** 4
91. 4	**92.** 2	**93.** 1	**94.** 1	**95.** 3	**96.** 1	**97.** 2	**98.** 4	**99.** 2	**100.** 4
101. 3	**102.** 3	**103.** 1	**104.** 3	**105.** 3	**106.** 2	**107.** 4	**108.** 4	**109.** 2	**110.** 2
111. 4	**112.** 3	**113.** 1	**114.** 3	**115.** 1	**116.** 1	**117.** 2	**118.** 1	**119.** 4	**120.** 1
121. 4	**122.** 4	**123.** 2	**124.** 1	**125.** 1	**126.** 4	**127.** 4	**128.** 1	**129.** 3	**130.** 1
131. 1	**132.** 2	**133.** 4	**134.** 2	**135.** 1	**136.** 4	**137.** 2	**138.** 2	**139.** 1	**140.** 4
141. 2	**142.** 3	**143.** 1	**144.** 4	**145.** 3	**146.** 2	**147.** 3	**148.** 4	**149.** 1	**150.** 2
151. 3	**152.** 3	**153.** 2	**154.** 4	**155.** 1	**156.** 4	**157.** 4	**158.** 3	**159.** 1	**160.** 2
161. 2	**162.** 1	**163.** 3	**164.** 2	**165.** 2	**166.** 4	**167.** 3	**168.** 3	**169.** 1	**170.** 4
171. 4	**172.** 4	**173.** 3	**174.** 3	**175.** 4	**176.** 3	**177.** 2	**178.** 4	**179.** 1	**180.** 1
181. 4	**182.** 2	**183.** 1	**184.** 3	**185.** 4	**186.** 1	**187.** 2	**188.** 4	**189.** 4	**190.** 4
191. 1	**192.** 4	**193.** 4	**194.** 3	**195.** 4	**196.** 4	**197.** 3	**198.** 4	**199.** 2	**200.** 3

UNIT-6

PLANT PHYSIOLOGY

1. C_4 rice has been developed by transforming rice with
(1) PCPC
(2) PPDK
(3) GS
(4) Both PEPC and PPDK

2. Stress signalling is mediated by
(1) ABA
(2) GA
(3) Both the above
(4) None of the above

3. Which among the following enzymes is not a component of nitrogen assimilation complex ?
(1) Nitrate reductase
(2) Glutamate synthase
(3) Lactate dehydrogenase
(4) Glutamine synthetase

4. *nif* genes which encode the nitrogenase complex and other enzymes involve
(1) ammonification
(2) nitrogen fixation
(3) nitrification
(4) denitrificatrion

5. ABA is a
(1) growth promoter
(2) stress hormone
(3) protein
(4) polyamine

6. Break down of Gibberellic acid is mediated by
(1) GA_{20} Oxidase
(2) GA_2 Oxidase
(3) GA_3 Oxidase
(4) Kaurene Oxidase

7. Isopentenyl transferase is an enzyme involved in
(1) cytokinin synthesis
(2) auxin synthesis
(3) proline synthesis
(4) purine synthesis

8. Slender (*slr*) mutant in rice is due to mutation in
(1) GA signaling
(2) ABA signalling
(3) Auxin signaling
(4) Cytokinin signaling

9. Biological nitrogen fixation occurs when atmospheric nitrogen is converted into
(1) ammonia
(2) nitrate
(3) nitrite
(4) nitrogen dioxide

10. The first product of photosynthesis in C_3 plants is
(1) glycerate 3 phosphate
(2) malate
(3) glycerate 1,3 bisphosphate
(4) phosphoenol pyruvate

11. Respiratory Quotient is given by
(1) moles of CO_2 produced / moles of O_2 consumed
(2) moles of O_2 consumed / moles of CO_2 produced
(3) moles of biomass produced / moles of O_2 consumed
(4) moles of biomass produced/ moles of CO_2 produced

12. LEAs are classified as
(1) shoot development proteins
(2) seed storage proteins
(3) mutant derived proteins
(4) leaf development proteins

13. Engineering plants using chitinase gene leads to development of
(1) viral resistance
(2) bruchid resistance
(3) bacterial resistance
(4) cold tolerance

14. Grain number (*Gn1*) in rice is regulated by
(1) OsMADS1
(2) cytokinin oxidase
(3) gibberellin oxidase
(4) histidine kinase

15. Glyphosate-resistant gene *gox* is isolated from
(1) *Arthrobacter* sp
(2) *Achromobacter* sp
(3) *Bacillus* sp
(4) *Streptomyces* sp

16. The gene responsible for dwarfing character in rice is
(1) Tift 23A
(2) Norin 10
(3) Dee-geo-woo-gen
(4) Opaque 2

17. Which of the following plants contain the largest genome?
(1) *Arabidopsis thaliana*
(2) *Fritillaria assyriaca*
(3) *Zea mays*
(4) *Triticum dicoccum*

18. A hybrid between species followed by polyploidy or chromosome doubling is known as
(1) Autopolyploid
(2) Aneuploid
(3) Haploid
(4) Allopolyploid

19. The ploidy level of zygote: endosperm: maternal tissue ratio in a well developed seed is
(1) 1:1:1 (2) 2:3:2
(3) 1:3:1 (4) 1:2:1

20. ABA catabolism is mediated by
(1) ABA-8' carboxylase
(2) ABA-8' hydroxylase
(3) ABA-8' aminotransferase
(4) ABA-8' oxygenase

21. Nodulating genes in rhizobium are influenced by the presence of which one of the following in the roots?
(1) flavones
(2) lignin
(3) tannins
(4) cellulose

22. Which of the following is responsible for the protection of target molecules from reactive oxygen species?
(1) Halliwell-Asada pathway
(2) Calvin cycle
(3) Krebs cycle
(4) Pentose phosphate pathway

23. Which of the following enzymes is not responsible for dissipation of hydrogen peroxide?
(1) Ascorbate peroxidase
(2) Catalase
(3) Guaiacol peroxidase
(4) Superoxide dismutase

24. Among the following reporter genes which is the best that can be used for studying gene expression in a real time manner in plants?

(1) Luciferase
(2) GUS
(3) Green Fluorescent Protein
(4) Chloramphenicol Acetyl Transferase

25. Nitrogen use efficiency of the plants can be regulated by overexpressing which of the following genes?

(1) BZip
(2) Dof
(3) Leucine zipper
(4) Zinc finger

26. The herbicide that kills plants by blocking the photosynthetic electron flow of photosystem I is

(1) Diuron (2) Paraquat
(3) Glyphosate (4) Atrazine

27. In submerged plants the root tip stimulates the activity of

(1) ACC synthase
(2) ACC oxidase
(3) ACC synthase and ACC oxidase
(4) ACC kinase

28. In a microbial system, how are true and apparent growth yields related ?

(1) True growth yield is more than apparent growth yield
(2) True growth yield is less than apparent growth yield
(3) True growth yield is equal to apparent growth yield
(4) True growth yield and apparent growth yield are not related at all

29. Which of the following plant hormones is synthesized from an amino acid precursor?

(1) Ethylene (2) Gibberllic acid
(3) Cytokinin (4) Abscisic acid

30. Which of the following chemicals prevents precocious germination, promotes embryo maturation and increases dessication tolerance of somatic embryos?

(1) Abscisic acid
(2) Silver nitrate
(3) 2,4-Dichloro phenoxy acetic acid
(4) Sodium chloride

31. Which of the following is formed without fertilization in diplosporic apomicts?

(1) Embryo (2) Seed
(3) Endosperm (4) Embryo sac

32. Which one of the following phytohormones is produced under water-deficit and plays an important role in tolerance against drought?

(1) Abscisic acid (2) Cytokinin
(3) Ethylene (4) Gibberellin

33. Phytoalexins are chemical compounds that are produced by plants. The following statements refer to these compounds

1. These compounds are stress related factors produced in plants due to physical chemical or microbial stress
2. A large number of secondary metabolites belong to this class
3. Some genes are derepressed because of these factors

Of the statements given above

(1) Only 1 is true
(2) Only 1 and 2 are true
(3) Only 1 and 3 are true
(4) All are true

34. Barley RIP gene was demonstrated to provide protection in tobacco plant against

(1) *Rhizoctonia solani*
(2) *Alternaria* blight
(3) *Phytophthora infestans*
(4) Tobacco Mosaic Virus

35. CO_2 acceptor in C_3 plants is

(1) Phosphoglyceric acid
(2) Ribulose monophosphate
(3) Ribulose 1,5- bisphosphate
(4) Phosphoenol pyruvate

36. Why haploids are preferred for plant breeding experiments?
(1) Dominant characters are expressed
(2) Recessive characters are expressed
(3) Induction of mutation is easy
(4) Incomplete dominance is expressed

37. Among the following which enzyme is present in C_4 rice mesophyll cells?
(1) PEPC and PPDK
(2) PEPC and RUBISCO
(3) PEPC and NADP-ME
(4) PEPC and NAD-ME

38. Active form of Gibberellin in plants is
(1) GA1 (2) GA3
(3) GA15 (4) GA9

39. AAO gene family in plants is involved in
(1) ABA biosynthesis
(2) GA biosynthesis
(3) Auxin biosynthesis
(4) Ethylene biosynthesis

40. Oilseed mustard overexpressing 18:1 delta 12 desaturase will show increased levels of the following fatty acid in the seed oil
(1) oleic acid (2) linoleic acid
(3) stearic acid (4) erucic acid

41. The technique of estimation of chlorophyll in water bodies through remote sensing satellites uses electromagnetic radiations in the approximate wavelength range
(1) 1 to 100 cm
(2) 10^{-5} to 10^{-4} cm
(3) 10^{-9} to 10^{-12} cm
(4) 1 to 10 nm

42. A type of apomixis in which embryo sac develops from vegetative cells of the ovule is called
(1) apospory (2) apogamy
(3) diplospory (4) polyembryony

43. Banana bunchy top disease is transmitted by
(1) *Ferrisia virgate*
(2) *Aphis gossypii*
(3) *Pentalonia nigronervosa*
(4) *Thrips tabaci*

44. The chemical nature of GA_3 is
(1) Phenolic (2) Terpene
(3) Purine (4) Indole

45. Rice grains are deficient in
(1) Lysine (2) Glycine
(3) Isoleucine (4) Alanine

46. Lysimeter is used in the measurement of
(1) Light (2) Transpiration
(3) Lysine content (4) Water potential

47. Aerenchyma formation is related to which of the following hormones?
(1) ABA (2) Ethylene
(3) Cytokinin (4) Auxin

48. Barnase has----------activity
(1) Dnase (2) RNase
(3) Protease (4) Restriction

49. ________is used as a bacteriostat in the *Agrobacterium* mediated plant transformation experiments
(1) Kanamycin (2) Hygromycin
(3) Cefotaxime (4) Ampicillin

50. Variation in clonally reproducing crop arises from
(1) Genetic recombination
(2) Chromosomal segregation
(3) Alternative splicing
(4) Mutation

51. Seeds germinating in dark beneath the surface of the soil undergo
(1) Skotomorphogenesis
(2) Photomorphogenesis
(3) Embryogenesis
(4) Dessication

52. The substrate for photorespiration is
(1) Glycine
(2) Phospho Glycolic Acid
(3) Glycolic acid
(4) Phospho-glyceric acid

53. Leaf abscission is a phenomenon regulated by the amount of auxin and ethylene. Which one of the following statements is correct regarding the amount of auxin and ethylene during leaf abscission?

(1) Reduction in ethylene and increase in auxin
(2) Increase in both ethylene and auxin
(3) Reduction in auxin and increase in ethylene
(4) Decrease in both ethylene and auxin

54. Sciophytes are plants that prefer to grow in

(1) Sun
(2) Shade
(3) Cold temperature
(4) Water

55. Which one of the following is a micronutrient of plants?

(1) Mn (2) P
(3) Ca (4) Mg

56. Which one of the following receptors perceives blue light in plants?

(1) Cryptochrome
(2) Phytochrome
(3) Phototropin
(4) Photopsin

57. Photon capturing prosthetic group moiety of chlorophyll molecule is known as

(1) Mn Porphyrin
(2) Mg Porphyrin
(3) Fe Porphyrin
(4) Mo Porphyrin

58. Among the six types of plant cation channels identified, the shaker channels have been most thoroughly characterized. The shaker channels are highly selective to which one of the following cations?

(1) Na^+
(2) K^+
(3) Ca^{2+}
(4) Mg^{2+}

59. The process of oxidation of plastohydroquinone (PQH_2) molecule during light reaction of photosynthesis takes place in which of the following major protein complexes of the thylakoid membrane?

(1) ATP Synthase
(2) Photosystem I
(3) Photosystem II
(4) Cytochrome b_6f

60. In a metabolic engineering experiment the flux from primary carbon metabolism was diverted towards Methyl Erythritol Phosphate (MEP) pathway. Which one of the following secondary metabolites will be maximally synthesized?

(1) Alkaloids (2) Phenolics
(3) Terpenes (4) Glucosinolates

61. Which one of the following sequence of events is the first to take place during systemic production of protease inhibitors in young tomato leaves following insect feeding?

(1) Prosystemin is synthesized
(2) Systemin is produced
(3) Jasmonic acid is transported through phloem to other parts of plants
(4) Systemin is released from damaged cells into apoplast

62. Which of the following is not a typical feature of hypersensitive response observed in plants during the attack of invading microbes?

(1) Cells immediately surrounding the infection site die rapidly
(2) Often preceded by rapid accumulation of reactive oxygen species
(3) Results in increase in net photosynthesis rate
(4) A rapid spike of nitric oxide production accompanies the oxidative burst in the infected leaves

63. Which one of the following is true for the polymer-trapping model of sugar diffusion in plant cells?
(1) should be less concentrated in the mesophyll than in intermediary cells
(2) Sucrose should be more concentrated in the mesophyll than in the intermediary cells
(3) Sucrose concentration should be equal in mesophyll and intermediary cells
(4) Sucrose should be hydrolysed

64. What will happen to the growth of a vertically growing root if half of the root cap is removed?
(1) It will bend towards the side with remaining half of the root cap
(2) It will bend opposite to the side with remaining half of the root cap
(3) The root growth will be inhibited
(4) The root will keep growing gravitropically

65. Which one of the following statements is not correct regarding ethylene signaling in plants?
(1) Ethylene binding inactivates ethylene receptors
(2) In the absence of ethylene the receptors are inactive and suppress the ethylene response
(3) Disrupted ethylene receptors are inactive in the presence or absence of ethylene leading to constitutive ethylene response
(4) Missense mutation at binding site makes receptor insensitive to ethylene and could shut off the ethylene response

66. The precursor amino acid for secondary metabolite nicotine obtained from *Nicotiana tobaccum* is?
(1) Tyrosine
(2) Ornithine
(3) Leucine
(4) Tryptophan

67. The major plant hormone for Systemic Acquired Resistance and production of pathogenesis related protein is?
(1) Jasmonic acid
(2) Absiccic acid
(3) Benzoic acid
(4) Salisylic acid

68. The alternate oxidase system in plant is induced in presence of?
(1) Heat stress and salt stress
(2) Salisylic acid and increase in reduced pool of ubiquinone
(3) Cyanide in soil
(4) Severe nitrogen starvation conditions

69. Which of the following would you expect to see as products of the dark reaction of photosynthesis?
(1) Carbon dioxide, NADPH and ATP
(2) NADPH, Pi and ATP
(3) Carbon dioxide, ATP and glucose
(4) Glucose, $NADP^+$ and ADP

70. What is a property of C_4 plants, but not C_3 plants?
(1) Initially fix carbon dioxide in mesophyll cells as 4-carbon long compounds
(2) Initially fix carbon dioxide in bundle sheath cells as 3-carbon long compounds
(3) Initially fix carbon dioxide in mesophyll cells as 3-carbon long compounds
(4) Initially fix carbon dioxide in bundle sheath cells as 4-carbon long compounds

71. Which statement about photosynthesis in C_4 plants is not true?
(1) The first product of carbon dioxide fixation is a compound with 4 carbon atoms
(2) C_4 photosynthesis is an adaptation for plants living in hot, arid climates
(3) Carbon dioxide is initially fixed in mesophyll cells, but the Calvin cycle is active in bundle sheath cells in leaves of C_4 plants
(4) Less ATP is used overall for sugar biosynthesis in C_4 than in C_3 plants

72. During photosynthetic electron transport, the interior compartment of the thylakoid membranes becomes
(1) enriched in ATP
(2) more acidic than the stroma
(3) basic compared to the reaction centers
(4) the site for glucose synthesis

73. The overall source of energy for photosynthesis is
(1) energy of electron transport in the thylakoid membrane
(2) energy from the hydrolysis of ATP
(3) light energy from the sun
(4) energy from the reduced electron carriers NADH and $FADH_2$

74. Aquaporins occur in plants plasma membranes and membranes of the vacuole. The secondary structure in such proteins is
(1) Alpha helix
(2) Beta sheets
(3) Beta barrel
(4) Random coils

75. Which of the following is not a prosthetic group for nitrate reductase in plants
(1) FAD (2) NAD
(3) MoCo (4) Haeme-Fe

76. The enzyme nitrogenase that fixes atmospheric nitrogen in the root nodules of leguminous plants comprises two sub-units, the composition of each respectively being
(1) Fe-protein and Fe-Mo-protein
(2) Mo-protein and Fe-protein
(3) Mg-protein and Fe-Mo-protein
(4) Fe-protein and Fe-protein

77. During senescence of a plant
(1) the plant displays sequential senescence of the leaves
(2) metabolites like carbohydrates, proteins etc are transported from young leaves to the ageing leaves
(3) metabolites like carbohydrates, proteins etc are transported out of the ageing leaves to the young leaves
(4) secondary plant products accumulate in the tissues and organs during ageing

78. Which one of the following statements is correct with respect to the synthesis of starch and sucrose?
(1) Starch is synthesized in the mitochondria and sucrose in the cytosol
(2) Starch is synthesized in the cytosol and sucrose in the chloroplasts
(3) Starch is synthesized in the chloroplast and sucrose in the mitochondria
(4) Starch is synthesized in the chloroplast and sucrose in the cytosol

79. The enzyme that catalyzes the conversion of atmospheric nitrogen into ammonia is
(1) rhizobium
(2) nitrogenase
(3) megapascal
(4) rubisco

80. Mycorrhizae are
(1) nutrients required by plants in relatively small amounts
(2) plants such as mistletoe that parasitize other plants
(3) cells that control evaporation of water from leaves
(4) associations of roots with beneficial fungi

81. Once a flower is pollinated, changes occur that make it less attractive to insects. Its petals, for example, shrivel and fall off. Pollination must
(1) increase the output of cytokinins in the flower
(2) block the flow of auxins from the roots
(3) trigger the release of ethylene in the flower
(4) increase the formation of phytochrome, which sets the biological clock

82. Which hormone is called as stress hormone
(1) Abscisic Acid (2) Salisylic Acid
(3) Ethylene (4) Jasmonic Acid

83. Stomata opens when guard cells-
(1) sense an increase in CO_2 in the air spaces of leaf
(2) flap open because decrease in turgor pressure
(3) become more turgid because of an influx of K^+, followed by entry of water
(4) close aquaporins, preventing water uptake

84. The movement of sap from a sugar source to sugar sink
(1) occurs through apoplast of sieve tube members
(2) may translocate sugars from the breakdown of stored starch in a root tip to developing shoots
(3) is similar to flow of xylem sap in depending on tension or negative pressure
(4) is an active process where sucose is mainly utilized as energy source

85. The productivity of crop declines when leaves begin to wilt mainly because-
(1) the chloroplast of wilting leaves decomposes
(2) flaccid mesophyll cells are incapable of photosynthesis
(3) stomata close, preventing CO_2 from entering the leaf
(4) Photolysis of water cannot occur

86. Micronutrients are needed in very small amount because-
(1) most of them are mobile in plants
(2) most function as cofactor for enzymes
(3) most are supplied in large quantities in seed
(4) only the growing region of plant require them

87. Mycorrhizae enhance plant nutrition mainly by
(1) absorbing water and minerals through fungal hyphae
(2) converting atmospheric nitrogen to ammonia
(3) enabling the roots to parasitize neighboring plants stimulating the development of root hairs
(4) stimulating the development of root hairs

88. Carnivorous plant mainly compensate for soil that has relatively low content of-
(1) K^+ (2) Nitrogen
(3) PO_4^{--} (4) Ca^{++}

89. A mineral deficiency is likely to affect older leaves more then younger leaves if
(1) the mineral is micronutrient
(2) mineral is mobile within plant
(3) mineral is required for chlorophyll synthesis
(4) Deficiency persists for longer time

90. In plants IAA causes cell elongation due to
(1) increase in pH of Apoplast
(2) increase in pH of cytoplasm
(3) decrease in pH of Apoplast
(4) increase in pH of cytoplasm

91. A hormone that controls closure of stomata in response to water stress is-
(1) Abscissic acid
(2) Gibberellins
(3) Proline
(4) Ethylene

92. Which hormone is related with α-amylase activity in germinating seeds
(1) GA (2) Kinetin
(3) Auxin (4) ABA

93. Major role of carotenoids in plant is
(1) Antenna molecule
(2) Provide color to leaves and fruits
(3) Antioxidant
(4) Photo-protection

94. Among the following which enzyme is not activated by light
(1) Rubisco Kinase
(2) Fructose-1,6-bisphosphatase
(3) NADP: glyceraldehyde-3-phosphate dehydrogenase
(4) Hexoisomerase

95. Phytochromes are bluish chromoproteins. The phytochrome chromophores are called as
(1) phycobillins
(2) caretenoids
(3) xanthophylls
(4) porphyrins

96. Maximum water absorption by root takes place at
(1) meristimatic zone
(2) root cap
(3) elongation zone
(4) maturation zone

97. Which of the following protein-pigment complex can change its lateral position in thylakoid membrane as per need of electron flow?
(1) Cyt b6-f (2) PS-II
(3) LHC-II (4) PS-I

98. Which statement is false about photosystems in plants?
(1) two reaction centers are arranged in tandem (parallel) and connected by an electron transport chain containing plastoquionone and cytochromes-b6 and -f
(2) PS II a very strong oxidant is generated for oxidation of water
(3) PS I a very strong reductant for reduction of NADP
(4) Complete energy of the light absorbed by the two photosystems is used for chemical work

99. The proton motive force of a proton gradient is eliminated by uncouplers. Among the following which is a potential uncoupler for photophosphorylation?
(1) Carbonylcyanide-p-trifluormethoxy-phenylhydrazone (FCCP)
(2) Valinomycin
(3) Gramicidine
(4) All of the above

100. Nitrifying bacteria are examples of
(1) photoautotrophs
(2) photoheterotrophs
(3) chemoautotrophs
(4) chemoheterotrophs

101. Dissimilatory sulfate reducing bacteria utilizes
(1) sulfur as a final electron acceptor
(2) sulfur as a final electron donor
(3) H_2S as a final acceptor
(4) H_2S as a final donor

102. C_4 plants and CAM plants are similar in their photosynthetic adaptations in which of these respects?
(1) In both cases, the stomata normally close during the day
(2) Both types of plants make their sugar without the C_3 (Calvin) cycle
(3) In both cases, an enzyme other than Rubisco carries out the first step in carbon fixation
(4) Both types of plants make most of their sugar in the dark

103. A certain short-day plant flowers when days are less than 12 hours long. Which of the following would cause it to flower?
(1) 9-hour night and 15-hour day with 1 minute of darkness after 7 hr
(2) an 8-hr day and 16-hour night with a flash of white light after 8 hr
(3) 13-hour night and 11-hour day with 1 min of darkness after 6 hr
(4) a 12-hour day and 12-hour night with a flash of red light after 6 hr

104. When Pr molecules are exposed to red light, most of them absorb it and are converted to Pfr, but some of the Pfr also absorbs the red light and is converted back to Pr because both Pr and Pfr absorb red light. Thus the proportion of phytochrome in the Pfr form after saturating irradiation by red light is only about 85%. Similarly, the very small amount of far-red light absorbed by Pr makes it impossible to convert Pfr entirely to Pr by broad spectrum far-red light. Instead, an equilibrium of 97% Pr and 3% Pfr is achieved. This equilibrium is termed the

(1) Photo-equilibrium
(2) Photo-reversibility
(3) Photo-stationary state
(4) Photo-inhibition

105. Which of the following statements is correct?

(1) Legumes need rhizobial bacteria to provide them with nitrogen
(2) Legumes infected with rhizobial bacteria are more competitive against grasses
(3) There is a precise match between species of legume and species of Rhizobium
(4) Rhizobium can only grow and reproduce in the presence of a legume

106. An unusual amino acid is found in a fruit when it is unripe, that is a potent inhibitor of acyl-CoA dehydrogenases. If this fruit is eaten when unripe, it causes sickness that may be due to

(1) utilization of reserve body fat as fuel
(2) nonutilization of blood glucose as fuel
(3) nonutilization of reserve body fat as fuel
(4) inhibition of glycolysis

107. Heterocyst is the

(1) specialized structure for nitrogen fixation some cyanobacteria
(2) feeding organ of pathogenic fungi
(3) spore obtained from interspecies cross of fungi
(4) swollen tissue due to nematode infection

108. Respiratory organ for lichens is

(1) soredia
(2) haustoria
(3) cephalodia
(4) cyphella

109. In electron transport chain

(1) complex I and III are mobile while cytochrome-C and coenzyme-Q are immobile
(2) complex I and III are immobile while cytoehrome-e and coenzyme-Q are mobile
(3) complex I and cytochrome-c are mobile while complex III and coenzyme-Q are immobile
(4) Complex I and cytochrome-c are immobile while complex m and coenzyme-Q are mobile

110. Which of the following micronutrient deficiencies will have immediate effect in nitrogen assimilation?

(1) Molybdenum
(2) Boron
(3) Copper
(4) Zinc

111. An algal mutant was deficient in plastocyanin upon illumination. Its plastoquinone pool will be

(1) in oxidized state
(2) in reduced state
(3) ubiquinone state
(4) degraded

112. Anemophily is a form of

(1) pollination where pollen is carried by wind
(2) pollination where pollen is carried by herbivorous animals
(3) locomotion of plants directed by wind movement
(4) locomotion of plants by the direction of sunlight

113. Sodium/Proton exchanger protein used for raising plants show better ion homeostasis capability. This protein is localized in

(1) any location in the plasma membrane

(2) tonoplast exclusively

(3) inner mitochondrial membrane

(4) lysosomal membrane

114. 'Shooty' tumors are caused by the infection of

(1) *Agrobacterium rhizogenes*

(2) *Agrobacterium tumefaciens* wild type strain

(3) *Agrobacterium tumefaciens* with mutation in auxin biosynthesis gene

(4) *Agrobacterium tumefaciens* with mutation in cytokinin biosynthesis gene

115. Choose the correct sentence/statement for Crassulacean Acid Metabolism

(1) It causes more water loss than C_4 cycle for per molecule of carbon synthesis

(2) It is a process where first stable intermediate is malate/malic acid

(3) C_4 acid formation is both temporally and spatially separated from Calvin cycle

(4) Reduces CO_2 concentration in chloroplast

116. The hormone responsible for leaf epinasty (curvature) in flooding-sensitive plants such as tomato is

(1) abscisic acid

(2) auxin

(3) ethylene

(4) gibbrellic acid

117. 'Nodulins' are

(1) bacterial genes of NOD operon involved in nodule formation

(2) bacterial proteins that are required for symbiosis with plant

(3) alkaloids released by plant that attract symbiotic bacteria

(4) nodule-specific plant proteins

118. During cytokinesis in plant cells, the cell plate is formed by the fusion of vesicles derived from which of the following?

(1) Microtubules

(2) The Golgi complex

(3) The plasma membrane

(4) The contractile ring

119. Which of the following is required for the functioning of the Calvin cycle?

(1) Light

(2) Darkness

(3) ATP alone

(4) ATP and NADPH

120. Which of the following organisms may be used as sensitive indicators of air pollution?

(1) Members of the Chlorophyta

(2) Lichens

(3) Photosynthetic bacteria

(4) Slime molds

121. The cyanide-resistant respiration in plants results in production of

(1) heat only

(2) heat and excess of ATP

(3) heat and limited amount of ATP

(4) death of plant due to ATP exhaust

122. Illumination of an algal mutant lacking cytochrome-f results in

(1) accumulation of plastoquinol

(2) accumulation of reduced ferredoxin

(3) accumulation of NADPH

(4) accumulation of reduced plastocyanine

123. Ferredoxin is a

(1) copper-containing protein

(2) heme-Fe protein

(3) non-heme Fe-S protein

(4) flavin- containing protein

124. In C_4 plants; the rate of photosynthesis is higher than C_3 plants because
(1) PEP carboxylase is the carboxylating enzyme in bundle-sheath cells
(2) a lot of CO_2 is present in bundle-sheath chloroplasts due to malate decarboxylation
(3) Rubisco of C_4 plants is mutated so that it does not interact with oxygen,
(4) C_4 plants have more chlorophyll than C_3 plant

125. In summer, trees often absorb water in night
(1) because rate of transpiration in trees is still very high during night
(2) because trees compensate for the loss of water from the parenchymatous cells that occurs during day
(3) because rate of absorption of water increases due to negative root pressure during night
(4) due to balanced rate of absorption and transpiration during night

126. A porphyrin-containing protein involved in nitrogen fixation in several legumes is
(1) flavodoxin (2) nitrogenase
(3) thioredoxin (4) leghaemoglobin

127. Cryptochromes and phytochromes have overlapping functions. Screening for cryptochrome mutants should be done under
(1) blue light
(2) white light
(3) red + blue light
(4) red light

128. A mutant plant compared to the wild type accumulates higher amount of saturated fatty acids in membrane lipids. The growth of the mutant plants will be affected more under
(1) altered growth temperature
(2) arid environment
(3) water-logged condition
(4) high humidity condition

129. Glyoxysomes are likely to be found more in
(1) wheat embryo
(2) growing leaf of sugarcane
(3) root hairs
(4) endosperm of germinating castor seed

130. Which statement about plant hormones is not true?
(1) Plant hormones include at least one form that is a gas
(2) Each naturally occurring hormone has a specific chemical structure
(3) Hormones are among the principal internal regulators of plant growth
(4) All known growth regulators of plant activities are chemicals produced naturally in plants

131. Which of the following ions are affected by gibberellins and is also involved in signal transduction?
(1) H^+ (2) Ca^{2+}
(3) Na^+ (4) Mg^{2+}

132. Which plant hormone was discovered in infected rice seedlings that grew extremely tall and slender?
(1) Abscisic acid (2) Auxin
(3) Cytokinin (4) Gibberellin

133. How does auxin actually work?
(1) In the presence of light, it increases the turgor pressure of the cell and accelerates growth
(2) In the presence of light, it increases the formation of hydrolytic enzymes that release energy needed for growth purposes
(3) In the presence of light, it moves toward the light source and activates a protein that binds to DNA and starts enzyme production
(4) In the presence of unidirectional light, it moves to the shady side and activates an ATP-driven proton pump that results in weakened cell walls and eventual elongation

134. How does gibberellin actually work?

(1) In the presence of light, it increases the turgor pressure of the cell and accelerates growth

(2) In the presence of light, it increases the formation of additional DNA that codes for additional plant structures

(3) In the presence of unidirectional light, it moves to the shady side and activates an ATP-driven proton pump that results in weakened cell walls and eventual elongation

(4) It attaches to a plasma membrane receptor and results in gene activation to produce the enzyme amylase that releases sugars for additional cell growth, division, and elongation

135. The cofactor/prosthetic group associated with oxidation-reduction reactions is

(1) NADH

(2) Zn^{2+}

(3) Pyridoxal phosphate

(4) Coenzyme A

136. A photosystem contains which of the following?

(1) One photochemical reaction centre and multiple antennae

(2) Multiple photochemical reaction centres and one antenna

(3) One photochemical reaction centre and one antenna

(4) Multiple photochemical reaction centres and multiple antennae

137. Which of the following reactions is not true about the C_4 pathway in plants?

(1) CO_2 enters Calvin cycle by condensing with Ribulose 1,5-bisphosphate

(2) Oxaloacetate formation is catalyzed by PEP carboxylase

(3) Oxaloacetate is rapidly converted to malate in mesophyll chloroplasts

(4) Malate is decarboxylated to release CO_2 in the cytosol of mesophyll cells

138. Which of the following regulates the phytochrome action?

(1) PFr/Ptotal (2) Total PR

(3) Total PFr (4) PFr in dark

139. Which of the following about succulent species such as the cacti is incorrect

(1) have CAM pathway

(2) true drought tolerant

(3) high water use efficiency

(4) true drought avoider

140. Which of the following would you expect to see as products of the dark reaction of photosynthesis?

(1) Carbon dioxide, NADPH and ATP

(2) NADPH, Pi and ATP

(3) Carbon dioxide, ATP and glucose

(4) Glucose, $NADP^+$ and ADP

141. Plant cells that commonly do not contain nucleus are

(1) sieve tube cells

(2) companion cells

(3) guard cells

(4) epithelial cells

142. Which statement about photosynthesis in C_4 plants is not true?

(1) the first product of carbon dioxide fixation is a compound with 4 carbon atoms

(2) C_4 photosynthesis is an adaptation for plants living in hot, arid climates

(3) carbon dioxide is initially fixed in mesophyll cells, but the Calvin cycle is active in bundle sheath cells in leaves of C_4 plants

(4) less ATP is used overall for sugar biosynthesis in C_4 than in C_3 plants

143. During photosynthetic electron transport, the interior compartment of the thylakoid membranes becomes

(1) enriched in ATP

(2) more acidic than the stroma

(3) basic compared to the reaction centers

(4) the site for glucose synthesis

144. Major pathway for terpene biosynthesis is
(1) Shikimic acid pathway
(2) MEP pathway
(3) Aceto lactate synthase pathway
(4) C_4 pathway

145. Which of the following statements about cyclic photophosphorylation is incorrect?
(1) it doesn't involve NADPH formation
(2) it is activated when ATP is limiting
(3) it does not generate O_2
(4) it involves a substrate-level phosphorylation

146. At depths of ocean, due to the green window of photosynthesis, only green light is available, as the light of the other wavelengths is absorbed by green algae living in the upper regions of the water. But then also blue green algae are able to carry out their photosynthesis in such a dim or green light because they have?
(1) Chlorophyll 'a'
(2) Phycobilisomes
(3) Xanthophylls
(4) Single but efficient photosystem

147. Which statement about the Mn center of photosystem II is incorrect?
(1) the Mn center has four possible oxidation states
(2) electrons are transferred from the Mn center to P_{680}
(3) a tyrosine residue on the D1 protein is an intermediate in electron transfer
(4) the O_2 released by the Mn center comes from the oxidation of water

148. Paraquat a herbicide act as
(1) blocking electron flow between Q_A & Q_B
(2) blocking electron flow between bf_6-PSI
(3) photolysis of water
(4) act by accepting electrons from the early acceptors of photosystem I and then reacting with oxygen to form superoxide, O_2^-

149. Emerson with his experiments was able to show that when the pulses had a very low light intensity, the amount of oxygen formed increased proportionally with the light intensity. From this it was calculated that the release of one molecule of oxygen had a minimum quantum requirement of about
(1) two photons
(2) four photons
(3) six photons
(4) eight photons

150. Which of the following is not a component of antenna molecule in higher plants?
(1) Chlorophyll b
(2) Carotenes
(3) Xantophylls
(4) Phycobilins

151. Which organelle occupies maximum volume of a plant cell?
(1) Vacuole (2) Chloroplast
(3) Cytosol (4) Mitochondria

152. Plant cell wall is composed of
(1) carbohydrates only
(2) carbohydrates and proteins
(3) carbohydrates, lignin and proteins
(4) none of the above

153. Which of the following is not a function of a vacuole?
(1) It maintain cell turgor
(2) Recycles those cellular constituents that are defective or no longer required
(3) Function as waste deposit of xenobiotics such as herbicides
(4) Involved in photorespiration and glyoxylate cycle

154. In leaves grown in the dark (etiolated plants), the plastids have a yellowish color and are termed as
(1) proplastid (2) etioplast
(3) leucoplast (4) choloroplast

155. Regarding plants in general
(1) the sporophyte is haploid and the gametophyte is diploid
(2) both sporophyte and gametophyte are diploid
(3) the sporophyte is diploid and the gametophyte is haploid
(4) both sporophyte and gametophyte are haploid

156. A plant tissue with spiral thickening in cell walls capable of absorbing water is known as
(1) epidermis (2) hypodermis
(3) cork (4) velamen

157. Reserpine, a drug used for treating mental disorders and high blood pressure, is extracted from the roots of which of the following plant species?
(1) *Ferula* (2) *Rauwolfia*
(3) *Digitalis* (4) *Atropa*

158. Under what condition photorespiration is favoured?
(1) O_2 concentration is low and CO_2 concentration is high
(2) CO_2 concentration is low and O_2 concentration is high
(3) Under low-light conditions
(4) Under nutrient-deficient conditions

159. For sulfur assimilation, the sulfate molecules that are absorbed by root cells bind to
(1) glutathione (2) ATP
(3) coenzyme A (4) sulfate reductase

160. Which plant hormone, when applied exogenously, would inhibit the coleoptile elongation?
(1) Auxin (2) Cytokinin
(3) Gibberellin (4) Abscisic acid

161. The middle lamella is mainly composed of
(1) calcium pectate
(2) cellulose and hemicellulose
(3) suberin
(4) glycolipids

162. Leghaemoglobin in symbiotic nodules helps
(1) by providing energy to the growing symbionts
(2) in nitrogen reallocation between symbionts and plants
(3) by oxygen homeostasis
(4) by stabilizing redox potential of host cells

163. Systemic Acquired Resistance (SAR) activation leads to
(1) a marked reduction in disease symptoms after subsequent infection by many different pathogen species
(2) a marked reduction in disease symptoms after subsequent infection by the same pathogen species only
(3) no effect on reduction is disease symptom as the affected cells are all dead by the time it occurs
(4) a situation where only there is temporary reduction in disease symptom during the same infection cycle but no relief in current cycle

164. Breads, fruits and dried foods are most commonly spoilt by which group of organisms?
(1) *Acetobacter* (2) Fungi
(3) *Pseudomonas* (4) *Saccharomyces*

165. *Fusicoccum amygdali* produces the toxin fusicoccin which targets
(1) Na^+/H^+ antiporter located at the plasma membrane
(2) Na^+ /H^+ anti porter located at the tonoplast
(3) H^+-ATPase localized on the plasma membrane
(4) H^+-ATPase localized on the tonoplast

166. Water activity (a_w) value below which most spoilage bacteria cannot grow is
(1) 0.3 (2) 0.5
(3) 0.7 (4) 0.9

167. Casparian strip is the
(1) waxy outer coat of some grass seeds
(2) stretch of islands in Pacific ocean that is unique for its ecosystem
(3) hydrophobic barrier in the root endodermis
(4) dorsal cell layer in developing *C. elegans* embryo

168. Which is not a correct statement about chloroplasts?
(1) Granal and stromal thylakoids are membrane components of the thylakoids
(2) The membranes that are staked in the thylakoid are called stroma
(3) Photo-harvesting apparatus are embedded in the thylakoid membrane
(4) Like mitochondrial membrane, chloroplast membranes are also capable of generating ATP using proton motif force

169. Systemic Acquired Resistance (SAR) is a phenomenon wherein a plant partly infected by a pathogen develops resistance against further infections in other parts. The statement that is not correct regarding SAR mediated protection is
(1) Its effect is long lasting
(2) It is observed even in the new leaves that emerge after the induction of SAR
(3) It is a genetic response
(4) It is specific to the pathogen that induces it

170. Which one of the following is not involved in maintaining oxygen homeostasis in nitrogen fixing nodules?
(1) Leghaemoglobin
(2) Cytochrome oxidase
(3) ATP dependent potassium pump
(4) Dinitrogenase reductase

171. What are DELLA domain proteins?
(1) Proteins involved in cell-cell signaling
(2) Repressor of GA signaling
(3) Homeotic proteins
(4) Auxin receptors

172. Which one of the following statements correctly describes the role of vernalization?
(1) It reduces apical dominance and leads multiple branching
(2) It causes genetic modification that leads to early flowering
(3) It delays flowering and allows maximum vegetative growth leading to higher yield
(4) It causes epigenetic modification that promotes early flowering

173. When critical night length is broken by giving light flashes?
(1) A flash of red light can induce flowering in short-day plant
(2) A flash of far-red light followed by a flash of red light will induce flowering in short -day plant
(3) A flash of far-red light followed by a flash of red light will induce flowering in long-day plant
(4) A flash of red light followed by a flash of far-red light will induce flowering in long-day plant

174. The ring structure of chlorophyll contains which one of the following?
(1) Fe^{++} (2) Cu^{++}
(3) Mg^{++} (4) Mn^{++}

175. Which of the following mechanisms is used by the viral movement proteins to facilitate virus movement within plants?
(1) Rupturing cell wall
(2) Increasing size exclusion limit of plasmodesmata
(3) Rupturing plasmodesmata by exerting mechanical force
(4) Degrading cell wall using secretory enzymes

176. Which of the following is the plant hormone involved in seed and bud dormancy and in response of plants to stress?
(1) Abscisic acid (2) Cytokinin
(3) Auxins (4) Gibberellins

177. Most of DNA binding protein binds to DNA by particular motif to modulate gene expression. Genes which are under regulation of gibberllic acid have GRE where, GREB binds. The motif in GREB is
(1) Leucine Zipper
(2) bZIP
(3) Zinc finger
(4) Homeodomain

178. The donor and acceptor of electron for cytb/f6 are
(1) PQ and UQ
(2) UQ and Cyt C
(3) UQ and PC
(4) PQ and PC

179. The chemical salicyl hydrooxamic acid inhibits
(1) Cytochrome oxidase
(2) Alternate oxidase
(3) ATP synthese
(4) NADH-dehyrogenase

180. Phytochromes involved in red/far-red response are dimeric chromoproteins. The two sub-units of phytochromes are linked with
(1) PAS domain
(2) PTB domain
(3) Hinge region
(4) Kinase domain

181. Which of the following do not occur during seed development?
(1) Accumulation of storage proteins
(2) Synthesis of LEA proteins
(3) Desiccation
(4) Synthesis of Gibberlic acid

182. Jasmonic acid and its methylated derivative, methyl jasmonate, play major roles in plant defense against insect herbivores and microbial pathogens. Which of the following molecules is not known to be induced by the jasmonate signaling pathway?
(1) Proteinase inhibitors
(2) Chitinase
(3) Oxoacids
(4) Salicylic acid

183. Consider the following table regarding the photoperiods and flowering habits of plants and identify the correct statement

Plant Species	Photoperiod		
	8 hr of light	14 hr of light	20 hr of light
Xanthium	Flowers	Does not flower	Does not flower
Hypocyanus	Does not flower	Flowers	Flowers
Sunflower	Flowers	Flowers	Flowers

(1) *Xanthium* is short-day plant; *Hyoscyamus* is long-day plant; Sunflower neutral plant
(2) *Xanthium* is long-day plant; *Hyoscyamus* is short-day plant; Sunflower is day neutral plant
(3) *Xanthium* is short-day plant; *Hyoscyamus* is day-neutral plant; Sunflower is long day plant
(4) *Xanthium* is long-day plant; *Hyoscyamus* is day-neutral plant; Sunflower is short day plant

184. The CO_2 that is released during photorespiration originates from
(1) glycine decarboxylation in mitochondria
(2) ribulose bisphosphate in chloroplasts
(3) glyoxylate decarboxylation in peroxisomes
(4) serine decarboxylation in mitochondria

185. The precursor for abscissic acid biosynthesis is
(1) carotenoid
(2) chlorophyll
(3) dormin
(4) tryptophan

186. GA induced starch degradation in endosperm essentially requires gene expression, protein synthesis and protein transport. In an experiment to understand this, GFP tagged protein is seen to be moving from cytoplasm into the nucleus before alpha amylase could be detected. What do you think this GFP is tagged with?

(1) It is in fusion with a gene regulator
(2) It is in fusion with a signaling kinase
(3) It is in fusion with alpha-amylase
(4) It is tagged with a cell wall protein

187. To increase the shelf-life of tomato, which of the following genes should be silenced?

(1) ACC synthase
(2) EPSP synthase
(3) PEP carboxylase
(4) ABC transporter

188. The amino acid which help in osmo-regulation during stress condition is-

(1) Proline (2) α-Alanine
(3) Tyrosine (4) All of the above

189. Jasmonic acid and it methyl esters are synthesized form unsaturated fatty acids linolenic acid and mediates in insect and disease resistance like salicylic acid. Which is incorrect about jasmonic acid

(1) It activates number of genes encoding for proteins with antifungal properties
(2) Activates the gene for Salicyclic acid
(3) Function as secondary messenger
(4) Plays important role in Induced systemic response

190. Which symbiotic association helps plants in mineral and water absorption?

(1) Lichens (2) Root nodule
(3) Mycorryzae (4) Lichens

191. Non-O_2-evolving (anoxygenic) organisms, such as the purple photosynthetic bacteria of the genera *Rhodobacter* and *Rhodopseudomonas*, contain

(1) only a single photosystem
(2) both photosystem
(3) no photosytems
(4) three photosystems

192. Evidence from spectral and ESR studies indicates that ________ acts as an early acceptor in photosystem II, followed by a complex of two plastoquinones in close proximity to an iron atom

(1) Plastoquinone (2) Ferrodoxin
(3) Pheophytin (4) Plastocyanin

193. Which of the following statements about thylakoids is not correct?

(1) The thylakoid membranes contain chlorophyll pigments
(2) The thylakoid membranes contain the photosystems
(3) The thylakoid membranes contain the Calvin cycle enzymes
(4) The thylakoid membranes contain the electron transport machinery

194. Which of the following statements about the mechanism of the light-dependent reactions of photosynthesis is correct?

(1) Electrons from photosystem I reduce NADPH
(2) Electrons from photosystem I reduce pheophytin
(3) Electrons from NADPH revert photosystem II back to the ground state
(4) Ferredoxin-NADP reductase reduces $NADP^+$ to NADPH

195. Which of the following statements about cyclic photophosphorylation is correct?

(1) Cyclic photophosphorylation utilizes excess ATP
(2) Cyclic photophosphorylation reduces NADP+ to NADPH
(3) Cyclic photophosphorylation occurs in the cytochrome bf complex and utilizes electrons from photosystem I
(4) Cyclic photophosphorylation utilizes electrons from photosystem II

196. **Which of the following statements about the mechanism of the Calvin cycle is correct?**

(1) The Calvin cycle is a metabolic pathway by which plants convert CO_2 into pentose sugars as the main product

(2) The Calvin cycle enzymes are more reactive if there is a decrease in light intensity

(3) In the Calvin cycle rubisco adds CO_2 to 3-phosphoglycerate

(4) The Calvin cycle is a metabolic pathway by which plants convert CO_2 and water into carbohydrates

197. **Which reaction in photosynthesis is carried out by 'Rubisco' or ribulose 1-5 bisphosphate carboxylase?**

(1) Conversion of 3 phosphoglycerate into glyceraldehyde 3 phosphate

(2) Utilisation of CO_2 to produce 3 phosphoglycerate

(3) Conversion of glyceraldehyde 3 phosphate into ribulose 5 phosphate

(4) Carboxylation of phosphoenol pyruvate to oxaloacetate

198. **Which of the following statements about chlorophyll is correct?**

(1) Chlorophyll molecules make up the electron transport chains in photosynthesis

(2) Chlorophyll molecules are found in the chloroplast stroma

(3) Chlorophyll molecules appear green because they reflect light in the red and blue ranges

(4) Chlorophyll molecules are the photosynthetic pigments in the thylakoid membrane

199. **Which of the following statements about photosystem I in chloroplasts is not correct?**

(1) Photosystem I is activated by light independently from Photosystem II

(2) Plastocyanin reduces photooxidised P700 in PSI

(3) Plastocyanin reduces photooxidised P680 in PSI

(4) Electrons ejected from P700 in photosystem I are replaced with electrons from water

200. **Which of the following statements about C_4 plants are correct?**

(1) C_4 plants minimize the oxygenase activity of rubisco by fixing CO_2 into PEP (phosphoenolpyruvate)

(2) C_4 plants minimize the oxygenase activity of RUBISCO by fixing CO_2 phosphoglycerate

(3) C_3 plants are more efficient in photosynthesis than are C_4 plants

(4) C_4 plants mainly occur in areas of high temperature and light intensity

ANSWER SHEET

Unit 6: Plant Physiology

1. 4	**2.** 1	**3.** 3	**4.** 2	**5.** 2	**6.** 2	**7.** 1	**8.** 1	**9.** 1	**10.** 1
11. 1	**12.** 2	**13.** 2	**14.** 2	**15.** 2	**16.** 3	**17.** 2	**18.** 4	**19.** 2	**20.** 2
21. 1	**22.** 1	**23.** 4	**24.** 3	**25.** 2	**26.** 2	**27.** 3	**28.** 1	**29.** 1	**30.** 1
31. 1	**32.** 1	**33.** 3	**34.** 2	**35.** 3	**36.** 2	**37.** 1	**38.** 1	**39.** 1	**40.** 2

41. 2	**42.** 1	**43.** 3	**44.** 2	**45.** 1	**46.** 2	**47.** 2	**48.** 2	**49.** 2	**50.** 4
51. 1	**52.** 2	**53.** 3	**54.** 2	**55.** 1	**56.** 3	**57.** 2	**58.** 2	**59.** 4	**60.** 3
61. 1	**62.** 3	**63.** 2	**64.** 1	**65.** 2	**66.** 2	**67.** 4	**68.** 2	**69.** 4	**70.** 1
71. 4	**72.** 2	**73.** 3	**74.** 1	**75.** 2	**76.** 1	**77.** 3	**78.** 4	**79.** 2	**80.** 4
81. 3	**82.** 1	**83.** 3	**84.** 2	**85.** 3	**86.** 2	**87.** 1	**88.** 2	**89.** 2	**90.** 3
91. 1	**92.** 1	**93.** 4	**94.** 4	**95.** 1	**96.** 4	**97.** 3	**98.** 4	**99.** 4	**100.** 3
101. 1	**102.** 3	**103.** 3	**104.** 4	**105.** 3	**106.** 3	**107.** 1	**108.** 1	**109.** 2	**110.** 1
111. 2	**112.** 1	**113.** 2	**114.** 3	**115.** 2	**116.** 3	**117.** 4	**118.** 2	**119.** 4	**120.** 2
121. 3	**122.** 1	**123.** 3	**124.** 2	**125.** 3	**126.** 4	**127.** 1	**128.** 1	**129.** 4	**130.** 4
131. 2	**132.** 4	**133.** 4	**134.** 4	**135.** 1	**136.** 1	**137.** 4	**138.** 1	**139.** 2	**140.** 4
141. 1	**142.** 4	**143.** 2	**144.** 2	**145.** 4	**146.** 2	**147.** 1	**148.** 4	**149.** 4	**150.** 4
151. 1	**152.** 3	**153.** 4	**154.** 2	**155.** 3	**156.** 4	**157.** 2	**158.** 2	**159.** 2	**160.** 4
161. 1	**162.** 3	**163.** 1	**164.** 2	**165.** 3	**166.** 3	**167.** 3	**168.** 2	**169.** 4	**170.** 2
171. 2	**172.** 4	**173.** 3	**174.** 3	**175.** 2	**176.** 1	**177.** 4	**178.** 4	**179.** 2	**180.** 3
181. 4	**182.** 4	**183.** 1	**184.** 1	**185.** 1	**186.** 2	**187.** 1	**188.** 1	**189.** 2	**190.** 3
191. 2	**192.** 3	**193.** 3	**194.** 4	**195.** 3	**196.** 4	**197.** 2	**198.** 4	**199.** 3	**200.** 4

UNIT-7

ANIMAL PHYSIOLOGY

1. Superoxide dismutase is an important enzyme for maintenance of red blood cells and is defective in some neurodegenerative diseases. What does this enzyme do?
   (1) catalyzes the conversion of O_2^{-1} to H_2O_2 and O_2
   (2) creates superoxides by oxidizing heme
   (3) converts H_2O_2 to water and O_2
   (4) removes H_2O_2 by oxidizing glutathione and producing water
2. Repolarization after an action potential occurs
   (1) through the opening of Na^+ channels
   (2) through the opening of Na^+ and closure of K^+ channels
   (3) through the closure of Na^+ and opening of K^+ channels
   (4) through the opening of Na^+ and K^+ channels
3. The Bohr effect in hemoglobin refers to
   (1) the effect of pH on hemoglobin and myoglobin
   (2) higher pH found in actively metabolizing tissues
   (3) increased affinity for O_2 at lower pH
   (4) reduced affinity for O at lower pH
4. The rate of impulse conduction in a nerve depends on
   (1) axon diameter and axon length
   (2) axon length and number of dendrites
   (3) axon diameter and thickness of myelination
   (4) myelination and nuclear size
5. Pain sensation is a subjective and conscious feeling. However, although the autonomic organs do not get represented in the cerebral cortex, one feels pain in those parts as well. The reason is
   (1) those parts receive less blood supply
   (2) increased pH in those parts
   (3) the pain is referred to other parts of the body
   (4) those organs are not superficially located
6. Which one of the following is not a function of glia?
   (1) providing support to the neural tissue
   (2) conduction of electrical signal
   (3) myelination of neurons
   (4) help in neuronal growth
7. At chemical synapse, communication between two neurons is
   (1) physical process
   (2) chemical process
   (3) physico-chemico-physical process
   (4) physico-chemico-mechanical process
8. Retrograde transport may be used for
   (1) nerve path tracing
   (2) determining nerve fiber diameter
   (3) determining soma size
   (4) estimating number of dendrites
9. Which of these electrodes will be preferred for intracellular potential recording?
   (1) glass capillary electrode
   (2) steel micro-electrode
   (3) copper micro-electrode
   (4) solid glass electrode

10. Which of the following types of neurons may be identified using Tyrosine hydroxylase immunostaining ?
(1) Cholinergic (2) GABA-ergic
(3) Glutamateric (4) Aminergic

11. Nerve is a bundle of fibres. In vertebrates it contains
(1) many myelinated axons of different diameters as well as large number of unmyelinated fibres
(2) many unmyelinated fibres as well as large number of myelinated axons of same diameter
(3) only myelinated axons of same diameter
(4) only unmyelinated axons of different diameter

12. The arrival of the action potential in a presynaptic knob initiates quantal release of neurotransmitter by influx of Ca ions, which may act by
(1) Changing the viscosity of cytoplasm and thereby increasing the Brownian movements of synaptic vesicles
(2) Activating calcium-calmodulin protein kinase
(3) Dephosphorylating clathrin
(4) Phosphorylation of dynamin

13. Which one of the following changes will occur when the external K^+ ion concentration in the neuromuscular junction of skeletal muscles is increased?
(1) The rate of occurrence of MEPP is increased
(2) The amplitude of MEPP is increased
(3) The rate of occurrence of MEPP is decreased
(4) The amplitude of MEPP is decreased

14. Which of the statements is true for matured human RBC? It
(1) divides once a day
(2) does not divide
(3) divides every 120 days
(4) divides under stressful condition

15. The intervention by which a gadget e.g. an electrode may be accurately guided to a predefined region deep inside the brain is known as
(1) stereoscopy
(2) stereotaxic surgery
(3) craniotomy
(4) laparoscopy

16. The main pacemaker for endogenous rhythms (Circadian rhythms) is the
(1) Zeitgeber
(2) Suprachasmatic nucleus
(3) Optic chiasm
(4) Core body temperature

17. Which of the following types of neurons is primarily lost in Parkinson's disease?
(1) Dopaminergic neurons in the substantia nigra
(2) Cholinergic neurons in the brain stem
(3) Noradrenergic neurons in the cerebellum
(4) GABA-ergic neurons in cortex

18. Apart from gas transport Hemoglobin plays an important role in
(1) red cell morphology
(2) blood buffering
(3) globin synthesis
(4) bone marrow regeneration

19. Diabetes insipidus is caused due to insufficient level of
(1) insulin (2) ADH
(3) thyroxine (4) TSH

20. In Parkinson's disease there is a predominant loss of
(1) dopaminergic neurons in the substantia nigra
(2) cholinergic neurons in the brain stem
(3) noradrenergic neurons in the locus coeruleus
(4) GABA-ergic neurons in the cortex

21. Calcium present in which of the following spaces take part in the release of neurotransmitter?
(1) Vesicles at the presynaptic terminal
(2) Extracellular space
(3) Intracellular space
(4) Presynaptic terminal in free form

22. Maximum concentration of dopaminergic neurons is present in
(1) locus coeruleus
(2) red nucleus
(3) substantia Nigra
(4) mammillary body

23. In Alzheimer's disease there is predominant loss of which type of neurons?
(1) Cholinergic (2) Cholinoceptive
(3) Noradrenergic (4) Noradrenoceptive

24. Under stress condition which of the following pairs of organs plays as haemopoetic organ other than bone marrow?
(1) Both liver and lymph node
(2) Both spleen and liver
(3) Both lymphnode and thymus
(4) Both spleen and thymus

25. Which one of the following is correct when Na ion concentration in the bathing fluid of a squid giant axon is less than that present in sea water, but the osmotic pressure is maintained by substituting it with choline chloride?
(1) Hyperpolarization in membrane potential
(2)The amplitude of the action generated in the nerve fibre will be higher than normal action potential in sea water
(3) Depolarization in membrane potential
(4)The amplitude of the action potential generated in the nerve fibre will be lower than normal action potential in sea water

26. When Tetrodotoxin is administered in bathing medium of a nerve fibre during voltage clamp experiments the membrane current may show which one of the following changes?
(1) Absence of outward membrane current
(2) Absence of inward membrane current
(3) Higher amplitude of outward membrane current
(4) Higher amplitude of inward membrane current

27. Primary colours of vision are
(1) red, black and yellow
(2) black, white and green
(3) orange, yellow and blue
(4) red, blue and green

28. Locus coeruleus in the brain possesses maximum concentration of
(1) cholinergic neurons
(2) noradrenergic neurons
(3) astrocytes
(4) microglia

29. The rate of impulse conduction in a nerve depends on
(1) axon diameter and axon length
(2) axon length and number of dendrites
(3) axon diameter and thickness of myelination
(4) myelination and nuclear size

30. Melatonin is secreted in the
(1) day time before noon
(2) day time after noon period
(3) just before sun set
(4) dark period

31. Heart rate would increase by the application of
(1) acetylcholine (2) adrenaline
(3) cold saline (4) cold glucose solution

32. Posture maintenance is mainly controlled by the
(1) cerebrum (2) cerebellum
(3) hypothalamus (4) mammilary bodies

33. In albino Wistar rats the red colour of the blood is due to

(1) lack of pigmentation
(2) absence of porphyrin ring in the heme of haemoglobin
(3) oxidized state of the iron in the heme
(4) reduced state of the iron in the heme

34. Tyrosine hydroxylase immunopositive neurons are

(1) only noradrenergic
(2) only dopaminergic
(3) either dopaminergic or noradrenergic
(4) only serotonergic

35. In adult neurogenesis, which of the brain areas has been suggested to play a role in periodic clearance of outdated hippocampal memory traces?

(1) cortex (2) brainstem
(3) dentate gyrus (4) hypothalamus

36. In a biological system, amplification of signal is a fundamental regulatory principle. Which one of the following is not an amplification system?

(1) Blood clotting
(2) Complement activation
(3) Transmembrane receptor-mediated gene expression
(4) Ciliary movement in cochlea

37. Xth nerve is an example of

(1) mixed cranial nerve
(2) sensory cranial nerve
(3) spinal nerve
(4) motor nerve

38. The intervention, by which a specific point deep inside the brain may be accurately targeted by an object e.g., an electrode, is known as

(1) stereoscopy (2) stereotaxic surgery
(3) craniotomy (4) laparoscopy

39. Which of the following neurotransmitters containing neurons is maximally present in the dorsal raphe ?

(1) Dopaminergic (2) Adrenergic
(3) Serotonergic (4) Cholinergic

40. Cerebellar damage would primarily lead to

(1) difficulty in smelling
(2) postural disturbance
(3) loss of taste
(4) memory loss

41. In case of nerve impulse propagation between neurons, the first site of fatigue is at

(1) axon
(2) electrical synapse
(3) chemical synapse
(4) dendrite

42. Na^+-K^+ ATPase exchanges Na^+ and K^+ across cell membrane. The enzyme is a

(1) tetramer and consumes two ATP molecules in every cycle
(2) dimer and consumes two ATP molecules in every cycle
(3) monomer and consumes one ATP molecule in every cycle
(4) tetramer and consumes one ATP molecule in every cycle

43. Which of the following types of neurons is predominantly lost in Narcolepsy?

(1) Cholinergic (2) Orexinergic
(3) Noradrenergic (4) Histaminergic

44. Retrograde transport may be used for

(1) nerve path tracing
(2) determining nerve fiber diameter
(3) determining soma size
(4) estimating number of dendrites

45. The conscious state of an individual may be best understood by studying ones

(1) electromyogram
(2) electrocardiogram
(3) electroretinogram
(4) electroencephalogram

46. Which of the following electrodes will be preferred for recording intracellular potential?
(1) glass capillary electrode
(2) steel micro-electrode
(3) copper micro-electrode
(4) solid glass electrode

47. For recording fast physiological response e.g., action potential in neurons, one needs a
(1) Cathode Ray Oscilloscope
(2) Polygraph
(3) Spectrophotometer
(4) Confocal microscope

48. In vertebrates, nerve bundle usually contains
(1) many myelinated axons of different diameters as well as large number of unmyelinated fibres
(2) many unmyelinated fibres as well as large number of myelinated axons of same diameter
(3) only myelinated axons of same diameter
(4) only unmyelinated axons of different diameter

49. Mark the correct statement for a normal living excitable cell at rest. The ionic concentrations across the cell membrane are such that
(1) they are at equi-potential
(2) the intracellular potential is positive relative to that of the extracellular
(3) the intracellular potential is negative relative to that of the extracellular
(4) the intracellular potential is positive relative to that of the intracellular potential of another cell

50. Sleeping sickness is caused by
(1) *Plasmodium vivax*
(2) *Leishmania donovani*
(3) *Trypanosoma cruzi*
(4) *Entamoeba histolytica*

51. Which of the following sets of cranial nerves falls under parasympathetic system?
(1) I, IV, V and X
(2) III, VII, IX and X
(3) II, VIII, IX, XI
(4) VI, XII, I and IV

52. The demyelination of the central nervous system white matter produced by the canine distemper virus is an example of
(1) Fat necrosis
(2) Coagulation necrosis
(3) Zenker's necrosis
(4) Liquefactive necrosis

53. The discoloration of tissue by iron sulphide after somatic cell death is referred to as
(1) Hypostatic congestion
(2) Imbibition with hemoglobin
(3) Imbibition with bile
(4) Pseudomelanosis

54. The specific condition that occurs subsequent to the inhalation of carbon is referred to as
(1) Anthracosis (2) Pneumoconiosis
(3) Siderosis (4) Acanthosis

55. Severe deficiency of which of the following vitamins leads to hemolytic anemia in animals ?
(1) Vit A (2) Vit E
(3) Vit D (4) Vit K

56. Which of the following chemotherapeutic drugs has neurotoxicity?
(1) Vincristine (2) Cyclophosphamide
(3) Anthracyclines (4) Adriamycin

57. The high concentration of myoglobin in marine mammals allows them to
(1) respire at a lower rate
(2) remain underwater for longer periods of time
(3) reduce the effects of osmotic pressure due to extra oxygen load in their body
(4) with stand large changes in CO_2 concentration in muscle

58. Conjugated bilirubin is
(1) transported in blood bound to serum albumin
(2) reduced in a deficiency of a UDP-glucoronosyl transferase
(3) less soluble in aqueous solution than in conjugated form
(4) reduced in serum in biliary tract obstruction

59. The colloidal pressure of vertebrate blood is mainly due to
(1) Neutrophil (2) Albumin
(3) Fibrinogen (4) Globulin

60. The anemia with increase in size of RBC with reduced haemoblobin concentration is termed as
(1) Microcytic hypochromic anemia
(2) Normocytic hypochromic anemia
(3) Microcytic normochromic anemia
(4) Piglet anemia

61. Hemoglobin shows sigmoidal curve for oxygen saturation. What is the shape of curve for myoglobin oxygen-binding ?
(1) Linear (2) Hyperbolic
(3) Sigmoidal (4) Bell shape

62. The action potential results from
(1) decrease in negative charge inside the nerve fibre
(2) increase in positive charge outside the nerve fibre
(3) opening of voltage-gated sodium channels
(4) activation of the sodium-potassium pump

63. The changes in the electrical potential of a neuron that constitute the action potential occur in the following order
(1) depolarization-> resting potential –> hyperpolarization –> resting potential
(2) resting potential –> depolarization-> hyperpolarization – > resting potential
(3) resting potential -> hyperpolarization-> resting potential-> resting potential
(4) resting potential –> hyperpolarization -> resting potential -> depolarization

64. Suppose you were to treat a normal mammalian cell with a substance that inhibits the Na-K ATPase, what would be the most immediate effect upon the cell ?
(1) There would be no change at all
(2) The cell's osmotic balance would be disrupted and the cell would begin to swell
(3) The cell membrane potential would immediately drop to zero
(4) The cell would very quickly run out of ATP

65. Which of the following sequences describes the passage of an action potential in the neuron?
(1) Axon, cell body, dendrite, synaptic cleft
(2) synaptic cleft, axon, dendrite, cell body
(3) dendrite, synaptic cleft, cell body, axon
(4) dendrite, cell body, axon, synaptic cleft

66. Vagus nerve is a
(1) sensory nerve
(2) sensory-motor mixed nerve
(3) motor nerve
(4) lumbar nerve

67. Nerve bundles in vertebrates are likely to contain
(1) many myelinated axons of different diameters as well as a large number of unmyelinated fibres
(2) many unmyelinated fibres as well as a larger number of myelinated axons of same diameters
(3) only myelinated axons of same diameter
(4) only unmyelinated axons of different diameter

68. A tissue was responding when treated with a chemical for a brief period. However, when the treatment was continued for a longer time, the response stopped. After washing and leaving for some time, the tissue started responding to the same chemical at the same dose. The reason for the reduced response is likely to be due to
(1) increased apoptosis of the treated cells
(2) increased necrosis of the treated cells
(3) fixation of the treated cells
(4) desensitization/down-regulation of the receptors on the treated cells

69. Which one of the following cells will migrate after injury in the central nervous system?
(1) Microglia (2) Oligodendrocytes
(3) Astrocytes (4) Ependymal cells

70. Paralytic shellfish poisoning (PSP) is caused by the consumption of molluscan shellfish contaminated with
(1) Brevetoxins (2) Domoic Acid
(3) Saxitoxins (4) Okadaic acid

71. Birds that live in marine environments and thus lack access to fresh drinking water
(1) osmoregulate without using a transport epithelium for this purpose
(2) drink seawater and secrete excess ions through their kidneys only
(3) drink seawater and secrete excess ions mainly through their nasal salt glands
(4) have plasma that is isoosmotic to ocean water

72. A human who has no access to fresh water but is forced to drink seawater instead
(1) will thrive under such conditions, as long as he has lived at the ocean most of his life
(2) will excrete more water molecules than taken in, because of the high load of ion ingestion
(3) will develop structural changes in the kidneys to accommodate the salt overload
(4) will find that drinking saltwater satiates his thirst

73. Urea is produced in the
(1) liver from NH_3 and CO_2
(2) liver from glycogen
(3) kidneys from glucose
(4) kidneys from glycerol and fatty acids

74. Ammonia is likely to be the primary nitrogenous waste in living conditions that include
(1) lots of fresh water flowing across the gills of a fish
(2) lots of seawater, such as a bird living in a marine environment
(3) lots of seawater, such as a marine mammal (e.g., a polar bear)
(4) a terrestrial environment, such as that supporting crickets

75. The nitrogenous waste that requires the most energy to produce is
(1) ammonia
(2) ammonium
(3) urea
(4) uric acid

76. In animals, nitrogenous wastes are produced mostly from the catabolism of
(1) starch and cellulose
(2) triglycerides and steroids
(3) proteins and nucleic acids
(4) phospholipids and glycolipids

77. Choose a pair that correctly associates the mechanism for osmoregulation or nitrogen removal with the appropriate animal
(1) Metanephridium → flatworm
(2) Malpighian tubul → frog
(3) Flame bulb → snake
(4) Direct cellular exchange → marine invertebrate

78. The filtrate in the proximal convoluted tubule of the human does not normally include
(1) ions
(2) glucose
(3) plasma proteins
(4) amino acids

79. If ATP production in a human kidney was suddenly halted, urine production would
(1) come to a complete halt
(2) decrease, and the urine would be hypoosmotic compared to plasma
(3) increase, and the urine would be isoosmotic compared to plasma
(4) increase, and the urine would be hyperosmotic compared to plasma

80. Compared to wetland mammals, water conservation in mammals of arid regions is enhanced by having more
(1) Juxtamedullary nephrons
(2) Bowman's capsules
(3) Ureters
(4) Podocytes

81. When stimulated by aldosterone, the reabsorption of Na^+ is increased along
(1) the loop of Henle
(2) the collecting duct
(3) Bowman's capsule
(4) the distal tubule

82. Increased ADH secretion is likely after
(1) drinking lots of pure water
(2) sweating-induced dehydration increases plasma osmolarity
(3) ingestion of ethanol (drinking alcoholic drinks)
(4) eating a small sugary snack

83. After blood flow is artificially reduced at one kidney, you would expect that kidney to secrete more of the hormone known as
(1) erythropoietin
(2) angiotensinogen
(3) renin
(4) antidiuretic hormone

84. After drinking alcoholic beverages, increased urine excretion is the result of
(1) increased aldosterone production
(2) increased blood pressure
(3) inhibited secretion of antidiuretic hormone (ADH)
(4) increased reabsorption of water in the proximal tubule

85. ADH and RAAS work together in maintaining osmoregulatory homeostasis through which of the following ways?
(1) ADH regulates the osmolarity of the blood and RAAS regulates the volume of the blood
(2) ADH regulates the osmolarity of the blood by altering renal reabsorption of water, and RAAS maintains the osmolarity of the blood by stimulating Na^+ reabsorption
(3) ADH and RAAS work antagonistically; ADH stimulates water reabsorption during dehydration and RAAS causes increased excretion of water when it is in excess in body fluids
(4) Both stimulate the adrenal gland to secrete aldosterone, which increases both blood volume and pressure via its receptors in the urinary bladder

86. African lungfish, which are often found in small stagnant pools of fresh water, produce urea as a nitrogenous waste. What is the advantage of this adaptation?
(1) Urea takes less energy to synthesize than ammonia
(2) Small stagnant pools do not provide enough water to dilute the toxic ammonia
(3) The highly toxic urea makes the pool uninhabitable to potential competitors
(4) Urea forms an insoluble precipitate

87. The muscle spindle is
(1) An actin-myosin complex
(2) A troponin-tropomyosin complex
(3) Axons wound around muscle fibers
(4) A group of dendrite-encircled muscle fibers

88. Statocysts contain cells that are
(1) mechanoreceptors which function in orientation to gravity
(2) chemoreceptors used in selecting migration routes
(3) photoreceptors used in setting biological rhythms
(4) thermoreceptors used in prey detection

89. The pathway leading to the perception of sound by mammals begins with the
(1) hair cells of the organ of Corti, which rests on the basilar membrane, coming in contact with the tectorial membrane
(2) hair cells of the organ of Corti, which rests on the tympanic membrane, coming in contact with the tectorial membrane
(3) hair cells of the organ of Corti, which rests on the tectorial membrane, coming in contact with the basilar membrane
(4) hair cells of the organ of Corti coming in contact with the tectorial membrane as a result of fluid waves in the cochlea causing vibrations in the round window

90. The perceived pitch of a sound depends on
(1) which part of the tympanic membrane is being vibrated by sound waves
(2) which part of the oval window produces waves in the cochlear fluid
(3) which region of the basilar membrane was set in motion
(4) whether or not the sound moves the incus, malleus, and stapes

91. Compared to viewing a distant object, viewing an object held within 5 cm of the eye requires a lens that
(1) has been flattened, as a result of contraction of the ciliary muscles
(2) has been made more spherical, as a result of contraction of the ciliary muscles
(3) has been flattened, as a result of relaxation of the ciliary muscles
(4) has been made more spherical, as a result of relaxation of the ciliary muscles

92. Rods exposed to light will
(1) depolarize due to the opening of sodium channels
(2) hyperpolarize due to the closing of sodium channels
(3) depolarize due to the opening of potassium channels
(4) hyperpolarize due to the closing of potassium channels

93. An injury to the occipital lobe will likely impair function of the
(1) primary visual cortex
(2) thalamus
(3) optic chiasma
(4) sense of taste

94. The contraction of skeletal muscles is based on
(1) actin filaments coiling up to become shorter
(2) myosin filaments coiling up to become shorter
(3) actin and myosin filaments both coiling up to become shorter
(4) myosin cross-bridges binding to actin and then flexing

95. Myasthenia gravis is a form of muscle paralysis in which
(1) motor neurons lose their myelination and the ability to rapidly fire action potentials
(2) acetylcholine receptors are destroyed by an overactive immune system
(3) ATP production becomes uncoupled from mitochondrial electron transport
(4) the spinal cord is infected with a virus that attacks muscle stretch receptors

96. Most of the ATP supplies for a skeletal muscle undergoing 1 hour of sustained exercise come from
(1) creatine phosphate
(2) glycolysis
(3) substrate phosphorylation
(4) oxidative phosphorylation

97. The muscles of a recently deceased human can remain in a contracted state, termed *rigor mortis*, for several hours, due to the lack of
(1) phosphorylated myosin
(2) ATP needed to break actin-myosin bonds
(3) calcium ions needed to bind to troponin
(4) oxygen supplies needed for myoglobin

98. Skeletal, cardiac, and smooth muscle all have
(1) A bands and I bands
(2) transverse tubules
(3) gap junctions
(4) thick and thin filaments

99. Calcium ions regulate contraction of smooth muscle cells by binding to
(1) troponin. (2) tropomyosin
(3) actin. (4) calmodulin

100. Action potentials in the heart move from one contractile cell to the next via
(1) chemical synapses using acetylcholine
(2) chemical synapses using norepinephrine
(3) electrical synapses using gap junctions
(4) myelinated motor neurons

101. If a person loses a large amount of water in a short period of time, he or she may die from dehydration. ADH can help reduce water loss through its interaction with its target cells in the
(1) anterior pituitary
(2) posterior pituitary
(3) adrenal gland
(4) kidney

102. Oxytocin and antidiuretic hormone are synthesized in the
(1) hypothalamus
(2) adenohypophysis
(3) anterior pituitary
(4) posterior pituitary

103. Which endocrine disorder is correctly matched with the malfunctioning gland?
(1) diabetes insipidus and the posterior pituitary gland
(2) giantism and the posterior pituitary gland
(3) goiter and the adrenal medulla
(4) diabetes mellitus and the parathyroid glands

104. A disease that destroys the adrenal cortex should lead to an increase in the plasma levels of
(1) glucocorticoid hormones
(2) epinephrine
(3) adrenocorticotropic hormone (ACTH)
(4) glucose

105. Winter hibernation and spring reproduction in bears are cued by seasonal changes in the secretion of
(1) melatonin from the pineal gland
(2) melatonin from the hypothalamus
(3) thyroxine from the anterior pituitary gland
(4) acetylcholine from the pineal gland

106. Animals that migrate great distances would obtain the greatest energetic benefit of storing chemical energy as
(1) proteins. (2) fats
(3) carbohydrates. (4) amino acids

107. Certain nutrients are considered "essential" in the diets of some animals because
(1) only those animals use those nutrients
(2) the nutrients are subunits of important polymers
(3) these animals are not able to synthesize these nutrients
(4) the nutrients are necessary coenzymes

108. A general rule relating the capacity of a specific animal's digestive system to provide adequate access to substrates for biosynthesis of cellular components, as well as fuel molecules needed for ATP production, is that the animal should have access to

(1) a high-protein, low-carbohydrate diet

(2) a diet low in lipids and high in protein

(3) a low-calorie diet with a large intake of fluids, especially water

(4) a diet that matches the "food pyramid" for the species

109. After ingestion by humans, the first category of macromolecules to be chemically digested by enzymes in the mouth is

(1) proteins

(2) carbohydrates

(3) cholesterol and other lipids

(4) nucleic acids

110. Upon activation by stomach acidity, the secretions of the parietal cells

(1) initiate the digestion of protein in the stomach

(2) initiate the mechanical digestion of lipids in the stomach

(3) initiate the chemical digestion of lipids in the stomach

(4) include pepsinogen

111. The absorption of fats differs from that of carbohydrates in that the

(1) processing of fats does not require any digestive enzymes, whereas the processing of carbohydrates does

(2) fat absorption occurs in the stomach, whereas carbohydrates are absorbed from the small intestine

(3) carbohydrates need to be emulsified before they can be digested, whereas fats do not

(4) most absorbed fat first enters the lymphatic system, whereas carbohydrates directly enter the blood

112. For a nondiabetic person, the glucose concentration in this part of the vasculature varies more than in any other part

(1) abdominal artery

(2) coronary arteries

(3) pulmonary veins

(4) hepatic portal vessel

113. Because adult lampreys attach onto the surface of large fish for long periods of time to feed on body fluids, they can accomplish nutritional balance without need for a

(1) liver. (2) pancreas

(3) intestine. (4) stomach

114. A significant contribution of intestinal bacteria to human nutrition is the benefit of bacterial

(1) production of vitamins A and C

(2) generation of gases needed for elimination

(3) absorption of organic materials

(4) production of vitamin K

115. Cattle are able to survive on a diet consisting almost entirely of plant material because

(1) they have cellulose-digesting, symbiotic microorganisms in chambers of their stomachs

(2) cattle, like rabbits, re-ingest their feces

(3) they manufacture all 15 amino acids out of sugars in the liver

(4) cattle saliva has enzymes capable of digesting cellulose

116. PKU (phenylketonuria) is a hereditary condition in which infants and young children who ingest the amino acid phenylalanine risk serious neurological damage. However, the risk of damage can be substantially reduced by the severe restriction of phenylalanine in the diet. Which of the following is the nutritional concept that forms the basis for this preventive treatment?

(1) Enzymatic hydrolysis

(2) Essential nutrients

(3) Symbiosis

(4) Dehydration synthesis

117. Circulatory systems in molluscs
(1) are open in all species of molluscs
(2) are closed in all species of molluscs
(3) are open in species of large-sized molluscs and are closed in species of small-sized molluscs
(4) are open in species of small-sized molluscs and are closed in species of large-sized molluscs

118. A significant increase in the amount of interstitial fluid surrounding the capillary beds of a human's lungs will cause
(1) an increase in the amount of carbon dioxide moving from the blood to the lungs
(2) an increase in the amount of oxygen moving from the lungs into the blood
(3) a decrease in the amount of oxygen moving from the lungs into the blood
(4) an increase of pressure that would cause the capillary beds to burst

119. The only vertebrates in which blood flows directly from respiratory organs to body tissues without first returning to the heart are the
(1) amphibians. (2) birds
(3) fishes (4) mammals

120. Which of the following develops the greatest pressure on the blood in the mammalian aorta?
(1) Systole of the left atrium
(2) Diastole of the right ventricle
(3) Systole of the left ventricle
(4) Diastole of the right atrium

121. After several weeks of exercise, a human athlete's resting heart rate is typically lower than before because
(1) the body needs less oxygen than before
(2) the body temperature has increased
(3) the stroke volume has increased
(4) the cardiac output has decreased

122. A patient with a blood pressure of 120/75, a pulse rate of 40 beats/minute, a stroke volume of 70 mL/beat, and a respiratory rate of 25 breaths/minute will have a cardiac output of
(1) 500 mL/minute
(2) 1,000 mL/minute
(3) 1,750 mL/minute
(4) 2,800 mL/minute

123. Among the following choices, which organism likely has the highest systolic pressure?
(1) mouse (2) rabbit
(3) human (4) giraffe

124. The velocity of blood flow is the lowest in capillaries because
(1) the total cross-sectional area of the capillaries is greater than the total cross-sectional area of the arteries or any other part of the circulatory system
(2) the capillaries are far from the heart, and blood flow slows as distance from the heart increases
(3) the diastolic blood pressure is too low to deliver blood to the capillaries at a high flow rate
(4) the systemic capillaries are supplied by the left ventricle, which has a lower cardiac output than the right ventricle

125. If, during protein starvation, the osmotic pressure on the venous side of capillary beds drops below the hydrostatic pressure, then
(1) hemoglobin will not release oxygen
(2) fluids will tend to accumulate in tissues
(3) the pH of the interstitial fluids will increase
(4) most carbon dioxide will be bound to hemoglobin and carried away from tissues

126. A species that has a normal resting systolic blood pressure of >260 mm Hg is likely to be

(1) an animal that is small and compact, without the need to pump blood very far from the heart

(2) an animal with abundant lipid storage

(3) a species that has very wide diameter veins

(4) an animal that has a very long distance between its heart and its brain

127. Dialysis patients, who will have blood withdrawn, dialyzed, then replaced, are always weighed when they enter the facility and then weighed carefully again before they leave, because

(1) even small changes in body weight may signify changes in blood volume and therefore blood pressure

(2) many people who have dialysis are diabetic and must control their weight carefully

(3) dialysis removes blood proteins and these weigh more than other blood components

(4) dialysis is likely to cause edema and such swelling must be controlled

128. Dissolved proteins in human plasma include which of the following?

I. Fibrinogen

II. Hemoglobin

III. Immunoglobulin

(1) I only (2) II only

(3) I and III only (4) II and III only

129. Cyanide poisons mitochondria by blocking the final step in the electron transport chain. Human red blood cells placed in an isotonic solution containing cyanide are likely to

(1) retain the normal cell shape, but the mitochondria will be poisoned

(2) be unaffected

(3) switch to anaerobic metabolism

(4) become unable to carry oxygen

130. A normal event in the process of blood clotting is the

(1) production of erythropoietin

(2) conversion of fibrin to fibrinogen

(3) activation of prothrombin to thrombin

(4) increase in platelets

131. When the air in a testing chamber is specially mixed so that its oxygen content is 10% and its overall air pressure is 400 mm Hg, then PO_2 is

(1) 400 mm Hg. (2) 82 mm Hg

(3) 40 mm Hg (4) 21 mm Hg

132. Countercurrent exchange is evident in

(1) the flow of water across the gills of a fish and that of blood within those gills

(2) the flow of blood in the dorsal vessel of an insect and that of air within its tracheae

(3) the flow of air within the primary bronchi of a human and that of blood within the pulmonary veins

(4) the flow of water across the skin of a frog and that of blood within the ventricle of its heart

133. Carbon dioxide levels in the blood and cerebrospinal fluid affect its pH. This enables the organism to sense a disturbance in gas levels as

(1) the brain directly measures and monitors carbon dioxide and causes breathing changes accordingly

(2) the medulla oblongata, which is in contact with cerebrospinal fluid, monitors pH and uses this measure to control breathing

(3) the brain alters the pH of the cerebrospinal fluid to force the animal to retain more or less carbon dioxide

(4) stretch receptors in the lungs cause the medulla oblongata to speed up or slow breathing

134. The Bohr shift on the oxygen-hemoglobin dissociation curve is produced by changes in
(1) the partial pressure of oxygen
(2) the partial pressure of carbon monoxide
(3) hemoglobin concentration
(4) pH

135. Most of the carbon dioxide produced by humans is
(1) converted to bicarbonate ions by an enzyme in red blood cells
(2) bound to hemoglobin
(3) transported in the erythrocytes as carbonic acid
(4) simply dissolved in the plasma

136. The hemocyanin of arthropods and molluscs differ from the hemoglobin of mammals in that
(1) the oxygen dissociation curve for hemocyanin is linear
(2) hemocyanin carries appreciably more carbon dioxide
(3) hemocyanin has protein coupled to copper rather than iron
(4) the protein of hemocyanin is not bound to metal

137. Which of the following reactions prevails in red blood cells traveling through alveolar capillaries? (Hb = hemoglobin)
(1) $Hb + 4\ O_2 \rightarrow Hb(O_2)_4$
(2) $Hb(O_2)_4 \rightarrow Hb + 4\ O_2$
(3) $CO_2 + H_2O \rightarrow H_2CO_3$
(4) $H_2CO_3 \rightarrow H^+ + HCO_3^-$

138. When acetylcholine is broken down by acetylcholinesterase in the synaptic cleft, what happens to the choline?
(1) It is destroyed
(2) It is made into new acetylcholinesterase
(3) It is reabsorbed at the presynaptic terminal and combined with more acetic acid to make acetylcholine
(4) It is taken up at the post-synaptic terminal

139. Which one of the following statements is true?
(1) Gastric acid is secreted by parietal cells of the gastric glands in response to hormonal stimulation
(2) Most of the acid and pepsinogen secreted by the stomach occurs during the intestinal phase of gastric secretion
(3) Gastric secretion does not begin until food enters the stomach
(4) Secretin secreted by the duodenum stimulates gastric secretion

140. In muscles, ionic strength of G-actin is increased due to the presence of
(1) aluminium (2) magnesium
(3) zinc (4) iron

141. During a forceful exhalation, which of the following muscles would be activated?
(1) The diaphragm
(2) External intercostals
(3) Internal intercostals
(4) 1 and 2

142. Imagine that a person has been given a mysterious drug that causes the rate of oxygen delivery to tissues to be reduced. Assuming the pO_2 of the alveoli and of the tissues are normal, and do not change after the drug is given, which of the following might be possible effects of the drug that would explain this pattern?
(1) It left-shifts the oxygen affinity curve, and/or causes the formation of more carbamino compounds
(2) It right-shifts the oxygen affinity curve, and/or causes the formation of more carbamino compounds
(3) It left-shifts the oxygen affinity curve, and/or damages hemoglobin so its total carrying capacity is reduced
(4) It right-shifts the oxygen affinity curve, and/or damages hemoglobin so its total carrying capacity is reduced

143. One major advantage of the filtration-reabsorption method of urine production is that
(1) it is effective at getting rid of a wide variety of toxins and wastes that might be found in the blood
(2) it is very energy efficient, using little ATP to process the urine
(3) only a small quantity of water and solute needs to be reabsorbed from the nephron
(4) all the urea that initially enters the nephron leaves in the final urine

144. If a person had his adrenal cortex damaged, so that he could no longer release aldosterone, which of the following would be expected if he experienced low blood Na^+ concentrations?
(1) His liver would not release angiotensinogen
(2) He would not be able to release renin
(3) He would not be able to convert angiotensinogen into angiotensin I
(4) He would not experience increased Na^+ reabsorption in the distal tubule

145. One of the effects of the release of atrial natriuretic factor (ANF) is
(1) increased release of anti-diuretic hormone (ADH)
(2) increased water reabsorption in the collecting ducts
(3) increased reabsorption of Na^+
(4) increased vasodilation

146. Which of the following might be a problem for someone who secreted little or no stomach acid?
(1) Bacteria and other pathogens might not be killed as effectively
(2) Proteins might not be digested as effectively
(3) It might be harder to maintain a relatively neutral pH in the small intestine
(4) Two of the above might be a problem (1 and 2)

147. The release of digestive enzymes from the pancreas is controlled mainly by the hormone
(1) cholecystokinin
(2) enterokinase
(3) gastrin
(4) pro-carboxypeptidase

148. Bile salts are useful in the processing of fats in the small intestine because they
(1) cleave fatty acids off triglyceride fats
(2) act as a detergent to help break up large fat droplets
(3) carry fatty acids across the epithelial wall
(4) convert fatty acids back into triglyceride fats in the epithelial cells

149. The bacterial fauna found in our large intestines
(1) are just little parasites that eat our leftover nutrients and do us more harm than good
(2) don't really hurt us, but don't do us any good, either
(3) can provide some essential nutrients, and may have more profound health effects
(4) are absolutely essential to life – you couldn't live 24 hours without them

150. The liver stores glycogen primarily as
(1) a source of energy for its own needs
(2) a source of energy for the kidneys
(3) a source of energy for muscles
(4) a source of energy for the nervous system

151. The switch from glucose to fats/fatty acids as the primary source of fuel by most body cells during the post-absorptive period occurs in response to
(1) increasing blood insulin levels
(2) decreasing blood insulin levels
(3) increasing blood glucagon levels
(4) decreasing blood glucagon levels

152. Which of the following best describes the process of spermatogenesis?
(1) Spermatids undergo meiosis to form spermatocytes, which develop into spermatozoa that become fully mature after about two days
(2) Spermatids undergo meiosis to form spermatocytes, which develop into spermatozoa that become fully mature after about two months
(3) Spermatocytes undergo mitosis to form spermatids, which develop into spermatozoa that become fully mature after about two months
(4) Spermatocytes undergo meiosis to form spermatids, which develop into spermatozoa that become fully mature after about two months

153. During puberty in females, increasing levels of estrogen cause all of the following except
(1) Development of the breasts
(2) Deposition of fat on the buttocks and thighs rather than the abdomen
(3) The limitation of body hair growth to primarily the axillary and pubic regions
(4) Increased muscle mass

154. Which of the following best describes the cellular activity in a well-developed follicle in an ovary?
(1) Granulosa cells produce estrogen; theca cells support development of the oocyte
(2) Granulosa cells produce androgens; theca cells convert androgen to estrogen and support development of the oocyte
(3) Theca cells produce estrogen; granulosa cells support development of the oocyte
(4) Theca cells produce androgens; granulosa cells convert androgen to estrogen and support development of the oocyte

155. The growth of a new endometrial layer in the uterus every month is driven by increasing levels of
(1) estrogen. (2) luteinizing hormone
(3) oxytocin. (4) progesterone

156. The delay between fertilization of an oocyte by a sperm and implantation of the resulting blastocyst into the uterine wall is about
(1) seven minutes. (2) 30 minutes
(3) seven hours. (4) seven days

157. The reason that a woman's estrogen and progesterone levels plummet immediately after she gives birth is that
(1) her corpus luteum rapidly degenerates
(2) her levels of LH and FSH drop to almost nothing
(3) she begins lactating
(4) the placenta that was supplying these hormones has detached from her uterus

158. Which of the following fluids is normally high in K^+ and low in Na^+?
(1) Intracellular fluid (2) Interstitial fluid
(3) Blood plasma (4) All of the above

159. Which of the following does not represent an example of a positive feedback loop?
(1) The increasing strength of contraction seen in the uterus during parturition
(2) The erection of the penis in men
(3) The increasing levels of estrogen seen shortly before ovulation in women
(4) The increase in body temperature seen in response to an infection

160. If we see water moving from one side (side A) of an epithelial cell layer to the other (side B), which of the following would not by itself be a potential explanation for this movement?
(1) The overall concentration is higher on side B than side A
(2) The overall concentration is higher on side A than side B, but paracellular pockets of high concentration on side B are generated by epithelial cells
(3) The two sides are initially the same overall concentration, but a solute diffuses from side A to side B
(4) The two sides are the same overall concentration, but the pH is higher on side A than side B

161. Imagine a neuron with a relatively long axon covered in myelin but lacking any nodes of Ranvier (i.e., the myelin is completely continuous along the axon) Which of the following best describes the result?

(1) The axon would have better cable properties than if it lacked myelin, and an action potential would be able to propagate the length of the axon
(2) The axon would have better cable properties than if it lacked myelin, but an action potential would not be able to propagate the length of the axon
(3) The axon would have worse cable properties than if it lacked myelin, but an action potential would still be able to propagate the length of the axon
(4) The axon would have worse cable properties than if it lacked myelin, and an action potential would not be able to propagate the length of the axon

162. At a neural synapse, the release of neurotransmitter by the presynaptic neuron causes the postsynaptic neuron to hyperpolarize after a delay of about 100 ms, and it stays hyperpolarized for several seconds. This means that the receptors are most likely

(1) ionotropic receptors that cause Na^+ channels to open
(2) ionotropic receptors that cause Na^+ channels to close
(3) metabotropic receptors that cause Na^+ channels to open
(4) metabotropic receptors that cause Na^+ channels to close

163. The process of neuronal integration (including both spatial and temporal summation) takes information input from any number of presynaptic neurons and ends up encoding it as a single output, which is in the form of

(1) axonal contractions of different strengths
(2) axonal contractions of different frequencies
(3) action potentials of different amplitudes
(4) action potentials at different frequencies

164. Taste receptors for salt increase neurotransmitter release when

(1) Na^+ activates G-protein pathways that open Ca^{2+} channels in the receptor
(2) Na^+ enters through channel proteins, depolarizing the cell and opening Ca^{2+} channels
(3) Cl- activates G-protein pathways that open Ca^{2+} channels in the receptor
(4) Cl- exits through channel proteins, depolarizing the cell & opening Ca^{2+} channels

165. A problem that occasionally occurs with the semicircular canals is that a stray otolith will get into the canal, and end up pushing against the cupula of one of the canals. When this happens, the person is likely to feel that as though

(1) her head is spinning
(2) she is rapidly accelerating
(3) she is falling straight down
(4) her inner ear smells the scent of rose flowers on a spring morning

166. A person is listening to a pure tone (single frequency) sound that is initially fairly quiet and then becomes louder. The reason the person is able to detect the increase in the volume of the sound is

(1) a different region along the length of the cochlea vibrates
(2) the frequency of the vibration in the sensitive region of the cochlea increases
(3) the amplitude of the vibrations of the sensitive region of the cochlea increases
(4) the cochlea increases in overall length

167. Which of the following statements about rod photoreceptors is false?

(1) The molecule that detects light is located in the lamellae
(2) Rhodopsin changes form when struck by a photon
(3) Transducin is inactivated when a photoreceptor is exposed to light
(4) Levels of cGMP control the state (open or closed) of Na+ channels

168. All hormones
(1) are produced by endocrine glands
(2) are lipid-soluble molecules
(3) are carried to target cells in the blood
(4) are protein molecules

169. In a positive-feedback system where hormone A alters the amount of protein X
(1) an increase in A always produces an increase in X
(2) an increase in X always produces a decrease in A
(3) a decrease in A always produces an increase in X
(4) a decrease in X always causes a decrease in A

170. Testosterone is an example of a chemical signal that affects the very cells that synthesize it, the neighboring cells in the testis, along with distant cells outside the gonads. Thus, testosterone is an example of
(1) an autocrine signal
(2) a paracrine signal
(3) an endocrine signal
(4) an autocrine signal, a paracrine signal, and an endocrine signal

171. Which category of signal exerts its effects on target cells by binding to membrane-bound receptor proteins?
(1) Neurohormones. (2) Estrogens
(3) Androgens (4) Vitamin D

172. A paracrine signal that relaxes smooth muscle cells is
(1) nitric oxide (2) vitamin D
(3) testosterone. (4) cortisol

173. Prostaglandins are local regulators whose chemical structure is derived from
(1) oligosaccharides
(2) fatty acids
(3) steroids
(4) amino acids

174. Ecdysone is a(n)
(1) peptide hormone
(2) amine derivative
(3) steroid hormone
(4) iodinated protein hormone made by the thyroid gland

175. Aspirin and ibuprofen both
(1) inhibit the synthesis of prostaglandins
(2) inhibit the release of nitric oxide, a potent vasodilator
(3) activate the paracrine signaling pathways that form blood clots
(4) stimulate the release of oxytocin

176. A cell with membrane-bound proteins that selectively bind a specific hormone is called that hormone's
(1) secretory cell
(2) plasma cell
(3) endocrine cell
(4) target cell

177. The reason that the steroid hormone aldosterone affects only a small number of cells in the body is that
(1) only its target cells get exposed to aldosterone
(2) only its target cells contain aldosterone receptors
(3) it is unable to enter nontarget cells
(4) nontarget cells destroy aldosterone before it can produce any effect

178. Different body cells can respond differently to the same peptide hormones because
(1) different target cells have different sets of genes
(2) each cell converts that hormone to a different metabolite
(3) a target cell's response is determined by the components of its signal transduction pathways
(4) the circulatory system regulates responses to hormones by routing the hormones to specific targets

179. Insect hormones and their receptors
(1) act independently of each other
(2) are a focus in pest-control research
(3) utilize cell-surface receptors only
(4) are active independently of environmental cues

180. Endocrine glands that are sources of steroid hormones
(1) secrete the steroids through ducts into the blood
(2) store those hormones in membrane-bound vesicles
(3) have a very short latency between steroid synthesis and steroid release
(4) are all controlled by the autonomic nervous system

181. A chemical signal operating in a paracrine manner is one that
(1) is active at a neuronal synapse
(2) affects only the cells that synthesize the paracrine signal
(3) requires transport in the blood before it can act on its target cells
(4) evokes responses from all parts of the vascular system

182. Hormone X activates the cAMP second messenger system in its target cells. The greatest response by a cell would come from
(1) applying a molecule of hormone X to the extracellular fluid surrounding the cell
(2) injecting a molecule of hormone X into the cytoplasm of the cell
(3) applying a molecule of cAMP to the extracellular fluid surrounding the cell
(4) injecting a molecule of cAMP into the cytoplasm of the cell

183. When a steroid hormone and a peptide hormone exert similar effects on a population of target cells, then
(1) the steroid and peptide hormones must use the same biochemical mechanisms
(2) the steroid and peptide hormones must bind to the same receptor protein
(3) the steroid hormones affect the synthesis of effector proteins, whereas peptide hormones activate effector proteins already present in the cell
(4) the steroid hormones affect the activity of certain proteins within the cell, whereas peptide hormones directly affect the processing of mRNA

184. For hormones that homeostatically regulate cellular functions,
(1) negative feedback typically regulates hormone secretion
(2) the circulating level of a hormone is held constant through a series of positive feedback loops
(3) both lipid-soluble hormones and water-soluble hormones bind to intracellular protein receptors
(4) endocrine organs release their contents into the bloodstream via specialized ducts

185. Nitric oxide and epinephrine
(1) both function as neurotransmitters
(2) both function as steroid hormones
(3) are both involved in the "fight-or-flight" response
(4) bind the same receptors

186. Suppose that substance X is secreted by one cell, travels via interstitial fluid to a neighboring cell, and produces an effect on that cell. All of the following terms could describe this substance except
(1) paracrine signal (2) neurotransmitter
(3) prostaglandin. (4) pheromone

187. Hormones that promote homeostasis
(1) are not found as members of antagonistic signaling mechanisms
(2) travel to target cells without passage in the plasma
(3) utilize receptors that bind any hormone
(4) usually operate as part of a negative feedback system

188. During mammalian labor and delivery, the contraction of uterine muscles is enhanced by oxytocin. This is an example of
(1) a negative feedback system
(2) a hormone that acts in an antagonistic way with another hormone
(3) a hormone that is involved in a positive feedback loop
(4) signal transduction immediately changing gene expression in its target cells

189. During short-term starvation, most available fuel molecules are catabolized to provide energy for metabolism rather than being used as building blocks for growth and repair, a trade-off that is hormonally regulated by
(1) acetylcholine. (2) glucagon
(3) oxytocin. (4) antidiuretic hormone

190. Based on their effects, which pair below would not be expected to be active at the same time and place?
(1) Prostaglandin F and nitric oxide
(2) Growth hormone and insulin-like growth factors
(3) Endocrine and exocrine glands
(4) Hormones and target cells

191. The interrelationships between the endocrine and the nervous systems are especially apparent in
(1) a neuron in the spinal cord
(2) a steroid-producing cell in the adrenal cortex
(3) a neurosecretory cell in the hypothalamus
(4) a brain cell in the cerebral cortex

192. The hypothalamus modulates hormone secretion by the anterior pituitary by means of
(1) peptide hormones
(2) steroid hormones
(3) electrical synapses
(4) neurotransmitters

193. Portal blood vessels connect two capillary beds found in the
(1) hypothalamus and thalamus
(2) anterior pituitary and posterior pituitary
(3) hypothalamus and anterior pituitary
(4) posterior pituitary and thyroid gland

194. If a person loses a large amount of water in a short period of time, he or she may die from dehydration. ADH can help reduce water loss through its interaction with its target cells in the
(1) anterior pituitary
(2) posterior pituitary
(3) adrenal gland
(4) kidney

195. The hypothalamus
(1) functions only as an endocrine target, by having lots of receptors on its cells
(2) functions only in neuronal transmission
(3) does not have any hormone receptors on its cells
(4) includes neurosecretory cells that terminate in the posterior pituitary

196. Prolactin stimulates mammary gland growth and development in mammals and regulates salt and water balance in freshwater fish. Many scientists think that this wide range of functions indicates that prolactin
(1) is a nonspecific hormone
(2) has a unique mechanism for eliciting its effects
(3) is an evolutionarily conserved hormone
(4) is derived from two separate sources

197. A product of the anterior pituitary gland that causes color changes in its target cells is
(1) FSH. (2) LH
(3) TSH. (4) MSH

198. To prevent insect pests from maturing into reproducing adults, pest controllers use synthetic agonists of
(1) ecdysone. (2) juvenile hormone
(3) oxytocin. (4) brain hormone

199. In a lactating mammal, the two hormones that promote milk synthesis and milk release, respectively, are
(1) prolactin and calcitonin
(2) prolactin and oxytocin
(3) follicle-stimulating hormone and luteinizing hormone
(4) luteinizing hormone and oxytocin

200. Endocrine structures derived from nervous tissue include the
(1) thymus and the thyroid
(2) ovaries and the testes
(3) liver and the pancreas
(4) posterior pituitary gland and the adrenal medulla

ANSWER SHEET

Unit 7: Animal Physiology

1. 1	**2.** 3	**3.** 4	**4.** 3	**5.** 3	**6.** 2	**7.** 3	**8.** 1	**9.** 1	**10.** 4
11. 1	**12.** 2	**13.** 1	**14.** 2	**15.** 2	**16.** 2	**17.** 1	**18.** 2	**19.** 2	**20.** 1
21. 2	**22.** 3	**23.** 1	**24.** 2	**25.** 4	**26.** 2	**27.** 4	**28.** 2	**29.** 3	**30.** 4
31. 2	**32.** 2	**33.** 3	**34.** 3	**35.** 3	**36.** 4	**37.** 1	**38.** 2	**39.** 3	**40.** 2
41. 3	**42.** 4	**43.** 2	**44.** 1	**45.** 4	**46.** 1	**47.** 1	**48.** 1	**49.** 3	**50.** 3
51. 2	**52.** 4	**53.** 4	**54.** 1	**55.** 2	**56.** 1	**57.** 2	**58.** 2	**59.** 2	**60.** 3
61. 2	**62.** 3	**63.** 1	**64.** 2	**65.** 4	**66.** 2	**67.** 1	**68.** 4	**69.** 1	**70.** 3
71. 3	**72.** 2	**73.** 1	**74.** 1	**75.** 4	**76.** 3	**77.** 3	**78.** 3	**79.** 3	**80.** 1
81. 4	**82.** 2	**83.** 1	**84.** 3	**85.** 2	**86.** 2	**87.** 4	**88.** 1	**89.** 1	**90.** 3
91. 2	**92.** 2	**93.** 1	**94.** 4	**95.** 2	**96.** 4	**97.** 2	**98.** 4	**99.** 4	**100.** 3
101. 4	**102.** 1	**103.** 1	**104.** 3	**105.** 1	**106.** 2	**107.** 3	**108.** 4	**109.** 2	**110.** 1
111. 4	**112.** 4	**113.** 4	**114.** 4	**115.** 1	**116.** 2	**117.** 4	**118.** 3	**119.** 3	**120.** 3
121. 3	**122.** 4	**123.** 4	**124.** 1	**125.** 2	**126.** 4	**127.** 1	**128.** 3	**129.** 2	**130.** 3
131. 3	**132.** 1	**133.** 2	**134.** 4	**135.** 1	**136.** 3	**137.** 1	**138.** 3	**139.** 1	**140.** 2
141. 4	**142.** 1	**143.** 1	**144.** 4	**145.** 4	**146.** 4	**147.** 1	**148.** 2	**149.** 3	**150.** 4
151. 3	**152.** 4	**153.** 4	**154.** 2	**155.** 1	**156.** 4	**157.** 4	**158.** 1	**159.** 1	**160.** 2
161. 2	**162.** 3	**163.** 4	**164.** 2	**165.** 1	**166.** 1	**167.** 1	**168.** 3	**169.** 1	**170.** 4
171. 1	**172.** 1	**173.** 2	**174.** 3	**175.** 1	**176.** 4	**177.** 2	**178.** 3	**179.** 2	**180.** 3
181. 1	**182.** 1	**183.** 3	**184.** 1	**185.** 1	**186.** 4	**187.** 4	**188.** 3	**189.** 2	**190.** 1
191. 3	**192.** 1	**193.** 3	**194.** 4	**195.** 4	**196.** 3	**197.** 4	**198.** 2	**199.** 2	**200.** 4

UNIT - 8

GENETICS

1. Virus-mediated transfer of cellular genetic material from one bacterial cell to another by means of virus particles is called
(1) transduction (2) transposition
(3) transformation (4) transfection

2. Which of the following statements is not true for Hfr strains of *E. coli* ?
(1) F factor is integrated in the genome
(2) Chromosomal markers are transferred from donor to recipient
(3) They act as donors in the cross
(4) Progeny of the cross always becomes F +

3. Which of the following organisms contains a 11 bp specific DNA sequence that acts as self recognition sequence in natural transformation?
(1) *Streptococcus pneumonia*
(2) *Bacillus subtilis*
(3) *E. coli*
(4) *Hemophilus infuenzae*

4. A woman has a color blind father but husband with normal vision. What are the chances for their sons and daughters to be color blind?
(1) ½ for sons; ½ for daughters
(2) ¼ for sons; ¾ for daughters
(3) ½ for sons; zero for daughters
(4) Zero for sons; ¾ for daughters

5. Cystic fibrosis is due to
(1) defective chloride channel
(2) defective LDL receptor
(3) High levels of HDL
(4) increased dopamine

6. Sickle-cell anemia is an example of Single Nucleotide Polymorphism (SNP) of
(1) A to T mutation
(2) T to A mutation
(3) G to C mutation
(4) C to G mutation

7. Hemophilia A, a common X-linked bleeding disorder is caused due to lack of function of a gene for
(1) Factor VIII (2) Factor VII
(3) Platelets (4) Fibrinogen

8. A cross between two true breeding lines one with dark blue flowers and one with bright white flowers produces F1 offspring that are light blue. When the F1 progeny are selfed a 1:2:1 ratio of dark blue to light blue to white flowers is observed) What genetic phenomenon is associated with these results?
(1) Epistasis
(2) Incomplete dominance
(3) Co-dominance
(4) Inbreeding depression

9. Mutations which occur in vegetative parts during growth which do not go on to form gametes can be classified as
(1) auxotrophic mutations
(2) somatic mutations
(3) morphological mutations
(4) oncogenes

10. *Arabidopsis* is advantageous for plant genetic research because
(1) it is commercially important as a food crop
(2) it is having longer life cycle
(3) it is a small plant with a small genome size which can be raised inexpensively
(4) it is a close relative of corn and results with this species can be applied to problems in corn

11. The gene which is suppressed by another nonallelic gene through interaction is known as
(1) Epistatic (2) Incomplete
(3) Hypostatic (4) Homologs

12. A mapping method for identifying markers linked to a trait of interest in a natural population is called
(1) Linkage mapping
(2) Association mapping
(3) Transcriptome mapping
(4) Chromosome walking

13. In meiosis
(1) Chromosomes separate in meiosis I and chromatids separate in meiosis II
(2) Chromosomes separate in meiosis II and chromatids separate in meiosis I
(3) Chromosomes separate in both meiosis I and II
(4) Chromatids separate in both meiosis I and II

14. Galactosemia is a recessive human disease that is treatable by restricting lactose and glucose in the diet. A couple is heterozugous for the galactosemia gene. If the couple has 4 children, what is the probability that none of the four will have galactosemia?
(1) 1/16 (2) 9/16
(3) 1/256 (4) 81/256

15. When Hfr strain of *E. coli* is crossed with F- strain, recombinants obtained are
(1) always F+ (2) always HFr+
(3) rarely F+ (4) rarely HFr+

16. Which of the following bacteria is not naturally competent?
(1) *Bacillus subtilis*
(2) *E. coli*
(3) *Streptococcus peumoniae*
(4) *Hemophilus influeanzae*

17. Hemophilia A and Hemophilia B have nearly identical phenotypes, but they result from mutations in different genes on the X chromosome. This is an example of
(1) Locus heterogeneity
(2) Epistatic interaction
(3) Double heterozygosity
(4) Variable expressivity

18. Molecular analysis is performed on the three copies of chromosome 21 in a child with Down's syndrome using markers of DNA polymorphism for which both parents are heterozygous for different alleles. Two of the chromosomes (#21) have the same mother's alleles. Based on this information, when did the nondisjunction event most likely occur?
(1) Maternal meiosis I
(2) Maternal meiosis II
(3) Paternal meiosis I
(4) Paternal meosis II

19. Which of the following statements correctly indicates the relative position of two genes in higher plants?
(1) The closer two genes are to each other on a chromosome, the higher the frequency of recombination between them
(2) The more distant two genes are to each other on chromosome, the higher the frequency of recombination between them
(3) If two genes are located on different chromosomes then there will be high frequency of recombination between them
(4) If two genes are located on different chromosomes then there will be low frequency of recombination between them

20. In human, pointed eyebrows are dominant to smooth eyebrows and "widow's peak" frontal hairline is dominant to continuous hairline. What phenotypic ratio would you expect in the offspring from a marriage between an individual heterozygous for both the genes and an individual homozygous recessive for both the genes?

(1) 9 : 3: 3: 1 (2) 9: 7
(3) 1: 1 (4) 1: 1: 1: 1

21. If a man of blood group AB marries a woman of blood group A whose father was of blood group O, to what different blood groups can this man and woman expect their children to belong?

(1) A, AB, B (2) A, AB
(3) AB, O (4) A, O, B

22. Which of these techniques is used by geneticists to map human genes?

(1) Diploid cross
(2) Tetrad analysis
(3) Somatic cell hybridization
(4) Complementation analysis

23. A woman who is heterozygous for both phenylketonuria mutation and for X-linked hemophilia mutation has a child with a phenotypically normal man who is also heterozygous for a phenylketonuria mutation. What is the probability that the child will be affected by both ?

(1) 1 /8 (2) 1 /4
(3) 1/16 (4) 3/8

24. In a family a female child is diagnosed with a known genetic disorder. She has four unaffected brothers and sisters. Neither parent nor any of the four biological grand-parents of the affected child had the disease. The most likely genetic explanation is that the disease is inherited as ________disease

(1) autosomal dominant
(2) autosomal recessive
(3) X-linked recessive
(4) X-linked dominant

25. The genotypes of a husband and wife are I^AI^A X I^AI^O. Among the blood types of their children,how many different genotypes and phenotypes are possible ?

(1) 2 genotypes, 1 phenotypes
(2) 3 genotypes, 4 phenotypes
(3) 4 genotypes, 4 phenotypes
(4) 4 genotypes, 3 phenotypes

26. In a conjugation experiment, the order in which the markers from Hfr donor appeared in F- recipient was : *mal* (10 min) *thi* (22 min) *met* (17 min) *trp* (57 min) *thr* (33 min) What will be the order of markers on the chromosome?

(1) *mal-thi-met-trp-thr*
(2) *trp-thr-thi-met-mal*
(3) *mal-met-thi-thr-trp*
(4) *thr-trp-mal-met-thi*

27. A female rat homozygous for a recessive X-linked mutation is mated to a male with wild type phenotype. The phenotypes of the F1 progeny will be

(1) all wild type
(2) 50% mutant irrespective of sex
(3) all females wild type & all males mutant
(4) all males wild type & all females mutant

28. Red green colour blindness is X-linked in human. If a male is red green colour blind and both parents have normal colour vision, which of the male's grandparents is most likely to be red green colour blind ?

(1) maternal grandmother
(2) maternal grandfather
(3) paternal grandmother
(4) paternal grandfather

29. Which of the following is incorrect with respect to modification of Mendelian dihybrid ratio?

(1) Complementary gene interaction : 9:7
(2) Recessive epistasis: 9:3:4
(3) Dominant epistasis: 12:3:1
(4) Additive gene interaction: 10:6

30. The number of phenotypes in the F_2 of the dihybrid will be

(1) 2 (2) 3
(3) 4 (4) 8

31. Which of the following media is best suited for the selective growth of *E. coli* with genotype: Str+ his- leu- lys- ?

(1) Minimal medium with thiamine, histidine, leucine and lysine
(2) Luria Agar
(3) Minimal medium with thiamine and streptomycin
(4) Minimal medium with thiamine, histidine, leucine, lysine and streptomycin

32. If two heterozygous individuals suffering from an autosomal dominant disorder marry, what is the occurrence risk for this disorder in their offspring ?

(1) 100% (2) 75%
(3) 50% (4) 25%

33. A man who is affected with hemophilia A marries a woman who is a carrier of this disorder. What proportion among the following of this couple's daughters will be affected and what proportion of the daughters will be carriers ?

(1) 0.75; 0.25 (2) 0.25; 0.75
(3) 0 ; 1 (4) 0.5 ; 0.5

34. The character regulated by a holandric gene

(1) passes from the father through the son to the grandson
(2) passes from the father through the daughter to the grandsons
(3) passes from the mother through the daughter to the granddaughter
(4) passes from the mother through the daughter to the grandson

35. Which of the following statements about heritability (H2) is true?

(1) It is a measure of level of gene linkage
(2) It is a measure of inbreeding
(3) It is a measure of heterozygotes in a population
(4) It is a measure of the proportion of variation which is contributed by genetic factors

36. A threshold trait is one which is

(1) expressed on a continuous scale
(2) present in a few discrete classes but is influenced by environmental factors
(3) caused by only a single gene, with no environmental influence
(4) associated with superior survival of the heterozygote

37. A homeotic mutation is one which

(1) is present only in one form in an individual
(2) results in developmental block of any tissue-specific gene expression
(3) substitutes one body part for another during development
(4) results from transposon mediated mutagenesis

38. A woman who is a heterozygous carrier of an X-linked recessive disease gene mates with a phenotypically normal male. The disease gene has a penetrance of 80%. On an average what proportion among the following of this couple's sons will be affected with the disorder ?

(1) 0.8 (2) 0.4
(3) 0.2 (4) 0.5

39. In the incomplete dominance of a monohybrid, the number of phenotypes in the F_2 will be

(1) 1 (2) 2
(3) 3 (4) 4

40. The genetic map for three genes A, B and C is as follows: A-B = 10 map units, B-C = 5 map units and A-C = 15 map units. In an individual of genotype AbC/aBc, the percentage of gametes expected to be of ABC is

(1) 0.5 (2) 5
(3) 15 (4) 50

41. Mendel's law of segregation, as applied to the behavior of chromosomes during cell division means that

(1) Alleles of a gene separate from each other when homologous chromosomes separate in meiosis

(2) Alleles of a gene separate from each other when chromatids separate in meiosis II

(3) Alleles of a gene separate from each other when homologs separate in meiosis I, or when chromatids separate in meiosis II if there is a cross over between the gene and the centromere

(4) Alleles of a gene separate from each other when chromatids separate in meiosis I, or when homologs separate in meiosis II if there is a cross over between the gene and the centromere

42. With reference to lac operon which one of the following merodiploids will show a constitutive expression of β-galactosidase?

(1) $I^-O^+Z^+Y^-$/ F' I^+ $O^+Z^-Y^+$

(2) $I^-O^CZ^+Y^-$ / F' $I^+O^+Z^-Y^+$

(3) $I^-O^+Z^+Y^-$ / F' $I^+O^CZ^-Y^+$

(4) $I^-O^CZ^-Y^-$ / F' $I^+O^+Z^+Y^+$

43. In *E. coli* four Hfr stains donate the following genetic markers, shown in the order donated

Strain 1: A D C E Q

Strain 2: A K M T Y

Strain 3: C E Q W X

Strain 4: M T Y X W

All these HFr strains are derived from the same F^+ strains. What is the order of these markers on the circular chromosome of the original F^+?

(1) A D C E Q W X Y T M K A

(2) Y T M K A D C E Q W X Y

(3) D C E Q A K M T Y X W D

(4) C E Q W X Y T M A K D C

44. Which of the following statements is true?

(1) There is a lesser probability for a crossover to occur between 2 genes farther apart from the genes nearer to each other

(2) There is a greater probability for a crossover to occur between 2 genes farther apart from the genes nearer to each other

(3) Probability of crossover between 2 genes is not related to the distance between them

(4) Maximum frequency of recombination that can result from crossing over between linked genes is 100%

45. Formation of the following pattern in an ordered octad is due to what event during meiosis? Pattern: AAaaaaAA

(1) No crossing over and normal segregation in meiosis I

(2) Crossing over and normal segregation during meiosis I

(3) Crossing over and segregation during meiosis II

(4) No crossing over and segregation during meiosis II

46. You set up a mapping cross involving two genes in Aspergillus and determine the following percentages of tetrad types: parental ditype 51%, nonparentalditype 49%. What can you say about your genes?

(1) They are closely linked

(2) They are unlinked

(3) They are lethal

(4) They are not expressed in the phenotype

47. You set up a mapping cross involving your favorite gene (YFG) in Neurospora and determine the following octad types: 4:4 arrangement 83; 2:4:2 arrangement 7; 2:2:2:2 arrangement 10. What is the distance between your gene and the centromere?

(1) 8.5 cM (2) 10cM

(3) 17 cM (4) 83cM

48. Which is actual evidence supporting the endosymbiotic theory for the origin of eukaryotes?
(1) Fossil evidence suggest early eukaryotes ate proteobacteria
(2) Free-living mitochondria still exist in some environments
(3) Mitochondria and chloroplasts have their own DNA and have 60S ribosomes
(4) Only prokaryotes are autotrophic

49. For obtaining recombinants in conjugation of Hfr x F- cells, which one of the events is necessary following conjugation?
(1) Survival of donor cells
(2) Survival of recipient cells
(3) F- cells get converted into Hfr
(4) Cells should not be disturbed

50. Bacterial cell with two copies of some genes, one on the bacterial chromosome and one on the newly introduced F plasmid are termed as
(1) Mesozygotes (2) Diploid
(3) Merozygotes (4) Sex duction

51. Which of the following statements about familial hypercholesterolemia is false?
(1) Familial hypercholesteremia is caused by a defect in membrane transport
(2) Familial hypercholesteremia can cause early heart attack and stroke
(3) Sufferers from this disease have high cholesterol levels in their liver cells
(4) Familial hypercholesterolemia can be due to a single point mutation

52. What offspring would be expected if the two genes A and B are 10 map units apart and the heterozygote has trans (repulsive) arrangement of alleles on chromosome?
(1) 25% of each possible combination of traits: A and B, A and b, a and B, and a and b
(2) 45% of the offspring will exhibit A and b, 45% will exhibit a and B, 5% will exhibit A and B, and 5% will exhibit a and b
(3) 40% of the offspring will exhibit A and B, 40% will exhibit a and b, 10% will exhibit A and b, and 10% will exhibit a and B
(4) 45% of the offspring will exhibit A and B, 45% will exhibit a and b, 5% will exhibit A and b, and 5% will exhibit a and B

53. How many map units is a recombination frequency of 5 percent equal to?
(1) 2.5 centimorgans
(2) 10 centimorgans
(3) 5 centisturtevants
(4) 5 centimorgans

54. A linkage map
(1) orders genes on a chromosome based on recombination frequencies
(2) can only be constructed for sex chromosomes
(3) orders genes on a chromosome based on their location with respect to a stained band
(4) shows the actual ordering and spacing of genes on a chromosome

55. A male bee is
(1) X Y
(2) diploid
(3) Z W
(4) haploid

56. What is the probability that a male will inherit an X-linked recessive gene from his father?
(1) 0 (2) 25 %
(3) 50 % (4) 100 %

57. Which of the following human genetic disorders is sex linked?
(1) hemophilia
(2) PKU
(3) cystic fibrosis
(4) achondroplasia

58. A genetic defect in humans results in the absence of sweat glands in the skin. Some men have this defect all over their bodies, but in women it is usually expressed in a peculiar way. A woman with this defect typically has small patches of skin with sweat glands and other patches where sweat glands are lacking. This pattern suggests the phenotypic effect of

(1) X-linked defect

(2) chromosome inactivation

(3) RNA splicing

(4) Genomic imprinting

59. Assuming equal sex ratio, what is probability that a sib ship of four children consists entirely of boys-

(1) 25 % (2) 12. 5 %

(3) 6.25 % (4) 3.125 %

60. The segregation of Mendelian factors generally takes place during-

(1) Meiosis I (2) Meiosis II

(3) Mitosis (4) Inter phase

61. The calculation of χ^2 can be represented in expression as

(1) $\chi^2 = \Sigma \dfrac{\text{Observed} - \text{Expected}}{\text{Expected}}$

(2) $\chi^2 = \Sigma \dfrac{(\text{Observed} - \text{Expected})^2}{\text{Expected}^2}$

(3) $\chi^2 = \Sigma \dfrac{\text{Observed}^2 - \text{Expected}^2}{\text{Expected}^2}$

(4) $\chi^2 = \Sigma \dfrac{(\text{Observed} - \text{Expected})^2}{\text{Expected}^2}$

62. In single factor crosses between two true breeding strains of tree frog, a scientist discovers that plain green tree frogs and tree frogs with a red 'racing stripe' always produce offspring that have red racing stripes. Why?

(1) Stripes are dominant to plain green coloration

(2) Stripes must allow the frog to hide more effectively in the leaves

(3) Tree frogs with stripes are often eaten by birds

(4) Tree frogs with stripes must be more common than plain green ones

63. The dihybrid F_2-ratio 9:7 represents

(1) Dominance

(2) Law of independent assortment

(3) epistasis

(4) Co-dominance

64. Frequency of crossing over is directly proportional to-

(1) Length of chromosomes

(2) Number of genes

(3) Width of chromosomes

(4) Distance between genes

65. When a cross is made between dominant long and full endosperm (*LLSS*) and recessive short and shrunkeb endosperm (*llss*) The progeny obtained 98.3% were long and full endosperm and short and shrunken, rest 1.7% were long shrunken endosperm and short full endosperm. Such a ratio was observed due to

(1) Epitasis (2) Linkage

(3) Co-dominance (4) Over dominance

66. The ratio of observed number of double crossover to expected number of cross over is termed as

(1) Coefficient of coincidence

(2) Coefficient of interference

(3) Hitch hiking

(4) Linkage suppression

67. The general term for an *E. coli* strain requiring amino acid supplements for growth is termed a/an

(1) oligotroph

(2) auxotroph

(3) prototroph

(4) lysotroph

68. What proportion of offspring of a male carrying an X-linked dominant trait will be affected?

(1) ½ sons and ½ daughters
(2) All sons and no daughters
(3) All daughters and no sons
(4) ¼ daughters and ½ sons

69. Gene transfer in bacteria by transformation has the following characteristic

(1) A majority of the donor genes are transferred
(2) It involves a plasmid
(3) It depends on phage infection of the recipient cell
(4) It can be carried out using free DNA extracted from the donor

70. Two strains with the auxotrophies met-bio-leu+ and met+bio+leu- were mixed for a mating experiment in bacteria. Bacterial strain selected on a complete medium in this experiment would have which of the following markers?

(1) $met^- bio^- leu^-$ (2) $met^+ bio^- leu^-$
(3) $met^+ bio^+ leu^+$ (4) $met^+ bio^+ leu^+$

71. What do we mean when we use the terms monohybrid cross and dihybrid cross?

(1) A monohybrid cross involves a single parent, whereas a dihybrid cross involves two parents
(2) A monohybrid cross produces a single progeny, whereas a dihybrid cross produces two progeny
(3) A dihybrid cross involves organisms that are heterozygous for two characters and a monohybrid cross involves only one
(4) A monohybrid cross is performed for one generation, whereas a dihybrid cross is performed for two generations

72. Why did the F_1 offspring of Mendel's classic pea cross always look like one of the two parental varieties?

(1) No genes interacted to produce the parental phenotype
(2) Each allele affected phenotypic expression
(3) The traits blended together during fertilization
(4) One phenotype was completely dominant over another

73. What was the most significant conclusion that Gregor Mendel drew from his experiments with pea plants?

(1) There is considerable genetic variation in garden peas
(2) Traits are inherited in discrete units, and are not the results of "blending"
(3) Recessive genes occur more frequently in the F_1 generation than do dominant ones
(4) Genes are composed of DNA

74. How many unique gametes could be produced through independent assortment by an individual with the genotype *AaBbCCDdEE*?

(1) 4 (2) 8
(3) 16 (4) 32

75. The individual with genotype *AaBbCCDdEE* can make many kinds of gametes. Which of the following is the major reason?

(1) segregation of maternal and paternal alleles
(2) recurrent mutations forming new alleles
(3) crossing over during prophase I
(4) different possible alignments of chromosomes

76. Why did Mendel continue some of his experiments to the F_2 or F_3 generation?

(1) to obtain a larger number of offspring on which to base statistics
(2) to observe whether or not a recessive trait would reappear
(3) to observe whether or not the dominant trait would reappear
(4) to distinguish which alleles were segregating

77. Which of the following differentiates between independent assortment and segregation?

(1) The law of independent assortment requires describing two or more genes relative to one another

(2) The law of segregation requires describing two or more genes relative to one another

(3) The law of segregation requires having two or more generations to describe

(4) The law of independent assortment is accounted for by observations of prophase I

78. Two plants are crossed, resulting in offspring with a 3:1 ratio for a particular trait. What does this suggest?

(1) that the parents were true-breeding for contrasting traits

(2) that the trait shows incomplete dominance

(3) that a blending of traits has occurred

(4) that the parents were both heterozygous for a single trait

79. A sexually reproducing animal has two unlinked genes, one for head shape (*H*) and one for tail length (*T*) Its genotype is *HhTt*. Which of the following genotypes is possible in a gamete from this organism?

(1) *tt* (2) *Hh*

(3) *HhTt* (4) *HT*

80. When crossing an organism that is homozygous recessive for a single trait with a heterozygote, what is the chance of producing an offspring with the homozygous recessive phenotype?

(1) 0% (2) 25%

(3) 50% (4) 75%

81. Mendel accounted for the observation that traits which had disappeared in the F_1 generation reappeared in the F_2 generation by proposing that

(1) new mutations were frequently generated in the F_2 progeny, "reinventing" traits that had been lost in the F_1

(2) the mechanism controlling the appearance of traits was different between the F_1 and the F_2 plants

(3) traits can be dominant or recessive, and the recessive traits were obscured by the dominant ones in the F_1

(4) the traits were lost in the F_1 due to dominance of the parental traits

82. The fact that all seven of the pea plant traits studied by Mendel obeyed the principle of independent assortment most probably indicates which of the following?

(1) None of the traits obeyed the law of segregation

(2) The diploid number of chromosomes in the pea plants was 7

(3) All of the genes controlling the traits were located on the same chromosome

(4) All of the genes controlling the traits behaved as if they were on different chromosomes

83. Mendel was able to draw his ideas of segregation and independent assortment because of the influence of which of the following?

(1) His reading and discussion of Darwin's *Origin of Species*

(2) The understanding of particulate inheritance he learned from renowned scientists of his time

(3) His discussions of heredity with his colleagues at major universities

(4) His reading of the scientific literature current in the field

84. Mendel's observation of the segregation of alleles in gamete formation has its basis in which of the following phases of cell division?

(1) prophase I of meiosis

(2) anaphase II of meiosis

(3) metaphase I of meiosis

(4) anaphase I of meiosis

85. Mendel's second law of independent assortment has its basis in which of the following events of meiosis I?

(1) synapsis of homologous chromosomes
(2) crossing over
(3) alignment of tetrads at the equator
(4) separation of homologs at anaphase

86. Black fur in mice ((2) is dominant to brown fur ((2) Short tails (*T*) are dominant to long tails (*t*) What fraction of the progeny of crosses *BbTt* × *BBtt* will be expected to have black fur and long tails?

(1) 1/16 (2) 3/16
(3) 3/8 (4) 1/2

87. In certain plants, tall is dominant to short. If a heterozygous plant is crossed with a homozygous tall plant, what is the probability that the offspring will be short?

(1) 1 (2) 1/2
(3) 1/4 (4) 0

88. In the cross *AaBbCc* × *AaBbCc*, what is the probability of producing the genotype *AABBCC*?

(1) 1/4 (2) 1/8
(3) 1/16 (4) 1/64

89. Given the parents *AABBCc* × *AabbCc*, assume simple dominance for each trait and independent assortment. What proportion of the progeny will be expected to phenotypically resemble the first parent?

(1) 1/4 (2) 1/8
(3) 3/4. (4) 3/8

90. Which of the following is the best statement of the use of the addition rule of probability?

(1) the probability that two or more independent events will both occur
(2) the probability that two or more independent events will both occur in the offspring of one set of parents
(3) the probability that either one of two independent events will occur
(4) the probability of producing two or more heterozygous offspring

91. Which of the following calculations require that you utilize the addition rule?

(1) Calculate the probability of black offspring from the cross *AaBb* × *AaBb*, when *B* is the symbol for black
(2) Calculate the probability of children with both cystic fibrosis and polydactyly when parents are each heterozygous for both genes
(3) Calculate the probability of each of four children having cystic fibrosis if the parents are both heterozygous
(4) Calculate the probability of a child having either sickle-cell anemia or cystic fibrosis if parents are each heterozygous for both genes

92. In cattle, roan coat color (mixed red and white hairs) occurs in the heterozygous (*Rr*) offspring of red (*RR*) and white (*rr*) homozygotes. Which of the following crosses would produce offspring in the ratio of 1 red:2 roan:1 white?

(1) red × white
(2) roan × roan
(3) white × roan
(4) red × roan

93. Which of the following describes the ability of a single gene to have multiple phenotypic effects?

(1) incomplete dominance
(2) multiple alleles
(3) pleiotropy
(4) epistasis

94. Cystic fibrosis affects the lungs, the pancreas, the digestive system, and other organs, resulting in symptoms ranging from breathing difficulties to recurrent infections. Which of the following terms best describes this?

(1) incomplete dominance
(2) multiple alleles
(3) pleiotropy
(4) epistasis

95. Which of the following is an example of polygenic inheritance?
(1) pink flowers in snapdragons
(2) the ABO blood group in humans
(3) Huntington's disease in humans
(4) skin pigmentation in humans

96. *Hydrangea* plants of the same genotype are planted in a large flower garden. Some of the plants produce blue flowers and others pink flowers. This can be best explained by which of the following?
(1) the knowledge that multiple alleles are involved
(2) the allele for blue hydrangea being completely dominant
(3) the alleles being codominant
(4) environmental factors such as soil pH

97. Which of the following provides an example of epistasis?
(1) Recessive genotypes for each of two genes (*aabb*) results in an albino corn snake
(2) The allele *b17* produces a dominant phenotype, although *b1* through *b16* do not
(3) In rabbits and many other mammals, one genotype (*cc*) prevents any fur color from developing
(4) In *Drosophila* (fruit flies), white eyes can be due to an X-linked gene or to a combination of other genes

98. Most genes have many more than two alleles. However, which of the following is also true?
(1) At least one allele for a gene always produces a dominant phenotype
(2) Most of the alleles will never be found in a live-born organism
(3) All of the alleles but one will produce harmful effects if homozygous
(4) There may still be only two phenotypes for the trait

99. How could you best predict the maximum number of alleles for a single gene whose polypeptide product is known?
(1) Search the population for all phenotypic variants of this polypeptide
(2) Count the number of amino acids in the polypeptide
(3) Mate all known genotypes and collect all possible offspring different from the parents
(4) Count the number of DNA nucleotides that are in the code for the polypeptides

100. An ideal procedure for fetal testing in humans would have which of the following features?
(1) the procedure that can be performed at the earliest time in the pregnancy
(2) lowest risk procedure that would provide the most reliable information
(3) the procedure that can test for the greatest number of traits at once
(4) a procedure that provides a three-dimensional image of the fetus

101. A scientist discovers a DNA-based test for one allele of a particular gene. This and only this allele, if homozygous, produces an effect that results in death at or about the time of birth. Of the following, which is the best use of this discovery?
(1) Screen all newborns of an at-risk population
(2) Design a test for identifying heterozygous carriers of the allele
(3) Introduce a normal allele into deficient newborns
(4) Follow the segregation of the allele during meiosis

102. An obstetrician knows that one of her patients is a pregnant woman whose fetus is at risk for a serious disorder that is detectable biochemically in fetal cells. The obstetrician would most reasonably offer which of the following procedures to her patient?
(1) CVS (2) ultrasound imaging
(3) amniocentesis. (4) blood transfusion

103. The frequency of heterozygosity for the sickle-cell anemia allele is unusually high, presumably because this reduces the frequency of malaria. Such a relationship is related to which of the following?

(1) Mendel's law of independent assortment
(2) Mendel's law of segregation
(3) Darwin's explanation of natural selection
(4) Darwin's observations of competition

104. Phenylketonuria (PKU) is a recessive human disorder in which an individual cannot appropriately metabolize a particular amino acid. The amino acid is not otherwise produced by humans. Therefore, the most efficient and effective treatment is which of the following?

(1) Feed them the substrate that can be metabolized into this amino acid
(2) Transfuse the patients with blood from unaffected donors
(3) Regulate the diet of the affected persons to severely limit the uptake of the amino acid
(4) Feed the patients the missing enzymes in a regular cycle, such as twice per week

105. Hutchinson-Gilford progeria is an exceedingly rare human genetic disorder in which there is very early senility and death, usually from coronary artery disease, at an average age of approximately 13. Patients, who look very old even as children, do not live to reproduce. Which of the following represents the most likely assumption?

(1) All cases must occur in relatives; therefore, there must be only one mutant allele
(2) Successive generations of a family will continue to have more and more cases over time
(3) The disorder may be due to mutation in a single protein-coding gene
(4) Each patient will have had at least one affected family member in a previous generation

106. One of two major forms of a human condition called neurofibromatosis (NF 1) is inherited as a dominant gene, although it may range from mildly to very severely expressed. If a young child is the first in her family to be diagnosed, which of the following is the best explanation?

(1) The mother carries the gene but does not express it at all
(2) One of the parents has very mild expression of the gene
(3) The condition skipped a generation in the family
(4) The child has a different allele of the gene than the parents

107. In a particular plant, leaf color is controlled by gene locus *D*. Plants with at least one allele *D* have dark green leaves, and plants with the homozygous recessive *dd* genotype have light green leaves. A true-breeding dark-leaved plant is crossed with a light-leaved one, and the F_1 offspring is allowed to self-pollinate. The predicted outcome of the F_2 is diagrammed in the Punnett square shown in Figure, where 1, 2, 3, and 4 represent the genotypes corresponding to each box within the square

	D	*d*
D	1	2
d	3	4

Which of the boxes correspond to plants with a heterozygous genotype?

(1) 1 only
(2) 1 and 2
(3) 2 and 3
(4) 2, 3 and 4

108. 108. The following questions refer to the pedigree chart in Figure for a family, some of whose members exhibit the dominant trait, *W*. Affected individuals are indicated by a dark square or circle

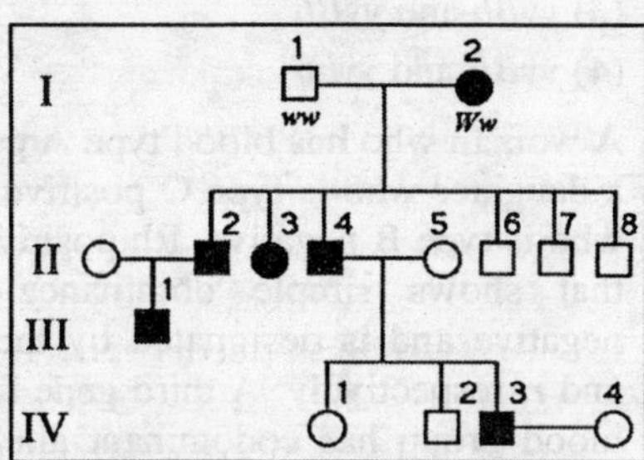

What is the likelihood that the progeny of IV-3 and IV-4 will have the trait?

(1) 0% (2) 25%
(3) 50%. (4) 75%

109. Use the following pedigree (Figure) for a family in which dark-shaded symbols represent individuals with one of the two major types of colon cancer. Numbers under the symbols are the individual's age at the time of diagnosis

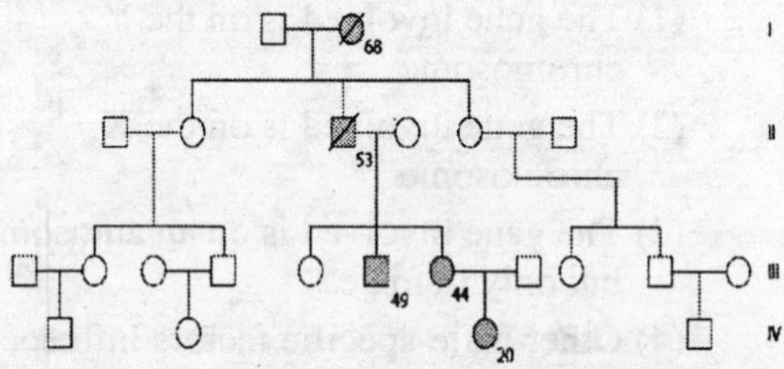

In each generation of this family after generation I, the age at diagnosis is significantly lower than would be found in nonfamilial (sporadi(3) cases of this cancer (~ 63 years) What is the most likely reason?

(1) Members of this family know to be checked for colon cancer early in life
(2) Hereditary (or familial) cases of this cancer typically occur at earlier ages than do nonfamilial forms
(3) This is pure chance; it would not be expected if you were to look at a different family
(4) This cancer requires mutations in more than this one gene

110. Two true-breeding stocks of pea plants are crossed. One parent has red, axial flowers and the other has white, terminal flowers; all F_1 individuals have red, axial flowers. The genes for flower color and location assort independently. Among the F_2 offspring, what is the probability of plants with white axial flowers?

(1) 9/16 (2) 1/16
(3) 3/16. (4) 1/8

111. Labrador retrievers are black, brown, or yellow. In a cross of a black female with a brown male, results can be either all black puppies, 1/2 black to 1/2 brown puppies, or 3/4 black to 1/4 yellow puppies. How many genes must be responsible for these coat colors in Labrador retrievers?

(1) 1 (2) 2
(3) 3 (4) 4

112. Radish flowers may be red, purple, or white. A cross between a red-flowered plant and a white-flowered plant yields all-purple offspring. The part of the radish we eat may be oval or long, with long being the dominant characteristic. In the F_2 generation of the above cross, which of the following phenotypic ratios would be expected?

(1) 9:3:3:1 (2) 9:4:3
(3) 1:1:1:1 (4) 6:3:3:2:1:1

113. *Drosophila* (fruit flies) usually have long wings (+) but mutations in two different genes can result in bent wings (*bt*) or vestigial wings (*vg*) If a homozygous bent wing fly is mated with a homozygous vestigial wing fly, which of the following offspring would you expect?

(1) all *+bt +vg* heterozygotes
(2) 1/2 bent and 1/2 vestigial flies
(3) all homozygous + flies
(4) 3/4 bent to 1/4 vestigial ratio

114. Tallness (*T*) in snapdragons is dominant to dwarfness (*t*), while red (*R*) flower color is dominant to white (*r*) The heterozygous condition results in pink (*Rr*) flower color. If snapdragons are heterozygous for height as well as for flower color, a mating between them will result in what ratio?

(1) 9:3:3:1 (2) 6:3:3:2:1:1
(3) 1:2:1 (4) 27:9:9:9:3:3:3:1

115. Skin color in a certain species of fish is inherited via a single gene with four different alleles. One fish of this type has alleles 1 and 3 (S_1S_3) and its mate has alleles 2 and 4 (S_2S_4) If each allele confers a unit of color darkness such that S_1 has one unit, S_2 has two units, and so on, then what proportion of their offspring would be expected to have five units of color?

(1) 1/4 (2) 1/5
(3) 1/8 (4) 1/2

116. Gene *S* controls the sharpness of spines in a type of cactus. Cactuses with the dominant allele, *S*, have sharp spines, whereas homozygous recessive *ss* cactuses have dull spines. At the same time, a second gene, *N*, determines whether or not cactuses have spines. Homozygous recessive *nn* cactuses have no spines at all. If doubly heterozygous *SsNn* cactuses were allowed to self-pollinate, the F_2 would segregate in which of the following ratios?

(1) 3 sharp-spined:1 spineless
(2) 1 sharp-spined:2 dull-spined:1 spineless
(3) 1 sharp-spined:1 dull-spined:1 spineless
(4) 9 sharp-spined:3 dull-spined:4 spineless

117. Feather color in budgies is determined by two different genes, *Y* and *B*, one for pigment on the outside and one for the inside of the feather. *YYBB*, *YyBB*, or *YYBb* is green; *yyBB* or *yyBb* is blue; *YYbb* or *Yybb* is yellow; and *yybb* is white. Two blue budgies were crossed. Over the years, they produced 22 offspring, 5 of which were white. What are the most likely genotypes for the two blue budgies?

(1) *yyBB* and *yyBB*
(2) *yyBB* and *yyBb*
(3) *yyBb* and *yyBb*
(4) *yyBB* and *yybb*

118. A woman who has blood type A positive has a daughter who is type O positive and a son who is type B negative. Rh positive is a trait that shows simple dominance over Rh negative and is designated by the alleles *R* and *r*, respectively. A third gene for the MN blood group has codominant alleles *M* and *N*. Which of the following is a possible partial genotype for the son?

(1) I^BI^B (2) $I^B I^A$
(3) *ii* (4) $I^B i$

119. When Thomas Hunt Morgan crossed his red-eyed F_1 generation flies to each other, the F_2 generation included both red- and white-eyed flies. Remarkably, all the white-eyed flies were male. What was the explanation for this result?

(1) The gene involved is on the Y chromosome
(2) The gene involved is on the X chromosome
(3) The gene involved is on an autosome, but only in males
(4) Other male-specific factors influence eye color in flies

120. Sturtevant provided genetic evidence for the existence of four pairs of chromosomes in *Drosophila* in which of these ways?

(1) There are four major functional classes of genes in *Drosophila*
(2) *Drosophila* genes cluster into four distinct groups of linked genes
(3) The overall number of genes in *Drosophila* is a multiple of four
(4) The entire *Drosophila* genome has approximately 400 map units

121. Which of the following is the meaning of the chromosome theory of inheritance as expressed in the early 20th century?

(1) Individuals inherit particular chromosomes attached to genes

(2) Mendelian genes are at specific loci on the chromosome and in turn segregate during meiosis

(3) Homologous chromosomes give rise to some genes and crossover chromosomes to other genes

(4) No more than a single pair of chromosomes can be found in a healthy normal cell

122. Thomas Hunt Morgan's choice of *Drosophila melanogaster* has been proven to be useful even today. Which of the following has/have continued to make it a most useful species?

I. its four pairs of chromosomes

II. a very large number of visible as well as biochemically mutant phenotypes

III. easy and inexpensive maintenance

IV. short generation time and large number of offspring

(1) I and IV only

(2) II and III only

(3) I, II, and III only

(4) I, II, III, IV, and V

123. A woman is found to have 47 chromosomes, including three X chromosomes. Which of the following describes her expected phenotype?

(1) Masculine characteristics such as facial hair

(2) Enlarged genital structures

(3) Excessive emotional instability

(4) Normal female

124. Males are more often affected by sex-linked traits than females because

(1) male hormones such as testosterone often alter the effects of mutations on the X chromosome

(2) female hormones such as estrogen often compensate for the effects of mutations on the X chromosome

(3) X chromosomes in males generally have more mutations than X chromosomes in females

(4) males are hemizygous for the X chromosome

125. SRY is best described in which of the following ways?

(1) A gene present on the X chromosome that triggers female development

(2) An autosomal gene that is required for the expression of genes on the Y chromosome

(3) A gene region present on the Y chromosome that triggers male development

(4) An autosomal gene that is required for the expression of genes on the X chromosome

126. In cats, black fur color is caused by an X-linked allele; the other allele at this locus causes orange color. The heterozygote is tortoiseshell. What kinds of offspring would you expect from the cross of a black female and an orange male?

(1) tortoiseshell females; tortoiseshell males

(2) black females; orange males

(3) orange females; orange males

(4) tortoiseshell females; black males

127. Red-green color blindness is a sex-linked recessive trait in humans. Two people with normal color vision have a color-blind son. What are the genotypes of the parents?

(1) $X^c X^c$ and $X^c Y$

(2) $X^c X^c$ and $X^C Y$

(3) $X^C X^C$ and $X^c Y$

(4) $X^C X^c$ and $X^C Y$

128. Cinnabar eyes is a sex-linked recessive characteristic in fruit flies. If a female having cinnabar eyes is crossed with a wild-type male, what percentage of the F_1 males will have cinnabar eyes?

(1) 0%
(2) 25%
(3) 50%
(4) 100%

129. Calico cats are female because

(1) the males die during embryonic development
(2) a male inherits only one of the two X-linked genes controlling hair color
(3) the Y chromosome has a gene blocking orange coloration
(4) only females can have Barr bodies

130. In birds, sex is determined by a *ZW* chromosome scheme. Males are *ZZ* and females are *ZW*. A recessive lethal allele that causes death of the embryo is sometimes present on the *Z* chromosome in pigeons. What would be the sex ratio in the offspring of a cross between a male that is heterozygous for the lethal allele and a normal female?

(1) 2:1 male to female
(2) 1:2 male to female
(3) 1:1 male to female
(4) 4:3 male to female

131. Sex determination in mammals is due to the SRY region of the Y chromosome. An abnormality of this region could allow which of the following to have a male phenotype?

(1) Turner syndrome, 45, X
(2) Translocation of SRY to an autosome of a 46, XX individual
(3) A person with an extra X chromosome
(4) A person with one normal and one shortened (deleted) X

132. In humans, clear gender differentiation occurs, not at fertilization, but after the second month of gestation. What is the first event of this differentiation?

(1) Formation of testosterone in male embryos
(2) Formation of estrogens in female embryos
(3) Anatomical differentiation of a penis in male embryos
(4) Activation of SRY in male embryos and masculinization of the gonads

133. Duchenne muscular dystrophy (DMD) is caused by a gene on the human X chromosome. The patients have muscles that weaken over time because they have absent or decreased dystrophin, a muscle protein. They rarely live past their 20s. How likely is it for a woman to have this condition?

(1) Women can never have this condition
(2) One-half of the daughters of an affected man could have this condition
(3) One-fourth of the children of an affected father and a carrier mother could have this condition
(4) Very rarely would a woman have this condition; the condition would be due to a chromosome error

134. Women (and all female mammals) have one active X chromosome per cell instead of two. What causes this?

(1) Modification of the XIST gene so that it is active only on one X chromosome, which then becomes inactive
(2) Activation of the Barr gene on one of the two X chromosomes that then inactivates
(3) Crossover between the XIST gene on one X chromosome and a related gene on an autosome
(4) Inactivation of the XIST gene on the X chromosome derived from the male parent

135. Which of the following statements is true of linkage?

(1) The closer two genes are on a chromosome, the lower the probability that a crossover will occur between them

(2) The observed frequency of recombination of two genes that are far apart from each other has a maximum value of 100%

(3) All of the traits that Mendel studied–seed color, pod shape, flower color, and others–are due to genes linked on the same chromosome

(4) Linked genes are found on different chromosomes

136. How would one explain a testcross involving F_1 dihybrid flies in which more parental-type offspring than recombinant-type offspring are produced?

(1) The two genes are closely linked on the same chromosome

(2) The two genes are linked but on different chromosomes

(3) Recombination did not occur in the cell during meiosis

(4) The testcross was improperly performed

137. What does a frequency of recombination of 50% indicate?

(1) The two genes are likely to be located on different chromosomes

(2) All of the offspring have combinations of traits that match one of the two parents

(3) The genes are located on sex chromosomes

(4) Abnormal meiosis has occurred

138. What is the reason that linked genes are inherited together?

(1) They are located close together on the same chromosome

(2) The number of genes in a cell is greater than the number of chromosomes

(3) Chromosomes are unbreakable

(4) Alleles are paired together during meiosis

139. Three genes at three loci are being mapped in a particular species. Each has two phenotypes, one of which is markedly different from the wild type. The unusual allele of the first gene is inherited with either of the others about 50% of the time. However, the unusual alleles of the other two genes are inherited together 14.4% of the time. Which of the following describes what is happening?

(1) The genes are showing independent assortment

(2) The three genes are linked

(3) The first gene is linked but the other two are not

(4) The first gene is assorting independently from the other two that are linked

140. The centimorgan (cM) is a unit named in honor of Thomas Hunt Morgan. To what is it equal?

(1) the physical distance between two linked genes

(2) 1% frequency of recombination between two genes

(3) 1 nanometer of distance between two genes

(4) the distance between a pair of homologous chromosomes

141. Recombination between linked genes comes about for what reason?

(1) Mutation on one homolog is different from that on the other homolog

(2) Independent assortment sometimes fails because Mendel had not calculated appropriately

(3) When genes are linked they always "travel" together at anaphase

(4) Crossovers between these genes result in chromosomal exchange

142. Why does recombination between linked genes continue to occur?

(1) Recombination is a requirement for independent assortment
(2) Recombination must occur or genes will not assort independently
(3) New allele combinations are acted upon by natural selection
(4) The forces on the cell during meiosis II always result in recombination

143. Map units on a linkage map cannot be relied upon to calculate physical distances on a chromosome for which of the following reasons?

(1) The frequency of crossing over varies along the length of the chromosome
(2) The relationship between recombination frequency and map units is different in every individual
(3) Physical distances between genes change during the course of the cell cycle
(4) The gene order on the chromosomes is slightly different in every individual

144. Which of the following two genes are closest on a genetic map of *Drosophila*?

(1) *b* and *vg* (2) *vg* and *cn*
(3) *rb* and *cn* (4) *b* and *rb*

145. If nondisjunction occurs in meiosis II during gametogenesis, what will be the result at the completion of meiosis?

(1) All the gametes will be diploid
(2) Half of the gametes will be $n + 1$, and half will be $n - 1$
(3) 1/4 of the gametes will be $n + 1$, 1/4 will be $n - 1$, and 1/2 will be n
(4) There will be three extra gametes

146. One possible result of chromosomal breakage is for a fragment to join a nonhomologous chromosome. What is this alteration called?

(1) Deletion (2) Transversion
(3) Inversion (4) Translocation

147. A nonreciprocal crossover causes which of the following products?

(1) Deletion only
(2) Duplication only
(3) Nondisjunction
(4) Deletion and duplication

148. In humans, male-pattern baldness is controlled by an autosomal gene that occurs in two allelic forms. Allele *Hn* determines nonbaldness, and allele *Hb* determines pattern baldness. In males, because of the presence of testosterone, allele *Hb* is dominant over *Hn*. If a man and woman both with genotype *HnHb* have a son, what is the chance that he will eventually be bald?

(1) 0% (2) 25%
(3) 75%. (4) 50%

149. Of the following human aneuploidies, which is the one that generally has the most severe impact on the health of the individual?

(1) 47, +21
(2) 47, XXY
(3) 47, XXX
(4) 47, XYY

150. A phenotypically normal prospective couple seeks genetic counseling because the man knows that he has a translocation of a portion of his chromosome 4 that has been exchanged with a portion of his chromosome 12. Although he is normal because his translocation is balanced, he and his wife want to know the probability that his sperm will be abnormal. What is your prognosis regarding his sperm?

(1) 1/4 will be normal, 1/4 will have the translocation, and 1/2 will have duplications and deletions
(2) All will carry the same translocation as the father
(3) None will carry the translocation since abnormal sperm will die
(4) His sperm will be sterile and the couple might consider adoption

151. Abnormal chromosomes are frequently found in malignant tumors. Errors such as translocations may place a gene in close proximity to different control regions. Which of the following might then occur to make the cancer worse?

(1) An increase in nondisjunction

(2) Expression of inappropriate gene products

(3) A decrease in mitotic frequency

(4) Death of the cancer cells in the tumor

152. An inversion in a human chromosome often results in no demonstrable phenotypic effect in the individual. What else may occur?

(1) There may be deletions later in life

(2) Some abnormal gametes may be formed

(3) There is an increased frequency of mutation

(4) All inverted chromosomes are deleted

153. What is the source of the extra chromosome 21 in an individual with Down syndrome?

(1) Nondisjunction in the mother only

(2) Nondisjunction in the father only

(3) Duplication of the chromosome

(4) Nondisjunction or translocation in either parent

154. Down syndrome has a frequency in the U.S. population of ~1/700 live births. In which of the following groups would you expect this frequency to be significantly higher?

(1) people in Latin or South America

(2) the Inuit and other peoples in very cold habitats

(3) people living in equatorial areas of the world

(4) No groups have such higher frequency

155. A couple has a child with Down syndrome. The mother is 39 years old at the time of delivery. Which of the following is the most probable cause of the child's condition?

(1) The woman inherited this tendency from her parents

(2) One member of the couple carried a translocation

(3) One member of the couple underwent nondisjunction in somatic cell production

(4) One member of the couple underwent nondisjunction in gamete production

156. In 1956 Tijo and Levan first successfully counted human chromosomes. What is the reason it took so many years to do so?

(1) Watson and Crick's structure of DNA was not done until 1953

(2) Chromosomes were piled up on top of one another in the nucleus

(3) Chromosomes were not distinguishable during interphase

(4) Chromosomes were piled up on top of one another in the nucleus, chromosomes were not distinguishable during interphase, and a method had not yet been devised to halt mitosis at metaphase

157. At which phase(s) is it preferable to obtain chromosomes to prepare a karyotype?

(1) Early prophase

(2) Late telophase

(3) Anaphase

(4) Late prophase or metaphase

158. What is a syndrome?

(1) A characteristic facial appearance

(2) A group of traits, all of which must be present if an aneuploidy is to be diagnosed

(3) A group of traits typically found in conjunction with a particular chromosomal aberration or gene mutation

(4) A characteristic trait usually given the discoverer's name

159. Which of the following is known as a Philadelphia chromosome?

(1) A human chromosome 22 that has had a specific translocation

(2) A human chromosome 9 that is found only in one type of cancer

(3) An animal chromosome found primarily in the mid-Atlantic area of the United States

(4) An imprinted chromosome that always comes from the mother

160. At what point in cell division is a chromosome lost so that, after fertilization with a normal gamete, the result is an embryo with 45, X?

I. An error in anaphase I

II. An error in anaphase II

III. An error of the first postfertilization mitosis

IV. An error in pairing

(1) I or II only

(2) II or IV only

(3) III or IV only

(4) I, II, III, or IV

161. Which of the following is true of aneuploidies in general?

(1) A monosomy is more frequent than a trisomy

(2) 45 X is the only known human live-born monosomy

(3) Some human aneuploidies have selective advantage in some environments

(4) Of all human aneuploidies, only Down syndrome is associated with mental retardation

162. A gene is considered to be non-Mendelian in its inheritance pattern if it seems to "violate" Mendel's laws. Which of the following would be considered Mendelian?

(1) A gene whose expression varies depending on the gender of the transmitting parent

(2) A gene derived solely from maternal inheritance

(3) A gene transmitted via the cytoplasm or cytoplasmic structures

(4) A gene transmitted to males from the maternal line and from fathers to daughters

163. Genomic imprinting is generally due to the addition of methyl ($-CH_3$) groups to C nucleotides in order to silence a given gene. If this depends on the sex of the parent who transmits the gene, which of the following must be true?

(1) Methylation of C is permanent in a gene

(2) Genes required for early development stages must not be imprinted

(3) Methylation of this kind must occur more in males than in females

(4) Methylation must be reversible in ovarian and testicular cells

164. Correns described that the inheritance of variegated color on the leaves of certain plants was determined by the maternal parent only. What phenomenon does this describe?

(1) Mitochondrial inheritance

(2) Chloroplast inheritance

(3) Genomic imprinting

(4) Infectious inheritance

165. Mitochondrial DNA is primarily involved in coding for proteins needed for electron transport. Therefore, in which body systems would you expect most mitochondrial gene mutations to be exhibited?

(1) The immune system and the blood

(2) The excretory and respiratory systems

(3) The skin and senses

(4) The nervous and muscular systems

166. A certain kind of snail can have a right-handed direction of shell coiling ((4) or left-handed coiling ((4) If direction of coiling is due to a protein deposited by the mother in the egg cytoplasm, then a *Dd* egg-producing snail and a *dd* sperm-producing snail will have offspring of which genotype(s) and phenotype(s)?

(1) 1/2 *Dd*:1/2 *dd*; all right coiling
(2) All *Dd*; all right coiling
(3) 1/2 *Dd*:1/2 *dd*; half right and half left coiling
(4) All *Dd*; all left coiling

167. Which of the following produces a Mendelian pattern of inheritance?

(1) Genomic imprinting
(2) A mitochondrial gene mutation
(3) A chloroplast gene mutation
(4) A trait acted upon by many genes

168. Suppose that a gene on human chromosome 18 can be imprinted in a given pattern in a female parent but not in a male parent. A couple in whom each maternal meiosis is followed by imprinting of this gene have children. What can we expect as a likely outcome?

(1) All sons but no daughters will bear their mother's imprinting pattern
(2) All daughters but no sons will bear their mother's imprinting pattern
(3) All sons and daughters will have a 50% chance of receiving the mother's imprinting pattern
(4) All the children will bear their mother's imprinting pattern but only daughters will then pass it down

169. The following is a map of four genes on a chromosome

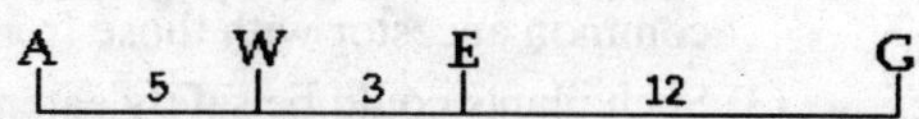

Between which two genes would you expect the highest frequency of recombination?

(1) *A* and *W* (2) *W* and *E*
(3) *E* and *G* (4) *A* and *G*

170. In a series of mapping experiments, the recombination frequencies for four different linked genes of *Drosophila* were determined as shown in Figure. What is the order of these genes on a chromosome map?

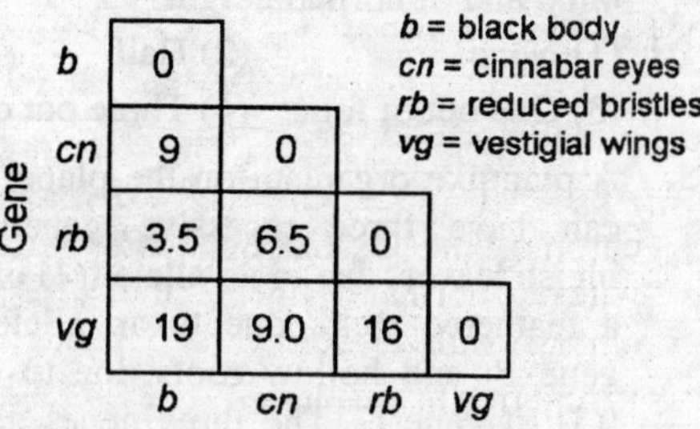

The numbers in the boxes are the recombination frequencies in between the genes (in percent).

(1) *rb-cn-vg-b* (2) *vg-b-rb-cn*
(3) *cn-rb-b-vg* (4) *b-rb-cn-vg*

171. The pedigree in figure shows the transmission of a trait in a particular family. Based on this pattern of transmission, the trait is most likely

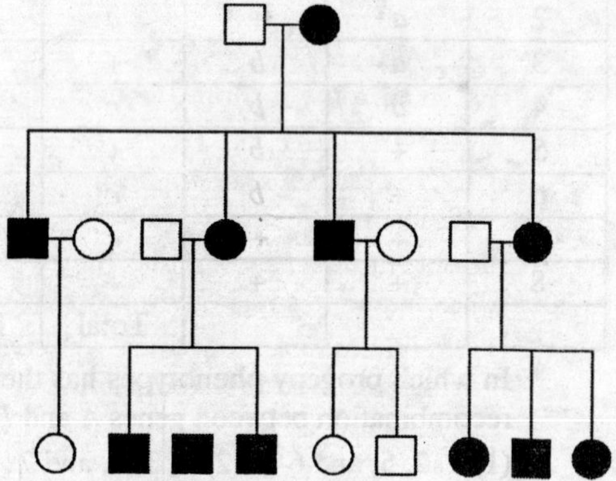

(1) mitochondrial
(2) autosomal recessive
(3) sex-linked dominant
(4) sex-linked recessive

172. A man who is an achondroplastic dwarf with normal vision marries a color-blind woman of normal height. The man's father was 6 feet tall, and both the woman's parents were of average height. Achondroplastic dwarfism is autosomal dominant, and red-green color blindness is X-linked recessive. What proportion of their sons would be color-blind and of normal height?

(1) None (2) Half
(3) One out of four (4) Three out of four

173. A plantlike organism on the planet Pandora can have three recessive genetic traits: bluish leaves, due to an allele ((*1*) of gene *A*; a feathered stem, due to an allele ((*2*) of gene *B*; and hollow roots due to an allele ((*3*) of gene *C*. The three genes are linked and recombine as follows: A geneticist did a testcross with an organism that had been found to be heterozygous for the three recessive traits and she was able to identify progeny of the following phenotypic distribution (+ = wild type) :

Phenotypes	Leaves	Stems	Roots	Number
1	*a*	+	+	14
2	*a*	+	*c*	0
3	*a*	*b*	+	32
4	*a*	*b*	*c*	440
5	+	*b*	+	0
6	+	*b*	*c*	16
7	+	+	*c*	28
8	+	+	+	470
			Total	1,000

In which progeny phenotypes has there been recombination between genes *A* and *B*?

(1) 1, 2, 5, and 6 (2) 1, 3, 6, and 7
(3) 2, 4, 5, and 8 (4) 2, 3, 5, and 7

174. The greatest distance among the three genes is between *a* and *c*. What does this mean?

(1) Gene *a* is closest to *b*
(2) Genes are in the order: *a–b–c*
(3) Gene *a* is not recombining with *c*
(4) Gene *a* is between *b* and *c*

175. Humanoids on the newly explored planet Brin (in a hypothetical galaxy in ~50 years from the present) have a gene structure similar to our own, but many very different plants and animals. One species of a small birdlike animal has an extremely variable tail length, which is a highly polymorphic trait. Geneticists have come to realize that there are eight separate genes for tail length per haploid genome, with each gene having two alleles. One allele for each gene (*a*, *b*, and so on) increases the length by 1 cm, whereas the other allele (a_2, b_2, and so on) increases it by 0.5 cm. One bird was analyzed & found to have the following genotype

$a_1a_1b_2b_2c_1c_2d_1d_2e_2e_2f_1f_2g_1g_1h_1h_2$

What is the length of its tail?

(1) 6 cm (2) 8 cm
(3) 12 cm. (4) 24 cm

176. One species of green plant, with frondlike leaves, a spine-coated stem, and purple cup-shaped flowers, is found to be self-pollinating. Which of the following is true of this species?

(1) The species must be haploid
(2) Its reproduction is asexual
(3) All members of the species have the same genotype
(4) Some of the seeds would have true-breeding traits

177. If the environmental parameters, such as temperature, humidity, atmosphere, sunlight, and so on, are mostly Earthlike, which of the following do you expect of its types of leaves, stems, and flowers?

(1) The genes for them would have originated on Earth
(2) Genes for these traits would have a common ancestor with those from Earth
(3) Such plants could be safely eaten by humans
(4) Phenotypes would be selected for or against by these environmental factors

178. Marfan syndrome in humans is caused by an abnormality of the connective tissue protein fibrillin. Patients are usually very tall and thin, with long spindly fingers, curvature of the spine, sometimes weakened arterial walls, and sometimes ocular problems, such as lens dislocation. Which of the following would you conclude about Marfan syndrome from this information?

(1) It is recessive
(2) It is dominant
(3) It has a late age of onset (> 60)
(4) It is pleiotropic

179. Sometimes the effect of gene interaction is that one gene masks (hides) the effect of another gene at a different locus, a phenomenon known as

(1) Epistasis
(2) Incomplete penetrance
(3) Variable expressivity
(4) Anticipation

180. Test in which individuals homozygous for different mutations are crossed, can be used to determine if the mutations occur at the same locus or at different loci is

(1) Test cross
(2) Reciprocal cross
(3) Complementation
(4) Back cross

181. Pattern baldness in humans is a sex-influenced trait that is autosomal dominant in males and recessive in females. Jack has a full head of hair. Jill also has a full head of hair, but her mother is bald (In women, pattern baldness is usually expressed as a thinning of the hair) If Jack and Jill marry, what proportion of their children are expected to be bald?

(1) 0 % (2) 25 %
(3) 50 % (4) 75 %

182. The ratio of observed double crossovers to expected double crossovers is termed as

(1) Coefficient of coincidence
(2) Coefficient of interference
(3) Coefficient of recombination
(4) Coefficient of linkage

183. Physical-mapping methods determine the physical locations of genes on chromosomes and which do not include

(1) deletion mapping
(2) somatic-cell hybridization
(3) in situ hybridization
(4) linkage mapping

184. Bacterial strains that are unable to synthesize the amino acid leucine will not grow on minimal medium but will grow on medium to which leucine has been added are termed as

(1) Protrophs (2) Heterotrophs
(3) Auxotrophs (4) Autotrophs

185. Plasmids that are capable of either freely replicating or integrating into the bacterial chromosomes are termed as

(1) Col plasmids
(2) Prophage
(3) Episomes
(4) Replicative plasmids

186. Bacterial viruses (bacteriophages) carry DNA from one bacterium to another. Once inside the bacterium, the newly introduced DNA may undergo recombination with the bacterial chromosome. Such an mode of parasexual reproduction is termed as

(1) Conjugation (2) Transformation
(3) Transduction (4) Lysogeny

187. During conjugation between an F′lac cell and an F - cell, the F plasmid is transferred to the F - cell, which means that any genes on the F plasmid, including those from the bacterial chromosome, may be transferred to F - recipient cells. This process is called

(1) Transformation
(2) Conjuagation
(3) Transduction
(4) Sex duction

188. Conjugation can be used to map bacterial genes by mixing ________ cells that differ in genotype and interrupting conjugation at regular intervals. The amount of time required for individual genes to be transferred from the one bacteria to the other indicates the relative positions of the genes on the bacterial chromosome

(1) F+ X F- (2) hfr X F-
(3) F' X F- (4) F' X hfr

189. Which of the following would refer to the production of a functional RNA by insertion or alteration of nucleotides in an RNA after it has been synthesized?

(1) Alternative RNA splicing
(2) RNA interference
(3) RNA editing
(4) miRNA splicing

190. Replicative transposons increases their copy number by replicating its copies and inserting into new position while keeping its copy at original place. The enzyme involved in this process are

(1) Transposase
(2) Transposase and Integrase
(3) Transposase and Resolvase
(4) Transposase, Resolvase and Integrase

191. A cell undergoing meiosis produces four daughter cells, two of which are aneuploids, while two are haploid. This can occur due to

(1) Non-dis-junction during first meiotic division only
(2) Non-dis-junctin during second meiotic division only
(3) Non-dis-junction during either first or second meiotic divisions
(4) Non-dis-junction during both first and second meiotic divisions

192. What proportion of offspring of a male suffering from an X-linked recessive trait and carrier female, will be affected?

(1) 1/2 sons and 1/2 daughters
(2) All sons and no daughters
(3) All daughters and no sons
(4) 3/4 daughters and 1/4 sons

193. If a protein coding gene has an 'amber mutation' and if you want to overcome the effect of this mutation on the protein, which of the second mutations will you introduce into another gene?

(1) Reading frame mutation
(2) Suppressor tRNA
(3) Stop codon mutation
(4) Start codon mutation

194. The nucleotide sequence 5' AUG AGC CUC AGG UAA 3' would be translated in the nucleus and mitochondria of a mammalian cell as

(1) Met-Ser-Leu-Arg in both the nucleus and mitochondria
(2) Met-Ser-Leu in both the nucleus and mitochondria
(3) Met-Ser-Leu in the nucleus and Met-Ser-Leu-Arg in the mitochondria
(4) Met-Ser-Leu in the mitochondria and Met-Ser-Leu-Arg in the nucleus

195. Tallness (*T*) in snapdragons is dominant to dwarfness (*t*), while red (*R*) flower color is dominant to white (*r*) The heterozygous condition results in pink (*Rr*) flower color. If snapdragons are heterozygous for height as well as for flower color, a mating between them will result in what ratio?

(1) 9:3:3:1 (2) 6:3:3:2:1:1
(3) 1:2:1 (4) 27:9:9:9:3:3:3:1

196. You set up the following dihybrid mapping cross in fruit flies

P e^+e roro cqcq x ee ro+ro$^+$ cq$^+$cq+. After backcrossing F1 males to eerorocqcq females, you get the following results: gray body, rough eyes, claret eyes 576; ebony body, smooth, red eyes 564. How can you explain this result?

(1) The genes are not linked
(2) Crossing over does not occur in male Drosophila
(3) These genes are X linked
(4) These genes are autosomal and showing independent assortment

197. The DNA in a eukaryotic chromosome is many times longer than the chromosome itself. How many times longer is the DNA in a human chromosome than the length of the chromosome?

(1) 10 x (2) 100 x
(3) 10,000 x (4) 1,000 x

198. Which of the following statements is not correct?

(1) The enzyme that retroviruses and retrotransposons use to synthesise DNA is known as reverse transcriptase
(2) Transposons occur only in eukaryotes
(3) Transposons are responsible for most of the repetitive DNA in human cells
(4) Transposons have no known direct function but they may have benefits in evolution

199. A large part of the human genome has in the past been regarded as 'junk' DNA. Which of the following statements about 'junk' DNA is correct according to our current understanding?

(1) About 1.6% of the human genome has no known function and has been described as 'junk' DNA
(2) Bacteria have no junk DNA
(3) There are no genes in junk DNA
(4) Micro RNA genes do not code for proteins

200. Which of the following statements about split genes is correct?

(1) The coding sequence in human genes is interrupted by spacer sequences called exons
(2) Human genes can be split by up to 500 spacer sequences
(3) Most genes in bacteria are split
(4) The biggest human gene is bigger than the *E. coli* genome

ANSWER SHEET

Unit 8: Genetics

1. 1	**2.** 4	**3.** 4	**4.** 3	**5.** 1	**6.** 1	**7.** 1	**8.** 2	**9.** 2	**10.** 3
11. 3	**12.** 2	**13.** 1	**14.** 4	**15.** 3	**16.** 2	**17.** 1	**18.** 2	**19.** 2	**20.** 4
21. 1	**22.** 3	**23.** 3	**24.** 2	**25.** 1	**26.** 3	**27.** 3	**28.** 2	**29.** 4	**30.** 3
31. 4	**32.** 2	**33.** 4	**34.** 1	**35.** 4	**36.** 2	**37.** 3	**38.** 2	**39.** 3	**40.** 2
41. 3	**42.** 2	**43.** 2	**44.** 2	**45.** 3	**46.** 2	**47.** 1	**48.** 3	**49.** 2	**50.** 3
51. 3	**52.** 2	**53.** 4	**54.** 1	**55.** 4	**56.** 1	**57.** 1	**58.** 2	**59.** 3	**60.** 1
61. 4	**62.** 1	**63.** 3	**64.** 4	**65.** 2	**66.** 1	**67.** 2	**68.** 3	**69.** 4	**70.** 4
71. 3	**72.** 4	**73.** 2	**74.** 2	**75.** 4	**76.** 2	**77.** 1	**78.** 4	**79.** 4	**80.** 3
81. 3	**82.** 4	**83.** 4	**84.** 4	**85.** 3	**86.** 4	**87.** 4	**88.** 4	**89.** 3	**90.** 3
91. 4	**92.** 2	**93.** 3	**94.** 3	**95.** 4	**96.** 4	**97.** 3	**98.** 4	**99.** 4	**100.** 1
101. 2	**102.** 3	**103.** 3	**104.** 3	**105.** 3	**106.** 2	**107.** 3	**108.** 3	**109.** 2	**110.** 3
111. 2	**112.** 4	**113.** 1	**114.** 2	**115.** 4	**116.** 4	**117.** 3	**118.** 4	**119.** 2	**120.** 2
121. 2	**122.** 4	**123.** 4	**124.** 4	**125.** 3	**126.** 4	**127.** 4	**128.** 4	**129.** 2	**130.** 1

131. 2	**132.** 4	**133.** 4	**134.** 1	**135.** 1	**136.** 1	**137.** 1	**138.** 1	**139.** 4	**140.** 2
141. 4	**142.** 3	**143.** 1	**144.** 4	**145.** 3	**146.** 4	**147.** 4	**148.** 3	**149.** 1	**150.** 1
151. 2	**152.** 3	**153.** 4	**154.** 4	**155.** 4	**156.** 4	**157.** 4	**158.** 3	**159.** 1	**160.** 4
161. 2	**162.** 4	**163.** 4	**164.** 2	**165.** 4	**166.** 1	**167.** 4	**168.** 4	**169.** 4	**170.** 4
171. 1	**172.** 2	**173.** 1	**174.** 4	**175.** 3	**176.** 1	**177.** 2	**178.** 4	**179.** 1	**180.** 3
181. 2	**182.** 1	**183.** 4	**184.** 3	**185.** 3	**186.** 3	**187.** 4	**188.** 2	**189.** 3	**190.** 3
191. 2	**192.** 1	**193.** 2	**194.** 4	**195.** 2	**196.** 2	**197.** 3	**198.** 2	**199.** 4	**200.** 2

UNIT-9

DIVERSITY AMONG LIFE FORMS

1. The origin of tetradotoxin in mollusks is
(1) endogeneous
(2) exogeneous
(3) symbiotic microorganisms
(4) all the above

2. Which of the following commercially available cancer drug is obtained from *Catherenthus roseus*?
(1) Bleomycin (2) AraC
(3) Cisplatin (4) Vinblastin

3. How many species of animals are there?
(1) 5 million (2) 60 million
(3) 30 million
(4) somewhere between (1) and (3)

4. Rank the following animal groups from greatest to least (left to right) in the number of described species: Mammalia (mammals), Aves (Birds), Mollusca (clams, snails, etc), and Insecta
(1) Mollusca, Aves, Insecta, Mammalia
(2) Insecta, Mollusca, Aves, Mammalia
(3) Insecta, Aves, Mammalia, Mollusca
(4) Mammalia, Aves, Insecta, Mollusca

5. Heparin is a
(1) lipopolysaccharide
(2) glycated lipopolysaccharide
(3) sulphated polysaccharide
(4) sulphated lipopolysaccharide

6. Eutrophication in coastal water results in the following phenomenon
(1) red tide (2) diurnal tide
(3) mixed tide (4) neap tide

7. Barophiles are capable of growth up to
(1) 100 to 200 atm (2) 700 to 800 atm
(3) 300 to 600 atm (4) 1 to 100 atm

8. The most abundant group of organisms inhabiting hydrothermal vents are
(1) Sulphate reducing bacteria
(2) Chemoautotrophic sulphur bacteria
(3) Sulphur oxidizing chemolithotrophs
(4) Nitrifiers

9. Deep sea hydrothermal vents are habitats where the primary producers are
(1) Organotrophic bacteria
(2) Chemolithotrophic bacteria
(3) Chemoorganotrophs
(4) Methylotrophs

10. Giant tube worms receive their nutrition
(1) through unusual haemoglobins which bind H_2S as well as O_2, transport to the trophosome and release to bacterial symbionts
(2) from methanotrophic symbionts living in symbiotic association
(3) from thermophilic prokaryotes which reside in smoker chimneys
(4) through normal haemoglobin

11. The lux gene from *Vibrio fischeri* has been used to make glowing tobacco plants. This gene is involved in
(1) Bioluminescence
(2) Photosynthesis
(3) Phosphorescence
(4) Fluorescence

12. Hydrocolloids extracted from seaweeds have attained commercial significance specially as
(1) Detergents
(2) Biofuels
(3) Food additives
(4) Laboratory chemicals

13. Which theory proposes that the number of species at a site is determined by the balancing of rates of immigration of species to that site with the local extinction of species already present?
(1) Stability-Time Hypothesis
(2) Pleistocene Forest Refugia Hypothesis
(3) Equilibrium Theory of Island Biogeography
(4) Periodic Extinction Hypothesis

14. Nutraceuticals available in the market from marine sources are largely from
(1) bacteria (2) fungi
(3) diatoms (4) macro algae

15. ___________ are being cultured and harvested as a source of biofuels
(1) anaerobic bacteria
(2) luminescent bacteria
(3) non-methanogenic bacteria
(4) sea weeds

16. *Epulopiscium fishelsoni* whose length ranges from 200-500 μm belongs to the microbial group
(1) Microalgae (2) Fungi
(3) Protozoa (4) Bacteria

17. Agar-agar is a polymer of
(1) Glucose (2) Sulphated sugar
(3) Pectin (4) Protein

18. The most uncommon characteristic of marine microorganisms is
(1) They require low nutrients
(2) They are slow growing
(3) 95% are gram negative
(4) They do not exhibit pleomorphism

19. Cleistogamy occurs in
(1) Rice (2) Barley
(3) Maize (4) Pearl millet

20. An example of a motile microorganism is
(1) Brucella (2) Lactobacillus
(3) Shigella (4) Pseudomonas

21. Archea is considered as a separate group from bacteria and eukaryotes, based on
(1) genome sequence
(2) 16S rRNA gene sequence
(3) 23S rRNA gene sequence
(4) EFTu sequence

22. Homology modelling can be used to predict the 3D structure of only
(1) paralogs (2) orthologs
(3) xenologs (4) homologs

23. The autochthonous probiotic bacteria used in aquaculture are isolated from
(1) microbial flora associated with seaweeds
(2) the gastrointestinal tract of aquaculture animals
(3) the sediments, especially from the intertidal region
(4) the microbial flora associated with mangrove plants

24. Foraminiferans and radiolarians are
(1) non-photosynthetic protists
(2) photosynthetic protists
(3) microscopic bacteria
(4) biogenic sediments

25. With regard to ocean waters, which one of the following is not a depth-wise division?
(1) Epipelagic (2) Mesopelagic
(3) Abyssopelgic (4) Neritopelagic

26. Organisms which can be used for producing silicon like component for use in the field of nanotechnology are
(1) Diatoms (2) Rhabdovirus
(3) *Gracilaria corticata*
(4) *Sargassum tennerimum*

27. Marine bacteria that can grow over a wide range of temperature are referred to as
(1) thermophiles (2) thermotolerants
(3) stenothermals (4) eurythermals

28. In polar oceans, the main factor affecting the phytoplankton growth is
(1) depletion of nutrients in water
(2) vertical migration of nutrients
(3) shortage of sunlight
(4) depletion of phosphates

29. Organisms which reproduce in sea water and live as adults in fresh water are called
(1) catadromous (2) anadromous
(3) migratory (4) epipelagic

30. Periods of mass extinction as evidenced in the fossil record may have been caused by
(1) Periodic asteroid or comet impacts leading to a nuclear winter like scenario
(2) Plate tectonics and the fusion of continents reducing the availability and provinciality (isolation) of suitable habitat
(3) Episodes of widespread disease
(4) (1) and (2)

31. Organisms that are plankton in the juvenile stage, but nekton or benthos in the adult stage are called
(1) meroplankton (2) macroplankton
(3) holoplankton (4) picoplankton

32. A giant bacterium measuring up to 0.75 mm and referred to as the "Sulfur Pearl" is
(1) *Thioploca* sp
(2) *Epulopiscium fishelsoni*
(3) *Thiomargarita nambiensis*
(4) *Beggiatoa* sp

33. During protoplast isolation from *Gracilaria corticata*, which one of the following is added as an osmoticum?
(1) glucose (2) mannose
(3) mannitol (4) fructose

34. The first group of organisms that colonize the hydrothermal vents are
(1) tube worms
(2) chemolithotrophic bacteria
(3) chemoautotrophic sulfur bacteria
(4) crabs

35. What are zooxanthallae?
(1) Deep sea dwelling brightly pigmented fish
(2) Algae living in corals
(3) A species of crab
(4) *Xanthomonas*-infected zooplankton

36. Which of the following statements is not true for giant tube worms observed at hydrothermal vents?
(1) Digestive tract of tube worms produces combination of thermostable proteases and polysaccharases
(2) The tube worms obtained their nutrients from symbiotic chemolithotropic bacteria
(3) The tube worms have the fastest growth rate compared to any known marine invertebrates
(4) The hemoglobin present in tube worm binds both H_2S and O_2

37. The bacterial pathogen which is most detrimental to shrimp aquaculture is
(1) *Vibrio* sp
(2) *Pseudomonas* sp
(3) *Flavobacterium* sp
(4) *Micrococcus* sp

38. Which of the following is not an example of the evolution of a key innovation (a new trait) allowing occupation of a new adaptive zone (habitat or lifestyle)
(1) Evolution of donut eating by police officers
(2) Invasion of the land by plants & animals
(3) Evolution of wings in insects, birds, and bats
(4) The evolution of plant –eating among insects

39. Which of the following traits do not help distinguish animals from other forms of life?
(1) The presence of DNA in the cell nucleus
(2) The presence of two types of tissues: nervous tissues for impulse conduction and muscle tissue for movement
(3) Cell walls that have structural support
(4) 2 and 3

40. Arrange the following taxonomic categories in their hierarchical order from highest to lowest (left to right): Genus, Family, Class, Order, Phylum
(1) Phylum, Order, Class, Genus, Family
(2) Class, Phylum, Order, Family, Genus
(3) Order, Phylum, Class, Family, Genus
(4) Phylum, Class, Order, Family, Genus

41. Which one of the following protozoans is transmitted by ingestion of tick?
(1) *Haemoproteus columbae*
(2) *Ehrlichia canis*
(3) *Hepatozoon canis*
(4) *Histomonas meleagridis*

42. Bovine group A rotavirus contains
(1) ss RNA (2) ds RNA
(3) ss DNA (4) ds DNA

43. Among the following diagrams which taxon represented by the shaded region represents a monophyletic group?

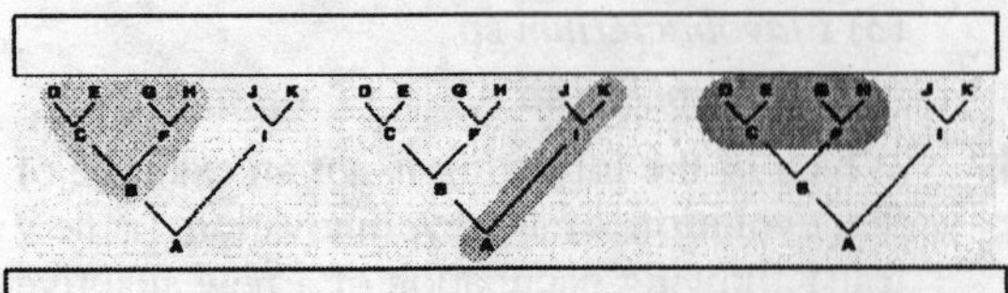

(1) Taxon A and B (2) Taxon B and C
(3) Taxon C only (4) Taxon A only

44. Scrapie is caused by
(1) Fungal protein
(2) Bacterial protein
(3) Plant lipoprotein
(4) Prion

45. Human intestinal flora cannot digest
(1) Cellulose (2) Lignin
(3) Pectin (4) Starch

46. *Xenopsylla cheopis* is the vector for
(1) Indian tick typus (2) Epidemic typus
(3) Plague (4) Kala azar

47. In animal kingdom, the group amniota includes
(1) birds and reptiles
(2) birds and mammals
(3) reptiles and mammals
(4) reptiles, birds and mammals

48. In evolutionary biology, 'biological species' concept is based on
(1) geographical isolation and karyotypic difference
(2) morphological differentiation only
(3) ecological differentiation coupled with morphological differentiation
(4) primarily reproductive isolation

49. Higher version of BLOSUM can be used to detect
(1) Closely related sequences
(2) Distantly related sequences
(3) Unrelated sequences
(4) Partially related sequences

50. TBlastX matches a DNA query sequence, translated into all six reading frames, against a DNA database with
(1) No gaps allowed
(2) Gaps allowed
(3) Gaps depending on the input sequence
(4) Gaps depending on the database

51. Changing which of the following Blast parameters would tend to yield fewer search results?
(1) Turning off the low complexity filter
(2) Changing the expected value from 1-10
(3) Raising the threshold value
(4) Changing the scoring matrix from PAM30 to PAM70

52. Cameleons are
(1) Coral reef associated fish that change colour depending on the color of corals
(2) Chimeric protein consisting of calmodulin and mutant GFP
(3) Fluorescent proteins isolated from *Demospongiae* that depict different color fluorescence depending on the presence of metal ions
(4) Deep sea spotted Rat fish which puffs up mimicking a puffer fish to ward off predators

53. Scombrotoxic fish poisoning is due to
(1) histamine and histamine-like products generated in fin-fishes due to contamination by members of *Enterobacteriaceae* family
(2) toxins produced by *Noctiluca* that resides commonly in fishes
(3) toxin generated by *Shigella flexneri*
(4) toxin generated by *Clostridium botulinum*

54. Beta-glucan, a structural element in fungal cell walls has been used in fish aquaculture
(1) as an immunostimulant
(2) to stimulate natural bacterial flora for bioremediation of waste matter
(3) for inhibiting bacterial cell wall synthesis and thus preventing their growth
(4) as a bioflocculant for precipitaton of waste in aquaculture ponds

55. "Crabyon" an anti-smell, non-allergic soft fibre used in sports- wear is a blend of
(1) Crab shells and rayon
(2) Modified chitosans
(3) Crustacean shells and calcium
(4) Chitin and Cellulose

56. Giant bacteria are normally associated with sulphur compounds however the only species so far reported to store calcite in their cells is
(1) *Thiomargarita* sp
(2) *Epulopiscium* sp
(3) *Achromatium* sp
(4) *Thioploca* sp

57. Each specimen of a type series other than the haplotype
(1) Neotype (2) Syntype
(3) Paratype (4) Holotype

58. Ancestral shared traits
(1) Apomorphic traits
(2) Sympatric
(3) Plesiomorphic traits
(4) Synapomorphic traits

59. The naming of taxonomic groups is based on specimen submitted by author himself is
(1) Holotype (2) Lectotype
(4) Neotype (4) Nomenclature

60. Sister species evolve in segregated habitats across a narrow contact zone - little mixing in spite of proximity
(1) Paratype (2) Sympatric
(3) Topotype (4) Parapatric

61. Not all descendents of a common ancestor are included
(1) Paraphyletic group
(2) Polyphyletic group
(3) Monophyletic group
(4) Paralectotype

62. Which of the following is correct explanation for paraphyltetic group?
(1) It contains various unrelated organisms
(2) It includes most recent ancestors
(3) it includes all representative of clade but ancestors are not included
(4) It includes all common ancestors but not all of its decedents

63. Which of the following microbe have been extensively utilized in organic farming?
(1) *Rhyzopus*
(2) *Mucor*
(3) *Fusarium*
(4) *Trichorderma viride*

64. Biodiesal is obtained from
(1) *Jatropa curcus*
(2) *Calotropis procera*
(3) *Prosopis cinereria*
(4) *Catharanthus roseus*

65. During evolution if a single species give rise to many descendents of different phylogenetic taxons, it is termed as
(1) clade (2) cline
(3) cluster (4) clone

66. Biological species concept cannot be applied on
(1) Parthenogenic species
(2) Sympatric species
(3) Species producing viable hybrids
(4) Species of aquatic ecosystem

67. Increase in amount of the following is not a consequence of sewage effluents in river system
(1) enhanced biodiversity
(2) phosphate level
(3) biological Oxygen Demand
(4) cynaobacterial density

68. Chordate pharyngeal slits appear to have functioned first as
(1) the digestive system's opening
(2) suspension-feeding devices
(3) components of the jaw
(4) gill slits for respiration

69. Which of the following statements would be least acceptable to most zoologists?
(1) The extant lancelets are contemporaries, not ancestors, of vertebrates
(2) The first fossils resembling lancelets appeared in the fossil record around 530 million years ago
(3) Recent work in molecular systematics supports the hypothesis that lancelets are the most recent common ancestor of all vertebrates
(4) The extant lancelets are the immediate ancestors of the fishes

70. Which extant chordates are postulated to be *most* like the earliest chordates in appearance?
(1) lancelets. (2) adult tunicates
(3) amphibians (4) reptiles

71. What do all craniates have that earlier chordates did not have?
(1) brain
(2) vertebrae
(3) cartilaginous pipe surrounding notochord
(4) partial or complete skull

72. Lampreys differ from hagfishes in
(1) lacking jaws
(2) having a cranium
(3) having pharyngeal clefts that develop into pharyngeal slits
(4) having a notochord that is surrounded by a tube of cartilage

73. The feeding mode of the extinct conodonts was
(1) herbivory. (2) suspension feeding
(3) predation (4) filter feeding

74. The earliest known mineralized structures in vertebrates are associated with which function?
(1) reproduction (2) feeding
(3) locomotion (4) defense

75. The lamprey species whose larvae live in freshwater streams, but whose adults live most of their lives in seawater, are similar in this respect to certain species of
(1) chondrichthyans
(2) actinopterygians
(3) lungfishes
(4) coelacanths

76. Jaws first occurred in which extant group of fishes?
(1) Lampreys
(2) Chondrichthyans
(3) Ray-finned fishes
(4) Lungfishes

77. Which of these might have been observed in the common ancestor of chondrichthyans and osteichthyans?
(1) a mineralized, bony skeleton
(2) opercula
(3) bony fin rays
(4) a spiral valve intestine

78. To which of the following are the scales of chondrichthyans most closely related in a structural sense?
(1) Osteichthyan scales
(2) Reptilian scales
(3) Mammalian scales
(4) Chondrichthyan teeth

79. Which of these statements accurately describes a similarity between sharks and ray-finned fishes?
(1) The skin is typically covered by flattened bony scales
(2) They are equally able to exchange gases with the environment while stationary
(3) They are highly maneuverable due to their flexibility
(4) They have a lateral line that is sensitive to changes in water pressure

80. Which group's members have had both lungs and gills during their adult lives?
(1) sharks, skates, and rays
(2) lungfishes
(3) lancelets
(4) amphibians

81. There is evidence that ray-finned fishes evolved
(1) in response to a crisis that wiped out the chondrichthyans
(2) directly from lampreys and hagfish
(3) early in the Cambrian period
(4) the swim bladder from a lung

82. The swim bladder of ray-finned fishes
(1) was probably modified from simple lungs of chondrichthyans
(2) developed into lungs in saltwater fishes
(3) first appeared in sharks
(4) provides for regulation of buoyancy

83. Which of the following belongs to the lobe-fin clade?
(1) chondrichthyans (2) ray-finned fishes
(3) lampreys (4) tetrapods

84. A trend first observed in the evolution of the earliest tetrapods was
(1) the appearance of jaws
(2) the appearance of bony vertebrae
(3) feet with digits
(4) the mineralization of the endoskeleton

85. What should be true of fossils of the earliest tetrapods?
(1) They should show evidence of internal fertilization
(2) They should show evidence of having produced shelled eggs
(3) They should indicate limited adaptation to life on land
(4) They should be transitional forms with the fossils of chondrichthyans that lived at the same time

86. Which of these are amniotes?
(1) amphibians
(2) fishes
(3) egg-laying mammals
(4) eichinodermates

87. Why is the amniotic egg considered an important evolutionary breakthrough?
(1) It has a shell that increases gas exchange
(2) It allows deposition of eggs in a terrestrial environment
(3) It prolongs embryonic development
(4) It provides insulation to conserve heat

88. Which era is known as the "age of reptiles"?
(1) Cenozoic (2) Mesozoic
(3) Paleozoic (4) Devonian

89. Which of these characteristics added most to vertebrate success in relatively dry environments?
(1) the shelled, amniotic egg
(2) the ability to maintain a constant body temperature
(3) two pairs of appendages
(4) bony scales

90. Which of the following is characteristic of most extant reptiles and most extant mammals?
(1) ectothermy (2) diaphragm
(3) shelled eggs (4) keratinized skin

91. At the end of which era did most dinosaurs and pterosaurs become extinct?
(1) Cretaceous. (2) Permian
(3) Devonian (4) Ordovician

92. Which of the following are the only extant animals that descended directly from dinosaurs?
(1) lizards (2) crocodiles
(3) snakes (4) birds

93. Examination of the fossils of *Archaeopteryx* reveals that, in common with extant birds, it had
(1) a long tail containing vertebrae
(2) feathers
(3) teeth
(4) Three of the options listed are correct

94. Why is the discovery of the fossil *Archaeopteryx* significant? It supports the
(1) phylogenetic relatedness of birds and reptiles
(2) contention that birds are much older than we originally thought
(3) claim that some dinosaurs had feathers well before birds had evolved
(4) idea that the first birds were ratites

95. Among extant vertebrates, a sheet of muscle called the diaphragm is not found
(1) birds (2) monotremes
(3) marsupials. (4) placentals

96. Which of these would a paleontologist b most likely to do in order to determin whether a fossil represents a reptile or mammal?
(1) Look for the presence of milk-producing glands
(2) Look for the mammalian characteristics of a four-chambered heart and a diaphragm
(3) Examine the teeth
(4) Use molecular analysis to look for the protein keratin

97. Which of the following represents th strongest evidence that two of the thre middle ear bones of mammals ar homologous to certain reptilian jawbones?
(1) They are similar in size to the reptilian jawbones
(2) They are similar in shape to the reptilian jawbones
(3) The mammalian jaw has fewer bones than does the reptilian jaw
(4) These bones can be observed to move from the developing jaw to the developing middle ear in mammalian embryos

98. Female birds lay their eggs, thereby facilitating flight by reducing weight. Which "strategy" seems most likely for female bats to use to achieve the same goal?
(1) lay shelled eggs
(2) limit litters to a single embryo
(3) refrain from flying throughout pregnancy (about 6 weeks long)
(4) give birth to underdeveloped young, and subsequently carry them in a pouch that has teats

99. In which vertebrates is fertilization exclusively internal?
(1) chondrichthyans, osteichthyans, and mammals
(2) amphibians, mammals, and reptiles
(3) chondrichthyans, osteichthyans, and reptiles
(4) reptiles and mammals

00. Their scales most closely resemble teeth in both structure and origin
(1) amphibians (2) nonbird reptiles
(3) chondrichthyans (4) mammals

01. Internal fertilization, leathery amniotic egg, and skin that resists drying are characteristics of
(1) amphibians. (2) nonbird reptiles
(3) chondrichthyans. (4) mammals

02. Which of the following are the most abundant and diverse of the extant vertebrates?
(1) ray-finned fishes. (2) birds
(3) amphibians (4) nonbird reptiles

103. What is the single unique characteristic that distinguishes extant birds from other extant vertebrates?
(1) endothermy (2) feathers
(3) an amniotic egg (4) flight

104. Which of these traits is most strongly associated with the adoption of bipedalism?
(1) fingerprints
(2) enhanced depth perception
(3) shortened hind limbs
(4) repositioning of foramen magnum

105. Which of the following statements about human evolution is correct?
(1) Modern humans are the only human species to have evolved on Earth
(2) Human ancestors were virtually identical to extant chimpanzees
(3) Human evolution has occurred within an unbranched lineage
(4) Fossil evidence indicates that early anthropoids were arboreal and cat-sized

106. Which of the following are considered apes?
(1) Lorises
(2) New World monkeys
(3) Old World monkeys
(4) Orangutans

107. Which of these species was the first to have been adapted for long-distance bipedalism?
(1) *Australopithecus garhi*
(2) *Homo erectus*
(3) *H. ergaster*
(4) *H. habilis*

108. Which of these species was apparently the first to craft stone tools?
(1) *Australopithecus garhi*
(2) *H. erectus*
(3) *H. ergaster*
(4) *H. habilis*

109. Which of these species was the first to have some members migrate out of Africa?
(1) *Australopithecus garhi*
(2) *H. erectus*
(3) *H. ergaster*
(4) *H. habilis*

110. Which of these species is currently thought to have coexisted (at the same time and places) with *H. neanderthalensis*?
(1) *H. erectus* (2) *H. ergaster*
(3) *H. habilis* (4) *H. sapiens*

111. Which of these species had members who moved out of Africa?
(1) *H. erectus*
(2) Both *H. sapiens* and *H. erectus*
(3) *H. habilis*
(4) *H. sapiens*

112. With which of the following statements would a biologist be most inclined to agree?
(1) Humans and apes represent divergent lines of evolution from a common ancestor
(2) Humans evolved directly from Old World monkeys
(3) Humans represent the pinnacle of evolution and have escaped from being affected by natural selection
(4) Humans evolved from chimpanzees

113. Which of these statements about human evolution is correct?
(1) The ancestors of *Homo sapiens* were chimpanzees
(2) Human evolution has proceeded in an orderly fashion from an ancestral anthropoid to *Homo sapiens*
(3) The evolution of upright posture and enlarged brain occurred simultaneously
(4) Different species of the genus *Homo* have coexisted at various times throughout hominin evolution

114. The oldest fossil remains of *Homo sapiens* found so far date from about
(1) 6 million years ago
(2) 1.6 million years ago
(3) 195,000 years ago
(4) 60,000 years ago

115. Which of the following statements is correct in regard to *Homo erectus*?
(1) Their fossils are not limited to Africa
(2) On average, *H. erectus* had a smaller brain than *H. habilis*
(3) *H. erectus* had a level of sexual dimorphism less than that of modern humans
(4) *H. erectus* was not known to use tools

116. Which of the following is the most inclusive (most general) group, all of whose members have foramina magna centrally positioned in the base of the cranium?
(1) Hominoids (2) *Homo*
(3) Anthropoids (4) Hominins

117. Which of the following is the most inclusive (most general) group, all of whose members have fully opposable thumbs?
(1) Apes (2) *Homo*
(3) Anthropoids. (4) Hominins

118. Which of the following is the most specific group that includes both the Old World monkeys and the New World monkeys?
(1) apes (2) *Homo*
(3) anthropoids. (4) hominins

119. At least one of these has been found in all species of eumetazoan animals studied thus far
(1) *Hox* (2) *Dlx*
(3) *Otx* (4) *FOXP2*

120. Which of the following is a cluster of genes coding for transcription factors involved in the evolution of innovations in early vertebrate nervous systems and vertebrae?
(1) *Hox* (2) *Dlx*
(3) *Otx* (4) *FOXP2*

121. Vertebrates and tunicates share
(1) jaws adapted for feeding
(2) a high degree of cephalization
(3) the formation of structures from the neural crest
(4) a notochord and a dorsal, hollow nerve cord

122. Some animals that lived 530 million years ago resembled lancelets but had a brain and a skull. These animals may represent
(1) the first chordates
(2) a "missing link" between urochordates and cephalochordates
(3) early craniates
(4) marsupials

123. Which of the following could be considered the most recent common ancestor of living tetrapods?
(1) A sturdy-finned, shallow-water lobe-fin whose appendages had skeletal supports similar to those of terrestrial vertebrates
(2) An armored, jawed placoderm with two pairs of appendages
(3) An early ray-finned fish that developed bony skeletal supports in its paired fins
(4) A salamander that had legs supported by a bony skeleton but moved with the side-to-side bending typical of fishes

124. Unlike eutherians, both monotremes and marsupials
(1) lack nipples
(2) have some embryonic development outside the uterus
(3) lay eggs
(4) are found in Australia and Africa

125. Which clade does not include humans?
(1) synapsids (2) lobe-fins
(3) diapsids. (4) craniates

126. The legless condition that is observed in several groups of extant reptiles is the result of
(1) their common ancestor having been legless
(2) a shared adaptation to an arboreal (living in trees) lifestyle
(3) several instances of the legless condition arising independently of each other
(4) individual lizards adapting to a fossorial (living in burrows) lifestyle during their lifetimes

127. The various taxonomic levels (namely, genera, classes, etc.) of the hierarchical classification system differ from each other on the basis of
(1) how widely the organisms assigned to each are distributed throughout the environment
(2) their inclusiveness
(3) the relative genome sizes of the organisms assigned to each
(4) morphological characters that are applicable to all organisms

128. Linnaeus was a "fixist" who believed that species remained fixed in the form in which they had been created. Linnaeus would have been uncomfortable with
(1) classifying organisms using the morpho-species concept
(2) the scientific discipline known as taxonomy
(3) phylogenies
(4) nested, ever-more inclusive categories of organisms

129. Which of the following is (are) problematic when the goal is to construct phylogenies that accurately reflect evolutionary history?
(1) polyphyletic taxa (2) paraphyletic taxa
(3) monophyletic taxa (4) 1 and 2 both

130. Which individual would make the *worst* systematist? One who is uncomfortable with the
(1) Linnaean system of classification
(2) notion of hypothetical phylogenies
(3) PhyloCode method of classification
(4) notion of permanent polytomies

131. The term homoplasy is most applicable to which of the following features?
(1) the legless condition found in various lineages of extant lizards
(2) the five-digit condition of human hands and bat wings
(3) the β hemoglobin genes of mice and of humans
(4) the fur that covers Australian moles and North American moles

132. If, someday, an archaean cell is discovered whose rRNA sequence is more similar to that of humans than the sequence of mouse rRNA is to that of humans, the best explanation for this apparent discrepancy would be
(1) homology
(2) homoplasy
(3) common ancestry
(4) retro-evolution by humans

133. The best classification system is that which most closely
(1) unites organisms that possess similar morphologies
(2) conforms to traditional, Linnaean taxonomic practices
(3) reflects evolutionary history
(4) reflects the basic separation of prokaryotes from eukaryotes

134. Which of the following pairs are the best examples of homologous structures?
(1) bones in the bat wing and bones in the human forelimb
(2) owl wing and hornet wing
(3) bat wing and bird wing
(4) eyelessness in the Australian mole and eyelessness in the North American mole

135. Some molecular data place the giant panda in the bear family (Ursidae) but place the lesser panda in the raccoon family (Procyonidae) Consequently, the morphological similarities of these two species are probably due to
(1) inheritance of acquired characteristics
(2) sexual selection
(3) inheritance of shared derived characters
(4) possession of analogous structures

136. The importance of computers and of computer software to modern cladistics is most closely linked to advances in
(1) light microscopy
(2) radiometric dating
(3) fossil discovery techniques
(4) molecular genetics

137. Which mutation should least require realignment of homologous regions of a gene that is common to several related species?
(1) three-base insertion
(2) one-base substitution
(3) four-base insertion
(4) one-base deletion

138. Which of the following is true of all horizontally oriented phylogenetic trees, where time advances to the right?
(1) Each branch point represents a point in absolute time
(2) Organisms represented at the base of such trees are descendants of those represented at higher levels
(3) The fewer branch points that occur between two taxa, the more divergent their DNA sequences should be
(4) The common ancestor represented by the rightmost branch point existed more recently in time than the common ancestors represented at branch points located to the left

139. When using a cladistic approach to systematics, which of the following is considered most important for classification?
(1) shared primitive characters
(2) analogous primitive characters
(3) shared derived characters
(4) the number of homoplasies

140. Cladograms (a type of phylogenetic tree) constructed from evidence from molecular systematics are based on similarities in
(1) morphology
(2) the pattern of embryological development
(3) biochemical pathways
(4) mutations to homologous genes

141. There is some evidence that reptiles called cynodonts may have had whisker-like hairs around their mouths. If true, then what can be properly said of hair?
(1) It is a shared derived character of mammals, even if cynodonts continue to be classified as reptiles
(2) It is a shared derived character of the amniote clade, and not of the mammal clade
(3) It is a shared ancestral character of the amniote clade, but only if cynodonts are reclassified as mammals
(4) It is a shared derived character of the mammals, but only if cynodonts are reclassified as mammals

142. A researcher wants to determine the genetic relatedness of several breeds of dog (*Canis lupus familiaris*) The researcher should compare homologous sequences of this type of biochemical ________ which can be described as ________

(1) fatty acids; highly conserved
(2) lipids; poorly conserved
(3) proteins; highly conserved
(4) nucleic acids, poorly conserved

143. Concerning growth in genome size over evolutionary time, which of these is *least* associated with the others?

(1) orthologous genes
(2) gene duplications
(3) paralogous genes
(4) gene families

144. Nucleic acid sequences that undergo few changes over the course of evolutionary time are said to be conserved. Conserved sequences of nucleic acids

(1) are found in the most crucial portions of proteins
(2) include all mitochondrial DNA
(3) are abundant in ribosomes
(4) are proportionately more common in eukaryotic introns than in eukaryotic exons

145. Species that are not closely related and that do not share many anatomical similarities can still be placed together on the same phylogenetic tree by comparing their

(1) plasmids
(2) mitochondrial genomes
(3) homologous genes that are poorly conserved
(4) homologous genes that are highly conserved

146. Which kind of DNA should provide the best molecular clock for determining the evolutionary relatedness of several species whose common ancestor became extinct billions of years ago?

(1) that coding for ribosomal RNA
(2) intronic DNA belonging to a gene whose product performs a crucial function
(3) paralogous DNA that has lost its function (i.e., no longer codes for functional gene product)
(4) mitochondrial DNA

147. A phylogenetic tree constructed using sequence differences in mitochondrial DNA would be most valid for discerning the evolutionary relatedness of

(1) archaeans and bacteria
(2) fungi and animals
(3) chimpanzees and humans
(4) sharks and dolphins

148. Which statement represents the best explanation for the observation that the nuclear DNA of wolves and domestic dogs has a very high degree of sequence homology?

(1) Dogs and wolves have very similar morphologies
(2) Dogs and wolves belong to the same order
(3) Dogs and wolves are both members of the order Carnivora
(4) Dogs and wolves shared a common ancestor very recently

149. Paralogous genes that have lost the function of coding for any functional gene product are known as "pseudogenes." Which of these is a valid prediction regarding the fate of pseudogenes over evolutionary time?

(1) They will be preserved by natural selection
(2) They will be highly conserved
(3) They will ultimately regain their original function
(4) They will have relatively high mutation rates

150. The most important feature that permits a gene to act as a molecular clock is
(1) having a large number of base pairs
(2) having a larger proportion of exonic DNA than of intronic DNA
(3) having a reliable average rate of mutation
(4) its recent origin by a gene-duplication event

151. Neutral theory proposes that
(1) significant proportion of mutations are not acted upon by natural selection
(2) most mutations of highly conserved DNA sequences should have no functional effect
(3) DNA is less susceptible to mutation when it codes for amino acid sequences whose side groups (or R groups) have a neutral pH
(4) DNA is less susceptible to mutation when it codes for amino acid sequences whose side groups (or R groups) have a neutral electrical charge

152. When it acts upon a gene, which of the following processes consequently makes that gene an accurate molecular clock?
(1) transcription
(2) directional natural selection
(3) mutation
(4) proofreading

153. Which of these would, if it had acted upon a gene, prevent this gene from acting as a reliable molecular clock?
(1) neutral mutations
(2) genetic drift
(3) mutations within introns
(4) natural selection

154. What kind of evidence has recently made it necessary to assign the prokaryotes to either of two different domains, rather than assigning all prokaryotes to the same kingdom?
(1) molecular. (2) behavioral
(3) nutritional (4) anatomical

155. What important criterion was used in the late 1960s to distinguish between the three multicellular eukaryotic kingdoms of the five-kingdom classification system?
(1) the number of cells present in individual organisms
(2) the geological stratum in which fossils first appear
(3) the nutritional modes they employ
(4) the biogeographic province where each first appears

156. Members of which kingdom have cell walls and are all heterotrophic?
(1) Plantae (2) Fungi
(3) Animalia (4) Protista

157. Which kingdom has been replaced with two domains?
(1) Plantae (2) Fungi
(3) Animalia (4) Monera

158. Which eukaryotic kingdom is polyphyletic, and therefore unacceptable, based on cladistics?
(1) Plantae (2) Fungi
(3) Animalia (4) Protista

159. Which eukaryotic kingdom includes members that are the result of endosymbi oses that included an ancient proteobact-erium and an ancient cyanobacterium?
(1) Plantae. (2) Fungi
(3) Animalia (4) Protista

160. In a comparison of birds and mammals, having four limbs is
(1) a shared ancestral character
(2) a shared derived character
(3) a character useful for distinguishing birds from mammals
(4) an example of analogy rather than homology

161. To apply parsimony to constructing a phylogenetic tree,
(1) choose the tree that assumes all evolutionary changes are equally probable
(2) choose the tree in which the branch points are based on as many shared derived characters as possible
(3) base phylogenetic trees only on the fossil record, as this provides the simplest explanation for evolution
(4) choose the tree that represents the fewest evolutionary changes, either in DNA sequences or morphology

162. Which of the following is *not* evidence that charophytes are the closest algal relatives of plants?
(1) similar sperm structure
(2) the presence of chloroplasts
(3) similarities in cell wall formation during cell division
(4) genetic similarities in chloroplasts

163. Which of the following characteristics of plants is absent in their closest relatives, the charophyte algae?
(1) chlorophyll *b*
(2) cellulose in cell walls
(3) formation of a cell plate during cytokinesis
(4) alternation of multicellular generations

164. In plants, which of the following are produced by meiosis?
(1) haploid sporophytes
(2) haploid gametes
(3) diploid gametes
(4) haploid spores

165. Microphylls are found in which plant group?
(1) mosses (2) liverworts
(3) lycophytes. (4) ferns

166. Which of the following is a land plant that has flagellated sperm and a sporophyte-dominated life cycle?
(1) fern. (2) moss
(3) liverwort (4) charophyte

167. Suppose an efficient conducting system evolved in a moss that could transport water and other materials as high as a tall tree. Which of the following statements about "trees" of such a species would not be true?
(1) Fertilization would probably be more difficult
(2) Spore dispersal distances would probably increase
(3) Females could produce only one archegonium
(4) Unless its body parts were strengthened, such a "tree" would probably flop over

168. Among the following which groups do not obey Linneus Binomial Nomenclature?
(1) Virus (2) Bacteria
(3) Archae (4) fungi

169. The structural integrity of bacteria is to peptidoglycan as the structural integrity of plant spores is to
(1) lignin (2) cellulose
(3) secondary compounds
(4) sporopollenin

170. All of the following are common to both charophytes and land plants except
(1) sporopollenin. (2) lignin
(3) chlorophyll *a*. (4) cellulose

171. Which taxon is essentially equivalent to the "embryophytes"?
(1) Viridiplantae (2) Plantae
(3) Pterophyta (4) Bryophyta

172. What is true of charophytes?
(1) They are the ancestors of green algae
(2) They are examples of seedless vascular plants
(3) They are the closest living algal relatives of land plants
(4) They share some features in common with land plants, namely spores surrounded by

173. The following are all adaptations to life on land except

(1) rings of cellulose-synthesizing complexes
(2) cuticles
(3) tracheids
(4) reduced gametophyte generation

174. Considering that the mature sporophytes of true mosses get their nutrition from the gametophytes on which they grow, and considering these generations as individual plants, what is true of the relationship between true moss sporophytes and gametophytes?

(1) Sporophytes are endosymbionts of gametophytes
(2) Sporophytes are mutualists of gametophytes
(3) Sporophytes are commensalists of gametophytes
(4) Sporophytes are parasites of gametophytes

175. Where in an angiosperm would you find a megasporangium?

(1) in the style of a flower
(2) inside the tip of a pollen tube
(3) enclosed in the stigma of a flower
(4) within an ovule contained within an ovary of a flower

176. A fruit is most commonly

(1) a mature ovary
(2) a thickened style
(3) an enlarged ovule
(4) a modified root

177. With respect to angiosperms, which of the following is incorrectly paired with its chromosome count?

(1) egg–*n*
(2) megaspore–2*n*
(3) microspore–*n*
(4) zygote–2*n*

178. Which of the following is not a characteristic that distinguishes gymnosperms and angiosperms from other plants?

(1) alternation of generations
(2) ovules
(3) integuments
(4) pollen

179. All fungi share which of the following characteristics?

(1) symbiotic (2) heterotrophic
(3) flagellated (4) pathogenic

180. Which feature seen in chytrids supports the hypothesis that they diverged earliest in fungal evolution?

(1) the absence of chitin within the cell wall
(2) coenocytic hyphae
(3) flagellated spores
(4) formation of resistant zygosporangia

181. Which of the following cells or structures are associated with asexual reproduction in fungi?

(1) ascospores (2) basidiospores
(3) zygosporangia (4) conidiophores

182. The photosynthetic symbiont of a lichen is often

(1) a moss. (2) a green alga
(3) a brown alga. (4) an ascomycete

183. Among the organisms listed here, which are thought to be the closest relatives of fungi?

(1) animals. (2) vascular plants
(3) mosses (4) brown algae

184. The adaptive advantage associated with the filamentous nature of fungal mycelia is primarily related to

(1) the ability to form haustoria and parasitize other organisms
(2) avoiding sexual reproduction until the environment changes
(3) the potential to inhabit almost all terrestrial habitats
(4) an extensive surface area well suited for invasive growth and absorptive nutrition

185. ICBN stands for
(1) International Code for Botanical Nomenclature
(2) International Code for Biological Nomenclature
(3) International Code for Botanical Naming
(4) International Code for Binomial Nomenclature

186. If haustoria from the fungal partner were to appear within the photosynthetic partner of a lichen, and if the growth rate of the photosynthetic partner consequently slowed substantially, then this would support the claim that
(1) algae and cyanobacteria are autotrophic
(2) lichens are not purely mutualistic relationships
(3) algae require maximal contact with the fungal partner in order to grow at optimal rates
(4) fungi get all of the nutrition they need via the "leakiness" of photosynthetic partners

187. When pathogenic fungi are found growing on the roots of grape vines, grape farmers sometimes respond by covering the ground around their vines with plastic sheeting and pumping a gaseous fungicide into the soil. The most important concern of grape farmers who engage in this practice should be that the
(1) fungicide might also kill the native yeasts residing on the surfaces of the grapes
(2) lichens growing on the vines' branches are not harmed
(3) fungicide might also kill mycorrhizae
(4) sheeting is transparent so that photosynthesis can continue

188. Gymnosperms and angiosperms have the following in common except
(1) seeds and ovules
(2) pollen
(3) vascular tissue
(4) ovaries

189. Plastids that are surrounded by more than two membranes are evidence of
(1) evolution from mitochondria
(2) fusion of plastids
(3) origin of the plastids from archaea
(4) secondary endosymbiosis

190. Biologists suspect that endosymbiosis gave rise to mitochondria before plastids partly because
(1) the products of photosynthesis could not be metabolized without mitochondrial enzymes
(2) all eukaryotes have mitochondria (or their remnants), whereas many eukaryotes do not have plastids
(3) mitochondrial DNA is less similar to prokaryotic DNA than is plastid DNA
(4) without mitochondrial CO_2 production, photosynthesis could not occur

191. Which group is incorrectly paired with its description?
(1) rhizarians: morphologically diverse group defined by DNA similarities
(2) diatoms: important producers in aquatic communities
(3) red algae: acquired plastids by secondary endosymbiosis
(4) apicomplexans: parasites with intricate life cycles

192. Which protists are in the same eukaryotic supergroup as land plants?
(1) green algae
(2) dinoflagellates
(3) red algae
(4) both green algae and red algae

193. In life cycles with an alternation of generations, multicellular haploid forms alternate with
(1) unicellular haploid forms
(2) unicellular diploid forms
(3) multicellular haploid forms
(4) multicellular diploid forms

194. Genetic variation in bacterial populations cannot result from
(1) transduction (2) transformation
(3) conjugation (4) meiosis

195. Photoautotrophs use
(1) light as an energy source and CO_2 as a carbon source
(2) light as an energy source and methane as a carbon source
(3) N_2 as an energy source and CO_2 as a carbon source
(4) CO_2 as both an energy source and a carbon source

196. Which of the following statements is *not* true?
(1) Archaea and bacteria have different membrane lipids
(2) Both archaea and bacteria generally lack membrane-enclosed organelles
(3) The cell walls of archaea lack peptidoglycan
(4) Only bacteria have histones associated with DNA

197. Which of the following involves metabolic cooperation among prokaryotic cells?
(1) binary fission
(2) endospore formation
(3) endotoxin release
(4) biofilms

198. Bacteria perform each of the following ecological roles. Which role typically does not involve a symbiosis?
(1) skin commensalist
(2) decomposer
(3) aggregates with methane-consuming archaea
(4) gut mutualist

199. Plantlike photosynthesis that releases O_2 occurs in
(1) cyanobacteria (2) chlamydias
(3) archaea. (4) actinomycetes

200. Which statement about the domain Archaea is true?
(1) Genetic prospecting has recently revealed the existence of many previously unknown archaean species
(2) No archaeans can reduce CO_2 to methane
(3) The genomes of archaeans are unique, containing no genes that originated within bacteria
(4) No archaeans can inhabit solutions that are nearly 30% salt

---

ANSWER SHEET

Unit 9: Diversity Among Life Forms

---

1. 3	**2.** 4	**3.** 4	**4.** 2	**5.** 3	**6.** 1	**7.** 3	**8.** 3	**9.** 3	**10.** 1
11. 1	**12.** 3	**13.** 3	**14.** 4	**15.** 4	**16.** 4	**17.** 2	**18.** 4	**19.** 2	**20.** 4
21. 2	**22.** 2	**23.** 2	**24.** 1	**25.** 4	**26.** 1	**27.** 4	**28.** 3	**29.** 1	**30.** 4
31. 1	**32.** 3	**33.** 3	**34.** 2	**35.** 2	**36.** 1	**37.** 2	**38.** 1	**39.** 1	**40.** 4
41. 3	**42.** 2	**43.** 4	**44.** 4	**45.** 1	**46.** 3	**47.** 3	**48.** 4	**49.** 3	**50.** 3
51. 3	**52.** 1	**53.** 1	**54.** 1	**55.** 4	**56.** 1	**57.** 3	**58.** 3	**59.** 1	**60.** 4

61. 1 **62.** 4 **63.** 4 **64.** 1 **65.** 1 **66.** 1 **67.** 1 **68.** 2 **69.** 4 **70.** 1
71. 4 **72.** 4 **73.** 3 **74.** 2 **75.** 2 **76.** 2 **77.** 1 **78.** 4 **79.** 4 **80.** 2
81. 1 **82.** 4 **83.** 4 **84.** 3 **85.** 3 **86.** 3 **87.** 2 **88.** 2 **89.** 1 **90.** 4
91. 1 **92.** 4 **93.** 2 **94.** 1 **95.** 1 **96.** 3 **97.** 4 **98.** 2 **99.** 4 **100.** 3
101. 3 **102.** 1 **103.** 2 **104.** 4 **105.** 4 **106.** 4 **107.** 3 **108.** 1 **109.** 2 **110.** 4
111. 2 **112.** 1 **113.** 4 **114.** 3 **115.** 1 **116.** 4 **117.** 3 **118.** 3 **119.** 1 **120.** 2
121. 4 **122.** 3 **123.** 1 **124.** 2 **125.** 3 **126.** 3 **127.** 2 **128.** 3 **129.** 4 **130.** 2
131. 1 **132.** 2 **133.** 3 **134.** 1 **135.** 4 **136.** 4 **137.** 2 **138.** 4 **139.** 3 **140.** 4
141. 4 **142.** 4 **143.** 1 **144.** 3 **145.** 4 **146.** 1 **147.** 3 **148.** 4 **149.** 4 **150.** 3
151. 1 **152.** 3 **153.** 4 **154.** 1 **155.** 3 **156.** 2 **157.** 4 **158.** 4 **159.** 1 **160.** 1
161. 4 **162.** 2 **163.** 4 **164.** 4 **165.** 3 **166.** 1 **167.** 3 **168.** 1 **169.** 4 **170.** 2
171. 2 **172.** 3 **173.** 1 **174.** 4 **175.** 4 **176.** 1 **177.** 2 **178.** 1 **179.** 2 **180.** 3
181. 4 **182.** 2 **183.** 1 **184.** 4 **185.** 1 **186.** 2 **187.** 3 **188.** 4 **189.** 4 **190.** 2
191. 3 **192.** 4 **193.** 4 **194.** 1 **195.** 1 **196.** 4 **197.** 4 **198.** 2 **199.** 1 **200.** 1

UNIT-10

ECOLOGY

1. Most animals in coral reefs are brightly colored, especially red, while those in the deeper ocean are not. This is because
   (1) The deep ocean does not contain pigments that animals can absorb for their colouration
   (2) Longer wavelength of lights are attenuated with water depth
   (3) Shorter wavelength of lights are attenuated with water depth
   (4) The coral reef uses photosynthetic organisms to produce green colour against which other colours, such as red, are clearly visible

2. Two students were testing the amount of fertilizers that would best promote the growth of strawberries in a garden. Which of the following could be an avoidable source of experimental error?
   (1) Length of the study
   (2) Variation in the plants
   (3) Variable amount of watering
   (4) Fertilization during the study

3. A blue whale stranded on a beach will suffocate because
   (1) It needs to breathe water
   (2) It envaginated gills will dry out
   (3) It will not be able to contract its diaphragm
   (4) It will get air bubbles in its blood stream

4. The dawn chorus of singing birds is a very striking and widespread phenomenon. Which of the following is unlikely to be a plausible explanation for the evolution of the dawn chorus of birds?
   (1) Increasing light levels at dawn trigger singing activity in birds
   (2) By singing at dawn, birds avoid acoustic interference from other sympatric callers such as cicadas and frogs
   (3) By calling at dawn, the bird community can maximize its fitness relative to the frog
   (4) Atmospheric conditions for transmission of birdsongs are optimal at dawn

5. Which of the following is not a critically endangered animal species of India?
   (1) *Gyps indicus*
   (2) *Hemitragus hylocrius*
   (3) *Panthera pardus*
   (4) *Ardeotis nigriceps*

6. Which of the following is not an endangered plant species?
   (1) *Ceropegia mahabalei*
   (2) *Butea monosperma*
   (3) *Fimbristylis hirsutifolia*
   (4) *Ischaemum jayachandranii*

7. If the milk is left open, lactose is fermented first to produce acid. This is followed by protelytic bacteria which increases the pH. Ultimately milk fats are degraded to produce rancidity. This is an example of
   (1) ecological succession
   (2) Antagonism
   (3) interference competition
   (4) Microevolution

8. Based on per molecule, which of the following gas has the most powerful greenhouse effect?

(1) CO_2 (2) CH_4
(3) N_2O (4) CFCs

9. Which of the following organism is widely used as a bio-control in organic farming?

(1) *Rhyzobium tropicii*
(2) *Trichorderma viridis*
(3) *Fusarium oxysporum*
(4) *Nostoc muscorum*

10. Following figure shows McArthur and Wilson's equilibrium model of biota on a single island

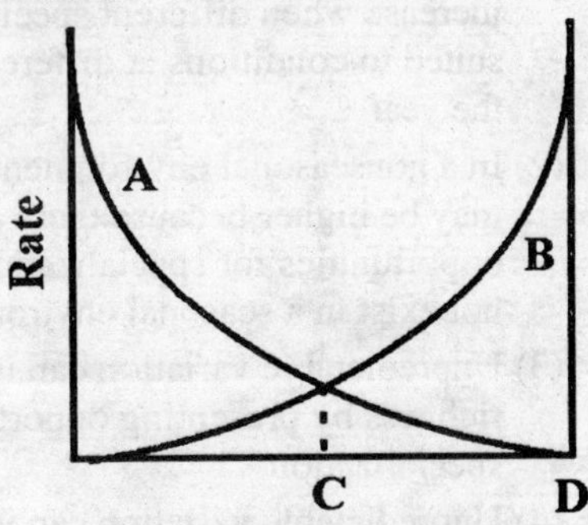

In this figure, terms A, B, C and D in order are

(1) extinction. immigration, equilibrium number of species, size of species pool
(2) immigration. extinction, equilibrium number of species, size of species pool
(3) extinction. immigration, size of species pool, equilibrium number of species
(4) immigration, extinction, size of species pool, equilibrium number of species

11. Which of the following characteristic of an climax community?

(1) Wide niche
(2) Low species diversity
(3) High community production
(4) open mineral cycling

12. An elephant is considered to have a slow life-history pattern. Which of these characteristics would be matched to an elephant?

(1) Unstable environment
(2) Long lifespan
(3) Reproduce early
(4) Mature quickly

13. Which of these is not an example of a terrestrial ecosystem?

(1) Human skin (2) Garden plot
(3) Rotting log (4) Volcano site

14. In the 1960's myxomatosis was 'accidentally' released killing approximately 90% of rabbits in the UK. What happened in the food chain below?

Carrots → Rabbits → Foxes

(1) The number of carrots increased
(2) The number of foxes increased
(3) The number of carrots decreased and the number of foxes increased
(4) The number of carrots increased and the number of foxes decreased

15. Rabbits are important grazers, maintaining some important grassland habitats. If another catastrophe were to reduce rabbit populations what would happen?

(1) Important grassland butterflies would be lost
(2) There would be more grass for the surviving rabbits
(3) The grassland would be invaded by trees and shrubs
(4) All of the above

16. Which of the following statements about pesticides is true?

(1) Spraying pesticides only affects the farmer's crops
(2) There do not affect food chains
(3) Without pesticides farmers could not grow crops
(4) Pesticides can end up in rivers and streams

17. Many pesticides are stored in an animal's body fat. If a pesticide got into a lake, which organism would have the highest level? Microscopic plant → Micro animal → Fish → Osprey
(1) Osprey
(2) Fish
(3) Microscopic animals
(4) Microscopic plants

18. Fertilizers can be washed into rivers by the rain. This can cause
(1) Bioaccumulation
(2) Eutrophication
(3) Biodegradation
(4) Biomagnification

19. The maximum possible rate of increase of population of a given species under ideal conditions would depend upon its
(1) biotic potential
(2) potential natality
(3) carrying capacity
(4) biomass

20. The total number of species of insects in two large fields is the same. Within each of these fields are two smaller patches. In Field 1, the two patches share very few species. In the other, the two patches share most species. Which of the following statements is true?
(1) The individual patches in field I have more species than in field 2
(2) The individual patches in field 2 have more species than in field 1
(3) The individual patches in both fields have the same number of species
(4) There is high species turnover between fields 1 and 2

21. Which of the following is not an explanation for higher species richness in a community?
(1) Species are more specialized in their use of resources
(2) Species have more overlap in their use of resources
(3) A greater range of resources is available
(4) Some species are better at utilizing resources

22. In which of the following places would you expect the highest number of plant species per hectare of forest?
(1) Corbett National Park
(2) Ranthambore National Park
(3) Mudumalai National Park
(4) Silent Valley National Park

23. Which of the following has not been proposed as an effect of climatic variation on species richness?
(1) In a seasonal environment, richness may increase when different species are suited to conditions at different times of the year
(2) In a nonseasonal environment, richness may be higher because there are opportunities for specialization that do not exist in a seasonal environment
(3) Unpredictable variation can increase richness by presenting opportunities for specialization
(4) Unpredictable variation can increase richness by preventing competitive exclusion

24. In a species-area relationship
(1) the number of species in a sample increases as the area covered by that sample increases
(2) the number of species in an area of habitat increases as the area of that habitat increases
(3) the type of species observed in a habitat alters as the area of habitat alters
(4) species diversity increases linearly with island size

25. Species richness tends to decrease with increasing altitude. Which of the following is not believed to play some part in this trend?
(1) Higher-altitude areas tend to be younger, in evolutionary terms
(2) Higher-altitude areas tend to occupy relatively small areas
(3) Productivity declines with altitude
(4) High-altitude habitats tend to be isolated from one another

26. One of the criteria that conservationists typically use to identify habitats for conservation is high species diversity. This may not always be a good criterion because
(1) Low diversity areas are common
(2) Some common habitats have high diversity
(3) Some unique habitats have low diversity
(4) High diversity areas are unique

27. Conservationists argue over whether reserves should be set aside in single Large blocks or several small blocks. Single large reserves may be better because larger areas support mare species. However, small areas may have advantages as well. One of the arguments proposed in favour of several small patches concerns
(1) Habitat loss
(2) Global climate change
(3) Minimum viable populations
(4) Disease

28. Climate change, habitat loss, ultraviolet radiation, fungal disease, over exploitation and invasive have been implicated in the decline of
(1) Frogs (2) Vultures
(3) Tigers (4) Snakes

29. Which of the following biomes is characterized by a wet, mild winter and a warm, dry summer?
(1) Taiga (2) Tundra
(3) Chaparral
(4) Temperate deciduous forest

30. Which of the following is not a prediction of MacArthur and Wilson's theory of Island Biogeography?
(1) The number of species on an island should eventually become roughly constant through time
(2) Large islands should support more species than small islands
(3) Species number should decline with the increasing remoteness of an island
(4) The rate of species turnover should increase with island area

31. Which life history trait has not generally been recognized as decreasing a species sensitivity to extinction?
(1) Lack of dispersal ability
(2) Lower trophic status
(3) Low longevity
(4) High fecundity

32. Which is not one of the main threat categories used for threatened species?
(1) Endangered (2) Vulnerable
(3) Rare (4) Imperiled

33. Which factor currently threatens most species with extinction?
(1) Habitat loss
(2) Exploitation
(3) Introduced species
(4) All of the above

34. In which taxa is the percentage of endangered species highest?
(1) Amphibians (2) Reptiles
(3) Birds (4) Fish

35. The habitat with the highest percentage of threatened mammals is
(1) Grassland
(2) Freshwater
(3) Tropical forest
(4) Oceanic island

36. Which of the following statements regarding alpha diversity is correct?
(1) Alpha diversity is represented by the number of species in a specified area
(2) it decreases with the total number of individuals encompassed and thus with the increase in the area sampled and the productivity per unit area
(3) It is less on remote islands and increases as one moves towards the equator
(4) It is diversity between two habitats with diverse environment gradients

37. Which of the following statements regarding biodiversity is/are correct?
(1) Beta diversity is represented by the turnover of species across space
(2) Beta diversity refers to the degree to which species composition change along an environmental gradient
(3) Gamma diversity is a species turn-over rate with distance between sites of similar habitat or with expanding geographical area
(4) All of the above

38. Which of the following habitats show the highest diversity of living species?
(1) Grassland (2) Temperate forest
(3) Desert (4) Tropical rainforest

39. The richness of species in an ecosystem is termed as
(1) Genetic diversity
(2) Species diversity
(3) Community diversity
(4) All of the above

40. Indian rhinoceros is being protected from extinction in
(1) Kaziranga National Park
(2) Jim Corbet
(3) Silent Valley
(4) Kanha

41. The maximum water pollution is caused by-
(1) Detergents (2) Pesticides
(2) Industrial wastes (4) Ammonia

42. Which of the following is not a major shift in species richness observable in the fossil record?
(1) The Cambrian explosion
(2) The Permian decline of invertebrate families
(3) The flowering of angiosperms in the Cretaceous
(4) The Devonian decimation of large mammals

43. Lyme disease, caused by *Borrelia burgdorferi*, can be passed directly from deer to humans when ticks simply transfer infected blood from one to the other. Borrelia may be said, therefore, to have
(1) alternative hosts and a vector
(2) alternative and alternate hosts
(3) alternate hosts and a vector
(4) neither alternate nor alternative hosts

44. Which of the following statements is correct?
(1) Legumes need rhizobial bacteria to provide them with nitrogen
(2) Legumes infected with rhizobial bacteria are more competitive against grasses
(3) There is a precise match between species of legume and species of Rhizobium
(4) Rhizobium can only grow and reproduce in the presence of a legume

45. Assume that an animal generates heat at a rate proportional to its volume and can radiate heat at a rate proportional to its body surface area. Which of these animals would be best at maintaining its body temperature in a cold climate?
(1) Mouse
(2) Bear
(3) Rabbit
(4) Fox

46. Recently, a new species of frog was discovered in the Western Ghats whose closest relatives are in Seychelles and Madagascar. This proves that

(1) India was connected to these islands in the past

(2) Frogs are good at dispersing across oceans

(3) Frogs are Living fossils

(4) India was a part of Gondwanaland

47. Consider a population that shows exponential growth over time: Nt = No e^{rt} where No is initial population size at time = 0, Nt is population size at time t, r is the intrinsic rate of increase, and e is the constant 2.7. In such a population

(1) Population growth rate is highest when the population is small and declines over time

(2) Population growth rate is smallest when the population is small, increases with time and reaches an asymptote

(3) Population growth rate is smallest when the population is small and increases with time

(4) Population growth rate is constant

48. Batesian mimicry is an example of

(1) assortative mating

(2) heterozygote advantage

(3) stabilizing selection

(4) frequency-dependent selection

49. A forest patch has 100 deer on Sunday, May 1, 2007. Hunting season opens on May 8,2007. Hunters kill 20 deer every Sunday. Each deer eats 1 kg of grass per day. If none of these populations are reproducing, how much grass would there need to be for the hunters to wipe out the deer population before the grass is wiped out?

(1) 700 kg

(2) 1.4 tonnes

(3) 2.1 tones

(4) 1.8 tonnes

50. Bird species A and B both nest in marshes. Species A is dominant over B when they are found together. Species B will occupy the entire marsh if Species A is absent, but if Species A is present, it will exclude Species B from the marsh center, and restrict Species B to the edge of the marsh. For Species B, the entire marsh would represent its

(1) realized niche

(2) habitat

(3) range

(4) fundamental niche

51. Marsupials are found in

(1) Australia and South America

(2) Australia alone

(3) Australia and Madagascar

(4) Australia and Africa

52. Which of the following traits is not characteristic of all vascular plants?

(1) Presence of seeds

(2) Alteration of generations

(3) Presence of xylem and phloem

(4) Presence of lignin in cell walls

53. Due to global warming, species of plants and animals can generally be expected to migrate from

(1) higher altitudes to lower altitudes

(2) lower altitudes to higher altitudes

(3) west to East along a given latitude

(4) east to West along a given latitude

54. A number of viruses have RNA genomes. In positive-strand viruses,

(1) the RNA genome codes directly for viral protein

(2) a complementary copy of the genome is translated to form viral proteins

(3) both the strands can be utilized for protein synthesis

(4) none of the strands have information for proteins

55. Myelination of axons results in

(1) An increase in the frequency of firing of action potential

(2) A decrease in the frequency of firing of action potentials

(3) A decrease in the velocity of conduction of action potentials

(4) An increase in the velocity of conduction of action potentials

56. Mullerian mimicry is when

(1) A harmless species mimics a dangerous species

(2) A dangerous species mimics a harmless species

(3) One harmless species mimics another

(4) One dangerous species mimics another

57. Two resources may be described as antagonistic when

(1) Consuming both kills the animal

(2) One can replace the other in an animal's diet

(3) The animal requires proportionately more resources when both are consumed together

(4) The animal cannot consume both simultaneously

58. Under which scenario two species could co-exist in same niche violating the Gauss exclusion principle

(1) High predator pressure or disease

(2) Limited food awailibility

(3) Limited space

(4) Functioning under same environment gradients

59. In population A, each woman bears exactly two female children, born at maternal ages 20 and 25. Population B is the same, except that the children are born when the mother is 25 and 30. Assume that these are the only differences between A and B. When the sizes of the two populations are monitored over time, we will find that

(1) Both are stationary

(2) Both grow at the same rate

(3) B grows faster than A

(4) A grows faster than B

60. Which of the following is least likely to explain greater bird species diversity in the terrestrial tropics along a latitudinal gradient?

(1) Higher plant diversity

(2) Greater area

(3) Higher temperatures

(4) Higher rainfall

61. The number of individuals or biomass at each trophic level is usually depicted as an ecological pyramid. Sometimes, inverted pyramids such as produced by a host parasite system, may be observed. These can be seen in

(1) Pyramids of numbers and biomass, but never energy

(2) Pyramids of biomass and energy, but never numbers

(3) Pyramids of energy and numbers, but never biomass

(4) All kinds of pyramids

62. Ears of different moth species are found on different parts of the body, including the thorax, wings, abdomen and proboscis. This suggests that

(1) Moths are evolving more rapidly than other insect groups

(2) Ears in moths arose as a single event in their phylogenetic history

(3) Ears in moths arose multiple times in their phylogenetic history

(4) The structure of moth ears is different from those of other insect groups

63. The roots of many aquatic plants have special structures that project above the surface of the water. For example, *cypress* trees (which grow in swamps) have knees that extend upward above water level. Which of the following is the most logical function of these structures?

(1) Obtaining carbon dioxide for photosynthesis
(2) Nitrogen fixation
(3) Obtaining oxygen for the roots
(4) Absorbing trace minerals

64. You carefully study populations of two very similar field mice, one from Rajasthan and one from Gujarat. You want to know whether the populations belong to the species or to two different species. You could most confidently decide this if you could

(1) Show that the ranges of the two mice overlap without hybridization occurring
(2) Bring the two types of mice into the laboratory to see if they will mate
(3) Demonstrate that the natural ranges of the two types of mice are entirely allopatric
(4) Show that the Rajasthan mice live in wetter habitats than the Gujarat mice

65. During a study on fish in a pond, 200 fish are caught and marked on the first day. On the second day of sampling, 150 fish are caught, of which 50 are already found to be marked. What is the estimate of the total number of fish in the pond?

(1) 600 (2) 300
(3) 150 (4) 1200

66. A plant ecologist has to study the diversity of plants in a forest with a strong gradient of rainfall from North to South. A good sampling design would be

(1) Random placement of quadrats in the forest
(2) Placement of quadrats from North to South
(3) Placement of quadrats from East to West
(4) Enumeration of all individuals in the forest

67. From our understanding of energy flow through various trophic levels of an ecosystem, which of the following statement is correct?

(1) About 50% of the energy in one trophic level is passed on to the next level
(2) Primary consumers have the highest ecological efficiency
(3) Only a negligible amount of energy is lost along each link in the food chain
(4) Producing meat is an inefficient way of obtaining energy trapped by photosynthesis

68. In a simple ecological system with two competitor species A and B competing over a single resource, theory predicts that stable co-existence of the two species is most likely when

(1) interspecific competition is stronger than intraspecific competition in species A
(2) intraspecific competition is weaker than interspecific competition in species B
(3) intraspecific competition is stronger than interspecific competition in both species
(4) interspecific competition is stronger than intraspecific competition in both species

69. When the net primary productivity (NPP) of grassland is plotted as a function of the mean annual rainfall (R), which of the following functions would best describe the relationship between them?

(1) Linear
(2) Hyperbolic
(3) Negative exponential
(4) Positive exponential

70. Which of the following life history adaptations is least expected when predation pressure on a fish species that grows continuously throughout its lifespan is focused on large size classes?

(1) Allocate more resources preferentially to early reproductive effort than to growth

(2) Allocate more resources preferentially to growth than to early reproductive effort

(3) Early age at maturity

(4) Produce many small offspring in relatively few reproductive seasons

71. Which of the following is likely to be heard at the West distance from source from the caller (given that all calls are produced with the same loudness)?

(1) The alarm call of a langur

(2) The chirping of a cricket

(3) The song of a magpie robin

(4) The clicks of a bat in a roost

72. A weed growing in a lake doubles in area every week. If it takes a month for it to cover the entire surface of the lake, how long does it take to cover 1/4th of the surface?

(1) 0.5 weeks

(2) 1.0 weeks

(3) 1.5 weeks

(4) 2.0 weeks

73. If the polar ice melts entirely because of global warming, the rate at which the earth rotates around its own axis should

(1) slow down

(2) speed up

(3) vary cyclically

(4) remain unchanged

74. Which of the following is a wrong explanation for why temperate and Arctic lakes never freeze completely solid?

(1) The hydrogen bonds in ice make ice less dense than liquid water

(2) Water expands when it solidifies

(3) Water is denser as a solid than as a liquid

(4) Floating ice insulates the liquid water below

75. If temperatures increase by more than 40°C due to global warming the most likely impact on forests in the tropics would be

(1) stimulated growth of trees resulting in increased biomass

(2) increase in the diversity of trees in tropical forests

(3) increase in respiration over photosynthesis resulting in decreased biomass

(4) migration of temperate tree species to the tropics

76. Semelparous reproduction, where the organism produces all its offspring in a single reproductive event, is favoured when

(1) juvenile survival is high relative to adult survival

(2) adult survival is high relative to juvenile survival

(3) juvenile survival increases with size at birth

(4) does not depend on survival rates

77. A metapopulation structure is most likely to prevent the regional extinction of a species when

(1) the dynamics of the local populations are asynchronous

(2) the dynamics of the local populations are synchronous

(3) individuals move extensively between local populations so that they behave like one large population

(4) local populations are isolated and closed

78. You are given individuals of three species of fish. Of these, species A is covered with multi-coloured spots all over the body, species 0 is black with very large eyes and species C is brown with reduced eyes. The predicated habitats of the three species (A, 0 and C) would be
(1) Coral reefs, benthos and estuaries respectively
(2) Coral reefs, estuaries and benthos respectively
(3) Benthos, estuaries and coral reefs respectively
(4) Estuaries, benthos and coral reefs respectively

79. Resources are not evenly distributed in time and space. The type of resource distribution in time that most closely resembles a patchily distributed resource in space is
(1) scarce resource
(2) common resource
(3) perennial resource
(4) fluctuating resource

80. Which of the following should not be described as 'migration'?
(1) The movement of young male rodents away from the territories in which they were born
(2) The movement of crabs up and down the shore with the tides
(3) The practice of transhumance, moving flocks of sheep and goats up to mountain pastures in summer and down again in the fall
(4) The movement of planktonic plants to the upper layers of the water in lakes during the day and to the lower, nutrient-rich depths at night

81. The logistic equation cannot be written
(1) dN/dt = r N{K - (N/K)}
(2) dN/dt = r N{[K - N]/K}
(3) dN/dt = r N{1 - (N/K)}
(4) dN/dt (1/N) = r - (r/K) N

82. Which of the following is not an adverse effect on plants of living in soils of low pH?
(1) Calcium toxicity
(2) Reduced nitrogen fixation
(3) Increased aluminium solubility
(4) Reduced uptake of phosphorus

83. The association of algae and fungi to form lichens is an example of
(1) a mutualistic symbiosis
(2) a commensal relationship
(3) obligatory parasitism
(4) facultative parasitism

84. In community successions, if they run their full course, the number of species
(1) first increases (because of rising productivity) but eventually decreases (as productivity declines)
(2) first increases (because of colonization) but eventually decreases (because of predation)
(3) first increases (because of increasing spatial heterogeneity) but eventually decreases (because of competition)
(4) first increases (because of colonization) but eventually decreases (because of competition)

85. Which of the following statements is correct?
(1) Microparasites are distinguishable by being much smaller than their hosts
(2) Microparasites are typical of animal rather than plant hosts
(3) The dynamics of macroparasites are usually followed indirectly, by monitoring infected hosts
(4) Microparasites usually pass through several generations in the same host individual

86. The life cycle of many species of bamboo may be described as
(1) iteroparous-annual
(2) semelparous-annual
(3) iteroparous-perennial
(4) semelparous-perennial

87. In a Type 2 survivorship curve
(1) the number of survivors remains same
(2) the risk of mortality remains constant
(3) a plot of number against age is a straight line
(4) the line has a slope of +(1)

88. The term dispersal refers to
(1) the way individuals spread away from each other
(2) the mass directional movement of large numbers of a species from one location to another
(3) the distribution of organisms in space
(4) the process in which individuals leave a population

89. Which of the following is not correct?
(1) At a carrying capacity per capita birth rates and death rates are equal
(2) At a carrying capacity a population has no general tendency either to increase or to decrease
(3) The carrying capacity is a concept that emerges from the logistic equation
(4) No real population is characterized by a single carrying capacity

90. Which of the following is not a typical characteristic of an r-selected species
(1) large size
(2) many, small offspring
(3) a large allocation of resources to reproduction
(4) early reproductive maturity

91. If a species is a keystone predator, then its removal from a community should
(1) decrease population size of the predator's preferred prey
(2) decrease species diversity in the prey community
(3) decrease productivity of the predator's preferred prey
(4) increase species diversity in the prey community

92. All of the following may be considered as sites for secondary succession except
(1) an abandoned agricultural field
(2) a bare rock shelf on a mountainside
(3) a forest gap resulting from a tree fall
(4) a hurricane-damaged forest

93. The members of same plant species but adapted to different ecological conditions are termed as
(1) Ecotype (2) Ecophene
(3) Ecospecies (4) Ecotone

94. Acid rain damage depends on the buffering capacity of the soils in a given region. Damage has been greatest where the soil layer is
(1) thin and contains little calcium and magnesium
(2) thin and contains abundant calcium and magnesium
(3) thin and contains abundant calcium but little magnesium
(4) thick and contains abundant calcium and magnesium

95. Nitrogen is a major environmental pollutant of groundwater in many parts of the world. What is the major cause of this nitrogen accumulation in groundwater?
(1) Excessive use of nitrogen fertilizers
(2) Destruction of nitrifying bacteria, which results in soil nitrogen remaining as NH_4^+
(3) Excess soil moisture, which causes ammonification or denitrification
(4) Lack of moisture in the topsoil, which attracts nitrogen to the subsurface and groundwater

96. A community with low species diversity and high dominance is said to be
(1) unproductive but stable
(2) productive but unstable
(3) neither productive nor stable
(4) both productive and stable

97. Small body size often allows animals to exploit desert habitats because it
(1) decreases the ratio of surface area to volume and hence the tendency to lose water
(2) promotes heat gain
(3) enhances thermal inertia
(4) enables exploitation of favorable microhabitats

98. The amount of energy entering a food chain depends on the
(1) direction of energy flow in the system
(2) efficiency of energy recycling in the system
(3) biomass of heterotrophs and their efficiency in transforming plant tissue into animal tissue
(4) biomass of autotrophs and their efficiency in transforming solar energy into chemical energy

99. If average family size for the human population could be immediately reduced to two children per couple, the population would
(1) remain constant
(2) begin an immediate decline
(3) continue to increase for 20 to 40 years
(4) continue to increase indefinitely

100. When exposed to a cold environment for several weeks, many small mammals dramatically increase their capacity for heat production primarily by means of
(1) decreased insulation
(2) increased insulation
(3) shivering thermogenesis
(4) non-shivering thermogenesis

101. A broad-based, pyramid shaped age structure is characteristic of a population that is
(1) growing rapidly
(2) at carrying capacity
(3) stable
(4) limited by density-dependent factors

102. An ecologist would suspect population is growing rapidly if it
(1) contains many more pre-reproductive than reproductive individuals
(2) is near its carrying capacity
(3) is limited only by density-dependent factors
(4) shows a clumped pattern of dispersion

103. Which of these equations represents zero population growth?
(1) Birthrate - Death rate = 0
(2) Birthrate - Death rate > 0
(3) Birthrate + Death rate < 0
(4) Birthrate + Death rate = 0

104. Life-history characteristics not associated with r-selected organisms include
(1) rapid reproductive rates, short generation times, and large body size
(2) repeated reproduction, few progeny, and large body size
(3) inhabiting early successional state communities, rapid maturation rates, and numerous large offspring
(4) inhabiting climax communities, many small offspring, and short life span

105. A population of rabbits experiences an increase over time. Which of these statements about that population is true?
(1) The population may face increased competition due to resource depletion
(2) The population my face decreased competition due to resource depletion
(3) The population may experience exponential growth due to resource depletion
(4) The population may experience straight-line growth due to resource depletion

106. A population is correctly defined as having which of the following characteristics?

I. inhabiting the same general area

II. belonging to the same species

III. possessing a constant and uniform density and dispersion

(1) I only (2) III only
(3) I and II only. (4) II and III only

107. An ecologist recorded 12 white-tailed deer, *Odocoileus virginianus*, per square mile in one woodlot and 20 per square mile in another woodlot. What was the ecologist comparing?

(1) density
(2) dispersion
(3) carrying capacity
(4) cohorts

108. During the spring, you are studying the mice that live in a field near your home. The population density is high, but you realize that you rarely observe any reproductive female mice. This most likely indicates

(1) that there is selective predation on female mice
(2) that female mice die before reproducing
(3) that this habitat is a good place for mice to reproduce
(4) that you are observing immigrant mice

109. Uniform spacing patterns in plants such as the creosote bush are most often associated with

(1) chance
(2) patterns of high humidity
(3) the random distribution of seeds
(4) competitive interaction between individuals of the same population

110. Which of the following groups would be most likely to exhibit uniform dispersion?

(1) Red squirrels, who actively defend territories
(2) Cattails, which grow primarily at edges of lakes and streams
(3) Dwarf mistletoes, which parasitize particular species of forest tree
(4) Moths, in a city at night

111. To construct a reproductive table for a sexual species, one needs to

(1) assess sperm viability for the males in the population
(2) keep track of all of the offspring of a cohort
(3) keep track of the females in a cohort
(4) keep track of all of the offspring of the females in a cohort

112. Which of the following examples would most accurately measure the density of the population being studied?

(1) Counting the number of prairie dog burrows per hectare
(2) Counting the number of times a 1 kilometer transect is intersected by tracks of red squirrels after a snowfall
(3) Counting the number of coyote droppings per hectare
(4) Counting the number of zebras from airplane census observations

113. Which of the following assumptions have to be made regarding the capture-recapture estimate of population size?

I. Marked and unmarked individuals have the same probability of being trapped

II. The marked individuals have thoroughly mixed with the population after being marked

III. No individuals have entered or left the population by immigration or emigration, and no individuals have been added by birth or eliminated by death during the course of the estimate

(1) I only
(2) II only
(3) I and II only
(4) I, II and III

114. Long-term studies of Belding's ground squirrels show that immigrants move nearly 2 km from where they are born and become 1%-8% of the males and 0.7%-6% of the females in other populations. On an evolutionary scale, why is this significant?

(1) These immigrants make up for the deaths of individuals, keeping the other populations' size stable
(2) Young reproductive males tend to stay in their home population and are not driven out by other territorial males
(3) These immigrants provide a source of genetic diversity for the other population
(4) Those individuals that emigrate to these new populations are looking for less crowded conditions with more resources

115. Which of the following sets of measurements is the most useful when studying populations?

(1) Density, dispersion, and demographics of a population
(2) Gene frequency over time and the ratio of reproductive to nonreproductive individuals
(3) Annual precipitation averages and mean annual temperatures
(4) Minimum and maximum amounts of precipitation & annual temperature extreme

116. Which of the following scenarios would provide the most legitimate data on population density?

(1) Count the number of nests of a particular species of songbird and multiply this by a factor that extrapolates these data to actual animals
(2) Count the number of pine trees in several randomly selected 10 m x 10 m plots and extrapolate this number to the fraction of the study area these plots represent
(3) Use the mark-and-recapture method to estimate the size of the population
(4) Calculate the difference between all of the immigrants and emigrants to see if the population is growing or shrinking

117. Which of the following is the best example of uniform distribution?

(1) Bees collecting pollen in a wildflower meadow
(2) Snails in an intertidal zone at low tide
(3) Territorial songbirds in a mature forest during mating season
(4) Mushrooms growing on the floor of an old growth forest

118. Which of the following choices would most likely promote random distribution?

(1) Territorial species
(2) Species that secrete chemicals to attract or inhibit other individuals
(3) Flocking and schooling behaviors
(4) Homogeneous chemical and physical factors in the environment

119. Which of the following best defines a cohort?

(1) A group of individuals that inhabits a small isolated region within the range for the species
(2) All of the individuals that are annually added to a population by birth and immigration
(3) The reproductive males and females within the population
(4) A group of the individuals from the same age group, from birth until they are all dead

120. Why do some invertebrates, such as lobsters, show a "stair-step" survivorship curve?

(1) Many invertebrates mate and produce offspring on multiyear cycles
(2) Within a species of invertebrates, younger individuals have a higher survivorship than older individuals
(3) Many invertebrates molt in order to grow, and they are vulnerable to predation during their "soft shell" stage
(4) Many invertebrate species have population cycles that go up and down according to the frequency of sunspots

121. Which of the following is the most important assumption for the capture-recapture method to estimate the size of wildlife populations?

(1) All females in the population have the same litter size

(2) More individuals emigrate from, as opposed to immigrate into, a population

(3) Over 50% of the marked individuals need to be trapped during the recapture phase

(4) Marked individuals have the same probability of being recaptured as unmarked individuals during the recapture phase

122. A population of ground squirrels has an annual per capita birth rate of 0.06 and an annual per capita death rate of 0.02. Calculate an estimate of the number of individuals added to (or lost from) a population of 1,000 individuals in one year

(1) 120 individuals added

(2) 40 individuals added

(3) 20 individuals added

(4) 400 individuals added

123. Exponential growth of a population is represented by dN/dt =

(1) $\frac{rN}{K}$ (2) rN

(3) $rN(K+N)$ (4) $rN\frac{(K-N)}{K}$

124. Starting from a single individual, what is the size of a population of bacteria that reproduce by binary fission every 20 minutes at the end of a 2-hour time period? (Assume unlimited resources and no mortality)

(1) 64 (2) 18

(3) 128 (4) 512

125. Which of the following is the equation for zero population growth (ZPG)?

(1) $R = b - m$

(2) $dN/dt = rN$

(3) $dN/dt = r_{max} N (K - N)/K$

(4) $dN/dt = r_{max} N$

126. In 2008, the population of Jodhpur was approximately 4,275,000 people. If the birth rate was 14 births for every 1,000 people, approximately how many births occurred in Jodhpur in 2008?

(1) 6,000 (2) 42,275

(3) 60,000. (4) 140,000

127. Consider two forests: one is an undisturbed old-growth forest, while the other has recently been logged. In which forest are species likely to experience exponential growth, and why?

(1) Old growth, because of stable conditions that would favor exponential growth of all species in the forest

(2) Old growth, because each of the species is well established and can produce many offspring

(3) Logged, because the disturbed forest affords more resources for increased specific populations to grow

(4) Logged, because the various populations are stimulated to a higher reproductive potential

128. Evidence shows that some grasses benefit from being grazed. Which of the following terms would best describe this plant-herbivore interaction?

(1) Mutualism. (2) Commensalism

(3) Parasitism (4) Competition

129. As N approaches K for a certain population, which of the following is predicted by the logistic equation?

(1) The growth rate will not change

(2) The growth rate will approach zero

(3) The population will show an Allee effect

(4) The population will increase exponentially

130. In models of logistic population growth,

(1) the population growth rate slows dramatically as *N* approaches *K*

(2) new individuals are added to the population most rapidly at the beginning of the population's growth

(3) new individuals are added to the population as *N* approaches *K*

(4) only density-dependent factors affect the rate of population growth

131. The Allee effect is used to describe a population that

(1) has become so small that it will have difficulty surviving and reproducing

(2) has become so large that it will have difficulty surviving and reproducing

(3) is viable and stable at its carrying capacity

(4) has exceeded its carrying capacity

132. Carrying capacity (K) is

(1) seldom reached by marine producers and consumers because of the vast resources of the ocean

(2) the maximum population size that a particular environment can support

(3) fixed for most species over most of their range most of the time

(4) determined by density and dispersion data

133. Which of the following causes populations to shift most quickly from an exponential to a logistic population growth?

(1) increased birth rate

(2) removal of predators

(3) decreased death rate

(4) competition for resources

134. Which of the following statements about the evolution of life histories is correct?

(1) Stable environments with limited resources favor *r*-selected populations

(2) *K*-selected populations are most often found in environments where density-independent factors are important regulators of population size

(3) Most populations have both *r*- and *K*-selected characteristics that vary under different environmental conditions

(4) The reproductive efforts of *r*-selected populations are directed at producing just a few offspring with good competitive abilities

135. Natural selection involves energetic trade-offs between

(1) choosing how many offspring to produce over the course of a lifetime and how long to live

(2) producing large numbers of gametes when employing internal fertilization versus fewer numbers of gametes when employing external fertilization

(3) the emigration of individuals when they are no longer reproductively capable or committing suicide

(4) high survival rates of offspring and the cost of parental care

136. The three basic variables that make up the life history of an organism are

(1) life expectancy, birth rate, and death rate

(2) number of reproductive females in the population, age structure of the population, and life expectancy

(3) age when reproduction begins, how often reproduction occurs, and how many offspring are produced per reproductive episode

(4) how often reproduction occurs, life expectancy of females in the population, and number of offspring per reproductive episode

137. Pacific salmon and annual plants are excellent examples of

(1) cohort disintegration

(2) dispersion

(3) iteroparous reproduction

(4) semelparous reproduction

138. Which of the following pairs of reproductive strategies is consistent with energetic trade-off and reproductive success?

(1) Pioneer species of plants produce many very small, highly airborne seeds, whereas large elephants that are very good parents produce many offspring

(2) Female rabbits that suffer high predation rates may produce several litters per breeding season, and coconuts produce few fruits, but most survive when they encounter proper growing conditions

(3) Species that have to broadcast to distant habitats tend to produce seeds with heavy protective seed coats, and animals that are caring parents produce fewer offspring with lower infant mortality

(4) Free-living insects lay thousands of eggs and provide no parental care, whereas flowers take good care of their seeds until they are ready to germinate

139. Which of the following is characteristic of *K*-selected populations?

(1) Offspring with good chances of survival

(2) Many offspring per reproductive episode

(3) Small offspring

(4) A high intrinsic rate of increase

140. Which variables define the ecological life history of a species?

(1) The age at which reproduction begins, frequency of reproduction, and the number of offspring for each reproductive episode

(2) The ratio of females to males, the length of the breeding season, and the number of offspring for each reproductive episode

(3) The number of offspring produced over a lifetime by a breeding pair and the survivability of the offspring

(4) Timing breeding sessions with optimal environmental conditions and the number of offspring produced during each breeding session

141. Which pattern of reproduction is correctly paired with a species?

(1) Iteroparity: Pacific salmon

(2) Iteroparity: elephant

(3) Semelparity: oak tree

(4) Semelparity: rabbit

142. Often the growth cycle of one population has an effect on the cycle of another. As moose populations increase, for example, wolf populations also increase. Thus, if we are considering the logistic equation for the wolf population,

$$dN/dt = rN\frac{(K-N)}{K}$$

which of the factors accounts for the effect on the moose population?

(1) r

(2) N

(3) rN

(4) K

143. In which of the following situations would you expect to find the largest number of *K*-selected individuals?

(1) A recently abandoned agricultural field in Ohio

(2) The sand dune communities of south Lake Michigan

(3) The flora and fauna of a coral reef in the Caribbean

(4) South Florida after a hurricane

144. Which of the following is most likely to contribute to density-dependent regulation of populations?

(1) The removal of toxic waste by decomposers

(2) Intraspecific competition for nutrients

(3) Earthquakes

(4) Floods

145. Why do populations grow more slowly as they approach their carrying capacity?

(1) Density-dependent factors lead to fewer births and increased mortality

(2) Density-independent factors lead to fewer births and increased mortality

(3) Hormonal changes promote higher death rates in crowded populations

(4) Individuals voluntarily stop mating so that overcrowding does not occur

146. A population of white-footed mice becomes severely overpopulated in a habitat that has been disturbed by human activity. Sometimes intrinsic factors cause the population to increase in mortality and lower reproduction rates to occur in reaction to the stress of overpopulation. Which of the following is an example of intrinsic population control?

(1) Owl populations frequent the area more often because of increased hunting success

(2) Females undergo hormonal changes that delay sexual maturation and many individuals suffer depressed immune systems and die due to the stress of overpopulation

(3) Clumped dispersion of the population leads to increased spread of disease and parasites, resulting in a population crash

(4) All of the resources (food and shelter) are used up by overpopulation and much of the population dies of exposure and/or starvation

147. Why is territoriality an adaptive behavior for songbirds maintaining populations at or near their carrying capacity?

(1) Songbirds expend a tremendous amount of energy defending territories so that they spend less time feeding their young and fledgling mortality increases

(2) Only the fittest males defend territories and they attract the fittest females so the best genes are conveyed to the next generation

(3) Songbird males defend territories commensurate with the size from which they can derive adequate resources for themselves, their mate, and their chicks

(4) Many individuals are killed in the agonistic behaviors that go along with territorial defense

148. Which of the following could be a density-independent factor limiting human population growth?

(1) social pressure for birth control

(2) earthquakes

(3) plagues

(4) famines

149. An ecological footprint is a construct that is useful

(1) for a person living in a developed nation to consider to make better choices when using global food and energy resources

(2) for a person living in a developing country to see how much of the world's resources are left for him/her

(3) in converting human foods' meat biomass to plant biomass

(4) in making predictions about the global carrying capacity of humans

150. Which of the following was the most significant limiting factor in human population growth in the 20th century?

(1) Famine

(2) Non-HIV disease

(3) HIV

(4) Clean water

151. Which statement is true with regard to human population growth?

(1) It is at a zero reproduction rate

(2) Its rate of increase continues to grow at an exponential rate

(3) Its rate of growth is slowing

(4) Its rate of growth is increasing

152. Which curve best describes survivorship in marine molluscs?

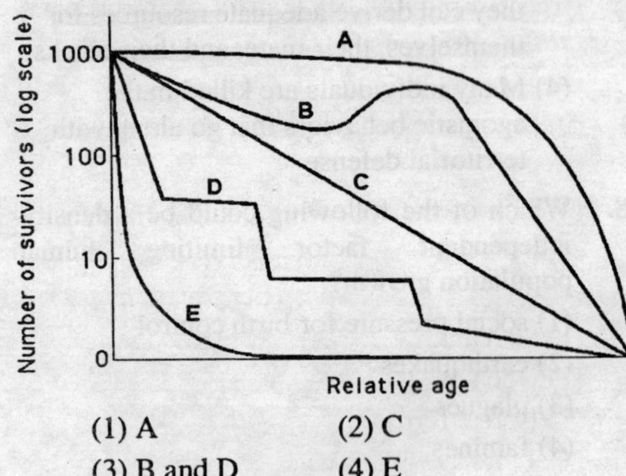

(1) A (2) C
(3) B and D (4) E

153. In ecosystems, why is the term *cycling* used to describe material transfer, whereas the term *flow* is used for energy exchange?

(1) Materials are repeatedly used, but energy flows through and out of ecosystems
(2) Both material and energy are recycled and are then transferred to other ecosystems as in a flow
(3) Materials are cycled into ecosystems from other ecosystems, but energy constantly flows within the ecosystem
(4) Both material and energy flow in a never-ending stream within an ecosystem

154. Which statement most accurately describes how matter and energy are used in ecosystems?

(1) Matter is cycled through ecosystems; energy is not
(2) Energy is cycled through ecosystems; matter is not
(3) Energy can be converted into matter; matter cannot be converted into energy
(4) Matter can be converted into energy; energy cannot be converted into matter

155. The law of conservation of matter states that matter cannot be created, yet matter is sometimes gained or lost to an ecosystem. What is the reason for this seeming contradiction?

(1) Chemoautotrophic organisms can convert matter to energy
(2) Matter can be moved in/out of an ecosystem from/to another ecosystem
(3) Photosynthetic organisms convert solar energy to sugars
(4) Detrivores convert matter to energy

156. Photosynthetic organisms are unique to most ecosystems because they

(1) synthesize organic compounds they obtain from decaying heterotrophs
(2) synthesize inorganic compounds from organic compounds
(3) use light energy to synthesize organic compounds
(4) use chemical energy to synthesize organic compounds

157. A cow's herbivorous diet indicates that it is a(n)

(1) primary consumer
(2) secondary consumer
(3) decomposer
(4) autotroph

158. To recycle nutrients, an ecosystem must have, at a minimum

(1) producers
(2) producers and decomposers
(3) producers, primary consumers, and decomposers
(4) producers, primary consumers, secondary consumers, and decomposers

159. If the sun were to suddenly stop providing energy to Earth, most ecosystems would vanish. Which of the following ecosystems would likely survive the longest after this hypothetical disaster?

(1) Tropical rain forest
(2) Tundra
(3) Benthic ocean
(4) Grassland

60. Which of the following is true of detrivores?

(1) They recycle chemical elements directly back to primary consumers

(2) They synthesize organic molecules that are used by primary producers

(3) They convert organic materials from all trophic levels to inorganic compounds usable by primary producers

(4) They secrete enzymes that convert the organic molecules of detritus into CO_2 and H_2O

161. The major role of detrivores in ecosystems is to

(1) provide a nutritional resource for heterotrophs

(2) recycle chemical nutrients to a form capable of being used by autotrophs

(3) prevent the buildup of the organic remains of organisms, feces, and so on

(4) return energy lost to the ecosystem by other organisms

162. Approximately 1% of the solar radiation that strikes a plant is converted into the chemical bond energy of sugars. Why is this amount so low?

(1) Approximately 99% of the solar radiation is converted to heat energy

(2) Only 1% of the wavelengths of visible light are absorbed by photosynthetic pigments

(3) Most solar energy strikes water and land surfaces

(4) Approximately 99% of the solar radiation is reflected

163. Subtraction of which of the following will convert gross primary productivity into net primary productivity?

(1) The energy contained in the standing crop

(2) The energy used by heterotrophs in respiration

(3) The energy used by autotrophs in respiration

(4) The energy fixed by photosynthesis

164. Which of these ecosystems accounts for the largest amount of Earth's net primary productivity?

(1) Tundra

(2) Salt marsh

(3) Open ocean

(4) Tropical rain forest

165. Which of these ecosystems has the highest net primary productivity per square meter?

(1) Savanna

(2) Open ocean

(3) Boreal forest

(4) Tropical rain forest

166. Which data is most useful to measure primary productivity in a terrestrial ecosystem?

(1) Temperature readings

(2) Potential evapotranspiration

(3) Intensity of solar radiation

(4) Annual precipitation

167. Which of the following is a true statement regarding mineral nutrients in soils and their implication for primary productivity?

(1) Globally, phosphorous availability is most limiting to primary productivity

(2) Adding a nonlimiting nutrient will stimulate primary productivity

(3) Adding more of a limiting nutrient will increase primary productivity, indefinitely

(4) Phosphorous is sometimes unavailable to producers due to leaching

168. The total biomass of photosynthetic autotrophs present in an ecosystem is known as

(1) gross primary productivity

(2) standing crop

(3) net primary productivity

(4) secondary productivity

169. How is it that the open ocean produces the highest net primary productivity of Earth's ecosystems, yet net primary productivity per square meter is relatively low?

(1) Oceans contain greater concentrations of nutrients compared to other ecosystems
(2) Oceans receive a lesser amount of solar energy per unit area
(3) Oceans have the largest area of all the ecosystems on Earth
(4) Ocean ecosystems have less species diversity

170. Why is net primary production (NPP) a more useful measurement to an ecosystem ecologist than gross primary production (GPP)?

(1) NPP can be expressed in energy/unit of area/unit of time
(2) NPP can be expressed in terms of carbon fixed by photosynthesis for an entire ecosystem
(3) NPP represents the stored chemical energy that is available to consumers in the ecosystem
(4) NPP is the same as the standing crop

171. How is Net Ecosystem Production (NEP) typically estimated in ecosystems?

(1) The ratio of producers to consumers
(2) The amount of heat energy released by the ecosystem
(3) The net flux of CO_2 or O_2 in or out of an ecosystem
(4) The rate of decomposition by detrivores

172. Aquatic primary productivity is most limited by which of the following?

(1) Light and nutrient availability
(2) Predation by primary consumers
(3) Increased pressure with depth
(4) Pollution

173. Aquatic ecosystems are least likely to be limited by which of the following nutrients?

(1) Nitrogen (2) Carbon
(3) Phosphorus (4) Iron

174. What is the primary limiting factor for aquatic productivity?

(1) Pressure
(2) Lack of nutrients
(3) Light availability
(4) Herbivores

175. Which of the following ecosystems would likely have a larger net primary productivity/hectare and why?

(1) Open ocean because of the total biomass of photosynthetic autotrophs
(2) Grassland because of the small standing crop biomass that results from consumption by herbivores and rapid decomposition
(3) Tropical rain forest because of the massive standing crop biomass and species diversity
(4) Cave due to the lack of photosynthetic autotrophs

176. How is it that satellites can detect differences in primary productivity on Earth?

(1) Photosynthetic organisms absorb more visible light in the 350-750 wavelengths
(2) Satellite instruments can detect reflectance patterns of the photosynthetic organisms of different ecosystems
(3) Sensitive satellite instruments can measure the amount of NADPH produced in the summative light reactions of different ecosystems
(4) Satellites detect differences by comparing the wavelengths of light captured and reflected by photoautotrophs to the amount of light reaching different ecosystems

177. Which of the following lists of organisms is ranked in correct order from lowest to highest percent in production efficiency?

(1) Mammals, fish, insects
(2) Insects, fish, mammals
(3) Fish, insects, mammals
(4) Insects, mammals, fish

178. The amount of chemical energy in consumers' food that is converted to their own new biomass during a given time period is known as which of the following?
(1) Biomass
(2) Secondary production
(3) Biomagnification
(4) Primary production

179. What is secondary production?
(1) Energy converted by secondary consumers from primary consumers
(2) Solar energy that is converted to chemical energy by photosynthesis
(3) Food that is converted to new biomass by consumers
(4) Energy that is not used by consumers for growth and reproduction

180. How does inefficient transfer of energy among trophic levels result in the typically high endangerment status of many top-level predators?
(1) Top-level predators are destined to have small populations that are sparsely distributed
(2) Predators have relatively large population sizes
(3) Predators are more disease-prone than animals at lower trophic levels
(4) Predators have short life spans and short reproductive periods

181. Trophic efficiency is
(1) The ratio of net secondary production to assimilation of primary production
(2) The percentage of production transferred from one trophic level to the next
(3) A measure of how nutrients are cycled from one trophic level to the next
(4) Usually greater than production efficiencies

182. Owls eat rats, mice, shrews, and small birds. Assume that, over a period of time, an owl consumes 5,000 J of animal material. The owl loses 2,300 J in feces and owl pellets and uses 2,500 J for cellular respiration. What is the primary efficiency of this owl?
(1) 0.02% (2) 1%
(3) 4% (4) 10%

183. Why does a vegetarian leave a smaller ecological footprint than an omnivore?
(1) Fewer animals are slaughtered for human consumption
(2) There is an excess of plant biomass in all terrestrial ecosystems
(3) Vegetarians need to ingest less chemical energy than omnivores
(4) Eating meat is an inefficient way of acquiring photosynthetic productivity

184. For most terrestrial ecosystems, pyramids of numbers, biomass, and energy are essentially the same- they have a broad base and a narrow top. The primary reason for this pattern is that
(1) secondary consumers and top carnivores require less energy than producers
(2) at each step, energy is lost from the system as a result of keeping the organisms alive
(3) as matter passes through ecosystems, some of it is lost to the environment
(4) biomagnification of toxic materials limits the secondary consumers and top carnivores

185. Which of the following is primarily responsible for limiting the number of trophic levels in most ecosystems?
(1) Many primary and higher-order consumers are opportunistic feeders
(2) Decomposers compete with higher-order consumers for nutrients and energy
(3) Nutrient cycles involve both abiotic and biotic components of ecosystems
(4) Energy transfer between tropic levels is in almost all cases less than 20% efficient

186. Which trophic level is most vulnerable to extinction?
(1) Producer level
(2) Primary consumer level
(3) Secondary consumer level
(4) Tertiary consumer level

187. Which of the following statements is consistent with the principle of competitive exclusion?
(1) Bird species generally do not compete for nesting sites
(2) The random distribution of one competing species will have a positive impact on the population growth of the other competing species
(3) Two species with the same fundamental niche will exclude other competing species
(4) Even a slight reproductive advantage will eventually lead to the elimination of the less well adapted of two competing species

188. According to the competitive exclusion principle, two species cannot continue to occupy the same
(1) habitat
(2) niche
(3) territory
(4) range

189. Which of the following best describes resource partitioning?
(1) Competitive exclusion that results in the success of the superior species
(2) Sight variations in niche that allow similar species to coexist
(3) Two species that can coevolve to share identical niches
(4) Differential resource utilization that results in a decrease in community species diversity

190. As you study two closely related predatory insect species, the two-spot and the three-spot avenger beetles, you notice that each species seeks prey at dawn in areas without the other species. However, where their ranges overlap, the two-spot avenger beetle hunts at night and the three-spot hunts in the morning. When you bring them into the laboratory and isolate the two different species, you discover that the offspring of both species are found to be nocturnal. You have discovered an example of
(1) mutualism
(2) character displacement
(3) Batesian mimicry
(4) resource partitioning

191. Resource partitioning would be most likely to occur between
(1) sympatric populations of a predator and its prey
(2) sympatric populations of species with similar ecological niches
(3) sympatric populations of a flowering plant and its specialized insect pollinator
(4) allopatric populations of the same animal species

192. Which of the following is an example of cryptic coloration?
(1) Bands on a coral snake
(2) Brown or gray color of tree bark
(3) Markings of a viceroy butterfly's wings
(4) A "walking stick" insect that resembles a twig

193. Which of the following is an example of Müllerian mimicry?
(1) Two species of unpalatable butterfly that have the same color pattern
(2) A day-flying hawkmoth that looks like a wasp
(3) A chameleon that changes its color to look like a dead leaf
(4) Two species of rattlesnakes that both rattle their tails

194. Which of the following is an example of Batesian mimicry?
(1) An insect that resembles a twig
(2) A butterfly that resembles a leaf
(3) A nonvenomous snake that looks like a venomous snake
(4) A fawn with fur coloring that camouflages it in the forest environment

195. Which of the following would be most significant in understanding the structure of an ecological community?
(1) Determining how many species are present overall
(2) Determining which particular species are present
(3) Determining the kinds of interactions that occur among organisms of different species
(4) Determining how many species are present overall, which particular species are present, the kinds of interactions that occur among organisms of different species, and the relative abundance of species

196. White-breasted nuthatches and Downy woodpeckers both eat insects that hide in the furrows of bark in hardwood trees. The Downy woodpecker searches for insects by hunting from the bottom of the tree trunk to the top, whereas the white-breasted nuthatch searches from the top of the trunk down. These hunting behaviors best illustrate which of the following ecological concepts?
(1) Competitive exclusion
(2) Resource partitioning
(3) Character displacement
(4) Keystone species

197. Prairie dogs once covered the expanses of the Great Plains. Their grazing made the grass more nutritious for the huge herds of bison, and they were preyed upon by a variety of snakes, raptors, and mammals. In fact, the black-footed ferret (now endangered) specialized in prairie dog predation. Today, increases in housing and agricultural developments have eradicated many prairie dog towns. Which of the following statements about prairie dogs is true?
(1) Their realized niche has expanded
(2) They have a competitive relationship with bison
(3) They are probably a poor candidate for keystone species
(4) Their fundamental niche has expanded

198. How might an ecologist test whether a species is occupying its realized or its fundamental niche?
(1) Study the temperature range and humidity requirements of the species
(2) Observe if the niche size changes after the addition of nutritional resources to the habitat
(3) Observe if the niche size changes after the introduction of a similar non-native species
(4) Remove a competitor species to see if the species expands its range

199. According to bottom-up and top-down control models of community organization, which of the following expressions would imply that an increase in the size of a carnivore (C) population would negatively impact on its prey (P) population, but not vice versa?
(1) $P \leftarrow C$. (2) $P \rightarrow C$
(3) $C \leftrightarrow P$ (4) $P \leftarrow C \rightarrow P$

200. Ecosystem stability would be more when
(1) there is linear food chains composed mainly of grazers
(2) there are more number of carnivores
(3) there is less then 1% energy flow among trophic level
(4) food web is complex and multiples species interacting among themselves

ANSWER SHEET

Unit 10: Ecology

1. 3 **2.** 3 **3.** 3 **4.** 3 **5.** 3 **6.** 2 **7.** 1 **8.** 4 **9.** 2 **10.** 2

11. 3 **12.** 2 **13.** 1 **14.** 4 **15.** 4 **16.** 4 **17.** 1 **18.** 2 **19.** 1 **20.** 2

21. 2 **22.** 4 **23.** 3 **24.** 2 **25.** 1 **26.** 3 **27.** 3 **28.** 1 **29.** 3 **30.** 4

31. 2 **32.** 4 **33.** 4 **34.** 1 **35.** 3 **36.** 1 **37.** 1 **38.** 4 **39.** 4 **40.** 1

41. 1 **42.** 1 **43.** 1 **44.** 3 **45.** 2 **46.** 1 **47.** 3 **48.** 1 **49.** 3 **50.** 4

51. 1 **52.** 1 **53.** 2 **54.** 1 **55.** 4 **56.** 4 **57.** 3 **58.** 1 **59.** 4 **60.** 3

61. 1 **62.** 3 **63.** 3 **64.** 1 **65.** 1 **66.** 1 **67.** 4 **68.** 3 **69.** 2 **70.** 2

71. 1 **72.** 4 **73.** 1 **74.** 3 **75.** 3 **76.** 2 **77.** 3 **78.** 1 **79.** 1 **80.** 2

81. 1 **82.** 3 **83.** 1 **84.** 4 **85.** 1 **86.** 4 **87.** 2 **88.** 3 **89.** 4 **90.** 1

91. 2 **92.** 2 **93.** 1 **94.** 1 **95.** 1 **96.** 2 **97.** 4 **98.** 4 **99.** 3 **100.** 4

101. 1 **102.** 1 **103.** 1 **104.** 2 **105.** 1 **106.** 3 **107.** 1 **108.** 4 **109.** 4 **110.** 1

111. 3 **112.** 1 **113.** 4 **114.** 3 **115.** 1 **116.** 2 **117.** 3 **118.** 4 **119.** 4 **120.** 3

121. 4 **122.** 2 **123.** 2 **124.** 1 **125.** 1 **126.** 3 **127.** 3 **128.** 1 **129.** 2 **130.** 3

131. 2 **132.** 2 **133.** 4 **134.** 3 **135.** 4 **136.** 3 **137.** 4 **138.** 2 **139.** 1 **140.** 1

141. 2 **142.** 4 **143.** 3 **144.** 2 **145.** 1 **146.** 2 **147.** 3 **148.** 2 **149.** 1 **150.** 4

151. 3 **152.** 4 **153.** 1 **154.** 1 **155.** 2 **156.** 3 **157.** 1 **158.** 2 **159.** 3 **160.** 3

161. 2 **162.** 2 **163.** 3 **164.** 3 **165.** 4 **166.** 4 **167.** 4 **168.** 2 **169.** 3 **170.** 3

171. 3 **172.** 1 **173.** 1 **174.** 2 **175.** 2 **176.** 4 **177.** 1 **178.** 2 **179.** 3 **180.** 1

181. 2 **182.** 3 **183.** 4 **184.** 2 **185.** 4 **186.** 4 **187.** 4 **188.** 2 **189.** 2 **190.** 4

191. 1 **192.** 4 **193.** 1 **194.** 3 **195.** 4 **196.** 2 **197.** 4 **198.** 4 **199.** 1 **200.** 4

UNIT-11

EVOLUTION

1. The extreme eusocial behavior in honey bees where sterile female helps queen is rearing offsprings is explained as an example of
   (1) kin selection
   (2) group selection
   (3) sexual selection
   (4) natural selection

2. The Hardy-Wienberg principle comes from considering what happens when Medelian genes act on population. The model predicts that there will be no change in allele frequencies when
   (1) migration into the population occurs at a steady rate
   (2) reproduction alone is the factor effecting the population
   (3) a rare new mutation is associated with a sharp increase in fitness
   (4) natural selection process is at work

3. Among the following events in history of life
   a. prokaryotic cell
   b. eukaryotic cell
   c. natural selection
   d. organic molecules
   e. self replicating molecule
   Which is the correct chronological order?
   (1) d→e→c→a→b
   (2) d→e→a→b→c
   (3) e→d→a→c→b
   (4) d→e→a→c→b

4. Sexual selection results in variation in the reproductive success of males, often due to female choice with particular phenotypes. This type of sexual selection is because
   (1) males cannot compete with other males
   (2) cost of breeding is higher for females as compared to males
   (3) inappropriate mating results in a similar reduction in fitness of females and males
   (4) males are a limiting resource for females

5. A monophyletic group
   (1) contains unrelated organisms
   (2) includes the most recent common ancestors but not all of its decedents
   (3) includes all the representatives of a clade but not the most recent common ancestor
   (4) contains all the representative of a clade and most recent common ancestor

6. Which of the following is expected have highest basal metabolic rate?
   (1) Rat (2) Cat
   (3) Dog (4) Elephant

7. Imagine that you observe an increase in the frequency of brown coloration genes and a decrease in the frequency of green coloration genes in a beetle population. Any combination of the mechanisms of microevolution might be responsible for the pattern, but which is least likely be true?
   (1) Natural selection or differential reproduction
   (2) Gene flow or migration
   (3) Mutation/ genetic drift
   (4) Random matting

8. Which of the following is not a post-zygotic isolating mechanism?
(1) F_1 hybrid break down
(2) Developmental/Segregational hybrid sterility
(3) F_2 breakdown
(4) Gametic isolation

9. N. Eldredge and S. J. Gould suggested that speciation itself, and the morphological changes accompanying speciation, occur rapidly, separated by long periods of time when little change occurs (stasis) They called their model
(1) Phyletic gradualism
(2) Cladogenesis
(3) Neutral hypothesis
(4) Punctuated equilibrium

10. If a polymorphism arises within a parasitic species that allows an individual with a certain genotype to adapt to a new host, this genotype may be the forerunner of a new species. If the parasite not only feeds on the new host but also mates on the new host, a barrier to gene flow arises, although the parasite may be surrounded by other members of its species with the original genotype. Such a mode of speciation is termed as
(1) Sympatric speciation
(2) Allopatric speciation
(3) Parapatric speciation
(4) Peripatric speciation

11. Often the identical mutation is found in many children born to the same parents. Neither parent carries the mutation. A likely explanation is that the mutation occurred in
(1) the germ line of both parents after meiosis
(2) the soma of both parents
(3) the germ line of one parent before meiosis
(4) all the affected zygotes

12. A rare X-linked disorder is present in the human population among males at a frequency of 1 in 10. The proportion of females who will suffer from same disease is about
(1) 1 in 2 (2) 1 in 10
(3) 1 in 20 (4) 1 in 100

13. The DNA from which taxon would be the best to use to prove the connection between India and Seychelles in geological time?
(1) Bats (2) Birds
(3) Frogs (4) Sharks

14. The number of copies of an autosomal gene in a population of diploid species of N individuals is 2N. Assuming that the gender ratio in the population equals one, the number of copies of a gene on the X chromosome in a population of N individuals will be
(1) 2N (2) 1.5N
(3) N (4) 0.5N

15. Two identical DNA sequences of 100 base pairs length accumulate nucleotide substitutions at the same rate of one change per year or 0.01 change/site/year. The percent nucleotide difference between these two sequences after infinite time will be
(1) 100 (2) 75
(3) 50 (4) 25

16. Which of the following trees shows the right evolutionary relationship between bird, tortoise, crocodile, lizard and mammal?

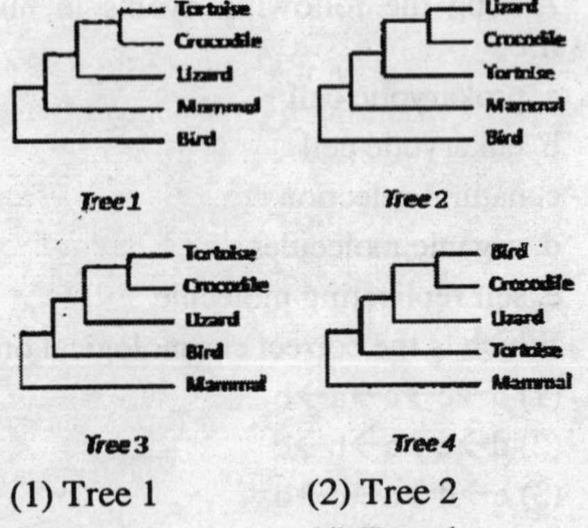

(1) Tree 1 (2) Tree 2
(3) Tree 3 (4) Tree 4

17. DNA sequences of the mitochondrial cytochrome b gene from two species of birds were compared and found to have 4 differences between them. Nevertheless when the cytochrome b protein sequences were compared they were identical in the two species. This is because the differences at the DNA sequence were most probably

(1) due to insertions

(2) at the second position of the codon

(3) at the third position of the codon

(4) due to deletions

18. Which cellular features do you expect to find in neuronal pathways mediating escape responses in animals?

(1) Long, small-diameter axons

(2) Long, large-diameter axons

(3) Short, small diameter axons

(4) Short, large-diameter axons

19. A monkey is briefly presented with a mirror and allowed to see its own image in it. A red spot is then applied to its forehead, a few minute later, it is again presented with the mirror in which its own image is visible. The monkey then wipes the spot off its forehead. For which ability does this experiment provide supporting evidence?

(1) Visual memory

(2) Long term memory

(3) Self awareness

(4) Deception

20. For a behavior to be considered altruistic by an evolutionary biologist, which of the following conditions should be demonstrated?

(1) The behavior should have no cost to the donor and increased the individual fitness of the recipient

(2) The behavior should increases the individual fitness of donor and increases the individual fitness of the recipient

(3) The behavior should decreases the individual fitness of the donar and increases the individual fitness of the recipient

(4) The behavior should increase the individual fitness of the recipient; the effect on the individual fitness of the donor is not relevant

21. The effective population size of a population of elephants is

(1) Larger than the absolute population size

(2) Equal to the total number of males and females

(3) Between the total number of adults and total population sixe

(4) Smaller than the total population size

22. In a Drosophila population in Hardy-Wienberg equilibrium, 9% of the flies have white eyes. If white eyes are caused by a recessive allele at one locus, what is the expected frequency of the wild type allele in that population?

(1) 70 % (2) 30 %

(3) 49 % (4) 21 %

23. The genetic relatedness between you and your grandfather's brother is

(1) 0.1 (2) 0.125

(3) 0.25 (4) 0.5

24. Which of the following is least likely to cause reduction of genetic variations in a population?

(1) Drift

(2) Inbreeding

(3) Non-random mating

(4) Immigration

25. In a population, the genotype frequencies at a locus with two alleles A1 and A2 are A1A1:0.25, A1A2:0.50 and A2A2:0.25. After 5 generations of selection the frequency of the A1 allele goes up by 20 %. Assuming Hardy-Weinberg equilibrium for the fifth generation, what is the percent increase in the A1A1 genotype?

(1) 20 % (2) 25 %

(3) 36 % (4) 44 %

26. The Red Queen evolutionary dynamics refers to which of the following types of selection processes?
(1) Coevolution
(2) Convergent evolution
(3) Adaptive radiation
(4) Disruptive selection

27. When the operational sex ratio in a population switches from being highly male-biased to highly female-biased, mating systems theory predicts a corresponding switch from
(1) monogamy to polygyny
(2) polygyny to polyandry
(3) polyandry to polygyny
(4) polyandry to mongoamy

28. In carrier pigeons there is rare inherited condition that caused the death of the chicks before hatching. In order for this disease to be passed from generation to generation there must be parent birds that
(1) are heterozygous for the disease
(2) have the disease themselves
(3) produce new mutations for this disease
(4) are closely interbred

29. The relative abundance of individuals in a community can be described by a model known as the "broken stick" where the stick can be broken in different ways to produce a distribution of pieces that represents the proportion of species in the community. A method that will produce very unequal pieces is
(1) Random breakage of the stick
(2) Uniform breakage of the stick
(3) Sequential breakage of the stick where the larges piece is further sub divided
(4) Sequential breakage of the stick where the smallest piece is further subdivided

30. The hardy-weinberg equation for genotype frequency for X-linked allele in a diploid population would be
(1) $p^2 + 2pq + q^2 = 1$
(2) $(p + q)^2 = 1$
(2) $(p + q + x + z) = 1$
(4) $p^2 + 2pq + q^2 + p + q = 1$

31. Hardy Weinberg law could be extended for three allele at a locus. Which of the collect representation for genotype frequency for three alleles in a population?
(1) $p^2 + 2pq + q^2 = 1$
(2) $(p + q)^2 = 1$
(2) $(p + q)^3 = 1$
(4) $(p + q + r)^2 = 1$

32. Two closely related insect species are found on the same tree. If they mate, they produce sterile offspring. One of the species is active in the early evening, and the other in the late evening. The two species present a case of
(1) co-evolution
(2) reproductive isolation
(3) competition
(4) allopatric speciation

33. The bee orchid mimics a female bee and attracts male bees for
(1) pollination
(2) copulation
(3) seed dispersal
(4) scaring away herbivores

34. A certain Drosophila mutant grows faster in a medium containing chemical X. The wild type flies were accidently grown in a medium containing X, and it was observed that 4% of the flies were faster growing type. What percentage of the population is expected to be homozygous wild type if the genes are in Hardy-Weinberg equilibrium?
(1) 8%
(2) 16%
(3) 32%
(4) 64%

35. Isolated islands such as the Hawaiian and Galapagos islands have a higher proportion of closely related endemic species occupying different niches than the mainland. Which of the following statements is not a plausible explanation for the above observation?

(1) Very few species from the mainland colonize these islands

(2) Species that colonize these islands eventually radiate into multiple species

(3) Isolated islands often have numerous unoccupied niches

(4) These islands have higher rates of species extinction

36. Ears of different moth species are found on different parts of the body, including the thorax, wings, abdomen and proboscis. This suggests that

(1) moths are evolving more rapidly than other insect groups

(2) ears in moths arose as a single event in their phylogenetic history

(3) ears in moths arose multiple times in their phylogenetic history

(4) the structure of moth ears is different from those of other insect groups

37. You carefully study populations of two very similar field mice, one from Rajasthan and one from Gujarat. You want to know whether the populations belong to the species or to two different species. You could most confidently decide this if you could

(1) show that the ranges of the two mice overlap without hybridization occurring

(2) bring the two types of mice into the laboratory to see if they wil1 mate

(3) demonstrate that the natural ranges of the two types of mice are entirely allopatric

(4) show that the Rajasthan mice live in wetter habitats than the Gujarat mice

38. Which of the following is not a major shift in species richness observable in the fossil record?

(1) The Cambrian explosion

(2) The Permian decline of invertebrate families

(3) The flowering of angiosperms in the Cretaceous

(4) The Devonian decimation of large mammals

39. Charles Darwin discussed all of the following except

(1) natural selection tends to remove those organisms that are poorly adapted to their environments

(2) individuals in a population compete with one another for limited resources

(3) gene mutations are the source of variation for evolution

(4) organisms tend to produce more offspring than can survive in each generation

40. Branching evolution in which different populations of a species become reproductively isolated from each other by adapting to different ecological niches and eventually become separate species is termed as

(1) orthogenesis

(2) convergent evolution

(3) speciation

(4) adaptive radiation

41. Without genetic variation, some of the basic mechanisms of evolutionary change cannot operate. Which of the following will not be responsible for bring changes in gene pool

(1) mutations

(2) gene flow

(3) sexual reproduction/genetic shuffling

(4) random matting and homogeneous environmental conditions

42. Which organisms may be regarded as relics from an early period in the evolution of photosynthesis?
(1) Purple bacteria and green sulfur bacteria
(2) Cyanobacteria
(3) Algae
(4) Primitive plants

43. Which of the following best describes vestigial structures?
(1) Vestigial structures are similar structurally and functionally to comparable structures in other organisms
(2) Vestigial structures are small structures that are not present in close relatives
(3) Vestigial structures are well-developed structures that have a different function in close relatives
(4) Vestigial structures are functionless or rudimentary homologs of characters that are functional in close relatives

44. Homologous characters are
(1) similar in function, but not in structure
(2) features that show underlying structural similarity even if their superficial structure is different
(3) features that appear superficially similar even if their underlying structure is different
(4) known only from gross anatomy, not from any other biological feature

45. Uniformitarianism means that __________
(1) geological processes (like the deposition of sediments at beaches) have operated at the same rate throughout the Earth's history
(2) the Earth is very young
(3) extinctions cannot happen
(4) geological processes operating now operated the same way in the past

46. Which of the following attributes does not relate to homoplasy?
(1) Especially common in DNA sequence data sets
(2) Often due to reversal or convergent evolution
(3) "Noise" in data sets used to construct phylogeny
(4) Especially common in SINE and LINE data sets

47. A growing body of literature suggests that mutations are surprisingly frequent on a per genome/per generation basis; most are __________ or __________
(1) neutral/slightly beneficial
(2) neutral/slightly deleterious
(3) slightly deletrious/extremely deleterious
(4) slightly beneficial/highly beneficial

48. The Hardy-Weinberg formula allows geneticists to determine a population's gene frequencies from direct observation of it
(1) gene pool
(2) collective genotype
(3) collective phenotype
(4) DNA

49. An organism is capable of oxidizing H_2S and using the energy obtained from the reaction to reduce carbon dioxide. Which type of metabolism does this organism possess?
(1) Photolithotrophic autotroph
(2) Chemoorganotrophic heterotroph
(3) Photoorganotrophic heterotroph
(4) Chemolithotrophic autotroph

50. The students of school are tested for the tasting of PTC. The ability to taste PTC is dominant (T) over the inability to taste it (t) Among 1242 students, 67.8% were found to be tasters. What are the expected frequency of two alleles?
(1) T=0.700 and t=0.300
(2) T=0.680 and t=0.320
(3) T= 0.433 and t=0.567
(4) T=0.564 and t=0.436

51. The neutral theory of molecular evolution was proposed by
(1) Ronald Fisher (2) Sewall Wright
(3) Motoo Kimura (4) Gregor Mendel

52. In which species concept morphological and molecular characters are used to distinguish species?
(1) Genetic (2) Ecological
(3) Biological (4) Phylogenetic

53. Two nucleotide sequences found in different species are almost exactly same. This suggests that these species
(1) are evolving into the same species
(2) contain identical genome
(3) may share same ancestors
(4) have same number of mutations

54. Given the phylogeny below, which one of the following statement is correct?

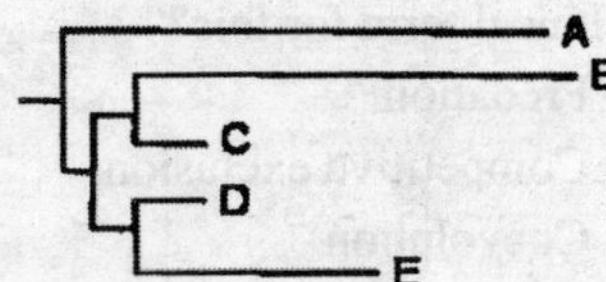

(1) Species A and B are sister taxa
(2) Species C is more closely related to D than B
(3) Species D is more closely related to C than E
(4) Species D and E share an immediate common ancestor

55. Which of the following was not present on earth before life evolved?
(1) Hydrogen (2) Ammonia
(3) Oxygen (4) Methane

56. During which geological period did the greatest diversification of life on earth occur?
(1) Permian (2) Jurassic
(3) Cambrian (4) Ordovician

57. The change in a trait with time as a result of natural selection is shown below. This type of natural selection is

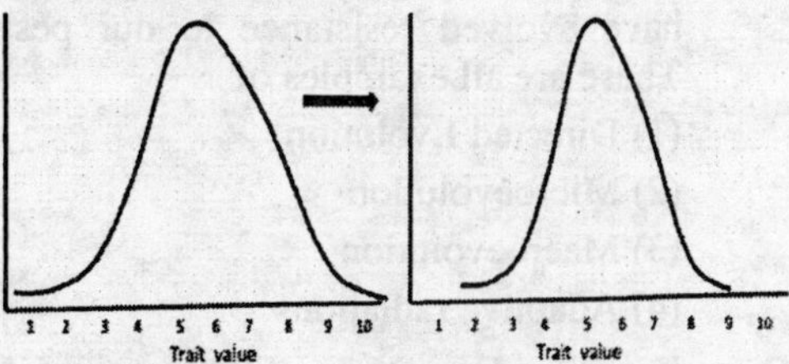

(1) directional. (2) disruptive
(3) stabilizing. (4) random

58. At present, the relationship between human and monkey can be stated as
(1) they both have common ancestors
(2) human have evolved from monkey
(3) both have very distinct phylogeny
(4) relationship can not be established

59. Imagine that in one generation, two brown beetles happened to have four offspring survive to reproduce. Several green beetles were killed when someone stepped on them and had no offspring. The next generation would have a few more brown beetles than the previous generation but just by chance. These chance changes from generation to generation are known
(1) natural selection
(2) sexual selection
(3) genetic drift
(4) over-dominance

60. Since all cells in our body contain DNA, there are lots of places for mutations to occur; however, not all mutations matter for evolution. Which of the following mutations are most likely to favor natural selection?
(1) Synonymous substitution (Neutral mutation)
(2) Non-synonymous substitution (Directed mutation)
(3) Non-sense mutations
(4) Silent mutations

61. House sparrows have adapted to the climate of North America, mosquitoes have evolved in response to global warming, and insects have evolved resistance to our pesticides. These are all examples of

(1) Directed Evolution
(2) Microevolution
(3) Macroevolution
(4) Adaptive radiations

62. Fossil evidence suggest that terrestrial plants evolved around

(1) 25 million years ago
(2) 50 million years ago
(3) 100 million years ago
(4) 500 million years ago

63. One reason that amoeba cannot be as large as a whale is that

(1) there are simply is not enough food in the oceans to support such a large amoeba
(2) the volume would be much too small for its surface area
(3) the surface area would be too small to hold that much volume
(4) the amount of surface area per unit volume would be too small for sufficient gas exchange

64. The speciation in which a population splits into two geographically isolated populations experience dissimilar selective pressure and genetic drift is known as

(1) sympatric speciation
(2) parapatric speciation
(3) peripatric speciation
(4) allopatric speciation

65. Which of the following is not an explanation for the preponderance of sexual reproduction?

(1) Increases genetic variability
(2) Winning in the arm races with parasites
(3) Avoidance of inbreeding
(4) Adaptation to heterogeneous environments

66. Which of the following insects have male helpers?

(1) Honey bees (2) Ants
(3) Termites (4) Wasps

67. What is required for sympatric speciation?

(1) Geographical isolation
(2) Inbreeding
(3) Rapid environmental fluctuation
(4) Reproductive isolation

68. Which of the following assumption support the Hardy-Weinberg Equilibrium?

(1) Presence of Natural Selection
(2) Random mating
(3) Genetic Drift
(4) Assortive mating

69. Two species or populations are competing for the exact same resources, and one will eventually exclude the other. What is the technical term for this?

(1) Predation
(2) Competitive exclusion
(3) Coevolution
(4) Mutualism

70. About 80% of moth species possess ears that are tuned exclusively to ultrasound. Only about 1% of all moth species produce ultrasonic matting calls. This suggests that hearing in moths

(1) is a vestigial trait
(2) evolved primarily in response to the predation pressure imposed by bats
(3) evolved as a result of sexual selection
(4) evolved as a mechanism to prevent inter-specific hybridization

71. Which of the following resting prey organisms would a pit viper find most difficult to detect at long range in total darkness?

(1) A mouse
(2) A warbler
(3) A cuckoo nestling
(4) A frog

72. A northern population of frog species that is sympatric with three other frog species was found to have lower call frequencies than a southern population of the same frog species. This is an example of
(1) Adaptive radiation
(2) Allopatric speciation
(3) Character displacement
(4) Convergence

73. Successful artificial selection, such as that practiced by plant and animal breeders to modify their crops and livestock, depends on which of the following?
(1) Breeders must be able to selectively breed the individuals with the most desirable traits
(2) Some individuals must have more desirable traits than others
(3) When individuals with the most desirable traits are bred, their traits must be passed from parents to offspring
(4) All of these are necessary for artificial selection to take place

74. Which is the best definition of Darwinian fitness?
(1) The ability of an individual to survive and reproduce in any potential environment, compared to other individuals
(2) The ability of a species to survive over time, compared to other species
(3) The ability of an individual to survive and reproduce in a certain environment, compared to other individuals
(4) The ability of a population to survive in any environment, compared to other populations

75. Which of the following statements is true?
(1) Natural selection generally leads to perfection of existing traits
(2) Natural selection acts directly on phenotypes, and only indirectly on genotypes.
(3) Natural selection cannot be progressive because of its random nature
(4) Natural selection usually results in an immediate increase in population variation

76. Which of the following is a true statement regarding pre-adaptations?
(1) A trait that evolved for one function, but that coincidentally is also suited for a new, different function
(2) A trait that is currently evolving for a function that will be needed in the future
(3) A trait that is imperfect due to genetic or developmental constraints
(4) A trait that is in the initial stages of evolving for a certain function, but has not yet been perfected

77. The creationist "argument from design" is based on the idea that complex structures and organisms can't arise by chance. The argument fails because, although ________ is random, ________ is not; it is directed in the sense of increasing adaptation
(1) mutation/selection
(2) evolution/selection
(3) selection/mutation
(4) adaptation/mutation

78. Which is actual evidence supporting the endosymbiotic theory for the origin of eukaryotes?
(1) Fossil evidence suggest early eukaryotes ate proteobacteria
(2) Free-living mitochondria still exist in some environments
(3) Mitochondria and chloroplasts have their own DNA and have 70S ribosomes
(4) Only prokaryotes are autotrophic

79. The degree of genetic relatedness between the offspring and their parents is
(1) higher than that between sister & brother
(2) lower than that between sister & brother
(3) the same as that between sister & brother
(4) dependent on the number of siblings

80. Which of the following is changing the fastest over evolutionary time?
(1) The amount of intergenic DNA
(2) The order of genes on chromosomes
(3) Microsatellites
(4) Exon DNA sequences

81. An isolated human population, with approximately equal number of blue eyed and brown eyed individuals, was killed due to earthquake. Only a few brown eyed people remained to form the next generation. This kind of change in the gene pool is called
(1) Hardy- Weinberg equilibrium
(2) blocked gene flow
(3) bottleneck effect
(4) founder effect

82. A pair of genes in two organisms of different species which are similar and they are strongly predicated to have the same function is known as
(1) homologous genes
(2) orthologous genes
(3) paralogous genes
(4) isoforms

83. During protein evolution the region of protein most prone to mutation is
(1) functional domain
(2) structurally conserved domain
(3) connective loops
(4) hydrophobic domain

84. Which of the following obtains energy from the oxidation of inorganic or organic chemicals?
(1) Chemotroph (2) Lithotroph
(3) Autotroph (4) Heterotroph

85. A set of closely related genes or genetic markers that are inherited as a single unit is
(1) cistron (2) gene families
(3) haplotype (4) haploid

86. In a population of 200 individuals which is at equilibrium, the frequency of one of the alleles under study is 0.11. What is the expected frequency of heterozygous individual?
(1) 0.89 (2) 0.0979
(3) 0.1958 (4) 0.842

87. Increased genetic diversity following extended time in a tissue culture is a problem called
(1) gene alteration
(2) temporal modification
(3) somaclonal variation
(4) culture shock

88. Balanced genetic polymorphism occurs when there is selection against
(1) heterozygotes
(2) all genotypes
(3) all homozygotes
(4) only homozygous recessive

89. "Living fossils" are the
(1) organisms which have become fossilized in recent geological era
(2) organisma that have become fossilized recently and all the structural proteins are not yet completely denatured
(3) ancient organisms persisting to modern times without further morphological evolution
(4) ancient organisms persisting to modern times with further morphological evolution

90. In animal kingdom, the group amniota includes
(1) birds and reptiles
(2) birds and mammals
(3) reptiles and mammals
(4) reptiles, birds and mammals

91. A lithoautotroph
(1) derives energy from reduced compounds of mineral origin
(2) can be bacterial or eukaryotic cell
(3) is always an extremophile
(4) requires light for producing

92. Cystic fibrosis is a recessive condition that affects about 1 in 2,500 babies in the Caucasian population. The frequency of heterozygotes or carriers of cystic fibrosis is
(1) 1 in 12500 (2) 1 in 25
(3) 1 in 625 (4) 1 in 125

93. Which of the following would cause deviation from the Hardy-Weinberg equilibrium?
(1) Small population size
(2) Random mating
(3) Lack of selection pressure
(4) No mutation

94. Mating between individuals which are closely related by ancestry is called
(1) Genetic assortative mating
(2) Genetic disassortative mating
(3) Random mating
(4) Poly cross

95. Genes related through descent from a common ancestral gene are called
(1) Orthologous (2) Homologous
(3) Heterologous (4) Paralogous

96. Catastrophism, meaning the regular occurrence of geological or meteorological disturbances (catastrophes), was Cuvier's attempt to explain the existence of
(1) evolution
(2) the fossil record
(3) uniformitarianism
(4) the origin of new species

97. With what other idea of his time was Cuvier's theory of catastrophism most in conflict?
(1) Gradualism
(2) The fixity of species
(3) Island biogeography
(4) Uniformitarianism

98. During a study session about evolution, one of your fellow students remarks, "The giraffe stretched its neck while reaching for higher leaves; its offspring inherited longer necks as a result." Which statement is most likely to be helpful in correcting this student's misconception?
(1) Characteristics acquired during an organism's life are generally not passed on through genes
(2) Spontaneous mutations can result in the appearance of new traits
(3) Only favorable adaptations have survival value
(4) Disuse of an organ may lead to its eventual disappearance

99. In the mid-1900s, the Soviet geneticist Lysenko believed that his winter wheat plants, exposed to ever-colder temperatures, would eventually give rise to ever more cold-tolerant winter wheat. Lysenko's attempts in this regard were most in agreement with the ideas of
(1) Cuvier
(2) Hutton
(3) Lamarck
(4) Darwin

100. Charles Darwin was the first person to propose
(1) that evolution occurs
(2) a mechanism for how evolution occurs
(3) that Earth is older than a few thousand years
(4) a mechanism for evolution that was supported by evidence

101. Which of these conditions should completely prevent the occurrence of natural selection in a population over time?
(1) All variation between individuals is due only to environmental factors
(2) The environment is changing at a relatively slow rate
(3) The population size is large
(4) The population lives in a habitat where there are no competing species present

102. Natural selection is based on all of the following *except*

(1) genetic variation exists within populations and populations tend to produce more individuals than the environment can support

(2) the best-adapted individuals tend to leave the most offspring

(3) individuals who survive longer tend to leave more offspring than those who die young

(4) individuals adapt to their environments and, thereby, evolve

103. Which of the following represents an idea that Darwin learned from the writings of Thomas Malthus?

(1) Technological innovation in agricultural practices will permit exponential growth of the human population into the foreseeable future

(2) Populations tend to increase at a faster rate than their food supply normally allows

(3) Earth changed over the years through a series of catastrophic upheavals

(4) The environment is responsible for natural selection

104. Given a population that contains genetic variation, what is the correct sequence of the following events, under the influence of natural selection?

1. Well-adapted individuals leave more offspring than do poorly adapted individuals

2. A change occurs in the environment

3. Genetic frequencies within the population change

4. Poorly adapted individuals have decreased survivorship

(1) 2 → 4 → 1 → 3

(2) 4 → 2 → 1 → 3

(3) 4 → 1 → 2 → 3

(4) 4 → 2 → 3 → 1

105. A biologist studied a population of squirrels for 15 years. During that time, the population was never fewer than 30 squirrels and never more than 45. Her data showed that over half of the squirrels born did not survive to reproduce, because of both competition for food and predation. In a single generation, 90% of the squirrels that were born lived to reproduce, and the population increased to 80. Which inference(s) about this population might not be true?

(1) The amount of available food may have increased

(2) The parental generation of squirrels developed better eyesight due to improved diet; the subsequent squirrel generation inherited better eyesight

(3) The squirrels of subsequent generations should show greater levels of genetic variation than previous generations, because squirrels that would not have survived in the past will now survive

(4) This is probably a good example of natural selection

106. Which of the following must exist in a population before natural selection can act upon that population?

(1) Genetic variation among individuals

(2) Variation among individuals caused by environmental factors

(3) Sexual reproduction

(4) Random matting

107. Which of Darwin's ideas had the strongest connection to Darwin having read Malthus's essay on human population growth?

(1) Descent with modification

(2) Variation among individuals in a population

(3) Struggle for existence

(4) The ability of related species to be conceptualized in "tree thinking"

108. If Darwin had been aware of genes, and of their typical mode of transmission to subsequent generations, with which statement would he most likely have been in agreement?

(1) If natural selection can change one gene's frequency in a population over the course of generations then, given enough time and enough genes, natural selection can cause sufficient genetic change to produce new species from old ones

(2) If an individual's somatic cell genes change during its lifetime, making it more fit, then it will be able to pass these genes on to its offspring

(3) If an individual acquires new genes by engulfing, or being infected by, another organism, then a new genetic species will be the result

(4) A single mutation in a single gene in a single gamete will, if perpetuated, produce a new species within just two generations

109. The role that humans play in artificial selection is to

(1) determine who lives and who dies

(2) create the genetic variants, which nature then selects

(3) choose which organisms breed, and which do not

(4) train organisms to breed more successfully

110. Currently, two extant elephant species (X & Y) are placed in the genus *Loxodonta*, and a third species (Z) is placed in the genus *Elephas*. Thus, which statement should be true?

(1) Species X and Y are not related to species Z

(2) Species X and Y share a greater number of homologies with each other than either does with species Z

(3) Species X and Y share a common ancestor that is still extant (in other words, not yet extinct)

(4) Species X and Y are the result of artificial selection from an ancestral species Z

111. The rise of methicillin-resistant *Staphylococcus aureus* (MRSA) can be considered to be an example of artificial selection because

(1) humans purposefully raise MRSA in large fermenters in an attempt to make the bacteria ever-more resistant

(2) *S. aureus* is cultivated by humans to replenish the soil with nutrients

(3) humans synthesize methicillin and create environments in which bacteria frequently come into contact with methicillin

(4) humans are becoming resistant to bacteria by taking methicillin

112. In a hypothetical environment, fishes called pike-cichlids are visual predators of algae-eating fish (in other words, they locate their prey by sight) If a population of algae-eaters experiences predation pressure from pike-cichlids, which of the following is least likely to be observed in the algae-eater population over the course of many generations?

(1) Selection for drab coloration of the algae-eaters

(2) Selection for nocturnal algae-eaters (active only at night)

(3) Selection for larger female algae-eaters, bearing broods composed of more, and larger, young

(4) Selection for algae-eaters that become sexually mature at smaller overall body sizes

113. DDT was once considered a "silver bullet" that would permanently eradicate insect pests. Today, instead, DDT is largely useless against many insects. Which of these would have been required for this pest eradication effort to be successful in the long run?

(1) Larger doses of DDT should have been applied

(2) All habitats should have received applications of DDT at about the same time

(3) The frequency of DDT application should have been higher

(4) None of the individual insects should have possessed genomes that made them resistant to DDT

114. If the bacterium *Staphylococcus aureus* experiences a cost for maintaining one or more antibiotic-resistance genes, then what should happen in environments from which antibiotics are missing?

(1) These genes should continue to be maintained in case the antibiotics ever appear

(2) These bacteria should be outcompeted and replaced by bacteria that have lost these genes

(3) The bacteria should try to make the cost worthwhile by locating, and migrating to, microenvironments where traces of antibiotics are present

(4) The bacteria should start making and secreting their own antibiotics

115. Of the following anatomical structures, which is homologous to the bones in the wing of a bird?

(1) Cartilage in the dorsal fin of a shark

(2) Bones in the hind limb of a kangaroo

(3) Chitinous struts in the wing of a butterfly

(4) Bones in the flipper of a whale

116. If two modern organisms are distantly related in an evolutionary sense, then one should expect that

(1) they live in very different habitats

(2) they should share fewer homologous structures than two more closely related organisms

(3) their chromosomes should be very similar

(4) they shared a common ancestor relatively recently

117. Over evolutionary time, many cave-dwelling organisms have lost their eyes. Tapeworms have lost their digestive systems. Whales have lost their hind limbs. How can natural selection account for these losses?

(1) Natural selection cannot account for losses, only for innovations

(2) Natural selection accounts for these losses by the principle of use and disuse

(3) Under particular circumstances that persisted for long periods, each of these structures presented greater costs than benefits

(4) The ancestors of these organisms experienced harmful mutations that forced them to find new habitats that these species had not previously used

118. Which of the following pieces of evidence most strongly supports the common origin of all life on Earth?

(1) All organisms require energy

(2) All organisms use essentially the same genetic code

(3) All organisms reproduce

(4) All organisms show heritable variation

119. Logically, which of these should cast the most doubt on the relationships depicted by an evolutionary tree?

(1) None of the organisms depicted by the tree ate the same foods

(2) Some of the organisms depicted by the tree had lived in different habitats

(3) The skeletal remains of the organisms depicted by the tree were incomplete (in other words, some bones were missing)

(4) Relationships between DNA sequences among the species did not match relationships between skeletal patterns

120. Which of the following statements most detracts from the claim that the human appendix is a completely vestigial organ?

(1) The appendix can be surgically removed with no immediate ill effects

(2) The appendix might have been larger in fossil hominids

(3) The appendix has a substantial amount of defensive lymphatic tissue

(4) Individuals with a larger-than-average appendix leave fewer offspring than those with a below-average-sized appendix

121. Members of two different species possess a similar-looking structure that they use in a similar fashion to perform the same function. Which information would best help distinguish between an explanation based on homology versus one based on convergent evolution?
(1) The two species live at great distance from each other
(2) The two species share many proteins in common, and the nucleotide sequences that code for these proteins are almost identical
(3) The sizes of the structures in adult members of both species are similar in size
(4) Both species are well adapted to their particular environments

122. Both ancestral birds and ancestral mammals shared a common ancestor that was terrestrial. Today, penguins (which are birds) and seals (which are mammals) have forelimbs adapted for swimming. What term best describes the relationship of the bones in the forelimbs of penguins & seals, and what term best describes the flippers of penguins and seals?
(1) homologous; homologous
(2) analogous; homologous
(3) homologous; analogous
(4) analogous; analogous

123. What must be true of any organ that is described as vestigial?
(1) It must be analogous to some feature in an ancestor
(2) It must be homologous to some feature in an ancestor
(3) It must be both homologous and analogous to some feature in an ancestor
(4) It need be neither homologous nor analogous to some feature in an ancestor

124. What is true of pseudogenes?
(1) They are composed of RNA, rather than DNA
(2) They are the same things as introns
(3) They are unrelated genes that code for the same gene product
(4) They are vestigial genes

125. It has been observed that organisms on islands are different from, but closely related to, similar forms found on the nearest continent. This is taken as evidence that
(1) island forms and mainland forms descended from common ancestors
(2) common environments are inhabited by the same organisms
(3) the islands were originally part of the continent
(4) the island forms and mainland forms are converging

126. If one wanted to find the largest number of endemic species, one should visit which of the following geological features (assuming each has existed for several millions of years)?
(1) An isolated ocean island in the tropics
(2) An extensive mountain range
(3) A midcontinental grassland with extreme climatic conditions
(4) A shallow estuary on a warm-water coast

127. A high degree of endemism is most likely in environments that are
(1) easily reached and heterogeneous
(2) isolated and heterogeneous
(3) isolated and homogeneous
(4) isolated and extremely cold

128. Which of the following is *not* an observation or inference on which natural selection is based?
(1) There is heritable variation among individuals
(2) Poorly adapted individuals never produce offspring
(3) Species produce more offspring than the environment can support
(4) Individuals whose characteristics are best suited to the environment generally leave more offspring than those whose characteristics are less well suited

129. Within six months of effectively using methicillin to treat *S. aureus* infections in a community, all new infections were caused by MRSA. How can this result best be explained?

(1) *S. aureus* can resist vaccines
(2) A patient must have become infected with MRSA from another community
(3) In response to the drug, *S. aureus* began making drug-resistant versions of the protein targeted by the drug
(4) Some drug-resistant bacteria were present at the start of treatment, and natural selection increased their frequency

130. The upper forelimbs of humans and bats have fairly similar skeletal structures, whereas the corresponding bones in whales have very different shapes and proportions. However, genetic data suggest that all three kinds of organisms diverged from a common ancestor at about the same time. Which of the following is the most likely explanation for these data?

(1) Humans and bats evolved by natural selection, and whales evolved by Lamarckian mechanisms
(2) Forelimb evolution was adaptive in people and bats, but not in whales
(3) Natural selection in an aquatic environment resulted in significant changes to whale forelimb anatomy
(4) Genes mutate faster in whales than in humans or bats

131. DNA sequences in many human genes are very similar to the sequences of corresponding genes in chimpanzees. The most likely explanation for this result is that

(1) humans and chimpanzees share a relatively recent common ancestor
(2) humans evolved from chimpanzees
(3) chimpanzees evolved from humans
(4) convergent evolution led to the DNA similarities

132. During an individual organism's lifetime, which of these is most likely to help the organism respond properly to changes in its environment?

(1) Microevolution
(2) Change in allele or gene frequency
(3) Change in gene expression
(4) Change in average heterozygosity

133. If, on average, 46% of the loci in a species' gene pool are heterozygous, then the average homozygosity of the species should be

(1) 23%
(2) 46%
(3) 54%
(4) There is not enough information to say

134. Which of these variables is likely to undergo the largest change in value as the result of a mutation that introduces a brand-new allele into a population's gene pool at a locus that had formerly been fixed?

(1) Average heterozygosity
(2) Nucleotide variability
(3) Geographic variability
(4) Average number of loci

135. Which statement about the beak size of finches on the island of Daphne Major during prolonged drought is true?

(1) Each bird evolved a deeper, stronger beak as the drought persisted
(2) Each bird's survival was strongly influenced by the depth and strength of its beak as the drought persisted
(3) Each bird that survived the drought produced only offspring with deeper, stronger beaks than seen in the previous generation
(4) The frequency of the strong-beak alleles increased in each bird as the drought persisted

136. Although each of the following has a better chance of influencing gene frequencies in small populations than in large populations, which one most consistently requires a small population as a precondition for its occurrence?

(1) Mutation
(2) Nonrandom mating
(3) Genetic drift
(4) Natural selection

137. In modern terminology, diversity is understood to be a result of genetic variation. Which of the following is a recognized source of variation for evolution?

(1) Mistakes in translation of structural genes
(2) Mistakes in protein folding
(3) Rampant changes to the dictionary of the genetic code
(4) Recombination by crossing over in meiosis

138. A trend toward the decrease in the size of plants on the slopes of mountains as altitudes increase is an example of

(1) a cline
(2) a bottleneck
(3) relative fitness
(4) genetic drift

139. The higher the proportion of loci that are "fixed" in a population, the lower is that population's

(1) nucleotide variability only
(2) genetic polyploidy only
(3) average heterozygosity only
(4) nucleotide variability and average heterozygosity only

140. Which statement about variation is true?

(1) All phenotypic variation is the result of genotypic variation
(2) All genetic variation produces phenotypic variation
(3) All nucleotide variability results in neutral variation
(4) All new alleles are the result of nucleotide variability

141. Rank the following one-base point mutations (from most likely to least likely) with respect to their likelihood of affecting the structure of the corresponding polypeptide

1. insertion mutation deep within an intron
2. substitution mutation at the third position of an exonic codon
3. substitution mutation at the second position of an exonic codon
4. deletion mutation within the first exon of the gene

(1) 1, 2, 3, 4
(2) 4, 3, 2, 1
(3) 2, 1, 4, 3
(4) 3, 1, 4, 2

142. Most invertebrates have a cluster of ten similar *Hox* genes, all located on the same chromosome. Most vertebrates have four such clusters of *Hox* genes, located on four nonhomologous chromosomes. The process that could have potentially contributed to the cluster's presence on more than one chromosome was ________

(1) binary fission
(2) translation
(3) gene duplication
(4) nondisjunction

143. Which of the following is a true statement concerning genetic variation?

(1) It is created by the direct action of natural selection
(2) It arises in response to changes in the environment
(3) It must be present in a population before natural selection can act upon the population
(4) It tends to be reduced by the processes involved when diploid organisms produce gametes

144. Whenever diploid populations are in Hardy-Weinberg equilibrium at a particular locus
(1) the allele's frequency should not change from one generation to the next, but its representation in homozygous and heterozygous genotypes may change
(2) natural selection, gene flow, and genetic drift are acting equally to change an allele's frequency
(3) this means that, at this locus, two alleles are present in equal proportions
(4) the population itself is not evolving, but individuals within the population may be evolving

145. In the formula for determining a population's genotype frequencies, the 2 in the term *2pq* is necessary because
(1) the population is diploid
(2) heterozygotes can come about in two ways
(3) the population is doubling in number
(4) heterozygotes have two alleles

146. In the formula for determining a population's genotype frequencies, the *pq* in the term *2pq* is necessary because
(1) the population is diploid
(2) heterozygotes can come about in two ways
(3) the population is doubling in number
(4) heterozygotes have two alleles

147. Hardy-Weinberg equilibrium must occur in populations wherein
(1) an allele remains fixed
(2) no genetic variation exists
(3) natural selection is operating
(4) random drift is operating

148. In a Hardy-Weinberg population with two alleles, *A* and *a*, that are in equilibrium, the frequency of the allele *a* is 0.3. What is the percentage of the population that is homozygous for this allele?
(1) 0.09 (2) 0.49
(3) 0.9 (4) 9.0

149. In a Hardy-Weinberg population with two alleles, *A* and *a*, that are in equilibrium, the frequency of allele *a* is 0.2. What is the percentage of the population that is heterozygous for this allele?
(1) 0.2 (2) 2.0
(3) 4.0 (4) 32.0

150. In a Hardy-Weinberg population with two alleles, *A* and *a*, that are in equilibrium, the frequency of allele *a* is 0.1. What is the frequency of individuals with *AA* genotype?
(1) 0.20
(2) 0.32
(3) 0.42
(4) 0.81

151. You sample a population of butterflies and find that 56% are heterozygous at a particular locus. What should be the frequency of the recessive allele in this population?
(1) 0.07
(2) 0.08
(3) 0.70
(4) Cannot be predicted from given information

152. In peas, a gene controls flower color such that *R* = purple and *r* = white. In an isolated pea patch, there are 36 purple-flowering plants and 64 white-flowering plants. Assuming Hardy-Weinberg equilibrium, what is the value of *q* for this population?
(1) 0.36
(2) 0.64
(3) 0.75
(4) 0.80

153. Over time, the movement of people on Earth has steadily increased. This has altered the course of human evolution by increasing
(1) nonrandom mating
(2) geographic isolation
(3) genetic drift
(4) gene flow

154. Swine are vulnerable to infection by bird flu virus and human flu virus, which can both be present in an individual pig at the same time. When this occurs, it is possible for genes from bird flu virus and human flu virus to be combined, thereby producing a genetically distinctive virus, which can subsequently cause widespread disease. The production of new types of flu virus in the manner described above is most similar to the phenomenon of

(1) bottleneck effect
(2) founder effect
(3) natural selection
(4) gene flow

155. If the original finches that had been blown over to the Galápagos from South America had already been genetically different from the parental population of South American finches, even before adapting to the Galápagos, this would have been an example of

(1) Natural selection
(2) bottleneck effect
(3) Sexual selection
(4) Genetic drift by founder effect

156. What is true of natural selection?

(1) Natural selection is a random process
(2) Natural selection creates beneficial mutations
(3) The only way to eliminate harmful mutations is through natural selection
(4) Mutations occur at random; natural selection can preserve and distribute beneficial mutations

157. The restriction enzymes of bacteria protect the bacteria from successful attack by bacteriophages, whose genomes can be degraded by the restriction enzymes. The bacterial genomes are not vulnerable to these restriction enzymes because bacterial DNA is methylated. This situation selects for bacteriophages whose genomes are also methylated. As new strains of resistant bacteriophages become more prevalent, this in turn selects for bacteria whose genomes are not methylated and whose restriction enzymes instead degrade methylated DNA. The outcome of the conflict between bacteria and bacteriophage at any point in time results from

(1) frequency-dependent selection
(2) evolutionary imbalance
(3) heterozygote advantage
(4) neutral variation

158. The restriction enzymes of bacteria protect the bacteria from successful attack by bacteriophages, whose genomes can be degraded by the restriction enzymes. The bacterial genomes are not vulnerable to these restriction enzymes because bacterial DNA is methylated. This situation selects for bacteriophages whose genomes are also methylated. As new strains of resistant bacteriophages become more prevalent, this in turn selects for bacteria whose genomes are not methylated and whose restriction enzymes instead degrade methylated DNA. Over the course of evolutionary time, what should occur?

(1) Methylated DNA should become fixed in the gene pools of bacterial species
(2) Nonmethylated DNA should become fixed in the gene pools of bacteriophages
(3) Methylated DNA should become fixed in the gene pools of bacteriophages
(4) Methylated and nonmethylated strains should be maintained among both bacteria and bacteriophages, with ratios that vary over time

159. Arrange the following from most general (i.e., most inclusive) to most specific (i.e., least inclusive)

1 natural selection 2. microevolution
3. intrasexual selection
4. evolution 5. sexual selection

(1) 4, 1, 2, 3, 5 (2) 4, 2, 1, 3, 5
(3) 4, 2, 1, 5, 3 (4) 1, 4, 2, 5, 3

160. Sexual dimorphism is most often a result of
(1) pansexual selection
(2) stabilizing selection
(3) intrasexual selection
(4) intersexual selection

161. In the wild, male house finches (*Carpodus mexicanus*) vary considerably in the amount of red pigmentation in their head and throat feathers, with colors ranging from pale yellow to bright red. These colors come from carotenoid pigments that are found in the birds' diets; no vertebrates are known to synthesize carotenoid pigments. Thus, the brighter red the male's feathers are, the more successful he has been at acquiring the red carotenoid pigment by his food-gathering efforts (all other factors being equal) During breeding season, one should expect female house finches to prefer to mate with males with the brightest red feathers. Which of the following is true of this situation?
(1) Alleles that promote more efficient acquisition of carotenoid-containing foods by males should increase over the course of generations
(2) Alleles that promote more effective deposition of carotenoid pigments in the feathers of males should increase over the course of generations
(3) There should be directional selection for bright red feathers in males
(4) All of the above

162. During breeding season, one should expect female house finches to prefer to mate with males with the brightest red feathers. Which of the following terms are appropriately applied to this situation?
(1) Sexual selection
(2) Mate choice
(3) Intersexual selection
(4) All of the above

163. Adult male humans generally have deeper voices than do adult female humans, which is the direct result of higher levels of testosterone causing growth of the larynx. If the fossil records of apes and humans alike show a trend toward decreasing larynx size in adult females and increasing larynx size in adult males, then
(1) sexual dimorphism was developing over time in these species
(2) intrasexual selection seems to have occurred
(3) stabilizing selection was occurring in these species concerning larynx size
(4) selection was acting more directly upon genotype than upon phenotype

164. Adult male humans generally have deeper voices than do adult female humans, which is the direct result of higher levels of testosterone causing growth of the larynx. If one excludes the involvement of gender in the situation, then the pattern that is apparent in the fossil record is most similar to one that should be expected from
(1) pansexual selection
(2) directional selection
(3) disruptive selection
(4) stabilizing selection

165. Which of the following statements best summarizes evolution as it is viewed today?
(1) It represents the result of selection for acquired characteristics
(2) It is synonymous with the process of gene flow
(3) It is the descent of humans from the present-day great apes
(4) It is the differential survival and reproduction of the most-fit phenotypes

166. Which of the following is most likely to produce an African butterfly species in the wild whose members have one of two strikingly different color patterns?
(1) Artificial selection
(2) Directional selection
(3) Stabilizing selection
(4) Disruptive selection

167. Most Swiss starlings produce four to five eggs in each clutch. Starlings producing fewer, or more, than this have reduced fitness. Which of the following terms best describes this situation?

(1) Artificial selection
(2) Directional selection
(3) Stabilizing selection
(4) Disruptive selection

168. The recessive allele that causes phenylketonuria (PKU) is harmful, except when an infant's diet lacks the amino acid phenylalanine. What maintains the presence of this harmful allele in a population's gene pool?

(1) Heterozygote advantage
(2) Stabilizing selection
(3) Diploidy
(4) Balancing selection

169. Heterozygote advantage should be most closely linked to which of the following?

(1) Sexual selection
(2) Stabilizing selection
(3) Random selection
(4) Directional selection

170. In seedcracker finches from Cameroon, small- and large-billed birds specialize in cracking soft and hard seeds, respectively. If long-term climatic change resulted in all seeds becoming hard, what type of selection would then operate on the finch population?

(1) Disruptive selection
(2) Directional selection
(3) Stabilizing selection
(4) No selection would operate because the population is in Hardy-Weinberg equilibrium

171. When imbalances occur in the sex ratio of sexual species that have two sexes (i.e., other than a 50:50 ratio), the members of the minority sex often receive a greater proportion of care and resources from parents than do the offspring of the majority sex. This is most clearly an example of

(1) sexual selection
(2) disruptive selection
(3) balancing selection
(4) frequency-dependent selection

172. The same gene that causes various coat patterns in wild and domesticated cats also causes the cross-eyed condition in these cats, the cross-eyed condition being slightly maladaptive. In a hypothetical environment, the coat pattern that is associated with crossed eyes is highly adaptive, with the result that both the coat pattern and the cross-eyed condition increase in a feline population over time. Which statement is supported by these observations?

(1) Evolution is progressive and tends toward a more perfect population
(2) Phenotype is often the result of compromise
(3) Natural selection reduces the frequency of maladaptive genes in populations over the course of time
(4) Polygenic inheritance is generally maladaptive, and should become less common in future generations

173. A proficient engineer can easily design skeletal structures that are more functional than those currently found in the forelimbs of such diverse mammals as horses, whales, and bats. The actual forelimbs of these mammals do not seem to be optimally arranged because

(1) natural selection has not had sufficient time to create the optimal design in each case, but will do so given enough time
(2) in many cases, phenotype is not merely determined by genotype, but by the environment as well
(3) though we may not consider the fit between the current skeletal arrangements and their functions excellent, we should not doubt that natural selection ultimately produces the best design
(4) natural selection is generally limited to modifying structures that were present in previous generations and in previous species

174. There are those who claim that the theory of evolution cannot be true because the apes, which are supposed to be closely related to humans, do not likewise share the same large brains, capacity for complicated speech, and tool-making capability. They reason that if these features are generally beneficial, then the apes should have evolved them as well. Which of these provides the best argument against this misconception?

(1) Advantageous alleles do not arise on demand
(2) A population's evolution is limited by historical constraints
(3) Adaptations are often compromises
(4) Evolution can be influenced by environmental change

175. In a very large population, a quantitative trait has the following distribution pattern

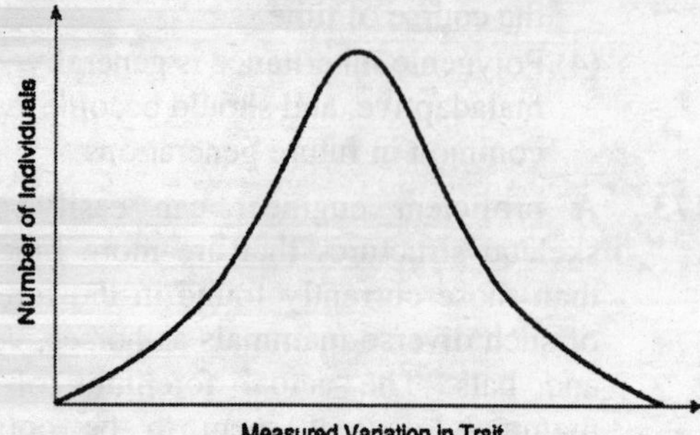

What is true of the trait whose frequency distribution in a large population appears in the previous figure? It has probably undergone

(1) directional selection
(2) stabilizing selection
(3) disruptive selection
(4) normal selection

176. HIV's genome of RNA includes the code for reverse transcriptase (RT), an enzyme that acts early in infection to synthesize a DNA genome off of an RNA template. The HIV genome also codes for protease (PR), an enzyme that acts later in infection by cutting long viral polyproteins into smaller, functional proteins. Both RT and PR represent potential targets for antiretroviral drugs. Drugs called nucleoside analogs (N(1) act against RT, whereas drugs called protease inhibitors (PI) act against PR. Which of the following represents the treatment option most likely to avoid the production of drug-resistant HIV (assuming no drug interactions or side effects)?

(1) Using a series of NAs, one at a time, and changed about once a week
(2) Using a single PI, but slowly increasing the dosage over the course of a week
(3) Using high doses of NA and a PI at the same time for a period not to exceed one day
(4) Using moderate doses of NA and two different PIs at the same time for several months

177. A large population of laboratory animals has been allowed to breed randomly for a number of generations. After several generations, 25% of the animals display a recessive trait (*aa*), the same percentage as at the beginning of the breeding program. The rest of the animals show the dominant phenotype, with heterozygotes indistinguishable from the homozygous dominants. What is the most reasonable conclusion that can be drawn from the fact that the frequency of the recessive trait (*aa*) has not changed over time?

(1) The population is undergoing genetic drift
(2) The two phenotypes are about equally adaptive under laboratory conditions
(3) The genotype *AA* is lethal
(4) There has been a high rate of mutation of allele *A* to allele *a*

178. You are studying three populations of birds. Population A has ten birds, of which one is brown (a recessive trait) and nine are red. Population B has 100 birds, of which ten are brown. Population C has 30 birds, and three of them are brown. In which population would it be least likely that an accident would significantly alter the frequency of the brown allele?

(1) Population A

(2) Population B

(3) Population C

(4) They are all the same

179. In those parts of equatorial Africa where the malaria parasite is most common, the sickle-cell allele constitutes 20% of the β hemoglobin alleles in the human gene pool. What should be the proportion of heterozygous individuals in populations that live here?

(1) 0.04 (2) 0.16

(3) 0.20 (4) 0.32

180. If the sickle-cell allele is recessive, what proportion of the population should be susceptible to sickle-cell anemia under typical conditions?

(1) 0.04 (2) 0.16

(3) 0.20 (4) 0.32

181. With respect to the sickle-cell allele, what should be true of the β hemoglobin locus in U.S. populations of African-Americans whose ancestors were from equatorial Africa?

1. The average heterozygosity at this locus should be decreasing over time
2. There is an increasing heterozygote advantage at this locus
3. Diploidy is helping to preserve the sickle-cell allele at this locus
4. Frequency-dependent selection is helping to preserve the sickle-cell allele at this locus

(1) 1 only (2) 1 and 3

(3) 2 and 3 (4) 1, 2, and 3

182. The sickle-cell allele is pleiotropic (i.e., it affects more than one phenotypic trait) Specifically, this allele affects oxygen delivery to tissues and affects one's susceptibility to malaria. Under conditions of low atmospheric oxygen availability, individuals heterozygous for this allele can experience life-threatening sickle-cell "crises." Such individuals remain less susceptible to malaria. Thus, pleiotropic genes/alleles such as this can help explain why

(1) new advantageous alleles do not arise on demand

(2) evolution is limited by historical constraints

(3) adaptations are often compromises

(4) chance events can affect the evolutionary history of populations

183. In the year 2500, five male space colonists and five female space colonists (all unrelated to each other) settle on an uninhabited Earthlike planet in the Andromeda galaxy. The colonists and their offspring randomly mate for generations. All ten of the original colonists had free earlobes, and two were heterozygous for that trait. The allele for free earlobes is dominant to the allele for attached earlobes. Which of these is closest to the allele frequency in the founding population?

(1) 0.1 *a*, 0.9 *A* (2) 0.2 *a*, 0.8 *A*

(3) 0.5 *a*, 0.5 *A* (4) 0.8 *a*, 0.2 *A*

184. If one assumes that Hardy-Weinberg equilibrium applies to the population of colonists on this planet, about how many people will have attached earlobes when the planet's population reaches 10,000?

(1) 100 (2) 400

(3) 800 (4) 1,000

185. If four of the original colonists died before they produced offspring, the ratios of genotypes could be quite different in the subsequent generations. This would be an example of

(1) diploidy (2) gene flow

(3) genetic drift (4) disruptive selection

186. You are maintaining a small population of fruit flies in the laboratory by transferring the flies to a new culture bottle after each generation. After several generations, you notice that the viability of the flies has decreased greatly. Recognizing that small population size is likely to be linked to decreased viability, the best way to reverse this trend is to

(1) cross your flies with flies from another lab
(2) reduce the number of flies that you transfer at each generation
(3) transfer only the largest flies
(4) change the temperature at which you rear the flies

187. The volcano is currently dormant, but in a hypothetical future scenario, satellite cones at the base of Mt. Kilimanjaro spew sulfurous gases and lava, destroying all life located between the base and 6,000 feet above sea level. As a result of this catastrophe, how should the frequency of the sickle-cell allele change in the remnant human population that survives above 6,000 feet, and which phenomenon accounts for this change in allele frequency?

(1) Decreases; disruptive selection
(2) Increases; genetic drift
(3) Decreases; gene flow
(4) Decreases; bottleneck effect

188. Swine are vulnerable to infection by bird flu virus and human flu virus, which can both be present in an individual pig at the same time. When this occurs, it is possible for genes from bird flu virus and human flu virus to be combined. If the human flu virus contributes a gene for Tamiflu resistance (Tamiflu is an antiviral drug) to the new virus, and if the new virus is introduced to an environment lacking Tamiflu, then what is most likely to occur?

(1) The new virus will maintain its Tamiflu-resistance gene, just in case of future exposure to Tamiflu
(2) The Tamiflu-resistance gene will undergo mutations that convert it into a gene that has a useful function in this environment
(3) If the Tamiflu-resistance gene involves a cost, it will experience directional selection leading to reduction in its frequency
(4) If the Tamiflu-resistance gene confers no benefit in the current environment, and has no cost, the virus will become dormant until Tamiflu is present

189. Which of the following factors weaken(s) the hypothesis of abiotic synthesis of organic monomers in early Earth's atmosphere?

1. The relatively short time between intense meteor bombardment and the appearance of the first life-forms
2. The lack of experimental evidence that organic monomers can form by abiotic synthesis
3. Uncertainty about which gases comprised early Earth's atmosphere

(1) 1 only
(2) 2 only
(3) 1 and 2
(4) 1 and 3

190. How were conditions on the early Earth of more than 3 billion years ago different from those on today's Earth?

(1) Only early Earth was intensely bombarded by large space debris
(2) Only early Earth had an oxidizing atmosphere
(3) Less ultraviolet radiation penetrated early Earth's atmosphere
(4) Early Earth's atmosphere had significant quantities of ozone

191. Which of the following is the correct sequence of events in the origin of life?

I. Formation of protobionts

II. Synthesis of organic monomers

III. Synthesis of organic polymers

IV. Formation of DNA-based genetic systems

(1) I, II, III, IV (2) I, III, II, IV
(3) II, III, I, IV (4) II, III, IV, I

192. Which of the following is a defining characteristic that all protobionts had in common?

(1) The ability to synthesize enzymes
(2) A surrounding membrane or membrane-like structure
(3) RNA genes
(4) The ability to replicate RNA

193. The synthesis of new DNA requires the prior existence of oligonucleotides to serve as primers. On Earth, these primers are small RNA molecules. This latter observation is evidence in support of the hypothesized existence of

(1) a snowball Earth
(2) earlier genetic systems than those based on DNA
(3) the abiotic synthesis of organic monomers
(4) the delivery of organic matter to Earth by meteors and comets

194. If natural selection in a particular environment favored genetic systems that permitted the production of daughter "cells" that were genetically dissimilar from the mother "cells," then one should expect selection for which of the following?

I. Polynucleotide polymerase with low mismatch error rates

II. Polynucleotide polymerases without proofreading capability

III. Batteries of efficient polynucleotide repair enzymes

IV. Polynucleotide polymerases with proof reading capability

V. polynucleotide polymerases with high mismatch error rates

(1) I only (2) I and IV
(3) I, III, and IV (4) II and V

195. If the half-life of carbon-14 is about 5,730 years, then a fossil that has one-sixteenth the normal proportion of carbon-14 to carbon-12 should be about how many years old?

(1) 1,400 (2) 2,800
(3) 11,200 (4) 22,900

196. Approximately how far back in time does the fossil record extend?

(1) 3.5 million yrs (2) 5.0 million yrs
(3) 3.5 billion yrs (4) 5.0 billion yrs

197. What is true of the fossil record of mammalian origins?

(1) It is a good example of punctuated equilibrium
(2) It shows that mammals and birds evolved from the same kind of dinosaur
(3) It includes transitional forms with progressively specialized teeth
(4) It indicates that mammals and dinosaurs did not overlap in geologic time

198. If a fossil is encased in a stratum of sedimentary rock without any strata of igneous rock (for example, lava, volcanic ash) nearby, then it should be

(1) easy to determine the absolute age of the fossil, because the radioisotopes in the sediments will not have been "reset" by the heat of the igneous rocks
(2) easy to determine the absolute age of the fossil, because the igneous rocks will not have physically obstructed the deposition of sediment of a single age next to the fossil
(3) easy to determine, as long as there is enough metamorphic rock nearby
(4) difficult to determine the absolute age of the fossil, because radiometric dating of sedimentary rock is less accurate than that of igneous rock

199. An early consequence of the release of oxygen gas by plant and bacterial photosynthesis was to
(1) generate intense lightning storms
(2) change the atmosphere from oxidizing to reducing
(3) make it easier to maintain reduced molecules
(4) cause iron in ocean water and terrestrial rocks to rust (oxidize)

200. What is true of the Cambrian explosion?
(1) There are no fossils in geological strata that are older than the Cambrian explosion
(2) Only the fossils of microorganisms are found in geological strata older than the Cambrian explosion
(3) The Cambrian explosion is evidence for the instantaneous creation of life on Earth
(4) Recent evidence supports the contention that the Cambrian explosion may not have been as "explosive" as was once thought

ANSWER SHEET

Unit 11: Evolution

1. 1	**2.** 2	**3.** 1	**4.** 2	**5.** 4	**6.** 1	**7.** 4	**8.** 4	**9.** 4	**10.** 1
11. 3	**12.** 4	**13.** 3	**14.** 2	**15.** 2	**16.** 4	**17.** 3	**18.** 4	**19.** 1	**20.** 3
21. 4	**22.** 1	**23.** 2	**24.** 4	**25.** 4	**26.** 1	**27.** 3	**28.** 1	**29.** 4	**30.** 4
31. 4	**32.** 2	**33.** 1	**34.** 4	**35.** 4	**36.** 3	**37.** 1	**38.** 1	**39.** 3	**40.** 4
41. 4	**42.** 1	**43.** 4	**44.** 2	**45.** 4	**46.** 1	**47.** 2	**48.** 1	**49.** 4	**50.** 3
51. 3	**52.** 4	**53.** 3	**54.** 4	**55.** 3	**56.** 4	**57.** 4	**58.** 2	**59.** 3	**60.** 2
61. 2	**62.** 4	**63.** 4	**64.** 4	**65.** 3	**66.** 4	**67.** 4	**68.** 2	**69.** 2	**70.** 2
71. 4	**72.** 3	**73.** 4	**74.** 1	**75.** 2	**76.** 1	**77.** 1	**78.** 3	**79.** 3	**80.** 3
81. 3	**82.** 2	**83.** 3	**84.** 2	**85.** 3	**86.** 3	**87.** 3	**88.** 3	**89.** 3	**90.** 4
91. 1	**92.** 4	**93.** 1	**94.** 1	**95.** 2	**96.** 2	**97.** 4	**98.** 1	**99.** 3	**100.** 4
101. 1	**102.** 4	**103.** 2	**104.** 1	**105.** 2	**106.** 1	**107.** 3	**108.** 1	**109.** 3	**110.** 2
111. 3	**112.** 3	**113.** 4	**114.** 2	**115.** 4	**116.** 2	**117.** 3	**118.** 2	**119.** 4	**120.** 3
121. 2	**122.** 3	**123.** 2	**124.** 4	**125.** 1	**126.** 1	**127.** 2	**128.** 2	**129.** 4	**130.** 3
131. 1	**132.** 3	**133.** 3	**134.** 1	**135.** 2	**136.** 3	**137.** 4	**138.** 3	**139.** 4	**140.** 4
141. 2	**142.** 2	**143.** 3	**144.** 1	**145.** 2	**146.** 4	**147.** 1	**148.** 4	**149.** 4	**150.** 4
151. 4	**152.** 4	**153.** 4	**154.** 4	**155.** 4	**156.** 4	**157.** 1	**158.** 4	**159.** 3	**160.** 3
161. 4	**162.** 4	**163.** 1	**164.** 3	**165.** 4	**166.** 4	**167.** 3	**168.** 3	**169.** 2	**170.** 2
171. 4	**172.** 2	**173.** 4	**174.** 1	**175.** 2	**176.** 4	**177.** 2	**178.** 2	**179.** 4	**180.** 1
181. 2	**182.** 3	**183.** 1	**184.** 1	**185.** 3	**186.** 1	**187.** 4	**188.** 3	**189.** 4	**190.** 1
191. 3	**192.** 3	**193.** 2	**194.** 4	**195.** 4	**196.** 3	**197.** 3	**198.** 4	**199.** 4	**200.** 4

UNIT-12

BIOTECHNOLOGY

1. Scientist involved with "Golden Rice" technology is
   (1) Norman Borlaug
   (2) I. Potrykus
   (3) M.S. Swaminathan
   (4) G.S. Khush
2. Haploid production by anther culture was first demonstrated by
   (1) Bhojwani
   (2) Guha and Maheshwari
   (3) Murashige and Skoog
   (4) Cocking
3. During batch fermentation lowest specific growth rate is achieved during
   (1) Exponential phase
   (2) Lag and stationary phase
   (3) When cells are growing at their fastest pace
   (4) Throughout the fermentation
4. Addition of salt to a culture medium only allows the salt-tolerant bacteria to grow. This is an example of a
   (1) Complex media
   (2) Chemically defined media
   (3) Selective media
   (4) Differential media
5. Identify the hormone combination that induces shoot development in vitro
   (1) no auxin + average cytokinin
   (2) high auxin + no cytokinin
   (3) high auxin + low cytokinin
   (4) low auxin + high cytokinin
6. Biodiesel is produced by
   (1) Transesterification
   (2) Fermentation
   (3) High pressure oxidation
   (4) Esterification
7. The first commercially released GM crop in India is
   (1) Cotton expressing *cryIAb* gene
   (2) Brinjal expressing *cry1Ab* gene
   (3) Corn expressing *cryIAb* gene
   (4) Cotton expressing *cryIAc* gene
8. Vitrification of cultured explants caused by
   (1) low light irradiance, high temperature and intensive sterilization
   (2) high auxin, low temperature and high light irradiance
   (3) higher agar, high nutrients and low pH
   (4) high pH, low temperature and high micronutrient concentration
9. Which of the following objectives can not be achieved through use of cybrids ?
   (1) Transfer of cytoplasmic male sterility
   (2) Recombination of cytoplasmic genes with nuclear gene of another species
   (3) Introgression of Chromosome segment
   (4) Development of true hybrid line
10. The most preferred choice for development of hybrid plants from a male sterile line would be
   (1) Pollen culture (2) Anther culture
   (3) Ovary culture (4) Meristem culture

11. The transplastomic lines bear no risk of gene escape through pollens because
(1) Pollens degenerate before fertilization
(2) Transformed mitochondrial DNA is lost during pollen maturation
(3) Transformed chloroplast DNA is lost during pollen maturation
(4) Transformed genomic DNA is maternally inherited

12. The most dominant trait incorporated in transgenic plants worldwide is
(1) Insect resistance
(2) yield
(3) nutritional quality
(4) herbicide tolerance

13. Clean gene technology in developing transgenic plants means
(1) transgenic plants without marker genes
(2) transgenic plants with provision of removing marker gene after transformation
(3) plant obtained with conventional breeding approach
(4) transgenic plants obtained through plastid transformation

14. Transformation method which avoids use of plant tissue culture technique is
(1) electroporation
(2) biolistic
(3) In planta
(4) Microinjection

15. Which of the following is an example of GURT?
(1) Hybridoma technology
(2) PCR technology
(3) Terminator technology
(4) Transgenic technology

16. Tobacco leaf discs are transferred with *Agrobacterium tumefaciens* strain containing binary vector (GUS as reporter gene) with selectable marker neo (kanamycin resistant gene) and then regenerated to the plants. The plants are kanamycin resistant but leaf tissues are negative to GUS assay. The explanation is
(1) the plants are transformed for both the genes but GUS gene is turned off
(2) the plants are transformed to only neo genes not the GUS genes
(3) the plants are not transformed at all, but the development of kanamycin resistance is due to somaclonal variation
(4) all of the above

17. Some of the genes from viruses introduced into plants in fully functional form often exhibit Mendelian inheritance, because
(1) the genes are stably integrated in chromosomes
(2) the genes are stably maintained in vectors
(3) the genes are co- expressed with chromosomal genes
(4) the genes are not interrupted by introns

18. The oxygen solubility in a bioreactor depends upon
(1) Agitation
(2) Aeration
(3) Both Agitation and Aeration
(4) Viscosity and surface tension

19. The *E. coli* concentration in a growth medium is 0.6 g/l. The respiration rate of *E.coli* at this condition in 0.6 g/l. h. What will be the oxygen uptake rate?
(1) 36 g/g.h
(2) 36 g/l.h
(3) 0.1 g/g.h
(4) 0.01 g/l h

20. Immobilization of microbial cells
(1) Increases apparent Ks
(2) Decreases apparent Ks
(3) Has no effect on Ks
(4) Increases cell's affinity for the substrate

21. The ratio of BOD/COD is approximately 0.5. When this ratio falls below 0.3, it signifies that it
(1) Contains large amount of microorganism
(2) Contains large amount of organic compounds
(3) Contains large amount of organic compounds that are not easily biodegradable
(4) Contains no organic compounds

22. In continuous sterilization process, fluid flows through the holding section where there is
(1) A positive axial dispersion value
(2) A negative axial dispersion value
(3) Axial dispersion is zero
(4) No relationship between axial dispersion and fluid flow

23. The physical significance of specific growth rate constant is
(1) Rate at which the organism is growing
(2) Fraction per unit growth of microorganism per unit time
(3) Grams of cell formed per grams cell mass per unit time
(4) Gram of cells formed per unit time

24. Anticancer vitamin is
(1) Retinol (2) Phylloquinone
(3) Thiamine (4) Pyridoxine

25. If the rate of product formation is approximately proportional to the rate of cell growth, then this pattern of product formation is referred to as
(1) Non-growth associated
(2) Growth associated
(3) Uncoupled
(4) Metabolically uncoupled

26. The addition of acid to maintain the pH at 2 –2.6 in the filtered fermentation broth before penicillin extraction is carried out to extract maximum amount of penicillin in solvent phase. The pH is adjusted to 2 – 2.6 because this helps
(1) In precipitation of proteins
(2) In maintaining penicillin in aqueous phase
(3) In maintaining penicillin in organic phase
(4) In reducing the bacterial contamination

27. Which of the following is the name of the satellite developed in 1999 to scan the oceans around the country?
(1) INSAT-2D
(2) INSAT-1B
(3) INSAT-2E
(4) Aryabhatta

28. Which one of the following algae is a wonder crop with about 70% protein even surpassing famous Soya bean in food value?
(1) Chlorella (2) Dunaliella
(3) Scenodesmus (4) Spirulina

29. β-lactoglobulin promoter is used for expression of gene in
(1) Liver (2) Spleen
(3) Mammary gland (4) Lymphnode

30. Which one of the following virus vectors has been used for development of blue-tongue virus particle like recombinant vaccine?
(1) AcNPV (2) BmNPV
(3) HSV (4) VSV

31. Some extremophiles produce extremozymes which have major industrial application as
(1) many industrial processes operate best at high temperatures
(2) they have high specificity and their ability to distinguish between chiral isomers enable them to function in extremes environment
(3) they are always produced by hyperthermophiles
(4) many bioprocesses operate at low pH

32. Transfer of T-DNA from Ti-plasmid into plant cells is mediated by
(1) MOB-gene (2) Nif gene
(3) Vir gene (4) Octopine gene

33. Asparaginase is used as an
(1) Anti-tumor agent
(2) Anti-tuberculosis agent
(3) Anti-malarial agent
(4) Anti-diabetic agent

34. The biosafety problem due to spread of transgenes from transgenic plants to its wild relatives can be avoided by
(1) developing transgenic plants with herbicide markers
(2) Posi-Tech selection using non-antibiotic markers like *pmi*
(3) developing transplastomic lines
(4) elimination of markers using Cre/*lox* system

35. The first commercially produced plant secondary metabolite using bioreactor technology is
(1) shikonin (2) colchicine
(3) cercosporin (4) cytokinin

36. Enucleated protoplast is called
(1) cybrid (2) tonoplast
(3) cytoplast (4) duplast

37. Doubled haploid lines can be generated by
(1) Protoplast fusion
(2) Transformation
(3) Anther culture
(4) RNAi technology

38. It was possible to engineer genes of interest between right and left border of T-DNA of *Agrobacterium* for plant transformation because
(1) T-DNA is nontoxic
(2) T-DNA is not required by *Agrobacterium*
(3) The Agrobacterial genes essential for mobilization of the T-DNA lie outside the TDNA
(4) T-DNA cannot function in plants

39. Integration of more than one copy of transgene is not desirable because
(1) It can make more transcript than needed
(2) It can cause toxicity
(3) It can cause mutation in the gene of interest
(4) It can induce gene silencing

40. Systemic acquired resistance
(1) is an enhanced resistance exhibited by uninfected plant tissue through a memory of previous infection
(2) is an resistance exhibited by plants towards any pathogen
(3) is a resistance acquired by a sensitive plant through breeding
(4) is a resistance exhibited by bacterial pathogens towards a large number of antibiotics

41. Which of the following reactor systems is generally used to generate microbial mutants?
(1) CSTR system (2) BSTR system
(3) PBR system (4) FBR system

42. In microbial fermentation, generally inhibitors
(1) are consumed by the microbes
(2) are incorporated in the synthesized molecule
(3) help in combating contamination
(4) help in the production of desired products

43. By a single experiment how would you know that the stationary phase in a batch culture is due to substrate depletion or toxin accumulation?
(1) By adding highly concentrated substrate in the stationary phase
(2) By adding more substrate in the stationary phase
(3) By diluting the broth by sterilized water in the stationary phase
(4) By extracting toxin from the broth in the stationary phase

44. The average value of dissolved oxygen concentration in water is

(1) 10 mg/l (2) 100 mg/l

(3) 160 mg/l (4) 1 mg/l

45. A method commonly used to determine the level of damage of animal cells in a culture medium is to monitor

(1) the concentration of lactate in the medium

(2) the level of lactate oxidase activity in the medium

(3) the level of lactate dehydrogenase activity

(4) the level of laccase activity in the medium

46. Which of the following antibiotics is produced by chemical synthesis ?

(1) Penicillin (2) Streptomycin

(3) Tetracycline (4) Chloramphenicol

47. *Clostridium tetanomorphum* is known to produce

(1) Vitamin A (2) Vitamin B

(3) Vitamin C (4) Vitamin K

48. When organisms make toxic substances more toxic, the process is called

(1) Bioremediation

(2) Biomagnification

(3) Biotoxification

(4) Bioamplification

49. Gene therapy through stem cells may be done using

(1) lentiviral vector

(2) plasmid vector

(3) episomal vector

(4) baculovirus vector

50. Defect in the SCID mice may be cured by inserting

(1) ADA gene (2) SCID gene

(3) SCDA gene (4) DAA gene

51. Which one of the following viruses has been extensively used as expression vector for a number of foreign genes?

(1) Vaccinia virus (2) Rotavirus

(3) Rabies virus (4) Papilloma virus

52. Metabolic engineering of *E.coli* as a commercial source of the fuel ethanol involves alteration of its

(1) Carbohydrate catabolic pathways

(2) Fermentative pathways

(3) TCA cycle

(4) Ability to grow autotrophically

53. CpG islands and codon bias are tools used in eukaryotic genomics to

(1) identify open reading frames

(2) differentiate between eukaryotic and prokaryotic DNA sequences

(3) Look for DNA-binding domains

(4) determine STS

54. The overall cost of production of recombinant DNA products for human use, in general increases due to complications in

(1) Upstream processing

(2) Fermentation process

(3) Downstream processing

(4) Formulation process

55. Which of the following does not represent a valid amino acid sequence?

(1) EINSTEIN (2) CRICK

(3) FARADAY (4) WATSON

56. Which of the following proteins was used to create the first transgenic fish?

(1) Green Fluorescent protein

(2) Anti freezing protein

(3) Horseshoe peroxidase

(4) Myosin protein

57. Curacin A is a potent anti-tumor agent obtained from a marine

(1) Actinomyces (2) Cyanobacterium

(3) Aspergillus (4) Coral reef

58. Gene coding for luminescence in marine luminescent bacteria is

(1) luxR (2) toxR
(3) recA (4) Luciferase

59. Red fluorescent protein used in the development of transgenic ornamental fish was isolated from

(1) Star fish (2) Sponges
(3) Red algae (4) Corals

60. Morphogenetic competence in callus tissue is obtained by

(1) Vitrification (2) Somaclonal variation
(3) Habituation (4) Dedifferentiation

61. Which of the following is the best method for localization of protein during transition from vegetative to flowering stage?

(1) Place a reporter gene next to the promoter of the gene encoding the protein, and identify the cellular location of the reporter gene
(2) Use an antibody against the specific protein
(3) Separate the cellular compartments by centrifugation followed by screening with antibodies
(4) Tag the protein with fluorescent amino acids and identify the cellular location by fluorescent microscopy

62. Transgenic for terminator seed is due to a lethal gene along with two other genes. Which of the following is the product of the lethal gene?

(1) Recombinase
(2) Ribosomal inhibiting protein
(3) Repressor protein
(4) Protein for late embryogenesis

63. Addition of activated charcoal to plant tissue culture medium is done in order to

(1) Reduce phenolic leachates
(2) Increase osmolarity of medium
(3) Supplement vitamins
(4) Resist pH changes

64. Wild type *Agrobacterium tumefaciens* does not contain one of the following classes of genes in the T-DNA region. These are

(1) Auxin biosynthesis genes
(2) Opine synthesis genes
(3) Opine catabolism genes
(4) Cytokinin biosynthesis genes

65. One way of overcoming homologous gene silencing by a transgene is

(1) Using CaMV35S promoter to drive transgene expression
(2) Inserting transgene into plastid genome instead of nuclear genome
(3) Removing the antibiotic resistance conferring genes from transformed plants
(4) Placing an intron between the promoter and coding sequence of the transgene

66. Which of the following genes is a non-antibiotic, non-herbicidal marker that is used for positive selection in plant genetic transformation?

(1) Neomycin phosphotransferase II
(2) Dihydrofolate reductase
(3) Glyphosate oxidoreductase
(4) Mannose 6-phosphate isomerase

67. NCED (9-cis epoxycarotenoid dioxygenase) is involved in

(1) ABA biosynthesis
(2) GA biosynthesis
(3) Auxin biosynthesis
(4) Ethylene biosynthesis

68. The single-stranded nick in DNA during T-DNA transfer is initiated by

(1) Vir C (2) Vir B
(3) Vir D (4) Vir E

69. Target protein for Glyphosate (Roundup®) is

(1) Acetolactate synthase
(2) Glutamine synthase
(3) 5-Enolpyruvyl shikimate 3-phosphate synthase
(4) Glutamate dehydrogenase

70. Application of molecular biological technique for commercial production of recombinant products in plants is referred as
(1) Transgenic technology
(2) Biotech crops technology
(3) Molecular Farming
(4) Recombinant DNA technology

71. Which of the following is the seed specific promoter used in plant genetic engineering?
(1) CaMV35S promoter
(2) Ubiquitin promoter
(3) ABRE promoter
(4) Glutelin promoter

72. Which of the following objectives cannot be achieved by using cybrids?
(1) Transfer of cytoplasmic male sterility
(2) Recombination of cytoplasmic genes with nuclear gene of another species
(3) Introgression of Chromosome segment
(4) Development of true hybrid line

73. Amylase inhibitor gene is transferred to plants for providing resistance against
(1) Fungi (2) Viruses
(3) Insects (4) Bacteria

74. PHB biosynthetic pathway for bioplastic production involves manipulation of *pha*A, *pha*B and *pha*C genes of
(1) *Escherichia coli*
(2) *Aspergillus nidulans*
(3) *Alcaligenes eutrophus*
(4) *Bacillus megaterium*

75. Sodium alginate is used in
(1) Protoplast culture
(2) Artificial seed formation
(3) Cryopreservation
(4) Media as gelling agent

76. Producing biodiesel from algae is a process which is
(1) carbon negative
(2) carbon positive
(3) carbon and nitrogen positive
(4) carbon neutral

77. Which of the following strategies is not adopted to minimize secretion of acetate in a recombinant protein producing *E. coli* ?
(1) Converting acetyl CoA to acetoin
(2) Altering the rate of glucose transport inside the cell
(3) Blocking the acetate synthesis pathway
(4) Increasing the flux in pentose phosphate pathway

78. Which statement is correct in the biotransformation of prochiral ketones to the desired enantiomeric alcohol?
(1) The enantiomeric excess should be maximum
(2) The conversion should be maximum
(3) Both the conversion and the enantiomeric excess should be maximum
(4) The conversion should be maximum with lowest enantiomeric excess

79. Which one of the following promoters is not derived from *Agrobacterium*?
(1) CaMV 35S
(2) nos
(3) ipt
(4) virD

80. An example of a co-dominant marker is
(1) AFLP marker
(2) ISSR marker
(3) RAPD marker
(4) SSR marker

81. In the biological treatment of waste, bacteria such as species of *Acinetobacter* and *Zoogloea* are considered to play a key role in floc formation
(1) by the synthesis and secretion of polysaccharides
(2) since they are acid producers
(3) as they are slow-growing methanogens
(4) since they break down acetic acid into methane and CO_2

82. During primary sedimentation and biological treatment of waste water, vast quantities of sludge are generated which are assessed by sludge volume index (SVI), defined as the volume occupied by 1 g of sludge after settling for 30 min in a 1L Imoff cone which measures the

(1) sludge thickening
(2) sludge stabilization
(3) sludge dewatering
(4) rate of acid utilization

83. Biofilms interrupt the flow of ions and water to and from the substrate surface by acting as adiffusion barrier. The reduction of localised oxygen can accelerate the corrosion of a metallic substrate and is called microbially induced corrosion (MIC) An example of MIC is

(1) sulphides from SRB which cause the pitting of steel surfaces
(2) remains of old barnacle exoskeletons
(3) electrostatic interactions and Van der Waal's forces
(4) crosslinks using cysteine residues

84. *Acetobacter aceti* produced 7.5 g/l CH_3COOH from a medium containing initial 10 g/l C_2H_5OH when the residual C_2H_5OH concentration is 2 g/l. What will be the overall yield of CH_3COOH from C_2H_5OH compared to the theoretical yield?

(1) 70% of the theoretical yield
(2) 72% of the theoretical yield
(3) 84% of the theoretical yield
(4) 94% of the theoretical yield

85. Which of the following pairs of amino acids is responsible for feed-back inhibition of lysine biosynthesis in Corynebacteria?

(1) Lysine and methionine
(2) Lysine and leucine
(3) Lysine and threonine
(4) Lysine and isoleucine

86. A 1.5 ml of a bacterial culture consisting of 10^8/ml is used to inoculate 100 ml of medium where the cells reach a density of $5x10^7$ cells/ml. How many generations did the cells go through approximately?

(1) 1 (2) 5
(3) 10 (4) 15

87. Which one of the following requires back crossing?

(1) Generation of Recombinant Inbred Lines (RILs)
(2) Generation of Doubled Haploids (DH)
(3) Generation of F_2s
(4) Generation of Near isogenic Lines (NILs)

88. *nptII* gene imparts resistance to

(1) Ampicillin
(2) Hygromycin
(3) Kanamycin
(4) Chloramphenicol

89. *Agrobacterium* mediated transformation of monocots requires the use of ______ for the induction of vir genes

(1) Agarose
(2) Acetophenone
(3) Acetosyringone
(4) Cefotaxime

90. Which of the following techniques is used to obtain hybrids between two species with prefertilization barrier?

(1) Embryo rescue
(2) Protoplast fusion
(3) Ovary culture
(4) Embryo implantation

91. Which of the following is used most commonly in the plant tissue culture medium to induce multiple shoots ?

(1) Benzylaminopurine
(2) Naphthalene Acetic acid
(3) 2,4-Dichlorophenoxy acetic acid
(4) Indole butyric acid

92. Engineering plants using chitinase gene leads to development of

(1) Viral resistance

(2) Fungal resistance

(3) Bacterial resistance

(4) Cold tolerance

93. The first GM potato developed at CPRI, India for increasing protein content in tubers consists of genes from

(1) Chick pea (2) Pigeon pea

(3) Cabbage (4) *Amaranthus*

94. Hybridization between species followed by polyploidy is known as

(1) Autopolyploid (2) Allopolyploid

(3) Aneuploid

(4) Species differentiation

95. Source of dwarfing genes in wheat is

(1) Ganga 101 (2) Norin 10

(3) Dee-geo-woo-gen

(4) Sonalika

96. Form of IPR that helped India win Basmati case is

(1) Novelty (2) Trade mark

(3) Geographical indication

(4) Industrial design

97. The probable Geographic Origin of the following crops is called: Wheat and barley, flax, lentils, chickpea, figs, dates, grapes, olives, lettuce, onions, cabbage, carrots, cucumbers, melons and fruits and nuts

(1) South America

(2) Mesoamerica and North America

(3) The Fertile Crescent

(4) South-east Asia

98. Which of the following regions typically has the highest primary productivity per unit surface area of the ocean?

(1) zones of upwelling

(2) coastal water

(3) the centres of ocean gyres

(4) tropical waters

99. In the preparation of Golden rice, a few exogenous genes have been used in order to achieve the production of Vitamin A through recombinant DNA technology. This is a true example of

(1) Metabolic repression

(2) Biochemical engineering

(3) Metabolic extension

(4) Combinatorial biosynthesis

100. Which one of the following microorganisms can be effectively used as a biocontrol agent?

(1) *Bacillus thuringiensis*

(2) *Bacillus megaterium*

(3) *Aspergillus niger*

(4) *Pseudomonas putida*

101. Which one of the following gases contributes most to the Greenhouse Effect?

(1) Acetylene

(2) Carbon dioxide (CO_2)

(3) Carbon monoxide

(4) Nitrous oxide (N_2O)

102. The treatment of hepatocytes in *in vitro* culture with Ricin leads to

(1) Inhibition of endocytosis

(2) Inhibition of translation

(3) Inhibition of Transcription

(4) Inhibition of signal transduction

103. One of the major reasons of the failure of somatic gene therapy protocol can be assigned to

(1) Lack of appropriate gene delivery system

(2) Lack of technique in site specific integration of foreign gene into the chromosome of target cells

(3) Low level expression of foreign genes in target cells

(4) High level expression of foreign genes in target cells

104. Plants carrying a transgene for glyphosate resistance will survive a field application of glyphosate that kills weeds. Which of the following enzyme is inhibited by glyphosate?

(1) Phosphoglycerate

(2) Phosphoglycolatephosphatase

(3) Enolpyruvateshikimate -3 – phosphate synthase

(4) Glycolatedehydrogenase

105. The T-DNA of Agrobacterium must be excised from its circular plasmid for its transfer into plant cells. Which one of the following complexes of Vir proteins is responsible for DNA nick at the left and right border sequences?

(1) VirA/VirC (2) VirB6/VirB7

(3) VirD1/VirD2 (4) VirE2/VirG

106. Transgenic plants were developed with a gene conferring kanamycin resistance. Four independent T_o events were selfed and ~100 T_1 seeds were germinated on kanamycin. The number of kanamycin sensitive (Kan^S) and kanamycin resistant (Kan^R) progeny obtained in case is listed below

Transgenic Lane No.	Kan^R	Kan^S
One	95	7
Two	72	25
Three	51	47
Four	11	87

Identify the line where the transgene has integrated at a single locus

(1) Line One (2) Line Two

(3) Line Three (4) Line Four

107. GA2-oxidase cDNA from bean is overexpressed in wheat plant by genetic engineering. Which one of the following phenotypes correctly describes the resultant transgenic plant?

(1) The plant will be taller

(2) The plant will be shorter

(3) There will be no change in height

(4) The plant will not survive

108. Using transgenic technology male sterile plants are developed by

(1) Expressing barnase gene in tapetum

(2) Targeting barnase protein to tapetum

(3) Expressing barnase gene in tapetum and barstar gene in the rest of the plant

(4) Expressing barnase gene in tapetum and targeting barstar protein in the rest of the plant

109. One of the concerns in commercial usage of transgenic plants is the spread of transgene through pollen flow. Which one of the following methods can be used to circumvent this problem?

(1) Use of terminator technology

(2) Chloroplast transformation

(3) Developing male sterile lines

(4) Planting of refugia

110. Which one of the following can be used as a selection marker for developing transgenic plants?

(1) green fluorescent protein

(2) β-glucuronidase

(3) β-galactosidase

(4) hygromycin phosphotransferase

111. Direct DNA uptake by protoplasts can be stimulated by

(1) Polyethylene glycol

(2) Sucrose

(3) $CaCl_2$

(4) LiCl

112. Aroma in rice is due to

(1) Acetyl choline

(2) 4-benzyl pyrroline

(3) 2-ethyl pyrroline

(4) 2-acetyl-1-pyrroline

113. The male sterile cytoplasm in pearl millet is
(1) Milo (2) CK 60A
(3) Tift 23A (4) W.A

114. The pathogen against which the crop rotation can be adopted
(1) Soil invader (2) Soil inhabitant
(3) Soil-borne (4) Biotrophic

115. In a large scale high cell density fed batch reactor, which one of the following statements is true?
(1) Dissolved oxygen starvation occurs at the top and glucose starvation at bottom of the fermenter
(2) Dissolved oxygen starvation occurs at the bottom and glucose starvation at top of the fermenter
(3) Both dissolved oxygen and glucose starvation occurs at the top of the fermenter
(4) Both dissolved oxygen and glucose starvation occurs at the bottom of the fermenter

116. Environmental gene tags are
(1) 16S rRNA gene signatures from the environment
(2) Habitat-specific functional genes from microbial communities
(3) Non-essential genes from microbial communities
(4) Fluorescently tagged metagenome

117. Which one of the following is not true of biophotolysis?
(1) It is a photoheterotrophic process
(2) It is a photoautotrophic process
(3) H_2O is the electron donor in the process
(4) [FeFe]-hydorgenase catalyses the reaction

118. Which is not a characteristic of a plasmid used for production of DNA vaccines in fish?
(1) Promoter- and enhancer sequences
(2) Poly-adenylation sequence
(3) Transcriptional termination sequence
(4) Poly-T- tail

119. If you want to use a plant tissue culture as a chemical factory for vitamins, which of the following will you choose?
(1) Suspension cultures
(2) Callus cultures
(3) Organ cultures
(4) Anther/pollen cultures

120. In which one of the following fermentations an inhibitor is added to increase the productivity?
(1) Rifampian B fermentation
(2) Tetracycline fermentation
(3) Glutamic acid fermentation
(4) Citric acid fermentation

121. The stability of a recombinant protein can be enhanced by
(1) altering the C-terminal region of the protein
(2) exclusion of PEST sequences from the protein
(3) production of compound similar to detergents to prevent formation of inclusion bodies
(4) altering the N-terminus by adding leucine or phenyl alanine by genetic manipulation

122. Which of the following transgenic crops occupies the largest area in the world?
(1) Herbicide tolerant soybean
(2) Herbicide tolerant maize
(3) Insect resistant cotton
(4) Insect resistant potato

123. In order to develop iron-rich rice which of the following genes was used for creating genetically modified plants?
(1) Ferritin
(2) Phytic acid
(3) Phytic acid and Ferritin
(4) Transferrin and Ferritin

124. Viable seeds can be produced without fertilization of the egg in a process called
(1) Apospory
(2) Apomixis
(3) Parthenogenesis
(4) Meiosis

125. Marker-free plants can be developed by
(1) co-transformation
(2) insertion
(3) deletion
(4) inversion

126. Somatic embryogenesis is _____
(1) germ line cells developing into embryos
(2) non-germ line cells developing into embryos
(3) embryos developing from zygotes
(4) embryonic tissue becoming somatic

127. The precursors for the synthesis of nopalines in Ti plasmid is _____
(1) phosphoenolpyruvate + amino acid
(2) pyruvate + amino acid
(3) alpha-Ketoglutarate + amino acid
(4) oxalate + amino acid

128. Somatic embryos from cotyledon explant would develop in which of the following sequences?
(1) Globular, torpedo, heart, cotyledonary stage
(2) Globular, heart, torpedo & cotyledonary stage
(3) Cotyledonary, heart, globular & torpedo
(4) Cotyledonary, torpedo, heart & globular

129. The protein(s) which remains attached to the T-DNA during transfer to plant cells is/are
(1) Vir D2
(2) Vir E2
(3) Vir G
(4) Both (1) and (2)

130. In plant tissue culture, induction of roots and shoots is accomplished by
(1) using a tissue of certain minimum size
(2) using a particular auxin cytokinin ratio
(3) using a specific concentration of sucrose in the medium
(4) manipulating physical factors such as light, pH, temperature

131. Which of the following is not obtained from plant sources
(1) Nattokinase
(2) Papain
(3) Bromelain
(4) Dornase α

132. Glofish is
(1) a patented zebra fish which has been genetically modified with GFP
(2) a commercial name given to tuna fish created by cloning growth hormone gene
(3) an angler fish harboring bioluminescent bacteria
(4) a cutter-shark fish which catches its prey with the help of bioluminescent bacteria residing near the gills

133. DsRed is a
(1) red fluorescent protein observed in *Aequorea victoria*
(2) common name given to red tide observed on the coast of Taiwan
(3) red fluorescent protein isolated from coral *Discosoma* genus
(4) red bioluminescent bacteria seen in certain species of copepod

134. Which of the following is most closely identical to the formation of twins?
(1) cell cloning
(2) therapeutic cloning
(3) use of adult stem cells
(4) organismal cloning

135. In 1997, Dolly the sheep was cloned. Which of the following processes was used?

(1) use of mitochondrial DNA from adult female cells of another ewe

(2) replication and dedifferentiation of adult stem cells from sheep bone marrow

(3) separation of an early stage sheep blastula into separate cells, one of which was incubated in a surrogate ewe

(4) fusion of an adult cell's nucleus with an enucleated sheep egg, followed by incubation in a surrogate

136. Which of the following problems with animal cloning might result in premature death of the clones?

(1) use of pluripotent instead of totipotent stem cells

(2) use of nuclear DNA as well as mtDNA

(3) abnormal regulation due to variant methylation

(4) the indefinite replication of totipotent stem cells

137. Reproductive cloning of human embryos is generally considered unethical. However, on the subject of therapeutic cloning there is a wider divergence of opinion. Which of the following is a likely explanation?

(1) Use of adult stem cells is likely to produce more cell types than use of embryonic stem cells

(2) Cloning to produce embryonic stem cells may lead to great medical benefits for many human diseases

(3) Cloning to produce stem cells relies on a different initial procedure than reproductive cloning

(4) A clone that lives until the blastocyst stage does not yet have human DNA

138. Which of the following is true of embryonic stem cells but not of adult stem cells?

(1) They can differentiate into many cell types

(2) They make up the majority of cells of the tissue from which they are derived

(3) They can continue to replicate for an indefinite period

(4) They can provide enormous amounts of information about the process of gene regulation

139. A researcher is using adult stem cells and comparing them to other adult cells from the same tissue. Which of the following is a likely finding?

(1) The cells from the two sources exhibit different patterns of DNA methylation

(2) Adult stem cells have more DNA nucleotides than their counterparts

(3) The two kinds of cells have virtually identical gene expression patterns in microarrays

(4) The nonstem cells have fewer repressed genes

140. In animals, what is the difference between reproductive cloning and therapeutic cloning?

(1) Reproductive cloning uses totipotent cells, whereas therapeutic cloning does not

(2) Reproductive cloning uses embryonic stem cells, whereas therapeutic cloning does not

(3) Therapeutic cloning uses nuclei of adult cells transplanted into enucleated nonfertilized eggs

(4) Therapeutic cloning supplies cells for repair of diseased or injured organs

141. The first cloned cat, called Carbon Copy, was a calico, but she looked significantly different from her female parent. Why?

(1) The environment, as well as genetics, affects phenotypic variation

(2) Fur color genes in cats are influenced by differential acetylation patterns

(3) Cloned animals have been found to have a higher frequency of transposon activation

(4) X inactivation in the embryo is random and produces different patterns

142. In recent times, it has been shown that adult cells can be induced to become pluripotent stem cells (iPS) In order to make this conversion, what has been done to the adult cells?

(1) A retrovirus is used to introduce four specific regulatory genes
(2) The adult stem cells must be fused with embryonic cells
(3) Cytoplasm from embryonic cells is injected into the adult cells
(4) An adenovirus vector is used to transfer embryonic gene products into adult cells

143. Let us suppose that someone is successful at producing induced pluripotent stem cells (iPS) for replacement of pancreatic insulin-producing cells for people with type 1 diabetes. Which of the following could still be problems?

I. the possibility that, once introduced into the patient, the iPS cells produce nonpancreatic cells
II. the failure of the iPS cells to take up residence in the pancreas
III. the inability of the iPS cells to respond to appropriate regulatory signals

(1) I only
(2) II only
(3) III only
(4) I, II and III

144. Genetic engineering is being used by the pharmaceutical industry. Which of the following is not currently one of the uses?

(1) production of human insulin
(2) production of human growth hormone
(3) production of tissue plasminogen activator
(4) creation of products that will remove poisons from the human body

145. Genetically engineered plants

(1) are more difficult to engineer than animals
(2) include a transgenic rice plant that can help prevent vitamin A deficiency
(3) are being rapidly developed, but traditional plant breeding programs are still the only method used to develop new plants
(4) are able to fix nitrogen themselves

146. Scientists developed a set of guidelines to address the safety of DNA technology. Which of the following is one of the adopted safety measures?

(1) Microorganisms used in recombinant DNA experiments are genetically crippled to ensure that they cannot survive outside of the laboratory
(2) Genetically modified organisms are not allowed to be part of our food supply
(3) Transgenic plants are engineered so that the plant genes cannot hybridize
(4) Experiments involving HIV or other potentially dangerous viruses have been banned

147. One successful form of gene therapy has involved delivery of an allele for the enzyme adenosine deaminase AD(1) to bone marrow cells of a child with SCID, and delivery of these engineered cells back to the bone marrow of the affected child. What is one major reason for the success of this procedure as opposed to many other efforts at gene therapy?

(1) The engineered bone marrow cells from this patient can be used for any other SCID patient
(2) The ADA-introduced allele causes all other ADA-negative cells to die
(3) The engineered cells, when reintroduced into the patient, find their way back to the bone marrow
(4) No vector is required to introduce the allele into ADA-negative cells

148. Which of the following is one of the technical reasons why gene therapy is problematic?

(1) Most cells with an engineered gene do not produce gene product
(2) Most cells with engineered genes overwhelm other cells in a tissue
(3) Cells with transferred genes are unlikely to replicate
(4) Transferred genes may not have appropriately controlled activity

149. As genetic technology makes testing for a wide variety of genotypes possible, which of the following is likely to be an increasingly troublesome issue?

(1) use of genotype information to provide positive identification of criminals
(2) using technology to identify genes that cause criminal behaviors
(3) the need to legislate for the protection of the privacy of genetic information
(4) discrimination against certain racial groups because of major genetic differences

150. Novel groups of polyketides can be synthesized in *Actinomycetes* by

(1) gene rearrangement
(2) gene amplification
(3) combinatorial biosynthesis
(4) metabolic engineering

151. The process of phytoremediation where complexation and immobilization of toxin takes place within the soil is called

(1) phytoextraction
(2) phytodegradation
(3) phytovolatilization
(4) phytostabilization

152. *Coliforms* are used as indicator organisms of sewage pollution because they

(1) are pathogens
(2) ferment lactose
(3) are abundant in human intestines
(4) grow within 24 hours

153. Majority of the cereals are highly recalcitrant to Agrobacterium-mediated transformation, and so direct transformation methods have been developed to transform such plants. Which of the following direct transformation methods is applicable to intact plant tissues?

(1) Calcium chloride and PEG-mediated transformation
(2) Liposome-mediated transformation
(3) Electroporation
(4) Transformation using microprojectiles

154. Which of the following ecological locations has the greatest species diversity?

(1) tundra (2) deciduous forests
(3) tropics. (4) grasslands

155. Estimates of current rates of extinction

(1) indicate that we have reached a state of stable equilibrium in which speciation rates equal extinction rates
(2) suggest that one-half of all animal & plant species may be gone by the year 2100
(3) indicate that rates may be greater than the mass extinctions at the close of the Cretaceous period
(4) indicate that only 1% of all of the species that have ever lived on Earth are still alive

156. Extinction is a natural phenomenon. It is estimated that 99% of all species that ever lived are now extinct. Why then do we say that we are now experiencing an extinction (loss of biodiversity) crisis?

(1) Humans are ethically responsible for protecting endangered species
(2) Scientists have finally identified most of the species on Earth and are thus able to quantify the number of species becoming extinct
(3) The current rate of extinction is high and human activities threaten biodiversity at all levels
(4) Humans have greater medical needs than at any other time in history, and many potential medicinal compounds are being lost as plant species become extinct

157. Although extinction is a natural process, current extinctions are of concern to environmentalists because

(1) more animals than ever before are going extinct

(2) most current extinctions are caused by introduced species

(3) the rate of extinction is unusually high

(4) current extinction is primarily affecting plant diversity

158. Which of the following groups is most threatened by global extinctions?

(1) mammals (2) birds

(3) fish (4) amphibians

159. The most serious consequence of a decrease in global biodiversity would be the

(1) increase in global warming and thinning of the ozone layer

(2) potential loss of ecosystem services on which people depend

(3) increase in the abundance and diversity of edge-adapted species

(4) loss of source of genetic diversity to preserve endangered species

160. Which of the following is the most direct threat to biodiversity?

(1) Increased levels of atmospheric carbon dioxide

(2) The depletion of the ozone layer

(3) Overexploitation of selected species

(4) Habitat destruction

161. According to most conservation biologists, the single greatest threat to global biodiversity is

(1) global climate change resulting from a variety of human activities

(2) stratospheric ozone depletion

(3) overexploitation of certain species

(4) alteration or destruction of the physical habitat

162. What is the biological significance of genetic diversity between populations?

(1) Genes for adaptive traits to local conditions make microevolution possible

(2) The population that is most fit would survive by competitive exclusion

(3) Genetic diversity allows for species stability by preventing speciation

(4) Isolated populations become more fit

163. Overexploitation encourages extinction and is most likely to affect

(1) animals that occupy a broad ecological niche

(2) large animals with low intrinsic reproductive rates

(3) most organisms that live in the oceans

(4) terrestrial organisms more than aquatic organisms

164. Of the following ecosystem types, which have been impacted the most by humans?

(1) Wetland and riparian

(2) Open and benthic ocean

(3) Desert and high alpine

(4) Taiga and second-growth forests

165. Imagine a population of 1,000 small rodents. Of these, 300 are breeding females, 300 are breeding males, and 400 are nonbreeding juveniles. What is the effective population size?

(1) 1,000 (2) 1,200

(3) 600 (4) 400

166. If the sex ratio in a population is significantly different from 50:50, then which of the following will always be true?

(1) The population will enter the extinction vortex

(2) The genetic variation in the population will increase over time

(3) The genetic variation in the population will decrease over time

(4) The effective population size will be less than the actual population size

167. Relatively small geographic areas with high concentrations of endemic species and a large number of endangered and threatened species are known as

(1) endemic sinks
(2) critical communities
(3) biodiversity hot spots
(4) endemic metapopulations

168. How are movement corridors potentially harmful to certain species?

(1) They increase inbreeding
(2) They promote dispersion
(3) They spread disease and parasites
(4) They increase genetic diversity

169. Biodiversity hot spots are not necessarily the best choice for nature preserves because

(1) hot spots are situated in remote areas not accessible to wildlife viewers
(2) their ecological importance makes land purchase very expensive
(3) a hot spot for one group of organisms may not be a hot spot for another group
(4) hot spots are designated by abiotic factors present, not biotic factors

170. Which of the following is true about "hot spots"?

(1) One-third of all species on Earth occupy less than 1.5% of Earth's land area (hot spots)
(2) All of the plants and animals containing genes that may be useful to humankind are located in Earth's hot spots
(3) Around 75% of all of the undiscovered species of organisms live in ecological hot spots
(4) As conservation measures improve over the next ten years, hot spots will likely disappear

171. This causes an increase in the intensity of UV radiation reaching Earth

(1) depletion of ozone layer
(2) acid precipitation
(3) biological magnification
(4) greenhouse effect

172. This term refers to the reflecting and absorption of infrared radiation by atmospheric methane, carbon dioxide, and water

(1) depletion of ozone layer
(2) acid precipitation
(3) biological magnification
(4) greenhouse effect

173. This is caused by excessive nutrient runoff into aquatic ecosystems

(1) depletion of ozone layer
(2) acid precipitation
(3) biological magnification
(4) eutrophication

174. This causes extremely high levels of toxic chemicals in fish-eating birds

(1) eutrophication
(2) acid precipitation
(3) biological magnification
(4) greenhouse effect

175. Which of the following statements is true of protoplast fusion?

(1) It occurs when the second sperm nucleus fuses with the polar nuclei in the embryo sac
(2) It can be used to form new plant varieties by combining genomes from two plants
(3) It is used to develop gene banks to preserve genetic variability
(4) It is the method of test-tube cloning that produces whole plants from callus

176. One major potential benefit resulting from protoplast fusion experiments in agriculturally relevant species is

(1) there is an increased potential of hopeful mutants
(2) species that are normally sexually incompatible can sometimes be successfully fused producing offspring with traits of both parent species
(3) many new species can be produced in a short period of time
(4) deleterious traits are often removed by this process

177. The most immediate potential benefits of introducing genetically modified crops include
(1) increasing the amount of land suitable for agriculture
(2) overcoming genetic incompatibility
(3) increasing the frequency of self-pollination
(4) increasing crop yield

178. Biofuels are mainly produced by
(1) the breakdown of cell wall biopolymers into sugars that can be fermented
(2) plants that convert hemicellulose into gasoline
(3) the genetic engineering of ethanol-generating genes into plants
(4) transgenic crops that have cell walls containing ethylene

179. One disadvantage of monoculture is that
(1) the whole crop ripens at the same time
(2) genetic uniformity makes a crop more vulnerable to a new pest or disease
(3) it predominantly uses vegetative propagation
(4) most grain crops self-pollinate

180. Which of the following is a scientific concern related to creating genetically modified crops?
(1) Herbicide resistance may spread to weedy species
(2) Beneficial insects may be harmed by eating pest species
(3) Their adaptive advantages would allow them to overpower natural ecosystems
(4) The monetary costs of growing genetically modified plants are significantly greater than traditional breeding techniques

181. All of the following strategies are being pursued with the goal of preventing transgene escape from genetically modified crops except
(1) the engineering of male sterility into plants
(2) the genetic engineering of apomixis into transgenic crops
(3) the genetic engineering of trangenes into the chloroplast DNA
(4) hybridization of transgenic crop genes with related wild weeds

182. Which of the following is proving to be the most difficult task for plant breeders regardless of the techniques they use for crop improvement due to the complexity of the metabolic pathways involved?
(1) insect resistance
(2) nitrogen fixation
(3) herbicide resistance
(4) improved nutritional quality

183. Which of the following is a successful example of the commercial use of transgenic crops?
(1) inserting *Bt* toxin genes into cotton, maize, and potato
(2) developing plants that are resistant to ultraviolet light mutations
(3) producing plants that resist attack by large herbivores
(4) developing plants that produce all the essential amino acids for humans

184. Which one of the following statements about plant tissue culture is correct
(1) cells can be cultured only on solid medium
(2) cells can be cultured both on solid and liquid medium
(3) callus do not need hormones
(4) the cells of the callus cannot be subcultured

185. POMATO, is
(1) a transgenic plant
(2) a plant obtained through protoplast hybridization
(3) a plant obtained by organ culture
(4) a plant developed by plant breeding method

186. The production of a large number of genetically similar plants through plant tissue culture is called
(1) hybridoma technology
(2) recombinant DNA technology
(3) gene therapy
(4) Micropropagation

187. Genes have been transferred into animals with a view to obtain a large scale production of the proteins encoded by these genes in the milk, blood etc. This approach is also referred generally as
(1) In vitro culture
(2) molecular farming
(3) gene therapy
(4) hybridoma technology

188. With reference to biotechnology, microinjection is a method of
(1) injecting a solution of DNA into the nucleus of a cell
(2) injecting nutrients into a cell culture media
(3) injecting microbes into a cell culture media
(4) injecting medicine to human beings

189. Pluripotent cells derived from the early pre implantation of an embryo in mice are called
(1) stem cells
(2) organ culture
(3) somatic cell hybridization
(4) Hybridoma

190. The advantage with embryonic stem cells in producing transgenic animals is
(1) these cells are immortal
(2) these cells can be maintained and multiplied in vitro long enough to permit various manipulations for gene transfer
(3) both 1 and 2
(4) neither 1 nor 2

191. In nature, *Agrobacterium tumefaciens* mediated infection of plant cells leads to
(P) crown gall disease in plants
(Q) hairy root disease in plants
(R) transfer of T-DNA into the plant chromosome
(S) transfer of Ri-plasmid into the plant cell
(1) S only (2) P and R only
(3) Q and S only (4) Q only

192. Protein A, which has strong affinity to Fc region of immunoglobulin, is extracted from
(1) *Saccharomyces cerevisiae*
(2) *Staphylococcus aureus*
(3) *Streptococcus pyogenes*
(4) *Streptococcus sanguis*

193. an siRNA experiment, a double-stranded 21-mer oligo against the 3' UTR of a gene leads to complete loss of function of the gene in cell culture. For the gain of function, one should transfect the cells with plasmid containing
(1) 3' UTR of the gene
(2) 5' UTR of the gene
(3) coding region of the gene
(4) gene with the introns

194. Xeroderma pigmentosum is a human genetic disease that renders affected individuals highly sensitive to sunlight resulting in skin lesions and cancer. The genes involved *(XPA, XPC..)* are of
(1) nucleotide excision repair pathway
(2) double-strand break repair pathway
(3) non-homologous end joining repair pathway
(4) transcription-coupled repair pathway

195. Which of the following would be considered an example of bioremediation?
(1) adding nitrogen-fixing microorganisms to a degraded ecosystem to increase nitrogen availability
(2) using a bulldozer to regrade a strip mine
(3) dredging a river bottom to remove contaminated sediments
(4) reconfiguring the channel of a river

196. In secondary metabolism two distinct phases-trophophase and idiophase refer respectively to
(1) Growth and production phase
(2) Early and late phase
(3) Primary and secondary metabolism
(4) Lag phase and log phase

197. What will be the proportion of Resistant (R) to susceptible (S) plants in the progeny of a T_0 tobacco plant carrying kanamycin resistance gene when selfed and when crossed with an untransformed plant separately?
(1) 1:1 and 3:1 (2) 1:1 and 1:3
(3) 3:1 and 1:3 (4) 3:1 and 1:1

198. What is the correct sequence of events in Southern blotting?
(1) Hybridization of DNA fragments with a labelled probe sequence followed by separation by electrophoresis and then transfer to a membrane
(2) Separation of DNA fragments by electrophoresis followed by hybridization with a labelled probe sequence and then transfer to a membrane
(3) Separation of DNA fragments by electrophoresis followed by transfer to a membrane and then hybridization with a labelled probe sequence
(4) Transfer of DNA fragments to a membrane followed by separation by electrophoresis and then hybridization with a labelled probe sequence

199. Plasmid vectors used in cloning often contain a gene for the N-terminal 146 amino acids of the enzyme β-galactosidase. What is the purpose of including this gene in the vector?
(1) It allows selection of *E. coli* host cells that contain the plasmid
(2) It allows selection of *E. coli* host cells that contain plasmid in which the insert has been ligated
(3) It cleaves the insert to allow it to be ligated into the vector
(4) It enables the plasmid vector to replicate in *E. coli* host cells

200. In the creation of a knockout mouse using homologous recombination, a viral thymidine kinase gene is often included in the vector outside of the region of homology between the vector and targeted chromosome. What is the purpose of this?
(1) To allow negative selection of cells in which integration of the targeting sequence has occurred by homologous recombination
(2) To allow negative selection of cells in which integration of the targeting sequence has occurred by random insertion into the genome
(3) To allow positive selection of cells in which integration of the targeting sequence has occurred by homologous recombination
(4) To allow positive selection of cells in which integration of the targeting sequence has occurred by random insertion into the genome

ANSWERS HEET

Unit 12: Biotechnology

1. 2	**2.** 2	**3.** 2	**4.** 3	**5.** 4	**6.** 1	**7.** 1	**8.** 1	**9.** 4	**10.** 3
11. 3	**12.** 4	**13.** 2	**14.** 3	**15.** 3	**16.** 4	**17.** 1	**18.** 3	**19.** 4	**20.** 1

21. 3	**22.** 3	**23.** 3	**24.** 1	**25.** 2	**26.** 3	**27.** 3	**28.** 4	**29.** 3	**30.** 1
31. 2	**32.** 3	**33.** 1	**34.** 3	**35.** 1	**36.** 3	**37.** 3	**38.** 3	**39.** 4	**40.** 1
41. 1	**42.** 4	**43.** 1	**44.** 1	**45.** 3	**46.** 4	**47.** 2	**48.** 2	**49.** 1	**50.** 1
51. 4	**52.** 2	**53.** 1	**54.** 3	**55.** 4	**56.** 2	**57.** 2	**58.** 1	**59.** 4	**60.** 4
61. 2	**62.** 1	**63.** 1	**64.** 3	**65.** 4	**66.** 3	**67.** 1	**68.** 3	**69.** 3	**70.** 3
71. 1	**72.** 3	**73.** 2	**74.** 3	**75.** 2	**76.** 2	**77.** 4	**78.** 4	**79.** 1	**80.** 4
81. 1	**82.** 1	**83.** 1	**84.** 2	**85.** 3	**86.** 2	**87.** 4	**88.** 3	**89.** 3	**90.** 2
91. 1	**92.** 2	**93.** 4	**94.** 2	**95.** 2	**96.** 3	**97.** 3	**98.** 1	**99.** 3	**100.** 1
101. 2	**102.** 2	**103.** 2	**104.** 3	**105.** 3	**106.** 2	**107.** 2	**108.** 1	**109.** 2	**110.** 4
111. 1	**112.** 4	**113.** 3	**114.** 1	**115.** 1	**116.** 2	**117.** 1	**118.** 4	**119.** 1	**120.** 1
121. 2	**122.** 1	**123.** 3	**124.** 2	**125.** 1	**126.** 2	**127.** 3	**128.** 2	**129.** 4	**130.** 2
131. 4	**132.** 1	**133.** 3	**134.** 4	**135.** 4	**136.** 3	**137.** 2	**138.** 2	**139.** 1	**140.** 4
141. 4	**142.** 1	**143.** 4	**144.** 4	**145.** 3	**146.** 1	**147.** 3	**148.** 4	**149.** 3	**150.** 3
151. 2	**152.** 2	**153.** 4	**154.** 3	**155.** 3	**156.** 3	**157.** 3	**158.** 4	**159.** 2	**160.** 4
161. 4	**162.** 1	**163.** 2	**164.** 1	**165.** 3	**166.** 4	**167.** 3	**168.** 3	**169.** 3	**170.** 1
171. 1	**172.** 4	**173.** 4	**174.** 3	**175.** 2	**176.** 2	**177.** 4	**178.** 1	**179.** 2	**180.** 1
181. 4	**182.** 3	**183.** 1	**184.** 2	**185.** 2	**186.** 4	**187.** 2	**188.** 1	**189.** 1	**190.** 2
191. 2	**192.** 2	**193.** 2	**194.** 1	**195.** 1	**196.** 3	**197.** 4	**198.** 3	**199.** 2	**200.** 2

UNIT-13

APPLIED BIOTECHNOLOGY & BIOTECHNIQUES

1. The plasmid cloning vector pBR322 contains amp^R and tet^R genes that confer resistance to ampicillin and tetracycline, respectively. The tet^R gene contains a site for the restriction endonuclease *BamH*I. pBR322 is first cleaved with *BamH*I, added to a *BamH*I restriction fragment from a different DNA molecule and the resulting mixture is treated with DNA ligase and used to transform *E. coli* cells. Under these conditions, which one of the following statements is true?
   (1) Tetracycline can then be used to select for transformed *E. coli* carrying recombinant plasmids
   (2) Tetracycline can then be used to select for transformed *E. coli* carrying non-recombinant plasmids
   (3) Tetracycline can then be used to select for non-transformed *E. coli*
   (4) *E. coli* cells with recombinant plasmids will grow on both tetracycline and ampicillin

2. Which one of the following radioisotopes does not emit β rays?
   (1) ^{14}C (2) ^{3}H
   (3) ^{32}P (4) ^{125}I

3. In nuclear magnetic resonance (NMR) spectroscopy, the absorption spectra result by the absorption of one of the following electromagnetic radiation by the spinning nucleus
   (1) Ultraviolet waves
   (2) Infrared waves
   (3) Radiowaves (4) Microwaves

4. A ribonuclase solution gave an absorbance of 1.0 at 278 nm in a UV spectrometer using a 1 cm quartz cuvette. Given that the molar extinction coefficient of the enzyme at 278 nm is 102 $M^{-1}cm^{-1}$, the concentration of the enzyme would be
   (1) 1 mM (2) 20 mM
   (3) 10 mM (4) 100 mM

5. Two proteins have the same molecular mass as well as the same isoelectric point. The best way to separate them would be to use
   (1) Reverse phase chromatography
   (2) Gel filtration chromatography
   (3) Ion-exchange chromatography
   (4) Chromatofocusing

6. Which type of restriction enzymes does not require ATP?
   (1) Type I (2) Type II
   (3) Type III (4) Type IV

7. The Southern blotting technique is used for
   (1) the detection of RNA fragments on membranes by specific radioactive antibodies
   (2) the detection of DNA fragments on membranes by a radioactive DNA probe
   (3) the detection of proteins on membranes using a radioactive DNA probe
   (4) the detection of DNA fragments on membranes by specific radioactive antibodies

8. Carboxymethyl cellulose is
(1) a cation -exchange matrix
(2) a gel filtration matrix
(3) an anion-exchange matrix
(4) a plant cell wall constituent

9. A tetanus booster shot results in the increased production of
(1) tetanus-specific NK cells
(2) T cells that recognize tetanus toxoid but not tetanus toxin
(3) antibodies which neutralize tetanus toxin
(4) T-cells which kill *Clostridium tetani*

10. ESTs are obtained through
(1) Genomic DNA library
(2) cDNA library
(3) RT-PCR
(4) Chromosome walking

11. Targetted suppression of gene expression is achieved by
(1) T-DNA insertion
(2) EMS
(3) RNAi
(4) Gamma ray

12. A set of two or more overlapping DNA fragments that form a contiguous stretch of DNA is called
(1) contigs (2) BAC clones
(3) YAC clones (4) map

13. A vector can accept an insert of 20 kb size and the recombinant vector can be replicated in *E. coli.* In order to make complete gene library of *E. coli* by using this vector, minimum number of bacterial colonies which must be present should not be less than
(1) 1.1×10^3 (2) 4.1×10^3
(3) 2.1×10^4 (4) 1.5×10^4

14. The DNA of temperate phage P4 is linear, double stranded, 11.5 kb long and has cohesive ends. Digestion with BamH1 yields fragments 6.4, 4.1 and 1.0 kb in length. The partial digestion with the same enzyme yields fragments 10.5, 7.4, 6.4, 4.1, and 1.0 kb in length. Circular P4 DNA made with DNA ligase can be digested with BamH1 to yield fragments in the DNA. What is the order of fragments in DNA ?
(1) 6.4- 1.0- 4.1 (2) 4.1- 6.4- 7.4
(3) 1.0- 6.4- 4.1 (4) 4.1- 1.0- 6.4

15. The dependence of molecular weight of protein molecule to the distance traveled in denaturing gel electrophoresis is
(1) linear
(2) cubic
(3) logarithmic
(4) inversely related to the amount of denaturant

16. What is PROSITE?
(1) A database of protein structures
(2) A database of interacting proteins
(3) A database of protein motifs
(4) A search tool

17. Which is the best annotated database?
(1) Genbank (2) PDB
(3) Prodom (4) Swissprot

18. Protein sequence comparison is more sensitive than nucleic acid sequence comparison because
(1) proteins are functional
(2) proteins have definite three dimensional structures
(3) the protein alphabet has more letters than the nucleic acid
(4) codon bias

19. Which of the following is not a part of the methods used in single locus probe analysis of VNTR regions of human DNA?
(1) DNA extraction
(2) Restriction endonuclease digestion of DNA
(3) Gel electrophoresis
(4) Recombinant DNA

20. A covalently closed circular DNA containing a single promoter is mixed with RNA polymerases that open complexes form where 1 turn of DNA is unwound. Which of the following statements will be true? Please note that L = linking number, T = twist, and W = writhe

(1) L will decrease because T will decrease by 1

(2) L will increase because W will increase by 1

(3) L will stay the same because T and W will not change

(4) L will stay the same because an increase in W will cancel out the decrease in T

21. A solution of a protein whose sequence includes three tryptophan residues, no tyrosine residues, and no phenylalanine residues has an absorbance of 0.3 at 280 nm in a cell with a path length of 1 cm. Estimate the concentration of the protein in units of molarity. If the protein has a molecular mass of 100 kDa, estimate the concentration in units of milligrams of protein per milliliter of solution. (ε= 10000 M^{-1} cm^{-1})

(1) 10 μM, I mg/mL

(2) 30 μM, 1 mg/mL

(3) 30 μM, 3 mg/mL

(4) 10 μM, 3 mg/mL

22. Which of the following techniques can be utilized to measure the rate of diffusion of membrane proteins?

(1) Patching and capping

(2) Immunodiffusion

(3) Patch-clamp

(4) FRAP

23. Sedimentation constant (S) is the ratio of

(1) Rate of sedimentation to acceleration due to gravity

(2) Rate of sedimentation to angular acceleration

(3) Rate of sedimentation to relative acceleration

(4) Velocity of sedimentation to acceleration due to gravity

24. For 100% purity of peak recovery in chromatography, the desired resolution factor (Rs) is

(1) Rs=0.99 (2) Rs= 1

(3) Rs=1.5 (4) Rs= infinity

25. One of the these purification steps requires initial high ionic strength in sample

(1) Ion exchange chromatography

(2) Hydrophobic interaction chromatography

(3) Chromatofocusing

(4) Preparative chromatography

26. Arrange the following in hierarchical top to bottom order as is done in SCOP

(1) Classes, domains, superfamilies, folds, families

(2) domains, superfamilies, folds, families, classes

(3) superfamilies, folds, families, domains, classes

(4) Classes, folds, superfamilies, families, domains

27. A restriction endonuclease recognizes a 8 bp unbiased conserved sequence as its cleavage site. How many probable site(s) can be present in a 70 kb DNA fragment ?

(1) 2 (2) 1

(3) 4 (4) 6

28. A cDNA encoding an eukaryotic gene was ligated to an expression vector which was then introduced into *E. coli* for expression of protein. However, the experiment resulted in poor expression of inactive form of the protein, which could be due to

(1) Absence of capping at the 5'end of the transcript

(2) Absence of polyadenylation at the 3'end of the transcript

(3) Codon bias

(4) Lack of splicing machinery in *E. coli*

29. The best method to permeabilize yeast cells chemically is to use

(1) EDTA and Lysozyme
(2) β-(1,3) glucanase and protease
(3) α-(1,6) glucanase
(4) Alkaline hydroxylase

30. 5' RACE is often necessary to

(1) delete sequences from 5' end of the DNA strand
(2) label 5'end of DNA with a dye
(3) clone 5' region of genes from mRNA
(4) add sequences at 5' end to facilitate annealing of a specific primer

31. The E-value in a Blast search measures

(1) the probability that the search result is non-random
(2) the significance of the search result
(3) the probability that the search result is obtained randomly
(4) the reliability of the search

32. Calf thymus terminal deoxynucleotidyl transferase

(1) adds nucleotide to the 3'OH terminus of a DNA molecule
(2) adds nucleotide to the 5' P terminus of a DNA molecule
(3) removes nucleotide from the 3'OH terminus of a DNA molecule
(4) removes nucleotide from 5' P terminus of a DNA molecule

33. The enzyme of choice for converting DNA with 3'end overhang into a blunt ended one is

(1) Klenow fragment of DNA Polymerase I
(2) DNA Polymerase I holoenzyme
(3) T_4 DNA polymerase
(4) S1 nuclease

34. In which type of chromatography are ion-pairing agents used for elution?

(1) Hydrophobic Interaction Chromatography
(2) Reverse Phase Chromatography
(3) Ion Exchange Chromatography
(4) Immobilized Metal Anion Chromatography

35. You have two distantly related proteins. Which of the following sets is the best for comparing them?

(1) BLOSUM45 or PAM250
(2) BLOSUM45 or PAM1
(3) BLOSUM80 or PAM250
(4) BLOSUM80 or PAM1

36. X-ray crystallography is

P. based on the elastic scattering of X-rays
Q. non-destructive analytical techniques
R. used in determining the arrangement of atoms within a crystal
S. based on Bragg's law

(1) P and Q
(2) R and S
(3) P, Q and S
(4) P, Q, R and S

37. Which of the following techniques is used to locate disulfide bonds in a protein?

(1) The protein is first reduced and carboxy methylated
(2) The protein is cleaved by acid hydrolysis
(3) The peptides are separated by SDS polyacrylamide gel electrophoresis
(4) The peptides are separated by two dimensional electrophoresis with an intervening performic acid treatment

38. Concentration of a protein solution determined using its extinction coefficient resulted in a value of 1 mg/ml. Given that its molecular weight is 100 kDa, its concentration in molar units will be

(1) 10 μM
(2) 100 μM
(3) 1 mM
(4) 10 mM

39. Urea and Guanidinium chloride are potent protein denaturants. They work by
(1) binding to the hydrophobic groups in proteins
(2) breaking of disulphide bonds
(3) enhancing the hydrogen bonded network of water
(4) binding to the peptide bonds of proteins

40. A bottle contains 1 mCi of L-Phenylalanine ^{14}C (uniformly labelled) in 2.0 ml of solution. Specific activity of the labelled amino acid is given as 150mCi/m mole. The concentration of L-Phenylalanine in the solution can be calculated as
(1) 2.335×10^{-3} M
(2) 3.335×10^{-3} M
(3) 3.535×10^{-3} M
(4) 2.535×10^{-3} M

41. Which one of the following statements is not correct?
(1) PCR is more sensitive than branched DNA assays
(2) Branched DNA is more sensitive than Ligase chain reaction
(3) Linear amplification occurs in branched DNA
(4) PCR is extremely liable to contamination

42. Which method is used to generate clonally amplified DNA fragments to be used as templates in Pyrosequencing?
(1) Ligation PCR (2) Shot gun cloning
(3) BAC cloning (4) Emulsion PCR

43. In a ligation reaction, what amount of insert DNA of 750 bp is required to clone into 5000 bp vector with a concentration of 80 ng/ µl at a molar ratio of vector to insert 1:3?
(1) 12 ng (2) 36 ng
(3) 40 ng (4) 24 ng

44. The virus inserted in genome can be recognized by
(1) FISH (2) Northern blot
(3) Microarray (4) Southern blot

45. Unsynchronized signals in EEG are generated during
(1) Deep Sleep
(2) Active and non-quite
(3) Slow wave but quite sleep
(4) REM sleep

46. DNA from a host sample can be amplified by a process known as the polymerase chain reaction (PCR) Which of the following is required for PCR?
(1) Knowledge of the genetic sequence to be amplified
(2) A single nucleotide primer
(3) A universal probe to detect the amplified product
(4) A heat-sensitive DNA polymerase enzyme

47. A patch–clamp device is used to
(1) measure the strength of an electrochemical gradient
(2) study the properties of individual neurotransmitters
(3) infuse different kinds of ions into an axon
(4) study the properties of individual membrane channels

48. Which of the following enzymes is not used in pyrosequencing?
(1) DNA Polymerase
(2) Pyrophosphatase
(3) Luciferase
(4) ATP sulfurylase

49. A ribonuclease stock solution at pH 3.0 in 10 mM glycine-HCl buffer is diluted 20 fold with the buffer and resulting absorbance of solution is taken in a quartz cuvette of path length 1 cm was 0.38 at its wavelength maximum. If the extinction coefficient of the protein is 0.74 ml/mg.cm, the concentration of the stock protein solution would be
(1) 5 mg/ml (2) 10 mg/ml
(3) 20 mg/ml (4) 50 mg/ml

50. Which of the following techniques cannot be used to determine the molecular weight of a protein?
(1) UV absorption
(2) Viscocity
(3) Light scattering
(4) Sedimentation equilibrium

51. β-lactoglobulin which is a monomer at neutral pH is known to tetramerise at acidic pH of 2. Which one of the following techniques could be effectively employed to demonstrate the formation of a tetramer ?
(1) Native gel electrophoresis
(2) Anion exchange chromatography
(3) SDS-polyacrylamide gel electrophoresis
(4) Reverse phase chromatography

52. ELISPOT assay is used for measuring
(1) cytokine concentration in serum
(2) antibody titre in serum
(3) frequency of B cell responses
(4) frequency of T cell responses

53. Which of the following is the best method to determine the phospholipid asymmetry in a plasma membrane?
(1) Electron microscopy
(2) Fluorescence spectroscopy
(3) Lectin binding
(4) Thin layer chromatography

54. APS (Ammonium persulphate) is used in SDS-PAGE for
(1) preventing oxidation
(2) cross-linking
(3) its role as a catalyst
(4) free radical formation

55. Identify the mismatch
(1) Alkaline phosphatase : remove phosphate group present at 5' end of DNA
(2) DNA Polymerase I : nick translation
(3) S1 endonucleases : cleaves only single strand DNA
(4) DNase I : cleaves only double stranded DNA

56. In protein 'A' glutamic acid is replaced by glutamine to make protein 'B'. Which technique can resolve these two proteins?
(1) Isoelectric focusing
(2) Pulse field electrophoresis
(3) SDS-PAGE
(4) Gel filtration

57. The advantage of the Edman's reagent (phenyl isothiocyanate-PTH) over Sanger's reagent (fluorodinotrobenzene-FDNB) in peptide analysis is
(1) complete oxidation of all disulfides
(2) complete denaturation
(3) that the process can be repeated on the remaining peptide
(4) complete hydrolysis

58. In a gel filtration chromatography
(1) The large protein will be eluted first
(2) The small protein will be eluted first
(3) Both large and small will elute at the same time
(4) The small protein with high charge will be eluted first

59. Two dimensional gel electrophoresis is a technique for separating proteins
(1) Based on charge
(2) Based on mass
(3) Based on both charge and mass
(4) Based on its pI value

60. Exact mass and sequence of proteins and peptides can be measured by
(1) MALDI-TOF
(2) Proton NMR
(3) X-Ray
(4) Mass spectroscopy

61. Fluorescence microscopy is based on the ability of certain molecules to
(1) Continuously emit light of a constant wavelength
(2) Absorb light of many different wavelengths
(3) Absorb light of a given wavelength and then emit light of a longer wavelength
(4) Absorb light of a given wavelength and then emit light of a shorter wavelength

62. Which one of the following tools of recombinant DNA technology is incorrectly paired with its applications?
(1) Restriction endonuclease - production of DNA fragments for gene cloning
(2) DNA ligase - enzyme that cuts DNA, creating sticky ends
(3) DNA polymerase - copies DNA sequences in the polymerase chain reaction
(4) Reverse transcriptase - production of cDNA from mRNA

63. The coefficient of correlation r satisfies
(1) $0 < r < 1$ (2) $[r] > 1$
(3) $[r] = 1$ (4) $-1 < r < +1$

64. When urea-denatured inclusion bodies are renatured, protein aggregates can sometimes be found due to misfolding. For small proteins with no disulphide linkages and a single domain, the probability of misfolding is the highest when
(1) Urea is diluted stepwise
(2) Urea is diluted in a single step
(3) Glutathione(s) addition is not done
(4) pH of urea solution is adjusted to 7.0-7.5

65. The net charge of a protein may not be sufficient to determine whether a protein will bind to an ion exchanger. This is due to
(1) The presence of hydrophobic patches on the protein surface
(2) Heterogeneous spatial distribution of charged amino acids
(3) The presence of repeating motifs in some proteins
(4) The strong hydration potential of protein

66. In Sephadex G series (gel filtration) as the G number increases from 10, 15, 25….200, the pore size of the beads
(1) Decreases
(2) Increases
(3) remains constant
(4) varies from manufacturer to manufacturer

67. Increasing aeration at a particular rpm does not always lead to increase in oxygen mass transfer in a bioreactor. This is because
(1) As aeration increases, the bubble size increases
(2) As aeration increases, the bubble size decreases
(3) Agitator can get flooded leading to inconsistencies in bubble size
(4) Residence time of the bubble does not change

68. A hypothetical relaxed circular plasmid has 4500 bp. If for supercoiled form of this plasmid the twist is 440 and the writhe is -20, then the plasmid can be considered as a
(1) nicked circular plasmid
(2) positively supercoiled plasmid
(3) negatively supercoiled plasmid
(4) relaxed circular plasmid

69. Restriction enzymes produced by *E. coli*, do not cut self DNA because cells are
(1) RecA+ (2) Dam +
(3) RecA- (4) Dam -

70. Which one of the following microscopic techniques is best suited to visualize the topology and distribution of transmembrane protein of a cell membrane?
(1) Scanning electron microscopy
(2) Transmission electron microscopy
(3) Freeze-fracture electron microscopy
(4) Thin-section electron microscopy

71. A novel type II restriction enzyme has been isolated from a thermophilic bacteria. This restriction endonuclease recognizes 5'ATAANNNTTAT3' (N= any nucleotide) and cuts after third 'A' in the above sequence. What is the fate of DNA after restriction digestion ?

(1) A 3 nucleotide long 5′ overhang
(2) A 7 nucleotide long 5′ overhang
(3) A 4 nucleotide long 5′ overhang
(4) A 3 nucleotide long 3′ overhang

72. Pyrosequencing derives its name from the fact that

(1) the bases are detected by pyrolysis
(2) it uses enzyme apyrase to detect the bases
(3) it detects pyrophosphate released during base incorporation
(4) it generates pyrograms as output

73. Electrophoresis of a purified protein in SDS-PAGE in the presence of 2-marcaptoethanol yields two bands of 35 kDa and 45 kDa. However, in a gel filtration chromatography, the same protein elutes as 80 kDa. What conclusion can be drawn from the above observation?

(1) Protein is not purified to homogeneity
(2) Two bands generated in SDS-PAGE due to degradation
(3) Protein is a multimer
(4) Protein is a heterodimer

74. A BSA stock solution is diluted 10 folds with phosphate buffer. The absorbance of the solution in a quartz cuvette of pathlength 1 mm at 281.5 nm is 0.330. If the extinction coefficient of the protein is 0.66 ml/mg.cm, the concentration of the stock protein solution would be

(1) 5 mg/ml
(2) 20 mg/ml
(3) 33 mg/ml
(4) 50 mg/ml

75. Sodium dodecyl sulphate, an anionic detergent commonly used in SDS-Polyacrylamide gel electrophoresis, works in facilitating electrophoretic separation of a mixture of proteins by its ability to bind to the

(1) negatively charged amino acid side chains in proteins
(2) hydrophobic side chains in proteins
(3) positively charged amino acid side chains in proteins
(4) peptide group in proteins

76. RT-PCR reaction sequentially uses

(1) RNA dependent DNA polymerase and DNA dependent DNA polymerase
(2) RNA dependent DNA polymerase and DNA polymerase 1
(3) RNA polymerase and DNA dependent DNA polymerase
(4) RNA polymerase and DNA polymerase 1

77. The linear and circular forms of the same DNA molecule can be distinguished using

(1) Absorbance at 260 nm
(2) Endonuclease digestion
(3) Viscosity of the solution
(4) Exonuclease digestion

78. Protein-protein interaction can be evaluated by all of the following except

(1) Far-Western blotting
(2) Chromatin immunoprecipitation
(3) Yeast-two hybrid system
(4) Co-immunoprecipitation

79. In pET expression vectors, high level of expression of cloned gene is achieved using

(1) T_7 promoter (2) SP6 promoter
(3) λ–P_L promoter (4) Trp promoter

80. Which of the following techniques can be used to determine the alpha-amylase gene polymorphism?

(1) Southern blot (2) Slot blot
(3) Dot blot (4) Northern blot

81. Assume that you are trying to insert a gene into a plasmid. Someone gives you a preparation of genomic DNA that has been cut with restriction enzyme X. The gene you wish to insert has sites on both ends for cutting by restriction enzyme Y. You have a plasmid with a single site for Y, but not for X. Your strategy should be to

(1) insert the fragments cut with restriction enzyme X directly into the plasmid without cutting the plasmid

(2) cut the plasmid with restriction enzyme X and insert the fragments cut with restriction enzyme Y into the plasmid

(3) cut the DNA again with restriction enzyme Y and insert these fragments into the plasmid cut with the same enzyme

(4) cut the plasmid twice with restriction enzyme Y and ligate the two fragments onto the ends of the DNA fragments cut with restriction enzyme X

82. How does a bacterial cell protect its own DNA from restriction enzymes?

(1) by adding methyl groups to adenines and cytosines

(2) by using DNA ligase to seal the bacterial DNA into a closed circle

(3) by adding histones to protect the double-stranded DNA

(4) by forming "sticky ends" of bacterial DNA to prevent the enzyme from attaching

83. What is the most logical sequence of steps for splicing foreign DNA into a plasmid and inserting the plasmid into a bacterium?

I. Transform bacteria with a recombinant DNA molecule

II. Cut the plasmid DNA using restriction enzymes

III. Extract plasmid DNA from bacterial cells

IV. Hydrogen-bond the plasmid DNA to nonplasmid DNA fragments

V. Use ligase to seal plasmid DNA to nonplasmid DNA

(1) I, II, IV, III, V

(2) II, III, V, IV, I

(3) III, II, IV, V, I

(4) III, IV, V, I, II

84. A principal problem with inserting an unmodified mammalian gene into a BAC, and then getting that gene expressed in bacteria, is that

(1) prokaryotes use a different genetic code from that of eukaryotes

(2) bacteria translate polycistronic messages only

(3) bacteria cannot remove eukaryotic introns

(4) bacterial RNA polymerase cannot make RNA complementary to mammalian DNA

85. A gene that contains introns can be made shorter (but remain functional) for genetic engineering purposes by using

(1) RNA polymerase to transcribe the gene

(2) a restriction enzyme to cut the gene into shorter pieces

(3) reverse transcriptase to reconstruct the gene from its mRNA

(4) DNA polymerase to reconstruct the gene from its polypeptide product

86. Why are yeast cells frequently used as hosts for cloning?

(1) They easily form colonies

(2) They can remove exons from mRNA

(3) They do not have plasmids

(4) They are eukaryotic cells

87. The DNA fragments making up a genomic library are generally contained in

(1) BACs

(2) recombinant viral RNA

(3) individual wells

(4) DNA-RNA hybrids

88. Yeast artificial chromosomes contain which of the following elements?
(1) Centromeres only
(2) Telomeres only
(3) Origin of replication only
(4) Centromeres, telomeres, and an origin of replication

89. Which of the following best describes the complete sequence of steps occurring during every cycle of PCR?
1. The primers hybridize to the target DNA
2. The mixture is heated to a high temperature to denature the double-stranded target DNA
3. Fresh DNA polymerase is added
4. DNA polymerase extends the primers to make a copy of the target DNA
(1) 2, 1, 4 (2) 1, 3, 2, 4
(3) 3, 4, 1, 2 (4) 3, 4, 2

90. A researcher needs to clone a sequence of part of a eukaryotic genome in order to express the sequence and to modify the polypeptide product. She would be able to satisfy these requirements by using which of the following vectors?
(1) A bacterial plasmid
(2) BAC to accommodate the size of the sequence
(3) A modified bacteriophage
(4) A YAC with appropriate cellular enzymes

91. A student wishes to clone a sequence of DNA of ~200 kb. Which vector would be appropriate?
(1) A plasmid
(2) A typical bacteriophage
(3) A BAC
(4) A plant virus

92. Sequencing an entire genome, such as that of *C. elegans*, a nematode, is most important because
(1) it allows researchers to use the sequence to build a "better" nematode, which is resistant to disease
(2) it allows research on a group of organisms we do not usually care much about
(3) the nematode is a good animal model for trying out cures for viral illness
(4) a sequence that is found to have a particular function in the nematode is likely to have a closely related function in vertebrates

93. To introduce a particular piece of DNA into an animal cell, such as that of a mouse, you would find more probable success with which of the following methods?
(1) The shotgun approach
(2) Electroporation followed by recombination
(3) Introducing a plasmid into the cell
(4) Infecting the mouse cell with a Ti plasmid

94. The major advantage of using artificial chromosomes such as YACs and BACs for cloning genes is that
(1) Plasmids are unable to replicate in cells
(2) Only one copy of a plasmid can be present in any given cell, whereas many copies of a YAC or BAC can coexist in a single cell
(3) YACs and BACs can carry much larger DNA fragments than ordinary plasmids can
(4) YACs and BACs can be used to express proteins encoded by inserted genes, but plasmids cannot

95. Which of the following is used to make complementary DNA (cDNA) from RNA?
(1) Restriction enzymes
(2) Gene cloning
(3) DNA ligase
(4) Reverse transcriptase

96. Why is it so important to be able to amplify DNA fragments when studying genes?

(1) DNA fragments are too small to use individually

(2) A gene may represent only a millionth of the cell's DNA

(3) Restriction enzymes cut DNA into fragments that are too small

(4) A clone requires multiple copies of each gene per clone

97. *Pax-6* is a gene that is involved in eye formation in many invertebrates, such as *Drosophila*. *Pax-6* is found as well in vertebrates. A *Pax-6* gene from a mouse can be expressed in a fly and the protein (PAX-6) leads to a compound fly eye. This information suggests which of the following?

(1) *Pax-6* genes are identical in nucleotide sequence

(2) PAX-6 proteins have identical amino acid sequences

(3) *Pax-6* is highly conserved and shows shared evolutionary ancestry

(4) PAX-6 proteins are different for formation of different kinds of eyes

98. Why are BACs preferred today rather than bacteriophages for making genomic libraries?

(1) The BAC carries more DNA

(2) The BAC can carry entire genes and their regulatory elements

(3) Larger BACs are easier to store

(4) The BAC carries more DNA, the BAC can carry entire genes and their regulatory elements, and larger BACs are easier to store

99. The reason for using Taq polymerase for PCR is that

(1) it is heat stable and can withstand the temperature changes of the cycler

(2) only minute amounts are needed for each cycle of PCR

(3) it binds more readily than other polymerases to primer

(4) it has regions that are complementary to primers

100. Why might a laboratory be using dideoxy nucleotides?

(1) To separate DNA fragments

(2) To clone the breakpoints of cut DNA

(3) To produce cDNA from mRNA

(4) To sequence a DNA fragment

101. In order to identify a specific restriction fragment using a probe, what must be done?

(1) The fragments must be separated by electrophoresis

(2) The fragments must be treated with heat or chemicals to separate the strands of the double helix

(3) The probe must be hybridized with the fragment

(4) The fragments must be separated by electrophoresis, the fragments must be treated with heat or chemicals to separate the strands of the double helix, and the probe must be hybridized with the fragment

102. Which of the following modifications is least likely to alter the rate at which a DNA fragment moves through a gel during electrophoresis?

(1) Altering the nucleotide sequence of the DNA fragment

(2) Methylating the cytosine bases within the DNA fragment

(3) Increasing the length of the DNA fragment

(4) Decreasing the length of the DNA fragment

103. DNA fragments from a gel are transferred to a nitrocellulose paper during the procedure called Southern blotting. What is the purpose of transferring the DNA from a gel to a nitrocellulose paper?

(1) To attach the DNA fragments to a permanent substrate
(2) To separate the two complementary DNA strands
(3) To transfer only the DNA that is of interest
(4) To prepare the DNA for digestion with restriction enzymes

104. DNA microarrays have made a huge impact on genomic studies because they

(1) can be used to eliminate the function of any gene in the genome
(2) can be used to introduce entire genomes into bacterial cells
(3) allow the expression of many or even all of the genes in the genome to be compared at once
(4) allow physical maps of the genome to be assembled in a very short time

105. Which of the following describes the transfer of polypeptide sequences to a membrane to analyze gene expression?

(1) Southern blotting
(2) Northern blotting
(3) Western blotting
(4) Eastern blotting

106. Which of the following uses reverse transcriptase to make cDNA followed by amplification?

(1) Southern blotting
(2) Northern blotting
(3) Western blotting
(4) RT-PCR

107. RNAi methodology uses double-stranded pieces of RNA to trigger a breakdown or blocking of mRNA. For which of the following might it more possibly be useful?

(1) To raise the rate of production of a needed digestive enzyme
(2) To decrease the production from a harmful gain-of-function mutated gene
(3) To destroy an unwanted allele in a homozygous individual
(4) To form a knockout organism that will not pass the deleted sequence to its progeny

108. A researcher has used *in vitro* mutagenesis to mutate a cloned gene and then has reinserted this into a cell. In order to have the mutated sequence disable the function of the gene, what must then occur?

(1) Recombination resulting in replacement of the wild type with the mutated gene
(2) Use of a microarray to verify continued expression of the original gene
(3) Replication of the cloned gene using a bacterial plasmid
(4) Transcription of the cloned gene using a BAC

109. Which of the following techniques used to analyze gene function depends on the specificity of DNA base complementarity?

(1) Northern blotting
(2) use of RNAi
(3) *in vitro* mutagenesis
(4) *in situ* hybridization

110. Silencing of selected genes is often done using RNA interference (RNAi) Which of the following questions would not be answered with this process?

(1) What is the function of gene 432 in this species of annelid?
(2) What will happen in this insect's digestion if gene 173 is not able to be translated?
(3) Is gene HA292 responsible for this disorder in humans?
(4) Will the disabling of this gene in *Drosophila* and in a mouse cause similar results?

111. In large scale, genome-wide association studies in humans, correlation is sought between

(1) lengthy sequences that might be shared by most members of a population
(2) single nucleotide polymorphisms found only in persons with a particular disorder
(3) single nucleotide polymorphisms found in families with a particular introns sequence
(4) single nucleotide polymorphisms in two or more adjacent genes

112. For a particular microarray assay (DNA chip), cDNA has been made from the mRNAs of a dozen patients' breast tumor biopsies. The researchers will be looking for

(1) a particular gene that is amplified in all or most of the patient samples
(2) a pattern of fluorescence that indicates which cells are overproliferating
(3) a pattern shared among some or all of the samples that indicates gene expression differing from control samples
(4) a group of cDNAs that act differently from those on the rest of the grid

113. Which of the following tools of recombinant DNA technology is incorrectly paired with its use?

(1) Restriction enzyme: analysis of RFLPs
(2) DNA ligase: cutting DNA, creating sticky ends of restriction fragments
(3) DNA polymerase: polymerase chain reaction to amplify sections of DNA
(4) Reverse transcriptase: production of cDNA from mRNA

114. Plants are more readily manipulated by genetic engineering than are animals because

(1) Plant genes do not contain introns
(2) More vectors are available for transferring recombinant DNA into plant cells
(3) A somatic plant cell can often give rise to a complete plant
(4) Genes can be inserted into plant cells by microinjection

115. A paleontologist has recovered a bit of tissue from the 400-year-old preserved skin of an extinct dodo (a bird) To compare a specific region of the DNA from the sample with DNA from living birds, which of the following would be most useful for increasing the amount of dodo DNA available for testing?

(1) RFLP analysis
(2) Polymerase chain reaction (PCR)
(3) Electroporation
(4) Gel electrophoresis

116. DNA technology has many medical applications. Which of the following is *not* done routinely at present?

(1) Production of hormones for treating diabetes and dwarfism
(2) Production of microbes that can metabolize toxins
(3) Introduction of genetically engineered genes into human gametes
(4) Prenatal identification of genetic disease alleles

117. In recombinant DNA methods, the term vector can refer to

(1) the enzyme that cuts DNA into restriction fragments
(2) the sticky end of a DNA fragment
(3) a SNP marker
(4) a plasmid used to transfer DNA into a living cell

118. Which of the following would not be true of cDNA produced using human brain tissue as the starting material?

(1) It could be amplified by the polymerase chain reaction
(2) It could be used to create a complete genomic library
(3) It was produced from mRNA using reverse transcriptase
(4) It could be used as a probe to detect genes expressed in the brain

119. Expression of a cloned eukaryotic gene in a bacterial cell involves many challenges. The use of mRNA and reverse transcriptase is part of a strategy to solve the problem of
(1) post-transcriptional processing
(2) electroporation
(3) post-translational processing
(4) nucleic acid hybridization

120. Which of the following sequences in double-stranded DNA is most likely to be recognized as a cutting site for a restriction enzyme?
(1) AAGG
TTCC
(2) AGTC
TCAG
(3) GGCC
CCGG
(4) ACCA
TGGT

121. In order to most efficiently quantitate an antigen available at low concentration in a biological fluid, which of the following ELISA would be preferred?
(1) Coating the plate with the fluid containing the antigen, followed by antigen-specific antibody-enzyme conjugate
(2) A sandwich ELISA, capturing the antigen using an antibody specific for the antigen
(3) A sandwich ELISA, capturing the antigen using an antibody and detecting the antigen with an antibody, which is biotinylated
(4) Plating the biological fluid on an ELISA plate coated with the antigen, followed by regular steps of ELISA

122. An observation was being made under a compound microscope using light of 500 nm wave length. The refractive index of the observation medium was 1.5. The sinα of the lens used was 0.94. The limit of resolution (μm) of the microscope using this lens will be
(1) 0.216.
(2) 216
(3) 0.94
(4) 1.41

123. Which of the following statement about RFLP (Restriction Fragment Length Polymorphism) and AFLP (amplified fragment length polymorphism) is not correct?
(1) Southern hybridization is used only for RFLP
(2) AFLP is a co-dominant molecular marker
(3) AFLP can generate DNA profiles regardless of the complexity or origin of the DNA molecule
(4) only RFLP requires prior information about DNA sequence

124. Western assays used to test serum samples for the presence of antibodies to infectious agents, such as HIV, are particularly useful as diagnostic assays because
(1) they are more sensitive than ELISA
(2) antibodies specific for multiple antigenic epitopes can be detected
(3) they provide quantitative data for sample analysis
(4) they allow multiple samples to be tested simultaneously

125. The major difference between transgenic mice and knockout mice is that
(1) transgenic mice always employ the use of cloned genes derived from other species
(2) transgenic mice have foreign genes that integrate at targeted loci through homologous recombination
(3) transgenic mice have a functional foreign gene added to their genome
(4) knockout mice always have a unique phenotype

126. SCID mice have a genetic defect that prevents development of functional

(1) hematopoietic cells
(2) B cells and T cells
(3) T cells and NK cells
(4) pluripotential stem cells

127. Which of the following statements regarding B cell hybridomas is false?

(1) They are immortal cell lines that produce antibodies of a single specificity
(2) They are derived from B cells that are first cloned and grown in cell culture for short periods
(3) They contain a large nucleus formed by the fusion of two nuclei
(4) They can be used to manufacture diagnostic or therapeutic monoclonal antibodies

128. An ELISA designed to test for the presence of serum antibody for a new strain of pathogenic bacteria is under development. Initially, a monoclonal antibody specific for a single epitope of the organism was used both to sensitize the wells of the ELISA plate and as the enzyme-labeled detecting antibody in a conventional sandwich ELISA. The ELISA failed to detect the antigen despite the use of a wide range of antibody concentrations. What is the most probable cause of this problem?

(1) The antigen is too large
(2) The antibody has a low affinity for the antigen
(3) The monoclonal antibody used to sensitize the wells is blocking access of the epitope, thus when the same antibody is enzyme-labeled, it cannot bind to the antigen
(4) The enzyme-labeled antibody used should have been a different isotype than the sensitizing antibody

129. What are the experimental steps needed for screening an expression library for a clone encoding a protein X that has been isolated and purified?

(P) m-RNA isolation
(Q) Antibody preparation
(R) Cloning into an appropriate vector
(S) Western blotting on transfected plaques

(1) P, S (2) Q, S
(3) Q, R (4) R,S

130. DNA content of Caenorhabditis elegans was analysed and found to contain 1.0 x 10^8 bp. How many standard λ-phage vectors carrying 20kb DNA fragments or YACs carrying 250kb DNA fragments are theoretically required to constitute a complete *C. elegans* genomic library?

(1) 500 λ -phage vectors or 40 yeast clones
(2) 400 λ -phage vectors or 5000 yeast clones
(3) 5000 λ -phage vectors or 400 yeast clones
(4) 5 x 10^4 λ -phage vectors or 4000 yeast clones

131. How many λ -phage vectors / yeast clones should be prepared in order to ensure that every sequence is included in the library?

(1) 25 x 10^3 λ -phage vectors / 2000 yeast clones
(2) 20 x 10^3 λ -phage vectors / 1600 yeast clones
(3) 5 x 10^4 λ -phage vectors / 4000 yeast clones
(4) 10 x 10^4 λ -phage vectors / 10000 yeast clones

132. A mixture separated on HPLC gives clearly defined peaks whereas in a manually run column the peaks tend to blend into each other. This is primarily because

(1) the small size of the packing material used in the HPLC column
(2) the better control of flow in HPLC
(3) the use of high pressure in HPLC
(4) the use of better detection systems in HPLC

133. Electrophoresis of a purified protein called X in the presence of sodium dodecyl sulfate (SDS) and beta-mercaptoethanol shown a single band of 60 kDa. In gel filtrations experiment, protein X elutes between alcohol dehydrogenase (160 kDa) and beta amylase (190 kDa) How many identical subunits protein X is composed of?

(1) One (2) Two
(3) Three (4) Four

134. Three polypeptides (A, B, C) whose masses are 55 kDa, 50kDa and 75 kDa with pI of 6.5, 7.0 and 8.0, respectively, were subjected to standard reducing SDS-PAGE. The order of their separation from top to bottom would be

(1) A, B and C (2) B, A and C
(3) A, C and B (4) C, A and B

135. A solution containing Aspartic acid (pI=2.98), Glycine (pI=5.97), Threonine (pI=6.53) and Lysine (pI=9.74) in 50 mM citrate buffer pH 3.0 was applied to a cation exchange column equilibrated with the same buffer and fraction collected. The order of elution of these amino acids from the column is

(1) Lysine, Threonine, Leucine, Glycine, Aspartic acid
(2) Asprtic acid, Threonine, Glycine, Leucine, Lysine
(3) Aspartic acid, Glycine, Leucine, Threonine, Lysine
(4) Aspartic acid, Threonine, Leucine, Glycine, Lysine

136. Normalized Differential Vegetation Index (NDVI) in remote sensing refers to the following spectral band derivation

(1) Near IR – Red
(2) Red / Near Red
(3) (Near IR – Red) / (Near IR + Red)
(4) (Near IR – Red) / Red

137. An oligomeric protein consisting of identical subunits has a molecular weight of 100000. Each subunit is known to contain 0.28% of iron by weight. What is the maximum possible number of subunits in the protein? (Atomic weight of Iron = 56)

(1) 2 (2) 3
(3) 4 (4) 5

138. Assume the molar extinction coefficient (ε 340 nm) of a compound is $6 \times 10^3\ M^{-1}\ cm^{-1}$. If you are given a 1 mM solution of this substance, in order to obtain an absorbance 0.3 in a 1 cm path length cell, you will dilute it

(1) 20000 times. (2) 2000 times
(3) 200 times (4) 20 times

139. A researcher is trying to identify a specific protein within a mixture. He subjects the mixture to gel electrophoresis and then transfers the separation to nitrocellulose filters. The filters are incubated with antibody to the specific protein and the excess antibody is washed off. The antibody-protein complex is then incubated with a radiolabeled protein that binds to the antibody. Autoradiography is performed to detect the presence of the protein. This technique represents

(1) Northern blotting
(2) Southern blotting
(3) Southwestern blotting
(4) Western blotting

140. Which of the following most correctly describes a shotgun technique for sequencing a genome?

(1) Genetic mapping followed immediately by sequencing
(2) Physical mapping followed immediately by sequencing
(3) Cloning large genome fragments into very large vectors such as YACs, followed by sequencing
(4) Cloning several sizes of fragments into various size vectors, ordering the clones, and then sequencing them

141. The biggest problem with the shotgun technique is its tendency to underestimate the size of the genome. Which of the following might best account for this?

(1) skipping some of the clones to be sequenced
(2) missing some of the overlapping regions of the clones
(3) missing some duplicated sequences
(4) having some of the clones die during the experiment and therefore not be represented

142. What is metagenomics?

(1) Genomics as applied to a species that most typifies the average phenotype of its genus
(2) The sequence of one or two representative genes from several species
(3) The sequencing of only the most highly conserved genes in a lineage
(4) Sequencing DNA from a group of species from the same ecosystem

143. Which procedure is not required when the shotgun approach to sequencing is modified as sequencing by synthesis, in which many small fragments are sequenced simultaneously?

(1) Use of restriction enzymes
(2) Sequencing each fragment
(3) Cloning each fragment into a plasmid
(4) PCR amplification

144. Bioinformatics can be used to scan sequences for probable genes looking for start and stop sites for transcription and for translation, for probable splice sites, and for sequences known to be found in other known genes. Such sequences containing these elements are called

(1) expressed sequence tags
(2) cDNA
(3) multigene families
(4) proteomes

145. A microarray known as a GeneChip, with most now-known human protein coding sequences, has been developed to aid in the study of human cancer by first comparing two to three subsets of cancer subtypes. What kind of information might be gleaned from this GeneChip to aid in cancer prevention?

(1) Information about whether or not a patient has this type of cancer prior to treatment
(2) Evidence that might suggest how best to treat a person's cancer with chemotherapy
(3) Data that could alert patients to what kind of cancer they were likely to acquire
(4) Information about which parent might have provided a patient with cancer-causing genes

146. What is gene annotation in bioinformatics?

(1) Finding transcriptional start and stop sites, RNA splice sites, and ESTs
(2) Describing the functions of protein-coding genes
(3) Describing the functions of noncoding regions of the genome
(4) Matching the corresponding phenotypes of different species

147. Why is it unwise to try to relate an organism's complexity with its size or number of cells?

(1) A very large organism may be composed of very few cells or very few cell types
(2) A single-celled organism, such as a bacterium or a protist, still has to conduct all the complex life functions of a large multicellular organism
(3) A single-celled organism that is also eukaryotic, such as a yeast, still reproduces mitotically
(4) A simple organism can have a much larger genome

148. Fragments of DNA have been extracted from the remnants of extinct woolly mammoths, amplified, and sequenced. These can now be used to

(1) introduce into relatives, such as elephants, certain mammoth traits
(2) clone live woolly mammoths
(3) study the relationships among woolly mammoths and other wool-producers
(4) understand the evolutionary relationships among members of related taxa

149. If humans have 2,900 Mb, a specific member of the lily family has 120,000 Mb, and a yeast has ~13 Mb, why can't this data allow us to order their evolutionary significance?

(1) Size matters less than gene density
(2) Size does not compare to gene density
(3) Size does not vary with gene complexity
(4) Size is mostly due to "junk" DNA

150. Which of the following is a representation of gene density?

(1) Humans have 2,900 Mb per genome
(2) *C. elegans* has ~20,000 genes
(3) Humans have ~20,000 genes in 2,900 Mb
(4) Humans have 27,000 bp in introns

151. The comparison between the number of human genes and those of other animal species has led to many conclusions, including

(1) the density of the human genome is far higher than in most other animals
(2) the number of proteins expressed by the human genome is far more than the number of its genes
(3) most human DNA consists of genes for protein, tRNA, rRNA, and miRNA
(4) the genomes of other organisms are most significantly smaller than the human genome

152. What is the most probable explanation for the continued presence of pseudogenes in a genome such as our own?

(1) They are genes that had a function at one time, but that have lost their function because they have been translocated to a new location
(2) They are genes that have accumulated mutations to such a degree that they would code for different functional products if activated
(3) They are duplicates or near duplicates of functional genes but cannot function because they would provide inappropriate dosage of protein products
(4) They are genes that are not expressed, even though they have nearly identical sequences to expressed genes

153. What characteristic of short tandem repeat DNA makes it useful for DNA fingerprinting?

(1) The number of repeats varies widely from person to person or animal to animal
(2) The sequence of DNA that is repeated varies significantly from individual to individual
(3) The sequence variation is acted upon differently by natural selection in different environments
(4) Every racial and ethnic group has inherited different short tandem repeats

154. Alu elements account for about 10% of the human genome. What does this mean?

(1) Alu elements cannot be transcribed into RNA
(2) Alu elements evolved in very ancient times, before mammalian radiation
(3) Alu elements represent the result of transposition
(4) No Alu elements are found within individual genes

155. A multigene family is composed of
(1) multiple genes whose products must be coordinately expressed
(2) genes whose sequences are very similar and that probably arose by duplication
(3) the many tandem repeats such as those found in centromeres and telomeres
(4) a gene whose exons can be spliced in a number of different ways

156. Which of the following can be duplicated in a genome?
(1) DNA sequences above a minimum size only
(2) DNA sequences below a minimum size only
(3) Entire chromosomes only
(4) Sequences, chromosomes, or sets of chromosomes

157. In comparing the genomes of humans and those of other higher primates, it is seen that humans have a large metacentric pair we call chromosome 2 among our 46 chromosomes, whereas the other primates of this group have 48 chromosomes and any pair like the human chromosome 2 pair is not present; instead, the primate groups each have two pairs of midsize acrocentric chromosomes. What is the most likely explanation?
(1) The ancestral organism had 48 chromosomes and at some point a centric fusion event occurred and provided some selective advantage
(2) The ancestral organism had 46 chromosomes, but primates evolved when one of the pairs broke in half
(3) At some point in evolution, human ancestors and primate ancestors were able to mate and produce fertile offspring, making a new species
(4) Chromosome breakage results in additional centromeres being made in order for meiosis to proceed successfully

158. Unequal crossing over during prophase I can result in one sister chromosome with a deletion and another with a duplication. A mutated form of hemoglobin, known as hemoglobin Lepore, is known in the human population. Hemoglobin Lepore has a deleted set of amino acids. If it was caused by unequal crossing over, what would be an expected consequence?
(1) If it is still maintained in the human population, hemoglobin Lepore must be selected for in evolution
(2) There should also be persons born with, if not living long lives with, an anti-Lepore mutation or duplication
(3) Each of the genes in the hemoglobin gene family must show the same deletion
(4) The deleted gene must have undergone exon shuffling

159. When does exon shuffling occur?
(1) During splicing of DNA
(2) During mitotic recombination
(3) As an alternative splicing pattern in post-transcriptional processing
(4) As an alternative cleavage or modification post-translationally

160. What are genomic "hot spots"?
(1) The locations that correspond to most genetic diseases
(2) The areas of a genome that most often mutate due to environmental effects
(3) The locations that correspond to most genetic diseases, the areas of a genome that most often mutate due to environmental effects, and the locations that most often correspond with chromosomal breakpoints
(4) The locations that correspond to most genetic diseases and the locations that most often correspond with chromosomal breakpoints

161. In order to determine the probable function of a particular sequence of DNA in humans, what might be the most reasonable approach?

(1) Prepare a knockout mouse without a copy of this sequence and examine the mouse phenotype

(2) Genetically engineer a mouse with a copy of this sequence and examine its phenotype

(3) Look for a reasonably identical sequence in another species, prepare a knockout of this sequence in that species, and look for the consequences

(4) Prepare a genetically engineered bacterial culture with the sequence inserted & assess which new protein is synthesized

162. Homeotic genes contain a homeobox sequence that is highly conserved among very diverse species. The homeobox is the code for that domain of a protein that binds to DNA in a regulatory developmental process. Which of the following would you then expect?

(1) That homeotic genes are selectively expressed over developmental time

(2) That a homeobox-containing gene has to be a developmental regulator

(3) That homeoboxes cannot be expressed in nonhomeotic genes

(4) That all organisms must have homeotic genes

163. Two eukaryotic proteins have one domain in common but are otherwise very different. Which of the following processes is most likely to have contributed to this similarity?

(1) Gene duplication

(2) RNA splicing

(3) Exon shuffling

(4) Histone modification

164. When biologists wish to study the internal ultrastructure of cells, they can achieve the finest resolution by using

(1) a phase contrast light microscope

(2) a scanning electron microscope

(3) a transmission electronic microscope

(4) a confocal fluorescence microscope

165. The advantage of light microscopy over electron microscopy is that

(1) light microscopy provides for higher magnification than electron microscopy

(2) light microscopy provides for higher resolving power than electron microscopy

(3) light microscopy allows one to view dynamic processes in living cells

(4) light microscopy provides higher contrast than electron microscopy

166. A primary objective of cell fractionation is to

(1) view the structure of cell membranes

(2) sort cells based on their size and weight

(3) determine the size of various organelles

(4) separate the major organelles so that their particular functions can be determined

167. In the fractionation of homogenized cells using centrifugation, the primary factor that determines whether a specific cellular component ends up in the supernatant or the pellet is

(1) the relative solubility of the component

(2) the size and weight of the component

(3) the percentage of carbohydrates in the component

(4) the presence or absence of nucleic acids in the component

168. Which of the following correctly lists the order in which cellular components will be found in the pellet when homogenized cells are treated with increasingly rapid spins in a centrifuge?

(1) Ribosomes, nucleus, mitochondria

(2) Chloroplasts, ribosomes, vacuoles

(3) Nucleus, ribosomes, chloroplasts

(4) Nucleus, mitochondria, ribosomes

169. Green fluorescent protein (GFP) can be used to fluorescently label a specific protein in cells by genetically engineering cells to synthesize the target protein fused to GFP. What is the advantage of using GFP fusions to visualize specific proteins, instead of staining cells with fluorescently labeled probes that bind to the target protein?

(1) GFP fusions enable one to track changes in the location of the protein in living cells; staining usually requires preserved cells

(2) GFP fusions enable higher resolution than staining with fluorescent probes

(3) GFP permits the position of the protein in the cell more precisely than fluorescent probes

(4) GFP permits visualization of protein-protein interactions; fluorescent probes do not

170. What is the reason that a modern electron microscope (TEM) can resolve biological images to the subnanometer level, as opposed to tens of nanometers achievable for the best super-resolution light microscope?

(1) The focal length of the electron microscope is significantly longer

(2) Contrast is enhanced by staining with atoms of heavy metal

(3) Electron beams have much shorter wavelengths than visible light

(4) The electron microscope has a much greater ratio of image size to real size

171. What technique would be most appropriate to use to observe the movements of condensed chromosomes during cell division?

(1) Light microscopy

(2) Scanning electron microscopy

(3) Transmission electron microscopy

(4) Confocal fluorescence microscopy

172. Pyrosequencing derives its name from the fact that

(1) The bases are detected by pyrolysis

(2) It uses the enzyme apyrase to detect the bases

(3) It detects pyrophosphate released during nucleotide incorporation

(4) It generates pyrogram as output

173. Protein binding regions of DNA are generally identified by one of the following techniques?

(1) Finger printing

(2) Foot printing

(3) Western blotting

(4) MALDI-TOF

174. Which of the following repetitive motifs is responsible for the formation of triple-helix in collagen?

(1) Ala-X-Y (2) Gly-X-Y

(3) Cys-X-Y (4) Pro-X-Y

175. Phosphinothricin is an active ingredient of the herbicide 'BASTA'. It is an analogue of which of the following amino acids?

(1) L-Glutamic acid (2) L-Glutamine

(3) L-Aspartic acid (4) L-Asparagine

176. Over-expression of Adenosine-5- Phosphosulfate (APS) reductase and disruption of APS kinase in *Arabidopsis* is most likely to result in reduced formation of which of the following sulphur-containing compounds?

(1) Glucosinolates

(2) Sulfite

(3) Glutathione

(4) Thiosulfate

177. Which of the following statements about *vir* genes in Ti plasmid is incorrect?

(1) they are capable of functioning in *trans* orientation

(2) *vir* A and *vir* G are polycistronic with other *vir* genes

(3) *vir* A and *vir* G gene products regulate the expression of other *vir* genes

(4) *vir* A product probably acts as a chemoreceptor which senses the presence of acetosyringone

178. *Cot* values or DNA reassociation kinetics experiments tell us about

(1) DNA complexity
(2) Copy number of genes
(3) Genome size
(4) Ploidy level

179. In the maple syrup urine disease of human beings, the metabolic defect is related to

(1) Deficiency of the vitamin niacin
(2) Oxidative decarboxylation
(3) Transamination of an amino acid
(4) Synthesis of branched chain amino acids

180. What is the maximum Rf value for any molecule in paper chromatography?

(1) 0.1
(2) 1.0
(3) 10.0
(4) Depends on the composition of the sample loaded

181. The light emitting group in Green Fluorescent Protein (GFP) is a derivative of following three consecutive amino acids

(1) Serine, tyrosine and glycine
(2) Glycine, alanine and cysteine
(3) Phenylalanine, tryptophan and tyrosine
(4) Histidine, glutamic acid and phenylalanine

182. In addition to adjuvant, generation of anti-hapten antibodies will require

(1) addition of the hapten into the mice
(2) injection of mixture of hapten and proteins into mice
(3) injection of hapten covalently coupled with a protein into mice
(4) injection of a mixture of hapten and lipid into mice

183. A homogenate of liver cells is centrifuged at 100,000 g for 1 hours. Following this, the supernatant is separated from the pellet and the pellet is suspended in buffered medium. Assuming inclusion of substrates and cofactors, which of the following enzymatic activities can be measured in the supernatant?

(1) Succinate dehydrogenase
(2) Glyceraldehyde 3-phosphate dehydrogenase
(3) Glycogen synthase
(4) Aconitase

184. Human genome sequencing project involved the construction of genomic library in

(1) bacterial artificial chromosome
(2) pBR322
(3) bacteriophage
(4) pcDNA3.1

185. The nucleotide analogue used in DNA sequencing by chain termination method is

(1) 1',3'-dideoxy nucleoside triphosphate
(2) 2',3'-dideoxy nucleoside triphosphate
(2) 2',4'-dideoxy nucleoside triphosphate
(3) 2',5'-dideoxy nucleoside triphosphate

186. In nature, the horizontal gene transfer across bacteria is mediated by

(1) gene cloning followed by transformation
(2) conjugation and transformation
(3) conjugation only
(4) transformation only

187. A complete restriction digestion of a circular plasmid (5000bp) was carried out with *Hind*III, *Bam*HI and *Eco*RI individually. Restriction digestion yielded following fragments

Plasmid + *Hind*III→ 1200bp and 3800bp
Plasmid + *Bam*HI→5000bp
Plasmid + *Eco*RI→ 2500bp

The number of sites for *Eco*RI, *Bam*HI *and Hind*III present on this plasmid are

(1) *Eco*RI-2, *Bam*HI-1, *Hind*III-2
(2) *Eco*RI-1, *Bam*HI-1, *Hind*III-2
(3) *Eco*RI-3, *Bam*HI-1, *Hind*III-2
(4) *Eco*RI-2, *Bam*HI-2, *Hind*III-1

188. A humanized antibody is one in which the
(1) heavy and light chains are from human
(2) heavy chain is from human and light chain is from mouse
(3) light chain is from human and heavy chain is from mouse
(4) CDRs are from mouse, and the rest is from human

189. Dimethyl sulfoxide (DMSO) is used as a cryopreservant for mammalian cell cultures because
(1) it is an organic solvent
(2) it easily penetrates cells
(3) it protects cells by preventing crystallization of water
(4) it is also utilized as a nutrient

190. Nude mice refers to
(1) mice without skin
(2) mice without thymus
(3) knockout mice
(4) transgenic mice

191. Protein-DNA interactions *in vivo* can be studied by
(1) gel shift assay
(2) Southern hybridization
(3) chromatin immunoprecipitation assay
(4) fluorescence *in situ* hybridization assay

192. Synteny refers to
(1) gene duplication from a common ancestor
(2) a tree representation of related sequences
(3) the extent of similarity between two sequences
(4) local conservation of gene order

193. Which of the following statements are correct?
P. the complement fixation test is a relatively simple procedure used in serological microbiology
Q. in an agglutination test, a visible clump of particles appears to the observer
R. in the precipitation test, cells and large particles enter the reaction and clump together
S. the double-diffusion Oudin test utilizes the precipitation reaction as its basis
(1) P and Q
(2) Q and S
(3) P, Q, and R
(4) P, Q, R and S

194. The tertiary structure of protein is detected by
(1) X-ray diffraction/crystallographs
(2) spectrophotometry
(3) electrophoresis
(4) chromatography

195. UV-spectroscopy is not used to quantitate the compounds based on absorbance
(1) nucleic acids have absorbance peak at 260 nm
(2) NAD(P)H have absorbance peak at 340 nm
(3) aromatic amino acids have absorption maxima at about 280 nm
(4) sulfur containing amino acids have absorbance at 270 nm

196. Which of the following techniques yields the 3D information about soluble protein in its native state?
(1) automated Edman analysis
(2) circular dichroism spectroscopy
(3) magnetic resonance spectroscopy
(4) X-ray crystallography

197. Standard deviation of a data is 6. When each observation is increased by 1, then the standard deviation of new data is
(1) 5
(2) 7
(3) 6
(4) 8

198. A microorganism when viewed under a compound microscope with objective of 40X and eye piece of l0X magnification measured 400 mm in length. The same microorganism when viewed under a dissecting microscope with l0X lens, would measure

(1) 100 mm (2) 10 mm
(3) 40 mm (4) 400 mm

199. Dideoxynucleoside triphosphates (ddNTPs) are used in sequencing DNA because

(1) ddNTPs are fluorescent
(2) ddNTPs are incorporated very efficiently into DNA by DNA polymerase
(3) ddNTPs cannot be incorporated into DNA by DNA polymerase
(4) ddNTPs prevent further DNA synthesis once they are incorporated into the DNA sequence

200. Which of the following is earliest step of somatic embryo?

(1) Torpedo (2) Globulur
(3) Heart (4) Oval

ANSWER SHEET
Unit 13: Applied Biotechnology & Biotechniques

---

1. 2	**2.** 4	**3.** 3	**4.** 3	**5.** 1	**6.** 2	**7.** 2	**8.** 1	**9.** 3	**10.** 2
11. 3	**12.** 1	**13.** 1	**14.** 3	**15.** 3	**16.** 3	**17.** 4	**18.** 4	**19.** 4	**20.** 4
21. 1	**22.** 4	**23.** 2	**24.** 3	**25.** 2	**26.** 4	**27.** 2	**28.** 3	**29.** 2	**30.** 3
31. 2	**32.** 1	**33.** 2	**34.** 2	**35.** 1	**36.** 4	**37.** 4	**38.** 1	**39.** 4	**40.** 4
41. 1	**42.** 1	**43.** 1	**44.** 1	**45.** 3	**46.** 1	**47.** 4	**48.** 2	**49.** 2	**50.** 1
51. 1	**52.** 4	**53.** 2	**54.** 4	**55.** 4	**56.** 1	**57.** 3	**58.** 1	**59.** 3	**60.** 1
61. 3	**62.** 2	**63.** 4	**64.** 1	**65.** 2	**66.** 2	**67.** 3	**68.** 3	**69.** 2	**70.** 3
71. 2	**72.** 3	**73.** 4	**74.** 4	**75.** 2	**76.** 1	**77.** 4	**78.** 2	**79.** 1	**80.** 1
81. 3	**82.** 1	**83.** 3	**84.** 3	**85.** 3	**86.** 4	**87.** 1	**88.** 4	**89.** 4	**90.** 3
91. 3	**92.** 4	**93.** 4	**94.** 3	**95.** 4	**96.** 2	**97.** 3	**98.** 4	**99.** 1	**100.** 4
101. 4	**102.** 1	**103.** 1	**104.** 3	**105.** 3	**106.** 4	**107.** 2	**108.** 1	**109.** 3	**110.** 3
111. 2	**112.** 3	**113.** 2	**114.** 3	**115.** 2	**116.** 3	**117.** 4	**118.** 2	**119.** 1	**120.** 3
121. 3	**122.** 1	**123.** 2	**124.** 2	**125.** 1	**126.** 2	**127.** 2	**128.** 3	**129.** 2	**130.** 3
131. 3	**132.** 2	**133.** 3	**134.** 4	**135.** 3	**136.** 3	**137.** 4	**138.** 4	**139.** 4	**140.** 4
141. 3	**142.** 4	**143.** 3	**144.** 1	**145.** 3	**146.** 1	**147.** 4	**148.** 4	**149.** 3	**150.** 3
151. 2	**152.** 4	**153.** 1	**154.** 3	**155.** 2	**156.** 4	**157.** 1	**158.** 2	**159.** 3	**160.** 3
161. 3	**162.** 1	**163.** 3	**164.** 3	**165.** 3	**166.** 4	**167.** 2	**168.** 4	**169.** 1	**170.** 3
171. 1	**172.** 3	**173.** 3	**174.** 2	**175.** 1	**176.** 1	**177.** 4	**178.** 1	**179.** 4	**180.** 2
181. 1	**182.** 3	**183.** 2	**184.** 1	**185.** 2	**186.** 2	**187.** 1	**188.** 4	**189.** 3	**190.** 2
191. 3	**192.** 1	**193.** 4	**194.** 1	**195.** 4	**196.** 4	**197.** 3	**198.** 2	**199.** 4	**200.** 2

Part - B

UNIT-1

BIOCHEMISTRY

1. Given the following ΔH values, calculate ΔH for the reaction

$$6\ CO_2 + 6\ H_2O \rightarrow C_6H_{12}O_6 + 6\ O_2$$

	ΔH_f kJ‡/mol
$C_6H_{12}O_6$	– 1274.9
CO_2	– 393.3
H_2O	– 286.2
O_2	0

(1) 2802.1 kJ/mol (2)-2802.1 kJ/mol
(3) 594.5 kJ/mol (4) -594.5 kJ/mol

2. The enzyme phosphoglyceromutase catalyzes an intermediate step in glycolysis that converts two molecules of 3-phosphoglyceric acid to two molecules of 2-phosphoglyceric acid. The phosphorus atom simply shifts position on the molecule. The *Keq* for this reaction at physiological pH (7.4) and temperature (310 K) has been reported to be 5.0. What is the change in free energy?

(1) -4.1475 J/mol (2) -41.475 J/mol
(3) -414.75 J/mol (4) -4147.5 J/mol

3. The free-energy change of a reaction (Δ *G*) tells us if the reaction can occur spontaneously. Consider the following statement and find which of the following is true?

A. A reaction can occur spontaneously only if Δ G is negative. Such reactions are said to be exergonic

B. A system is at equilibrium and no net change can take place if Δ G is zero

C. A reaction cannot occur spontaneously if Δ G is positive. An input of free energy is required to drive such a reaction. These reactions are termed endergonic

D. The ΔG of a reaction is independent of the path (or molecular mechanism) of the transformation

E. The Δ G provides no information about the rate of a reaction

(1) A, B and C (2) A, C and D
(3) B, C, D, E (4) A, B, C, D and E

4. Which of the following does not apply to the standard change in reduction potential for an oxidation-reduction reaction $\Delta E^{0'}$

(1) is positive when $\Delta G^{0'}$ is negative

(2) is the sum of the standard reduction potentials of the individual oxidation or reduction steps in the reaction

(3) is negative when the reaction is spontaneous under standard condition

(4) is positive when electrons moves toward atoms or molecules with more positive reduction potential

5. During the fermentation of wine, a buffer system consisting of tartaric acid and potassium hydrogen tartrate is produced by a biochemical reaction. Assuming that at some time the concentration of potassium hydrogen tartrate is twice that of tartaric acid, calculate the pH of the wine. The pKa of tartaric acid is 2.96

(1) 2.96

(2) 2.66

(3) 3.26

(4) 4.96

6. What is the pH of a solution prepared by mixing 150 mL of 0.1 M HC1 with 300 mL of 0.1 M sodium acetate (NaOAc) and diluting the mixture of 1 L? (The pKa of acetic acid is 4.76)

(1) 4.76 (2) 5.06
(3) 4.66 (4) 5.76

7. Threonine constitutes 1.8 percent by weight of the amino acid content of insulin. Given that the molecular weight of threonine is 119 and there is single threonine molecule in insulin, what is the minimum molecular weight of insulin?

(1) 5600 dalton (2) 6611 dalton
(3) 1111 dalton (4) 4300 dalton

8. An amino-acid has one proton donating group in the side chain (R) The pK_{COOH}, pK_{NH2} and pK_R values for this amino-acid are 2.19, 9.67 and 4.25, respectively. Which one of the following statements about this amino-acid is correct?

(1) Majority of the molecules will have a net charge of -1 at pH of 7.0
(2) Majority of the molecules will have a net charge of 0 at pH of 4.25
(3) All the molecules will have a deprotonated R group at pH of 3.22
(4) During titration with a strong base, deprotonation will start with the R group

9. Water is an unusual compound whose physical properties set it apart from other liquids Consider the following statements

A. Water has a low dielectric constant. Water is therefore a non-polar solvent, with the consequence that electrically charged molecules or particles are easily separated in its presence

B. Water has a very high heat capacity; a large amount of heat is needed to raise its temperature each degree K. In biological systems, this is an advantage, acting as a thermal buffer that prevents the temperature of a cell undergoing moderate metabolic activity from rising significantly

C. Water has a low heat of vaporization unlike other liquids capable of hydrogen bonding, such as ethanol and acetic acid

D. Liquid water has a higher density than ice at standard temperature and pressure. There is a negative volume change upon freezing, which is why ice floats

E. Water has a high surface tension. This requires biological organisms to use detergent-like compounds to modify the surface tension. Lung surfactants are an example of these detergents

F. Water has a higher conductivity than might be expected on the basis of its ionization constant alone. The conductivity of ice is also high even though its ionization constant is 1000-fold lower. This has been attributed to the presence of an ordered structure which allows the electric charge to be carried by a modified "brigade" mechanism

Which of the following statement is not correct?

(1) D, E and F (2) A, B and C
(3) A, C and D (4) B, E and F

10. The pKa's of lysine are given below
pKa (COOH) = 2.2, pKa (α-NH_3^+J) = 9.0, pKa (ε-NH_3^+) = 10.0 The pI of lysine is

(1) 7.07 (2) 9.50
(3) 6.10 (4) 5.60

11. The molecular weight of a bacterial DNA molecule is 2.64 x 10^9. The average molecular weight of a nucleotide pair is 660. Assume that the average protein is made up of a chain of 400 amino acid residues. What is the maximum number of proteins that can be coded by the bacterial DNA molecule?

(1) 20000
(2) 6667
(3) 3333
(4) None of the above

12. Van der Waals forces are the result of induced electrical interactions between closely approaching atoms or molecules as their negatively-charged electron clouds fluctuate instantaneously in time. These fluctuations allow attractions to occur between the positively charged nuclei and the electrons of nearby atoms. The van der Waals interaction energy profile as a function of the distance, *r*, between the centers of two atoms is shown in diagram below. The energy can be calculated using the empirical equation $U = B/r^{12} - A/r^6$

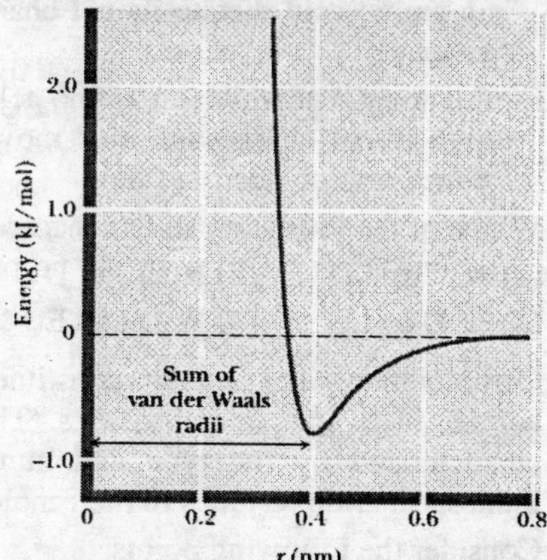

Consider the following statements

A. When two atoms approach each other so closely that their electron clouds interpenetrate, strong repulsion occurs. Such repulsive van der Waals forces follow an inverse 12th-power dependence on r ($1/r^{12}$)

B. Van der Waals interactions include dipole–dipole interactions, whose interaction energies decrease as $1/r^3$; dipole-induced dipole interactions, which fall off as $1/r^5$; and induced dipole-induced dipole interactions, often called London dispersion forces, which diminish as $1/r^6$

C. Between the repulsive and attractive domains lies a low point in the potential curve. This low point defines the distance known as the Van der Waals contact distance, which is the interatomic distance that results if only van der Waals forces hold two atoms together and at this distance the interaction is strongest

D. At best, Van der Waals interactions are weak and individually contribute 0.4 to 4.0 kcal/mol of stabilization energy

E. Van der Waals interactions are formed only in non-polar environments

Which of the following statement is true?

(1) A, B and C (2) B, C and D

(3) C, D and E (4) D, A and E

13. Indicate whether each of the following pairs of sugars consists of epimers?

A. D-glyceraldehyde and dihydroxyacetone

B. D-glucose and D-mannose

C. D-glucose and D-fructose

D. α-d-glucose and β-d-glucose

E. D-ribose and D-ribulose

F. D-galactose and D-glucose

(1) A, B, C (2) B, D, F

(3) E and F (4) B and F

14. The specific rotations of the α and β anomers of d-glucose are +112 degrees and +18.7 degrees, respectively. Specific rotation, $[\alpha]^d$, is defined as the observed rotation of light of wavelength 589 nm (the d line of a sodium lamp) passing through 10 cm of a 1 g ml^{-1} solution of a sample. When a crystalline sample of a-d glucopyranose is dissolved in water, the specific rotation decreases from 112 degrees to an equilibrium value of 52.7 degrees. On the basis of this result, what are the proportions of the α and β anomers at equilibrium? Assume that the concentration of the open-chain form is negligible

(1) α-glucose = 0.36, β-glucose = 0.64

(2) α-glucose = 0.64, β-glucose = 0.36

(3) α-glucose = 0.50, β-glucose = 0.50

(4) α-glucose = 0.18, β-glucose = 0.82

15. What is the charge on the oligonucleotide 5'pppGpGpApCpT 3' at neutral pH?

(1) -5 (2) -6
(3) -7 (4) -8

16. In an experiment involving planar bilayers, a solution of table salt (sodium and chloride ions in water) is added on the left side of the membrane while pure water is added on the right side. After 30 minutes the researchers test for the presence of ions on each side of the membrane by noting whether the solutions are capable of conducting electricity. The right side tests negative for ions. What can you conclude?

(1) The experiment failed
(2) The water somehow blocked the movement of ions across the membrane
(3) The left side would probably also test negative for ions
(4) Ions cannot cross planar bilayers

17. What is the pH when 100 mL of 0.1 *N* NaOH is added to 150 mL of 0.2 *M* CH_3COOH if p*Ka* for acetic acid = 4.76?

(1) 4.76 (2) 4.66
(3) 4.96 (4) 4.46

18. Consider the following statements regarding pH and Buffers

A. pH is the negative log of $[H^+]$. A low pH characterizes an acidic solution, and a high pH denotes a basic solution
B. The strength of weak acids is expressed by p*Ka*, the negative log of the acid dissociation constant
C. Strong acids have high p*Ka* values and weak acids have low p*Ka* values
D. Buffers resist a change in pH when protons are produced or consumed. Maximum buffering capacity occurs ± 1 pH unit on either side of p*Ka*. Physiologic buffers include bicarbonate, orthophosphate, and proteins

Which of the following statements are true?

(1) A and C (2) C and D
(3) A, B and D (4) A, B and C

19. Consider the following statements for amino acids

A. Both D-amino acids and non-α-amino acids occur in nature, but only D- α - amino acids are present in proteins
B. All amino acids possess at least two weakly acidic functional groups, $R\text{-}NH^{3+}$ and R-COOH
C. Many amino acids also possess additional weakly acidic functional groups such as -OH, -SH, guanidino, or imidazole groups
D The p*Ka* values of all functional groups of an amino acid dictate its net charge at a given pH
E. pI is the pH at which an amino acid bears highest net charge and thus moves in a direct current electrical field

Which of the following statements are true?

(1) A, C and D (2) B, C and D
(3) A, D and E (4) B, D and E

20. Carbohydrates are major constituents of animal food and animal tissues. They are characterized by the type and number of monosaccharide residues in their molecules

Consider the following points

A. Glucose is the most important carbohydrate in mammalian biochemistry because nearly all carbohydrate in food is converted to glucose for metabolism
B. Glucose have two stereoisomers because it contain a asymmetric carbon atoms
C. The disaccharides include maltose (glucosyl glucose), an intermediate in the digestion of starch; sucrose (glucosyl fructose), important as a dietary constituent containing fructose; and lactose (galactosyl glucose), in milk
D. Starch and glycogen are structural polymers of glucose in plants and animals respectively. Starch is the major fiber in the diet
E. Complex carbohydrates contain other sugar derivatives such as amino sugars, uronic acids, and sialic acids. They include proteoglycans and glycosamino

glycans, associated with structural elements of the tissues; and glycoproteins, proteins containing attached oligosaccharide chains. They are found in many situations including the cell membrane

Which of the following statement is not true?

(1) A, C and E (2) B, C and D
(3) C and E (4) B and D

21. Eicosanoids are formed from 20-carbon polyunsaturated fatty acids and make up an important group of physiologically and pharmacologically active compounds known as prostaglandins, thromboxanes, leukotrienes, and lipoxins. Which of the following statement is not true?

(1) Prostaglandins exist in virtually every mammalian tissue, acting as local hormones; they have important physiologic and pharmacologic activities. They are synthesized in vivo by cyclization of the center of the carbon chain of 20-carbon (eicosanoic) polyunsaturated fatty acids (eg, arachidonic acid) to form a cyclopentane ring

(2) thromboxanes, have the cyclopentane ring interrupted with an oxygen atom and play important role in blood clotting

(3) The leukotrienes are characterized by the presence of three or four conjugated double bonds, respectively. Leukotrienes cause bronchoconstriction as well as being potent proinflammatory agents and play a part in asthma

(4) Synthesis of all ecanosoids is inhibited by Aspirin (Methyl salisylate)

22. Among the following which is not an essential fatty acid?

(1) Cervonic acid (22:6; 4, 7, 10, 13, 16, 19)
(2) Timnodonic (20:5; 5, 8, 11, 14, 17)
(3) Arachidonic acid (20:4; 5, 8, 11, 14)
(4) Elaidic acid (18:1; 9)

23. Which of the following forms of cysteine predominates at pH 9?

(1) $H_3{}^+N-CH(COO^-)-CH_2SH$ (2) $H_3{}^+N-CH(COOH)-CH_2SH$

(3) $H_3{}^+N-CH(COO^-)-CH_2S^-$ (4) $H_2N-CH(COO^-)-CH_2S^-$

24. The form of aspartate depicted below predominates at which point on the titration curve?

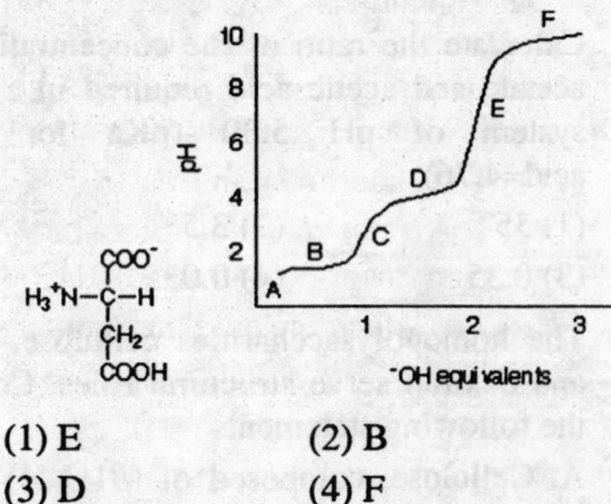

(1) E (2) B
(3) D (4) F

25. Of the following tripeptides, which is more likely to be soluble in a nonpolar solvent?

(1) Threonylleucylhistidine
(2) Glutamylalanylphenylalanine
(3) Valyltryptophanylglutamine
(4) Leucylalanyltrypthophan

26. Cellulose and starch are both polymers of glucose, yet most mammals can use starch, but not cellulose, as a source of fuel. Ruminents have successfully overcome this limitation and subsist on a diet of grasses. How have they accomplished this feat?

(1) They have cellulose as one of the digestive enzyme
(2) They have symbiotic microflora which digest cellulose & provide them glucose
(3) They have anerobic and alkaline environment in rumen which aids in digestion
(4) They have symbiotic bacteria in rumen which digest cellulose and provide them free fatty acids

27. Calculate the p*K*a of lactic acid, given that when the concentration of lactic acid is 0.01 M and the concentration of lactate is 0.087 M, the pH is 4.80

(1) 3.9 (2) 3.5
(3) 4.5 (4) 5.7

28. Calculate the pH of a mixture of 0.10 M acetic acid and 0.20 M sodium acetate. The p*K*a of acetic acid is 4.76

(1) 5.1 (2) 4.6
(3) 4.8 (4) 6.1

29. Calculate the ratio of the concentrations of acetate and acetic acid required in a buffer system of pH 5.30 (pKa for acetic acid=4.76)

(1) 35 (2) 3.5
(3) 0.35 (4) 0.035

30. The homopolysaccharides cellulose, chitin, and dextran serve structural roles. Consider the following statements

A. Cellulose, composed of (β1→ 4)-linked D-glucose residues, lends strength and rigidity to plant cell walls
B. Chitin, a polymer of (αl→ 4)-linked *N*-acetylglucosamine, strengthens the exoskeletons of arthropods
C. Dextran, a polymer of (α1→5 and α1→3) linked D-glucose forms an adhesive coat around certain bacteria
D. Bacterial and algal cell walls are strengthened by heteropolysaccharides—peptidoglycan in bacteria, agar in red algae. The repeating disaccharide in peptidoglycan is GlcNAc(αl→ 4)Mur2Ac; in agarose, it is D-Gal(α1→ 4)3,6-anhydro-L-Gal

Which of the following combination is correct?

(1) A and B (2) C and D
(3) A and C (4) B and D

31. A solution of one stereoisomer of a given monosaccharide rotates plane polarized light to the left (counter clockwise) and is called the levorotatory isomer, designated (-); the other stereoisomer rotates plane-polarized light to the same extent but to the right (clockwise) and is called the dextrorotatory isomer, designated (+) An equimolar mixture of the (+) and (-) forms

(1) does not rotate plane-polarized light
(2) generates circularaly polarized light
(3) does rotate plane polarized light
(4) optical rotation is ± 90°

32. The optical activity of a stereoisomer is expressed quantitatively by its optical rotation, the number of degrees by which plane-polarized light is rotated on passage through a given path length of a solution of the compound at a given concentration. The *specific rotation* $[\alpha]_b^{25°C}$ of an optically active compound is defined thus

$$[\alpha]_b^{25°C} = \frac{\text{observed optical rotation }(°)}{\text{optical path length (dm)} \times \text{concentration (g/mL)}}$$

The temperature and the wavelength of the light employed (usually the D line of sodium, 589 nm) must be specified in the definition. A freshly prepared solution of α-D-glucose shows a specific rotation of +112°. With time there is change in optical rotation. In contrast, a freshly prepared solution of β-D-glucose has a specific rotation of +19°. The rotation of this solution increases over time to the same equilibrium value as that shown by the α anomer. The equilibrium value is equal to

(1) 0°
(2) 52.5°
(3) -52.0°
(4) –40°

33. The enzyme invertase (now commonly called sucrase) is allowed to act on a 20% (0.2 g/mL) solution of sucrose until hydrolysis is complete. What will be the observed optical rotation of the solution in a

10 cm cell? (Ignore a possible small contribution from the enzyme)

(1) 2 (2) 4

(3) -2 (4) -4

34. Which of the following is not a role of carbohydrates in nature? Carbohydrates

(1) serve as energy stores in plants and animals

(2) are major structural components of mammalian tissues

(3) are constituents of nucleic acids

(4) are found in the structures of all the coenzymes

35. The pH of a buffer solution containing 0.1 mol L^{-1} acetic acid and 0.1 mol L^{-1} sodium acetate is 4.7. The pKa of acetic acid i s 4.7. What would be the pH value after adding 0.05 mol L^{-1} NaOH?

(1) 4.7 (2) 5.2

(3) 4.2 (4) 12.7

36. Which one of the following correctly pairs an amino acid with a valid chemical characteristic?

A. Glutamine: Contains a hydroxyl group in its side chain

B. Serine: Can form disulfide bonds

C. Cysteine : Contains the smallest side chain

D. Isoleucine: Is nearly always found buried in the center of proteins

E. Glycine: Contains an amide group in its side chain

(1) A and D (2) B and C

(3) C and D (4) Only D

37. Which of the following statements are true of water and/or hydrogen bonds?

(1) The oxygen atom in water has a partial positive charge

(2) The hydrogen atoms of water each bear a partial positive charge

(3) H-bonds can form only between water molecules

(4) H-bonds can form between different parts of polypeptide chains

(5) H-bonds are weaker in strength than van der Waals forces

(6) H-bonds are relatively weak compared to covalent bonds

(7) H-bonds in water contribute to its cohesiveness

(1) 3, 4, 5, 7 (2) 1, 3, 4, 6

(3) 2, 4, 6, 7 (4) 1, 5, 6, 7

38. The R-group of histidine has a pKa = 6.0. What percent of this group will exist as the conjugate base at pH = 5.5?

(1) 76% (2) 24%

(3) 97% (4) 67%

39. A mixture of Ala, Arg, and Asp in a pH 5.5 buffer was placed on a cation exchange column (the column is negatively charged) and eluted with the same buffer. What is the order of elution from the column?

Use these pKa values

terminal COOH – 2'

terminal NH_{3+} - 9

R - GROUPS

Amino - 10

COOH -3

(1) Arg, ala then asp

(2) Arg, asp then ala

(3) Asp, ala then arg

(4) Asp, arg then asp

40. Consider the following statement for amino acids

A. Amino acids can act as proton donors and acceptors

B. A conjugate acid/base pair is at its greatest buffering capacity when the pH equals its pI

C. The stronger the acid, the higher the Ka

D. The stronger the acid, the lower the pH

E. Amino acids are amphoteric

Which of the following are true?

(1) A, C, D (2) A, B, C, D

(3) B, C, D, E (4) A, C, D, E

41. Three buffers are made by combining a 1 M solution of acetic acid with a 1 M solution of sodium acetate in the ratios shown below.

	1 M acetic acid	1M sodium acetate
Buffer 1	10 mL	90 mL
Buffer 2	50 mL	50 mL
Buffer 3	90 mL	10 mL

Which of these statements is true of the resulting buffers?

(1) pH of buffer 1 < pH of buffer 2 < pH of buffer 3

(2) pH of buffer 1 = pH of buffer 2 = pH of buffer 3

(3) pH of buffer 1 > pH of buffer 2 > pH of buffer 3

(4) The problem cannot be solved without knowing the value of p*Ka*

42. A 1.0 M solution of a compound with 2 ionizable groups (pKa = 6.2 and 9.5; 100 mL total) has a pH of 6.8. If a biochemist adds 60 mL of 1.0 M HCl to this solution, the solution will change to pH

(1) 5.60

(2) 8.90

(3) 9.13

(4) The pH cannot be determined from this information

43. The standard free energy change ΔG^0 for reaction catalyzed by phosphoglucomutase Glucose-6-P ⇄ Glucose-1-P is +1.3 kcal mol^{-1} at 25°C. If initial concentration of glucose-1-P was used to carry out reaction, then the concentration of glucose-6-P and glucose-1-P at equilibrium would be

(1) 96 mM and 45 mM respectively

(2) 45 mM and 96 mM respectively

(3) 100 mM and 10 mM respectively

(4) 10 mM and 100 mM respectively

44. When a mixture of 3-phosphoglycerate and 2-phosphoglycerate is incubated at 25 °C with phosphoglycerate mutase until equilibrium is reached, the final mixture contains six times as much 2-phosphoglycerate as 3-phosphoglycerate. Which one of the following statements is most nearly correct, when applied to the reaction as written? (R = 8.315 J/mol·K; T = 298 K)

3-Phosphoglycerate→ 2-phosphoglycerate

(1) $\Delta G'^{\circ}$ is –4.44 kJ/mol

(2) $\Delta G'^{\circ}$ is zero

(3) $\Delta G'^{\circ}$is +12.7 kJ/mol

(4) $\Delta G'^{\circ}$is incalculably large and positive

45. For the following reaction, $\Delta G'^{\circ}$ = +29.7 kJ/mol

L-Malate + NAD^+ → oxaloacetate + NADH + H^+

The reaction as written

(1) can never occur in a cell

(2) can occur in a cell only if it is coupled to another reaction for which $\Delta G'^{\circ}$ is positive

(3) can occur only in a cell in which NADH is converted to NAD+ by electron transport

(4) may occur in cells at some concentrations of substrate and product

46. Equilibrium constant (K) of noncovalent interaction between two non-bonded atoms of two different groups was measured at 27°C. It was observed that K = 100 M^{-1}. The strength of this noncovalent interaction in terms of Gibbs free energy change is

(1) 2746 cal/mole

(2) -2746 cal/mole

(3) 247 cal/mole

(4) -247 cal/mole

47. If van der Waals interaction is described by the following relation

$$\Delta G_{van} = \frac{A}{r^{12}} - \frac{B}{r^6} + \frac{q_1 q_2}{r}$$

Where ΔG_{Van} is the free energy of the van der Waals interaction, A and B are constants, r is the distance between two nonbonded atoms 1 and 2 and q_1 and q_2 are partial charges on the dipoles 1 and 2. In this relation, the parameter A describes

(1) electron shell attraction
(2) electron shell repulsion
(3) dipole-dipole attraction
(4) dipole-dipole repulsion

48. The pH of blood of a healthy person is maintained at 7.40 ± 0.05, Assuming that this pH is maintained entirely by the bicarbonate buffer (pKa_1, and pKa_2 of carbonic acid are 6.1 and 10.3, respectively), the molar ratio of [bicarbonate]/ [carbonic acid] in the blood is

(1) 0.05 (2) 1
(3) 10 (4) 20

49. In a venous blood sample, the ratio of $[HPO_4^{-2}]/[H_2PO_4^-]=2$

$$CO_2(d) + H_2O \rightleftharpoons H_2CO_3 \underset{}{\overset{pK_a\ 6.1}{\rightleftharpoons}} HCO_3^- + H^+ \overset{pK_a\ 10.1}{\rightleftharpoons} CO_3^{-2} + H^+$$

$$H_3PO_4 \overset{pK_a\ 2.1}{\rightleftharpoons} H_2PO_4^- + H^+ \overset{pK_a\ 6.8}{\rightleftharpoons} HPO_4^{-2} + H^+ \overset{pK_a\ 12.3}{\rightleftharpoons} PO_4^{-3} + H^+$$

In this same blood sample, what would be the ratio of $[HCO_3^-]/[CO_2(d)]$?

(1) 20:1 (2) 15:1
(3) 10:1 (4) 5:1

50. A sample of urine from a patient with ketosis contains 28 mmoles of total acetoacetic acid-acetoacetate (i.e. $HA + A^- =$ 28 mmoles) The pKa for acetoacetic acid is 4.5; the pH of the urine is also 4.5. Assuming that Na^+ ions are excreted as a counter ion with equivalent amounts of the conjugate base (acetoacetate), what is the amount of Na^+ excreted?

(1) 7 mmoles (2) 14 mmoles
(3) 28 mmoles (4) 35 mmoles

51. The enzymes that catalyze reactions 1, 2, and 3, respectively are called

$$A + B \leftrightarrow A-B \quad (1)$$

$$\begin{matrix} X & Y \\ | & | \\ A & - & B \end{matrix} \leftrightarrow \begin{matrix} Y & X \\ | & | \\ A & - & B \end{matrix} \quad (2)$$

$$A-B+H_2O \leftrightarrow A-B+B-OH \quad (3)$$

(1) Ligase, transferase, oxidoreductase
(2) Ligase, isomerase, hydrolase
(3) Ligase, lyase, hydrolase
(4) Transferase, lyase, hydrolase

52. Which of the following are characteristics of the Watson-Crick DNA double helix?

A. The two polynucleotide chains are coiled about one another and about a common axis
B. Hydrogen bonds between A and C and between G and T help hold the two chains together
C. The helix makes one complete turn every 34Å because each base pair is rotated by 36° with respect to adjacent base pairs and is separated by 3.4Å from them along the helix axis
D. The purines and pyrimidines are on the inside of the helix and the phosphodiester linked backbones are on the outside
E. Base composition analyses of DNA duplexes isolated from many organisms show that the amounts of A and T are equal as are the amounts of G and C
F. The sequence in one strand of the helix varies independently of that in the other strand

(1) A, B, C and D
(2) A, C, D and E
(3) A, C, D, E and F
(4) A, B, C, D, E and F

53. A protein is a linear polymer of amino acids linked together by peptide bonds. Various, mostly non-covalent, interactions between amino acids in the linear sequence stabilize a specific folded three-dimensional structure (conformation) for each protein

A. The α helix, β strand and sheet, and turn are the most prevalent elements of protein secondary structure, which is stabilized by hydrogen bonds between atoms of the peptide backbone

B. Protein tertiary structure results from hydrophilic ionic interactions between polar side groups and hydrogen bonds between polar side groups that stabilize folding of the secondary structure into a compact overall arrangement, or conformation

C. Large proteins often contain distinct domains, independently folded regions of tertiary structure with characteristic structural or functional properties or both. The incorporation of domains as modules in different proteins in the course of evolution has generated diversity in protein structure and function

D. Quaternary structure encompasses the number and organization of subunits in multimeric proteins

Which of the following statement is correct in relation to protein folding and its higher level structures?

(1) A and B (2) B, C and D
(3) A, C and D (4) All of the above

54. A mixture of urease (pI = 5.1, mol. wt. 482,700), catalase (pI = 5.6, mol. wt. 247,500), lactoglobin (pI 5.2, mol. wt. 37,100) and hemoglobin (pI 6.9, mol. wt. 64,500) were applied to a size-exclusion (molecular sieve) chromatography column. What was their order of elution?

(1) Urease, lactoglobin, catalase, hemoglobin
(2) Hemoglobin, catalase, lactoglobin, urease
(3) Urease, catalase, hemoglobin, lactoglobin
(4) Lactoglobin, hemoglobin, catalase, urease

55. White and Anfinsen in their studies on the renaturation of ribonuclease provided evidence that the conformation of a protein resulted from its amino acid sequence. In their experiments what sequence of events led to renaturation of a functional ribonuclease protein and provided the basis for their conclusion?

(1) 1) Denaturation and reduction, 2) dialysis, 3) oxidation
(2) 1) Denaturation and reduction, 2) oxidation, 3) dialysis
(3) 1) Denaturation, 2) oxidation, 3) reduction, 4) dialysis
(4) 1) Oxidation, 2) dialysis, 3) denaturation, 4) reduction

56. What is the net charge pentapeptide, Cys-Cys-Phe-Arg-Ser, at pH 8?

Use these pKa values: terminal COOH – 4 ; terminal NH_3^+ - 9 ; R-GROUPS: Cys – 8 Arg – 12

(1) +1 (2) +2
(3) -1 (4) 0

57. In an experiment 5'-terminal phosphates of short segments of DNA were allowed to bind to a mica surface. After the mica-bound DNA was exposed to DNAse the resulting fragments were separated by electrophoresis in a polyacrylamide. When the experiment is carried out, we see clusters of DNA fragments of average sizes 10 and 11, 21, 31 and 32 base pairs and so forth. This experiments proves that the value of base pairs per turn in B-DNA is approximately most closest to

(1) 10 (2) 10.4
(3) 11 (4) 10.8

58. Early X-ray diffraction studies of DNA, which were carried out using concentrated solutions of DNA that had been drawn out into thin fibers, revealed three kinds of structures, the B, A and the Z forms of DNA

Consider the following statements

A. The B form, which is observed at high humidity, most closely corresponds to the average structure of DNA under physiological conditions. It has 10 base pairs per turn, and a wide major groove and a narrow minor groove

B. The A form, which is observed under experimental conditions of low humidity, has 11 base pairs per turn but absent in biological system

C. B-DNA does adopt the A structure in certain DNA-protein complexes. Also, the A form is similar to the structure that RNA adopts when double helical

D. DNA containing alternative purine and pyrimidine (CGCGCG...) residues can fold into left-handed as well as right-handed helices. The change to the syn position in the purine residues to alternating anti-syn conformations gives the backbone of left-handed DNA a zigzag look (hence its designation of Z DNA), which distinguishes it from right-handed forms

Which of the following statement is correct?

(1) A, B and C (2) B, C and D
(3) A, C and D (4) A, B and D

59. Hoogsteen AT and GC base pairs (named after their discoverer Karst Hoogsteen) have altered patterns of hydrogen bonding compared with Watson–Crick base pairs

Which of the following statement regarding Hoogsteen base pairing is incorrect

(1) In the Hoogsteen AT pair, the adenine base is rotated through 180° about the bond to the sugar

(2) The Hoogsteen GC pair only forms two hydrogen bonds, compared with three in the Watson–Crick GC pair

(3) Hoogsteen GC base pairs are stable at the neutral pH of cells (pH 7–8) One of the nitrogens on the cytosine must loose a hydrogen added to it for this type of base pair to form, and this de-protonation requires a lower pH (pH 4–5)

(4) Hoogsteen base pairs have gained importance recently because they are occasionally found in complexes of DNA with anticancer drugs and they show up in triple helices associated with genetic disease

60. Oligonucleotides of any desired sequence can be synthesized in sufficient quantity and pure enough to be studied by single-crystal X-ray. Structures have been determined for a number of synthetic oligonucleotides under various conditions. Studies have revealed the basic structure of three fundamental types of double helix and their features are shown in table below

	X	Y	Z
Orientation	**Right-handed**	**Right-handed**	**Left-handed**
Major groove	Deep and narrow	Moderate depth, wide	Very shallow, virtually nonexistent, sometimes called a "single groove"
Minor groove	Shallow and broad (superficial)	Moderate depth, narrow	Very deep and narrow
Bases/turn	11	10.5	12
Conditions	Low humidity (75%), high salt	High humidity (95%), low salt	High MgCl2 (>3M), NaCl, or ethanol In the presence of methylated cytosine: high humidity and low salt

Which of the following is correct combination?

(1) X→A-DNA; Y→B-DNA; Z→Z-DNA
(2) X→B-DNA; Y→Z-DNA; Z→Z-DNA
(3) X→Z-DNA; Y→B-DNA; Z→A-DNA
(4) X→A-DNA; Y→Z-DNA; Z→B-DNA

61. A red blood cell (RBC) requires a constant supply of NADPH and energy from the glucose entering the cell. Which of the following lists three enzymes necessary to meet the RBC's needs?

(1) Hexokinase, transketolase, fructose 1,6-bisphosphatase

(2) Phosphoglycerate mutase, transaldolase, phosphoenolpyruvate carboxykinase

(3) Phosphoglucomutase, glucose 6-phosphate dehydrogenase, glyceraldehyde 3-phosphate dehydrogenase
(4) Glucose 6-phosphate dehydrogenase, 6-phosphogluconate dehydrogenase, phosphoglycerate kinase

62. The relationship between the ratio of acid to base in a solution and its pH is described by the Henderson-Hasselbalch equation

pH = p*K* + log [base]/[acid]

The p*K* of acetic acid is 4.8. What is the approximate pH of an acetate solution containing 0.2 NI acetic acid and 2 NI acetate ion?

(1) 0.48 (2) 4.8
(3) 5.8 (4) 6.8

Question 63 and 64: A student was given a task of identifying the contents of five bottles of amino acids from which the labels had fallen off. Each of the original bottles contained one of the following: arginine, histidine, cysteine, proline, and tryptophan

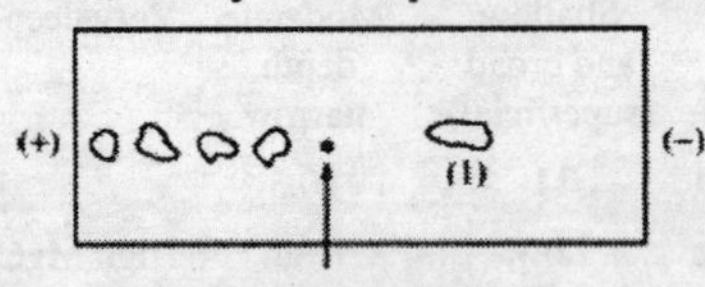

Point of Sample Application

A student was given a task of identifying the contents of five bottles of amino acids from which the labels had fallen off. Each of the original bottles contained one of the following: arginine, histidine, cysteine, proline, and tryptophan

63. Which of the following methods could be most readily employed to identify tryptophan?

(1) Electrophoresis (2) UV spectroscopy
(3) Gel filtration (4) Optical rotation

64. Each amino acid was subjected to paper electrophoresis in a pH 9.5 buffer, and then the amino acids were visualized by spraying the paper with ninhydrin. Which amino acid is at the point labeled (1) in the figure above?

(1) Arginine (2) Histidine
(3) Cysteine (4) Proline

65. Equation below shows that the free energy change for a reaction can be very different from the standard-state value if the concentrations of reactants and products differ significantly from unit activity (1 *M* for solutions) The effects can often be dramatic. Consider the hydrolysis of phosphocreatine

Phosphocreatine + H_2O → creatine + Pi

This reaction is strongly exergonic and $\Delta G°$ at 37°C is -42.8 kJ/mol. Physiological concentrations of phosphocreatine, creatine, and inorganic phosphate are normally between 1 m NI and 10 m NI. Assuming 1 m NI concentrations of reactants and products and using Equation above, the ΔG for the hydrolysis of phosphocreatine is

(1) -60.5 KJ/mol (2) -17.7 KJ/mol
(3) -25.1 KJ/mol (4) +25.1 KJ/mol

66. Consider a system consisting of an egg in an incubator. The white and yolk of the egg contain proteins, carbohydrates, and lipids. If fertilized, the egg is transformed from a single cell to a complex organism. Consider the developing chick as the system. The nutrients, egg shell, and outside world are the surroundings. In this irreversible process which of the following statement is not true in terms of the entropy changes in the system, surroundings, and universe

(1) Transformation of the single cell into a chick drastically reduces the entropy of the system (increases the order)
(2) Initially, the parts of the egg outside the embryo (within the surroundings) contain complex fuel molecules (a low-entropy condition)
(3) During incubation, some of these complex molecules are converted to large numbers of CO_2 and H_2O molecules (high entropy)
(4) The decrease in entropy of the surroundings is larger than the increase in entropy of the chick (the system) Thus, the entropy of the universe (the system + surroundings) decreases

67. 100 ml of 0.02 M acetic acid (pK_a= 4.76) is titrated with 0.02 N KOH. After adding some base to the acid solution, the observed pH is 2.76. At this pH degree of protonation is

(1) 0% (2) 10%
(3) 90% (4) 99%

68. Glycogen is synthesized from glucose-1-phosphate. To be incorporated into glycogen, glucose-1-phosphate is converted to a derivative of the nucleotide uridine disphosphate (UDP) The UDP serves as an excellent leaving group in the condensation reaction to form the glycogen polymer. The reaction is

Glucose-1-phosphate + UTP —> UDP-glucose + PPi

The the ΔG_O value for this reaction is approximately zero. If PPi (pyrophosphate) is hydrolyzed,

PPi + $H_2O \rightarrow$ 2 Pi

The loss in free energy (ΔG_o) is —33.5 kJ. Determine the $\Delta G°_{overall}$ value

(1) 1 (2) 0
(3) -33.5 (4) +33.5 kJ

69. The isoelectric pH of γ-globulin, human serum albumin (HSA), ribonuclease (RNAase) and hen egg lysozyme (HEL) are 6.6, 4.9, 7.8 and 11.0 respectively. The order in which these proteins will be eluted from a CM cellulose ion exchange column by an increasing salt gradient at pH 7.0 would be

(1) HSA, γ- globulin, RNAase and HEL
(2) HEL, RNAase, γ-globulin, HSA
(3) HEL, γ-globulin, RNAase, HSA
(4) HSA, RNAase, HEL, γ- globulin

70. Many integral membrane proteins are composed of a number of membrane-spanning segments, which form bundles of a helices packed closely together, often forming a membrane channel or pore. Each membrane-spanning sequence of most integral membrane proteins is an α helix composed of 18 to 20 amino acids. What is the width of the hydrocarbon core of the membrane?

(1) 10 A^0 (2) 20 A^0
(3) 30 A^0 (4) 40 A^0

71. The protein collagen, from skin and tendons, as unusual structure. Consider the following statements

A. It is composed of approximately 30 percent proline and hydroxyproline and 30 percent glycine
B. This protein has three chains, each with a conformation very similar to that of polyproline (left handed), are twisted about each other to make a triple helix
C. The three strands are hydrogen bonded to each other, through hydrogen bonds between the —NH of glycine residues and the —C=O groups of the other amino acids
D. All the three strands are coded by same gene
E. Non-covalent interactions mainly stabilize collagen structure

Which of the following statements are correct?

(1) A, B and C (2) A, C and D
(3) B, C and E (4) B, D and E

72. Hemoglobin A (the major, normal form in humans) has an isoelectric point of pH 6.9. The variant hemoglobin M has a glutamate residue in place of the normal valine at position 67 of the a chain. What effect will this substitution have on the electrophoretic behavior of the protein at pH 7.5?

(1) Mutant Hb will move toward cathode slower than normal Hb
(2) Mutant Hb will move toward cathode faster than normal Hb
(3) Mutant Hb will move toward anode slower than normal Hb
(4) Mutant Hb will move toward anode faster than normal Hb

73. Match the levels of protein structures in the left column with the appropriate descriptions in the right column

A. primary (1) association of protein subunits

B. secondary (2) overall folding of a single chain, can include α-helical and β sheet structures

C. tertiary (3) linear amino acid sequence

D. quaternary (4) repetitive arrangement of amino acids that are near each other in the linear sequence

(1) A-1 B-2 C-3 D-4
(2) A-3 B-4 C-2 D-1
(3) A-3 B-4 C-2 D-1
(4) A-4 B-3 C-2 D-1

74. Enzyme action can be influenced by the presence of inhibitors. Which of the following statements correctly matches the type of inhibitor with its effect on an enzyme

(1) Irreversible and Renders the enzyme permanently inactive
(2) Competitive and Inhibitor binds only to ES complex, only important when [S] high, V_{max} lower, K_m lower
(3) Noncompetitive and Can be overcome with high [S]; V_{max} unchanged, K_m higher
(4) Uncompetitive and Cannot be overcome with high [S]; V_{max} lower, but K_m unchanged

75. Ahlfors and Mansour studied the activity of purified sheep phosphofructokinase (PFK) as a function of the concentration of ATP in experiments that were carried out at a constant concentration of fructose 6-phosphate. Typical results are shown in Figure

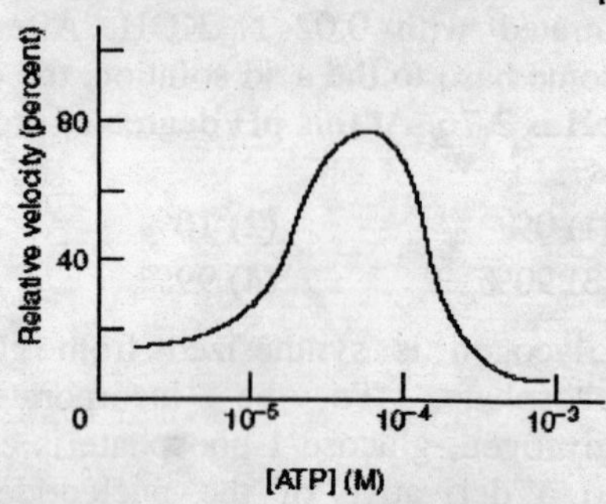

On basis of experiment following conclusion can be drawn to the role of PFK in the glycolytic pathway

A. The rate of the reaction catalyzed by PFK initially decreases with the ATP concentration because ATP is a inhibitor for the reaction; it binds at the active site of PFK with fructose 6-phosphate

B. At higher concentrations, ATP binds only at the active site; this alters the conformation of the enzyme and increases the level of its activity

C. When concentrations of ATP are relatively low, the activity of PFK is stimulated so that additional fructose 1,6-bisphosphate is made available for subsequent energy-generating reactions; when concentrations of ATP are higher and the demand of the cell for energy is lower, ATP inhibits PFK activity, thereby allowing glucose and other substrates to be utilized in other pathways

D. In many cells, ATP concentration is maintained at relatively high and constant levels, so that PFK is always subject to inhibition by ATP. Inhibition can be relieved by fructose 2,6-bisphosphate, which is synthesized when glucose is readily available

Which of the following conclusions are correct?

(1) A and C (2) B and D
(3) A and B (4) C and D

76. Cofactors can be subdivided into two groups: metals and small organic molecules. Cofactors that are small organic molecules are called coenzymes

Which of the following statement is true for coenzymes?

A. Co enzymes are often derived from water soluble vitamins, and if they are tightly bound to enzymes, they are called prosthetic groups

B. Loosely associated coenzymes are more like co-substrates because they bind to and are released from the enzyme just as substrates and products

C. Co-enzymes are always organic substance except proteins

D. Co-enzyme bind some other site else than active sites of enzyme

(1) A and B (2) B and C

(3) A, B and D (4) B, C and D

77. In the conversion of A into D in the following biochemical pathway, enzymes EA, EB, and EC have the *K* M values indicated under each enzyme. If all of the substrates and products are present at a concentration of 10^{-4} M, which step will be rate limiting

$$A \underset{}{\overset{E_A}{\rightleftharpoons}} B \underset{}{\overset{E_B}{\rightleftharpoons}} C \underset{}{\overset{E_C}{\rightleftharpoons}} D$$

$$K_M = \quad 10^{-2}M \quad 10^{-4}M \quad 10^{-4}M$$

(1) First step catalyzed by E_A

(2) Second step catalyzed by E_B

(3) Third step catalyzed by E_C

(4) None of them would be rate limiting

78. A reversible inhibitor of an enzyme was studied by kinetic analysis of the enzyme. When the inhibitor was present in the kinetic experiment at a concentration that achieved 50% inhibition of the activity measure in the normal activity assay (saturating substrate concentrations), the following was observed (the results of a control kinetic experiment done without inhibitor are also shown for comparison)

	No inhibitor present	With inhibitor present
K_M for substrate A	0.52 ± 0.11	0.55 ± 0.20
K_M for substrate B	2.2 ± 0.4	4.5 ± 0.5
V_{Max}	110 ± 5	112 ± 10

The inhibitor most likely is

(1) competitive with substrate A

(2) competitive with substrate B

(3) competitive with both substrates

(4) non-competitive

79. Regulated enzymes frequently exhibit allosteric activation and thus a higher order response to substrate concentration than non-regulated enzymes, which are commonly first-order. If enzyme "G" is first-order and enzyme "N" is allosterically activated exhibiting fourth-order kinetics

(1) enzyme N will be more sensitive to changes in substrate concentration than enzyme G

(2) enzyme G is said to be cooperative

(3) a graph of velocity versus substrate concentration for enzyme G will have the shape of a parabola

(4) enzyme G will require energy from ATP to produce a product, but not enzyme N

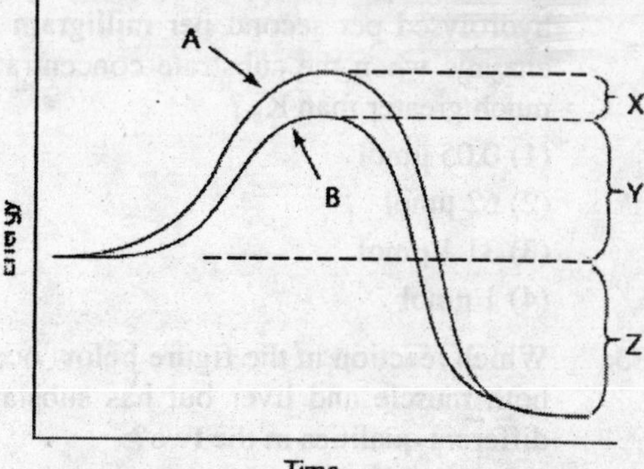

80. For the graph given above, the two curves describe the potential energy of substances during the progress of a chemical reaction. All of the following items could apply to this graph except

(1) Curve B could be showing the influence of an enzyme

(2) The sum of energy in the products of the reaction is less than the sum of energy in the reactants

(3) The activation energy of this reaction could be given by X + Y + Z

(4) This reaction graph could describe the reaction ATP → ADP + P*i*

81. Which of the following amino acids are least to be expected in α-helix of a protein?

(1) Glu, Ala, Leu, His, Met

(2) Gln, Trp, Val, Phe, Lys

(3) Pro, Gly, Tyr, Asn, Ile

(4) Asp, Thr, Ser, Arg, Cys

82. The hydrolysis of pyrophoshate to orthophosphate is important for several biosynthetic reactions. In *E. coli*, the molecular mass of the enzyme pyrophosphatase is 120 kD and it consists of six identical subunits. The enzyme activity is defined as the amount of enzyme that hydrolyzes 10 µmol of pyrophosphate in 15 minutes at 37^0 under standard assay condition. The purified enzyme has a V_{max} of 2800 units per milligram of the enzyme. How many moles of the substrate are hydrolysed per second per milligram of the enzyme when the substrate concentration is much greater than K_m?

(1) 0.05 µmol

(2) 62 µmol

(3) 31.1 µmol

(4) 1 µmol

83. Which reaction in the figure below occurs in both muscle and liver but has substantially different qualities in the two?

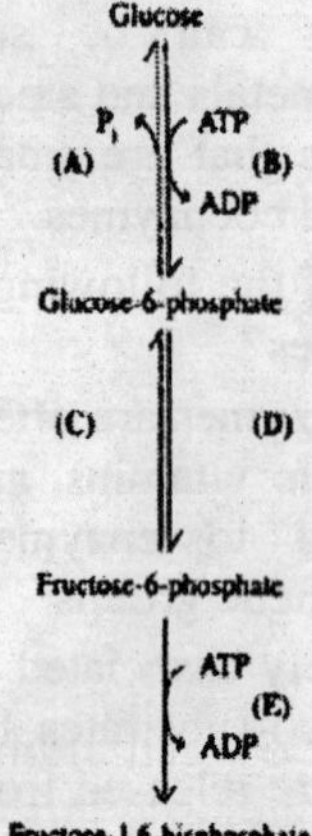

(1) Reaction A (2) Reaction B

(3) Reaction C (4) Reaction D and E

84. When blood glucose levels are low, glucagon is secreted. Which of the following are the effects of increased glucagon levels on glycolysis and related reactions in liver?

A. Phosphorylation of phosphofructokinase 2 and fructose bisphosphatase 2 occurs

B. Dephosphorylation of phosphofructokinase 2 and fructose bisphosphatase 2 occurs

C. Phosphofructokinase is activated

D. Phosphofructokinase is inhibited

E. Glycolysis is accelerated

F. Glycolysis is slowed down

(1) A, D and F (2) B, D and F

(3) A, C and E (4) B, C and E

85. Match hexokinase and glucokinase with the descriptions from the right column that are appropriate.

(a) hexokinase	(1) is found in the liver
(b) glucokinase	(2) is found in non-hepatic tissues
	(3) is specific for glucose
	(4) has a broad specificity for hexoses
	(5) requires ATP for reaction

(6) has a high Km for glucose

(7) is inhibited by glucose 6-phosphate

(1) (a) 1, 2, 4, 5, 7 (b) 1, 3, 5, 6
(2) (a) 1, 3, 4, 5, 7 (b) 1, 2, 5, 6
(3) (a) 1, 2, 5, 5, 7 (b) 1, 3, 4, 6
(4) (a) 1, 2, 4, 5, 6 (b) 1, 3, 5, 7

86. Oxidative phosphorylation forms ATP from ADP and Pi, using the energy released when molecular oxygen ultimately combines with hydrogen atoms to form water
Consider the following statements

A. The most of enzymes for oxidative phosphorylation are located in matrix of mitochondria

B. Hydrogen atoms derived from glycolysis, the Krebs cycle, and the breakdown of fatty acids are delivered, most bound to coenzymes, to the electron transport chain, which regenerates the hydrogen-free forms of the coenzymes NAD^+ and FAD by transferring the hydrogens to molecular oxygen to form water

C. The reactions of the electron transport chain produce a hydrogen-ion gradient across the inner mitochondrial membrane. The flow of hydrogen ions back across the membrane provides the energy for ATP synthesis

D. High amounts of reactive oxygen species, which can damage proteins, lipids, and nucleic acids, are formed during electron transport

Which of the following statement is correct?
(1) A and B
(2) B and C
(3) C and D
(4) A and C

87. The following is the sequence of reactions of gluconeogenesis from pyruvate to phosphoenolpyruvate

Pyruvate $\xrightarrow{A}$ oxaloacetate $\xrightarrow{B}$ malate $\xrightarrow{C}$

oxaloacetate $\xrightarrow{D}$ phosphoenolpyruvate

Match the capital letters indicating the reactions of the gluconeogenic pathway with the following statements

A. occurs in the mitochondria	E. requires ATP
B. occurs in the cytosol	F. requires GTP
C. produces CO_2	G. is regulated by acetyl CoA
D. consumes CO_2	H. requires a biotin cofactor

(1) (A) A, B (B) C, D (C) D (D) A (E) A (F) D (G) A (H) A
(2) (A) C, D (B) A, B (C) D (D) A (E) A (F) D (G) A (H) A
(3) (A) C, D (B) A, B (C) C (D) D (E) A (F) D (G) A (H) A
(4) (A) A, B (B) C, D (C) D (D) A (E) A (F) D (G) D (H) C

88. The following plot represents the Lineweaver-Burk equation of an enzymatic reaction both in the presence and the absence of inhibitor. Here, V is the velocity of reaction and S is the substrate concentration

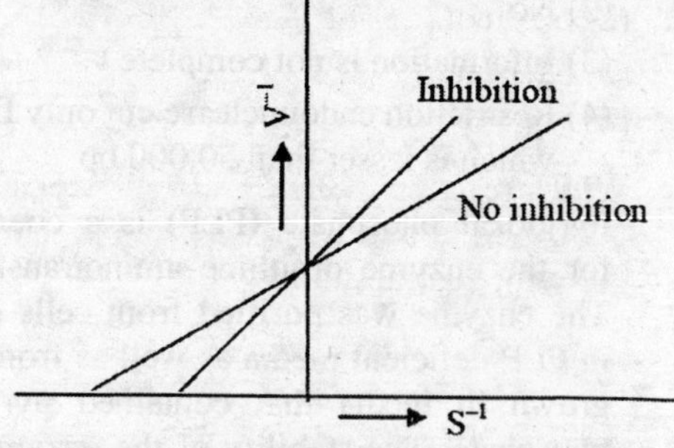

The nature of inhibition shown in the plot is
(1) Non-competitive
(2) Anti-competitive
(3) Competitive (4) Mixed type

89. Denaturation profiles of DNA are shown below

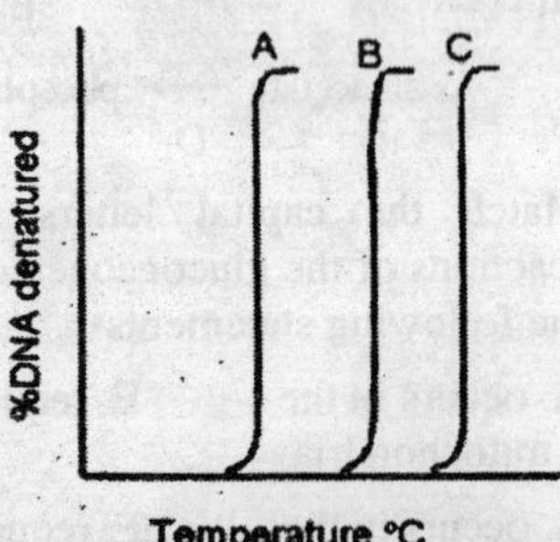

The differences in the profiles arise because

(1) the DNA is single stranded but of different sizes

(2) A + T content of A > B > C and the DNA are from complex genomes

(3) G + C content of C > B > A in DNA of comparable sizes isolated from simple genomes

(4) G + C content is identical but A + T content in A > B > C in DNA of comparable sizes isolated from simple genomes

90. A researcher has isolated a restriction endonuclease that cleaves at only one particular 10-base-pair site. Would this enzyme be useful in protecting cells from viral infections, given that a typical viral genome is 50,000 base pairs long?

(1) Yes

(2) No

(3) Information is not complete

(4) Restriction endonuclease cut only DNA which is lesser than 50,000 bp

91. Pyridoxal phosphate (PLP) is a coenzyme for the enzyme ornithine aminotransferase. The enzyme was purified from cells grown in PLP-deficient media as well as from cells grown in media that contained pyridoxal phosphate. The stability of the enzyme was then measured by incubating the enzyme at 37°C and assaying for the amount of enzyme activity remaining. The following results were obtained

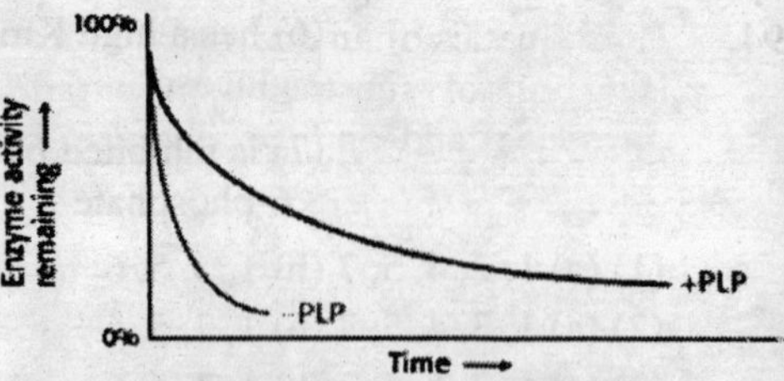

Why does the amount of enzyme from the PLP-deficient cells decline more rapidly?

(1) Due to denaturation of enzyme structure in presence of PLP

(2) The co-enzyme PLP apparently helps to stabilize the structure of enzyme

(3) PLP helps in catalysis and when enzyme is catalyzing reaction it can not be denatured

(4) 37°C temperature has same effect on denaturation of enzyme whether it is with or without PLP

92. Most uncatalyzed chemical reactions show an increased rate with increasing temperature. However, most biochemical reactions in a cell have a maximal rate around physiological temperature and show decreased rates at higher temperatures. This is because

(1) cells are disrupted by osmotic shock at higher temperatures

(2) intracellular catalysts become inhibited at low and high temperatures

(3) intracellular catalysts become denatured at high temperature

(4) the intracellular pH changes with temperature and this lowers the reaction rate

93. A twenty residue peptide composed of the sequence $(L\text{-}Ala)_{20}$ forms a right handed alpha helix in methanol. If every alternate L-Ala residue is replaced with D-Ala, the peptide will probably

(1) form a right handed alpha helix

(2) form a left handed alpha helix

(3) form ten residues of right handed helix followed by ten residues of a left handed helix

(4) not form a helix of any kind

94. The kinetics of an enzyme are measured as a function of substrate concentration in the presence and in the absence of 2 mM inhibitor (I)

[S] (μM)	Velocity (μmol/minute)	
	No inhibitor	Inhibitor
3	10.4	4.1
5	14.5	6.4
10	22.5	11.3
30	33.8	22.6
90	40.5	33.8

What type of inhibition is it?

(1) Irreversible inhibitor

(2) Competitive inhibitor

(3) Non-competitive inhibitor

(4) Un-competitive inhibitor

95. Referring to the structure of ATP as shown below, which of the following statements are true?

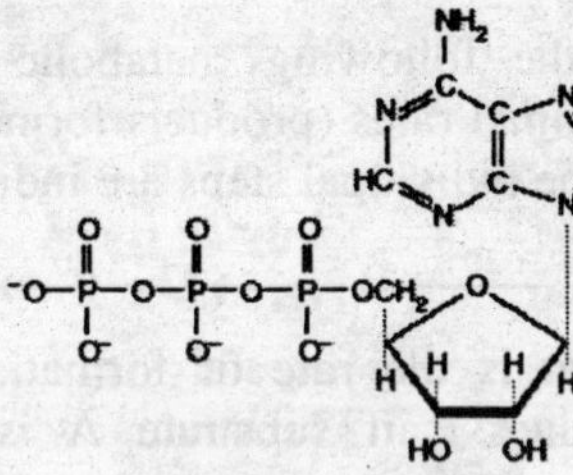

Adenosine triphosphate (ATP)

The structure of ATP

(1) contains a β-*N*-glycosidic linkage

(2) contains a pyranose ring

(3) exists in equilibrium with the open Fischer structure of the sugar

(4) contains a ketose sugar

96. Consider the following bonds/interaction in respect to biological system

A. Covalent Bonds

B. Hydrogen bonds

C. Ionic Interactions/Salt Bridge

D. Hydrophobic Interactions

Here are some of properties of such interactions

I. Attractions between peptide bonds forming the alpha helix structure of proteins and between polar amino acid side chains contributing to protein conformation; attractions between water molecules

II. Attractions between ionized groups in amino acid side chains contributing to protein conformation; attractions between ions in a salt

III. Attractions between nonpolar amino acids in proteins contributing to protein conformation; attractions between lipid molecules

IV. Most bonds linking atoms together to form molecules

Which of the following is correct match?

(1) A-I; B-II; C-III; D-IV

(2) A-IV; B-III; C-II; D-I

(3) A-IV; B-I; C-III; D-II

(4) A-IV; B-I; C-II; D-III

97. The rate of a chemical reaction (in other words, how many molecules of product are formed per unit time) can be determined by measuring the change in the concentration of reactants or products per unit of time. The faster the product concentration increases or the reactant concentration decreases, the greater the rate of the reaction. Four factors influence the reaction rate: reactant concentration, activation energy, temperature, and the presence of a catalyst

Consider the following statements

A. The lower the concentration of reactants, the slower the reaction simply because there are fewer molecules available to react. Conversely, the higher the concentration of reactants, the faster the reaction rate

B. The activation energy does not affect the difference in energy content between the reactants and final products since the activation energy is released when the products are formed

C. If the activation energy required for a reaction is small, then the probability of a given reactant molecule acquiring this amount of energy will be small, and the reaction rate will be slow. Thus, the lower the activation energy, the slower the rate of a chemical reaction

D. The higher the temperature, the faster molecules move and thus the greater their impact when they collide Therefore, one reason that increasing the temperature increases a reaction rate is that reactants have a better chance of acquiring sufficient activation energy from a collision

E. Since less activation energy is required, a reaction will proceed at a slower rate in the presence of a catalyst

Which of the following statements are correct?

(1) A, C and E
(2) A, B and D
(3) B, C and E
(4) B, C and D

98. Most of the chemical reactions in the body, if carried out in a test tube with only reactants and products present, would proceed at very low rates because they have high activation energies. In order to achieve the high reaction rates observed in living organisms, catalysts are required to lower the activation energies. These particular catalysts are called enzymes

Consider the following statements regarding enzymes

A. An enzyme undergoes no net chemical change as a consequence of the reaction it catalyzes

B. The binding of substrate to an enzyme's active site has all the characteristics—chemical specificity, affinity, competition, and saturation—of a ligand binding to a protein

C. An enzyme increases the rate of a chemical reaction and cause a reaction to occur that would not occur in its absence

D. An enzyme increases both the forward and reverse rates of a chemical reaction and thus does not change the chemical equilibrium that is finally reached. It only increases the rate at which equilibrium is achieved

E. An enzyme increases the activation energy of a reaction but does not alter the net amount of energy that is added to or released by the reactants in the course of the reaction

Which of the following statements are correct?

(1) A, B and C
(2) B, C and E
(3) A, B and D
(4) A, C and D

99. In the following metabolic pathway, the maximal rates (products formed per second) of the individual steps are indicated

$$A \xrightarrow{30} B \xrightarrow{5} C \xrightarrow{20} D \xrightarrow{40} E$$

What is the rate of formation of the end product E if substrate A is present at a saturating concentration?

(1) 40 (2) 95
(3) 23.75 (4) 5

100. Which one of the following statements is correct?

(1) All amino acids in biological system are chiral
(2) Only D form of amino acids exist in biological system
(3) All amino acids has only two buffering range
(4) D alanine and L alanine rotate plane polorized light in opposite direction.

ANSWER SHEET

Unit 1: Biochemistry

1. 1	**2.** 4	**3.** 4	**4.** 3	**5.** 3	**6.** 1	**7.** 2	**8.** 1	**9.** 3	**10.** 2
11. 3	**12.** 1	**13.** 4	**14.** 1	**15.** 4	**16.** 4	**17.** 4	**18.** 3	**19.** 2	**20.** 4
21. 4	**22.** 4	**23.** 3	**24.** 3	**25.** 4	**26.** 2	**27.** 1	**28.** 1	**29.** 2	**30.** 3
31. 1	**32.** 2	**33.** 3	**34.** 4	**35.** 2	**36.** 4	**37.** 3	**38.** 4	**39.** 3	**40.** 4
41. 3	**42.** 1	**43.** 3	**44.** 2	**45.** 4	**46.** 2	**47.** 2	**48.** 3	**49.** 3	**50.** 2
51. 2	**52.** 2	**53.** 3	**54.** 3	**55.** 1	**56.** 4	**57.** 2	**58.** 1	**59.** 3	**60.** 1
61. 4	**62.** 3	**63.** 2	**64.** 1	**65.** 1	**66.** 4	**67.** 4	**68.** 3	**69.** 1	**70.** 3
71. 1	**72.** 4	**73.** 2	**74.** 1	**75.** 3	**76.** 1	**77.** 1	**78.** 2	**79.** 1	**80.** 3
81. 3	**82.** 3	**83.** 2	**84.** 1	**85.** 1	**86.** 2	**87.** 1	**88.** 3	**89.** 3	**90.** 2
91. 2	**92.** 3	**93.** 4	**94.** 2	**95.** 1	**96.** 4	**97.** 2	**98.** 3	**99.** 4	**100.** 4

UNIT - 2

CELL BIOLOGY

1. Consider the following statement about plasma membrane

   I. Biological membranes usually contain both integral (transmembrane) and peripheral membrane proteins, which do not enter the hydrophobic core of the bilayer

   II. Most integral membrane proteins contain one or more membrane-spanning hydrophobic α helices and hydrophilic domains that extend from the cytosolic and exoplasmic faces of the membrane

   III. The porins, unlike other integral proteins, contain membrane spanning β sheets that form a barrel-like channel through the bilayer

   IV. Long-chain lipids attached to certain amino acids anchor some proteins always to inner membrane leaflet

   Among the following correct statements are

   (1) I and II only (2) II and III only

   (3) II, III and IV (4) I, II and III

2. The correct decreasing order of permeability through a lipid bilayer of the molecules/ions for Isoleucine, Tyrosine, O_2 and Na^+ is

   (1) O_2> Na +> Isoleucine> Tyrosine

   (2) O_2> Isoleucine> Tyrosine> Na^+

   (3) Isoleucine> Tyrosine> O_2> Na^+

   (4) Isoleucine> Tyrosine> $Na^+ > O_2$

3. Which of the following responses are not likely to be seen when a bacterial culture growing at 37°C is transferred to a culture room maintained at 25°C?

   A. Initial decrease in membrane fluidity

   B. Gradual replacement of shorter-chain fatty acids by longer-chain fatty acids in the membrane phospholipids

   C. Gradual replacement of stearate by oleate in the membrane phospholipids

   D. Enhanced rate of synthesis of unsaturated fatty acids

   E. Incorporation of more cholesterol into the membrane

   (1) A, C and D (2) B and E only

   (3) B, C and E (4) A and C only

4. The effects of temperature and lipid composition on membrane fluidity are often studied by using artificial membranes containing only one or a few kinds of lipids and no proteins. Assume that you and your lab partner have made the following artificial membranes

   Membrane A: Made entirely from phosphatidylcholine with saturated 16-carbon fatty acids

   Membrane B: Same as membrane A, except that each of the 16-carbon fatty acids has a single cis double bond

   Membrane C: Same as membrane A, except that each of the saturated fatty acids has only 14 carbon atoms

   After determining the transition temperatures of samples representing each of the membranes, you discover that your lab partner failed to record which membranes the samples correspond to. The three values you determined are –36°C, 23°C, and 41°C. Assign each of these transition temperatures to the correct artificial membrane

(1) Membrane A: 23°C, B: 41°C, C: -36°C
(2) Membrane A: 41°C, B: -36°C, C: 23°C
(3) Membrane A: -36°C, B: 23°C, C: 41°C
(4) Membrane A: 41°C, B: 23°C, C: -36°C

5. We know that exposed regions of membrane proteins can be labeled with ^{125}I by the lactoperoxidase (LP) reaction. Similarly, carbohydrate side chains of membrane glycoproteins can be labeled with ^{3}H by oxidation of galactose groups with galactose oxidase (GO) followed by reduction with tritiated borohydride ($^{3}H\ BH_4$)

Consider the following statements

A. When intact cells are incubated with LP in the presence of ^{125}I and the membrane proteins are then extracted and analyzed on SDS–polyacrylamide gels, several of the bands on the gel are found to be radioactive

B. When intact cells are incubated with GO and then reduced with $^{3}H\ BH_4$, several of the bands on the gel are found to be radioactive

C. All of the proteins of the plasma membrane that are known to contain carbohydrates cannot be labeled by the GO/$^{3}H\ BH_4$ method

D. None of the proteins of the erythrocyte plasma membrane that are known to be devoid of carbohydrate is labeled by the LP/^{125}I method

E. If the erythrocytes are ruptured before the labeling procedure, the LP procedure fails to label virtually all of the major membrane proteins

Noting that both LP and GO are too large to penetrate into the interior of an intact cell, explain which of the following observations would be made with intact erythrocytes

(1) C and E (2) A and D
(3) B and E (4) A and C

6. Bacterial mutants that are unable to synthesize fatty acids will incorporate them into their membranes when fatty acids are supplied in their growth medium. Suppose that each of two cultures contains a mixture of several types of straight-chain fatty acids, some saturated and some unsaturated, ranging in chain length from 10 to 20 carbon atoms. If one culture is maintained at 18ºC and the other is maintained at 40ºC over several generations, what differences in the composition of the cell membranes of the two cultures would you expect to observe?

(1) The bacteria grown at low temperature select fatty acids that will remain gel at a lower temperature in order to prevent their membranes from becoming too fluid

(2) The bacteria grown at the lower temperature will have incorporated a higher number of the longer fatty acids and a greater proportion of the saturated fatty acids. The membranes of bacteria grown at 40°C will have more short chain fatty acids and more that are unsaturated

(3) The bacteria grown at the higher temperature will have incorporated a higher number of the longer fatty acids and a greater proportion of the saturated fatty acids. The membranes of bacteria grown at 18°C will have more short chain fatty acids and more that are unsaturated

(4) The type of fatty acids in both kinds of bacteria will be almost same

7. Given two bilayer systems, one composed of phospholipids having saturated acyl chains 20 carbons in length and the other having acyl chains of the same length but with cis double bonds at C-5, C-8, C-11, and C-14, compare the effect of the acyl chains on T_m for each system

(1) T_m will remain same for phospholipids regardless of type of fatty acids

(2) Phospholipids having saturated fatty acids will have low T_m

(3) Phospholipids having unsaturated fatty acids will have high T_m

(4) T_m will be high for phospholipids having saturated fatty acids

8. The transition temperature for dipalmitoyl-phosphatidyl ethanolamine is 63ºC. Suppose you have a sample of this phospholipid in excess water at 50ºC, and you add cholesterol until it constitutes about 50% of the total lipid, by weight, in the sample. What would you expect when you attempt to determine the transition temperature for the mixture?

(1) Cholesterol will enhance the transition

(2) Cholesterol will allow sharp transition from gel to sol state

(3) Transition temperature will decrease by 10ºC

(4) Cholesterol will diminish or even to abolish the transition, by preventing the close packing of the fatty acyl chains that impart rigidity to the molecular assembly

9. Match the sodium (eel electric organ) and potassium (shaker) channels with the corresponding properties listed in the right column.

A.	Sodium channel	(1)	four 70-kd subunits
B.	Potassium channel	(2)	single 260-kd polypeptide chain
		(3)	fourfold symmetry
		(4)	tetrodotoxin binding site
		(5)	positively charged, voltage-sensing 54 helical segment
		(6)	ball-and-chain inactive mechanism
		(7)	3-A-diameter channel

(1) A: 2, 3, 4, 5, 6 B:1,3,5,6,7

(2) A: 1, 3, 4, 5, 7 B:2,3,5,6,7

(3) A: 1, 3, 4, 5, 6 B:2,3,4,6,7

(4) A: 2, 4, 5, 6, 7 B:1,2,3,4,5

10. Describe the characteristics in the right column either to active transport or to transport of ions or molecules through channels

A.	Active transport	(1) flux ~ 10^7 s^{-1}
B.	Transport through channels	(2) flux 3 x 10 to 2 x 10^3 s^{-1}
		(3) ions can flow from either side of the membrane
		(4) flux in a specific direction

(1) A:2,4 B:1,3

(2) A:1,2 B:3,4

(3) A:1,4 B:2,3

(4) A:2,3 B:1,4

Question 11-13

You are studying a yeast protein called transportin. Although the function of the protein remains to be determined, previous biochemical fractionation studies have demonstrated that transportin is present in a membrane-bound compartment

You have been given the task of defining the cellular compartment and also determining the pathway by which the protein gets there. Using a biochemical fractionation scheme that employs differential centrifugation followed by equilibrium density centrifugation, you've successfully separated various organelles into fractions. The identification of each organelle is determined by immunoblot using specific antibodies to the proteins indicated in panel a

You then introduce transportin into a mutant yeast strain that lacks endogenous transportin and induce its expression for 5 min allowing some protein to accumulate in the cell. At specific time points you then fractionate cells to assay the location of transportin in the cell (panel b)

A) equilibrium density gradient fractions

H 1 2 3 4 5 6 7 8 9 10 11 12 13 14 15 16 17 18 19 20

sec61
TIM23
GM120
GLP-1
Asg7p

B) equilibrium density gradient fractions

H 1 2 3 4 5 6 7 8 9 10 11 12 13 14 15 16 17 18 19 20

20 min
30 min
50 min
90 min

A) Immunoblot analysis of fractionated organelles. Organelle identity was determined by immunobloting for the following resident proteins: sec61 (endoplasmic reticulum), TIM23 (mitochondria), GM120 (Golgi apparatus), GLP-1 (lysosomes), Asg7p (plasma membrane). B) Immunoblot analysis of transportin following the fractionation of cells at each indicated time point.

11. Using panel a as a standard for organelle fractionation, indicate the pathway that transportin follows in the cell

(1) ER to golgi to lysosome to plasma membrane

(2) Cytoplasm to nucleus

(3) ER to lysosome to golgi

(4) ER to golgi to plasma membrane to lysosome

12. Is the protein co-translationally targeted into the membrane compartment or post-translationally targeted?

(1) Post- translationally

(2) Co-translationally

(3) Co-translationally and post-translatioally

(4) Cannot be interpreted

13. Which of the following events can induce a transient arrest in the translation of a secretory protein?

(1) Binding of a polysome to an ER receptor

(2) Binding of SRP to an N-terminal signal sequence

(3) Binding of snRNPs to the large ribosomal subunit

(4) Presence of a stop-transfer sequence in the polypeptide

14. Calculate the free-energy change for the transport of an uncharged species from a concentration of 5 mM outside a cell to a concentration of 150 mM inside. Assume that the temperature is 25°C

(1) +2.02 kcal/mol (2) +0.64 kcal/mol

(3) -1.38 kcal/mol (4) +3.40 kcal/mol

15. Calculate the free-energy change for the transport of an charged species from a concentration of 5 mM outside a cell to a concentration of 150 mM inside ion with a charge of +1 that is crossing a membrane with a potential of -60 mV (interior negative with respect to the exterior)

(1) +2.02 kcal/mol (2) +0.64 kcal/mol

(3) -1.38 kcal/mol (4) +3.40 kcal/mol

16. Mouse liver cells were homogenized and the homogenate subjected to equilibrium density-gradient centrifugation with sucrose gradients. Fractions obtained from these gradients were assayed for marker molecules (i.e., molecules that are limited to specific organelles) The results of these assays are shown in the figure

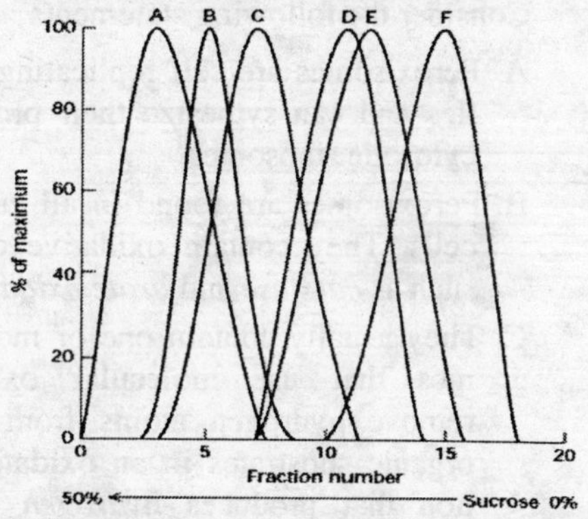

The marker molecules have the following functions: Cytochrome oxidase is an enzyme involved in the process by which ATP is formed in the complete aerobic degradation of glucose or fatty acids; ribosomal RNA forms part of the protein-synthesizing ribosomes; catalase catalyzes decomposition of hydrogen peroxide; acid phosphatase hydrolyzes monophosphoric esters at acid pH; cytidylyl transferase is involved in phospholipid biosynthesis; and amino acid permease aids in transport of amino acids across membranes. Which of the following is correct combination of organalles and curve A to F represented in figure above?

(1) A: ER B: Lysosome; C: Peroxisome; D:Plasma Membrane E:Ribsome; F: Mitochondria
(2) A: Mitochondria; B: Ribosome; C: Peroxisome; D: Lysosome; E: ER; F: Plasma membrane
(3) A: Mitochondria; B: Lysosome; C: Ribosome; D: ER; E: Plasma Membrane; F: Peroxisome
(4) A: Mitochondria; B: Ribosome; C: Peroxisome; D: Lysosome; E: Plasma membrane; F: ER

17. Peroxisomes are unusually diverse organelles and even in the different cells of a single organism may contain very different sets of enzymes

Consider the following statements

A. Peroxisomes are self replicating organelles and can sythesize their proteins by cytosolic ribosomes
B. Peroxisomes are found in all eukaryotic cells. They contain oxidative enzymes, such as *catalase* and *urate oxidase*
C. They usually contain one or more enzymes that use molecular oxygen to remove hydrogen atoms from specific organic substrates in an oxidative reaction that produces *hydrogen peroxide* (H_2O_2)
D. Catalase utilizes the H_2O_2 by the "peroxidative" reaction, particularly important in liver and kidney cells, to detoxify various toxic molecules that enter the bloodstream
E. β oxidation in yeast and plant cells occurs both in mitochondria and peroxisomes

Which of the following statement is correct?

(1) A, B and C (2) B, C and D
(3) C, D and E (4) A, D and E

18. All eukaryotic cells have an endoplasmic reticulum (ER) Its membrane typically constitutes more than half of the total membrane of an average animal cell

Consider the following statements

A. Import of proteins into the Rough ER always occurs co-translationally and have a signal peptide which is cleaved after entry into lumen of ER
B. Smooth ER is prominent in cells that specialize in lipid metabolism such as cells that synthesize steroid hormones from cholesterol
C. Smooth ER contains cytochrome P450 enzymes that catalyze a series of reactions to detoxify both lipid-soluble drugs and various harmful compounds produced by metabolism
D. Glycosphingolipids (glycolipids) and sphingomyelin are synthesized from ceramide in Smooth ER

Which of the following statement is correct?

(1) A and B (2) B and C
(3) C and D (4) A and D

19. Plant cells are surrounded by a rigid extracellular matrix in the form of a cell wall, which is responsible for many of the unique features of a plant's life-style

Consider the following statements

A. The cell wall is composed of tough cellulose microfibrils embedded in a highly cross-linked matrix of polysaccharides (mainly pectins and hemicellulose) and glycoproteins
B. A cortical array of microtubules can determine the orientation of newly deposited cellulose microfibrils, which in turn determines the manner in which the cell expands and therefore the cell's final shape and cell-division patterns
C. Organization of microtubules in plant cell is controlled, although it has been shown that they can rapidly reorient in response to extracellular stimuli, including low-molecular-weight plant growth

factors such as ethylene and gibberellic acid

D. Like other matrix macromolecules, which are made in the endoplasmic reticulum and Golgi apparatus and secreted, cellulose, is synthesized by a golgi-membrane-bound enzyme complex (cellulose synthase)

Which of the following statement is not correct?

(1) A and C (2) C and D
(3) Only C (4) Only D

20. You are studying the binding of proteins to the cytoplasmic face of cultured liver cells and have found a method that gives a good yield of inside-out vesicles from the plasma membrane. Unfortunately, your preparations are contaminated with variable amounts of right-side-out vesicles. Nothing you have tried avoids this contamination. Somebody suggests that you pass the vesicles over an affinity column made of lectin coupled to Sepharose beads. What is the rational of this suggestion?

(1) Right-side-out-vesicles will be lysed by lectin coupled to Sepharose beads
(2) Right-side-out-vesicles will simply bind to the lectin coupled Sepharose beads
(3) Lectin will bind to the carbohydrate residues present only on the inside out vesicles
(4) Lectin will bind to only glycoproteins and glycolipids present on the inside-out vesicles

21. Match the terms in Group I with terms in Group II

Group-I	**Group II**
P. Leucoplast	1. Protein modification and targeting
Q. Mitochondria	2. Microtubule organizing centre
R. Golgi Complex	3. Starch storage
S. Centriole	4. Kreb's cycle
	5. Glycogen storage
	6. Calvin cycle

(1) P-3, Q-4, R-2, S-1
(2) P-5, Q-4, R-6, S-5
(3) P-3, Q-6, R-4, S-5
(4) P-3, Q-4, R-1, S-2

22. In 1971 T. O. Diener of reported that potato spindle-tuber disease, which causes potatoes to become gnarled and cracked, is caused by an infectious agent consisting of a small circular RNA molecule that totally lacks a protein coat. Diener named the pathogen a viroid

Consider the following statements

A. The RNAs of viroids range in size from about 240 to 600 nucleotides, one tenth the size of the smaller viruses
B. Naked viroid RNA codes for limited number of proteins
C. Duplication of the viroid RNA within an infected cell utilizes the host's RNA polymerase I, an enzyme that normally transcribes the host's DNA into messenger RNAs
D. Viroids are thought to cause disease by interfering with the cell's normal path of gene expression. The effect on crops can be serious: a viroid disease called cadang-cadang has devastated the coconut palm groves of the Philippines, and another viroid has wreaked havoc on the chrysanthemum industry in the United States

Which of the following statement regarding viroids is correct?

(1) A and B (2) B and C
(3) C and D (4) D and A

23. A model depicting possible steps in the evolution of eukaryotic cells, including the origin of mitochondria and chloroplasts by endosymbiosis

Consider the following steps in origin of modern eukaryotic cell

A. A large anaerobic, heterotrophic prokaryote takes in a small aerobic prokaryote. Evidence strongly indicates that the engulfed prokaryote was an ancestor of modern-day Rickettsia, a group of bacteria that causes typhus and other diseases

B. The aerobic endosymbiont has evolved into a mitochondrion

C. a portion of the plasma membrane has invaginated and is seen in the process of evolving into a nuclear envelope and associated endoplasmic reticulum. The primitive eukaryote depicted gave rise to two major groups of eukaryotes

D. the primitive eukaryote evolves into nonphotosynthetic protist, fungal, and animal cells

E. the primitive eukaryote takes in a photosynthetic prokaryote, which will become an endosymbiont and evolve into a chloroplast

Which of the following is correct sequence of origin of modern eukaryotic cell?

(1) A→C→D→B→E

(2) A→B→C→D→E

(3) D→A→E→B→C

(4) D→A→B→C→E

24. Suppose you were a scientist and were studying a disease of tobacco crops that stunted the growth of the plants and mottled their leaves. You find that the sap from a diseased plant, when added to a healthy plant, is capable of transmitting the disease to that plant. You examine the sap in the best light microscopes of the period and see no evidence of bacteria. You force the sap through filters whose pores are so small that they retard the passage of the smallest known bacteria, yet the fluid that passes through the filters is still able to transmit the disease. Like Dimitri Ivanovsky, who conducted these experiments more than a hundred years ago, you would probably conclude that the infectious agent is

(1) Fungus (2) Virus

(3) Bacteria (4) Prions

25. A tenfold difference in concentration of a nonelectrolyte across a membrane at 25^{0}C generates a ΔG of -1.4 kcal/mol. Suppose the concentration gradient consisted of Na^+ ions, which were present at tenfold higher concentration outside the cell than in the cytoplasm. Because the voltage across the membrane of a cell is typically about -70 mV, the free-energy change for the movement of a mole of Na^+ ions into the cell under these conditions would be [Given F = 23.06 kcal/V . mol]

(1) -3.1 kcal/mol (2) -3100 kcal/mol

(3) -1.7 kcal/mol (4) -1700 kcal/mol

26. Viruses are useful in studying protein trafficking studies because they turn infected cells into factories for the production of viral proteins, which are carried like any other protein cargo through the biosynthetic pathway. When a cell is infected with modified vesicular stomatitis virus (VSV), massive amounts of the VSVG-GFP protein are produced in the endoplasmic reticulum (ER) The VSVG-GFP molecules then traffic through the Golgi complex and are transported to the plasma membrane of the infected cell where they are incorporated into viral envelopes. As in a radioactive pulse chase experiment, the use of a virus allows investigators to follow a relatively synchronous wave of protein-GFP movement, in this case represented by a wave of green fluorescence that begins soon after infection

Consider the following observation and conclusion

A. Virus with a mutant VSVG protein that is unable to leave the ER of infected cells that are grown at an elevated temperature (e.g., 40ºC) protein will accumulate in ER

B. When the temperature is lowered to 32ºC, the fluorescent GFP-VSVG protein that had accumulated in the ER moves synchronously to the Golgi complex, where various processing events occur, and then on to the plasma membrane

C. Mutants of this type that function normally at a elevated (non-permissive) temperature, but not at an normal (permissive) temperature are described as temperature-sensitive mutants

Which of the following are valid observation and conclusion?

(1) A and B (2) B and C
(3) A and C (4) A, B and C

27. It was a surprise to discover that misfolded proteins are not destroyed in the ER, but instead are transported into the cytosol by a process of "dislocation." Once in the cytosol, the oligosaccharide chains are removed, and the misfolded proteins are destroyed in proteasomes. The negative consequences of this is observed in most patients with cystic fibrosis, the plasma membrane of epithelial cells is lacking the abnormal protein encoded by the cystic fibrosis gene. In these cases, the mutant protein is destroyed by the quality control process of the ER and thus fails to reach the cell surface. This process which occurs in ER and cytosol that ensures that aberrant proteins are not transported to other parts of the cell known as

(1) ER-associated degradation (ERAD)
(2) Ubiquitinization
(3) Unfolded protein response
(4) Autophagy

28. Under certain circumstances, misfolded proteins can be generated in the ER at a rate faster than they can be exported to the cytoplasm. The accumulation of misfolded proteins, which is potentially lethal to cells, triggers a comprehensive "plan of action" within the cell known as the **Unfolded Protein Response (UPR)** The ER contains protein sensors that monitor the concentration of unfolded or misfolded proteins in the ER lumen. According to the prevailing model

A. the sensors are normally kept in an inactive state by molecular chaperones, particularly BiP. If circumstances should lead to an accumulation of misfolded proteins, the BiP molecules in the ER lumen are called into service as chaperones for the misfolded proteins, rendering them incapable of inhibiting the sensors

B. Activation of the sensors leads to a multitude of signals that are transmitted into both the nucleus and cytosol and result in the expression of hundreds of different genes whose encoded proteins have the potential to alleviate stressful conditions within the ER. These include genes that encode ER-based molecular chaperones that can help misfolded proteins reach the native state, proteins involved in the transport of the proteins out of the ER and proteins involved in the selective destruction of abnormal proteins

C. Phosphorylation of a key protein (eIFα) required for protein synthesis. This modification inhibits protein synthesis and decreases the flow of newly synthesized proteins into the ER. This gives the cell an opportunity to remove those proteins that are already present in the ER lumen

D. The UPR is more than a cell-survival mechanism; it also includes the activation of a pathway that leads to the death of the cell. It is presumed that the UPR provides a mechanism to relieve itself of the stressful conditions. If these corrective measures are not successful, the cell-death pathway is triggered and the cell is destroyed

Which of the following statement is correct regarding UPR?

(1) A, B and C (2) B, C and D
(3) A and D only (4) A, B, C and D

29. Fibroblast cells from either a control subject (closed circles) or a patient with homozygous FH (open circles) were grown in dishes containing fetal calf serum. On day 6 (which corresponds to 0 hours on the graph), the medium was replaced with fresh medium containing lipoprotein-deficient human plasma. At the indicated time, extracts were prepared, and HMG CoA reductase activity was measured

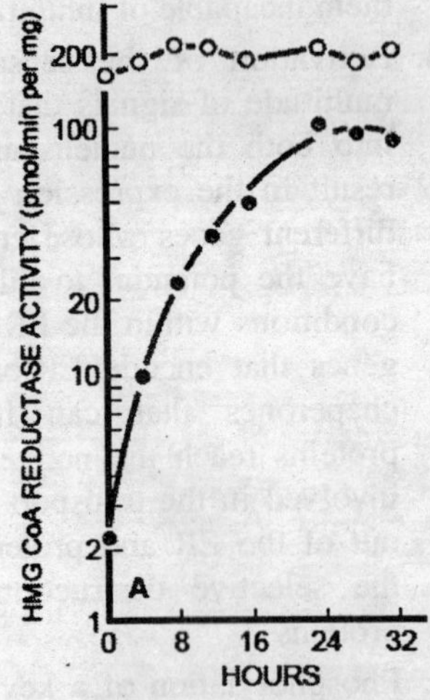

A. If we look at the control cells and cells from the Familial Hypercholesterolemia, it is apparent that at the beginning of the monitoring period the cells of patients have very little enzyme activity because the medium had contained enough cholesterol-containing lipoproteins that the cells did not need to synthesize their own

B. When the medium was changed to the lipoprotein-deficient plasma, the patient cells were no longer able to use cholesterol from the medium and thus increased the amount of enzyme within the cell

C. The cells from the Familial Hypercholesterolemia patients showed no response to either the presence or absence of lipoproteins in the medium

Which of the following is valid conclusion based on experimental observation?

(1) A and B (2) B and C
(3) A and C (4) A, B and C

30. The orientation of the Na^+-K^+ pump in cell membranes determines the side of the membrane where the various processes involved in the transport of Na^+ and K^+ will take place. Assign each of the steps or processes in the right column to the intracellular or extracellular side of the pump

(a) Intracellular side	(1) binding of cardiotonic steroids
(b) Extracellular side	(2) hydrolysis of ATP and phosphorylation of the pump
	(3) binding of K^+
	(4) binding of Na^+

(1) (a) 2,4 (b) 1,3
(2) (a) 1,2 (b) 3, 4
(3) (a) 1,4 (b) 2,3
(4) (a) 2,3 (b) 1,4

31. The amino acid sequence of a protein dictates its folding into a specific three-dimensional conformation, the native state

Consider the following statement about protein folding

A. Protein folding in vivo occurs with assistance from molecular chaperones (Hsp70 proteins), which bind to folded polypeptides and makes them stabile

B. Chaperonins, large complexes of Hsp60-like proteins, shelter some partly folded or misfolded proteins in a barrel-like cavity, providing additional time for proper folding

C. Viral proteins produced within infected cells, normal cytosolic proteins, and misfolded proteins are marked for destruction by the covalent addition of a polyubiquitin chain and then degraded within lysosome, large membranous bag within cell which has multiple proteases in their interiors

D. Neurodegenerative diseases like CJD and Alzheimer disease are caused by aggregates of proteins that are stably folded in an alternative conformation

Which of the following statement is true?

(1) A and B (2) B and D
(3) C and D (4) D and A

32. Match some of the known sorting signal that directs proteins to specific transport vesicles

Group I

A. Mannose-6-Phosphate (M6P)
B. Lys-Lys-X-X (KKXX)
C. Leu-Leu (LL)
D. Asn-Pro-X-Tyr (NPXY)
E. Lys-Asp-Glu-Leu (KDEL)

Group II

1. ER-resident luminal proteins
2. ER-resident membrane proteins
3. Plasma membrane proteins
4. Soluble lysosomal enzymes
5. LDL receptor in the plasma membrane

(1) A-4 B-1 C-5 D-3 E-1
(2) A-4 B-2 C-3 D-5 E-1
(3) A-4 B-1 C-5 D-3 E-2
(4) A-5 B-2 C-3 D-4 E-2

33. Match the enzymes in **Group I** with their corresponding activity in **Group II**

Group I	Group II
P. Flippase	1. Catalyzes the movement of any phospholipid across the lipid bilayer down its concentration gradient
Q. Floppase	2. Catalyzes the translocation of amino-phospholipids from the extracellular to the inner leaflet
R. Lipase	3. Catalyzes the translocation of membrane phospholipids from cytosolic to the extracellular leaflet
S. Scramblase	4. Degradation of phospholipids from the lipid bilayer including the inner and outer leaflets

(1) P-2, Q-3, R-4, S-1
(2) P-1, Q-2, R-3, S-4
(3) P-4, Q-1, R-2, S-3
(4) P-3, Q-2, R-4, S-1

34. Golgi apparatus is also termed as cellular post office, since it packages and transports cellular proteins across various organelles and outside the cell. In general, the Golgi is perinuclear in location and is closely associated with the endoplasmic reticulum. A chemical compound, Monensin inhibits all trafficking from Golgi. If Golgi is visualized by immunofluorescence microscopy after treatment with this compound, the Golgi will be

(1) absent (2) normal
(3) swollen (4) fragmented

35. Colchicine is a poison that binds to tubulin and prevents its assembly into microtubules; cytochalasins are compounds that bind to the ends of actin filaments and prevent their elongation. What effects do you think that these two substances would have on cell division in animal cells?

(1) Both will inhibit mitosis
(2) Only colchicines inhibit mitosis
(3) None of them will inhibit mitosis
(4) Both of them will inhibit cytokinesis

36. Nucleosomes are the structural unit of chromatin. In the formation of nucleosomes about 200 bp are wrapped around the histones. The nucleosomes (diameter about 11 nm) are folded into a second order structure to form a solenoid consisting of 6 nucleosomes per turn. Calculate the total packaging ratio of DNA at the level of solenoid. Further if the genome size of a eukaryotic microorganism is $2x\ 10^7$ bp and the size of its chromosome is 4 μm, what

will be the final packaging ratio of DNA in the chromosome?

(1) 40 and 1700 (2) 8 and 6800
(2) 40 and 6800 (4) 8 and 1700

37. In nerves, vesicles can move the length of an axon at a rate that far exceeds that which would be predicted for simple diffusion. Which of the following models best explains vesicular movement in these cells?

(1) Depolymerization of actin microfilaments attached to vesicles pulls the vesicles toward the site of depolymerization
(2) Vesicles are moved by alternate contraction and relaxation of actin-myosin "muscle" complexes
(3) Vesicles, by virtue of their net negative charge, are attracted to positively charged regions of the cell
(4) Vesicles are attached to the protein kinesin, which slides along microtubules by an ATP dependent process

38. Choose the statement that is not correct for the cytoskeletal protein actin

(1) α-actins are found in various types of muscles
(2) Polymerization of pure actin in vitro requires GTP
(3) Actin filaments have a slow growing minus end and a fast growing plus end
(4) Cytochalasins are the inhibitors for actin polymerization

39. The dependence of the rate of sucrose uptake with respect to sucrose concentration in plant cell was studied and data are shown in the following graph

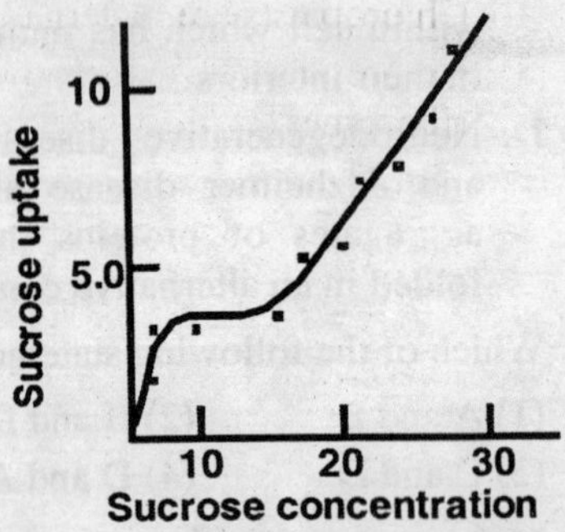

From the above data it can be inferred that

(1) the sucrose uptake is energy independent and no special carrier is involved
(2) the sucrose uptake is energy dependent and a special carrier is involved
(3) at lower concentration of sucrose the uptake of sucrose is energy dependent and carrier mediated
(4) at higher concentration of sucrose the uptake is energy dependent and carrier mediated

40. In bacterial cell division

A. MreB, similar to actin controls width of Bacteria
B. FtsZ, a tubulin homologue, helps in cytokinesis
C. Pur gene and min gene regulates chromosomal separation and binary fission
D. Spindle fibres (microtubules) takes chromosome to opposite poles

(1) A and B (2) C and D
(3) A, B and C (4) B, C and D

41. Plant chloroplasts are believed to have evolved from cyanobacteria-like progenitor through endosymbiosis. Which of the following statements support thi hypothesis?

I. Chloroplasts and cyanobacteria share similar photosynthetic pigments and thylakoid membranes
II. Cyanobacteria exhibit an oxygenic photosynthesis

III. Chloroplasts are maternally inherited
IV. Chloroplasts have their own DNA and ribosomes
V. Viable chloroplasts can be isolated from cells but cannot be cultured in vitro
VI. Prokaryotic genes express well in chloroplasts

(1) I, III, IV and V
(2) I, II, V and VI
(3) I, II, III and V
(4) II, III, IV and V

42. The components of poison ivy and poison oak that produce the characteristic itchy rash are catechols substituted with long-chain alkyl groups

$$\text{catechol ring with OH, OH and }(CH_2)_n\text{—}CH_3$$

$pK_a \approx 8$

If you were exposed to poison ivy, which of the treatments below would you apply to the affected area?

(1) Wash the area with cold water
(2) Wash the area with dilute vinegar or lemon juice
(3) Wash the area with soap and water
(4) Wash the area with soap, water, and baking soda (sodium bicarbonate)

43. Protein transport through the secretory pathway can be visualized by fluorescence microscopy of cells pproducing a GFP-tagged membrane protein. Cultured cells were transfected with a hybrid gene encoding the viral membrane glycoprotein VSV G protein linked to the gene for green fluorescent protein (GFP) A mutant version of the viral gene was used so that newly made hybrid protein (VSVG-GFP) is retained in the ER at 40ºC but is released for transport at 32ºC

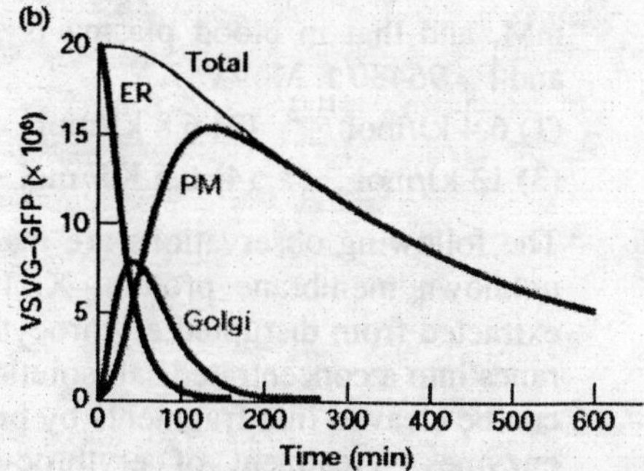

Figure shows plot of the levels of VSVG-GFP in the endoplasmic reticulum (ER), Golgi, and plasma membrane (PM) at different times after shift to lower temperature (The decrease in total fluorescence that occurs at later times probably results from slow inactivation of GFP fluorescence) The kinetics of transport from one organelle to another can be reconstructed by analysis of these data as follows

(1) Golgi→PM→ER
(2) PM→Golgi→ER
(3) ER→Golgi→PM
(4) Golgi→PM→Virus

44. Glucose uptake by epithelial cells of intestine is blocked by drug ovabain which inhibit Na^+-K^+ pump. Which of the following is correct mode of action of ovabain?

(1) Ovabain blocks Na^+ transport from intestinal lumen to epithelial cells
(2) Ovabain blocks Na^+ transport from intestinal lumen to interstial cells
(3) Ovabain blocks Na^+ transport from epithelial cells to basolateral side
(4) Ovabain blocks Na^+ transport from interstial cell to intestinal lumen

45. For a typical vertebrate cell with a membrane potential of -0.070 V (inside negative), what is the probable free-energy change for transporting 1 mol of Na^+ out of the cell and into the blood at 37ºC? Assume the concentration of Na^+ inside the cell is 12 mM, and that in blood plasma is 145 mM and F= 96480 J. Mol/V

(1) 6.4 kJ/mol (2) 6.8 kJ/mol
(3) 13 kJ/mol (4) 0.5 KJ/ mol

46. The following observations are made on an unknown membrane protein, X. It can be extracted from disrupted erythrocyte membranes into a concentrated salt solution, and it can be cleaved into fragments by proteolytic enzymes. Treatment of erythrocytes with proteolytic enzymes followed by disruption and extraction of membrane components yields intact X. However, treatment of erythrocyte "ghosts" (which consist of just plasma membranes, produced by disrupting the cells and washing out the hemoglobin) with proteolytic enzymes followed by disruption and extraction yields extensively fragmented X. What do these observations indicate about the location of X in the plasma membrane? Do the properties of X resemble those of an integral or peripheral membrane protein?

(1) Peripheral protein attached to extrinsic face of RBC membrane
(2) Peripheral protein attached to cytosolic face of RBC membrane
(3) Integral protein spanning across RBC membrane
(4) Lipid anchored protein associated with cytosolic face of RBC membrane

47. You are studying the uptake of L-leucine by epithelial cells of the mouse intestine. Measurements of the rate of uptake of L-leucine and several of its analogs, with and without Na^+ in the assay buffer, yield the results given in the table. What can you conclude about the properties of the leucine transporter?

	Upper in presence of Na^+		Uptake in absence of Na^+	
Substrate	V_{max}	K_t (mm)	V_{max}	K_t (mm)
L- Leucine	420	0.24	23	0.2
D- Leucine	310	4.7	5	4.7

(1) It is primary active transporter
(2) It is secondary active transporter
(3) It is carrier for facillated diffusion
(4) Transporter is not stereo specific

48. Consider the following statements regarding cell organelles

A. The nucleus, mitochondrion, and chloroplast are bounded by two bilayer membranes separated by an intermembrane space. All these organelles have evolved as per endosymbiosis theory
B. Endosomes internalize plasma-membrane proteins and soluble materials from the extracellular medium, and they sort them back to the membranes or to lysosomes for degradation and have acidic pH
C. Lysosomes have an acidic interior and contain various hydrolases that degrade worn-out or unneeded cellular components and some ingested materials
D. Liposomes are small organelles containing enzymes that oxidize various organic compounds without the production of ATP. By-products of oxidation are used in biosynthetic reactions

Which of the following statements are correct?

(1) A, B and C (2) B, C and D
(3) A and D (4) B and C

49. The cytosol is the internal aqueous medium of a cell exclusive of all organelles and the cytoskeleton. It contains numerous soluble enzymes responsible for much of the cell's metabolic activity. Three major types of protein filaments-actin filaments, microtubules, and intermediate filaments-make up the cytoskeleton

Consider the following statement?

A. Microfilaments are assembled from monomeric actin subunits; microtubules, from α,β-tubulin subunits; and intermediate filaments, from lamin subunits and other tissue specific proteins. In all animal and plant cells, the cytoskeleton provides structural stability for the cell and contributes to cell movement. Some bacteria have a primitive cytoskeleton

B. Tubulin bundles form the core of microvilli and other fingerlike projections of the plasma membrane

C. Cortical spectrin-tubulin networks are attached to the cell membrane by bivalent membrane-microfilament binding proteins such as ankyrin and band 4.1

D. Intermediate filaments are assembled into networks and bundles by various intermediate filament–binding proteins, which also cross-link intermediate filaments to the plasma and nuclear membranes, microtubules, and microfilaments

Which of the following statement is correct?

(1) A and D (2) A, B and C
(3) B, C and D (4) A, B, C and D

50. P-class pumps are composed of a catalytic α- subunit, which becomes phosphorylated as part of the transport cycle. A β - subunit, present in some of these pumps, may regulate transport. Which of the following is not a P-class of pump

(1) Na^+/K^+ pump on plasma membrane of higher eukaryotes

(2) H^+/K^+ pump on apical plasma membrane of mammalian stomach

(3) Ca^{2+} pump on sarcoplasmic reticulum membrane in muscle cells

(4) H^+ pump on lysosmal membranes and plasma membrane of osteoclasts and some kidney tubule cells

51. Ouabain is a specific inhibitor of the active transport of sodium ions out of the cell and is therefore a valuable tool in studies of membrane transport mechanisms. Which of the following processes in your own body would you expect to be sensitive to inhibition by ouabain?

A. Facilitated diffusion of glucose into a muscle cell

B. Active transport of dietary phenylala-nine across the intestinal mucosa

C. Uptake of potassium ions by red blood cells

D. Active uptake of lactose by the bacteria in your intestine

(1) A and B (2) B and C
(3) A and D (4) B and D

52. Name the organelle for which the following signal confers transport to

i. Amphipathic alpha helix

ii. Lysine rich, positively charged sequence located anywhere in protein

iii. Signal patch

iv. KDEL

v. Ser-Lys-Leu-COOH

(1) i-mitochondria, ii-nucleus, iii-lysosome, iv-ER (retention), v-peroxisome

(2) i-nucleus, ii- mitochondria, iii-lysosome, iv-ER (retention), v-peroxisome

(3) i- lysosome, ii-nucleus, iii- mitochondria, iv-ER (retention), v-peroxisome

(4) i- peroxisome, ii-nucleus, iii-lysosome, iv-ER (retention), v- mitochondria

53. Consider the following statements

A. Like the lumen of the ER, the interior of the nucleus is topologically equivalent to the outside of the cell

B. Membrane-bound and free ribosomes, which are structurally and functionally identical, differ only in the proteins they happen to be making at a particular time

C. To avoid the inevitable congestion that would occur if two-way traffic through a single pore were allowed, nuclear pore complexes are specialized so that some mediate import while others mediate export

D. Peroxisomes are found in only a few specialized types of eucaryotic cell

E. In multipass transmembrane proteins the odd-numbered transmembrane segments (counting from the N-terminus) act as start-transfer signals and

Which of the following are true?

(1) A and B (2) C and D
(3) D and E (4) B only

54. If the enzyme Di Hydro Folate Reductase (DHFR), which is normally located in the cytosol, is engineered to carry a mitochondrial targeting sequence at its N-terminus, it is efficiently imported into mitochondria. If the modified DHFR is first incubated with methotrexate, which binds tightly to the active site, the enzyme remains in the cytosol. How do you suppose that the binding of methotrexate interferes with mitochondrial import?

(1) The binding of methotrexate to the active site prevents N-terminal mitochondrial signal sequence to be recognized by its transporter
(2) The binding of methotrexate to the active site permanently denature enzyme and do not allow it to get translocated
(3) The binding of methotrexate to the active site locks the enzyme in its folded form, thus it is not able to be get translocated into mitochondria
(4) Binding of methotrexate bring allosteric change in enzyme and increases its molecular weight

55. Reassociation experiments revealed that eukaryotic DNA does not renature uniformly. These experiments are performed by fragmenting total DNA from an organism into pieces about 1000 base pairs long. The DNA is then melted into single strands and allowed to reform complementary base-paired double helices. Which class of DNA would be expected to have the most rapid rate of renaturation (reassociation)?

(1) Single-copy DNA
(2) Pseudogenes
(3) Simple sequence repeat DNA
(4) Small multigene families

56. Gene density in eukaryotic organisms is consistently lower and more variable than in prokaryotes. Compared to lower eukaryotes like the yeast *S. cerevisiae* (= 500 genes/Mb), the human genome has 50 fold lower gene density. The eukaryotic genes are also much larger in size. This is because

(1) eukaryotic genome has high proportion of repeats and satellite DNAs and genes have introns
(2) eukaryotic genome is diploid and has to cross over; also genes are polycistronic
(3) Prokaryotic genome has more non-coding DNA and interrupted genes
(4) Eukaryotic genome has telomeres and centromeres and genes are clustered as families

57. Phosphatidyl serine (PS) is mostly located in the inner bilayer of plasma membrane of red blood cells (RBCs) You have to prove this fact about PS by an experiment. You are provided with PS-specific lytic enzymes (PSE) and other reagents needed. Identify the correct sequence of experiments to be carried out to settle this issue

(1) RBCs → inside out vesicles→ PSE → Thin Layer Chromotography (TLC)
(2) RBCs → right side out vesicles → TLC →PSE
(3) RBCs → PSE → Inside out vesicles →TLC
(4) RBCs → PSE → TLC → Inside out vesicles

58. ATP-driven pumps hydrolyze ATP to ADP and phosphate and use the energy released to pump ions or solutes across a membrane. There are many classes of these pumps and representatives of each are found in all prokaryotic and eukaryotic cells. Which of the following statements about these pumps is not correct?

(1) P-type pumps are multipass transmembrane proteins which phosphorylate themselves during pumping and involve in ion transport

(2) F-type pumps normally use the H^+ gradient across the membrane to drive the synthesis of ATP

(3) V-type pumps normally use voltage gradient for transport of small molecules

(4) ABC transporters primarily pump small molecules across cell membrane

59. You create a N terminally GFP tagged version of the Interleukin -79 receptor. Using this construct, you then make a series of deletion mutants

A. A deletion of the signal sequence

B. A deletion of the transmembrane domain

C. A deletion of the fifteen C-terminal amino acids

However, you dropped the tubes before you labeled them! You have decided to transfect these mutants into cells to determine their fate

Match the above mutant with the localization patterns below

I. The GFP signal is in the cytosol of the cell

II. The GFP signal is in the ER lumen

III. The GFP signal is associated with the ER membrane

(1) I: A, II: B, III: C

(2) I: A, II: C, III: B

(3) I: C, II: B, III: A

(4) I: B, II: C, III: A

60. Cystic fibrosis (CF) trans membrane conductance regulator (CFTR) protein is known to be a cAMP- dependent Cl^- channel. CF patients (with mutant CFTR proteins) show reduced Cl^- permeability and as a result exhibit elevated Cl^- level in sweat. To prove this, CFTR proteins (both wild type and mutant) are inserted in a model membrane (liposome) and Cl^- transport is followed with radioactive Cl^-. It is known that topology of CFTR in membrane is important for its function. Despite no proteolytic degradation of denaturation of CFTR proteins, wild type CFTR failed to transport Cl^- in liposome Which of the following is the correct explanation of this?

(1) CFTR protein gets mutated during insertion in liposomes

(2) CFTR protein loses affinity with Cl- ions

(3) CFTR protein gets wrongly inserted in liposomes

(4) CFTR protein loses channel forming property in liposomes

61. Consider the following statement regarding membrane transport

A. Transport by transporters can be either active or passive, whereas transport by channels is always passive

B. Transporters saturate at high concentrations of the transported molecule when all their binding sites are occupied; channels, on the other hand, do not bind the ions they transport and thus the flux of ions through a channel does not saturate

C. The membrane potential arises from movements of charge that leave ion concentrations practically unaffected, causing only a very slight discrepancy in the number of positive and negative ions on the two sides of the membrane

Which of the following statement is true?

(1) A only (2) A and B

(3) B and C (4) A and C

62. Consider the transport of K^+ from the blood (where its concentration is about 4 mM) into an erythrocyte that contains 150 mM K+. The transmembrane potential is about 60 mV, inside negative relative to outside. The free-energy change for this transport process is: (These values may be of use to you: R = 8.315 J/mol.K; T = 298 K; 9 (Faraday constant) = 96,480 J/V; N = 6.022×10^{23}/mol)

(1) about 5 J/mol (2) about 15 J/mol

(3) about 5 kJ/mol (4) about 15 kJ/mol

63. You are following the intracellular sorting of an integral plasma membrane protein in a living cell, in culture. You have decided to probe this protein by metabolic labeling technique with ^{35}S-methionine (pulse-chase technique) After one cycle of division, the cells were treated with a potent inhibitor of protein biosynthesis and processed for subcellular fractionation. In which of the following fractions will you expect the immunoprecipitation with a specific antibody?

(1) Only cytoplasm
(2) Only plasma membrane
(3) Both endoplasmic reticulum and plasma membrane
(4) Only secretory vesicles and endoplasmic reticulum

64. The principal pathway for transport of lysosomal hydrolases from the trans Golgi network (pH 6.6) to the late endosomes (pH 6.0) and the recycling of M6P (mannose 6 phosphate) receptors back to the Goigi depends on the pH difference between those two compartments. From what you know about M6P receptor binding and recycling and the pathways for delivery of material to lysosomes, predict what would happen if the pH in late endosomes was raised to 6.6?

(1) M6P will bind to hydrolases but will not release the hydrolases in the late endosomes
(2) M6P will bind to hydrolases and will release the hydrolases in the late endosomes
(3) At higher endosomal pH, the receptor would not release the hydrolase and could not be recycled back to the trans Golgi network
(4) M6P will be degraded at higher pH

65. DNA was isolated from a newly discovered worm collected near a deep-sea vent in the Pacific Ocean. This DNA was sheared into pieces, heated to melting, and then cooled slowly. The amount of renaturation was measured with optical absorbance, and the following results were obtained

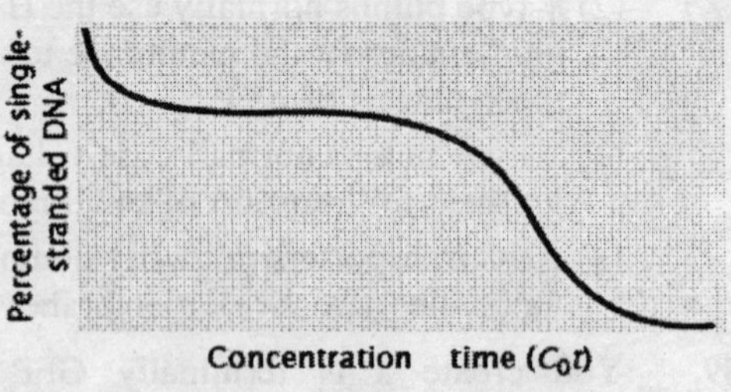

What conclusions can you draw about the type of sequences found in this DNA?

(1) Simple unique sequences
(2) Mixture of unique and repetitive sequences
(3) Highly repetitive sequences
(4) Moderately repetitive sequences

66. A renaturation reaction is carried out on the genomic DNA from three different bacterial species. Species I has a genome size of 2 X 10^6 bp, species II has a genome size of 1 X 10^8 bp, and species III has a genome size of 1 X 10^6 bp. Assume that the same total amount of DNA is used in each renaturation reaction and mark a *Cot* curve for each species, showing the relative positions of each species on the graph below

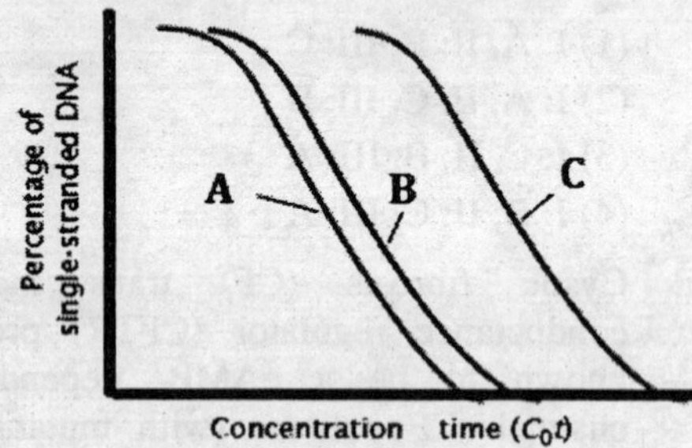

(1) Species I: A; Species II: B; Species III: C
(2) Species I: A; Species II: B; Species III: C
(3) Species I: B; Species II: C; Species III: A
(4) Species I: C; Species II: B; Species III: A

67. Table below represents co-translational and post-translational sorting signals for proteins

Organelle	Protein signal sequence	Receptor for signal
A	C-terminal Lys-Asp-Glu-Leu	HDEL- Receptor or ERD2
B	Internal Import Signal: Pro-Pro-Lys-Lys-Lys-Arg-Lys-Val Export Signal: Leu-Ala-Leu-Lys-Leu-Ala-Gly-Leu-Asp-Ile	Importin and Exportin
C	N-terminal Asn-Pro-Ile-Arg	BP80 or VSR-At1
D	C-Terminal Ser-Lys-Leu	SKL receptor

Identify the correct destination organelles

(1) A: Chloroplast B: Golgi, C: Peroxisome D: Nucleus

(2) A: ER B: Golgi C: Vacuole D: Peroxisome

(3) A: ER B: Golgi C: Nucleus D: Peroxisome

(4) A: ER B: Nucleus C: Vacuole D: Peroxisome

68. A special class of targeting recognition proteins, called SNAREs and Rabs, are involved in the selective fusion and fission of vesicles and tubules that serve as transporters between the compartments of the endomembrane system. Different SNAREs and Rabs are found associated with different membrane systems. SNAREs on the shuttle vesicle, called v-SNAREs, bind with SNAREs on the target membrane, called t-SNAREs. There are many more SNARE genes in plants than animals, and they may not only play a role in membrane fusion, but also membrane-membrane anchoring

Consider the following statements

A. The Rabs are small GTP-binding proteins that provide both energy and directionality to the process. Rabs are activated by specific guanine nucleotide exchange factors (GEFs) and inactivated by GTPase-activating proteins (GAPs)

B. Rab-GTP binds to the guanine nucleotide exchange protein or GEF on the donor organelle membrane, which causes GTP to be exchanged for GDP

C. Rab-GDP then attaches to the donor membrane via an exposed lipid tail

D. v-SNAREs and coat proteins accumulate at the same site, leading to vesicle formation, and the vesicle migrates to the target organelle. After the coat is released the vesicle docks with the complementary t-SNARE on the acceptor organelle membrane. The GTP bound to Rab is hydrolyzed to GDP, which releases Rab-GDP to the cytosol, where it recycles to the donor organelle. The vesicle fuses with its acceptor membrane. Important protein ATPase on the cytoplasmic leaflet of most endomembranes that is involved in breaking the bonds between the v- and t-SNAREs once fusion has occurred

Which of the following statements are correct?

(1) A and D (2) B and C

(3) A and B (4) C and D

69. Purified G-actin polymerization starts slowly, with monomers nucleating into oligomeric structures (the nucleation phase) Oligomeric "seeds" can then elongate during the elongation phase until the F–actin reaches a length dictated by the steady of state of G–actin association and disassociation with the F–actin, the steady-state phase. In the cell, actin filament formation is regulated by several actin-binding proteins (ABPs)

Consider the following statements

A. Profilin, which also charges ADP–actin with ATP, facilitates elongation

B. Formins also act at the plus end, directly adding G–actin while remaining attached to the plus end. They also bind to the sides of actin filaments and initiate growth there, producing bundles of microfilaments that are stabilized by other ABPs, such as fimbrin or villin

C. Actin filaments grow branches through the action of actin–related proteins 2 and 3, which form a complex (ARP 2/3) that binds to the sides of microfilaments and initiates growth at about a forty-five degree angle, while remaining at the filament junction

D. Actin length is dictated by the dissociation of G-actin from the plus and minus ends of F–actin, the protein, severin, can cut F–actin into smaller filaments. F–actin in the center of bundled actin cables is less susceptible to this common means of turnover in the cell

Which of the following statements are correct?

(1) A, B and C (2) A, C and D
(3) B, C and D (4) A, B, C and D

70. A 28 yr old male presents with anemia and low levels of measured erythropoietin due to a high rate of turnover. He is found to have a mutation in the erythropoietin gene resulting in an Asn → Asp amino acid substitution. Of the following options, what is the most likely cause of his condition?

(1) mislocalization due to a change in targeting signal
(2) lack of O-glycosylation
(3) improper protein hydroxylation
(4) lack of N-glycosylation

71. The proteins which take basic stain are known as the basic proteins. The most important basic proteins of the nucleus are nucleoprotamines and the nucleohistones. The nucleoprotamines are simple and basic proteins having very low molecular weight (about 4000 daltons) The most abundant amino acid of these proteins has the pH 10 to 11. The protamines usually remain bounded with the DNA molecules by the salt linkage. The protamines occur in the spermatozoa of the certain fishes. The nucleohistones have high molecular weight, i.e., 10,000 to 1,00,000 daltons. The histone proteins remain associated with the DNA by the ionic bonds and they occur in the nuclei of most organisms. According to the composition of the amino acids following types of histone proteins have been recognized, as

(1) Tyrosine, Threonine and Histidine
(2) Lysine, Alanine and Serine
(3) Arginine, Lysine and Histidine
(4) Proline, Serine and Threonine

72. Drugs like nigericin can artificially increase the electric potential across inner mitochondrial membranes relative to the matrix. Although nigericin treatment increases the rate and extent of translocation of matrix proteins, the rate of translocation reaches a plateau at a lower drug concentration than does the drug-induced change in membrane potential, as shown in Figure

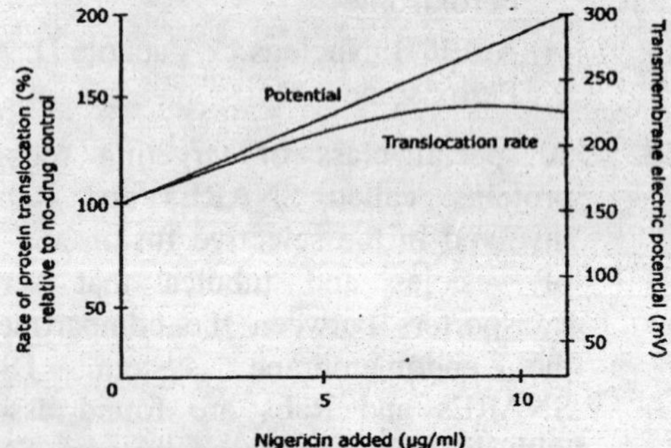

Explain these data based on your knowledge of the mechanisms in protein translocation to the mitochondrial matrix

A. Translocation of proteins across the inner mitochondrial membrane is dependent on the membrane potential

B. The rate of translocation is limited by the number of receptors on the outer membrane

C. Translocation of proteins across the inner mitochondrial membrane is independent of the membrane potential

(1) only A (2) A and B
(3) B and C (4) only C

73. Why the microsome obtained from sonication of inner membrane of mitochondria are preferred to investigate mechanism for electron transport chain and ATP synthesis as compare to intact mitochondria?

(1) It is easier to control the concentration of NAD, FAD, ADP etc in surrounding of microsome as compare to matrix of mitochondria

(2) The individual enzyme of inner membrane can be studied in microsomes

(3) Intact mitochondria do not survive outside cell

(4) Mitochondria cannot be purified

74. Cystic fibrosis is due to non-functional Cystic Fibrosis Transmembrane Conductance Regulator (CFTR) transporter for chlorine. To test this both wild type and mutated type CFRT proteins were embedded in membrane of liposome in absence of protease or denaturants. It was observed that neither wild type nor mutant transporter on liposome were able to uptake chlorine from surrounding. The possible reason is

(1) CFTR inserted in invert topology in membrane of liposome

(2) Wild type CFTR protein got mutated

(3) CFTR lost affinity for chlorine

(4) CFTR lost its functional conformation

75. You have identified a sequence (T-E-X-A-S) that you think might be a signal involved in importing proteins to the peroxisome. Which of the following experiments would test whether the peptide sequence (T-E-X-A-S) was sufficient for transport to the peroxisome (Remember the distinction between necessary and sufficient)

(1) Deleting the amino acids T-E-X-A-S from a peroxisomal protein and finding that the protein is no longer transported into the peroxisome

(2) Changing the amino acids to S-D-L-G-A and finding that the protein is no longer transported into the peroxisome

(3) Comparing the sequence of peroxisomal proteins and finding that the majority of them contain T-E-X-A-S in their amino acid sequence

(4) Adding the sequence T-E-X-A-S to a cytosolic protein and finding that the cytosolic protein is transported into the peroxisome

76. For many experiments, it is desirable to have a population of cells that are traversing the cell cycle synchronously. One of the first, and still often used, methods for synchronizing cells is the so-called double thymidine block. When high concentrations of thymidine are added to the culture fluid, cells in S phase stop DNA synthesis, though other cells are not affected. The excess thymidine blocks the enzyme ribonucleotide reductase, which is responsible for converting ribonucleotides into deoxyribonucleotides. When this enzyme is inhibited, the supply of deoxyribonucleotides falls and DNA synthesis stops. When the excess thymidine is removed by changing the medium, the supply of deoxyribonucleotides rises and DNA synthesis resumes normally

For a cell line with a 22-hour cell cycle divided so that M phase = 0.5 hour, G_1 phase = 10.5 hours, S phase = 7 hours, and G_2 phase = 4 hours, a typical protocol for synchronization by a double thymidine block would be as follows

1. At 0 hours (t= 0 hours) add excess thymidine
2. After 18 hours (t = 18 hours) remove excess thymidine
3. After an additional 10 hours (t= 28 hours) add excess thymidine
4. After an additional 16 hours (t = 44 hours) remove excess thymidine

At what point in the cell cycle is the cell population when the second thymidine block is removed?

(1) G_1 phase (2) S phase

(3) G_2 phase (4) M phase

77. You want to study the cell cycle in fish embryo cells, in which the cell cycle length is about 30 min. You examine the amount of M-cyclin in these cells and see that it varies over the cell cycle. You add a drug, called MI575, to cells, examine M-cyclin, and get the results shown in the figure below

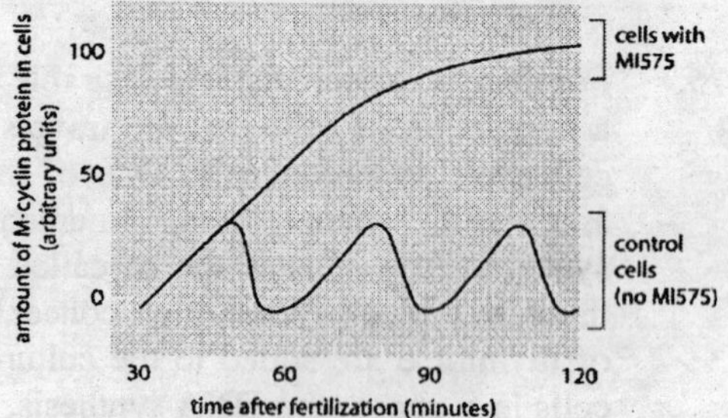

From these data, would you predict that cells treated with MI575 are arrested in the cell cycle at which stage?

(1) G_1 (2) G_2
(3) S (4) M

78. You use FACS to measure the DNA content of three samples: a cell preparation from mouse testis with all the testis cell types, isolated prophase I arrested mouse oocytes, and mammalian cell culture. Unfortunately you forgot to label the tubes

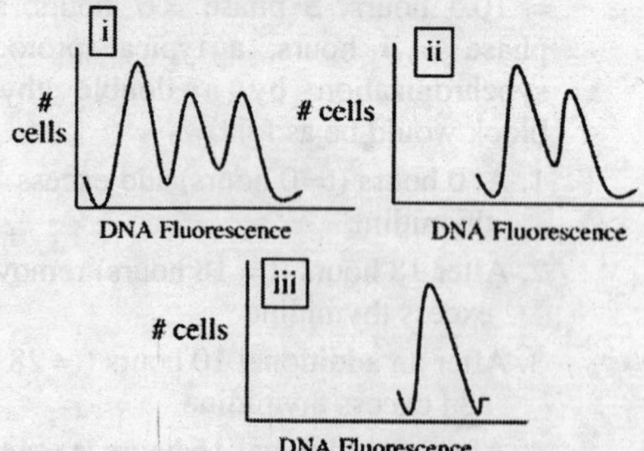

Which of the three FACS profiles below is most likely to correspond to each of the three samples?

(1) (i) Cell mainly in G1 stage; (ii) Cells in S-phase or G_2+M stage of cell cycle; (iii) Cells arrested at G_2+ M stage

(2) (i) Cells in S-phase or G_2+M stage of cell cycle (ii) Rapidly dividing unsynchronized cell line (iii) Cells arrested at G_2+ M stage

(3) (i) Rapidly dividing unsynchronized cell line (ii) Cells arrested at G_2+ M stage (iii) Cells in S-phase or G_2+M stage of cell cycle

(4) (i) Rapidly dividing unsynchronized cell line (ii) Cells in S-phase and G_2+M stage of cell cycle; (iii) Cells arrested at G_2+ M stage

79. Progression through the fission yeast cell cycle requires the phosphorylation and dephosphorylation of critical cdc2 residues. During G2, the cdc2 kinase interacts with a mitotic cyclin but remains inactive as the result of phosphorylation of a key tyrosine residue (Tyr 15 in fission yeast) by 'A' (step 1)

A separate kinase, called CAK, transfers a phosphate to another residue (Thr 161), which is required for cdc2 kinase activity later in the cell cycle. When the cell reaches a critical size, an enzyme called 'B' is activated, which removes the inhibitory phosphate on the Tyr 15 residue. The resulting activation of the cdc2 kinase drives the cell into mitosis (step 2) By the end of mitosis (step 3), the stimulatory phosphate group is removed from Thr 161 by another phosphatase. The free cyclin is subsequently degraded by 'C', and the cell begins another cycle

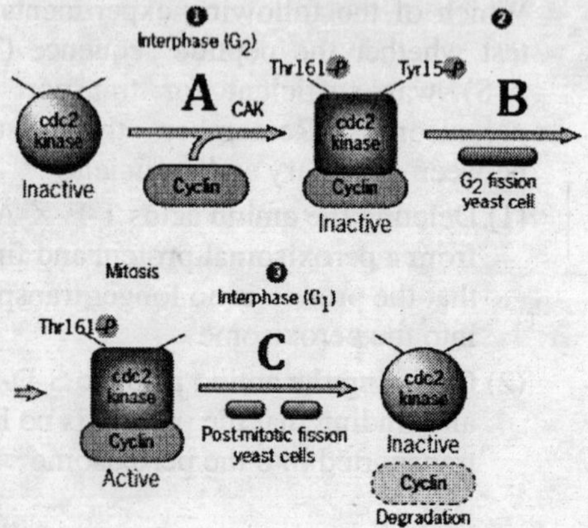

Identify A, B and C

(1) A: Wee 1, B: APC C: cdc25

(2) A: Cyclin B, B: Weel C: APC

(3) A: Cdc2. B: Wee 1 C: APC

(4) A: Weel, B: cdc25, C: APC

80. The roles of the various cyclin–Cdk complexes have been determined by a wide range of biochemical studies carried out on mammalian cells for more than two decades. Over the past few years, the roles of these proteins have been reexamined in knockout mice, with some surprising results. As expected, the phenotype of a particular knockout mouse depends on the gene that has been eliminated

Consider the following statements

A. Mice that are unable to synthesize Cdk1, cyclin B1, or cyclin A2, die as early embryos, suggesting that the proteins encoded by these three genes are essential for a normal cell cycle

B. A mouse embryo that lacks the genes encoding all four of the other cell cycle Cdks (namely, Cdks 2, 3, 4, and 6) is capable of developing to a stage with fully formed organs, although the animal does not survive to birth

C. As in yeast, Cdkl is the only Cdk required to drive a mammalian cell through all of the stages of the cell cycle. In other words, even though the other Cdks are normally expressed at specific times during the mammalian cell cycle, Cdk1 is able to "cover" for their absence, ensuring that all of the required substrates are phosphorylated at each stage of the cell cycle

D. Absence of one of these "nonessential" cyclins or Cdks typically results in distinct cell cycle abnormalities, at least in certain types of cells. Mice lacking a gene for cyclin D1, for example, are smaller than control animals, which stems from a reduction in the level of cell division throughout the body. In addition, cyclin D1-deficient animals display a particular lack of cell proliferation during development of the retina

E. Mice lacking Cdk4 develop without insulin-producing cells in their pancreas. Mice lacking Cdk2 appear to develop normally but exhibit specific defects during meiosis, which reinforces the important differences in the regulation of mitotic and meiotic divisions

Which of the following are correct observations?

(1) A, B, C and D

(2) B, C, D and E

(3) A, D and E

(4) A, B, C, D and E

81. Anaphase promoting complex, or APC, plays a key role in regulating events that occur during mitosis. The APC contains about a dozen core subunits, in addition to an "adaptor protein" that plays a key role in determining which proteins serve as the APC substrate. Two alternate versions of this adaptor protein—Cdc20 and Cdh1—play an important role in substrate selection during mitosis

APC complexes containing one or the other of these adaptors are known as APC^{Cdc20} or APC^{Cdh1} as shown in figure

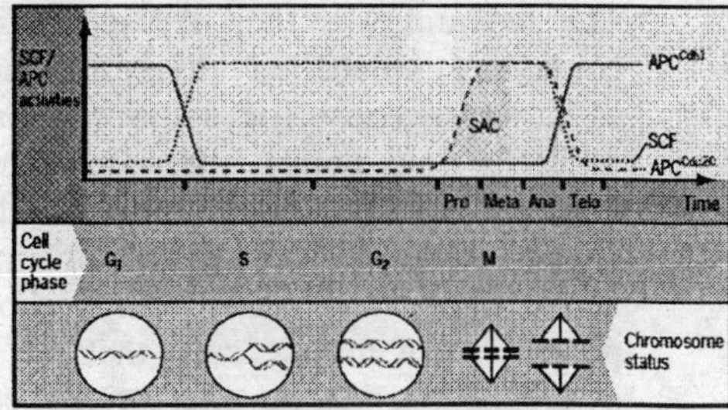

Consider the following statements

A. APC^{Cdh1} becomes activated prior to metaphase and ubiquitinates a major anaphase inhibitor called securing. The ubiquitination and destruction of securin at the end of metaphase release an active protease called separase. Separase then cleaves a key subunit of the cohesin molecules that hold sister chromatids together. Cleavage of cohesin triggers the separation of sister chromatids to mark the onset of anaphase

B. Near the end of mitosis, Cdh1 is inactivated, and the alternate adaptor, Cdc20, takes control of the APC's substrate selection. When Cdc20 is associated with the APC, the enzyme completes the ubiquitination of cyclin B

C. Destruction of the cyclin leads to a precipitous drop in activity of the mitotic Cdk (cyclin B–Cdk1) and progression of the cell out of mitosis and into the G1 phase of the next cell cycle

D. If the destruction of cyclin B is prevented with an inhibitor of the proteasome, cells remain arrested in a late stage of mitosis. If such cells that are arrested in mitosis are subsequently treated with a compound that inhibits the kinase activity of Cdk1, the cell will return to its normal activities and continue through mitosis and cytokinesis

Which of the following statement is correct?

(1) A and B (2) A and C
(3) C and D (4) B and D

82. In a given experiment the cells were labeled for 30 minutes with radioactive thymidine. The medium was then replaced with that containing unlabelled thymidine and the cells were grown for additional time. At different time points after replacement of medium the fraction of mitotic cells were analysed. Based on the results obtained, the above figure was drawn which shows the percentage of mitotic cells that are labeled as a function of time after brief incubation with radioactive thymidine

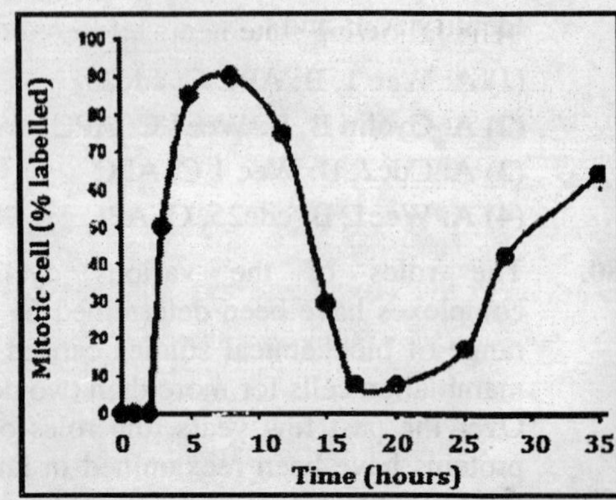

Considering the above experiment, the following statements were made

A. Cells in the S-phase of the cell cycle during the 30 minute labeling period contain radioactive DNA

B. It takes about 20 hours before the first labeled mitotic cells appear

C. The cells enter the second round of mitosis at t_{30} hours

D. The total length of the cell cycle is about 17 hours with G_1, being more than 8 hours

Which of the combination of above statements is correct?

(1) A and B (2) B and C
(3) C and D (4) A and D

83. Microbes produce either primary or secondary metabolites during fermentation. A metabolite production curve is shown below

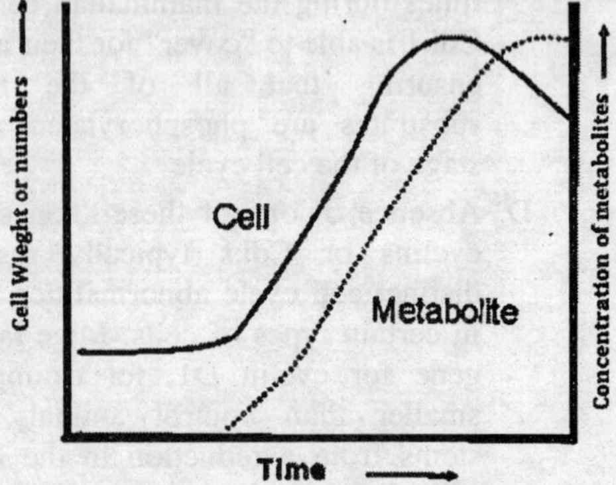

The following statements refer to the above figure

A. A primary metabolite has a production curve that lags behind the line showing cell growth

B. A primary metabolite is produced after the Trophophase is completed

C. A secondary metabolite is produced mainly during Idiophase

D. The curve shows the production of Penicillin from mold

Which of the above statements are correct?

(1) A and B (2) C and D

(3) A and C (4) B and D

84. Cell that grow and divide in a medium containing radioactive thymidine covalently incorporate the thymidine into their DNA during S phase. Consider a simple experiment in which cells are labelled by a brief (30 minutes) exposure to radioactive thymidine. The medium is then replaced with one containing unlabeled thymidine and the cells grow and divide for some additional time. At different time points after replacement of the medium, cells are examined under a microscope. Cells in mitosis are easy to recognize by their condensed chromosomes and the fraction of mitotic cells that have radioactive DNA can be estimated by autoradiography and plotted as a function of time after the thymidine labeling as in the figure below

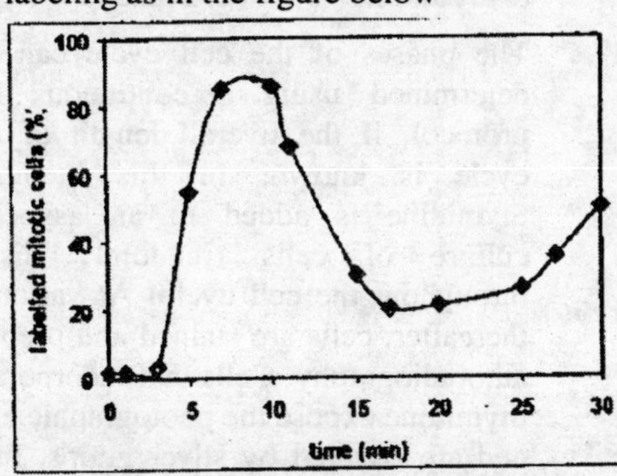

The rise and fall of the curve is because

(1) initial rise of the curve corresponds to cells that were just finishing DNA replication when radioactive thymidine was added (S phase)

(2) the peak of the curve corresponds to cells in M phase

(3) the rise in curve after 20 min corresponds to cells in apoptotic phase

(4) the fall in curve after 10 min indicates the cells exiting M phase

85. A rapidly growing bacterial species such as E. coli exhibits a typical phase of growth cycle in liquid nutrient broth (Iagphase → log phase → stationary phase →death phase) If a bacterial culture has starting density of 10^3 cells/ml has a lag time of 10 minutes and a generation time of 10 minutes, what will the cell density be at (cells/ml) 30 minutes?

(1) 6.0 X 10^3 (2) 2.0 X 10^3

(3) 3.0 X 10^3 (4) 4.0 X 10^3

86. Regulation of the cell cycle involves processes crucial to the survival of a cell, including the detection and repair of genetic damage as well as the prevention of uncontrolled cell division. The molecular events that control the cell cycle are ordered and directional; that is, each process occurs in a sequential fashion and it is impossible to "reverse" the cycle. Two key classes of regulatory molecules, cyclins and cyclin-dependent kinases (CDKs), determine a cell's progress through the cell cycle. Regulation of the eukaryotic cell cycle includes

(1) A G_0 checkpoint to decide whether there are sufficient deoxynucleotides for DNA replication to begin

(2) A G_1 checkpoint where the DNA is "evaluated" and repaired, if possible, before beginning DNA replication

(3) An S phase checkpoint where DNA replication is terminated if too much damage is encountered

(4) A G_2 checkpoint to determine whether all the chromosomes are condensed

87. If a small amount of a lipid with a fluorescent-labeled acyl chain is added to a membrane, the amount of acyl-chain movement can be determined by a technique called fluorescence anisotropy. The less acyl-chain motion in a membrane, the higher its fluorescence anisotropy. For example, the fluorescence anisotropy of a gel-state membrane is about 0.35, while that of a fluid-state membrane is about 0.10. Figure shows the fluorescence anisotropy of two preparations of bacterial membranes as a function of temperature. The membrane were isolated from two cultures of the same bacterial strain, one of which was grown at 37°C and the other at 22°C

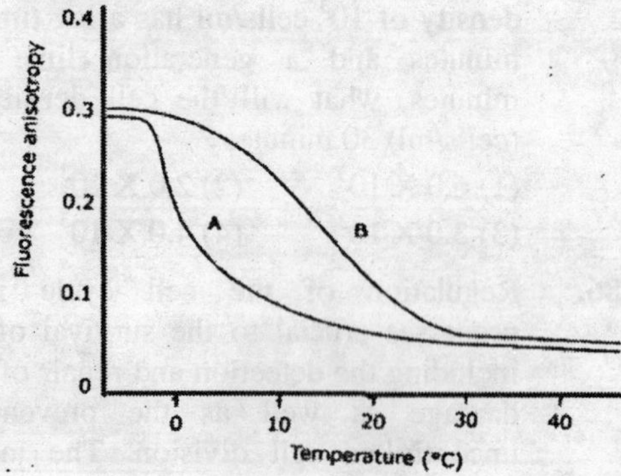

According to this graph, choose the *correct* statements

(1) Curve A represents membrane in the fluid state only at temperature above about 27°C

(2) Membrane lipids of the cells grown at 22°C have a lower transition temperature than the cells grown at 37°C

(3) Curve A represents the membranes of the bacteria grown at 37°C

(4) Membrane phospholipids of the cells grown at 22°C have a higher proportion of saturated fatty acyl side chains

88. The figure below is a graph of the reannealing (reforming of double-stranded molecules) of total genomic DNA from two different organisms. DNA was extracted from each organism, sheared to a uniform size, and then denatured by heating. An identical amount of each preparation of sheared, single-stranded DNA was allowed to reanneal under identical conditions of pH, buffer, ionic strength, and temperature. The reannealing curves shown were obtained by plotting the percentage of DNA remaining single-stranded versus the product of the total DNA concentration (C_O), in moles of nucleotide per liter, times the renaturation time (t), in seconds

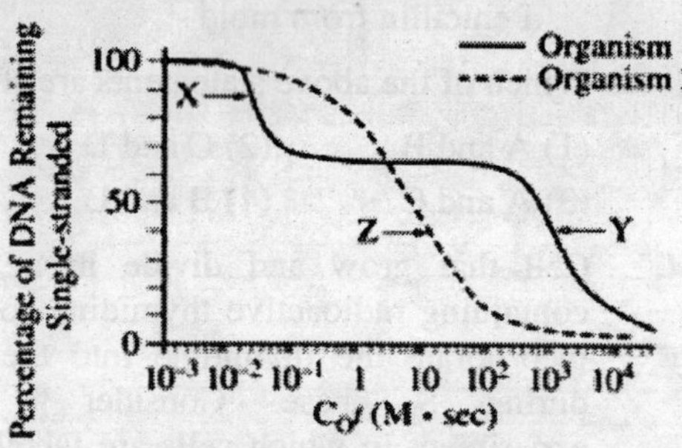

Consider the following conclusion

A. Organism 1 is prokaryotic and organism 2 is eukaryote

B. Organism 1 is eukaryotic and organism 2 is prokaryote

C. Y is more complex than either X or Z

D. X is more complex than either Y or Z

Which of the following is correct conclusion?

(1) A and C (2) B and D

(3) A and D (4) B and C

89. The phases of the cell cycle can also be determined using a continuous labeling protocol, if the overall length of the cell cycle is known. In this method, ^{3}H-thymidine is added to an asynchronous culture of cells (randomly distributed throughout the cell cycle) At various times thereafter, cells are stained and prepared for autoradiography. Cells that incorporated ^{3}H-thymidine expose the photographic emulsion and are covered by silver grains. In Figure A, the fraction of mitotic cells that are labeled is plotted as a function of time after addition of ^{3}H-thymidine. In Figure B, the average number of silver grains above mitotic cells is plotted as a function of time after addition of ^{3}H - thymidine

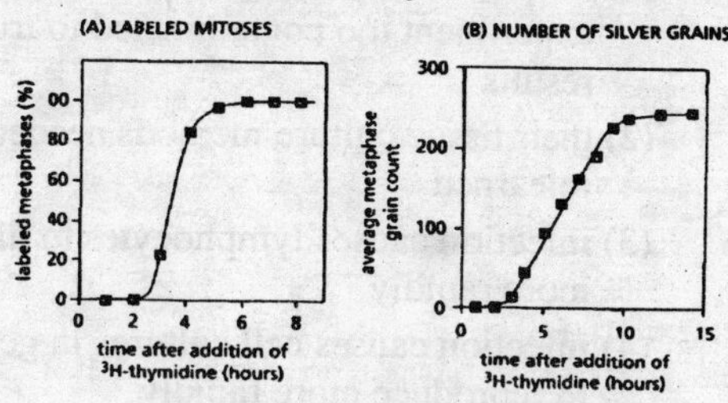

Figure Labeled mitotic cells as a function of time after addition of ^{3}H-thymidine
(A) Fraction of labeled mitotic cells.
(B) Average number of silver grains above labeled mitotic cells.

Assuming the M phase lasts 30 minutes, and that the overall length of the cell cycle was measured at 20 hours, deduce the duration of the G_1, S, and G_2 (in hours)

(1) G_1: 6.5 S: 10 G_2 : 3
(2) G_1:3 S: 6,5 G_2: 10
(3) G_1:18 S:1 G_2:1
(4) G_1:10 S:6.5 G_2:3

90. You have discovered an integral membrane protein that is used to transport the amino acid Tryptophan into a cell. If you measure the rate of transport of Tryptophan into the cell as a function of Tryptophan concentration outside the cell you obtain the following curve

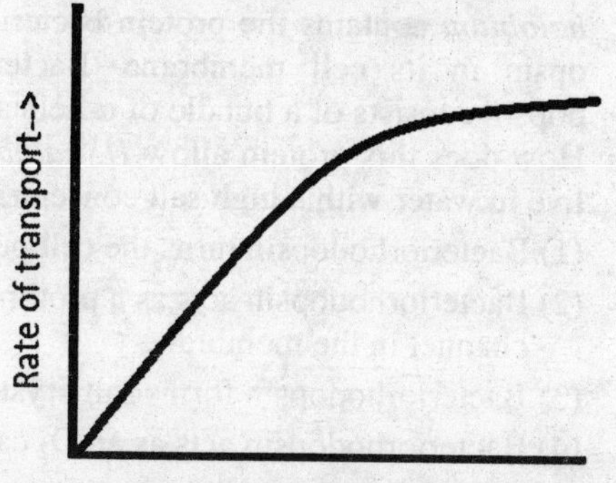

Therefore, you conclude that the Tryptophan transporter is a

(1) Ion channel
(2) Facilitative transporter
(3) Cotransporter
(4) Symporter

91. If you use immunofluorescence microscopy to examine the distribution of the cytoskeletal protein actin, you find it in the cytosol. You express a mutant actin that now ends in the sequence K-D-E-L. If you examine the distribution of mutant actin, where will you find the mutant actin?

(1) The mutant actin is in the cytosol
(2) The mutant actin is in the ER
(3) The mutant actin is in the Golgi
(4) The mutant actin is has been secreted from the cell

92. You have isolated a strain of mutant yeast that divides normally at 30°C but arrests in the cell cycle after S phase at 37°C. You have isolated its mitotic cyclin and mitotic Cdk and find that both are perfectly normal and form a normal MPF complex at both temperatures. Consider the following conditions

A. Inactivation of an enzyme that ubiquinates the MPF cyclin
B. Inactivation of a protein kinase that acts on the MPF Cdk kinase
C. Inactivation of a protein kinase that phosphorylates the MPF cyclin
D. Inactivation of a phosphatase that acts on the MPF Cdk kinase
E. The continuous activation of a phosphatase that removes all phosphate groups from MPF

Which of the following temperature-sensitive mutations could be responsible for the behavior of this strain of yeast?

(1) A, B and C
(2) B, D and E
(3) B, C and D
(4) A, C and E

93. You have cloned a cyclin gene from a mouse and want to find out where in the cell cycle the cyclin acts. To do this you introduce the cloned gene into mutant yeast cells, each of which is mutant in one of the cyclins. The experimental data are given in Table. Where in the mouse cell cycle does your cyclin act?

Yeast mutant cells	Phenotype of yeast cells without mouse cyclin gene	Phenotype of yeast with introduced mouse cycline gene
1.	Cells arrested in G_1 phase	Cells arrested in G_1 phase
2.	Cells arrested in G_2 phase	Cells arrested in G_2 phase
3.	Cells arrested in G_1 phase and smaller than normal	Cells grow to normal size progress through the cell cycle normally

(1) G1 checkpoint
(2) G2 checkpoint
(3) S checkpoint
(4) M checkpoint

94. Place the following events in the order in which they occur when cells stop dividing at a G1 checkpoint after the cells have been irradiated

A. Production of p21
B. DNA damage
C. Inactivation of cyclin-Cdk complex
D. Activation of p53

(1) A→B→C→D
(2) B→D→A→C
(3) B→A→D→C
(4) A→C-→D→A

95. Nucleotides can be radiolabeled before they are incorporated into newly forming DNA and can therefore be assayed to track their incorporation. In a set of experiments, a student-faculty research team used labeled T nucleotides and introduced these into the culture of dividing human cells at specific times. The research team used the setup to study the incorporation of labeled nucleotides into a culture of lymphocytes and found that the lymphocytes incorporated the labeled nucleotide at a significantly higher level after a pathogen was introduced into the culture. They concluded that

(1) the presence of the pathogen made the experiment too contaminated to trust the results
(2) their tissue culture methods needed to be relearned
(3) infection causes lymphocytes to divide more rapidly
(4) infection causes cell cultures in general to reproduce more rapidly

96. Which of the following statements about triosphosphate isomerase (TIM) is not true?

(1) The mechanism of action of TIM involves a conformational change in the structure of the enzyme that prevents escape of an activated intermediate
(2) The rate-limiting step in the reaction catalyzed by TIM is the release of the product glyceraldehyde 3-phosphate
(3) The kcat/KM ratio for the reaction catalyzed by TIM is close to the diffusion-controlled limit for a bimolecular reaction
(4) The isomerization of a hydrogen atom from one carbon atom to another in the TIM-catalyzed reaction is assisted by a base (the g-carboxyl of a glutamate residue) in the enzyme

97. The salt-tolerant bacteria *Halobacterium halobium* contains the protein bacteriorhodopsin in its cell membrane. Bacteriorhodopsin consists of a bundle of α-helical rods. How does this protein allow *H. halobium* to live in water with a high salt concentration?

(1) Bacteriorhodopsin turns the cell purple
(2) Bacteriorhodopsin acts as a proton channel in the membrane
(3) Bacteriorhodopsin forms salt crystals
(4) Bacteriorhodopsin acts as an O_2 carrier

98. Which of the following statements concerning microtubules are correct?

A. Microtubules are filaments composed of α-helical coiled-coil polypeptide chains

B. Microtubules contain α-tubulin and β-tubulin protomers that are disposed in a helical array around a hollow core to form a cylindrical structure

C. The cilia and flagella of eukaryotic cells contain nine microtubule doublets that surround a pair of microtubule singlets

D. The outer microtubules in an axoneme are linked together and to an ATPase called dynein

E. The powered movements of dynein in an axoneme shorten the structure

(1) A, B and C (2) B, C and D

(3) C, D and E (4) B, D and E

99. Cell fusion experiments reveal how constituents of the cytoplasm control the behavior of the nucleus. Consider the following fusion experiment

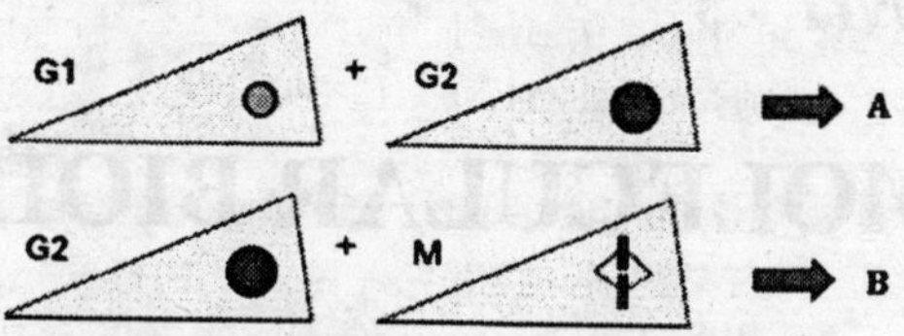

What will be outcome in A and B?

(1) A: Nuclei of G1 will remain in G1 and G2 will remain in G2
B: Both nuclei will move in M phase

(2) A: Both nuclei will move in G2 phase
B: Both nuclei will move in M phase

(3) A: Nuclei of G1 will remain in G1 and G2 will remain in G2
B: Nuclei of G2 will remain in G2 and M will remain in M

(4) A: Nuclei of G1 will remain in G1 and G2 will move to G1
B: Both nuclei will move in M phase

100. Lipid rafts are rich in

(1) Cholesterol and sphingolipid

(2) Phospholipid and sphinogolipid

(3) Cholesterol and Free fattyacids

(4) Free fattyacids and Phospholipid

ANSWER SHEET
Unit 2 : Cell Biology

1. 4	2. 2	3. 2	4. 2	5. 1	6. 3	7. 4	8. 4	9. 1	10. 1
11. 4	12. 2	13. 2	14. 1	15. 2	16. 2	17. 2	18. 2	19. 4	20. 2
21. 4	22. 4	23. 2	24. 2	25. 1	26. 1	27. 1	28. 4	29. 1	30. 1
31. 2	32. 2	33. 1	34. 3	35. 2	36. 1	37. 4	38. 2	39. 3	40. 3
41. 1	42. 4	43. 3	44. 3	45. 1	46. 2	47. 2	48. 4	49. 4	50. 4
51. 2	52. 1	53. 4	54. 3	55. 3	56. 1	57. 1	58. 3	59. 1	60. 3
61. 1	62. 4	63. 2	64. 1	65. 2	66. 3	67. 4	68. 1	69. 4	70. 4
71. 3	72. 2	73. 1	74. 1	75. 4	76. 2	77. 4	78. 4	79. 4	80. 4
81. 3	82. 4	83. 2	84. 1	85. 4	86. 2	87. 2	88. 4	89. 4	90. 2
91. 1	92. 2	93. 1	94. 2	95. 3	96. 3	97. 2	98. 4	99. 1	100. 1

UNIT - 3

MOLECULAR BIOLOGY

1. Presence of circular mRNAs for a specific protein in an eukaryotic cell reflects a rapid rate of synthesis of that protein. Following mechanisms are suggested

   A. eIF-4E, eIF-4G and PABP promote this process through 5'-3' interaction of mRNA

   B. ribosomes are less active in recognizing circular mRNA

   C. PABP - eIF-2 interaction promote this process

   D. ribosomes can reinitiate translation without being disassembled

   Which of the following is correct?

   (1) A and D (2) B and D

   (3) A and C (4) B and C

2. Some Human Hereditary Diseases and Cancers Associated with DNA-Repair Defects

   Group-I

   A. Hereditary nonpolyposis colorectal cancer

   B. Xeroderma Pigmentosum

   C. Hereditary breast cancer, BRCA-1 and BRCA-2 deficiency

   D. Fanconi anemia

   Group-II

   1. Repair of double-strand break by homologous recombination
   2. Nucleotide Excision repair
   3. DNA mismatch repair
   4. Defect in tyrosine kinase receptor

   Which of the following is correct match between group A and group B?

   (1) A-1; B-2; C-3; D-4

   (2) A-3; B-2; C-1; D-1

   (3) A-3; B-2; C-4; D-1

   (4) A-3; B-2; C-1; D-2

3. Consider the following statement regarding genome complexity for eukaryotes

   A. Among eukaryotes, there is a general trend for gene density to increase with increasing organism complexity

   B. increased gene density observed in eukaryotic cells due to increase in gene size and increases in the DNA between genes, called intergenic sequences

   C. About a 1/4 of the intergenic DNA is unique. These regions comprise many apparently nonfunctional relics, including nonfunctional mutant genes, gene fragments, and pseudogenes

   D. Average transcribed regions of a human gene is about 27 kb, whereas the average protein-coding region of a human gene is 1.3 kb

   E. Almost half of the human genome is composed of DNA sequences that are repeated many times in the genome. There are two general classes of repeated DNA: microsatellite DNA and genome-wide repeats

   Which of the following statement is correct?

   (1) A, B and C (2) B, C and D

   (3) C, D and E (4) A, C and E

4. Consider the following statement regarding DNA replication

A. DNA synthesis requires deoxynucleoside triphosphates and a primer: template junction. The template provides the ssDNA that will direct the addition of each complementary deoxynucleotide

B. The primer is complementary to the template have an exposed 5'phosphate adjacent to the single stranded region of the template that will be extended as new nucleotides are added

C. The phosphodiester bond is formed in an SN2 reaction in which the hydroxyl group at the 3' end of the primer strand attacks the α-phosphoryl group of the incoming nucleoside triphosphate

D. The leaving group for the DNA replication reaction is pyrophosphate, hydrolysis of pyrophosphate by pryro-phosphatase is the driving force for DNA synthesis

Which of the following statement are correct?

(1) A, B and C (2) B, C and D
(3) A, C and D (4) A, B and D

5. Consider the following statement regarding DNA polymerase

A. DNA polymerase uses a four different active site to catalyze the addition of the four deoxynucleoside triphosphates

B. rNTPs are incorporated at a rate that is more than 1,000-fold lower than dNTPs. This discrimination is mediated by the steric exclusion of rNTPs from the DNA polymerase active site. Nucleotide binding pocket is too small to allow the presence of a 2'OH on the incoming nucleotide

C. The palm domain of DNA polymerase is composed of a beta sheet and contains the primary elements of the catalytic site and two divalent metal ions (typically Mg^{+2} or Zn^{+2}) that alter the chemical environment around the correctly base-paired dNTP and the 3'OH of the primer

D. All bacterial DNA polymearase carry an associated 3'-->5' exonuclease proof reading activity

Which of the following are correct?

(1) A and B (2) B and C
(3) C and D (4) D and A

6. Consider the following statements regarding replication

A. Each DNA helicase moves along ssDNA in a defined direction. DNA helicases can have a polarity of either 5'-->3' or 3'->5'. This direction is always defined according to the strand of DNA bound. DNA helicase that functions on the lagging strand template of the replication fork, the polarity is 5'→ 3'. Helicases, uses energy in form of ATP hydrolysis

B. The negative supercoils introduced by the action of the DNA helicase are removed by topoisomerases II that act on the unreplicated double-stranded DNA in front of the replication fork

C. Sliding clamps (beta clamp/PCNA) increase DNA polymerase processivity and are opened and placed on DNA by clamp loaders which are coupled with ATP binding and hydrolysis

D. Nucleosome assembly factors, Okazaki fragment repair proteins, and other DNA repair proteins all interact with the same region of the sliding clamp as the clamp loader. Thus, sliding clamps are only removed from the DNA polymease once all the enzymes that interact with them have completed their function

Which of the following statement are correct?

(1) A, B and C (2) B, C and D
(3) A, C and D (4) A, B and D

7. Consider the following statements

A. DNA polymerase III (DNA Pol III) is the minor the enzyme involved in the replication of the bacterial chromosome which is least processive and fidel

B. DNA polymerase I (DNA Pol I) is specialized for the removal of the RNA primers that are used to initiate DNA synthesis using its 5' exonuclease activity. DNA Pol I is not highly processive, adding only 20-100 nucleotides per binding event

C. Eukaryotic cells also have multiple DNA polymerases, with a typical cell having more than 15. Of these, three are essential to duplicate the genome: DNA Pol δ, DNA Pol ε, and DNA Pol α/ primase

D. DNA Pol α/primase is specifically involved in initiating new DNA strands as well as removal of primer at end of replication

Which of the following are not true?

(1) A and B (2) A and D
(3) C and D (4) C and B

8. Consider the following factors regarding initiation of DNA replication

A. The *E. coli* initiator protein, DnaA. DnaA binds the repeated 9-mer elements in oriC and is regulated by ATP. When bound to ATP (but not ADP), DnaA also interacts with DNA in the region of the repeated 13-mer repeats of oriC. In E. coli DNA replication is regulated by DnaA-ATP Levels and SeqA

B. DnaA recruits a complex of two proteins: the DNA helicase, DnaG, and helicase loader DnaB

C. Eukaryotic chromosomes are replicated exactly once per cell cycle by phosphorylating two helicase loading proteins (Cdc6 and Cdtl) intially by S-CDK and later by M-CDK which are not able to inteact with Orc untill G1 stage

D. Bacteria and eukaryotic cells generally have multiple replicons from where replication can be initiated

Which of the following are not true?

(1) A and B (2) A and C
(3) C and D (4) D and B

9. Hybridization between mature mRNA and DNA can identify introns in eukaryotic genes. When mature mRNA produced by the insulin gene is hybridized with denatured chromosomal DNA, which of the following will most likely be observed?

(1) No hybridization would occur under any conditions
(2) Hybridization of mRNA would occur with random sections of chromosomal DNA
(3) Hybridization of mRNA with DNA would occur in a continuous stretch that is equal to the length of the mRNA
(4) Hybridization of mRNA with DNA would occur but with many single-stranded loops of DNA

10. Patients with hereditary non-polyposis colon cancer [HNPCC have genes with microsatellite instability, that is, many regions containing abnormal, small loops of unpaired DNA. This is a result of a mutation affecting

(1) mismatch repair
(2) transcription Coupled repair
(3) base excision repair
(4) tucleotide excision repair

11. An investigator incubates an unprocessed RNA molecule containing an intron in an extract absent of any protein. Correct splicing of the RNA is observed, but only when guanosine is present. Which is incorrect conclusion about the RNA molecule?

(1) It contains a ribozyme
(2) Self-splicing has occurred
(3) It contains a group II intron
(4) It contains a group I intron

12. Which of the following DNA repair mechanisms can repair double-stranded DNA breaks in eukaryotes?

A. Base excision repair
B. Homologous recombination
C. Mismatch excision repair

D. Nonhomologous end joining
E. Nucleotide excision repair

(1) A and B (2) D and E
(3) B and D (4) only B

13. Which one of the following statements correctly describes the splicing process undergone by most eukaryotic pre-mRNAs?

(1) In a two-step reaction, the spliceosome removes the exon as a lariat and joins - the two introns together
(2) Splicing requires conserved sequences, which are the 5'-splice site, the 3'-splice site, the branchpoint and the polypurine tract
(3) The U1 snRNP initially binds to the 5'-splice site, U2 to the branchpoint sequence and then the trisnRNP, U4, U5 and U6 can bind
(4) In the first step of splicing the G at the 3'-end of the intron is joined to the 2'-hydroxyl group of the A residue of the branchpoint sequence to create a lariat

14. The recent sequencing of the genomes of numerous organisms has provided a wealth of information concerning the makeup of chromosomal DNAs and how their characteristics have changed as organisms have increased in complexity

Consider the following statements

A. Prokaryotes mostly have single circular chromosomes, but they may have multiple chromosomes, linear chromosomes, or even both
B. The *E. coli* genome is composed almost entirely of genes. One critical element of the *E. coli* genome is not part of a gene is its origin of replication (*OriC*)
C. Gene density in eukaryotic organisms is consistently higher and less variable than in their prokaryotic counterparts
D. Among eukaryotes, there is still a general trend for gene density to increase with increasing organism complexity

Which of the following statement is correct?

(1) A and B only (2) C and D only
(3) B and C only (4) A, B and D

15. Rho factor is involved in termination of transcription in prokaryotes. Genetic manipulations indicate that Rho-dependent termination requires the presence of a specific recognition sequence on the newly synthesized RNA upstream of the termi-nation site. The recognition sequence must be on the nascent RNA rather than the DNA, as demonstrated by Rho's inability to terminate transcription in the presence of pancreatic RNAse. The essential features of this termination site have not been fully elucidated. Construction of synthetic termination sites indicates that it consists of 80 to 100 nts that lack a stable secondary structure and contain multiple regions that are rich in C and poor in G. Which of the following is/are suggested by the above observation?

(1) Rho factor attaches to nascent RNA at its recognition sequence and then migrates along the RNA in the $5^1 \rightarrow 3^1$ direction until it encounters an RNAP paused at the termination site
(2) Rho unwinds the RNA-DNA duplex forming the transcription bubble, thereby releasing the RNA transcript
(3) Rho factor attaches to the RNA at its recognition sequence while RNA is in the RNA-DNA hybrid condition
(4) There may be other factors and hence Rho factor does not need to unwind the RNA-DNA hybrid to release the transcript

16. DNA repair, synthesis and recombination are intimately connected and inter dependent. An apparent commonality between processes of DNA replication and repair in the enzymatically catalyzed synthesis of DNA polynucleotide segments, which can be assembled with preexisting poly-nucleotides, leading to repair or replication. Synthesis of these polynucleotide segments is catalyzed by a group of enzymes DNA-dependant DNA polymerases. In the case of

E. coli, DNA polymerase has been isolated in three distinct forms whereas five main types of polymerase have been isolated from mammalian cells. All the polymerases synthesize polynucleotides only in the 5'→3' direction. If polynucleotide chains could be elongated in 3→5' direction, the hypothetical growing 5' terminus, rather than the incoming nucleotide, would carry a triposphate that is unsuitable for further elongation. The 3' →5' exonuclease activity is not associated with all the polymerases and only present in

A. All *E. coli* DNA polymerases but not all mammalian polymerases
B. Pol I, Pol II, Pol III, Pol α, Pol β
C. Pol I, Pol II, Pol δ, Pol ε, Pol γ
D. Pol I, Pol II, Pol α, Pol ε

The correct statements are

(1) A and B
(2) A, B and C
(3) A and C
(4) A, C and D

17. When synthetic mRNA consisting of alternating A and C residues (ACACAC....) was translated in a bacterial extract, only one kind of polypeptide consisting of alternating threonine and histidine residues was made. When the base sequence was ...AACAACAAC.... three different polypeptides were formed : poiyasparagine, polythreonine and polyglutamine. Hence we may conclude that one codon for histidine is

(1) ACA (2) CAC
(3) AAC (4) CAA

18. During development and differentiation, there is a dynamic programme of differential expression of sets of genes. In bacteria, phage infections are among the simplest examples of developmental process. Typically, only a subset of the phage genome, offer referred to as immediate early genes, are expressed in the host immediately after phage infection. As time passes, early genes start to be expressed, and the immediate early genes and bacterial genes are turned off. In the final stage of phage infection, the early genes give way to late genes. One of the simplest way it is achieved is through

A. expression of cascade of σ factors
B. expression of new RNA polymerases
C. expression of different holoenzymes
D. expression of different transcription factors

The correct reasons are

(1) A and D
(2) A, C and D
(3) A, B and D
(4) A, B and C

19. A 10-year-old Caucasian girl is brought in by her parents for evaluation of a skin disorder. The child has many freckles on her face, arms, and legs. The parents were told by previous physicians that she suffers from Xeroderma pigmentosum and that they should limit her exposure to sun light. What is the most likely etiology of this disorder?

(1) Deficient DNA replication
(2) Impaired DNA repair by nucleotide excision
(3) Defective RNA transcription from TATA-less promoter
(4) Impaired ribosomal protein translation

20. The *E. coli* arabinose operon (*ara* gene) encodes three enzymes that convert the sugar arabinose into xylulose 5-phosphate, an intermediate in the pentose phosphate pathway. Which of the following mechanisms would most likely be used to produce the three different enzymes from a single RNA transcript?

(1) Three separate open reading frames on a single mRNA
(2) Alternative splicing leading to three different mRNAs
(3) Site-specific recombination generating three different versions of the *ara* gene
(4) Site-specific cleavage of a long polypeptide into three parts, each with distinct enzyme activity

21. Which of the following events would occur at the *E. coli* lac operon when the glucose concentration of the growth medium is low and the lactose concentration is high?

(1) The *lac* repressor protein will bind to the operator sequence

(2) Ribosomes will stall during translation of the first 13 amino acids

(3) The transcribed RNA will form a cAMP-dependent stem-loop structure to terminate transcription

(4) The catabolite gene activator protein (CAP), bound with cAMP, will stimulate binding of RNA polymerase to the promoter

22. A strain of *E. coli* has the following genotype with respect to the lac operon: $I^- O^+ Z^+ Y^- A^+$. This strain will

(1) produce permease only in the presence of lactose

(2) not produce beta galactosidase if lactose is present in the medium

(3) not produce permease irrespective of the presence or absence of lactose in the medium

(4) produce trans-acetylase only when induced

23. In a reaction containing dNTPs and DNA polymerase in buffer, on which of the following molecules DNA synthesis would occur

(1) a single-stranded closed circular DNA

(2) a double-stranded closed circular DNA

(3) a single-stranded closed circular DNA paired with a short oligonucleotide with a 3' terminal phosphate

(4) a single-stranded closed circular DNA paired with a short oligonucleotide with a 3' hydroxyl group

24. In cells that have not been previously exposed to lactose, how does lactose enter the cells to induce lac operon

(1) IPTG would induce the expression instead of lactose

(2) Lactose would enter by diffusion across the membrane without the need for permease

(3) Some-other transport protein facilitates the entry of lactose

(4) Basal level of expression of lac operon, leading to the low level synthesis of permease

25. Termination codons can be readily formed by single base changes from many other codons. Mutations of this type, termed nonsense mutations. Premature termination codons, are also commonly introduced into mRNAs during splicing. Cells possess an mRNA surveillance mechanism capable of detecting messages with premature termination codons. In most cases, mRNAs containing such mutations are translated only once before they are selectively destroyed and protects the cell from producing nonfunctional, shortened proteins. This surveillance mechanism is termed as

(1) Non-sense-mediated decay

(2) Non-stop mediated decay

(3) tm-RNA mediated release

(4) Spontaneous decay due to exon-junction complex

26. Suppose you mix the following components of protein synthesis in a test tube: amino acids from a rabbit; ribosomes from a dog; tRNAs from a mouse; mRNA from a chimpanzee; and necessary enzymes plus an energy source from a giraffe. If protein synthesis occurs, which animal's protein will be made?

(1) Rabbit

(2) Dog

(3) Mouse

(4) Chimpanzee

27. The diagram below shows the kinetics of β-galactosidase induction in *E. coli*

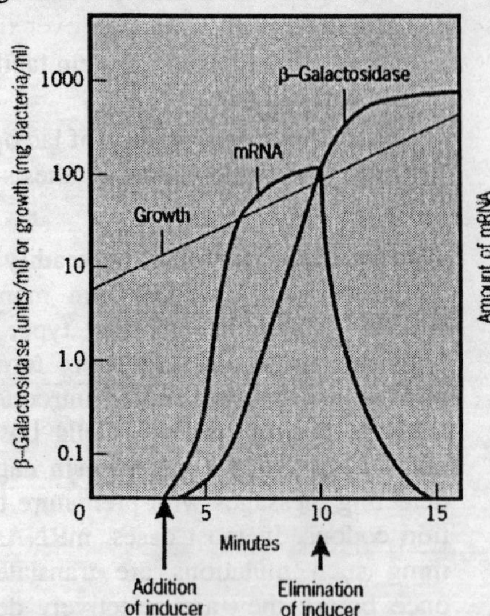

Which of the following statement is incorrect?

(1) When an appropriate inducer (a β-galactoside) is added, the production of mRNA for the enzyme β-galactosidase begins very rapidly, followed within a minute or so by the appearance of the enzyme, whose concentration increases rapidly

(2) Removal of the inducer leads to a precipitous drop in the level of the mRNA, which reflects its rapid degradation

(3) The amount of enzyme then levels off because new molecules are no longer synthesized

(4) Growth of bacteria after addition or removal of inducer is not affected at all

28. Describe the expected polypeptide composition resulting from a regular copolymer experiment, where the tetranucleotide CAUA was used. Given AUA, Ile; UAC, Tyr; ACA, Thr; CAU, His

(1) His- Thr-Tyr-Ile (2) He-His-Thr-Tyr

(3) Tyr-Ile-His-Thr (4) Thr-Tyr-Ile-His

29. Lac operon in *E. coli* is negatively regulated by lac repressor and positively by CAP protein. In an experiment following condition were observed for lac operon

(i) c-AMP levels were high

(ii) CRP/CAP was interacting with RNA polymerase

(iii) Repressor is bound with allolactose

Which of the following is most appropriate condition in experimental medium?

(1) Both glucose and lactose is present

(2) Both glucose and lactose is absent

(3) Glucose is present and lactose is absent

(4) Glucose is absent and lactose is present

30. When an uncharged transfer RNA finds its way into the A site of the ribosome, a likely event under amino acid starvation, it causes an idling reaction in the ribosome, which entails the production of the nucleotide guanosine tetraphosphate (5'-ppGpp-3') This is part of a control mechanism called the stringent response

Consider the following statement

A. A protein called the stringent factor, the product of the relA gene, produces guanosine tetraphosphate (ppGpp), originally called magic spot because of its sudden appearance on chromatograms (The gene is called rel from the relaxed mutant, which does not have the stringent response)

B. The stringent factor is associated with the ribosome, where ppGpp is synthesized, which is one of the structural proteins of the ribosome

C. The SpoT protein, the product of the spoT1 gene, breaks down ppGpp; thus, the concentration of ppGpp is regulated

D. The ppGpp interacts with RNA polymerase, causing an almost complete cessation of the transcription of ribosomal RNA; thus, no energy is wasted synthesizing ribosomes when translation is not possible

E. Many amino acid-synthesizing operons require ppGpp for transcription; ppGpp thus inhibits ribosomal RNA production and enhances the production of enzymes to synthesize amino acids, all when the cell is starved for amino acids

Which of the following statement is correct?

(1) A and D (2) B, C and E

(3) A, C, D and E (4) A, B, C and D

31. What are the phenotypes of the following partial diploids for the lac operon in *E. coli* in the presence and absence of lactose?

(F') $I^+ O^c P^+ Z^+ / I^+ O^+ P^- Z^+$ (Chromosome)

(1) Presence of lactose: On; Absence of lactose: off

(2) Presence of lactose: On; Absence of lactose: on

(3) Presence of lactose: Off; Absence of lactose: off

(4) Presence of lactose: off; Absence of lactose: on

32. Escherichia coli cells grown in a medium with lactose as the only carbon source are monitored for β- galactosidase activity over time with the results shown below

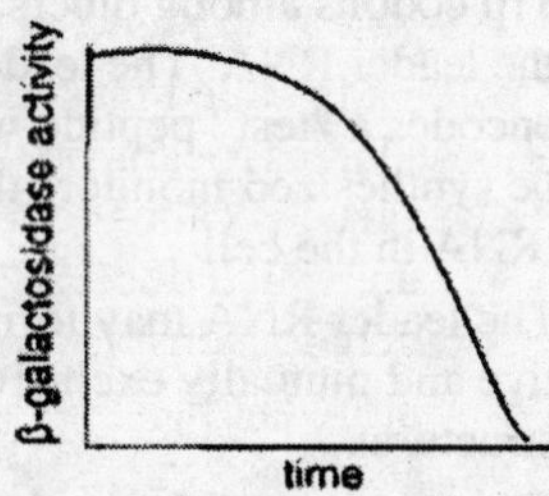

Which intracellular event would most likely be associated with the change in enzyme activity observed?

(1) Decreased synthesis of cytosolic cAMP

(2) Activation of a repressor protein by lactose

(3) Dissociation of a cAMP-CAP complex from a CAP-binding sequence

(4) Binding of a repressor protein to an operator sequence

33. A culture of E. coli is grown in a medium containing glucose and lactose. The expression of the lactose operon over time in the cells is shown in the graph below. Which statement best describes the change that occurred at point A?

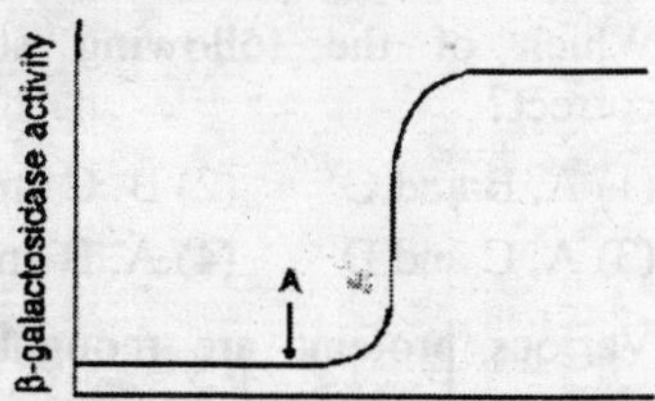

(1) Lactose was added to the culture

(2) cAMP concentration increased in the cells

(3) Repressor protein dissociated from the operator

(4) Repressor protein became bound to the operator

34. The elongating RNA polymerase is a processive machine that synthesizes and proofreads RNA. Consider the following statements

A. 5'→3' exonuclease activity

B. 3'→5' exonuclease activity

C. Pyrophosphorolytic editing

D. Hydrolytic editing

Which of the following strategies are used for proof reading by RNA polymerase?

(1) A and B (2) B and C

(3) C and D (4) B and D

35. Consider the following statements for eukaryotic transcription

A. The large subunit of eukaryotic RNA Pol II has a C-terminal domain (CTD) contains a series of repeats of the heptapeptide sequence: Tyr-Ser-Pro-Thr-Ser-Pro-Ser

B. TBP uses an extensive region of β- sheet to recognize the major groove of the TATA element

C. Within TFIIH are two subunits that function as ATPases (Helicase), and another that is a protein kinase, with roles in promoter melting and escape

D. TFIIS and hSPT5 are elongation factors--that is, factors that stimulate elongation

Which of the following statements are correct?

(1) A, B and C (2) B, C and D
(3) A, C and D (4) A, B and D

36. Various proteins are thought to stimulate elongation by eukaryotic RNA Pol II. Consider the following statements

A. Kinase P-TEFb, is recruited to polymerase by transcriptional activators, this protein phosphorylates the serine residue at position 2 of the CTD repeats

B. P-TEFb phosphorylates and thereby activates protein, called hSPT5, itself an elongation factor. hSPT5 recruits and stimulates the 5' capping enzyme

C. P-TEFb recruits TAT-SF1 elongation factor. TAT-SF1 recruits components of the splicing machinery

D. TFIIH stimulates the overall rate of elongation by limiting the length of time polymerase pauses when it encounters sequences that would otherwise tend to slow the enzyme's progress and also contributes to proofreading by polymerase

Which of the following statements are correct?

(1) A, B and C (2) B, C and D
(3) A, C and D (4) A, B and D

37. Consider the following statements

A. tRNAs are charged by the attachment of an amino acid to the 3' terminal adenosine nucleotide via a high-energy acyl linkage

B. aminoacyl t-RNA synthetases charge tRNAs in two steps. Step one is adenylylation in which the amino acid reacts with GTP and step two is tRNA charging in which the adenylylated amino acid, reacts with tRNA

C. Most organisms have 35 different tRNA, each aminoacyl tRNA synthetase attaches a single amino acid to only one tRNAs

D. tRNA Synthetases recognize acceptor arm & anticodon loop of cognate tRNAs

E. Some aminoacyl tRNA synthetases use an editing pocket besides catalytic pocket to charge tRNAs with high accuracy

Which of the following statements are correct?

(1) A, B and C (2) B, C and D
(3) A, D and E (4) A, B and E

38. Which of the following statements concerning the trp operon leader RNA, which has 162 nucleotides preceding the initiation codon of the first structural gene of the operon, are not correct?

(1) A deletion mutation in the DNA encoding the 3' region of the leader RNA gives rise to decreased levels of the biosynthetic enzymes forming Trp

(2) A short open reading frame, containing Trp codons among others, exists within the leader RNA. The leader RNA encodes a "test" peptide whose ability to be synthesized monitors the level of Trp-tRNA in the cell

(3) The leader RNA may form two alternative and mutually exclusive secondary structures

(4) The structure of the leader RNA in vivo depends on the position of the ribosomes translating it

39. When lactose is used as an inducer a lag occurs before the enzymes of the lactose operon are synthesized. With IPTG synthesis starts without a lag

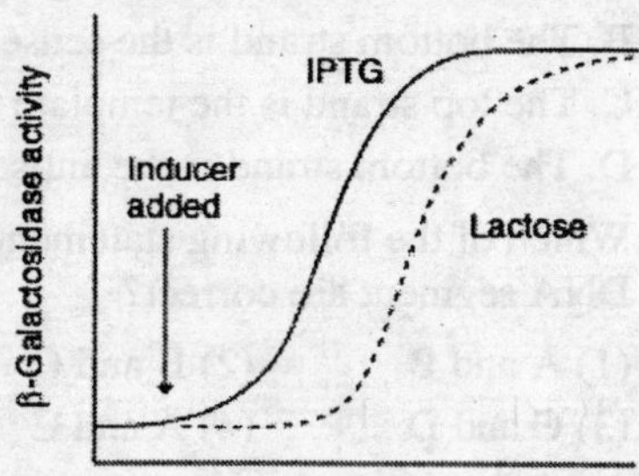

(1) The lag represents the time it takes for lactose to enter into bacteria. IPTG itself directly acts as an inducer by binding at membrane receceptor. Therefore, no lag is observed

(2) The lag represents the time it takes for lactose to be converted into 1,6-allolactose by residual β-galactosidase. IPTG itself directly acts as an inducer. Therefore, no lag is observed

(3) The lag represents the time it takes for lactose to be converted into 1,6-allolactose by residual β-galactosidase. IPTG itself directly acts as an repressor. Therefore, no lag is observed

(4) The lag represents the time it takes for IPTC to be converted into 1,6-allolactose by residual β-galactosidase. Lactose itself directly acts as an inducer. Therefore, no lag is observed

40. In order to prove that regulation by attenuation occurs in vivo, Charles Yanofsky and others studied tryptophan synthesis regulation in a series of *E. coli* mutants. Consider the following mutants

- Mutant A: mutations with decreased, but detectable, tryptophan-tRNA synthetase activity
- Mutant B: mutations in AUG or Shine-Dalgarno sequence of leader peptide sequence
- Mutant C: same as mutant B but with the leader peptide fully expressed on a plasmid
- Mutant D: mutations replacing the two Trp codons with Leu codons
- Mutant E: mutant E with a mutation in the Leu-tRNA synthetase gene
- Mutant F: mutant E that constitutively synthesizes leucine

Under which case expression of tryptophan synthesis genes would be observed even in absence of tryptophan

(1) Mutant B, C and F
(2) Mutant A, C and D
(3) Mutant B, D and F
(4) Mutant A, D and E

41. Consider the following statements regarding aminoacyl-tRNA synthetase reaction

A. ATP is a cofactor
B. GTP is a cofactor
C. The amino acid is attached to the 2′- or 3′-hydroxyl of the nucleotide cofactor (ATP)
D. The amino group of the amino acid is activated
E. A mixed anhydride bond is formed
F. An acyl ester bond is formed
G. An acyl thioester bond is formed
H. A phosphoamide (P–N) bond is formed

Which of the following statements about the aminoacyl-tRNA synthetase reaction are correct?

(1) A, B and C (2) C, D and E
(3) F, G and H (4) A, E and F

42. An experiment is carried out in which labeled amino acids are added to an in vitro translation system under the direction of a single mRNA species. Samples are withdrawn at different times, and the labeling patterns below are observed in the completed polypeptide chains. The dashes (-) represent unlabeled amino acids, X represents labeled amino acids, and A and B represent the ends of the intact protein

Time 1 (early) A - - - - - - - - - - - - - - - XXB

Time 2 A - - - - - - - - - - - - XXXXXB

Time 3 A - - - - - - - -- XXXXXXXXB

Time 4 (late) A - - - - XXXXXXXXXXXXB

Which of the following statements about these proteins are correct?

(1) The labeled amino acids are added in the B-to-A direction

(2) The labeled amino acids are added in the A-to-B direction

(3) B is the amino terminus of the protein

(4) A is the carboxy terminus of the protein

43. Which of the following statements about eukaryotic translation are correct?

A. A formylmethionyl-tRNA initiates each protein chain

B. It occurs on ribosomes containing one copy each of the 5S, 5.8S, 18S, and 28S rRNA molecules

C. The correct AUG codon for initiation is selected by the base-pairing of a region on the rRNA of the small ribosomal subunit with an mRNA sequence upstream from the translation start site

D. It is terminated by a release factor that recognizes stop codons and hydrolyzes GTP

E. It involves proteins that bind to the 5′ ends of mRNAs

F. It can be regulated by protein kinases

(1) A, B, C and D
(2) B, D, E and F
(3) C, D, E and F
(4) A, B, C and D

44. The sequence of a duplex DNA segment in a DNA molecule is

5′-ATCGCTTGTTCGGA-3′
3′-TAGCGAACAAGCCT-5′

When this segment serves as a template for *E. coli* RNA polymerase, it gives rise to RNA with the sequence 5′-CCGAACAAGCGAU-3′

Following conclusions can be drawn

A. The top strand is the coding strand

B. The bottom strand is the sense strand

C. The top strand is the template strand

D. The bottom strand is the antisense strand

Which of the following statements about the DNA segment are correct?

(1) A and B (2) B and C
(3) C and D (4) A and C

45. Consider the following statements for polyadenylation in eukaryotes

A. They are added as preformed polyriboadenylate segments to the 3′-ends of mRNA precursors by an RNA ligase activity

B. They are encoded by stretches of polydeoxythymidylate in the template strand of the gene

C. They are added by RNA polymerase II in a template-independent reaction using ATP as the sole nucleotide substrate

D. They are added by poly(A) polymerase using dATP as the sole nucleotide substrate

E. They are cleaved from eukaryotic mRNAs by a sequence-specific endoribonuclease that recognizes the RNA sequence AAUAAA

Which of the following statements about the poly(A) tails that are found on most eukaryotic mRNAs are correct?

(1) A and B (2) C and D
(3) D and E (4) none

46. Not all bacteria ingested by phagocytic cells are destroyed. In fact, some species hijack the phagocytic machinery to promote their own survival in the body

Consider some of the evasion mechanism which helps the bacteria to survive in lysosome

A. *Mycobacterium tuberculosis*, the agent responsible for tuberculosis, for example, is taken into the cytoplasm of a macrophage by phagocytosis, but the bacterium is able to inhibit the fusion of its phagosome with a lysosome

B. Even if the phagosome does become highly acidic, the bacterium is able to maintain its own physiological pH despite the lowered pH of its surrounding medium. The bacterium responsible for Q fever, *Coxiella burnetii*, becomes enclosed in a phagosome that does fuse with a lysosome, but neither the acidic environment nor the lysosomal enzymes can destroy the pathogen

C. *Listeria monocytogenes*, a bacterium that causes meningitis, produces proteins that destroy the integrity of the lysosomal membrane, allowing the bacterium to escape into the cell's cytosol

Which of the following statement is correct?

(1) A and B (2) B and C
(3) A and C (4) A, B and C

47. Which of the following responses are not likely to be seen when a bacterial culture growing at 37°C is transferred to a culture room maintained at 25°C?

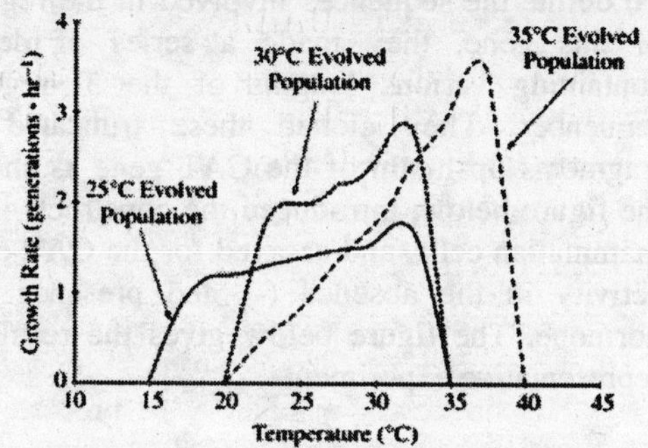

A. Initial decrease in membrane fluidity

B. Gradual replacement of shorter-chain fatty acids by longer-chain fatty acids in the membrane phospholipids

C. Gradual replacement of stearate by oleate in the membrane phospholipids

D. Enhanced rate of synthesis of unsaturated fatty acids

E. Incorporation of more cholesterol into the membrane

(1) A, C and D (2) B and E only
(3) B, C and E (4) A and C only

48. Match the descriptions in the column II with the type of splicing in the column I

Column I

A. group I splicing
B. group II splicing involved
C. nuclear mRNA splicing

Column II

(1) A 2′-5′phosphodiester bond is formed
(2) Nuclease and ligase activities are required
(3) Transesterification reactions break and form bonds
(4) A guanine nucleoside or nucleotide is required
(5) Spliceosomes are required
(6) Lariat intermediates are involved
(7) snRNPs are involved

(1) (A) 1, 3, 4 (B) 3, 5, 6 (C) 1, 3, 6, 7
(2) (A) 3, 4 (B) 1, 4, 6 (C) 2, 3, 5, 6, 7
(3) (A) 3, 4 (B) 1, 3, 6 (C) 1, 3, 5, 6, 7
(4) (A)1, 2, 4 (B) 1, 3, 6 (C) 1, 3, 5, 6

49. Consider the following statements reagarding DNA damage and repair

A. SOS repair removes errors that escape proofreading

B. Direct reversal of DNA damage is performed by photoreactivation by photolyase and methyltransferases against methylation at G

C. Base excision repair enzymes remove damaged bases by a base-flipping mechanism using DNA glycosylase

D. Nucleotide excision repair in *E. coli* is largely accomplished by four proteins:

UvrA, UvrB, UvrC, and UvrD and cleave damaged DNA on either side of the lesion

E. Transcription-coupled repair, involves recruitment of TF-IID to the stalled RNA polymerase of nucleotide excision repair proteins

Which of the following statements are correct?

(1) A, B and D (2) B, C and D
(3) A, D and E (4) B, C and E

50. In a eukaryotic gene, one of the introns is mutated at the 3' end as shown

Normal: 5'...TTTCCCACCCTTAG 3'
Mutant: 5'...TTTCCCACCCTTCG 3'

Which of the following processes is this mutation most likely to affect?

(1) Capping
(2) Hybridization
(3) Polyadenylation
(4) Splicing

51. Bacteriophage λ, has two modes in its life cycle, lytic and lysogenic. In the lysogenic mode, the expression of all the phage genes are repressed while the expression of repressor gene switches between on and off – position depending on the concentration of repressor. The following statements are made

A. Repressor may act both as a positive regulator and negative regulator
B. Expression of repressor gene, cI is independent of the expression of cll and cIII genes
C. Mutation of cI gene will cause it to it to form clear plaques on both wild type *E. coli* and *E. coli* (λ^+)
D. Mutation at operators, O_L and O_R will allow the phage to act as virulent phage

The correct statements are

(1) A and B (2) B and C
(3) C and D (4) D and A

52. Using established recombinant DNA technology, you insert a gene from a human liver cell into a bacterium. The bacterium then expresses a protein corresponding to the inserted DNA. To your dismay, you discover that the protein produced is useless and is found to contain many more amino acids than does the protein made by the eukaryotic cell. Assuming there is no mutation in the human gene, explain why this happened?

(1) Prokaryotes lack post translation modifications in proteins
(2) Prokaryotes lacks RNA editing mechanism
(3) Gene was fused with bacterial gene leading to formation of large protein
(4) Bacteria lacks mechanism for splicing mechanism of removal of introns

Answer questions 53-55 based on the information provided below

Researchers studying the regulation of a hormone-responsive gene isolated 750 base pairs of DNA immediately preceding the start site of transcription (+ 1) They demonstrated that if these sequences are cloned upstream of the bacterial chloramphenicol acetyltransferase (CAT) gene and the DNA then introduced into mammalian cells, CAT enzyme activity increases in response to hormone treatment. To define the sequences involved in the regulation of this gene, they made a series of deletions containing various lengths of the 5' regulatory sequences. They cloned these truncated DNA fragments upstream of the CAT gene as shown in the figure below, introduced the constructs into the mammalian cells, and assayed for the CAT enzyme activity in the absence (-) and presence (+) of hormone. The figure below gives the results of a representative experiment

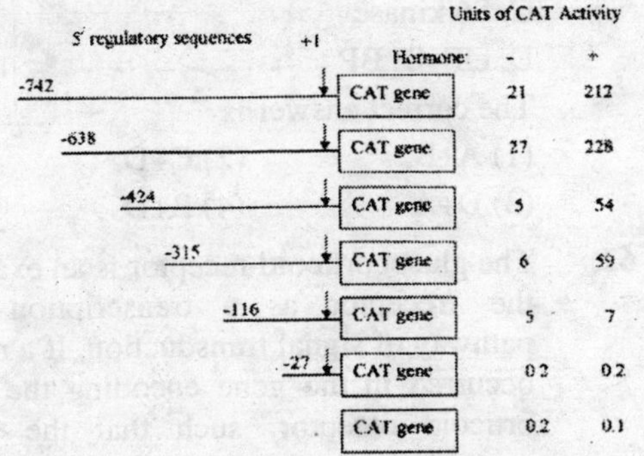

53. The maximal stimulation of CAT activity due to the addition of hormone is approximately

(1) 1-fold (2) 10-folds

(3) 50-folds (4) 100-folds

54. Assuming that there is a single hormone responsive regulatory element in the gene, that element is located between

(1) -742 and 638 (2) -638 and -424

(3) -424 and -315 (4) -315 and -116

55. A new construct was made that began at -742 and was identical to that shown in the figures except that the sequences between -424 and -315 were inverted. In this new construct, which of the following are closest to the expected CAT activities in the absence and presence, respectively, of hormones?

(1) 5 units/50 units

(2) 5 units/25 units

(3) 25 units/10 units

(4) 25 units/5 units

Answer questions 56-58 based on the information provided below

A transcription factor Y (TFY) is examined for its DNA binding ability and its expression in different cell lines. The specific DNA binding sequence has been determined and is used as a probe in an electromobility shift (EMS) assay

Specific antibody against TFY is used in Western blot analysis to check the subcellular localization of the protein. The EMS and Western blot results of TFY activity in muscle cells and nerve cells are shown below

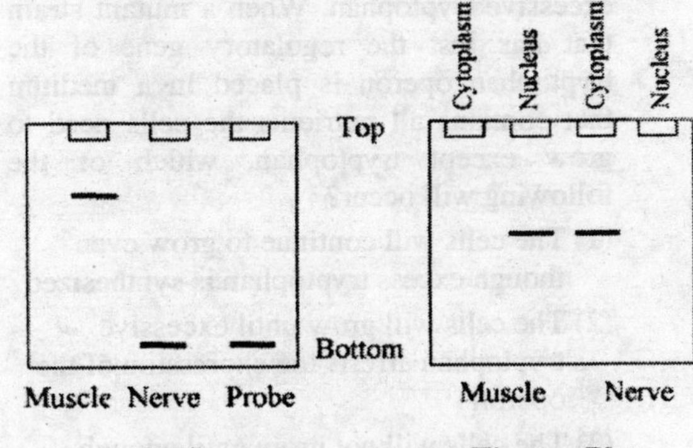

56. The EMS results suggest that

(1) TFY is expressed only in the muscle cells

(2) TFY binds DNA in the muscle cells, but not in nerve cells

(3) TFY binds DNA in the nerve cells, but not in muscle cells

(4) TFY is expressed only in the nerve cells

57. According to the results of the Western blot

(1) TFY binds with DNA in nerve cells

(2) TFY is expressed only in muscle cells

(3) TFY is expressed both in nerve and muscle cells

(4) TFY binds with DNA in muscle cells

58. It was observed that TFY activity is significantly blocked in the nerve cells in comparison to muscle cells. Based on these data, what could be most likely mechanism of regulation of TFY in nerve cells?

(1) Inhibition of transcription of TFY in the nucleus

(2) Inhibition of translation of TFY in cytoplasm

(3) Inhibition of translocation of TFY to the nucleus

(4) Inhibition of translocation of TFY to the cytoplasm

59. In *Escherichia coli*, the synthesis of tryptophan is controlled by the tryptophan operon that is repressed in the presence of excessive tryptophan. When a mutant strain that has lost the regulatory gene of the tryptophan operon is placed in a medium that contains all nutrients the cells need to grow except tryptophan, which of the following will occur?

(1) The cells will continue to grow even though excess tryptophan is synthesized
(2) The cells will grow until excessive tryptophan arrests the expression of the operon
(3) The cells will not grow until enough tryptophan has been synthesized to activate the co-repressor
(4) The cells will never grow unless tryptophan is added to the medium

60. Synthesis of an mRNA in mammalian cells was induced by estrogen, with the mRNA half-life of 10 min. If the cells were treated with estrogen and cycloheximide, the half-life was increased to 30 min. The effect of cycloheximide on the mRNA is likely to be at which of the following steps?

(1) Transcriptional
(2) Post-transcriptional
(3) Translational
(4) Post-translational

61. During mitogenic stimulation, cells proliferate at a higher rate and it is primarily determined by an enhanced rate of protein synthesis. Among other mechanisms, MAP kinase pathway of signal transduction is involved in this. Global protein synthesis may be regulated by many mechanisms involving various steps of protein synthesis, namely, initiation, elongation and termin-ation. Thus, many protein factors may be involved in the same. In the above process (mitogenic stimulation) the following factors are the portable targets

A. elF -2
B. eEF-1
C. S6 kinase
D. elF-4E BP

The correct answer is

(1) A+B (2) C+D
(3) D+A (4) B+D

62. The glucocorticoid receptor is an example of the "receptor as a transcription factor" pathway of signal transduction. If a mutation occurred in the gene encoding the glucoc-orticoid receptor, such that the receptor protein no longer could bind zinc atoms, which activity would be most directly affected?

(1) Ligand (hormone) binding
(2) DNA binding
(3) Transcriptional activation
(4) Nuclear localization

63. When bound to the glucocorticoid respon-sive element of the PEP carboxykinase gene, the transcriptional regulatory domain of the glucocorticoid receptor, bound with cortisol, can mediate which of the following?

(1) Recruit basal transcription factors to the core promoter (e.g. recruit TFII-D to the TATA box)
(2) Bring histone acetyl transferase to modify histones and unpack nucleosomes
(3) Recruit ATP-dependent chromatin-remodeling enzymes to increase accessibility for the transcription machinery including RNA polymerase II
(4) All the above

64. The core histories are among the most conserved eukaryotic proteins; therefore, the nucleosomes formed by these proteins are very similar in all eukaryotes. Eukaryotic cells commonly contain five abundant histones: H1, H2A, H2B, H3, and H4. Histories H2A, H2B, H3, and H4 are the core histones and form the protein core around which nucleosomal DNA is wrapped Consider the following statements

A. The Histone C-terminal tails stabilize DNA wrapping around the octamer and can be easily cleaved by trypsin treatment

B. Many DNA sequence-independent contacts is mediated by a large number of hydrogen bonds between the histones and the minor groove of DNA

C. H2A.z is a variant of H2A that is widely distributed in eukaryotic nucleosomes and is generally associated with transcribed regions of DNA. The presence of the H2A.z histone inhibits nucleosomes from forming repressive chromatin structures

D. In centromeric DNA chromosomal region, CENP-A replaces the histone H3 subunits in nucleosomes. These nucleosomes are incorporated into the kinetochore which mediates attachment of the chromosome to the mitotic spindle

Which of the following statement is correct?

(1) A, B and C (2) B, C and D
(3) A, C and D (4) A, B and D

65. Consider the following statement for chromosome organization

A. In the test tube, as salt concentrations are decreased, the addition of histone H_3 results in the nucleosomal DNA forming a 30-nm fiher. The histone C-terminal tails are required for the formation of the 30-nm fiber

B. The 30-nm fiber forms loops of 40-90 kb that are held together at their bases by a proteinacious structure referred to as the nuclear scaffold. Two classes of proteins that contribute to the nuclear scaffold have been identified. One of these is topoisomerase II and other are structural and maintenence protein

C. During the replication of the DNA, nucleosomes are transiently disassembled. Histone H_3-H_4 tetramers and H_2A-H_2B dimers are randomly distributed to one or the other daughter molecules. On average, each new DNA molecule receives half old and half new histones. This involves the function of specialized histone chaperones

D. A subset of nucleosomes is restricted to fixed positions in the genome and are said to be "positioned." Nucleosome positioning can be directed by DNA-binding proteins or particular DNA sequences (AT rich)

Which of the following statement are true?

(1) A, B and C (2) B, C and D
(3) A, C and D (4) A, B and D

66. The interaction of the DNA with the histones in the nucleosome is dynamic, allowing DNA-binding proteins intermittent access to the DNA

Consider the following statements

A. Histone H_1 is associated with transcriptionally active chromatin

B. Nucleosome remodeling complexes facilitate changes in nucleosome location or interaction with the DNA using the energy of ATP hydrolysis by "sliding" of the histone octamer along the DNA (SWI/SNF or ISWI or Mi2/NuRD) or the complete "transfer" of a histone octamer from one DNA molecule to another (SWI/SNF) or the "remodeling" of the nucleosome to allow increased access to the DNA (SWI/SNF)

C. A activator binding site near position 73 of the 147 base pairs tightly associated with a nucleosome is least frequently accessible, whereas binding sites near the ends (positions 1 or 147) of the nucleosomal DNA are most frequently accessible

D. Deacetylation of N-terminal tails is frequently associated with regions of active gone expression

Which of the following is correct?

(1) A and B (2) B and C
(3) C and D (4) D and A

67. Cells solve the end replication problem in a variety of ways

A. Terminally attached replication proteins which provide free 3'OH are found at the end of the linear chromosomes of certain species of bacteria and at the ends of the linear chromosomes of certain bacterial and animal viruses

B. Telomerase, a RNA dependent DNA polymerase, solves the end replication problem in higher animals by extending the 5' end (TTAGGG sequences) of the chromosome (called telomeres)

C. By extending the 5' ends of the telomere, telomerase leads to progressive loss of chromosome ends that conventional synthesis by the replication fork machinery would have not

Which of the following is correct?

(1) A only (2) B and C
(3) A and C (4) A, B and C

68. Which of the following statements is/are correct about leucine zipper motif

P. a DNA binding motif

Q. homodimer or heterodimer

R. present in transcription factors-Fos and Jun

S. present in steroid receptors

(1) P only (2) P, Q and R
(3) Q, R and S (4) P and S

69. Which of the following is/are true?

P. RNA polymerase needs general transcription factors to elongate the message

Q. tissue-specific transcription factors bind to TATA box

R. generalized transcription factors bind to proximal upstream of the promoters

S. C-terminus of RNA polymerase II is an important control point

(1) P only (2) R only
(3) Q and R (4) R and S

70. Researchers can learn which sites in the genome interact with a particular transcription factor. The following general strategies are often employed to make this determination

A. Deletion mapping

B. DNA foot printing

C. Genome wide location analysis

I. This strategy allows researchers to simultaneously monitor all of the sites within the genome that carry out a particular activity. To carry out this analysis, cells are isolated and then treated with formaldehyde which cross-links transcription factors to the DNA sites at which they were bound in the living cell. After this the cells chromatin is isolated, broken mechanically into small fragments, and treated with an antibody that binds to the transcription factor of interest induces the precipitation of chromatin fragments. Once this process of chromatin immunoprecipitation (or ChIP) has been carried out, the cross-links between the protein and DNA in the precipitate can be reversed and the segments of DNA can be purified

II. In this procedure, DNA molecules are prepared that contain deletions of various parts of the gene's promoter. Cells are then transfected with the altered DNA molecules; that is, they are caused to take up the DNA from the medium. Finally, the ability of the cells to transcribe the transfected DNA is determined

III. When a transcription factor binds to a DNA sequence, it protects that sequence from digestion by nucleases. Researchers take advantage of this property by isolating chromatin from cells and treating it with DNA-digesting enzymes, such as DNase I, that destroy those sections of the DNA that are not protected by bound transcription factors. Once the chromatin has been digested, the bound protein is removed, and the DNA sequences that had been protected are identified

Match the techniques with correct explanation

(1) A→I B→ II C→III

(2) A→II B→I C→III

(3) A→III B→II C→I

(4) A→II B→III C→I

71. Which of the following proteins will not transactivate transcription from the Gal4 upstream activator sequence (UASG)?

(1) A chimeric protein consisting of an SP1 binding domain and a Gal4 activation domain

(2) The Gal4 protein

(3) A chimeric protein consisting of a Gal4 DNA binding domain and an SP1 activation domain

(4) A chimeric protein consisting of a Gal4 DNA binding domain and a CTF1 activation domain

72. In mammals, males have one X chromosome and one Y chromosome and females have two X chromosomes. How is the expression of genes on the X chromosome regulated so that there is equal expression of genes on the X chromosome in males and females?

(1) One X chromosome in females is inactivated so that females only have a single X chromosome capable of transcription

(2) The genes on the X chromosome in males are transcribed twice as fast as in females

(3) All of the X chromosomes are inactivated so that no genes are expressed from the X chromosome in either males or females

(4) The Y chromosome contains balancing genes that help to raise the levels of mRNA produced by the X chromosome in males

73. One defense mechanism against viruses involves the generation of short interfering RNAs (siRNAs) that work through a RNA-induced silencing complex (RISC) in turning off the expression of certain genes. With respect to the mRNA corresponding to the gene whose expression needs to be decreased, which of the following represents an event mediated by RISC?

(1) covalent modification of histones to decrease accessibility of DNA template for transcription

(2) covalent modification of RNA polymerase II to lower its catalytic activity

(3) interference of translation of the mRNA by ribosomes

(4) N-glycosylation of the protein product to tag it for proteasomal degradation

74. Two patients both have a genetically inherited cause of anemia. The first patient has a mutation in codon 6 of the β-globin gene that replaces glutamate with valine. The second patient has a mutation in codon 6 of the β- globin gene that replaces glutamate with lysine. This would be considered an example of

(1) allelic heterogeneity

(2) syntenic genes

(3) locus heterogeneity

(4) homologous recombination

75. Your best friend is studying the influences of chromatin environment on the rate of transcription of a gene x. To test his hypothesis that local chromatin structure influences the rate of transcription of gene x, he moves gene x from its normal position on the chromosome to a position near the telomere. What effect on the rate of transcription of gene x does your friend expect to see?

(1) Gene x would be transcriptionally active

(2) Gene x would be transcriptionally silenced

(3) Gene x would be hyperactive

(4) Gene x would be expressed only during late prophase

76. Which of the following technique can be utilized for investigation of DNA (Promoter)-Protein interaction (Regulatory protein) under in-vivo conditions?

(1) RNAase protections Assay

(2) Gel Retardation Assay

(3) DNAse hypersensitivity Assay

(4) Chromatin Immunoprecipitation Assay

77. A patient at a specialty clinic is undergoing evaluation for liver disease. Although the cause of the liver dysfunction is not known, the doctor suspects that the patient has a deficiency of a liver enzyme resulting from a novel deletion of the gene for the enzyme. Which of the following tools of molecular genetics would be most useful for studying this gene?

(1) cDNA library

(2) Fluorescence-activated cell sorting (FACS)

(3) Genomic library

(4) Southern blotting

78. Consider the following statement regarding proteins involved in regulation of λ phage

A. cI : Repressor protein whose function favors lytic cycle

B. cII : Enhances transcription at the pI and pRE promoters

C. cIII : Inhibits the FtsH protease

D. cro: Antirepressor protein that favors lytic cycle

E. N: Antiterminator acting at nutR and nutL

F. rex : Protects bacterium from infection by T4 rII mutants

G. int : Integrase for prophage integration

H. Q: Antiterminator of early operon

I. FtsH : Bacterial protease that degrades cI protein

Which of the following is not correctly matched?

(1) A, H and I (2) C, D and I

(3) A, E and F (4) F, G and H

79. A hypothetical operon has a sequence of sites, Q R S T U, in the promoter region, but the exact location of the operator and promoter consensus sequences have not been identified. Various deletions of this operator region are isolated and mapped. Their locations appear as follows, with a "/" representing a deleted region. Deletions 3 and 4 are found to produce constitutive levels of RNA of the operon, and deletion 1 is found to never make RNA

Deletion	Q	R	S	T	U
1	//////				
2		//////			
3			//////		
4			//////////	//////////	
5					//////

Where are the operator and promoter consensus sequences probably located?

(1) Operator: S, T; Promoter: Q

(2) Operator: S, Promoter: Q

(3) Operator: Q; Promoter: S and T

(4) Operator: Q, R and S; Promoter: T and U

80. The formation and mechanism of action of siRNAs is shown below

A. the target RNA is cleaved at aspecific site by the Argonaute protein and subsequently degraded

B. the single-stranded guide RNA, in association with proteins of the RISC complex, binds to a target RNA that has a complementary sequence

C. the siRNA becomes associated with a protein complex, a pre-RISC, that contains an Argonaute protein (typically Ago2) capable of cleaving and removing the passenger strand of the siRNA duplex

D. both strands of a double-stranded RNA are cleaved by the endonuclease Dicer to form a small (21–23 nucleotide) siRNA, which has overhanging ends

What is the correct sequence for formation and functioning of siRNA?

(1) A→ B→C→D

(2) D→C→B→A

(3) B→D→C→A

(4) C→D→B→A

81. MicroRNAs are derived from single-stranded precursor RNAs that contain complementary sequences that allow them to fold back on themselves to form a double-stranded RNA with a stem-loop at one end. This pseudo dsRNA (or pri-miRNA) is cleaved near its terminal loop by a protein complex containing an endonuclease named Drosha to generate a pre-miRNA that has a 3' overhang at one end. The pre-miRNA is exported to the cytoplasm where it is cleaved by Dicer into a small duplex miRNA that has a 3' overhang at both ends

Consider the following statements

A. The double-stranded RNA becomes associated with a protein complex containing an Argonaute protein (typically Ago1), leading to the separation of the strands and removal of the passenger strand (called miRNA*)

B. The single-stranded guide miRNA then binds to a complementary region on an mRNA and inhibits translation of the message (or alternatively leads to destabilization and degradation of the mRNA

C. Unlike siRNAs, miRNAs that inhibit translation are only partially complementary to the target mRNA, hence the bulge

D. Most plant miRNAs and a handful of animal miRNAs are precisely complementary to the mRNA, or nearly so; in these cases, the outcome of the interaction tends to be cleavage of the mRNA

E. Certain class of miRNAs are derived from introns do not form long hairpin pri-miRNAs and do not require Drosha for processing

Which of the following statement is correct regarding miRNA?

(1) A, C and D

(2) B, D and E

(3) A, B and E

(4) A, B, C, D and E

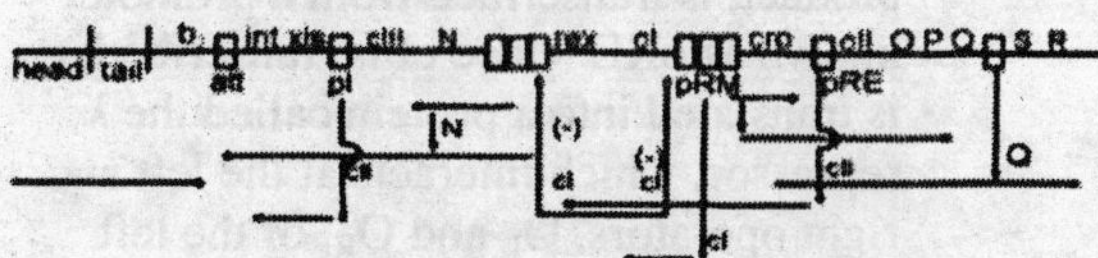

82. Bacteriophage λ genetic circuit may be represented as above. The control of gene expression occurred during the phage infection may be described as follows

A. N and Q protein act as antiterminator

B. CI acts as repressor

C. CII act as a retroregulator

D. CI act as positive and negative auto-regulator

Which one of the statement(s) is correct?

(1) A, B and C

(2) B, C and D

(3) A and D only

(4) A, B and D

83. Phage λ exhibits a complex system of controls of both early and late operons, as well as controls for the decision of lytic infection versus lysogenic integration. The λ genes are grouped into four operons: left, right, late, and repressor. The left and right operons contain the genes for DNA replication and recombination and phage integration. The late operon contains the genes that determine phage head and tail proteins and lysis of the host cell

Consider the following points for λ phage gene regulation

A. Transcription of lytic genes cannot continue until the protein product of the N gene is produced. This protein is called an antiterminator protein

B. The cIII-protein product inhibits a host cell protease, called FtsH, that would break down the cII-gene product

C. The cII protein binds at the promoters for cI transcription and for int transcription

D. The cI gene, with the aid of the cII-gene product, is transcribed from a promoter known as pRE. Once cI is transcribed, it is translated into a protein called the λ repressor, which interacts at the left and right operators, O_L and O_R, of the left and right operons and repress the genes for lytic cycle

Which of the following statement is correct?

(1) A, B and C (2) B, C and D

(3) A and D (4) A, B, C and D

Question 84-85

The anticodon of a tRNA is 5'-IGU-3'

84. Which amino acid would this tRNA insert during translation?

(1) Threonine, i.e., 5'-ACX-3'

(2) Alanine i.e 5'-GCU-3'

(3) Lysine or Arginine 5'-AC/AX-3'

(4) 1 and 2 both

85. Which codons would be read by this tRNA?

(1) ACT, ACC and ACA

(2) ACU, ACC and ACA

(3) ACU and ACC

(4) ACC and ACA

Questions 86-88

Five *E. coli* strains have been identified, each of which has a different mutation that disrupts the normal regulation of a particular operon

For each mutant strain, the mutation has been mapped to the promoter or the operator region; however, the exact sequence changes are not known for these mutations

It is known that the normal promoter/operator consists of a single binding site for a positively acting transcription factor located just upstream of the promoter itself

Short DNA fragments containing the promoter and the operator were sub-cloned from each of the five mutant strains and from the wild type, purified, and radiolabeled

These fragments were then incubated under conditions of DNA excess with either purified regulatory factor or RNA polymerase or with both polymerase and regulatory factor

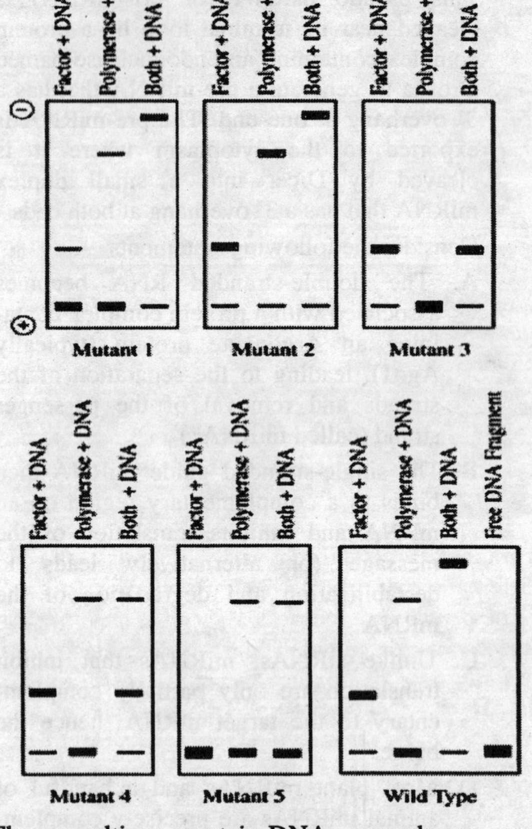

The resulting protein-DNA complexes were separated by electrophoresis, and the radioactive DNA fragments were detected by exposure to X-ray film, giving the results shown below. Electrophoresis is from top to bottom; the largest complexes run slowest

86. One of the mutations increases the affinity of the polymerase for the promoter. Transcription of the operon is not stimulated by the regulatory factor in this mutant. Which mutant is most likely to show this effect?

(1) Mutant 1 (2) Mutant 2
(3) Mutant 3 (4) Mutant 4

87. One of the mutations maps to the operator. Transcription of the operon is not stimulated by the regulatory factor in this mutant. Which mutant is most likely to show this effect?

(1) Mutant 1 (2) Mutant 2
(3) Mutant 4 (4) Mutant 5

88. One of the mutations is known to result from a small deletion between the operator and the promoter. The polymerase and the regulatory factor are each able to bind to the mutated DNA sequence, but are unable to form the three component complex. Transcription of the operon is not stimulated by the regulatory factor in this mutant. Which mutant shows the properties that might be expected for such a change?

(1) Mutant 5 (2) Mutant 2
(3) Mutant 3 (4) Mutant 4

Questions 89-92 refer to the following experiment

The graphs below represent various curves obtained in different in vitro bacterial translation experiments. The peptide product was assayed as acid-precipitable counts per minute (cpm) that resulted from incorporation of a mixture of radiolabeled amino acids. The arrows represent the times at which various substances were added to the systems

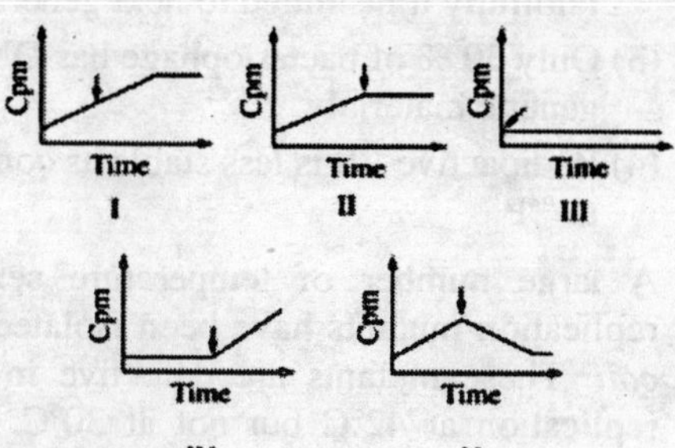

89. The addition of an antibiotic at the start of the assay results in the curve shown in graph III, while addition at a later time yields the curve shown in graph I. This antibiotic most likely blocks

(1) elongation only
(2) initiation only
(3) termination only
(4) both initiation and elongation

90. Addition of fmet-tRNA^f to a system lacking only this component is most likely to give the curve shown in graph

(1) I (2) II
(3) III (4) IV

91. Addition of a peptide that binds irreversibly to the A site of the ribosome will most likely give which of the following graphs?

(1) II only
(2) III only
(3) I and V only
(4) II and III only

92. Addition of which of the following would most likely give the curve seen in graph V?

(1) Ribonuclease
(2) Exopeptidase
(3) Deoxyribonuclease
(4) Chloramphenicol

93. Most eukaryotic pre-mRNAs are matured by which of the following modifications to their ends?

(1) Capping at the 3'-end, cleavage and polyadenylation at the 5'-end
(2) Addition of a GMP to the 5'-end, cleavage and polyadenylation to create the 3'-end
(3) Addition of a guanine residue to the 5'-end, cleavage and polyadenylation to create the 3'-end
(4) Addition of a GMP to the 5'-end, polyadenylation, then cleavage to create the 3'-end

94. The cell-free system was created using an extract from immature red blood cells from a rabbit, which are efficient factories for the synthesis of the α- and β-globin subunits of hemoglobin. The cell-free system was treated with a radioactive amino acid for a very few seconds (less than the time required to synthesize a complete globin chain) after which protein synthesis was immediately stopped. A brief radioactive labeling regime of this kind is known as a pulse or pulse-labeling. Next, globin chains that had completed their growth during the period of the pulse-labeling were separated from incomplete chains by gel electrophoresis. The full-length polypeptides were then treated with an enzyme, the protease trypsin, that cleaves proteins on particular sites in the polypetide chain, thereby generating a series of peptide fragments. In the final step of the experiment, the amount of radioactivity that had been incorporated into each peptide fragment was measured as shown in diagram

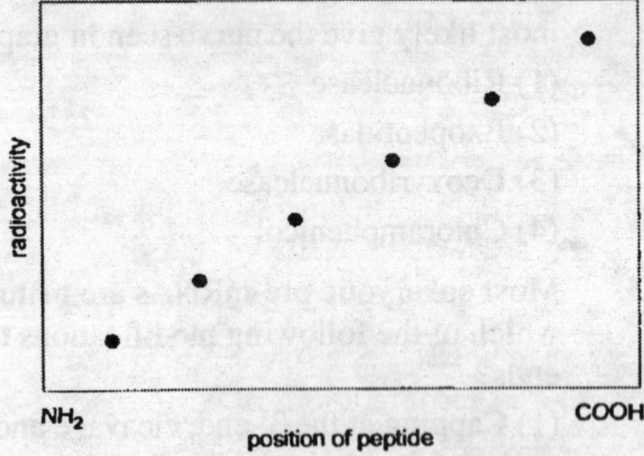

Which of the following interpretation is correct from above experiment?

(1) Radioactive labeling was lowest for peptides from the amino-terminal region of globin and greatest for peptides from the carboxyl-terminal region. Thus the direction of protein synthesis is from the amino-terminus to the carboxyl-terminus

(2) Radioactive labeling was lowest for peptides from the amino-terminal region of globin and greatest for peptides from the carboxyl-terminal region. Thus the direction of protein synthesis is from the carboxy-terminus to the amino-terminus

(3) Radioactive labeling was highest for peptides from the amino-terminal region of globin and greatest for peptides from the carboxyl-terminal region. Thus the direction of protein synthesis is from the amino-terminus to the carboxyl-terminus

(4) Radioactive labeling was highest for peptides from the carboxy-terminal region of globin and greatest for peptides from the amino-terminal region. Thus the direction of protein synthesis is from the carboxy-terminus to the amino-terminus

95. Bacteriophage T_4 attaches to its bacterial host and injects its DNA to initiate an infection that ultimately releases hundreds of progeny. In 1953 Hershey and Chase radiolabelled the DNA of bacteriophage T_4 with $^{32}PO_4^{3-}$ and the proteins with ^{35}S-methionine. They then mixed the labeled bacteriophage with bacterial and after brief time agitated the mixture previously in a blended to detach T_4 from bacteria. They then separated the phage from bacteria by centrifugation. They demonstrated that the bacteria contain 30% of ^{32}P label but virtually none of the ^{35}S label. When new bacteriophage were released from these bacteria, they were also found to contain ^{32}P but no ^{35}S. This experiment demonstrate that

(1) Protein may be genetic material which restrict only to one generation

(2) DNA is genetic material and it is faithfully transmitted to next generation

(3) Only 30 % of bacteriophage has DNA as genetic material

(4) Radioactive ^{35}S is less stable as compare to ^{32}P

96. A large number of temperature sensitive replication mutants have been isolated in *E. coli*. These mutants are defective in DNA replication at 42ºC but not at 30ºC. If the temperature of the medium is raised from

30ºC to 42ºC, these mutant stop DNA synthesis only after many minutes

Consider the following proteins in replication

A. Topoisomerase I

B. A replication initiator protein (DnaA)

C. Single stranded DNA binding protein

D. DNA helicase (DnaB)

E. DNA primase (DnaG)

F. DNA ligase

Predict which of the following protein, if temperature sensitive, would display a quick stop phenotype

(1) A, B, C, D and E

(2) B, C, D and E

(3) A, B,C and F

(4) D, E and F

97. Double-helical DNA can be denatured by conditions that disrupt hydrogen bonding and base stacking, resulting in the "melting" of the double helix into two single strands that separate from each other. No covalent bonds are broken in this process. Heat, alkaline pH, and chemicals such as formamide and urea are commonly used to denature DNA

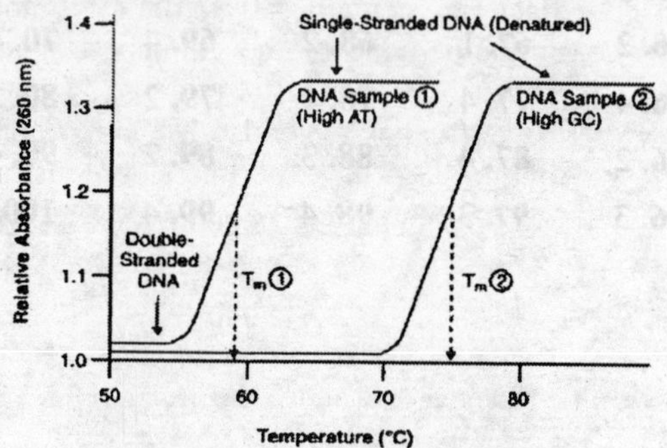

A melting curve for the denaturation of DNA by heat is shown in Figure above. In such a melting curve

A. Denaturation can be detected by the increase in absorbance of a DNA solution at a wavelength of 260 nm. This increase is known as the hypochromic effect

B. The melting temperature (Tm) is the temperature at which the molecule is half denatured. This represents the point at which enough heat energy is present to break half the hydrogen bonds holding the two strands together

C. A DNAsample with a high G-Ç content will have a high Tm because G-C base pairs have three hydrogen bonds, whereas A-T base pairs have only two

Which of the following statement is correct

(1) A and B (2) B and C

(3) A and C (4) A, B and C

98. It is now believed that a substantial proportion of the single nucleotide substitutions causing human genetic disease are due to misincorporation of bases during DNA replication. Which proofreading activity is critical in determining the accuracy of nuclear DNA replication and thus the base substitution mutation rate in human chromosomes?

(1) 3' to 5' polymerase activity of DNA polymerase α

(2) 3' to 5' exonuclease activity of DNA polymerase γ

(3) 5' to 3' polymerase activity of DNA polymerase III

(4) 3' to 5' exonuclease activity of DNA polymerase δ

99. The proliferation of cytotoxic T-cells is markedly impaired upon infection with a newly discovered human immunodeficiency virus, designated HIV- V. The defect has been traced to the expression of a viral-encoded enzyme that inactivates a host-cell nuclear protein required for DNA replication. Which protein is a potential substrate for the viral enzyme?

(1) TATA-box binding protein (TBP)

(2) Cap binding protein (CBP)

(3) Catabolite activator protein (CAP)

(4) Single-strand binding protein (SBP)

100. The immunoglobulin genes are expressed in B-cells, and the β-globin gene is expressed in red blood cells. What manipulations might be carried out to get the β-globin gene to be expressed in B-cells?

(1) The coding region of an immunoglobulin gene could be combined with the cis-regulatory region of the β-globin genes, and this artificial gene would then express β-globin in B-cells

(2) The cis-regulatory region of the β-globin gene could be inserted into B-cells, which would then express their own version of the β-globin gene

(3) The cis-regulatory region of an immunoglobulin gene could be combined with the coding region of the β-globin genes, and this artificial gene would then express β-globin in B-cells

(4) The transcription factors GATA-1 and GATA-2 (which activate the β-globin gene in erythrocytes) could be introduced into B-cells, which would then express their own copy of the β-globin gene

ANSWER SHEET

Unit 3 : Molecular Biology

1. 3	**2.** 2	**3.** 3	**4.** 3	**5.** 2	**6.** 3	**7.** 2	**8.** 4	**9.** 4	**10.** 1
11. 3	**12.** 3	**13.** 3	**14.** 1	**15.** 1	**16.** 3	**17.** 2	**18.** 2	**19.** 2	**20.** 1
21. 4	**22.** 3	**23.** 4	**24.** 4	**25.** 1	**26.** 4	**27.** 4	**28.** 1	**29.** 4	**30.** 3
31. 2	**32.** 4	**33.** 3	**34.** 3	**35.** 3	**36.** 1	**37.** 3	**38.** 1	**39.** 2	**40.** 1
41. 4	**42.** 2	**43.** 2	**44.** 2	**45.** 4	**46.** 4	**47.** 2	**48.** 3	**49.** 2	**50.** 4
51. 4	**52.** 4	**53.** 2	**54.** 4	**55.** 1	**56.** 2	**57.** 3	**58.** 3	**59.** 1	**60.** 3
61. 2	**62.** 2	**63.** 4	**64.** 2	**65.** 2	**66.** 2	**67.** 1	**68.** 2	**69.** 4	**70.** 4
71. 2	**72.** 1	**73.** 3	**74.** 1	**75.** 2	**76.** 4	**77.** 4	**78.** 1	**79.** 2	**80.** 2
81. 4	**82.** 4	**83.** 4	**84.** 1	**85.** 2	**86.** 2	**87.** 4	**88.** 3	**89.** 2	**90.** 4
91. 1	**92.** 2	**93.** 3	**94.** 1	**95.** 3	**96.** 3	**97.** 3	**98.** 4	**99.** 4	**100.** 3

UNIT-4

SIGNALLING, IMMUNOLOGY, CANCER

1. Two protein kinases, K1 and K2 function sequentially in regulating intracellular pathway in response to extracellular signal. The following observations are made
   (i) Response is observed even in the absence of extracellular signal when a mutation permanently activates K1
   (ii) Response is observed even in the absence of extracellular signal when K1 contains an activating mutation and K2 with inactivating mutation
   (iii) No response in the cells is detected even in the presence of extracellular signal when both kinases are inactivated by mutation

   Which one of the following is correct?
   (1) K1 activates K2
   (2) K2 activates K1
   (3) K1 inhibits K2
   (4) K2 inhibits K1

2. Conversion of proto-oncogene to oncogene may involve the following processes
   A. mutation in coding sequence
   B. gene amplification
   C. chromosome rearrangement
   D. mutation in non-coding sequence

   Which one is appropriate?
   (1) A, B and C (2) B, C and D
   (3) A, C and D (4) All

3. In an experiment erythrocytes containing epinephirine receptor but lacking adenylyl cyclase were fused with tumor cells containg adenylyl cyclase but lackinmg epinephirine receptors. Shortly after fusion, adenylyl cyclase can be stimulated by epinephrine. It indicates that
   (1) epinephrine receptors and adenylyl cyclase are same molecules
   (2) receptors derived from one cell can establish communication with adenylyl cyclase derived from another cell
   (3) different cells possesses different receptors for epinephirine
   (4) all cells will show similar response against epinephrine

4. Which of the following statement about the actions or targets of the second messengers of the phosphoinositide cascade is incorrect?
   (1) DAG activates protein kinase C (PKC)
   (2) Most of the effects of IP_3 and DAG are antagonistic
   (3) DAG increases the affinity of PKC for Ca^{2+}
   (4) PKC requires Ca^{2+} for its activity

5. The figure below illustrates the two-hit hypothesis for cancer. This hypothesis applies to _______, and the events shown in the bottom-half of the figure will result in the ______ of cancer

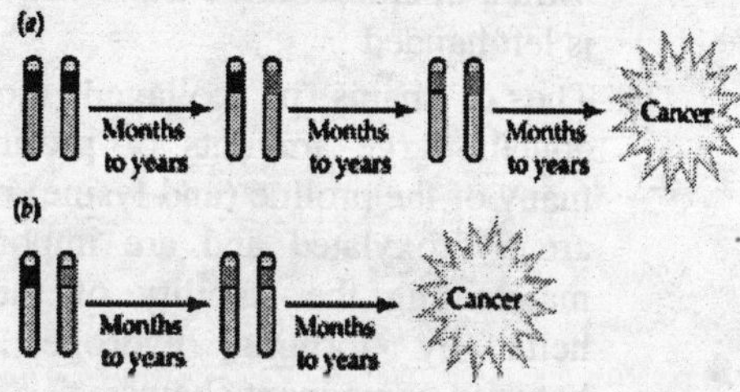

(1) Tumor suppressor genes only, inherited form
(2) Tumor suppressor genes only, sporadic form
(3) Oncogenes only, inherited form
(4) Oncogenes only, sporadic form

6. Steroid hormones alter gene expression by
(1) activating their receptors, making them capable of binding hormone response elements (HREs) in DNA
(2) having hormone response elements (HREs) that bind DNA
(3) activating their receptors, making their hormone response elements (HREs) capable of binding DNA
(4) binding hormone response elements (HREs) in DNA

7. Collagens comprise a family of fibrous glycoproteins that are present only in extracellular matrices. Collagens are found throughout the animal kingdom and are noted for their high tensile strength, that is, their resistance to pulling forces. It is estimated that a collagen fiber 1 mm in diameter is capable of suspending a weight of 10 kg without breaking

Consider the following statement regarding collagen

A. Collagen is the single most abundant protein in the human body
B. All collagen molecules are trimers consisting of three polypeptide chains, called α chains having right handed helicity
C. Along at least part of their length, the three polypeptide chains of a collagen molecule are wound around each other to form a unique, rodlike triple helix which is left handed
D. The α chains of collagen molecules contain large amounts of proline, and many of the proline (and lysine) residues are hydroxylated and are important in maintaining the stability of the triple helix by forming hydrogen bonds between component chains
E. Failure to hydroxylate collagen chains has serious consequences for the structure and function of connective tissues. This is evident from the symptoms of scurvy, a disease that results from a deficiency of vitamin C (ascorbic acid)

Which of the following statement is correct?

(1) A, B, and E (2) B, C and D
(3) A, D and E (4) A, B, C, D and E

8. Given abundance and widespread distribution of collagen, it is not surprising that abnormalities in fibrillar collagen formation can lead to serious disorders. Burns or traumatic injuries to internal organs can lead to an accumulation of scar tissue, which consists largely of fibrillar collagen

Consider the following statements

A. Mutations in genes encoding type I collagen can produce *osteogenesis imperfecta*, a potentially lethal condition characterized by extremely fragile bones, thin skin, and weak tendons
B. Mutations in genes encoding type II collagen alter the properties of cartilage tissue, causing dwarfism and skeletal deformities
C. Mutations in a number of other collagen genes can lead to a variety of distinct but related defects in collagen matrix structure (called *Ehler- Danlos syndromes*) Persons with one of these syndromes have hyperflexible joints and highly extensible skin
D. Not all collagens form fibrils. One of the nonfibrillar collagens is type IV, whose distribution is restricted to basement membranes. Mutations in type IV collagen genes have been identified in patients with *Alport syndrome*, an inherited kidney disease

Which of the following statement is correct?

(1) A, B and C (2) B, C, and D
(3) A, B and D (4) A, B, C and D

9. A human fibronectin molecule consists of two similar, but non-identical, polypeptides joined by a pair of disulfide bonds located near the C-termini. Each polypeptide is composed of a linear series of distinct modules, which are organized into several larger functional units, illustrated by the cylinders in this figure

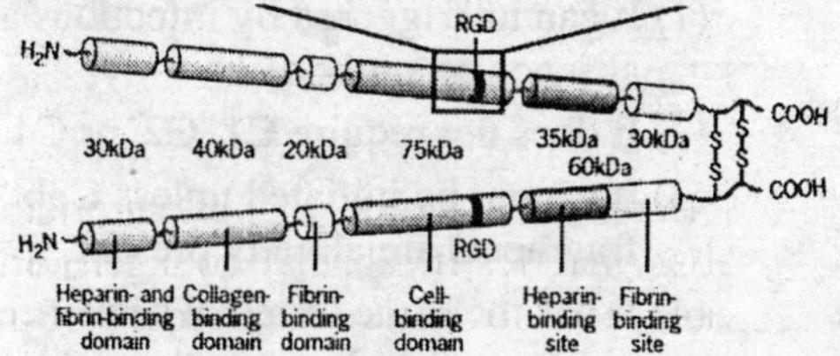

Each of these functional units contains one or more binding sites for either a specific component of the ECM or for the surface of cells

Consider the following statements

A. The cell-binding site of the polypeptide that contains the sequence arg-gly-asp, or RGD, binds specifically to a particular class of integral plasma membrane proteins selectin or cadherin that are involved in cell attachment and signal transduction

B. Neural crest cells can igrate out from a portion of the developing chick nervous system onto a glass culture dish that contains strips of fibronectin-coated surface alternating with strips of bare glass but the cells remain exclusively on the fibronectin coated surface

C. Cells that will reach the glass substratum tend to round up and lose their migratory capabilities

D. Mouse embryonic salivary gland that has been grown for 10 hours in culture with fibronectin divides into separate buds by a series of clefts whereas gland that has been grown for the same period of time in the presence of an anti-fibronectin antibody, which has completely inhibited cleft formation

Which of the following statement is correct?

(1) A, B and D (2) A, B and C
(3) B, C and D (4) A, C and D

10. Table below shows examples of physiologic processes mediated by GPCRs and heterotrimeric G Proteins

Stimulus	Receptor	Effector	Physiologic response
A. Epinephrine	Muscarinic receptor	Potassium channel	Slowing of pacemaker activity
B. Scrotonin	Serotonin receptor	Adenylyl cyclase	Behavior and learning in *Aplysia*
C. Light	Rhodopsin	cGMP phosphodiesterase	Visual excitation
D. IgE-antigen complex	Mast cell IgE receptor	Phospholipase C	Secretion
E. f-Met Peptide	Chemotactic receptor	Phospholipase C	Chemotaxis
F. Acetylcholine	β-Adrenergic receptor	Adenylyl cyclase	Glycogen breakdown

Which of the following is correct match

(1) A, B and C (2) C, D and E
(3) D, E and F (4) B, C, D and E

11. If a mouse is immunized with xenogeneic (human, rabbit, etc.) MHC molecules, it generates antibodies against many different epitopes present on the latter, including the epitopes specific to the MHC allele of the donor of these molecules ("polymorphic epitopes"), and the species-specific or "monomorphic" epitopes. Both categories of antibodies can be produced by hybridomas obtained after the immunization. Which of the following category of antibodies would you use to block the lysis of LCMV infected fibroblasts in the Doherty–Zinkernagel experiment?

(1) Hybridoma-derived antibodies
(2) Anti-polymorphic antibodies
(3) Anti-monomorphic antibodies
(4) Antibodies to the murine (H-2) allospecific determinants

12. A mouse is immunized with a protein antigen. After several days, macrophages from this mouse are put in culture together with the T cells from the same mouse. The antigen used for the immunization is added in the culture medium. What subpopulation of the murine T cells – will proliferate in this culture?

(1) CD4 cells
(2) CD8 cells
(3) Both CD4 and CD8
(4) Either CD4 or CD8

13. A fibroblast cell line transfected with an unknown gene has the following properties: it induces the growth and antibody secretion of B lymphocytes in culture and, in the presence of IL-4, stimulates the production of IgE. What is the gene that was used for transfection?

(1) CD40L (2) B7
(3) CD28 (4) CD19

14. A patient with rheumatic fever develops a sore throat from which beta-hemolytic streptococci are cultured. The patient is started on treatment with penicillin, and the sore throat resolves within several days. However, 7 days after initiation of penicillin therapy the patient develops a fever of 103°F, a generalized rash, and proteinuria. This most probably resulted from

(1) recurrence of the rheumatic fever
(2) an IgE response to penicillin
(3) an IgG-IgM response to penicillin
(4) a delayed hypersensitivity reaction to penicillin

15. Two dissimilar inbred strains of mice, A and B, are crossed to yield an F1 hybrid strain, AB. If a large dose of spleen cells from an adult A mouse is injected into an adult AB mouse, which one of the following is most likely to occur?

(1) The spleen cells will be destroyed
(2) The spleen cells will survive and will have no effect in the recipient
(3) The spleen cells will induce a graft-versus-host reaction in the recipient
(4) The spleen cells will survive and induce tolerance of strain A grafts in the recipient

16. Which one of the following is not true regarding the alternative complement pathway?

(1) It can be triggered by infectious agents in absence of antibody
(2) It does not require C1, C2, or C4
(3) It cannot be initiated unless C3b fragments are already present
(4) It has the same terminal sequence of events as the classic pathway

17. Regarding Th-1 and Th-2 cells, which one of the following is least accurate?

(1) Th-1 cells produce gamma interferon and promote cell-mediated immunity
(2) Th-2 cells produce interleukin-4 and -5 and promote antibody-mediated immunity
(3) Both Th-1 and Th-2 cells have both CD3 and CD4 proteins on their outer cell membrane
(4) Before naïve Th cells differentiate into Th-1 or Th-2 cells, they are double-positives; i.e., they produce both gamma interferon and interleukin-4

18. If a mouse is injected with Sheep Red Blood Cells (SRBC), it will develop a strong specific antibody response, which can be quantified in a hemolytic plaque assay. The number of plaques (sites of local hemolysis) in dishes with the immune mouse's lymphocytes and SRBC suspended in agar will be equal to the number of specific SRBC antibody-producing cells. If a mouse is injected with SRBC together with anti-SRBC IgG, the number of plaques will increase compared to the response of the mouse immunized with SRBC only. If, however, the mouse is injected with SRBC and anti-SRBC IgM, the plaque response will increase. The probable reason is

(1) the IgM binds complement (thus activating the B-cell co-receptor complex), but does not bind FcγRIIB. The IgG may also bind complement, but they simultaneously bind FcγRIIB and thus inhibit the signaling (antibody feedback)

(2) the IgG binds complement (thus activating the B-cell co-receptor complex), but does not bind FcgRIIB. The IgM may also bind complement, but they simultaneously bind FcμRIIB and thus inhibit the signaling (antibody feedback)

(3) Both IgG and IgM binds complement but does not bind to FcRIIB

(4) Both IgG and Igm binds complement and simultaneously bind FcRIIB and thus inhibits the signaling

19. The so-called "hyper-IgM syndrome" is an X-linked recessive disorder when patients do not express functional CD40 ligand. Which of the following effector functions of antibodies will be severely impaired in a boy suffering from this disease

A. neutralization of bacterial toxins

B. complement-dependent lysis of microbial particles

C. phagocytosis

D. ADCC

E. immediate hypersensitivity

(1) A, B and C (2) C, D and E

(3) B, D and E (4) B, C and E

20. Mouse bone marrow cells were fractionated to derive stem cell antigen-1^+ (Sca-1^+) cells. These cells were cultured with interleukin-3, or granulocyte-macrophage colony stimulating factor, or macrophage-colony stimulating factor, or granulocyte colony stimulating factor. Most numerous and varied colonies were obtained in the culture stimulated with

(1) interleukin-3

(2) granulocyte-macrophage colony stimulating factor

(3) macrophage-colony stimulating factor

(4) granulocyte-colony stimulating factor

21. Toll-like receptor 4 is associated with responsiveness to LPS, an endotoxin that causes lethal endotoxic shock. The mice deficient In Toll-like receptor 4 and BALB/c mice were injected with *Escherichia coli.* In addition, some BALB/b mice were also injected with the same bacteria alone or with anti-interleukin-10 (IL-10) antibody. The mice resistant to the lethal effect of the bacteria were

(1) BALB/b mice receiving the bacteria

(2) BALB/b mice receiving the bacteria and the anti-IL-10 antibody

(3) Mice deficient In Toll-like receptor

(4) BALB/c mice receiving the bacteria

22. Intracellular pathogens like *Mycobacteria, Salmonella, Leishmania* and *Listeria* survive in macrophages by modulating host cellular machinery. In order to study the fate of these intracellular pathogens in macrophages, cells were labelled with lysotracker Red arid infected with GFP-Iabelled organisms. After 2 hours at 37°C, cells were fixed, stained with anti-transferrin receptor antibody and probed with secondary antibody conjugated-blue dyes. Cells were viewed under confocal microscope

Observation: GFP-labelled *Mycobacteria, Salmonella* and *Listeria* were localized in the same compartment labelled with blue dyes; whereas GFP-*Leishmania* colocalize with red labelled compartment

Which of the following statement is true based on these observations?

(1) *Mycobacteria, Salmonella* and *Listeria* reside in the lysosomes

(2) *Leishmania* reside in lysosome like compartment

(3) *Leishmania* reside in a compartment which bears characteristics of early endocytic compartment

(4) *Mycobacteria, Salmonella* and *Listeria* lyse the phagosomal membrane and reside in cytosol

23. An autosomal recessive mutation resulting in **Severe Combined Immunodeficiency Disease (SCID)** developed spontaneously in a strain of mice called CB-17. These CB-17 SCID mice fail to develop mature T and B cells and consequently are severely compromised immunologically. This defect is due to a failure in V(D)J recombination. SCID mice must be housed in a sterile (germ-free) environment, because they cannot fight off microorganisms of even low pathogenicity. Consider the following statements for CB-17 mice

A. The absence of functional T and B cells enables these mice to accept foreign cells and grafts from other strains of mice or even from other species

B. When normal bone marrow cells are injected into SCID mice, normal T and B cells develop, and the mice are cured of their immunodeficiency

C. Because the mice lack mature T and B cells of their own, they do not reject the transplanted human tissue. The implanted human fetal liver contains immature lymphocytes (stem cells), which migrate to the implanted human tissues, where they mature into T and B cells, producing a SCID human mouse. Because the human lymphocytes are exposed to mouse antigens while they are still immature, they later recognize mouse cells as self and do not mount an immunologic response against the mouse host

Which of the following statement is correct?

(1) A and B (2) B and C
(3) A and C (4) A, B and C

24. Binding of TGF β, an antigrowth factor, causes activation of Smad transcription factors. In the absence of TGF β signaling due to either a receptor mutation or a SMAD mutation, cell proliferation and invasion of the surrounding extracellular matrix (ECM) increase

Which of the following phenomenon in this signaling pathway can lead to cancer?

(1) Loss-of-function mutations in receptors block TGF β signal
(2) Loss-of-function mutations in Smads block TGF β signal
(3) Increased production of p15 increases proliferation
(4) Decreased production of PAI-1 allows increased extracellular matrix degradation and, hence, metastasis

25. Figure below shows the time course of viral infection

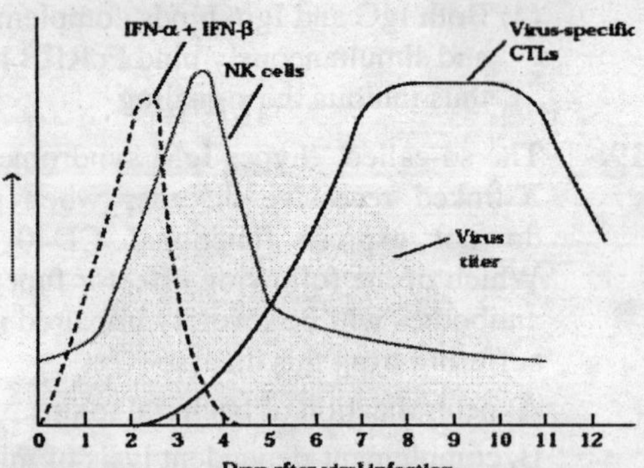

Which of the following statement is correct?

A. IFN-α and IFN-β (dashed curve) are released from virus-infected cells soon after infection

B. IFN-α and IFN-β cytokines are adaptive response which stimulate the viral specific NK cells, quickly leading to a rise in the NK-cell population (blue curve) from the basal level

C. NK cells help contain the infection during the period required for generation of CTLs (black curve)

D. Once the CTL population reaches a peak, the virus titer (blue area) rapidly decreases

(1) A, B and C
(2) B, C and D
(3) A, B and D
(4) A, B, C and D

26. Consider the following figure

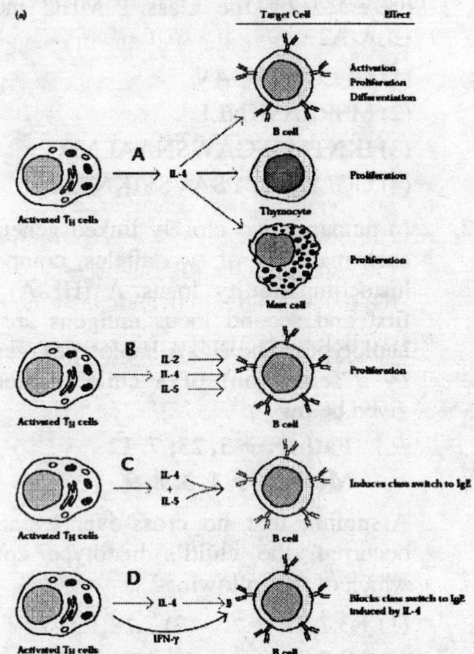

Which of the following is appropriate term for legend A, B, C and D?

(1) A: Plieotrophy B: Antagonism C: Synergy D: Redundancy

(2) A: Pleiotrophy B: Redundancy C: Synergy D: Antagonism

(3) A: Redundancy B: Pleiotrophy C: Synergy D: Antagonism

(4) A: Redundancy B: Antagonism C: Pleiotrophy D: Synergy

27. Four mice are immunized with antigen under the conditions listed below. In which case the induced serum antibodies will have high affinity and will be largely IgG

(1) A primary response to a low antigen dose

(2) A secondary response to a low antigen dose

(3) A primary response to a high antigen dose

(4) A secondary response to a high antigen dose

28. DNA was isolated from three sources: liver cells, pre-B lymphoma cells, and IgM-secreting myeloma cells. Each DNA sample was digested separately with the restriction enzymes *Bam*HI and *Eco*RI, which cleave germ-line heavy-chain and κ light-chain DNA as indicated in part (a) of the figure below. The digested samples were analyzed by Southern blotting using a radiolabeled $C_\mu 1$ probe with the *Bam*HI digests (blot #1) and a radiolabeled Cκ probe with the *Eco*RI digests (blot #2) The blot patterns are illustrated in part (b) of the figure

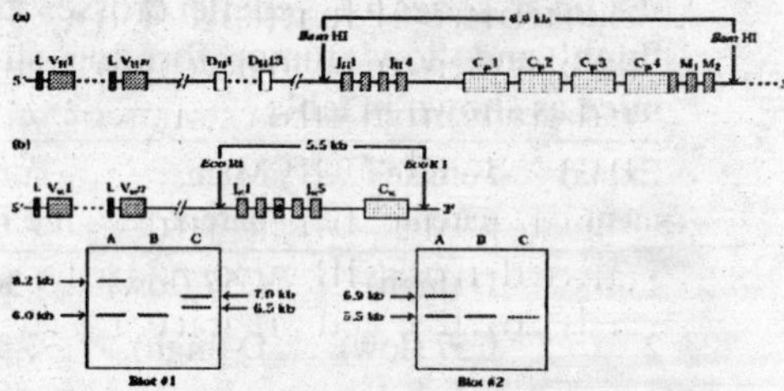

Based on this information the correct matching between DNA sample and organ from which it was isolated is?

(1) A: liver cells; B: pre-B lymphoma cells; C: IgM-secreting plasma cells

(2) A: IgM secreting plasma cells; B: liver cells; C: Pre-B lymphoma cells

(3) A: liver cells B: IgM-secreting plasma cells; C: pre-B lymphoma cells

(4) A: IgM-secreting plasma cells; B: pre-B lymphoma cells; C: Liver cells

29. Monoclonal antibodies (mAb) can be potentially used as therapeutic agents. The major advantage is that they can specifically target aberrant cells. However, there is a practical difficulty. Monoclonals are Raised in mouse and therefore it is expected that an immune reaction will develop if these are injected into humans. It is therefore, necessary to 'humanize' monoclonal antibody by

(1) expressing the genes for the mAb in cultured human cells and isolating the mAb from these cells

(2) replacing the Fv region of a mAb with one derived from a human IgG
(3) replacing C_L and C_H regions of the Mab with that obtained from a human IgG
(4) taking a human IgG and replacing the CDRs by those derived from the mouse mAb

30. Certain inbred strains of mice suffer tumors of breast at a relatively high frequency, whereas other inbred strain form breast tumors rarely, or not at all. To investigate the basis for their hereditary difference, you set up a series of genetic crosses between 'high' and 'low' tumor forming strains of mice as shown in table

Experi-ment	Female parent	Male parent	Tumors in F1 Females
1	D (high)	C57 (low)	36.1%
2	C57 (low)	D (high)	5.5%
3	A (high)	CBA (low)	86.3%
4	CBA (low)	A (high)	0.0%
5	Z (high)	I (low)	90.0%
6	I (low)	Z (high)	0.0%

You are amazed to find that high frequencies of tumors appear in Fl female mice only when their mothers were from the 'high' frequency strains. When you cross the Fl progeny generated in an experiment to produce F2 mice, you find the same result: high frequencies of tumor appear in F2 female mice only when their grandmothers were from high frequency strain. On basis of above result the location of chromosomal mutation and the probable mode of inheritance is?

(1) Autosomal dominant
(2) Autosomal recessive
(3) X-linked dominant
(4) X-linked recessive

31. The amino acid sequence of a small protein is

MSRLASKNLIRSDHAGGLLQATYSAVSS-IKNTMSFGAWSNAALNDSRDA

Predict the most likely peptide to be presented by the class I MHC molecule HLA-A2

(1) GLLQATYSAV
(2) MSRLASKNLI
(3) IKNTMSFGAWSNAALND
(4) GGLLQATYSAVSSIKNTM

32. In humans, two closely linked genetic loci each made up of two alleles, compose the histocompatibility locus A (HL-A) Paired first and second locus antigens are called haplotypes. The HL-A haplotypes (separated by a semicolon) of a child's parents are given below

Father:→ 3, 25; 7, 12
Mother:→ 1, 3; 8, 9

Assuming that no cross-over events have occurred, the child's histotype could be which of the following?

(1) 1,3;7,8
(2) 7,12;1,3
(3) 3,3;7,9
(4) 1,25;7,12

33. The graph below shows the sequential alteration in the type and amount of antibody produced after an immunization (Inoculation of antigen occurs at two different times, as indicated by the arrows. Curve A and curve B each represent a distinct type of antibody

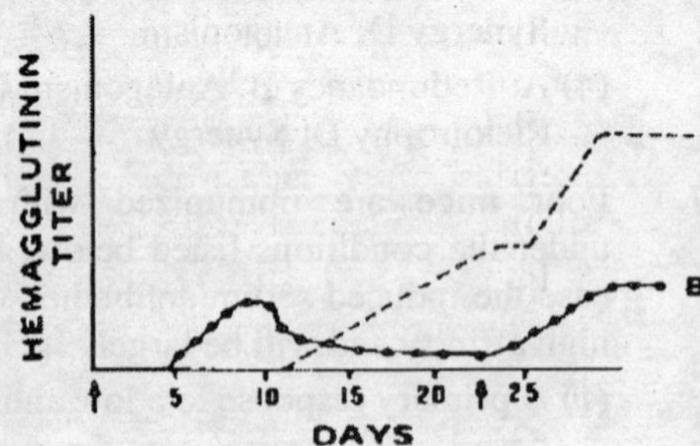

The class of immunoglobulin represented by curve B has which of the following characteristics?

(1) An estimated molecular weight of 150,000

(2) A composition of four peptide chains connected by disulfide links

(3) An appearance in neonates at approximately the third month of life

(4) The human ABO isoagglutinin

34. Complement-fixation (CF) testing is an important serological tool. One has to understand the conditions under which complement is bound and RBCs are lysed. Anti-*Mycoplasma* antibody + complement + hemolysin-sensitized red blood cells (RBC)+ anti-RBC antibody results in

(1) Complement is bound, red blood cells are lysed

(2) Complement is bound, red blood cells are not lysed

(3) Complement is not bound, red blood cells are lysed

(4) Complement is not bound, red blood cells are not lysed

35. One good example of this sort of "molecular hijacking" of normal cell adhesion processes occurs when pathogenic bacteria enter the digestive tract (such bacteria are called *enteropathogenic* bacteria; *entero* comes from the Greek word for "intestine") Such bacteria are responsible for several types of food poisoning, and their combined effects have a substantial impact on public health. Although some pathogenic bacteria can use multiple methods for gaining entry into the gut, in several well-studied cases, they attach to cell adhesion molecules such as integrins or cadherins. Consider the following statement

A. Infection by bacteria *Yersinia pseudotuberculosis* uses a 986-amino acid outer membrane protein, called *invasin,* to penetrate mammalian cells. The cellular receptors for invasin are integrins

B. Bacterium, *Shigella flexnerii,* also attaches to an integrin (α5β1) via proteins on its surface

C. *Listeria monocytogenes* expresses a protein called *internalin A* on its surface that can bind to E-cadherin on the surface of cells in the gut

Which of the following statements are correct?

(1) A and B (2) B and C

(3) A and C (4) A, B and C

Question 36-38

Sample No.	Lymphocytes/ml	Polymorpohonuclear Cells (PMN)/ml
1	14×10^6	3×10^6
2	8×10^6	5×10^6
3	55×10^6	4×10^6
4	30×10^6	35×10^6
5	11×10^6	5×10^6
6	52×10^6	6×10^6
7	71×10^6	6×10^6
8	12×10^6	3×10^6
9	92×10^6	4×10^6
10	49×10^6	3×10^6

Bcl-2 is a mitochondrial surface protein whose expression is associated with rescue from apoptosis (cell disintegration into particles that are then phagocytosed) Mice that are transgenic for bcl-2 gene (a proto oncogene) have been mated with non-transgenic mice. The transfene is inserted in such a way that allows it to be co-expressed with surface immunoglobulin. Blood samples from offsprings of the transgenic versus mating are analyzed and the result of the lymphocytes and polymorphonuclear cells is shown

36. The offspring most likely to have received the transgene are

(1) 3, 4, 6, 7, 9 and 10

(2) 1, 2, 5 and 8

(3) 4 only

(4) 3,6,7,9 and 10

(5) 2, 5 and 7

37. The elevation in mouse number 4's lymphocytes and polymorphonucleocytes is most likely due to
(1) bcl-2 trangene expression
(2) An error in count
(3) Malignancy
(4) Adverse drug reaction

38. If the proto-oncogene c-myc (associated with proliferation of cells) was inserted as a trangene rather than bcl-2 (same location), and the same blood counts were obtained from same backcross of transgenic to nontransgenic mice, which offspring would be most likely to have received the trangene
(1) 3, 4, 6, 7, 9 and 10
(2) 3, 6, 7, 9 and 10
(3) 4 only
(4) 1, 2, 5 and 8

39. Which of the following statements regarding oncogenic retroviruses is incorrect?
(1) They usually cause tumors in most species in which they can cause a productive infection
(2) Most infected permissive cells become tumor cells
(3) Integration of viral DNA into host DNA is obligatory for the production of progeny viruses
(4) All oncogenes of these viruses are essential for the production of progeny virions

40. Injection of an extract from the lotus leaves in a rabbit induces immune response. The characterization reveals that some of the antibodies can selectively recognize a protozoan antigen. Also the infection assays reveal that the rabbit becomes immune to the protozoan infection. Based on these observations, it is claimed that lotus leaves have an alkaloid that induces production of anti parasitic antibodies. Which of the following observations will weaken this claim substantially?
(1) The pond water- from where the lotus leaves were taken- shows very high parasitic count
(2) Lotus leaves do not have alkaloids that have been previously reported to have anti parasitic activity
(3) When the lotus leaves samples from 50 different ponds were screened, only 10% showed increased antibody production when injected in rabbits
(4) When the un-injected rabbits -grown in the laboratories were screened, 0.5% showed antibodies against the parasites

41. Most Gram-negative bacteria also possess a number of secretion systems. These include Type I-Type V secretion pathways. The Type I pathway, is relatively simple, consisting of just three proteins, and targets proteins with a specific 60-amino acid secretion signal at the carboxy terminus. Type II mechanism, is dependent on the GSP for transport of proteins to the periplasm and then uses a specific multiprotein complex to transport the protein across the outer membrane. The Type III pathway, which has been likened to a molecular syringe and which is used to inject proteins directly into the cytosol of eukaryotic cells. The Type IV secretory apparatus bears many similarities with the conjugal plasmid transfer systems and can be used by a bacterium to introduce proteins into eukaryotic cells. Type V secretion systems are similar, in that they are dependent on the Sec machinery for transport to the periplasm, but the translocated protein has a specialized C-terminal sequence which is able to insert into the outer membrane to form a pore. The N-terminal sequence of the same protein is able to pass through this pore, after which it is cleaved from the C-terminal sequence releasing it into the extracellular environment. These proteins are often called 'autotransporters' because they mediate transport through the outer membrane themselves and do not require additional machinery

Which of the following are not dependent on the GSP and are, therefore, termed Sec-independent

(1) Type I, II and III
(2) Type II, III, IV and V
(3) Type I, III and IV
(4) Type II and V

42. Tight junctions are apically placed junctions in the epithelial cells. These junctions are selectively permeable and regulate the water and electrolyte transport across the epithelium. The Trans-epithelial Electrical Resistance (TER) measurement reflects how permeable an epithelium is. The TER is inversely correlated with the epithelial permeability. Thus lower TER values are indicative of leakiness of the tight junctions. It is known that one of the cholera toxins called Zot increases the leakiness of the tight junctions, contributing to symptoms of cholera. A pharmaceutical company is screening for bacterial toxins, which can enhance the epithelial permeability. Their aim is to use the best of these toxins as an agent to enhance the transport of an anticancer drug across the intestinal epithelium. It is known that the transport of the drug increases when TER is below 5000 ohms. The effects of these 4 toxins (Figure A) and toxin+drug (Figure B) on TER were tested on cultured epithelial cells. The treatment started after 3 hours and the toxins and drugs were washed out after 4 hours of incubation

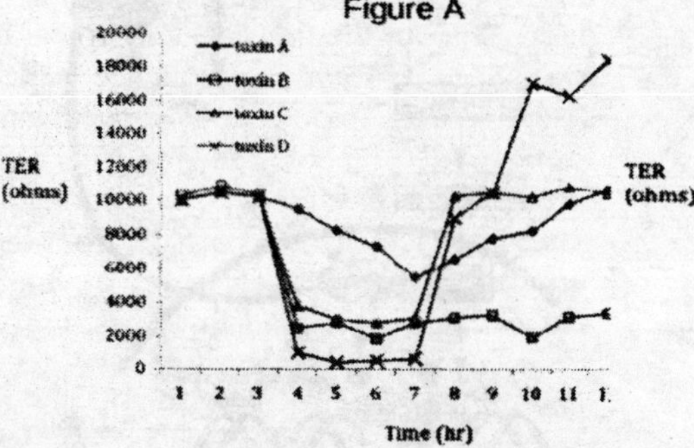

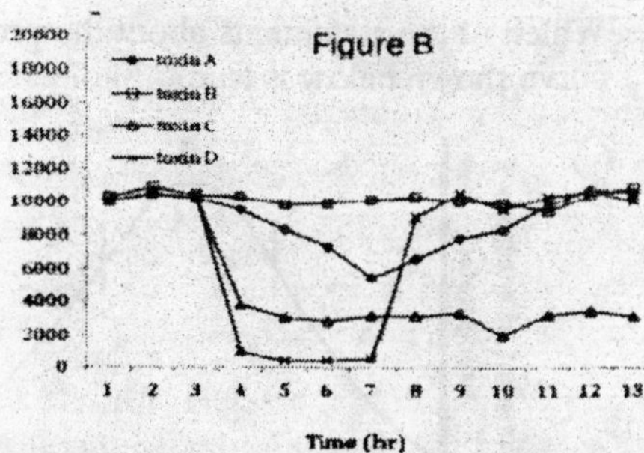

Which of these would you recommend for use in combination with the drug for clinical trials?

(1) Toxin B (2) Toxin A
(3) Toxin C (4) Toxin D

43. Polyomavirus (a DNA virus) causes tumors in "nude mice" (nude mice do not have a thymus, because of a genetic defect) but not in normal mice. The BEST interpretation is that

(1) macrophages are required to reject polyomavirus-induced tumors
(2) natural killer cells can reject polyomavirus-induced tumors without help from T lymphocytes
(3) T lymphocytes play an important role in the rejection of polyomavirus-induced tumors
(4) B lymphocytes play no role in rejection of polyomavirus-induced tumors

44. Which is the safest and most effective way to develop a vaccine that would protect the recipient if exposed to the natural toxin of *Vibrio cholerae*, the bacterium that causes cholera

(1) use purified natural toxin as a vaccine to elicit an immune response
(2) use live toxin-bearing Vibrio cholerae as a vaccine to elicit an immune response
(3) use a genetically engineered version of the cholera toxin with an altered carbohydrate-binding site
(4) use a genetically engineered version of the cholera toxin lacking the carbohydrate-binding site as a vaccine to elicit an immune response

45. Which of the statements about the precipitin curve shown below is true?

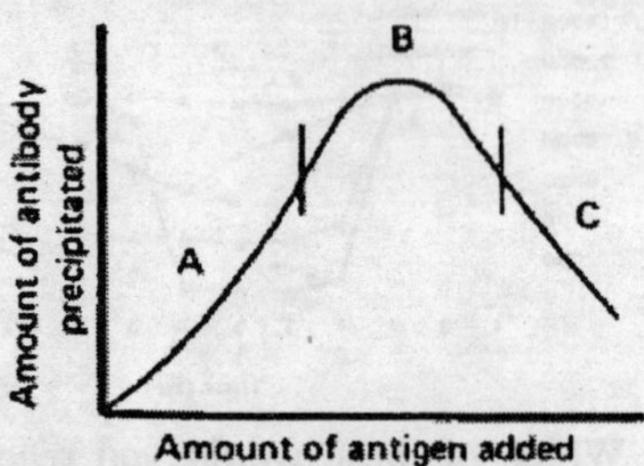

(1) In a multispecific system, a solution in zone B would have only an excess of antigen in the supernatant

(2) In a monospecific system, a solution in zone B would contain only reacted antibody and antige n

(3) A solution in zone A would be expected to have unreacted precipitable antigen in the supernatant

(4) A solution in zone C would be expected to have both antigen and antibody in excess

46. The table shows effect of transfection of DNA taken from malignant cells to normal cell lines

Source of DNA	*Recipient Cells*	*Number of Malignant Foci Found on 12 Plates*
Malignant Cell Lines		
MC5-5-0	NIH3T3 (normal fibroblasts)	48
MCA16	"	5
MB66 MCA ad 36	"	8
MB66 MCA ACL 6	"	0
MB66 MCA ACL 13	"	0
Normal Cell Lines		
NIH3T3	"	1
C3H10T1/2	"	0

On the basis of above data, following conclusion can be drawn

A. The DNA isolated from some (but not all) malignant cell lines could transform normal mouse cells, which then proliferated and produced malignant foci

B. From these experiments, it is clear why the DNA from two of the malignant cell lines, MB66 MCA ACL 6 and MB66 MCA ACL 13, could not transform the normal mouse cells. One possibility is that malignancies in these cell lines are caused by dominant oncogenes

C. Recessive genes would be able to transform normal mouse cells that already contain the normal (nonmalignant) dominant allele

Which of the following are valid conclusions?

(1) A and B (2) B and C
(3) A only (4) A, B and C

47. You are working with a type of fibroblast that normally responds to epidermal growth factor by increasing its rate of growth and division, and to epinephrine by lowering its rate of growth and division. You have determined that both of these responses require the MAP kinase pathway, and that EGF acts by means of an RTK and epinephrine by means of a G protein-coupled receptor. Suppose you identified a mutant strain of these cells that can still respond to EGF but is no longer inhibited by epinephrine. You suspect that the mutation is affecting the cross-talk between two pathways ?

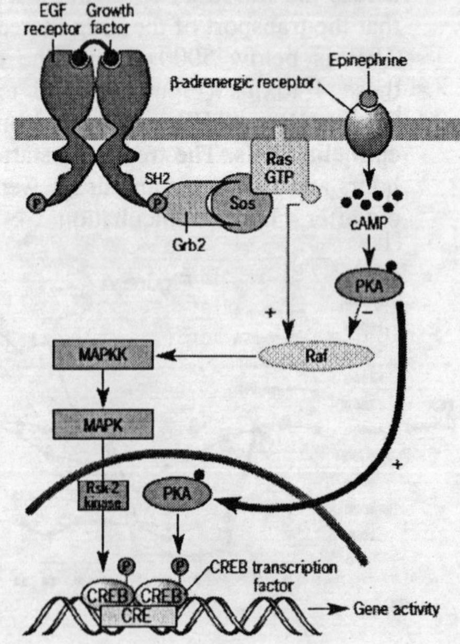

Which component in this figure might be affected by such a mutation?

(1) Raf (2) MAPKK
(3) PKA (4) CREB

48. Cancer is thought to be caused by the accumulation of two or more "hits" - that is, two or more mutations that affect cell proliferation and survival in the same cell. Many of these oncogenic mutations are dominant: one mutant copy of the gene is sufficient to change the proliferative properties of a cell. Consider the following events

A. a mutation that increases the number of copies of a transcriptional activator of cyclin A

B. a nonsense mutation located shortly after the beginning of translation in a gene that encodes a growth-factor receptor

C. a mutation that increases the level of FasL

D. a mutation that disrupts the active site of a cytoplasmic tyrosine-specific protein kinase

E. a translocation that joins a gene encoding an inhibitor of apoptosis to an active transcribed gene

Which of the following general types of mutations have the potential to create dominant oncogenes?

(1) B and C (2) A and E
(3) B, C and D (4) A, D and E

49. The *rb* gene encodes a protein that inhibits E2F, a transcription factor that activates several genes involved in cell division. Mutations in *rb* are associated with certain forms of cancer, such as retinoblastoma. Under each of the following conditions, would you expect cancer to occur?

(1) One copy of *rb* is defective; both copies of *E2F* are functional

(2) Both copies of rb are defective; both copies of *E2F* are functional

(3) Both copies of *rb* are defective; both copies of *E2F* is defective

(4) One copy of *rb* and both copies *E2F* are defective

50. A patient undergoes liver transplantation and during the course of post-operative treatment, becomes susceptible to infection. The patient can be treated in two different modes and can have alternative outcomes. Which of the following statements is correct?

(1) Treatment with immunostimulatory drugs reducing the infection but rejecting the transplant

(2) Treatment with immunostimulatory drugs reducing the infection and retaining the transplant

(3) Treatment with antibiotics reducing the infection but retaining the transplant

(4) Treatment with antibiotics reducing the infection but rejecting the transplant

51. The percent saturation of a ligand binding site depends upon two factors

A. the concentration of unbound ligand in the solution, and

B. the affinity of the binding site for the ligand

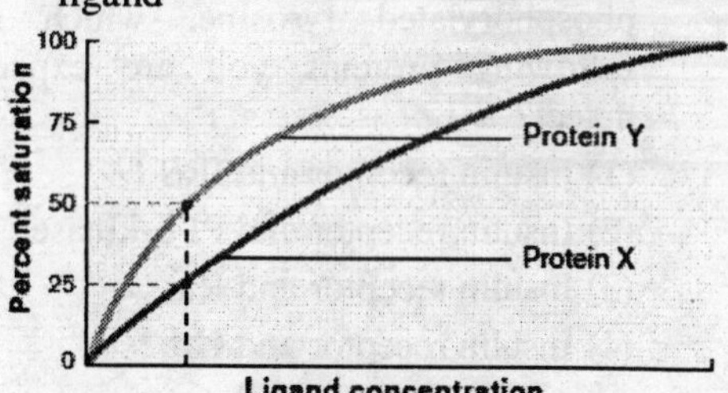

Which of the following is not valid conclusion based on graph shown above

(1) The percent saturation of binding sites increases with increasing ligand concentration until all the sites become occupied

(2) Assuming that the ligand is a molecule that exerts a biological effect when it is bound to a protein, the magnitude of the effect would also increase with increasing numbers of bound ligands until all the binding sites were occupied

(3) Once ligand binding site is 100 % saturated further increases in ligand concentration would produce no further effect since there would be no additional sites to be occupied

(4) Protein X shown above has high affinity ligand binding sites as compare to protein Y

52. In prokaryotes, environmental sensing frequently involves regulatory proteins (two-component systems) that sense and respond to changes in surroundings

I. Protein phosphorylaton

II. Transcriptional regulation

III. Membrane proteins

These two-component systems may involve which of the following?

(1) I only (2) II only
(3) III only (4) I, II and III

53. If you treat insulin-responsive mammalian cells in culture and immunoprecipitate with the antibody against insulin receptor and detect the proteins bound to the receptor by Western blotting using an antibody against phosphorylated tyrosine, which of the following proteins you are expected to detect?

(1) Insulin receptor and IRS-l

(2) Insulin receptor and PI-3-kinase

(3) Insulin receptor and Ras

(4) Insulin receptor and Grb2

54. Mutations in genes that encode the subunits of heterotrimeric G proteins can also lead to inherited disorders. This is illustrated by a report on male patient suffering from a rare combination of endocrine disorders: precocious puberty and hypoparathyroidism. Patients were found to contain a single amino acid substitution in one of the Gα isoforms. The alteration in amino acid sequence caused two effects on the mutant G protein. At temperatures below normal body temperature (33^{0}C), the mutant G protein remained in the active state, even in the absence of a bound ligand. In contrast, at normal body temperatures (37°C), the mutant G protein was inactive, both in the presence and absence of bound Ligand. The testes, which are housed outside of the body's core, have a lower temperature than the body's visceral organs (33°C vs. 37°C)

Following are statements drawn for mutation in Gα subunits

A. Normally, the endocrine cells of the testes initiate testosterone production at the time of puberty in response to the pituitary hormone LH, which begins to be produced at that time. The circulating LH binds to LH receptors on the surface of the testicular cells, inducing the synthesis of cAMP and subsequent production of the male sex hormone. The mutants will show normal phenotype

B. The testicular cells of the patients bearing the G protein mutation were stimulated to synthesize cAMP in the absence of the LH ligand, leading to premature synthesis of testosterone and precocious puberty

C. The mutation in this same Gα subunit in the cells of the parathyroid glands, which function at a temperature of 37°C, caused the G protein to remain inactive. As a result, the cells of the parathyroid gland could not respond to stimuli that would normally cause them to secrete parathyroid hormone, leading to the condition of hypoparathyroidism

D. The mutation in same Gα subunit in the cells of the parathyroid glands will have no effects at normal body temperature since receptor is able to dimerize at lower body temperature

Which of the conclusion are correct?

(1) A and D (2) B and C
(3) A and C (4) B and D

55. Intracellular receptors were identified and have been proposed to mediate actions of steroid hormones. In an experiment to study various actions of estrogen, it was covalently coupled to albumin, a protein of 68 kD. When this protein coupled hormone was added to estrogen responsive cells, a measureable response was observed. This indicates that

(1) estrogen receptor are not found inside the cell
(2) estrogen receptor are also present on the cell membrane
(3) estrogen coupled protein is taken by the cell by diffusion
(4) estrogen coupled protein is cleaved by the cells and free estrogen is taken by cells by passive diffusion

56. Consider the following statement regarding cell-cell/cell-matrix interactions

A. Adherens junction are forms cell-cell interaction connecting actin filament of one cell to another
B. Desmosomes are cell to matrix interaction connecting actin filament of one cell to ECM
C. Gap junctions are cylinders of connexin protein and acts as open channel for small hydrophilic molecules from one cell to another
D. Tight junctions are cell-cell interaction which seals the two cells
E. Hemi-desmosome are cell-matrix interaction connecting intermediate filament of cell to ECM

Which of the following statement are not correct?

(1) B only (2) B and C
(3) C and D (4) D and E

57. Cancer cells were either injected into the liquid-filled anterior chamber of a rabbit's eye, where there are no blood vessels, or they were placed directly on the iris. Tumor cells in the anterior chamber, nourished solely by diffusion, remain alive but stop growing before the tumor reaches a millimeter in diameter. In contrast, blood vessels quickly infiltrate the cancer cells implanted on the iris, allowing the tumors to grow to thousands of times their original mass. The valid conclusion which can be drawn from above experiments is

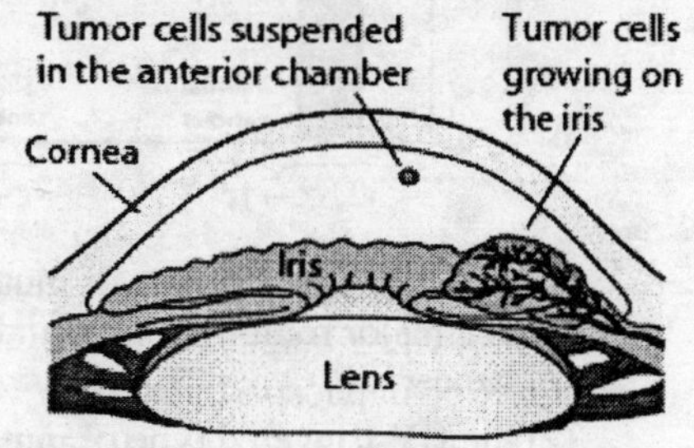

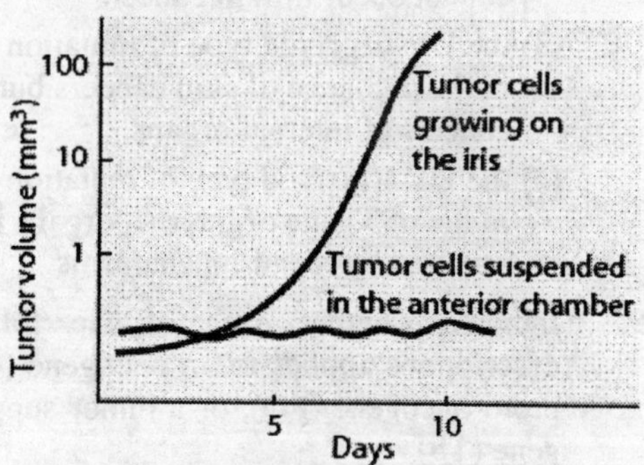

(1) cancer cell need a solid support for growth
(2) cancer cell required some growth factors for growth which are awailable in Iris
(3) angiogenesis is essential for survival and proliferation of cancer cells
(4) cancer cell are not always malignant

58. The graph above reports the incidence of two types of p53 mutations in skin cancer and internal cancers. The two bars on the left represent the frequency of CC→TT mutations, which are triggered by UV radiation. The two bars on the right represent the frequency of CST mutations not located at dipyrimidine sites, which are not caused by UV radiation. On basis of above information which is correct conclusion

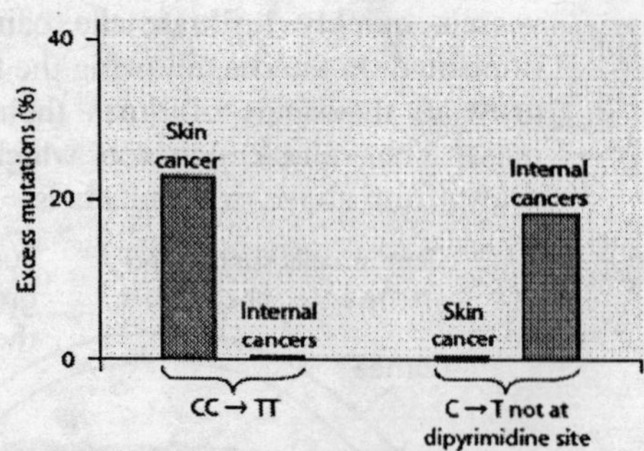

(1) the UV-triggered type of mutation in p53 is a major reason for skin and internal cancer

(2) the UV-triggered type of mutation in p53 cannot occur in skin cancers

(3) the UV-triggered type of mutation occurs in the p53 gene of skin cancers but not in cancers of internal organs

(4) the UV-triggered type of mutation occurs in the p53 gene of internal organs is repaired by repair mechanisms

59. Indicate whether each of the following descriptions applies to an oncogene (OG), a proto-oncogene (PO), or a tumor suppressor gene (TS)

A, A type of gene found in normal cells

B. A gene that could code for a normal growth factor

C. A type of gene found only in cancer cells

D. A gene whose absence can cause cancer

E. A type of gene found in normal cells and cancer cells

(1) OG: C PO: A, B, E TS: A, D

(2) OG: C, E PO: B, C, D TS: B, D, E

(3) OG: B, C, E PO: C, E TS: B, D, E

(4) OG: B, C, E PO: B, D, E TS: B, C, E

60. A signalling pathway I have identified has three proteins A, B, C that work to allow function X, When A is absent, X does not happen. The absence of C or B individually does not substantially affect function X. When C and B are both defective function X fails but can be restored if excess A is provided. Which of the following statements is false?

(1) Protein A functions after proteins C and B

(2) Protein A requires proteins C and B

(3) Protein A requires protein C or B to function

(4) Proteins C and B do not influence each others' function

61. Consider the following statement regarding oncogenes and tumor suppressor genes

A. Oncogenes are genes whose presence can cause cancer. Oncogenes always are brought into cells by viruses, rarely they arise from normal cellular genes (proto-oncogenes)

B. Many of the proteins produced by oncogenes are signaling pathway components, such as growth factors, receptors, plasma membrane GTP-binding proteins, nonreceptor protein kinases, transcription factors, and cell cycle and apoptosis regulators Oncogenes code for abnormal forms or excessive quantities of such proteins, thereby leading to excessive cell proliferation and survival

C. Tumor suppressor genes are genes whose loss or inactivation can lead to cancer. Susceptibility to cancer is increased in people who inherit defective tumor suppressor genes

D. Common tumor suppressor genes include the RB gene, which produces a protein that restrains passage from G1 into S phase; the p53 gene, which produces a protein that prevents proliferation and triggers apoptosis in cells with damaged DNA

Which of the following statement are correct?

(1) A and B (2) C and D

(3) B, C and D (4) All of the above

62. Diagram below represents Wnt signaling pathway which plays important role in cell proliferation and differentiation during embryonic development

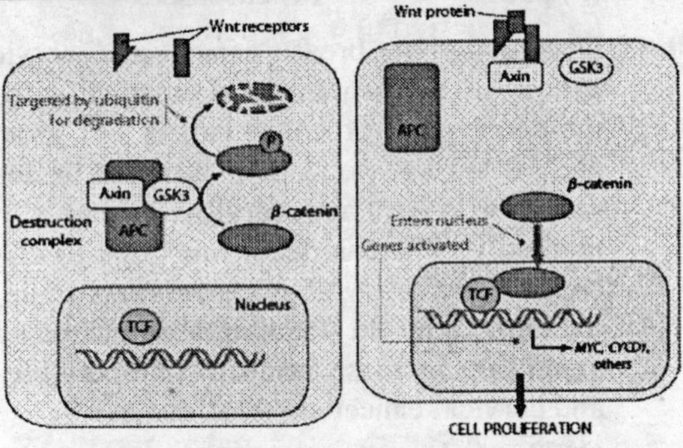

On the basis of above information which of the following condition can lead to uncontrolled growth?

(1) Loss-of function mutation in Wnt gene

(2) Loss of function mutation in TCF gene

(3) Loss-of-function mutations in the APC gene

(4) Loss of function mutation in myc or cycD1 gene

63. The chemical substance 2-acetylamino-fluorene (AAF) causes bladder cancer when injected into rats but not guinea pigs. If normal bladder cells obtained from rats and guinea pigs are grown in culture and exposed to AAF, neither are converted into cancer cells. How can you explain these findings?

(1) The rats are probably metabolizing AAF into a carcinogen (this is called 'activation'). The guinea pigs likely use a different enzyme to metabolyze AAF, and therefore 'fail' to activate it

(2) Detoxification process in the rats is probably occurring in the bladder. That is why when AAF is injected directly into the cell culture

(3) Detoxification of AAF in rat is absent while is present in cells of pig

(4) Detoxification and activation of AAF is absent both in pig and rats

64. Consider the following statement regarding signaling receptors and transduction process

A. The receptors involved in paracrine, synaptic, and endocrine signaling all have very high affinity for their respective signaling molecules

B. All small intracellular mediators (second messengers) are water soluble and diffuse freely through the cytosol

C. In the regulation of molecular switches, protein kinases and guanine nucleotide exchange factors (GEFs) always turn proteins on, whereas protein phosphatases and GTPase-activating proteins (GAPs) always turn proteins off

D. In contrast to the more direct signaling pathways used by nuclear receptors, catalytic cascades of intracellular mediators provide numerous opportunities for amplifying the responses to extracellular signals

Which of the following statements are correct?

(1) A and B (2) B and C

(3) C and D (4) Only D

65. Consider the following statement regarding cell communication

A. Binding of extracellular ligands to receptor tyrosine kinases activates the intracellular catalytic domain by propagating a conformational change across the lipid bilayer through a single transmembrane α helix

B. Protein tyrosine phosphatases display exquisite specificity for their substrates, unlike serine/threonine protein phosphatases, which have rather broad specificity

C. Even though plants and animals independently evolved multicellularity, they use virtually all the same signaling proteins and second messengers for cell–cell communication

Which of the following statements are not correct?

(1) A and B (2) B and C
(3) A and C (4) C only

66. Suppose that the circulating concentration of hormone is 10^{-10} M and the K_d for binding to its receptor is 10^{-8} M. What fraction of the receptors will have hormone bound?

(1) 1 % (2) 10 %
(3) 11 % (4) 99 %

67. Suppose that the circulating concentration of hormone is 10^{-10} M and the K_d for binding to its receptor is 10^{-8} M. If a meaningful physiological response occurs when 50% of the receptors have bound a hormone molecule, how much will the concentration of hormone have to rise to elicit a response?

(1) 10^{-10} M (2) 10^{-9} M
(3) 10^{-8} M (4) 10^{-5} M

68. Cells communicate in ways that resemble human communication. Decide which of the following forms of human communication are analogous to autocrine, paracrine, endocrine, and synaptic signaling by cells

A. A telephone conversation
B. Talking to people at a birthday party
C. A radio announcement
D. Talking to yourself

(1) autocrine: D, paracrine: B, endocrine: C synaptic: A
(2) autocrine: D, paracrine: A, endocrine: C synaptic: B
(3) autocrine: A, paracrine: B, endocrine: C synaptic: D
(4) autocrine: D, paracrine: C, endocrine: B synaptic: A

69. Cancer causing genes can be functionally classified into mainly three types
(i) genes that induce cellular proliferation
(ii) tumor suppressor genes
(iii) genes that regulate apoptotic pathway
Epstein-Barr virus that causes cancer by modulating apoptotic pathway, contains a gene having sequence homology with which of the following genes?

(1) bax (2) bcl-2
(3) p53 (4) caspase-3

70. Deaths due to breast cancer and cervical cancer in women differ dramatically from the same plot for colon cancer. At around age 50 the age-dependent increase in death rates for breast and cervical cancer slows markedly, whereas death rates due to colon cancer (and most other cancers) continue to increase. Why do you suppose that the age-dependent increase in death rates for breast and cervical cancer slows after age 50?

(1) Due to loss of regenerative tissue
(2) Lack of estrogen due to menopause
(3) Number of cells in colon increases after age of 50
(4) Immune system become weak after age of 50

71. Virtually all cancer treatments are designed to kill cancer cells, usually by inducing apoptosis. However, one particular cancer-acute promyelocltic leukemia (APL)- has been successfully treated with all-trans-retinoic acid, which causes the promyelocytes to differentiate into neutrophils. How might a change in the state of differentiation of APL cancer cells help the patient?

(1) Differentiated cells can be effectively removed by natural killer cells
(2) Differentiated cells stop dividing at arrest at G_0 state
(3) Differentiation leads to expression of novel antigen which can be recognized by CTL cells
(4) Retinoic acid induces expression of tumor suppressor genes in differentiated cells

72. In many signaling pathways, ligand binding to a receptor leads to activation of transcription factors in the cytosol, permitting them to translocate into the nucleus and stimulate (or occasionally repress) transcription of their target genes. Alternatively, receptor stimulation may lead to activation of cytosolic protein kinases that then translocate into the nucleus and regulate the activity of nuclear transcription factors. Some activated receptors, particularly certain G protein–coupled receptors, also can induce changes in the activity of preexisting proteins. Consider the following statements

A. **G protein-coupled Receptors:** Cytosolic domain with tyrosine kinase activity; activate cytosolic kinases (here MAP kinase) that translocate to the nucleus and activate nuclear transcription factors by phosphorylation

B. **TGFβ receptors:** Cytosolic domain with serine/threonine kinase activity; Activate Smad transcription factors in the cytosol by phosphorylation

C. **Cytokine receptors:** Associated with cytosolic JAK kinases; Activate cytosolic STAT transcription factors by phosphorylation

D. **Wnt receptors:** Palmitoylated Wnt ligand binds seven transmembrane protein receptor complex; Release an activated transcription factor from a multiprotein complex in the cytosol

Which of the following statements are correct?

(1) A, B and C (2) B, C and D
(3) A, C and D (4) A, B, C and D

73. Genetic studies demonstrated that TBP mutant cell extracts are deficient in transcription of genes from all three promoters viz. class I, II and III. Following statements describe characteristic features of TBP

A. TBP is considered as an universal basal transcription factor

B. TBP is not required for transcription of archaeal genes

C. TBP is involved in recognizing TATA box

D. TBP operates at all promoters regardless of their TATA content

Which of the following combinations is not correct?

(1) A and D (2) C and D
(3) B and D (4) A and C

74. Cancer is a fundamental aberration in cellular behavior, touching on many aspects of molecular cell biology. Most cell types of the body can give rise to malignant tumor (cancer) cells. Cancer cells usually arise from stem cells and other proliferating cells and bear more resemblance to these cells than to more mature differentiated cell types

Following statements describe characteristics of cancer cells

A. Cancer cells can multiply in the absence of at least some of the growth-promoting factors required for proliferation of normal cells and are resistant to signals that normally program cell death (apoptosis)

B. Cancer cells sometimes invade surrounding tissues, often breaking through the basal laminae that define the boundaries of tissues and spreading through the body to establish secondary areas of growth, a process called metastasis. Metastatic tumors often secrete proteases, which degrade the surrounding extracellular matrix

C. All tumors do not require angioge-nesis, the formation of new blood vessels, in order to grow to a large mass

D. The multi-hit model, which proposes that multiple mutations are needed to cause cancer, is consistent with the genetic homogeneity of cells from a given tumor, the observed increase in the incidence of human cancers with advancing age, and the cooperative effect of oncogenic transgenes on tumor formation in mice

E. Most oncogenic mutations occur in somatic cells and are carried to next generation

Which of the following statement is not correct?

(1) A and C (2) C, D and E
(3) A, B and C (4) C and E

75. Dominant gain-of-function mutations in proto-oncogenes and recessive loss-of-function mutations in tumor-suppressor genes are oncogenic. Among the proteins encoded by proto-oncogenes are growth-promoting signaling proteins and their receptors, signal-transduction proteins, transcription factors, and apoptotic proteins. Tumor-suppressor genes encode proteins that directly or indirectly slow progression through the cell cycle, checkpoint control proteins that arrest the cell cycle, components of growth-inhibiting signaling pathways, pro-apoptotic proteins, and DNA-repair enzymes. DNA microarray analysis can identify differences in gene expression between types of tumor cells that are indistinguishable by traditional criteria. Some tumor cells appear to be related to specific types of normal cells at certain stages of development based on similarities in their expression patterns. Consider the following statements

A. An activating mutation of one of the two alleles of a protooncogene is not sufficient to convert it in an oncogene. Mutation in both alleles can occur by point mutation, gene amplification, and gene translocation

B. Slow-acting retroviruses can cause cancer by integrating near a proto-oncogene in such a way that transcription of the cellular gene is activated continuously and inappropriately

C. The first human oncogene to be identified encodes a constitutively active form of Ras, a signal-transduction protein. This oncogene was isolated from a human bladder carcinoma. The first tumor-suppressor gene to be recognized, *RB*, is mutated in retinoblastoma and some other tumors

D. Inheritance of a single mutant allele of *RB* always leads to specific kind of cancer to develop, as is the case for many other tumor-suppressor genes (e.g., *APC* and *BRCA1*)

E. In individuals born heterozygous for a tumor-suppressor gene, a somatic cell can undergo loss of heterozygosity (LOH) by mitotic recombination, chromosome missegregation, mutation, or deletion

Which of the following statement is correct?

(1) A, B and C (2) A, B and E
(3) B, C and D (4) B, C and E

76. Indicate which of the following statements regarding haptens and carriers are true

A. Haptens are large protein molecules such as BSA

B. When a hapten-carrier complex containing multiple hapten molecules is injected into an animal, most of the induced antibodies are specific for the hapten

C. Carriers are needed only if one wants to elicit a cell-mediated response

D. It is necessary to immunize with a hapten-carrier complex in order to obtain antibodies directed against the hapten

E. Carriers include small molecules such as dinitrophenol and penicillenic acid (derived from penicillin)

(1) A, B and C (2) B, C and D
(3) B and D (4) A, C and E

77. For each pair of antigens listed below, indicate which is likely to be more immunogenic

A1: Native bovine serum albumin (BSA)

A2: Heat-denatured BSA

B1: Hen egg-white lysozyme (HEL)

B2: Hen collagen

C1: A protein with a molecular weight of 30,000

C2: protein with a molecular weight of 150,000

D1: BSA in Freund's complete adjuvant

D2 :BSA in Freund's incomplete adjuvant

(1) A1, B2, C1, D1 (2) A2, B1, C1, D2

(3) A1, B1, C2, D1 (4) A2, B2, C2, D2

78. Indicate whether each of the following statements is true?

A. A rabbit immunized with human IgG3 will produce antibody that reacts with all subclasses of IgG in humans

B. All immunoglobulin molecules on the surface of a given B cell have the same idiotype

C. All immunoglobulin molecules on the surface of a given B cell have the same isotype

D. All myeloma protein molecules derived from a single myeloma clone have the same idiotype and allotype

E. The heavy-chain variable region (V_H) is twice as long as the light-chain variable region (V_L)

(1) A, B and D (2) B, C and E

(3) A, D and E (4) C and E

79. Consider the following statements

A. Although IgA is the major antibody species that undergoes transcytosis, polymeric IgM, but not monomeric IgA, can also undergo transcytosis

B. The hypervariable regions make significant contact with the epitope

C. IgG functions more effectively than IgM in bacterial agglutination

D. Although monoclonal antibodies are often preferred for research and diagnostic purposes, both monoclonal and polyclonal antibodies are not highly specific

E. All isotypes are normally found in each individual of a species

Which of the following statements are false?

(1) C and D (2) A, B and E

(3) A, B and C (4) A, C and E

80. You are an energetic immunology student who has isolated protein X, which you believe is a new isotype of human immunoglobulin. You prepare rabbit antisera to whole human IgG, human λ chain, and human κ chain. Assuming protein X is, in fact, a new immunoglobulin isotype, to which of these antisera would it bind?

(1) λ chain

(2) κ chain

(3) both λ chain and κ chain

(4) whole human IgG and κ chain

81. Because immunoglobulin molecules possess antigenic determinants, they themselves can function as immunogens, inducing formation of antibody. For each of the following immunization scenarios, indicate whether anti-immunoglobulin antibodies would be formed to isotypic (IS), allotypic (AL), or idiotypic (ID) determinants

A. Anti-DNP antibodies produced in a BALB/c mouse are injected into a C57BL/6 mouse

B. Anti-BGG monoclonal antibodies from a BALB/c mouse are injected into another BALB/c mouse

C. Anti-BGG antibodies produced in a BALB/c mouse are injected into a rabbit

D. Anti-DNP antibodies produced in a BALB/c mouse are injected into an outbred mouse

E. Anti-BGG antibodies produced in a BALB/c mouse are injected into the same mouse

(1) IS: C, D AL: A ID: B

(2) IS: A, B AL: C ID: D

(3) IS: A, C AL: D, E ID: B

(4) IS: D, E AL: B ID: A, C

82. An investigator wanted to make a rabbit antiserum specific for mouse IgG. She injected a rabbit with purified mouse IgG and obtained an antiserum that reacted strongly with mouse IgG. To her dismay, however, the antiserum also reacted with each of the other mouse isotypes

(1) Because isotypes has same sequence in their constant region of heavy chain

(2) Because isotypes has minor variation in their constant region which are not sufficient to be distinguished

(3) Because antiserum would also contain antibodies to kappa and lambda light chain, which are present in all isotypes

(4) Because antibody generated against mouse IgG is polyclonal in nature which reacted with all isotypes

83. Epidermal growth factor (EGF) is needed for growth of almost all cells. EGF receptor is a transmembrane protein having an extracellular ligand-binding domain, a trans-membrane domain and a cytosolic domain of protein tyrosine kinase (PTK) Binding of EGF to the receptor activates PTK resulting in activation of transcription factors through intracellular transducers. In cell type A, much of the extra cellular ligand-binding domain is deleted by proteases such that cytosolic domain of PTK becomes constitutively active whereas cell type B is having normal EGF receptor. What will be the best-fit graph for the growth of the cultures of cell type A and B in complete medium in presence (+) and absence (-) of EGF?

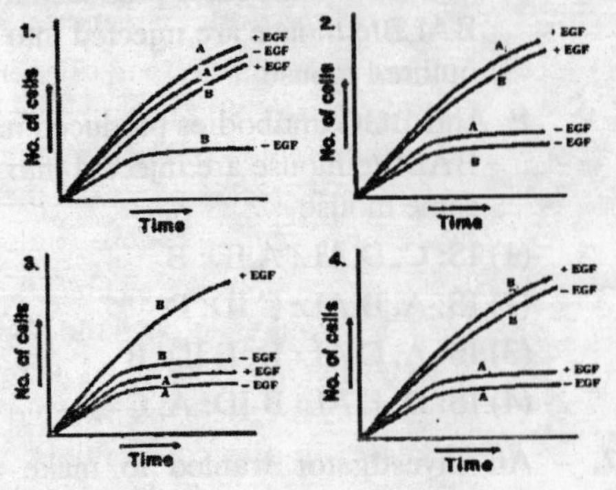

84. After successive surgery and chemotherapy, the tumor of a breast cancer patient subsided. However, after almost 5 years, the tumor relapsed in a more aggressive manner and did not respond to the conventional chemotherapy delivered earlier. The following postulations were made

A. Chemoresistant cells were persisting within the tumor even after therapy

B. A population of quiescent cells existed, which under favourable conditions, transformed to new tumor cells

C. High ABC (ATP-Binding Cassette) transporter expressing cells persisted in the breast during chemotherapy

D. Breast tumor cells which may have migrated to other tissues, returned to the breast immediately after chemotherapy was terminated

Which of the above combination of statements is true?

(1) A and C (2) A, B and C

(3) Only B (4) B and D

85. A particular type of cancer cell undergoes apoptosis by both extrinsic and intrinsic pathways when treated with a chemotherapeutic agent X. Caspase 8 and Caspase 9 are the initiator caspases associated with extrinsic and intrinsic pathways respectively. Now, if caspase 9 is silenced in the cancer cell by shRNA transfection, what will be the best fit graph for apoptosis scenario in the cancer cell when treated with agent X?

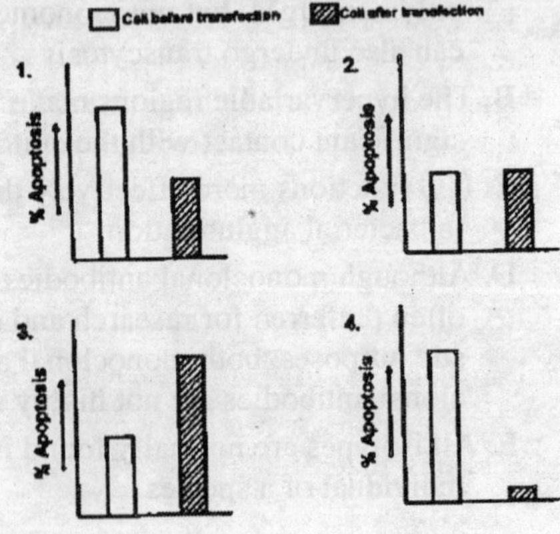

86. Integrity of membranes is maintained by association between different membrane proteins, which remains difficult to define. The association involving spectrin, ankyrin, band 3 and actin generates filamentous meshwork on the cytoplasmic surface of the red cell plasma membrane and has been investigated using antibodies that are specific to individual proteins. In this experiment, a mixture of two proteins is incubated together and then an antibody specific for one of them is added. The resulting antibody-protein complexes are then precipitated and analyzed

This technique applied to pair wise mixtures of spectrin, ankyrin, band 3 and actin yielded the following results:

Expt. No.	Protein mixture	Antibody Specificity	Proteins in pellet
1.	Band 3+ankyrin	Actin	Actin
2.	Band 3+spectrin	Spectrin	Spectrin
3.	Band 3+ankyrin	Ankyrin	Band 3+ankyrin
4.	Actin + spectrin	Spectrin	Actin + spectrin
5.	Actin + ankyrin	Ankyrin	ankyrin
6.	Spectrin+ankyrin	Spectrin	Spectrin+ankyrin

From the above data, the association between these proteins is deduced to be

(1) ankyrin-spectrin-actin-band 3
(2) spectrin-actin-ankyrin-band 3
(3) band 3-ankyrin-spectrin-actin
(4) actin-ankyrin-spectrin-band 3

87. Two antibiotics have different sites of action in a bacterial cell. The frequency of occurrence of resistance to these antibiotics used singly are 10^{-5} and 10^{-6}, respectively. When the antibiotics are used in combination the frequency of occurrence of resistance to both antibiotics will likely be

(1) 10^{-5} (2) 10^{-6}
(3) 10^{-11} (4) 10^{-30}

88. Consider the following statements

A. Basophils are non-phagocytic granulocytes that function by releasing pharmacologically active substances from their cytoplasmic granules which play a major role in certain allergic responses

B. Mast-cell precursors, which are formed in the bone marrow by hematopoiesis, are released into the blood as undifferentiated cells and enter the tissues including the skin, connective tissues of various organs. Mast cells, play an important role in the development of allergies

C. The follicular dendritic cell arise in bone marrow and has a major function in antigen-presenting

D. Neutrophils leave blood circulation and get mature into large phagocytotic cells termed as macrophage

Which of the following statements are correct?

(1) A and B (2) B and C
(3) C and D (4) A and D

89. The immunoglobulin genes are expressed in B-cells, and the β-globin gene is expressed in red blood cells. What manipulations might be carried out to get the β-globin gene to be expressed in B-cells?

(1) The coding region of an immunoglobulin gene could be combined with the cis-regulatory region of the β-globin genes, and this artificial gene would then express β-globin in B-cells
(2) The cis-regulatory region of the β-globin gene could be inserted into B-cells, which would then express their own version of the β-globin gene
(3) The cis-regulatory region of an immunoglobulin gene could be combined with the coding region of the β-globin genes, and this artificial gene would then express β-globin in B-cells
(4) The transcription factors GATA-1 and GATA-2 (which activate the β-globin gene in erythrocytes) could be introduced into B-cells, which would then express their own copy of the β-globin gene

90. Imagine that in an attempt to get the β-globin gene to be expressed in B-cells, the transcription factors (GATA-1 and GATA-2) known to control the β-globin gene in erythrocytes were introduced into B-cells, yet the β-globin gene was not expressed. What is a likely explanation for this result?

(1) The erythrocyte transcription factors are degraded rapidly in the B-cell, and so cannot act on the β-globin gene

(2) The β-globin gene in the B-cells is packaged into heterochromatin, and the erythrocyte transcription factors are insufficient to remodel that chromatin into an active state

(3) Erythrocyte transcription factors cannot interact with their target DNA sequences in B-cells, even though those sequences are present and accessible

(4) Even though the necessary erythrocyte-specific transcription factors are now present in the B-cells, the general transcription factors required for β-globin gene expression would not be present

91. In contrast to their similar brain abnormalities, new born mice deficient in Apaf1 or caspase-9 have distinctive abnormalities in their paws. Apaf1-deficient mice fail to eliminate the webs between their developing digits whereas caspase-9-deficient mice have normally formed digits (Figure) If Apaf1 and caspase-9 functions in the same apoptotic pathway, how is it possible for these deficient mice to differ in web cell apoptosis?

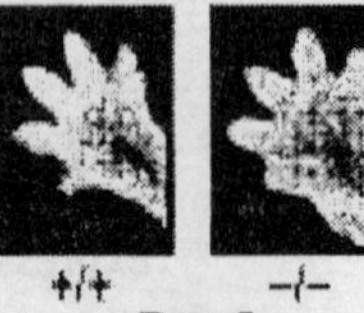

(1) Apaf1 promote apoptosis and caspase-9 is inhibitor of apoptosis

(2) Caspase-9 works upstream of Apaf-1

(3) Capase-9 and Apaf-1 are members of two different apoptotic cascades

(4) Capase-9 mutant are recessive to Apaf-1 mutants

92. Tumor-suppressor genes on their products can

P. act as dominant negative genes

Q. act as recessive genes

R. prevent cell-cycle progression

S. regulate apoptosis

(1) P and Q (2) Q and S
(3) P, Q and R (4) P, Q, R and S

93. Which of the following statement(s) is/are correct?

P. during the stimulation of B cells, a type of T cell called the suppresser T cell is involved

Q. the antibody identified as IgE functions in many allergic reaction

R. antibody-mediated immunity is dependent upon the activity of T lymphocytes

S. the largest antibody molecule, consisting of five monomer units, is IgD

(1) P and S (2) only Q
(3) P and R (4) P, Q, Rand S

94. Which of the following happens when RBC completes it lifespan of four months?

(1) Terminal sialic acid groups disappears

(2) Removal of sialic acid exposes underlying galactose residues

(3) Phosphotidyl serine is observed in outer leaflet of RBC membrane

(4) All of the above

95. Choose the incorrect statement

Toxin	Enzyme activity
(1) Botulinum toxin	Cleavage of SNARE
(2) Diptheria toxin	ADP ribosylation of eEF2
(3) Tetanus toxin	ADP ribosylation of G_s
(4) Shiga toxin	Inhibition of protein synthesis

96. What is globin gene switching?

(1) Globin gene switching refers to the switching on of the globin genes during erythrocyte differentiation

(2) Globin gene switching refers to the switching from α-globin to β-globin, so that α2β2 hemoglobin can be made

(3) Globin gene switching refers to the transition, during fetal development, from the production of hemoglobin containing ε-globin, to γ-globin, to β-globin

(4) Globin switching refers to a genetic rearrangement in the DNA that joins different segments of DNA, resulting first in ε-globin synthesis, then γ-globin synthesis, and finally β-globin synthesis

97. Opsonisation of a bacterium is a process by which specific antibody binds with the surface molecule of the bacteria. In an experimental condition, macrophage were infected with either WT: Mycobacteria or with opsonised: Mycobacteria for 2 hrs at 37°C. Subsequently, cells were washed and further incubated for 24 hrs at 37°C. Finally, bacterial load in macrophages were determined by colony forming unit (CFU) Which of the following observation is true?

(1) WT:Mycobacteria inhibits its transport to the lysosomes and survive in macrophages

(2) Opsonised: Mycobacteria inhibits its transport to the lysosomes and survive in macrophages

(3) WT: Mycobacteria are targeted to the lysosomes and killed in macrophages

(4) Opsonised: Mycobateria are targeted to the lysosomes and survive in macrophages

98. In an *in vitro* experiment using radio-labeled nucleotides, a researcher is trying to analyze the possible products or DNA replication by resolving the products using urea-polyacrylamide gel electrophoresis. In one experimental set up RNase H was added (Set 1), while in another set no RNase H was added (Set 2),

The possible observations of this experiment could be

A. There is no difference in the mobility of labelled DNA fragments between the Set I and Set 2

B. There is distinct difference in the mobility of the newly synthesized labeled DNA fragments between Set I and Set 2

C. The mobility of the newly synthesized labelled DNA fragments in case of Set 1 is faster as compared to the Set 2

D. The mobility of the newly synthesized labelled DNA fragments in case of Set 1 is slower as compared to the Set 2

Which of the following combinations represent correct observations?

(1) A and B (2) B and C
(3) A and D (4) B and D

99. The genome-wide comparative mRNA expression profile between normal and cancer cells can be studied using the microarray technology. The basic approach of this technology is

(1) hybridization of fluorescently labelled genomic DNA from human cells to normal and cancer cell cDNAs immobilized onto two glass slides

(2) hybridization of fluorescently labelled cDNAs from normal and cancer cells to a single glass slide containing immobilized clones of human genomic DNA

(3) hybridization of fluorescently labelled cDNAs from normal and cancer cells to a single glass slide containing immobilized clones of human cDNA

(4) hybridization of fluorescently labelled genomic DNA from normal and cancer cells to immobilized human cDNAs on a single glass slide

100. You are studying activated T helper cells in a mouse model system. T helper cells are a sub-group of lymphocytes that play an important role in establishing and maximizing the capabilities of the immune system without themselves having any cytotoxic or phagocytic activity. Your lab studies small molecules secreted by T helper cells that activate Natural Killer (NK) cells. You identify a small glycoprotein Interleukin-79 that is secreted by T helper cells, and seems to activate NK cells. This glycoprotein is particularly highly expressed in a subset of T helper cells which you have purified from primary mouse serum. Why isn't this primary cell line suitable for long term study?

(1) Primary cell lines are immortal

(2) Primary cell lines are resistant to apoptosis

(3) Primary cell line have finite life span

(4) Primary cell line cannot be cultured in absences of NK Cells

ANSWER SHEET

Unit 4 : Signalling, Immunology, Cancer

1. 2	**2.** 1	**3.** 2	**4.** 2	**5.** 1	**6.** 1	**7.** 3	**8.** 4	**9.** 3	**10.** 4
11. 4	**12.** 1	**13.** 1	**14.** 3	**15.** 3	**16.** 3	**17.** 4	**18.** 1	**19.** 2	**20.** 1
21. 2	**22.** 1	**23.** 4	**24.** 4	**25.** 4	**26.** 2	**27.** 4	**28.** 2	**29.** 4	**30.** 3
31. 1	**32.** 2	**33.** 4	**34.** 2	**35.** 4	**36.** 4	**37.** 2	**38.** 2	**39.** 4	**40.** 3
41. 3	**42.** 4	**43.** 3	**44.** 2	**45.** 2	**46.** 3	**47.** 3	**48.** 2	**49.** 2	**50.** 3
51. 4	**52.** 4	**53.** 1	**54.** 2	**55.** 4	**56.** 1	**57.** 3	**58.** 3	**59.** 1	**60.** 2
61. 3	**62.** 3	**63.** 1	**64.** 4	**65.** 4	**66.** 1	**67.** 3	**68.** 1	**69.** 2	**70.** 2
71. 2	**72.** 2	**73.** 3	**74.** 4	**75.** 4	**76.** 3	**77.** 3	**78.** 1	**79.** 1	**80.** 4
81. 1	**82.** 3	**83.** 1	**84.** 1	**85.** 2	**86.** 3	**87.** 3	**88.** 1	**89.** 3	**90.** 2
91. 3	**92.** 4	**93.** 2	**94.** 4	**95.** 3	**96.** 3	**97.** 1	**98.** 2	**99.** 3	**100.** 3

UNIT - 5

DEVELOPMENTAL BIOLOGY

1. The first few cell divisions in *C. elegans* produce six different *founder cells,* from embryonic stem cells, each with a separate fate as shown in Figure

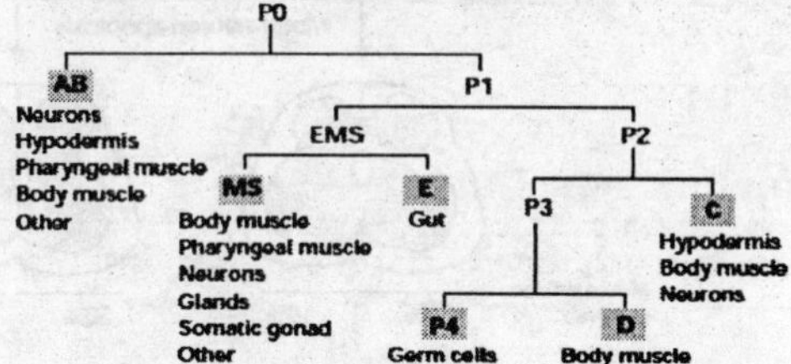

The initial division is asymmetric, giving rise to P1 and the AB founder cell. Further divisions in the P lineage form the other five founder cells. Some of the signals controlling division and fate asymmetry are known. For example, Wnt signals from the P2 precursor control the asymmetric division of the EMS cell into E and MS founder cells. Wnt signaling is also used in other asymmetric divisions in worms. Consider the following statements?

A. Embryonic stem cells (ES cells) are capable of giving rise to many kinds of differentiated cell types

B. During development, founder cells (AB/MS/E/C/D) generally gain potential; that is, they become progressively more potent in the number of different cell types they can form

C. Progenitor Cells (P1/P2/P3/P4) can produce more than one type of descendant cell, including in some cases a stem cell with a more-restricted potential to produce differentiated cell types under gradient of different morphogens

D. Founder cells have broad developmental potential and their prospective fate can be modified depending on the environment in which they are placed

Which of the following statement is false?

(1) A and D (2) B only

(3) B and D (4) C and D

2. Genetic studies in *C. elegans* defined an evolutionarily conserved apoptotic pathway with three major components: regulatory proteins, adapter proteins, and effector proteases called caspases in vertebrates

A. Once activated, apoptotic proteases cleave specific intracellular substrates leading to the demise of a cell. Adapter proteins (e.g., Apaf-1), which bind both regulatory proteins and caspases, are required for caspase activation

B. Pro-apoptotic regulator proteins (e.g., Bax, Bad) promote caspase activation, and anti-apoptotic regulators (e.g., Bcl-2) suppress activation

C. The Bcl-2 family contains pro-apoptotic; all are single-pass transmembrane proteins and engage in protein-protein interactions and are responsible for apoptosis

D. Binding of extracellular death signals, such as tumor necrosis factor and Fas ligand, to their receptors activates an associated protein (FADD) that in turn triggers the caspase cascade leading to cell murder

Which of the following statement are correct?

(1) Only A and D
(2) A, B and C
(3) A, B and D
(4) A, B, C and D

3. bHLH transcription factors have analogous functions in determination of precursor cells (i.e., A or C) and their subsequent differentiation into mature B or D cells. In both cases, the proteins encoded by the earliest-acting genes (*left*) are under both positive and negative control by other related proteins. Here, the cells A, B, C and D are

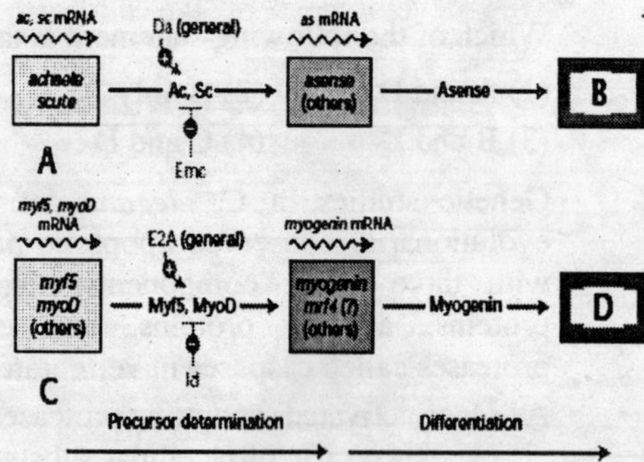

(1) A: Neuroblast, B: Neuron, C: Myoblast, D: Muscle
(2) A: Neural precursor cell, B: Neuron, C: Muscle precursor cell, D: Muscle
(3) A: Pre Neural Cell, B: Neuroblast, C: Myoblast, D: Muscle Precursor cell
(4) A: Mycoblast, B: Neuron, C: Neuroblast, D: Muscle

4. Removal of a limb bud from one side of a chick embryo at about 2.5 days results in a marked decrease in the number of motor neurons on the affected side. In an amputated embryo, normal numbers of motor neurons are generated on both sides (middle)

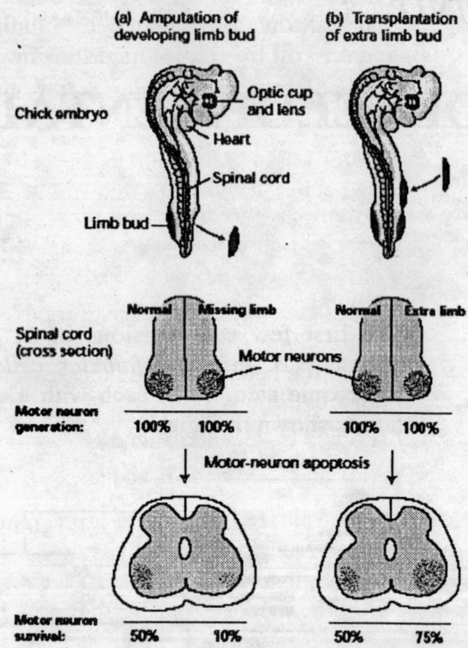

Later in development, many fewer motor neurons remain on the side of the spinal cord with the missing limb than on the normal side (bottom) Note that only about 50 percent of the motor neurons that originally are generated normally survive. (b) Transplantation of an extra limb bud into an early chick embryo produces the opposite effect, more motor neurons on the side with additional target tissue than on the normal side. This experiment concludes that

A. The survival of motor neurons depends on the size of the muscle target field they innervate
B. Neurons has finite number and cannot be varied
C. Neurons never enter into apoptosis if survival factor (NGF) is removed
D. Cell survival or Apoptosis is dependent on presence of NGF released from limb

Which of the following statement is correct?

(1) A and B
(2) B and C
(3) B, C and D
(4) A and D

5. Signal transduction initiated by binding of ligand on Toll like receptors results in

A. Activation of transcription factors such as NFkB

B. Promotes the expression of genes that contribute to inflammation which induces change in antigen presenting cells that make them more efficient in antigen presentation

C. causes the synthesis and export of intercellular signaling molecules that affect the behavior of leukocytes and other cells

(1) A only (2) A and B

(3) A, B and C (4) B and C

6. In the vertebrate eye precise arrangement of tissues is accomplished by one group of cells changing the behavior of an adjacent set of cells, thereby causing them to change their shape, mitotic rate, or fate. This kind of interaction at close range between two or more cells or tissues of different history and properties is called proximate interaction, or induction

A. Optic vesicle (presumptive retina) of *Xenopus laevis* is placed in an ectopic location (i.e., in a different place from where it normally forms) underneath the head ectoderm, it will induce that ectoderm to form lens tissue

B. Optic vesicle is inducer and even if placed beneath ectoderm in the flank or abdomen of the same organism, that ectoderm will be able to respond to form lens tissue

C. in the developing chick and mammalian eye, the Pax6 protein appears to be important in making the ectoderm competent to respond to the inductive signal from the optic vesicle

Which of the following statement is true in vertebrate eye development?

(1) Only C (2) A and B both

(3) A and C (4) A, B and C

7. When proteins synthesized by one cell can diffuse over small distances to induce changes in neighboring cells, the event is called a paracrine interaction, and the diffusible proteins are called paracrine factors or growth and differentiation factors (GDFs)

Consider the following statement about paracrine factors?

A. FGFs can activate a set of receptor tyrosine kinases called the fibroblast growth factor receptors (FGFRs) and kinase is activated, and it phosphorylates certain proteins within the responding cell. FGF2 is especially important in angiogenesis, and FGF8 is important for the development of the midbrain and limbs

B. Vertebrates have at least three homologues of the Drosophila *hedgehog* gene: *sonic hedgehog* (*shh*), *desert hedgehog* (*dhh*), and *indian hedgehog* (*ihh*) Desert hedgehog is expressed in the Sertoli cells of the testes, and mice homozygous for a null allele of *dhh* exhibit defective spermatogenesis. Indian hedgehog is expressed in the gut and in cartilage and is important in postnatal bone growth. Sonic hedgehog made by the notochord, it is processed so that only the amino-terminal two-thirds of the molecule is secreted. This peptide is responsible for patterning the neural tube such that motor neurons are formed from the ventral neurons and sensory neurons are formed from the dorsal neurons

C. The Wnts constitute a family of cysteine rich glycoproteins. Wnt1 appears to be active in inducing the dorsal cells of the somites to become muscle. Wnt proteins also are critical in establishing the polarity of insect and vertebrate limbs, and they are used in several steps of urogenital system development

D. The proteins encoded by TGF-β superfamily genes are processed such that the carboxy-terminal region contains the mature peptide. These peptides are dimerized into homodimers (with themselves) or heterodimers (with other TGF-β peptides) and are secreted from the cell. The TGF-β superfamily includes the TGF-β family, the activin family, the bone morphogenetic proteins (BMPs), the Vg1 family, and other proteins, including glial-derived neurotrophic factor (necessary for kidney and enteric neuron differentiation) and Müllerian inhibitory factor (which is involved in mammalian sex determination)

Which of the following statement is correct?

(1) A and C (2) B and D
(3) A, B and D (4) A, B, C and D

8. The anterior-posterior axis of the *Drosophila* embryo appears to be patterned before the nuclei even begin to function. The nurse cells of the ovary deposit mRNAs in the developing oocyte, and these mRNAs are apportioned to different regions of the cell

In particular, four maternal messenger RNAs are critical to the formation of the anterior-posterior axis which makes protein gradients in the early embryo

A. An anterior-to-posterior gradient of Bicoid protein
B. A posterior-to-anterior gradient of Hunchback protein
C. An anterior-to-posterior gradient of Nanos protein
D. A posterior-to-anterior gradient of Caudal protein

Which of the following statement is correct?

(1) A and D (2) B and C
(3) A, B and D (4) All of the above

9. Which of the following cannot be used for determination of tissue lineage of a given progenitor cell population in an animal?

(1) Marking progenitor cells with vital dye
(2) Transplanting equivalent progenitor cells from immunologically distinct but related organism
(3) Marking progenitor cells by genomic recombination coupled with reporter gene expression
(4) Marking progenitor cells by reporter gene expression under the control of a promoter – enhancer element specific for the given progenitor population

10. What would be the effect on the timing of somite formation, if a piece of the pre-somitic mesoderm of a chick embryo is rotated by 180° & reinserted in its original position?

(1) Antero-posterior development would be blocked at that point, and more posterior structures would fail to form
(2) The rotated mesoderm would degenerate and a gap would form in the antero-posterior pattern
(3) The rotated mesoderm will reverse the gradient of somite formation, so that somite formation posterior to the rotated mesoderm will begin in the posterior and move forward to meet the previously formed somites
(4) The timing of somite formation will be reversed in the rotated block only, proceeding from posterior to anterior, but the rest of the somites will form in a normal fashion, proceeding from anterior to posterior

11. Recently, it was reported that adult, differentiated cells can be made to behave as pluripotent stem cells by the introduction of a few genes, one of which was Oct4. What was the rationale for this experiment?

(1) Oct4 is a master switch gene that turns on the stem cell program
(2) Oct4 expression is one of the requisites for the maintenance of pluripotent embryonic stem cells in culture
(3) The Oct4 gene produces a cell-cell signaling protein that is used by stem cells to stimulate cell division
(4) Oct4 is only expressed in embryonic stem cells, and so its expression automatically reverses any differentiation that may have gone on in a cell

12. The role of maternal factors (such as Vg-1, Xwnt-1, and VegT) packaged into the vegetal region of the Xenopus oocyte is
   (1) in establishing the future dorso-ventral body axis
   (2) to allow the vegetal region to divide during cleavage
   (3) to make the vegetal region denser, so that the egg always floats with its animal pole upward
   (4) to specify the future animal-vegetal body axis

13. The development of drugs to block the activity of vascular endothelial growth factor (VEGF) has been applied to the control of the growth of cancers. What is the rationale for such a strategy?
   (1) VEGF acts as a growth factor to stimulate the growth of many solid tumors
   (2) Tumor cells produce VEGF to stimulate angiogenesis in their vicinity, and tumors cannot continue to grow without this supply of blood
   (3) Blocking VEGF prevents vasculogenesis in the embryo, and if done properly prevents cancer later in life
   (4) VEGF causes blood vessels to grow toward cancers, which kills the cancer due to too much blood

14. Is regeneration in *Hydra* similar to regeneration in vertebrates, or is it fundamentally different in some way?
   (1) Regeneration in *Hydra* occurs through a repatterning of existing cells by morphallaxis, unlike regeneration in vertebrates, which occurs through epimorphosis
   (2) Regeneration in *Hydra* occurs through the reinitiation of cell division in existing cells, followed by repatterning of those cells, very much like regeneration in vertebrates
   (3) Regeneration in *Hydra* occurs through the reinitiation of cell division in existing cells, followed by repatterning of those cells, which is totally different from regeneration in vertebrates
   (4) Regeneration in *Hydra* occurs through the reinitiation of cell division in existing cells, followed by repatterning of those cells, whereas regeneration in vertebrates occurs through a process of repatterning of existing cells

15. In a classic experiment to investigate cytokinesis, a glass bead was pressed into a fertilized sea urchin egg cell before it began embryonic cell divisions. Observation of the altered embryo suggested that
   (1) the position of cleavage furrows corresponded to the position of the metaphase plate of a mitotic spindle
   (2) defective contractile rings formed in the embryo that did not contain myosin filaments
   (3) cleavage furrows were positioned equidistant between two asters
   (4) formation of cleavage furrows was inhibited by the presence of the bead

Question 16-18

Figure (A) below represents the distribution of mRNA of different genes involved in Drosophila development. Figure (B) represents the distribution of their proteins in early cleavage embryo

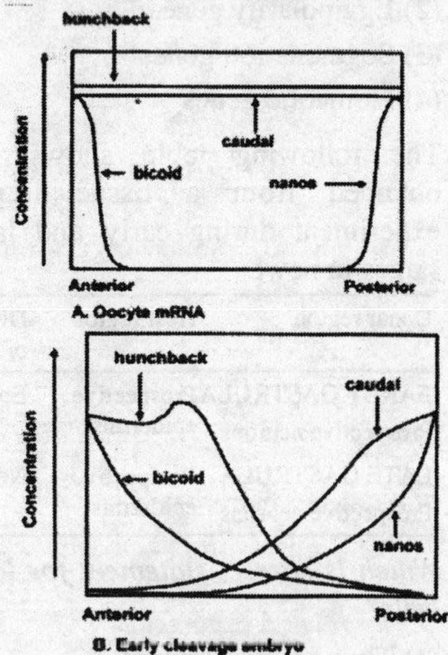

16. Where are these mRNAs transcribed?
(1) Anterior portion of embryo
(2) Posterior portion of embryo
(3) In oocyte prior to fertilization
(4) In different portion of embryo during cleavage

17. How is the distribution of mRNA in oocytes (Fig. A) converted into protein gradient as shown in Fig. B
(1) Expression of Caudal protein is suppressed by bicoid while expression of hunchback is suppressed by nanos at translational level
(2) Expression of Caudal protein is suppressed by bicoid while expression of hunchback is suppressed by nanos at transcriptional level
(3) Expression of bicoid protein is suppressed by caudal while expression of nanos is suppressed by hunchack at translational level
(4) Expression of nanos protein is suppressed by caudal while expression of bicoid is suppressed by hunchack at translational level

18. Caudal and hunchback genes are?
(1) Embryonic gene
(2) Egg polarity gene
(3) Segmenation genes
(4) Homeotic genes

19. The following table shows the results obtained from a tissue transplantation experiment during early and late gastrula stages in newt.

Donar region	Host region	Differentiation of donar tissue
EARLY GASTRULA Prospective neurons	Prospective epidermis	Epidermis
LATE GASTRULA Prospective neurons	Prospective epidermis	Neurons

Which is correct statement for late gastrula cells?
(1) They are pleuipotent
(2) They are totipotent
(3) They are determined or commited
(4) The development pattern in newt is mosaic at gastrula stage

Question 20-22

Experiment I

Embryonic Stage	Isolated Cell Type	Mesodermal Actin
32 cells	Crescent cells	–
	Animal pole cells	–
	Vegetal pole cells	–
128 cells	Crescent cells	–
	Animal pole cells	–
	Vegetal pole cells	–
512 cells	Crescent cells	+
	Animal pole cells	–
	Vegetal pole cells	–

After sperm entry, the cytoplasm of the amphibian egg rearranges to form an equatorial cresecent between the animal pole and vegetal pole. After cleavage, the crescent cells involulate during gastrulation to form mesoderm. Differentiated mesodermal cells produce a specific type of acitin. To investigate mesodermal differentiation, two experiments were preforemed. In experiment I, cells were isolated from the embryo at various stages and assayed for mesodermal actin. (+ indicates the presence of mesodermal actin)

Experiment 2

Embryonic Stage	Treatment	Mesodermal Actin		
		Animal Cells	Vegetal Cells	Crescent Cells
32 cells	Animal pole cells + vegetal pole cells	+	–	
	Vegetal pole cells + crescent cells		–	+
	Animal pole cells + crescent cells	–		–

In experiment 2, three different cell types from the 32-cell stage were isolated and recombined, as shown below. Later each cell type was assayed for mesodermal actin. The results are shown in the table below

20. As measured by the appearance of actin in experiment 1, mesoderm most likely differentiates
(1) before the 32-cell stage
(2) between the 32- and 128-cell stages
(3) at the 128 cell stage
(4) between 128- and 512-cell stages

21. Which of the following is the most logical conclusion based on the results of both experiments?

(1) Animal pole cells induce mesoderm

(2) Vegetal pole cells induce mesoderm

(3) Crescent cells induce mesoderm

(4) Animal pole cells self-differentiate as mesoderm

22. When isolated from 32-celled embryo, animal pole cells will produce an ectoderm-specific protein and isolated vegetal pole cells will produce an endoderm-specific protein. These observation and those of experiment 2 indicate that animal pole cells are competent to differentiate as

(1) mesoderm only

(2) ectoderm only

(3) endoderm only

(4) mesoderm and ectoderm

23. Manipulation of embryos often demonstrates that cell fate is influenced by the cell's environment. Cellular behavior may be influenced by intercellular signaling molecules (e.g., morphogens), direct cell-cell interactions (e.g., via cell-adhesion molecules) and extracellular matrix. A cell's fate is thus largely determined by its neighbors. Such an mode of development is termed as

(1) Conditional Development

(2) Mosaic Development

(3) Committed Development

(4) All of the above

24. At the early gastrula stage, the *Xenopus* (Ambhibian) embryo expresses genes for the important signaling molecules

I. Wnt8

II. Chordin, Noggin, Follistatin

III. BMP 2 and 4

IV. Dkk, Crescent, Frzb

These are used for patterning and morphogenesis at the gastrula/neurula stages

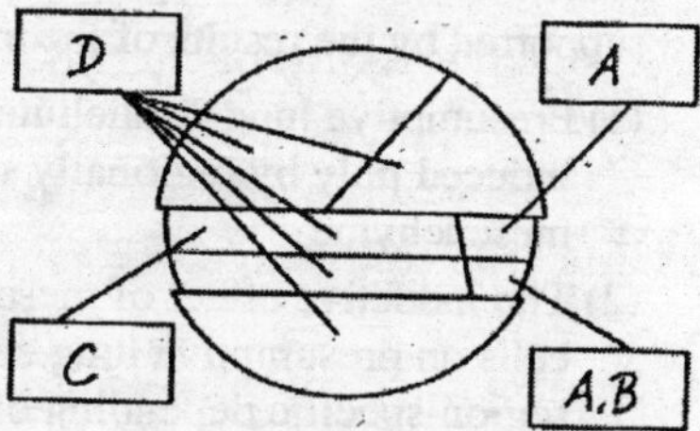

On early gastrula fate map shown above identify the genes for these signaling molecule which are expressed in box A, B, C and D

(1) A: I, B: IV, C: III, D: II

(2.) A: II, B: IV, C: I, D: III

(3) A: I, B: II, C: IV, D: III

(4) A: II, B: III, C: IV, D: I

25. Presumptive lung epithelium was removed from the endodermal tubes of a mammalian embryo and cultured in appropriate media both in the absence (1) and presence (2-6) of mesenchymal cells. The mesenchymal cells were chosen from specific locations in the embryo. The goal of the experiment was to ascertain of and in what ways patterns of epithelial differentiation varied in response to regional source of the mesenchymal inductive tissue. Results of the cell culture are shown in the diagram below

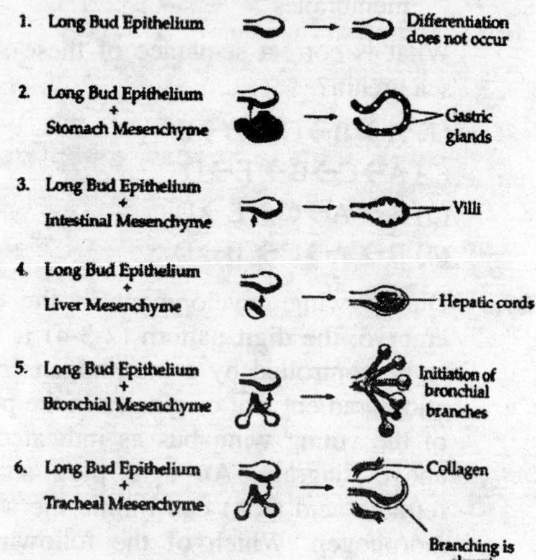

Which of the following statements is best supported by the results of the experiment?

(1) Presumptive lung epithelium will be induced pnly by regionally specific mesenchyme

(2) The inductive effect of mesenchymal cells on presumptive lung epithelium is region-specific depending on the source of mesenchyme

(3) Reciprocal induction occurs between regionally specific mesenchyme and presumptive lung epitherlium

(4) Epithelio-mesenchymal interactions are not a prerequisite for differentiation of presumptive lung epithelium

26. The interaction of sperm and egg generally proceeds according to five basic steps

A. The chemoattraction of the sperm to the egg by soluble molecules secreted by the egg

B. The exocytosis of the acrosomal vesicle to release its enzymes

C. The binding of the sperm to the extracellular envelope (vitelline layer or zona pellucida) of the egg

D. The passing of the sperm through this extracellular envelope

E. Fusion of egg and sperm cell plasma membranes

What is correct sequence of these steps in sea urchin?

(1) A→ B→C→ D→E

(2) A→C→B→ E→D

(3) B→ A→C→ E→D

(4) B→ A→C→ B→D

27. During wing development in the chicken embryo, the digit pattern (2-3-4) is thought to be controlled by a morphogen concentr-ation gradient that originates in the posterior of the young wing bus as indicated in the above diagram. An agar plug soaked in retinoic acid (RA) can mimic the action of morphogen. Which of the following digit patterns would be expected to result if an agar plug soaked in retinoic acid were placed in the anterior of developing wing bud?

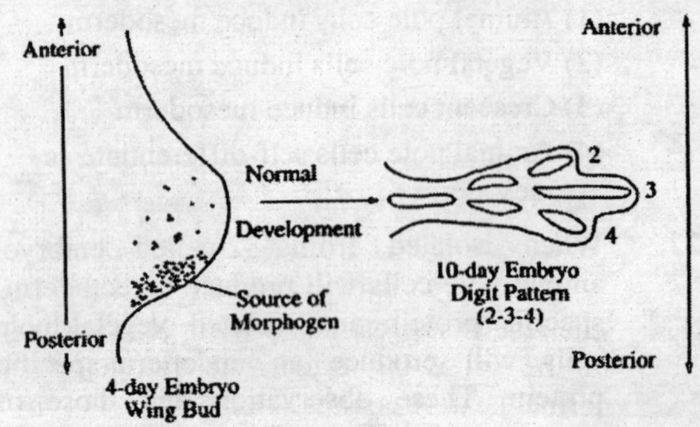

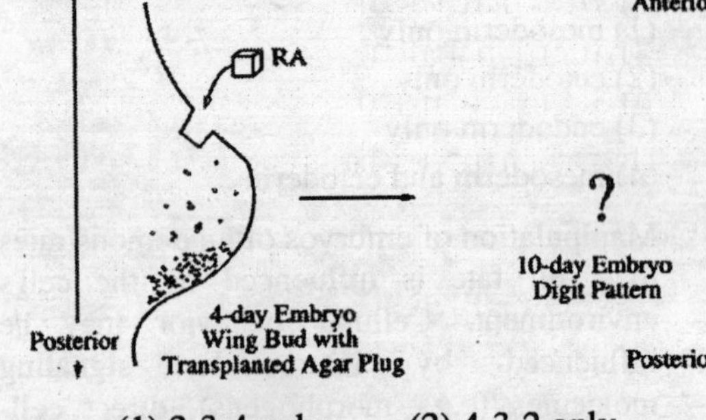

(1) 2-3-4 only
(2) 4-3-2 only
(3) 2-3-4-4-3-2
(4) 4-3-2-2-3-4

Questions 28-31

Protein X is a protein kinase encoded by mRNA X and is normally present in extracts of whole cleavage-stage frog embryos. Protein I, encoded by mRNA I, is an inactive mutant form of protein X and blocks the function of protein X present in the same cell by competing for binding. The results shown in the table below were obtained in a series of experiments in which all four cells in four-cell-stage embryos were injected as shown and allowed to develop to the early larval stage. Control embryos were pierced by a needle, but no material was injected. The larvae were scored for the presence or absence of eyes in the normal dorsal body axis and for the presence of eyes in the ventrally located secondary body axis that formed as a result of some of the treatments

Material Injected	Injection Location	Dorsal Body Axis with Eyes	Ventral Body Axis with Eyes
Control	Dorsal	100%	0%
mRNA X	Dorsal	0%	0%
mRNA I	Dorsal	100%	0%
Control	Ventral	100%	0%
mRNA X	Ventral	100%	0%
mRNA I	Ventral	100%	100%

28. The results suggest that protein X has which of the following effects on formation of normal head structures?

(1) Weak inhibition

(2) Weak stimulation

(3) Strong inhibition

(4) Strong stimulation

29. The data indicate that in at least some cells in the frog embryos, translation effectively occurs for

(1) mRNA X but not mRNA I

(2) mRNA I but not mRNA X

(3) neither mRNA X nor mRNA I

(4) both mRNA X and mRNA I

30. If translation of mRNA X occurs during cleavage, the results suggest that protein X would most likely be detectable at which location in untreated late-cleavage-stage frog embryos?

(1) Primarily in the dorsal cells of the embryos

(2) Primarily in the equatorial cells of the embryos

(3) Primarily in the ventral cells of the embryos

(4) In all of the cells of the embryos

31. In Mangold and Spemann's experiments pieces of dorsal lip tissue transplanted to the ventral surface of amphibian embryos induced the formation of a secondary body axis. Their results can be reconciled with the results shown above by assuming which of the following about the cells of the dorsal lip of the blastopore?

(1) They secrete protein X into the extracellular matrix of the embryo

(2) They destroy protein X in the extracellular matrix of the embryo

(3) They stimulate protein X synthesis or increase activity of protein X in neighboring cells in the embryo

(4) They inhibit protein X synthesis or decrease activity of protein X in neighboring cells in the embryo

32. Apoptosis is necessary not only for the proper spacing and orientation of neurons, but also for generating the middle ear space, the vaginal opening, and the spaces between our fingers and toes. Different tissues use different signals for apoptosis. One of the signals often used in vertebrates is bone morphogenetic protein 4 (BMP4)In other tissues, the cells are "programmed" to die, and they will remain alive only if some growth or differentiation factor is present to "rescue" them. This happens during the development of mammalian red blood cells

Consider the following statements-

A. While apoptosis-deficient nematodes deficient for CED-4 are viable (despite their having 15% more cells than wild-type worms), mice with loss-of function mutations for either caspase-3 or caspase-9 die around birth from massive cell overgrowth in the nervous system

B. Mice homozygous for targeted deletions of *Apaf-1* have severe craniofacial abnormalities, brain overgrowth, and webbing between their toes

C. In mammals, there is more than one pathway to apoptosis. The apoptosis of the lymphocytes, for instance, is not affected by the deletion of *Apaf-1* or *caspase-9*, and works by a separate pathway initiated by the CD95 protein

D. In vertebrate red blood cell development, the Stat5 transcription factor activated by erythropoietin functions by binding to the promoter of the *Bcl-X* gene, where it activates the synthesis of that anti-apoptosis protein

Which of the following statement is true for apoptosis?

(1) A, B and C (2) B and D only
(3) A, C and D (4) A, B, C and D

33. Rifampicin and actinomycin D are two antibiotics derived from the bacterium *Streptomyces*. Rifamycin binds to the b subunit of *E. coli* RNA polymerase and interferes with the formation of the first phosphodiester bond in the RNA chain. Actinomycin D binds to double-stranded DNA by intercalation (slipping between neighboring base pairs) When fertilized sea urchin eggs are treated with actinomycin D, they develop for many hours but eventually arrest as hollow balls of several hundred cells (called blastulae) The probable reason is

(1) Early cleavage division do not require proteins
(2) Embryonic gene expression occurs after mid blastula stage
(3) Maternal gene expression occurs after mid blastula stage
(4) Complete embryonic development is dependent on maternal encoded m-RNA

34. To study the role of proteins (e.g., transcription factors) that function in development is to microinject the mRNA that encodes a protein, or the purified protein itself, into an oocyte or embryo, and then determine how this affects the subsequent development of the embryo, larva, and adult. For example, if Bicoid protein is injected into the posterior region of an oocyte, the resulting embryo will develop into a larva that has anterior structures at both ends. Based on your understanding of the function of these developmental genes, what would be the predicted phenotype if the following proteins or mRNAs were injected into normal oocytes?

Experiments

A. *Nanos* mRNA injected into the anterior end of an oocyte
B. Antp protein injected into the posterior end of an embryo
C. *Toll* mRNA injected the dorsal side of an early embryo

Observation

1. The larva would develop with two posterior ends. The larva would not survive to the adult stage
2. The posterior end of the larva and adult fly would develop structures that were appropriate for thoracic segments. The adult fly may not survive
3. The dorsal side of the larva and adult fly would develop structures that were appropriate for the ventral side. The adult fly may not survive

(1) A: 3, B:2, C:1 (2) A:3, B:1, C:2
(3) A:1, B:3, C:2 (4) A:1, B:2, C:3

35. You have in your possession wild-type and bicoid mutant strains. You also have cloned cDNAs for nanos and bicoid. These plasmids are as follows

plasmid 1 full-length nanos cDNA
plasmid 2 full-length bicoid cDNA
plasmid 3 bicoid 5' UTR–bicoid ORF–nanos 3' UTR
plasmid 4 nanos 5' UTR-bicoid ORF–bicoid 3' UTR
plasmid 5 nanos 5' UTR-bicoid ORF–nanos 3'UTR

where, UTR = untranslated region and ORF= open reading frame (protein-coding region)

The plasmids are constructed so that you can generate a synthetic mRNA corresponding to the sequences described in each cDNA. Which of the following combination of plasmid will give normal embro having dorsal ventral axis?

(1) A and B only
(2) A and B or D and E only
(3) A and B or C and D or D and E
(4) A and B or A and D only

36. The gene bicoid exemplifies a maternal-effect gene that, when mutated, produces embryos that lack anterior structures but have the posterior structures duplicated. The Bicoid protein is a morphogen that establishes the domain of expression of the gap gene hunchback in a concentration dependent manner. The Bicoid protein is usually present in an anterior-to-posterior gradient with the highest concentration at the anterior end of the early embryo. Diagrammed below is the major domain of expression of hunchback in an embryo from a wildtype mother

What are the expected domains of hunchback expression in embryos from mothers with homozygous mutant for bicoid

(1) Increasing gradient from anterior to posterior end
(2) Decreasing gradient from posterior to anterior end
(3) Equally distributed through out embryo
(4) Totally absent from embryo

37. The development of the dorsal-ventral axis in vertebrates is thought to be controlled to a large extent by secreted inducing factors that are produced in a restricted part of the embryo and act at a distance. The initial event in dorsal-ventral patterning is a micro-tubule-directed rotation of the egg cortex The cortical rotation, which is completed before the first cleavage, modifies maternally deposited determinants of polarity (proteins or mRNAs) on the future dorsal side of the embryo. Irradiation of the vegetal hemisphere of the newly fertilized Xenopus embryo with ultraviolet (UV) light disrupts the microtubule array, and cortical rotation no longer occurs. The resulting embryo develops only ventral structures. A normal body axis can be restored to such UV-irradiated embryos in a number of ways

A. If UV-irradiated embryos are tipped during the first cell cycle, the force of gravity on the yolky vegetal hemisphere induces a cortical rotation, and such embryos develop normally
B. If dorsal vegetal blastomeres are transplanted into the irradiated embryo, an axis is restored
C. The cloning of a dorsalizing RNA (noggin) has an activity very similar to that of Xwnt8, and injected noggin mRNA can promote formation of a vegetal dorsalizing center (the Nieuwkoop center, which induces dorsal mesoderm, but whose own cells populate the yolky endoderm)

Which of the following statement is correct?

(1) A and B (2) A and C
(3) B and C (4) A, B and C

38. Which of the following mechanisms are true for Maintenance of the shoot meristems in adult Arabidopsis plants?

A. A transcription factor encoded by the SHOOT MERISTEMLESS gene is expressed in shoot meristem cells and maintains them in their undifferentiated state
B. A homeobox transcription factor encoded by the WUSCHEL gene is expressed in the organizing center and initiates a signal to the overlying cells to behave as stem cells
C. Shoot meristems cells secrete proteins encoded by the CLAVATA family that antagonize WUSCHEL expression, thereby restricting the size of the shoot meristems

(1) A and B (2) B and C
(3) A and C (4) A, B and C

39. Which of the following describes describe, in order, the pathway by which shoot meristems are turned into floral meristems in Arabidopsis by which day-length is translated into a signal that coverts a vegetative shoot meristem into a floral meristem?

A. The CONSTANS (CO) protein is degraded in the dark, but accumulates in the leaves as the days get longer

B. The CONSTANS protein is a transcription factor, which turns on expression of the FLOWERING LOCUS T (FT) gene

C. The FT mRNA travels to the shoot meristem, where it is translated; the protein then participates in activation of LEAFY and AP1

D. LEAFY and AP1 re-specify the shoot meristem into a floral meristem, capable of producing flowers

(1) A → B → C → D

(2) B → A → C → D

(3) C → A → B → D

(4) C → B → A → D

40. During vertebrate limb development, a specialised ectodermal structure, called Apical Ectodermal Ridge (AER), forms at the dorso-ventral ectodermal boundary at the distal tip of the developing limb bud. The following experimental facts about the AER is available

A. FGF 2, 4, and 8 are expressed in the AER

B. Removal of the AER causes cessation of limb growth

C. Removal of AER along with implantation of beads soaked in FGF 8 or

D. FGF 4 or FGF 2 protein rescues the AER removal phenotype and gives rise to normal limb

Which of the following statements cannot be made based on the above facts?

(1) FGF 2, 4, and 8 are secreted proteins

(2) FGF 2, 4, and 8 are necessary and sufficient for AER function

(3) FGF 2, 4, and 8 are sufficient for AER function

(4) FGF 2, 4, and 8 have largely redundant functions

41. During fertilization in amphibians, the fusion of egg and sperm plasma is preceeded by

A. release of enzymatic contents from the acrosomal vesile through exocytosis

B. binding and interaction of the sperm to vitelline membrane

C. chemoattraction of the sperm to the egg by soluble factors secreted by egg

D. passage of sperm through extracellular envelope

Which of the following is the correct sequence?

(1) A → B → C → D

(2) B → A → C → D

(3) C → A → B → D

(4) C → B → A → D

42. Which of the following statements is true about dorso-ventral patterning of drosophila embryo?

A. This is dictated by the location of the nurse cells

B. dorsal is the default fate

C. The whole process is regulated by preventing the entry of a transcription factor to the nucleus of dorsal cells

D. Homeobox containing genes play a critical role in this process

(1) A, B and C (2) A, B, C and D

(3) A and D (4) B and C

43. Which of the following is true about amphibian limb regeneration?

A. It requires a minimum number of functional nerves

B. The blastema of an amputated limb, if transplanted in the trunk region between two existing limbs in a host, will still give rise to a limb

C. The size of the regenerated limb is often grossly different from the original limb

(1) A (2) B and C

(3) A and C (4) A, B and C

44. Three classes of organ identity genes A, B and C are involved in specifying the four whorls of a flower. Sepals (se) are determined by the action of A class of genes, petals (pe) are determined by both A and B, stamens (st) by B and C, while carpels (ca) are determined by C alone. Mutations in these genes lead to homeotic transformation. Given below is the phenotype of floral whorls observed in wild type and mutants of floral organ identity genes

Whorl	1	2	3	4
Wild type	se	pe	st	ca
apetala 2 (ap2)	ca	st	st	ca
pistillata (pi)	se	se	ca	ca
agamous (ag)	se	pe	pe	se

To which class do the genes ap2, pi and Ag belongs

(1) Class A, B and C resp

(2) Class B, A and C respectively

(3) Class C, B and A resp

(4) Class B, C and A respectively

45. Two human disorders, Prader-Willi syndrome and Angelman syndrome, occur when the mother contributes a copy of chromosome 15 that contains a small deletion in a specific region. Why does this small deletion not behave as a recessive allele, that is, why is its loss not made up for by the good copy of the region on the paternally contributed chromosome 15?

(1) This portion of chromosome 15 is invariably mutated in males, so the father, too, contributes recessive alleles that are thus homozygous and perturb development

(2) The father's copy of chromosome 15 has genes in the region of the deletion that are imprinted, and thus inactive; in the absence of any active copies of these genes, development cannot proceed normally

(3) The genes in this portion of chromosome 15 are special in that they are required in two copies for normal development, and so the loss of one set does not allow normal development

(4) Two copies of every gene in the genome are required for development, so loss of one of the copies from this region perturbs development

46. How does the dose-dependence of retinoic acid treatment support the notion that a gradient of retinoic acid can act as a morphogen along the proximo-distal axis in the limb?

(1) Treatment with high levels of retinoic acid causes a proximal blastema to be respecified as a distal blastema, and only distal structures are regenerated

(2) Treatment with high levels of retinoic acid causes any blastema to form only distal structures

(3) Treatment with high levels of retinoic acid causes any blastema to be respecified as a proximal blastema, but it will form only proximal structures

(4) Treatment with high levels of retinoic acid causes a distal blastema to be respecified as a proximal blastema, and regeneration of a full limb may be initiated from proximal values

47. How is the ABC model for floral identity in *Arabidopsis* reminiscent of the models for homeotic gene function derived from studies in *Drosophila*?

(1) In both organisms, homeobox genes specify the identity of regions of the mature organism

(2) The key responsibility of the homeotic genes in both organisms is patterning of dorso-ventral asymmetry

(3) In both organisms, one homeotic gene is expressed at the two ends, a second expressed in to domains more central to that of the first, and a third expressed most centrally, thus contributing unambiguous identities to all regions of the organism

(4) In both organisms, the homeotic genes interact, so that it is often the combination of genes present that is critical in unambiguously specifying structures in the adult

48. Although the vertebrate body plan outwardly displays a mirror-image symmetry with regard to left and right, the internal organs are not symmetrical. How does this left-right asymmetry develop?

(1) The two cells of the Nieuwkoop center (at the 32-cell stage) signal each other and become either left or right

(2) Maternal factors that determine left and right are packaged into the egg, just as dorsalizing factors are

(3) Differential release of Ca^{2+} ions on the left side leads to expression of Nodal and Pitx2 on the left

(4) Gravity determines the left-right axis: that side which is down becomes right and that side which is up becomes left

49. What prevents the cells adjacent to the progenitor vulva cells in *C. elegans* from differentiating into vulva cells?

(1) These cells are too distant from the anchor cell to receive the LIN-3 signal

(2) These cells lack the ability to bind and respond to the LIN-3 signal

(3) These cells receive a signaling compound from a hypodermal cell that deactivates the LIN-3 signal

(4) These cells have condensed their chromatin in the regions responsive to the LIN-3 signal

50. Mutations in genes that control the early stages of development are often lethal. To circumvent this problem, developmental geneticists may try to isolate temperature-sensitive developmental mutants or *ts* alleles. If an embryo carries a *ts* allele, it will develop correctly at the permissive temperature (e.g., 25°C) but will fail to develop if incubated at the nonpermissive temperature (e.g., 30°C) In most cases, *ts* alleles have missense mutations that slightly alter the amino acid sequence of a protein, causing a change in its structure that prevents it from working properly at the nonpermissive temperature. *Ts* alleles are particularly useful because they can provide insight regarding the stage of development when the protein is necessary. Researchers can take groups of embryos that carry a *ts* allele and expose them to the permissive and nonpermissive temperature at different stages of development. In the experiment described next, embryos were divided into five groups and exposed to the permissive or nonpermissive temperature at different times after fertilization

Time After Fertilization (hours) Group:	1	2	3	4	5
0–1	25°C	25°C	25°C	25°C	25°C
1–2	25°C	30°C	25°C	25°C	25°C
2–3	25°C	25°C	30°C	25°C	25°C
3–4	25°C	25°C	25°C	30°C	25°C
4–5	25°C	25°C	25°C	25°C	30°C
5–6	25°C	25°C	25°C	25°C	25°C
SURVIVAL:	Yes	Yes	Yes	No	Yes

On basis of above experiment pinpoint the stage when the function of the protein encoded by this ts allele is critical?

(1) 1–2-hour stage of development

(2) 2–3-hour stage of development

(3) 3–4-hour stage of development

(4) 4–5-hour stage of development

51. Here are schematic diagrams of mutant larvae

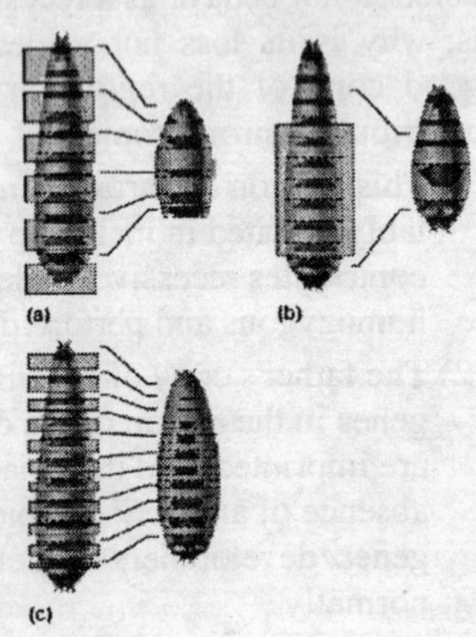

The left side of each pair shows a wild-type larva, with gray boxes showing the sections that are missing in the mutant larva. Which type of gene is defective in each larva?

(1) A: Pair-rule gene. B: Gap gene C: Segment polarity gene
(2) A: Pair-rule gene. B: Segment polarity gene C: Gap gene
(3) A:Segment polarity gene B: Pair-rule gene Gap gene C
(4) A: Gap gene B: Pair-rule gene C: Segment polarity gene

52. Some of the maternal gene products deposited in the oocyte of Drosophila, that are critical in the establishment of the antero-posterior, terminal, and dorso-ventral axes are described below

A. *Bicoid* mRNA is distributed in the anterior end of the oocyte and suppresses the formation of anterior structures
B. *Nanos* mRNA is localized to the anterior end and promotes the formation of anterior structures
C. The Torso receptor protein is found in the membrane and is activated by ligand binding at either end of the oocyte. It causes the formation of structures that are found only at the ends of the organism
D. The Toll receptor protein is activated by ligand binding at the ventral side of the embryo and establishes the dorso-ventral axis
E. The Torso and Toll receptor proteins are distributed throughout the plasma membranes of the oocyte and embryo, but they are activated by ligand binding only in the regions

Which of the following statements are correct?

(1) A and B (2) C, D and E
(3) Only C and D (4) Only C and E

53. If you observed fruit flies with the following developmental abnormalities, which of the following would you guess that a mutation has occurred in a homeotic gene?

A. Three abdominal segments were missing
B. One abdominal segment had legs
C. A fly with the correct number of segments had two additional thoracic segments and two fewer abdominal segments

(1) A and C (2) B and C
(3) A and B (4) A, B and C

54. Assuming that such a fly would be viable, what would be the sex of a fruit fly with the following chromosomal composition?

A. One X chromosome and two sets of autosomes
B. Two X chromosomes, one Y chromosome, and two sets of autosomes
C. Two X chromosomes and four sets of autosomes
D. Four X chromosomes, two Y chromosomes, and four sets of autosomes

(1) A: Male B: Female C: Male D: Female
(2) A: Male B: Male C: Female D: Female
(3) A: Male B: Female C: Female D: Male
(4) A: Female B: Male C: Male D: Female

55. Embryogenesis in plants requires specific gene expression. Analysis of *Arabidopsis* mutants that either fail to establish axial polarity or develop abnormally during embryogenesis has led to the identification of genes whose expression participates in tissue patterning during embryogenesis and at later stages of plant growth

Consider the following statements?

A. The *GNOM* gene helps establish apical–basal polarity and seedlings homozygous for mutations in the *GNOM* gene lack both roots and cotyledons
B. The *MONOPTEROS* gene is necessary for basal patterning and formation of the primary root. Plants homozygous for the *monopteros* mutation have a hypocotyl, a normal shoot apical meristem, and

cotyledons, but they lack the primary root

C. *SHORT ROOT* (*SHR*) and *SCARECROW* (*SCR*), are necessary for tissue differentiation and cell differentiation not only in the embryo, but also in both primary and secondary roots and in the hypocotyl. Mutants of *SHR* and *SCR* both produce roots with a single-celled layer of ground tissue

D. Root cap stem cells are formed from the hypophysis at the heart stage of embryogenesis, indicating that the root promeristem is established at least by this stage of embryogenesis. The expression of the *SHOOTMERISTEM-LESS* (*STM*) gene may be an early marker of root meristem identity. Mutants of the *SHOOTMERISTEMLESS* (*STM*) gene are defective in the formation of a functional embryonic root as well as lateral roots

E. The *HOBBIT* (*HBT*) gene is expressed specifically in the cells that will become the shoot apical meristem, and its expression in these cells is required for the formation of the shoot promeristem. *Arabidopsis* plants homozygous for a mutated, loss-of-function *HBT* gene do not form a shoot apical meristem, & instead all the cells in this region differentiate

Which of the following statements are correct?

(1) A, B and C (2) C, D and E
(3) A, B and D (4) A. B, C, D and E

Questions 56-57 refer to the following experiment

The marine coenocytic green alga *Caulerpa* grows in a unidirectional pattern. Photosynthetic fronds are produced on its upper surface and anchoring rhizoids on the lower surface (Figure I) When the same organism was inverted, new rhizoids were then initiated on the newly formed lower surface, while fronds were initiated on the newly formed upper surface (Figure II)

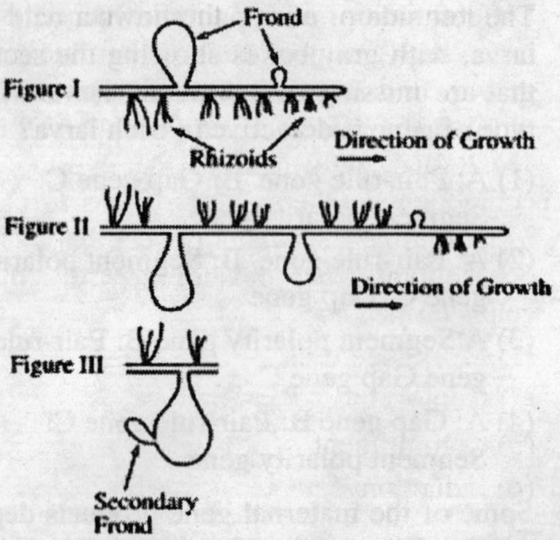

Figure I — Frond; Rhizoids; Direction of Growth

Figure II — Direction of Growth

Figure III — Secondary Frond

Secondary fronds can be initiated from the margins of the primary fronds. In another experiment, it was found that if the secondary fronds are initiated after the plant has been inverted, they will grow upward, even if the frond from which they are growing is oriented downward (Figure III)

56. Which of the following conclusions is best indicated by the additional experiment in Figure III ?

(1) Cellular gravity sensors are likely not responsible for the establishment of polarity
(2) Tip growth always occurs perpendicular to the pull of gravity
(3) Newly formed cytoplasm is fully differentiated and cannot respond to altered orientation
(4) The change in polarity is expressed only in newly formed cytoplasm

57. Which of the following procedures would best test whether the induction of polarity is regulated at the transcriptional level?

(1) Expose plants to a protein synthesis inhibitor, then invert
(2) Invert plants, then expose to a protein synthesis inhibitor
(3) Expose plants to an RNA synthesis inhibitor, then invert
(4) Invert plants, then expose to an RNA synthesis inhibitor

58. Differentiation under the influence of an organizer (group of cells that influence the growth and development of adjacent cells, through the effects of an evocator) or other "positional information" is an important component of theories of pattern formation in morphogenesis. This phenomenon where one cell influence development in adjacent cell is termed as

(1) Morphogenesis

(2) Competence

(3) Commitment

(4) Induction

59. The correct sequence of events during fertilization in mammals is

(1) capacitation, activation of egg, acrosomal reaction and entry of sperm in egg

(2) capacitation, acrosomal reaction, activation of egg and entry of sperm in egg

(3) acrosomal reaction, activation of egg, capacitation and entry of sperm in egg

(4) activation of egg, acrosomal reaction, entry of sperm in egg and capacitation

60. Match list-I (foetal membranes) with list-II (functions of foetal membranes) and select the correct answer using the codes given below the lists.

List-I	List-II
P. Yolk sac	1. elimination of metabolic wastes
Q. Chorion	2. shock absorption
R. Allantois	3. food storage and digestion
S. Amnion	4. exchange of substance between mother and embryo

Codes

(1) P-4, Q-3, R-l, 5-2

(2) P-4, Q-3, R-2, 5-1

(3) P-3, Q-4, R-2, 5-1

(4) P-3, Q-4, R-l, 5-2

61. Consider the following statements for apoptosis

P. it is mediated by an intracellular proteolytic cascade

Q. it is mediated by family of proteases called caspases

R. cytochrome c inhibitory role in apoptosis

S. gene Bcl-2 promote apoptosis

Which of the following is/are correct about apoptosis

(1) P and Q (2) Q and S

(3) P and S (4) P, Q, R and S

62. Consider the following proteins in reference of *Drosophila* development

P. torso

Q. bicoid

R. knirps

S. nanos

Which of the following maternal mRNAs encode proteins that directly influence early development in *Drosophila?*

(1) P and Q (2) Q and R

(3) P, Q and S (4) P, Q, R and S

63. These are some statements for homeodomain

P. proteins are a family of transcription factors

Q. characterized by a 60 amino acid domain

R. encoded by a 180 bp DNA sequence called Homeobox

S. highly conserved

Choose the correct statement(s) about homeodomain

(1) P, Q, R and S (2) Q, R and S

(3) P and Q only (4) R and S only

64. Consider the following egg patterns

P. Merobastic cleavage

Q. Mesolecithal egg

R. Discoidal cleavage

S. Centrolecithal egg

Which of the following is true about bird

(1) P and S (2) Q and S
(3) P and R (4) R and S

65. The correct sequence of expression of genes during the embryonic development of *Drosophila* is

(1) maternal effect genes → gap genes → pair rule genes → segment polarity genes
(2) maternal effect genes → pair rule genes → gap genes → segment polarity genes
(3) maternal effect genes → segment polarity genes → pair rule genes → gap genes
(4) gap genes → segment polarity genes → pair rule genes → maternal effect genes

66. These are some steps in fertilization in animals

P. sperm activation in female reproductive tract
Q. acrosome reaction
R. binding of sperm of zona pellucida
S. lysis of zona pellucida by sperm
T. fusion of sperm and egg membrane

Which of the following sequence of events are correct in mammalian fertilization?

(1) P, Q, R, S, T
(2) P, R, Q, S, T
(3) P, S, R, Q, T
(4) P, S, Q, R, P

67. Consider the cleavage pattern and representative group of animals

P. radial cleavage – echinoderms
Q. spiral cleavage - annelids
R. bilateral cleavage – tunicates
S. rotational cleavage - mammal

Which of the following is/are correct pair(s)?

(1) P and S (2) Q, R and S
(3) Q and R (4) P, Q, R and S

68. Consider the following statements regarding Dorsal-Ventral axis determination in *Drosophila?*

A. Gurken protein is released by follicle cells in dorsal anterior surface of egg which is recognized by Torpedo (Gurken receptor)
B. Overexpression of un-degradable cactus protein will lead to formation of dorsal side on both side
C. Dorsal is responsible for determining ventral side of embryo
D. Toll receptor are present though out egg membrane but they are activated by Spatzle protein only in ventral side

Which of the following combinations are correct?

(1) A and B (2) A and D
(3) B and C (4) C and D

69. Figure bellows represent summary of experiments demonstrating the effect of the apical ectodermal ridge (AER) on the underlying mesenchyme

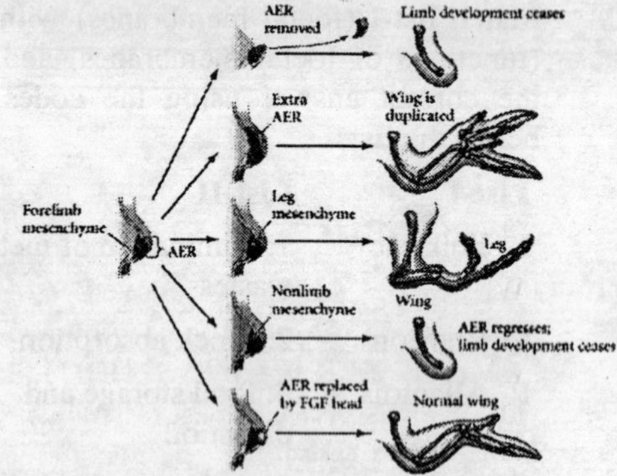

The following conclusions may be drawn

A. AER is sufficient and essential for progression of development
B. Leg mesenchymal cells are not determined they can form leg instead of forelimb when placed under AER of forelimb
C. Non-limb mesenchymal cells release FGF 10 but AER fails to formed due to lack of receptor Wnt3a

D. The role of AER can substituted with FGF 10

Which of the following conclusions are correct?

(1) A and B (2) B and D
(3) C and D (4) A and C

70. The following molecules play important roles in specifying cell fate

A. Sonic Hedgehog B. MyoD
C. Ephrin D. HoxC6
E. Chordin

Consider the following statements

I. Secreted Protein (dorsal ventral polarity in xenopus gastrulation)

II. Transcription Factor (vertebral axial identity)

III. Transmembrane Protein (Migration of Neural Crest cells and Sclerotome)

IV. Transcription Factor (muscle specification and differentiation)

V. Secreted Protein (Limb patterning, eye maintenance and taste bud development)

Which of the following is correct match?

(1) A: V B: IV C: II D: I E: III
(2) A: III B: II C: I D: IV E: V
(3) A: V B: IV C: III D: II E: I
(4) A: IV B: II C: I D: III E: V

71. When the AER is removed early, only proximal limb structures develop, while when the AER is removed later, more distal limb structures develop. Which of the following statement donot explain this observation?

(1) The cells that leave the progress zone early develop into proximal tissues

(2) The cells that leave the progress zone early develop into distil tissues

(3) There is a constant, absolute number of cells that die after AER removal, and when this happens early (small limb bud), the cell death wipes out the proximal and distal elements (they're all in a small area)

(4) When the AER is removed later, the limb but is much bigger and thus the same amount of cell death leads to the loss of only distal structures

72. In *Arabidopsis*, several mutations that affect SAM formation have been identified; shoot meristemless (STM), cup-shaped cotyledon1 (CUC1) CUC2 double mutants and wuschel (WUS) are defective in embryonic SAM formation

Consider the following statements

A. The WUS gene encodes a homeodomain protein of the KNOTTED1. KNOTTED1-like homeobox genes are thought to be involved in SAM formation and its maintenance

B. STM encodes a homeodomain protein of a novel class distinct from the KNOTTED1 class and is expressed in a group of cells underneath the stem cells of the WUS. STM is postulated to affect stem cell fate in a non-cell-autonomous manner

C. WUS and STM seem to act at different regulatory levels because the initiation of WUS expression is independent of STM activity, and because WUS is expressed in a small subdomain of the SAM

D. Cuc1 Cuc2 double mutant seedlings completely lack an embryonic SAM, and two cotyledons are fused to form one cup-shaped structure. However, cuc1 or cuc2 single mutant seedlings look normal except for occasional seedlings with fused cotyledons in one of their sides

E. CUC1 and CUC2 are functionally redundant genes that are required for embryonic SAM development and for keeping cotyledons and floral organs from fusing with each other. Accumulation of STM mRNA is blocked in cuc1 cuc2 double mutant embryos

Which of the following combinations are correct?

(1) A, B and D (2) A, D and E
(3) B, C and E (4) C, D and E

73. Match the terms that best describes the following statements

Terms

I. Resact
II. Bindin
III. Zona pellucida
IV. Ingression
V. *In situ* hybridization
VI. Targeted gene knock-out

Statements

A. a protective layer of glycoproteins, analogous to the vitelline envelope
B. a technique that can be used for loss-of-function studies in mouse
C. migration of individual cells into an embryo
D. a protein that ensures species specific binding between sperm membrane and vitelline envelope
E. a technique that detects mRNA expression of a specific gene
F. a species specific chemotaxic compound that attracts sperm

(1) A:II B:VI C:I D:III E:V F:IV
(2) A:III B:VI C:IV D:II E:V F:I
(3) A:I B:V C:IV D:II E:VI F:III
(4) A:III B:IV C:VI D:V E:II F:I

74. In an experiment, the cells that would normally become the middle segment of a *Drosophila* leg were removed from the leg forming area of the larva and were placed in the tip of the fly's antenna. Based on the "French flag" analogy for the operation of a gradient of positional information, which of the following statements is true?

(1) The transplanted cells are determined as leg cells and therefore would form a complete limb instead of an antenna
(2) The transplanted cells would intermingle with the cells present in the new environment and develop accordingly to give rise to an antenna
(3) The transplanted cells retain their committed status as leg cells and would develop to form a chimeric structure having proximal region made of antenna and the distal region ending in a complete leg
(4) The transplanted cells retain their committed status as leg cells, but respond to the positional information of their environment by becoming leg tip cells-i.e., claws

75. In the development of the nervous system, which is the correct order of events?

(1) First, cells must be specified to be neurons, second, these neurons must make connections with their targets, third, the neurons must migrate to their final position, and fourth, the neurons refine their synaptic connections
(2) First, cells must be specified to be neurons, second, the neurons must migrate to their final position, third, these neurons must make connections with their targets, and fourth, the neurons refine their synaptic connections
(3) First, cells must migrate to their final position, second, these cells must be specified to be neurons, third, the neurons must make connections with their targets, and fourth, the neurons refine their synaptic connections
(4) First, the prospective neurons must make connections with their targets, second, the neurons refine their synaptic connections, third, the neurons must migrate to their final position, and fourth, the cells are specified to become neurons

76. "Intercalary growth" in a regenerating amphibian limb means that if a distal blastema is grafted to a proximal stump,

(1) the distal blastema grows back, to regenerate proximal elements, then out, to regenerate distal structures

(2) the stump grows out to regenerate proximal structures, until the positional values of the blastema are produced, at which time the distal blastema takes over and completes outgrowth of distal structures

(3) the distal blastema grows proximally, intercalating intermediate structures between it and the stump, then growth stops, resulting in a regenerate lacking distal structures

(4) the stump grows out, intercalating intermediate structures between it and the distal blastema, then growth stops, resulting in a regenerate lacking distal structures

77. In clavata (clv) mutant of *Arabidopsis,* shoot apical meristem (SAM) size as well as expression of wuschel (wus) are increased whereas SAM size is reduced in wus mutant plants. With this information, choose the correct statement which best describes the function of wus and clv genes

(1) clv positively regulates wus expression and negatively regulates SAM size

(2) clv negatively regulates wus expression and wus positively regulates SAM size

(3) wus negatively regulates SAM size and clv expression

(4) wus and clv independently regulate SAM size

78. Mutations in *CONSTANS* (CO) of *Arabidopsis thaliana* results in late flowering phenotype. Transcript levels of CO were determined in long day and short day seedlings. Which of the following would likely represent the transcript profile of CO?

	Long-day seedlings	Short-day seedlings
(1)	▬▬ (thick)	—— (thin)
(2)	—— (thin)	▬▬ (thick)
(3)	—— (thin)	—— (thin)
(4)	▬▬ (thick)	▬▬ (thick)

79. A newly fertilized egg were treated either to cycloheximide or actinomycin D immediately, following fertilization. In presence of cycloheximide, the fertilized egg failed to divide while in presence of actinomycin D the egg divided several times to reach an early embryonic stage blastula. Data is also obtained on effect of differential treatment on protein synthesis (figure)

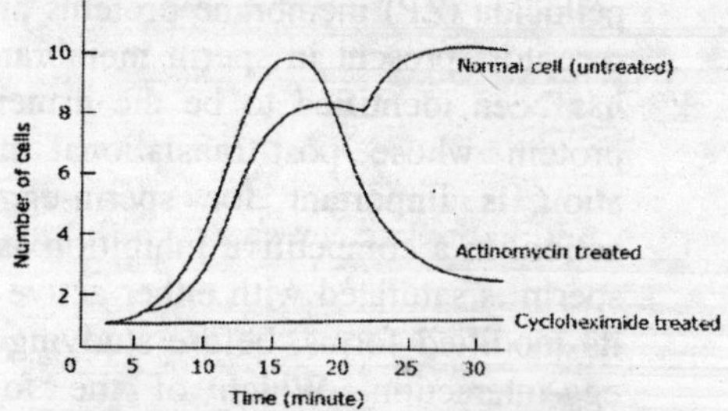

A. mRNA needed for earliest stages in sea urchin development must be present already

B. gene expression is not necessary for earliest of sea urchin development

C. gene expression is necessary for further progression of cells beyond early stage

D. from experiment it can be concluded that actinomycin D prevented cell cycle protein expression

Identify the correct conclusions

(1) A and B (2) A and C

(3) A, B and C (4) A, B, C and D

80. Segmentation genes in Drossophila are divided into three groups (gap, pair rule and segment polarity) based on their mutant phenotype

Below are some of the major genes expressed in a sequential, manner (with respect to the groups) affecting segmentation pattern

A. hairy → paired→ tailless → patched

B. hunchback → even-skipped → fushi tarazu → wingless

C. odd-skipped → giant → paired → wingless

D. tailless → hairy → fushi tarazu → gooseberry

Which of the above sequence(s) of genes expressed from early to late embryo is/are correct?

(1) D only (2) A and B
(3) C and B (4) B and D

81. During fertilization "in mammals, sperm-egg interaction is mediated by zona pellucida (ZP) membrane proteins and their receptors present in sperm membrane. ZP3 has been identified to be the principle ZP protein whose post-translational modification is important for sperm-egg interaction. In a competitive inhibition assay the sperm is saturated with either active ZP3 or its modified forms, before studying sperm-egg-interaction. Which of the following experiments will not inhibit sperm-egg-interaction

(1) Saturate sperm with ZP3 protein prior to use

(2) Deglycosylate the ZP3 protein and use it for saturation of sperm

(3) Phosphorylate the ZP3 protein and use it for saturation of sperm

(4) Dephosphorylate the ZP3 protein and use it for saturation of sperm

82. Consider the following features of messenger RNA

A. The mRNA in acts as a true 'messenger', being produced within the nucleus and migrating to the cytoplasm for translation to occur

B. Most mRNA is extremely short-lived; it typically has a half-life of a few minutes only, which may be less than the time required for producing or translating it. This can only be achieved by the coupling of transcription and translation

C. The ribosome in attaches to the 5' end of the mRNA and migrates until it reaches the start codon

D. Generally, genes contain introns and mRNA is often (but not always) polyadenylated, i.e. it has a run of adenine residues at the 3' end

Which of the following are characteristic feature of prokaryotic m-RNA?

(1) A only (2) A and C both
(3) B only (4) B, C and D

83. CUC (cup-shaped cotyledon) and STM (shoot meristem less) are two important regulators for shoot meristem (SAM) development and mutant of both the classes of genes fails to develop SAM. If artificially snoot meristem (SAM) is induced in CUC-mutant, the growth of the shoot takes place normally, whereas, in STM-mutant the artificially induced shoot meristem gets lost after a few rounds of cell division. What can be concluded from these observations?

(1) CUC gene is required for initiation and maintenance of SAM

(2) CUC gene is required only for maintenance of SAM

(3) STM gene is required for initiation and maintenance of SAM

(4) STM gene is required only for maintenance of SAM

84. During early cleavage of *Caenorabditis elegans* embryos, each asymmetrical division produces one founder cell which produces differentiated descendants and one stem cell. The very first cell division produces one anterior founder cell, namely AB and one posterior stem cell, namely P1. When these blastomeres are experimentally separated and allowed to proceed further with development, one could get the following possible outcomes

(1) P1 cell would develop autonomously while the AB would show conditional development
(2) P1 cells would show conditional development while AB would show autonomous development
(3) Both would show autonomous specification and result in mosaic development
(4) Both would show conditional specification and result in regulative development

85. In case of sea urchin, which of the following is the correct sequence of events taking place during the interaction of sperm and egg?

(1) Chemoattraction of sperm to the egg by soluble molecules secreted by the egg→ exocytosis of the sperm acrosomal vesicle to release its enzymes →binding of the sperm to the extracellular matrix of the egg →passage of sperm through this extracellular matrix →fusion of egg and sperm cell membranes
(2) Chemoattraction of sperm to the egg by soluble molecules secreted by the egg →binding of the sperm to the extracellular matrix of the egg → exocytosis of the sperm acrosomal vesicle to release its enzymes →passage of sperm through this extracellular matrix →fusion of egg and sperm cell membranes
(3) Chemoattraction of sperm to the egg by soluble molecules secreted by the egg →binding of the sperm to the extracellular matrix of the egg →passage of sperm through this extracellular matrix →exocytosis of the sperm acrosomal vesicle to release its enzymes →fusion of egg and sperm cell membranes
(4) Chemoattraction of sperm to the egg by soluble molecules secreted by the egg → passage of sperm through this extracellular matrix→binding of the sperm to the extracellular matrix of the egg→ exocytosis of the sperm acrosomal vesicle to release its enzymes →fusion of egg and sperm cell membranes

86. Injection of *noggin* mRNA into a 1-cell, UV-irradiated embryos of frog completely rescues dorsal development and allows the fonnation of a complete embryo. Some of the following statements (A-D) could possibly explain this observation

A. Noggin is a secreted protein which induces dorsal ectoderm to form neural tissue and it dorsalizes the mesoderm cells which would otherwise contribute to ventral mesoderm
B. Noggin binds directly to BMP4 and BMP2 thus preventing complex formation with their receptors
C. Noggin along with other molecules prevent BMP from binding to and inducing ectodenn and mesoderm cells near the organizer
D. Noggin is a secreted protein which induces the dorsal ectoderm to form the epidermis and it ventralizes the mesoderm cells which would otherwise contribute to dorsal mesoderm

Which of the above statements are correct?

(1) A, B and C (2) A and B
(3) B and C (4) A and D

87. Consider the following statements regarding functioning of pax6

A. Pax6 generates competence in optic vesicle and induces lens formation from head ectoderm
B. Pax6 generates competence in head ectoderm, so that it can respond to inducer from optic vesicle
C. Optic vesicle can induce lens formation of any part of head ectoderm which expresses pax6
D. Lens formation can also be induced by BMP and FGF besides pax6

Which of the following statement is correct?

(1) A and B (2) B and C
(3) C and D (4) B only

88. The type of cleavage in fertilized egg is decided by amount and distribution of yolk. Which of the following statement is correct?

(1) Centrolecithal egg → Holoblastic cleavage
(2) Telolecithal egg →Holoblastic cleavage
(3) Isolecithal egg → Holobalstic cleavage
(4) Isolecithal egg → Meroblastic cleavage

89. Consider the following statements regarding specification?

A. Potency of cell is equal to normal fate in mosaic development
B. Potency of cell is greater than normal fate in mosaic development
C. Potency of cell is equal to normal fate in regulative development
D. Potency of cell is greater than normal fate in regulative development

Which of the following statement is correct?

(1) A and B (2) A and D
(3) B and C (4) B and D

90. Consider the following statement regarding development of limb in amphibians

A. If AER is removed during limb elongation further development ceases
B. If hind limb mesenchymal cell is placed below fore limb AER, at the end hind limb structure would develop
C. If hind limb mesenchymal cell is placed below fore limb AER, at the end fore limb structure would develop
D. If extra AER is grafted in limb bud, supernumerary structure develops at distil end

Which of the following statement is correct?

(1) A and B (2) C and D
(3) A, B and D (4) A, C and D

91. Consider the following statement regarding plant development

A. Lateral roots develops from epidermal cells
B. Shoot axillary meristem develops from shoot apical meristem
C. Root cap cells are mainly dead cells
D. Lateral meristem and radial meristem are responsible for secondary growth in roots

Which of the following statements are correct?

(1) B and D (2) A and C
(3) B and C (4) A and D

92. *Dictyostelium* has been a wonderful experimental organism for developmental biologists because they are part time multicellular and initially identical cells are differentiated into one of two alternative cell types, spore and stalk

Consider the following statements

A. Aggregation is initiated as each of the cells begins to synthesize cGMP
B. When surgically removing the anterior part of the slug it abolish its ability to form a stalk
C. DIF appears to be necessary for stalk cell differentiation
D. The adenylyl cyclase expressed during aggregation, ACA is responsible for the synthesis of cAMP that is required for cell-cell signaling in early development. ACB is required for terminal differentiation. ACG, acts as an osmosensor and is involved in controlling spore germination

Which of the following is correct?

(1) A and C (2) B and D
(3) A and B (4) C and D

93. The sequence of expression of genes during embryonic development of *Drosophila* is

A	***hunchback, Krüppel, knirps, giant, tailless***
B	***engrailed, wingless, gooseberry, cubitus interruptus, patched, hedgehog, disheveled,***
C	***runt, hairy, fushi tarazu, even paired, odd paired, skipped, sloppy, paired, odd skipped***

(1) A→ B→C (2) A→ C→B
(3) B→ C→A (4) B→ A→C

94. The dorsal– ventral axis defines the back (dorsum) & belly (ventrum) of a *Drosophila* Consider the following statements

A. Dorsal protein remains in the cytoplasm; this side will become the dorsal surface. Along the other side, Dorsal protein is taken up into the nuclei; this side will become the ventral surface

B. The nuclear uptake of Dorsal protein is thought to be governed by a protein called Cactus, which binds to Dorsal protein and traps it in the cytoplasm

C. The presence of Toll, can alter Dorsal, allowing it to dissociate from Cactus and move into the nucleus

D. High nuclear concentration of Dorsal protein (as on the ventral side of the embryo) activates a gene called decapentaplegic, which causes mesoderm to develop

Which of the following is correct?

(1) A, B and C (2) B, C and D
(3) A, B and D (4) A, C and D

95. The development of the vulva in *C. elegans* offers several examples of induction on the cellular level. Consider the following statements

A. The anchor cell secretes the LIN-3 protein, a relative of epidermal growth factor (EGF) and the Boss protein

B. The six VPCs influenced by the anchor cell form an equivalence group

C. The LIN-3 protein is received by the LET-23 receptor tyrosine kinase on the VPCs, and the signal is transferred to the nucleus through the RTK-Ras pathway

D. The cell secreting LAG-2 becomes the ventral uterine precursor cell, while the cell receiving the signal through its LIN-12 protein becomes the gonadal anchor cell

Which of the following is correct?

(1) A, B and C (2) B, C and D
(3) A, B and D (4) A, C and D

96. Consider the following statements for eye development in *Xenopus*

A. Pax6 responsive for acquiring competence becomes expressed in the throughout surface ectoderm of the embryo

B. Competence to respond to the optic vesicle inducer can be conferred on ectodermal tissue by incubating it next to anterior neural plate tissue

C. The optic vesicle appears to secrete two paracrine factors, one of which may be BMP4, a protein that is received by the lens cells and induces the production of the Sox2 transcription factors. The other paracrine factor is *Fgf8*, signal that induces the appearance of the L-Maf transcription factor

D. The combination of Pax6, Sox2, and L-Maf in the ectoderm is needed for the production of the lens and the activation of lens-specific genes such as δ-crystallin

Which of the following is correct?

(1) A, B and C (2) B, C and D
(3) A, B and D (4) A, C and D

97. Consider the following statement for limb development in tetrapods

A. *Fgf10* immediately prior to limb formation restricted to lateral plate mesoderm where the limbs will form due to action of Wnt proteins-Wnt2b in the chick forelimb region and Wnt8c in the chick hindlimb region-which stabilize *Fgf10* expression at these sites

B. The gene encoding the Tbx5 transcription factor in mice is transcribed in the anterior lateral plate mesoderm and in the forelimbs while the genes encoding the transcription factors Tbx4 and Pitx1 are expressed in the posterior lateral plate mesoderm and in the hindlimbs

C. *Fgf8* is capable of inducing the AER in the competent ectoderm between the dorsal and ventral sides of the embryo

D. *Fgf8* from the lateral plate mesoderm acts by inducing a Wnt protein (Wnt3a in chicks; Wnt3 in humans and mice) in the prospective limb bud ectoderm. The Wnt protein acts through the canonical β-catenin pathway to induce *Fgf10* expression and form the AER. *Fgf10* stimulates mitosis in the mesenchyme cells beneath it and causes these cells to keep expressing *Fgf8*

Which of the following statement are correct

(1) B and C (2) B, C and D
(3) A and B (4) A, C and D

98. The proximal-distal growth and differentiation of the limb bud are made possible by a series of interactions between the AER and the limb bud mesenchyme directly (200 μm) beneath it. Consider the following statements

A. If the AER is removed at any time during limb development, further development of distal limb skeletal elements ceases

B. If an extra AER is grafted onto an existing limb bud, mirror images at the most distal end of the limb

C. If leg mesenchyme is placed directly beneath the wing AER, distal hind-limb structures develop at the end of the limb

D. If limb mesenchyme is replaced by non-limb mesenchyme beneath the AER, the AER progresses and new limb development start from ectopic position

Which of the following statements are correct

(1) A and C (2) B and D
(3) A and D (4) C and D

99. Thyroid hormones appear to play significant role in metamorphogenesis of amphibians by activating the transcription of some genes and repressing the transcription of others. Consider the following statements

A. Thyroid hormone receptors of two major types TRα and TRβ and binds to their specific sites on the chromatin only in presence of T_3 hormones

B. When T3 or T4 enters the cell, hormone-receptor complex is converted from a repressor to a strong transcriptional activator

C. T3 receptor forms a dimer with the retinoid receptor, retinoid receptor. This dimer binds T3 and can enter the nucleus to effect transcription

D. The hormone prolactin has been found to be responsible for the up-regulation of TRα and TRβ mRNAs

Which of the following statements are correct

(1) A and C (2) C and D
(3) B and C (4) B, C and D

100. The metamorphosis of insects appears to be regulated by effector hormones, which are controlled by neurohormones in the brain Consider the following statement

A. Insect molting and metamorphosis are controlled by two effector hormones: the steroid 20 hydroxyecdysone and the lipid juvenile hormone (JH)

B. The molting process is initiated in the brain, where neurosecretory cells release prothoracicotropic hormone (PTTH) in response to neural, hormonal, or environmental signals

C. PTTH is a peptide hormone stimulates the production of ecdysone by the prothoracic gland

D. An EcR protein forms an active molecule by pairing with an Ultraspiracle (Usp) protein, the homologue of the amphibian RXR that helps form the thyroid hormone receptor

Which of the following statements are correct

(1) A and C (2) C and D
(3) B and C (4) All of the above

ANSWER SHEET

Unit 5 : Developmental Biology

1. 3	**2.** 3	**3.** 1	**4.** 4	**5.** 3	**6.** 3	**7.** 4	**8.** 1	**9.** 4	**10.** 4
11. 2	**12.** 1	**13.** 2	**14.** 3	**15.** 1	**16.** 3	**17.** 1	**18.** 2	**19.** 3	**20.** 4
21. 2	**22.** 4	**23.** 1	**24.** 2	**25.** 2	**26.** 1	**27.** 4	**28.** 3	**29.** 4	**30.** 3
31. 4	**32.** 4	**33.** 3	**34.** 4	**35.** 4	**36.** 4	**37.** 4	**38.** 4	**39.** 1	**40.** 2
41. 3	**42.** 4	**43.** 4	**44.** 1	**45.** 2	**46.** 4	**47.** 4	**48.** 3	**49.** 2	**50.** 3
51. 1	**52.** 2	**53.** 2	**54.** 1	**55.** 1	**56.** 4	**57.** 3	**58.** 4	**59.** 2	**60.** 4
61. 1	**62.** 3	**63.** 1	**64.** 3	**65.** 1	**66.** 2	**67.** 4	**68.** 3	**69.** 4	**70.** 3
71. 2	**72.** 4	**73.** 2	**74.** 4	**75.** 2	**76.** 2	**77.** 2	**78.** 1	**79.** 3	**80.** 4
81. 2	**82.** 3	**83.** 4	**84.** 1	**85.** 1	**86.** 1	**87.** 2	**88.** 3	**89.** 2	**90.** 2
91. 1	**92.** 4	**93.** 2	**94.** 1	**95.** 1	**96.** 2	**97.** 3	**98.** 1	**99.** 3	**100.** 4

UNIT-6

PLANT PHYSIOLOGY

1. Figure below explains the regulation of respiratory rate by ADP during succinate oxidation in isolated mitochondria from mung bean (*Vigna radiata*) The numbers below the traces are the rates of oxygen uptake expressed as O_2 consumed (nmol min^{-1} mg $protein^{-1}$)

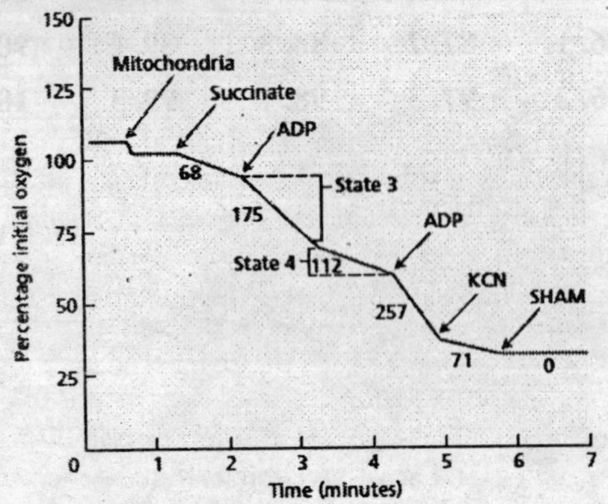

Which of the following conclusion is not true?

(1) Addition of succinate initiates mitochondrial electron transfer, which is measured with an oxygen electrode as the rate of oxygen reduction (to H_2O)

(2) Addition of cyanide inhibits electron flow through the main cytochrome pathway and only allows electron flow to oxygen through the alternative, cyanide-resistant pathway, which is subsequently inhibited by the addition of SHAM

(3) Addition of ADP inhibits electron transfer (state 3) by facilitating dissipation of the electrochemical proton gradient. The rate is further decreases after the second ADP addition because of activation of succinate dehydrogenase

(4) When all the ADP has been converted to ATP, electron transfer reverts to a lower rate (state 4)

2. The citric acid cycle reactions in plants are not all identical with those carried out by animal mitochondria

Consider the following statements

A. the step catalyzed by succinyl-CoA synthetase produces GTP in plants and ATP in animals

B. Significant activity of NAD^+ malic enzyme is found in the all plant mitochondria alternative pathways for the metabolism of PEP derived from glycolysis

C. Plant mitochondria have two pathways for oxidizing matrix $FADH_2$. Electron flow through complex II is sensitive to inhibition by rotenone and piericidin. In addition, plant mitochondria have a rotenone-resistant dehydrogenase for oxidation of $FADH_2$

D. Most plants have an "alternative" respiratory pathway for the reduction of oxygen which involves alternative oxidase that, unlike cytochrome *c* oxidase, is insensitive to inhibition by cyanide, azide, or carbon monoxide

Which of the following statement is correct?

(1) A and B (2) B and C
(3) B and D (4) A and C

3. The activity of RUBISCO is also regulated by light. Consider the following statements about regulation of RUBISCO activity

A. The ferredoxin–thioredoxin system reduces (sulphahydration) RUBISCO enzyme in the light. Upon reduction, RUBISCO enzymes are converted from an inactive to an active state. The activation process starts in the light by a reduction of ferredoxin by photosystem I

B. RUBISCO is activated when activator CO_2 reacts slowly with an uncharged ε-NH_2 group of lysine within the active site of the enzyme. The resulting carbamate derivative then rapidly binds Mg^{2+} to yield the activated complex

C. Two protons are released during the formation of the ternary complex rubisco–CO_2–Mg^{2+}, so activation is promoted by an increase in both pH and Mg^{2+} concentration. Thus, light-dependent stromal changes in pH and Mg^{2+} appear to facilitate the observed activation of RUBISCO by light

D. The binding of sugar phosphates, such as ribulose-1,5- bisphosphate, to rubisco helps in carbamylation. The sugar phosphates can be added by the enzyme RUBISCO activase, in a reaction that requires ATP. The primary role of RUBISCO activase is to de-accelerate the release of bound sugar phosphates, thus preparing RUBISCO for carbamylation

E. RUBISCO is also regulated by a natural sugar phosphate, carboxyarabinitol-1-phosphate, that closely resembles the six-carbon transition intermediate of the carboxylation reaction. This inhibitor is present at low concentrations in leaves of many species

Which of the following statement is correct?

(1) A, B and D (2) B, C and E
(3) A, B and E (4) C and D only

4. An important property of RUBISCO is its ability to catalyze both the carboxylation and the oxygenation of RuBP. Oxygenation is the primary reaction in a process known as photorespiration

A. $2\text{Phosphoglycolate} + 2H_2O \rightarrow 2\text{glycolate} + 2P_i$

B. $\text{Methylene-}H_4\text{-folate} + H_2O + \text{glycine} \rightarrow \text{serine} + H_4\text{-folate}$

C. $\text{Hydroxypyruvate} + NADH + H^+ \rightarrow \text{glycerate} + NAD^+$

D. $2\text{Glyoxylate} + 2\text{glutamate} \rightarrow 2\text{glycine} + 2\alpha\text{-ketoglutarate}$

Choose the location of above reactions [A-D] during photorespiration

(1) A: Peroxisome, B: Chloroplast, C: Mitochondria, D: Peroxisome

(2) A: Chloroplast, B: Mitochondria, C: Peroxisome, D: Peroxisome

(3) A: Chloroplast, B: Peroxisome, C: Mitochondria, D: Mitochondria

(4) A: Peroxisome, B: Mitochondria, C: Chloroplast, D: Mitochondria

5. The basic C_4 photosynthetic carbon cycle involves four stages in two different cell types. Consider the following statements

A. Fixation of CO_2 into a four-carbon acid occurs generally in a bundle sheath cell

B. Four-carbon acid is transported from the bundle sheath cell to mesophyll cell and is decarboxylated with the generation of a high CO_2 concentration in the mesophyll cell

C. The CO_2 released is fixed by rubisco and converted to carbohydrate by the Calvin cycle predominantly in bundle sheath cells

D. Residual three-carbon acid is regenerated into original CO_2 acceptor phosphoenolpyruvate in the chloroplast of mesophyll cells phosphoenolpyruvate which consumes two "high-energy" phosphate bonds: one in the reaction catalyzed by pyruvate –orthophosphate dikinase and another in the conversion of PPi to 2Pi catalyzed by pyrophosphatase

Which of the following statement are not true?

(1) A and B (2) A and C
(3) C and D (4) B and D

6. Consider the following statements about C_4 plants

A. The affinity of PEP carboxylase for its substrate, HCO^{3-}, is sufficiently high that the enzyme is saturated by HCO^{3-} in equlibrium with air levels of CO_2

B. The substrate PEP carboxylase is CO_2 whose structure resembles much with oxygen, so oxygen is a competitor in the reaction for PEPCO

C. This high activity of PEP carboxylase enables C_4 plants to reduce the stomatal aperture and thereby conserve water while fixing CO_2 at rates equal to or greater than those of C_3 plants

D. Beneficial feature of C_4 plant is the suppression of photorespiration resulting from the concentration of CO_2 in mesophyll cells where RUBISCO can do only carboxylase activity

Which of the following features enable C_4 plants to photosynthesize more efficiently at high temperatures than C_3 plants

Which of the following statement are true?

(1) A and B (2) A and C
(3) C and D (4) B and D

7. Figure below shows a response of photosynthesis to light in a C_3 plant

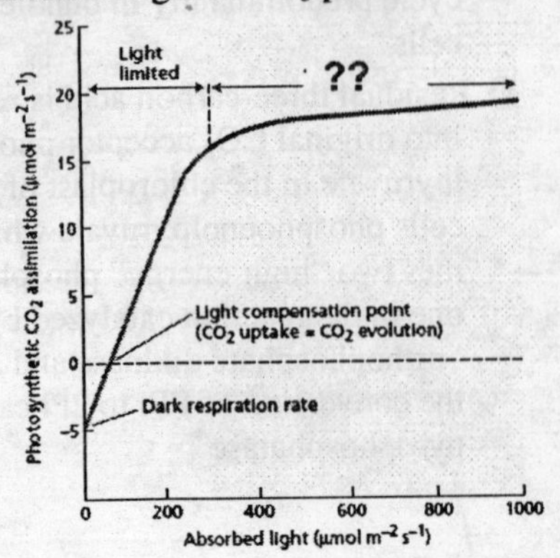

In darkness, respiration causes a net efflux of CO_2 from the plant. The light compensation point is reached when photosynthetic CO_2 assimilation equals the amount of CO_2 evolved by respiration. Increasing light above the light compensation point proportionally increases photosynthesis indicating that photosynthesis is limited by the rate of electron transport, which in turn is limited by the amount of available light. This portion of the curve is referred to as light-limited. Further increases in photosynthesis are eventually limited by the carboxylation capacity of RUBISCO or the metabolism of triose phosphates. This part of the curve is referred to as

(1) Saturation effect
(2) CO_2 limited
(3) Water limited
(4) Nutrient limited

8. Match list-I with list-II and select the correct answer using the codes given below the lists:

List-I	List-II
P. Richmond Lang effect	1. Reversal of red-drop
Q. Emerson effect	2. Inhibition of glycolysis by oxygen
R. Pasteur effect	3. Inhibition of photosynthesis by oxygen
S. Warburg effect	4. Senescence retardation in detached leaves

(1) P-l, Q-4, R-2, S-3
(2) P-4, Q-l, R-3, S-2
(3) P-l, Q-4, R-3, S-2
(4) P-4, Q-l, R-2, S-3

9. The figure shows light–response curves of photosynthetic carbon fixation in sun and shade plants. *Atriplex triangularis* (triangle orache) is a sun plant, and *Asarum caudatum* (a wild ginger) is a shade plant

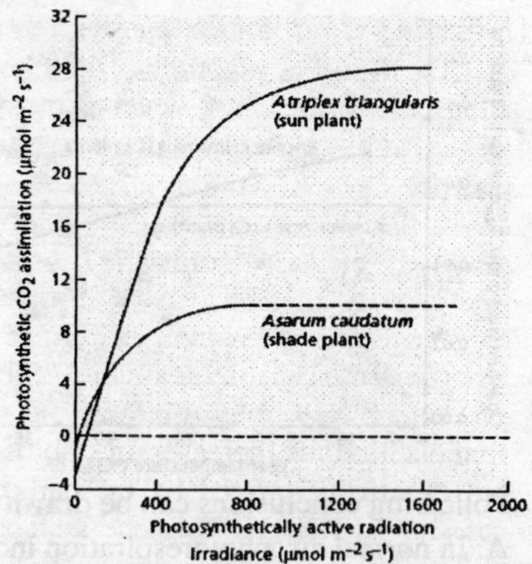

Which of the following is valid conclusion?

(1) Sun plants have a low light compensation point and have lower maximal photosynthetic rates than shade plants

(2) Shade plants have a low light compensation point and have lower maximal photosynthetic rates than sun plants

(3) Sun plants have a low light compensation point and have higher maximal photosynthetic rates than shade plants

(4) Shade plants have a high light compensation point and have higher maximal photosynthetic rates than sun plants

10. Figure below shows diurnal changes in xanthophyll content as a function of irradiance in sunflower (*Helianthus annuus*)

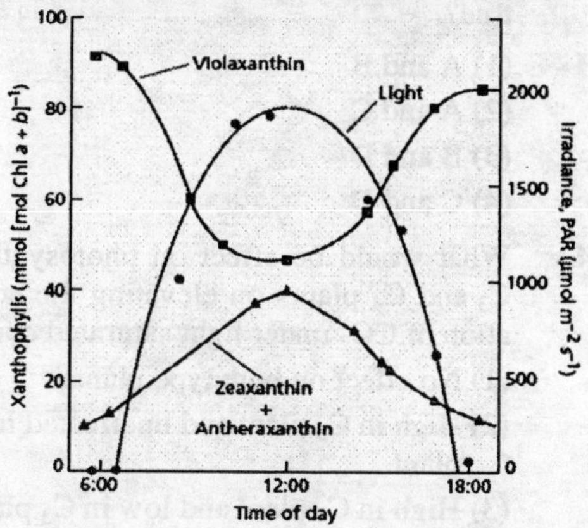

Which of the following is correct interpretation?

(1) As the amount of light incident to a leaf increases, a greater proportion of zeaxanthin is converted to antheraxanthin and violoxanthin, thereby dissipating excess excitation energy and protecting the photosynthetic apparatus

(2) As the amount of light incident to a leaf increases, a greater proportion of violaxanthin is converted to antheraxanthin and zeaxanthin, thereby dissipating excess excitation energy and protecting the photosynthetic apparatus

(3) As the amount of light incident to a leaf decreases, a greater proportion of violaxanthin is converted to antheraxanthin and zeaxanthin, thereby providing sufficient energy to photosynthetic apparatus for photosynthesis

(4) Violoxanthin is thermo labile while zeaxanthin and antheraxanthin are thremostable

11. The figure above shows changes in photosynthesis as a function of ambient intercellular CO_2 concentrations in plant A and B. It suggests that

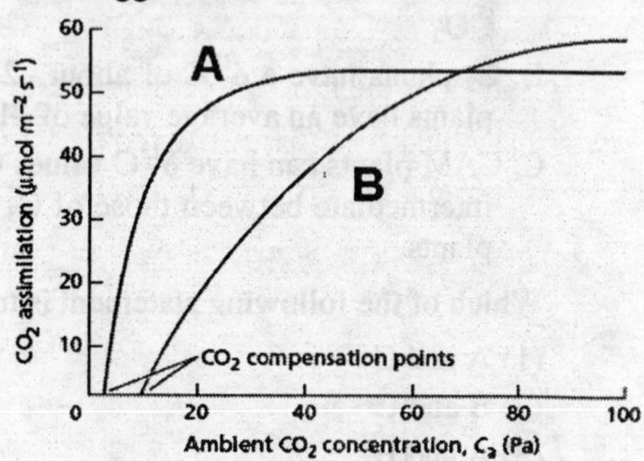

(1) Plant A is C_3 plant and Plant B is C_4 plant

(2) Plant A is C_4 plant and Plant B is C_3 plant

(3) Both are C_3 plants

(4) Both are C_4 plants

12. Atmospheric CO_2 contains the naturally occurring carbon isotopes ^{12}C, ^{13}C, and ^{14}C in the proportions 98.9%, 1.1%, and 10^{-10}%, respectively. $^{14}CO_2$ is present in such small quantities that it has no physiological relevance, but $^{13}CO_2$ is different. The chemical properties of $^{13}CO_2$ are identical to those of $^{12}CO_2$, but because of the slight difference in mass (2.3%)

Carbon isotope composition is measured by use of a mass spectrometer, which yields the following ratio

$$R = \frac{^{13}CO_2}{^{12}CO_2}$$

The isotope composition of plants, $\delta^{13}C$, is quantified on a per mil (–%) basis

$$\delta^{13}C\,\% = \left(\frac{R_{sample}}{R_{standard}} - 1\right) \times 1000$$

Consider the following statements

A. Most plants assimilate less $^{13}CO_2$ than $^{12}CO_2$. In other words, plants discriminate against the heavier isotope of carbon, and they have smaller ratios of ^{13}C to ^{12}C than are found in atmospheric CO_2

B. C_3 plants have a $\delta^{13}C$ of about –28%; C_4 plants have an average value of –14%

C. CAM plants can have $\delta^{13}C$ values that are intermediate between those of C_3 and C_4 plants

Which of the following statement is true?

(1) A and C
(2) B and C
(3) A and B
(4) A, B and C

13. Figure below explains the quantum yield of photosynthetic carbon fixation in a C_3 plant and in a C_4 plant as a function of leaf temperature

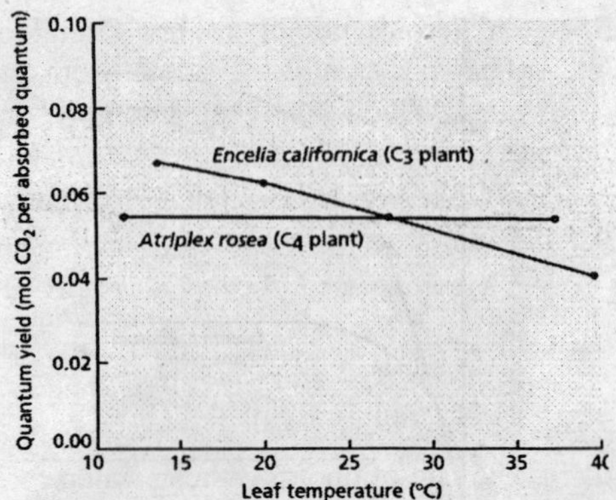

Following conclusions can be drawn

A. In normal air, photorespiration increases with temperature in C_3 plants, and the energy cost of net CO_2 fixation increases accordingly

B. The higher energy cost in C_3 is expressed in lower quantum yields at higher temperatures

C. Because of the CO_2 concentrating mechanisms of C_3 plants, photorespiration is low in these plants, and the quantum yield does not show a temperature dependence

D. At higher temperatures the quantum yield of C_3 plants is higher than that of C_4 plants, indicating that photosynthesis in C_4 plants is more efficient at lower temperatures

Which of the following statement is not true?

(1) A and B
(2) A and C
(3) B and D
(4) C and D

14. What would be effect on photosynthesis in C_3 and C_4 plants on elevating the concentration of CO_2 under light saturated condition?

(1) No effect on both type plants
(2) High in C_3 plant and unaffected in C_4 plant
(3) High in C_4 plant and low in C_3 plants
(4) High in both type plants

15. Thylokoid membrane has lateral asymmetrical positioning of photosytem in chloroplast. Which statement is correct?

(1) PS-I in non appressed portion and PS-II in appressed portion

(2) PS-II in non appressed portion and PS-I in appressed portion

(3) Both PS-I and PS-II in appressed portion

(4) Both PS-I and PS-II in non appressed portion of thylakoid

16. The rate of photosynthesis when red and far-red light are given together is greater than the sum of the rates when they are given apart. The enhancement effect provided essential evidence in favor of the concept that

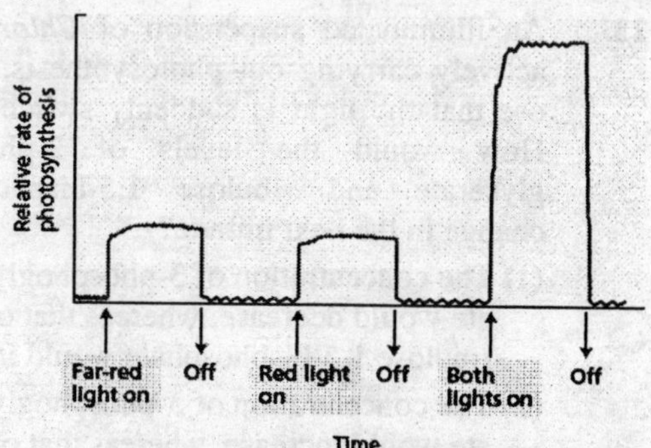

(1) Photosynthesis is carried out by two photochemical system working in sequence using same wavelength optima

(2) Photosynthesis is carried out by two photochemical systems working in tandem but with slightly different wavelength optima

(3) Photosynthesis is promoted during prolonged red or far red light wavelengths

(4) Far Red light promotes photosynthsis while red light inhibits photosynthesis

17. Chemical structure of dichlorophenyl-dimethylurea (DCMU) and methyl viologen (paraquat), two herbicides that block photosynthetic electron flow. DCMU is also known as diuron. Sites of action of the two herbicides has been marked with A and B. Which of the following is correct?

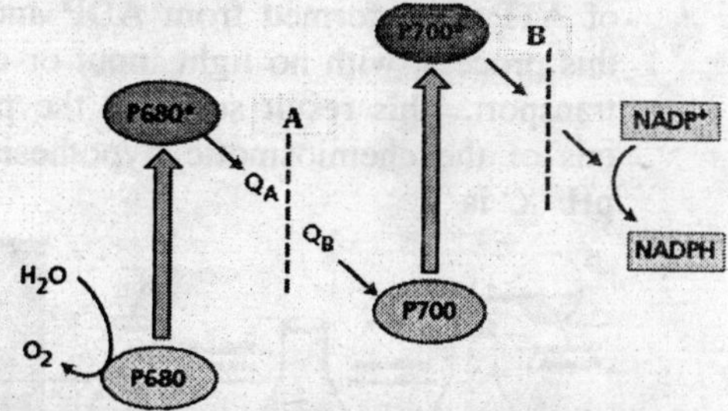

(1) A is Parquat and it blocks electron flow at the quinone acceptors of photosystem II, by competing for the binding site of plastoquinone while B is DCMU which acts by accepting electrons from the early acceptors of photosystem I

(2) A is is Paraquat which acts by accepting electrons from the early acceptors of photosystem II, while B is DCMU and it blocks electron flow at the quinone acceptors of photosystem I, by competing for the binding site of plastoquinone

(3) A is DCMU and it blocks electron flow at the quinone acceptors of photosystem II, by competing for the binding site of plastoquinone while B is Paraquat which acts by accepting electrons from the early acceptors of photosystem I

(4) A is DCMU and it blocks electron flow at the quinone acceptors of photosystem I, by competing for the binding site of plastoquinone and B is Parquat which acts by accepting electrons from the early acceptors of photosystem I

18. Some of the early evidence supporting a chemiosmotic mechanism of photosynthetic ATP formation was provided by an elegant experiment carried out by André Jagendorf and coworkers (Figure) They suspended chloroplast thylakoids in a pH 4 buffer, and the buffer diffused across the membrane, causing the interior, as well as the exterior, of the thylakoid to equilibrate at this acidic pH. They then rapidly transferred the thylakoids to a pH =X buffer, thereby creating a pH difference of 4 units across the thylakoid membrane. They found that large amounts of ATP were formed from ADP and Pi by this process, with no light input or electron transport. This result supports the predictions of the chemiosmotic hypothesis. Here pH 'X' is

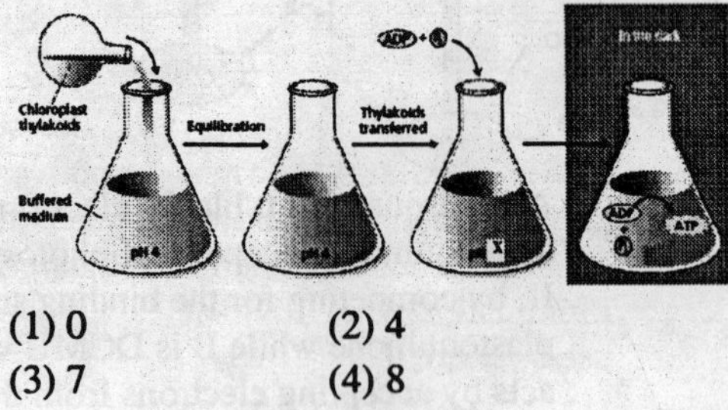

(1) 0 (2) 4
(3) 7 (4) 8

Question 19-21

Consider the relation between the energy of a photon and its wavelength

19. Some bacteria are able to harvest 1000-nm light. What is the energy (in kilocalories) of a mole (also called an einstein) of 1000-nm photons?
(1) 28.7 kcal $einstein^{-1}$
(2) 120 kcal $einstein^{-1}$
(3) 1.24 kcal $einstein^{-1}$
(4) 12 kcal $einstein^{-1}$

20. What is the maximum increase in redox potential that can be induced by a 1000-nm photon?
(1) 0.24 V
(2) 1.24 V
(3) 0.48 V
(4) 2.48 V

21. What is the minimum number of 1000-nm photons needed to form ATP from ADP and Pi? Assume a ΔG of 12 kcal mol^{-1} (50 kJ mol^{-1}) for the phosphorylation reaction
(1) 0.42
(2) 0.2
(3) 0.4
(4) 0.8

22. In 1939, Robert Hill discovered that chloroplasts evolve O_2 when they are illuminated in the presence of an artificial electron acceptor such as ferricyanide $[Fe^{3+}(CN)_6]^{3-}$. Ferricyanide is reduced to ferrocyanide $[Fe^{2+}(CN)_6]^{4-}$ in this process. No NADPH or reduced plastocyanin is produced. Propose a mechanism for the Hill reaction
(1) The electrons flow through photosystem I directly to ferricyanide
(2) The electrons flow through photosystem II to cyt f_6-f to PSI to ferricyanide
(3) The electrons flow through photosystem II to cyt b_6-f to ferricyanide
(4) The electrons flow through photosystem II directly to ferricyanide

23. An illuminated suspension of *Chlorella* is actively carrying out photosynthesis. Suppose that the light is suddenly switched off. How would the levels of 3-phosphoglycerate and ribulose 1,5-bisphosphate change in the next minute?
(1) The concentration of 3-phosphoglycerate would decrease, whereas that of ribulose 1,5-bisphosphate would increase
(2) The concentration of 3-phosphoglycerate would increase, whereas that of ribulose 1,5-bisphosphate would decrease
(3) The concentration of 3-phosphoglycerate would increase, whereas that of ribulose 1,5-bisphosphate would remain same
(4) The concentration of 3-phosphoglycerate would decrease, whereas that of ribulose 1,5-bisphosphate would remain same

24. An illuminated suspension of *Chlorella* is actively carrying out photosynthesis in the presence of 1% CO_2. The concentration of CO_2 is abruptly reduced to 0.003%. What effect would this reduction have on the levels of 3-phosphoglycerate and ribulose 1,5-bisphosphate in the next minute?

(1) The concentration of 3-phosphoglycerate would decrease, whereas that of ribulose 1,5-bisphosphate would increase
(2) The concentration of 3-phosphoglycerate would increase, whereas that of ribulose 1,5-bisphosphate would decrease
(3) The concentration of 3-phosphoglycerate would increase, whereas that of ribulose 1,5-bisphosphate would remain same
(4) The concentration of 3-phosphoglycerate would decrease, whereas that of ribulose 1,5-bisphosphate would remain same

Question 25-27

Graph A shows the photosynthetic activity of two species of plant, one a C_4 plant and the other a C_3 plant, as a function of leaf temperature

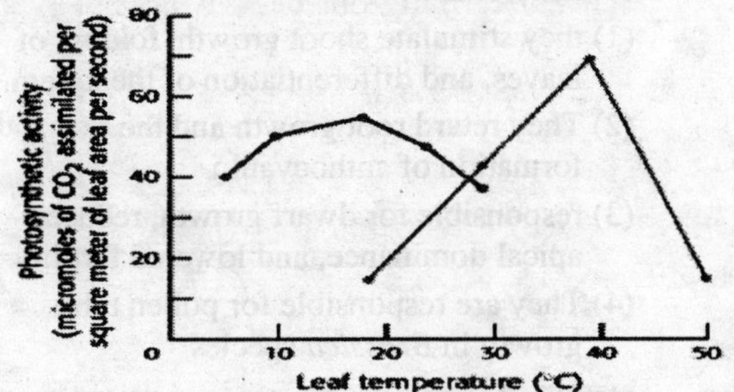

25. Which data were most likely generated by the C_4 plant and which by the C_3 plant?
(1) The curve on the right was generated by the C_4 plant
(2) The curve on the left was generated by the C_4 plant
(3) C_3 as well as C_4 plants can show both types of curve
(4) Nor C_3 neither C_4 can show such curves

26. Suggest some possible explanations for why the photosynthetic activity falls at higher temperatures
(1) Due to the oxygenase activity of RUBISCO in C_3 plants and denaturation of proteins in C_3 and C_4 plants
(2) Due to the carboxylase activity of RUBISCO in C_3 plants and denaturation of proteins in C_3 and C_4 plants
(3) Due to the oxygenase activity of PEPCO in C_4 plants and denaturation of proteins in C_3 plants
(4) Due to the carboxylase activity of PEPCO in C_4 plants and denaturation of proteins in C_3 and C_4 plants

27. Why does simultaneous illumination with 600 nm and 700 nm light support a higher, photosynthetic rate in higher plants than the sum of the photosynthetic rates supported by illumination with light of each wavelength used separately?
(1) Due to red drop
(2) Due to emmerson enhancement effect
(3) Intermediate wavelength (650 nm) is more suitable for photosynthesis
(4) Due to activation of RUBISCO

28. The figure above shows water potential of plants under various growing conditions, and sensitivity of various physiological processes to water potential. The intensity of the bar color corresponds to the magnitude of the process. For example, cell expansion decreases as water potential falls (becomes more negative) The plants with water potential in box are

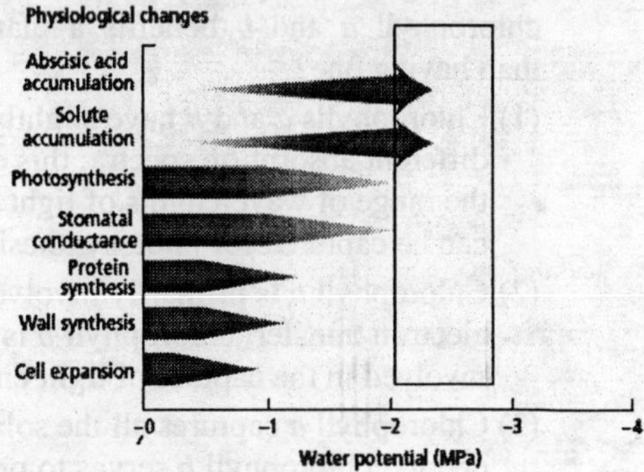

(1) Well-watered plants
(2) Plants under mild water stress
(3) Plants in arid/desert
(4) Hydrophytes

29. Almost a century and a half passed before the source of the O_2 released during photosynthesis was determined. One of the first uses of an isotopic tracer in biological research resulted in its identification. In these experiments, two groups of green plants were allowed to carry on photosynthesis. Plants in the first group were supplied with water containing the oxygen isotope ^{18}O and with CO_2 containing only the common oxygen isotope ^{16}O; plants in the second group were supplied with CO_2 labeled with ^{18}O and water containing only ^{16}O. When oxygen gas was collected from each group of plants and analyzed, it was found that O_2 containing ^{18}O was produced in abundance by the plants that had been given ^{18}O labeled water, but not by the plants given ^{18}O-labeled CO_2. These results showed that all the oxygen gas produced during photosynthesis comes from

(1) Water
(2) CO_2
(3) Both from water and CO_2
(4) Carbohydrates

30. Chlorophylls *a* and *b* are two pigments primarily responsible for the capture and processing of solar energy in the light reactions of photosynthesis. Which of the following best explains why having both chlorophyll *a* and *b* benefits a plant more than having one?

(1) Chlorophylls *a* and *b* have slightly different absorption spectra; this expands the range of wavelengths of light that can be captured for photosynthesis
(2) Chlorophyll *a* is primarily involved in electron transfer; chlorophyll *b* is mainly involved in the capture of light energy
(3) Chlorophyll *a* captures all the solar energy; chlorophyll *b* serves to protect chlorophyll *a* from overexcitation
(4) Chlorophyll *b* captures all the solar energy; since chlorophyll *a* is a precursor to chlorophyll *b*, a pool of chlorophyll *a* must be maintained

31. Glucose-1-P is converted to Fructose-6-P in two successive reactions

Glucose-1-P→ Glucose-6-P (ΔG_0 = -1.7 kcal/mol)

Glucose-6-P→ Fructose-6-P (ΔG_0 = -0. 4 kcal/mol)

What is ΔG_0 for the overall reactions?

(1) 1.3 kcal/mol (2) 6.8 kcal/mol
(3) 2.1 kcal/mol (4) -2.1 kcal/mol

32. Steroid phytohormones are the so-called brassinosteroids. The most effective member of the brassinosteroids is brassinolide. Brassinosteroids are synthesized from campesterol (a membrane lipid) Brassinosteroids regulate plant development. Which of the following is not a function of brassinosteroid?

(1) they stimulate shoot growth, folding of leaves, and differentiation of the xylem
(2) They retard root growth and the formation of anthocyanin
(3) responsible for dwarf growth, reduced apical dominance, and lowered fertility
(4) They are responsible for pollen tube growth in *Brassica* species

33. Fill in the blanks (a, b, c and d) in the following statements with a proper combination of m, n, o and p

Where in

m represents - longer

n represents - shorter

o represents -prevents

p represents - induces

Short day (SD) plants flower when night lengths are _____**a**____ than a critical dark period. Interruption of the dark period by a brief light treatment _______**b**_____ flowering in SD plants. Long day (LD) plants flower when night length is _______**c**______ than a critical period. Shortening of the night with a brief light treatment _______**d**_____ flowering in LD plants

	a	b	c	d		a	b	c	d
(1)	m	o	n	p	(2)	n	p	m	o
(3)	n	o	m	p	(4)	m	p	n	o

34. Mutations in Flowering locus T of *Arabidopsis thaliana* results in late flowering phenotype. Transcript levels of FLT were determined in long day and short day seedlings. Which of the following would likely represent the transcript profile of FLT?

	Long-day seedlings	**Short-day seedlings**
(1)		
(2)		
(3)		
(4)		

35. Consider the following statements regarding the salt tolerant plants

A. salt tolerant plants have salt glands at the surface of the leaves

B. accumulation of solute like sorbitol, glycine betaine, proline

C. salt tolerant plant overexpress SOS genes

D. salt tolerant plant selectively inhibit the movement of ions from soil to root

Which statement(s) is/are correct?

(1) A, B and C (2) B and C only

(3) A and D only (4) A, B, C and D

Question 36-37 refers to the following table

The photoperiodic response of three different types of plants were recoreded by observing flowering responses to varying length of light and dark. The photoreceptors responsible for flowering of plants is phytochrome. Phytochrome undergoes an allosteric change upon absorption of red light having specific wavelength. The P_R form of phytochrome absorbs red light of wavelength 660 nm; the P_{FR} form absorbs red light of wavelength 730 nm. P_{FR} spontaneously converts back to P_R in the absence of light. During daylight hours P_{FR} is more prevalent than P_R; during night time hours P_R is more prevalent

	16h 8h	8h 16h	8h 16h
Plant I	---	+++	---
Plant II	+++	+++	+++
Plant III	+++	---	+++

(+++ = flowering; --- = absence of flowering)
(□ = light; ■ = dark)

P_{FR} induces flowering in long day plants and inhibits flowering in short day plants

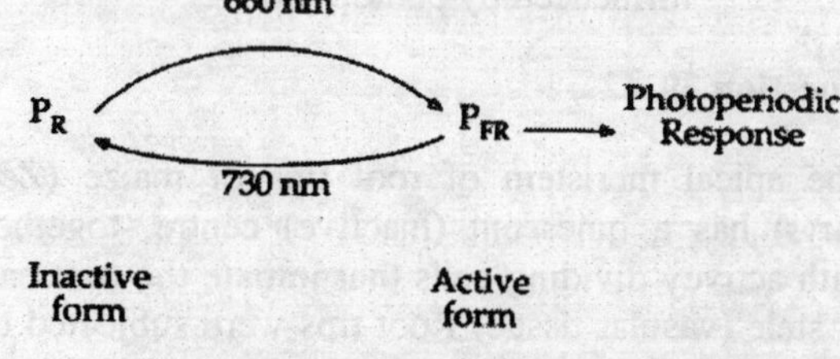

36. The critical factor in eliciting a flowering response in Plant II appears to be

(1) number of hours of day light

(2) number of hours of darkness with brief interrupted exposure to light

(3) number of hours of darkness uninterrupted by brief exposure to light

(4) Light and dark periods donot trigger flowering in these plants

37. Which of the following statement is the most reasonable conclusion concerning the influence of alternative forms of phytochrome on flowering by study plants?

(1) P_{FR} inhibits flowering in Plant I

(2) P_{FR} induces flowering in Plant I

(3) P_R inhibits flowering in Plant I

(4) P_{FR} inhibits flowering in Plant III

38. Certain seeds germinate when subjected to light of wavelength 660 nm. When briefly exposed to light of wavelength 660 nm followed by brief exposure to light of wavelength 730 nm, they fail to germinate. When briefly exposed to light of wavelength 660 nm followed by exposure to light of wavelength 730 nm followed by exposure to light of 660 nm, they germinate. The most likely interpretation of these results is that

(1) Exposure to light of wavelength 660 nm converts P_R to P_{FR}

(2) Exposure to light of wavelength 730 nm converts P_R to P_{FR}

(3) Germination in seeds displaying photoperiodism is influenced by the seed's final exposure to red light

(4) Germination in these seeds is not influenced by photoperiod

Question 39-42

The apical meristem of root tips of maize (*Zea mays*) has a quiescent (inactive) centre, together with activey dividing cells that initiate the root cap or stele (vasular tissue) Root tips were subjected to acute x-irradation at two dosages, which affected mitosis and cell division in three regions, as shown below

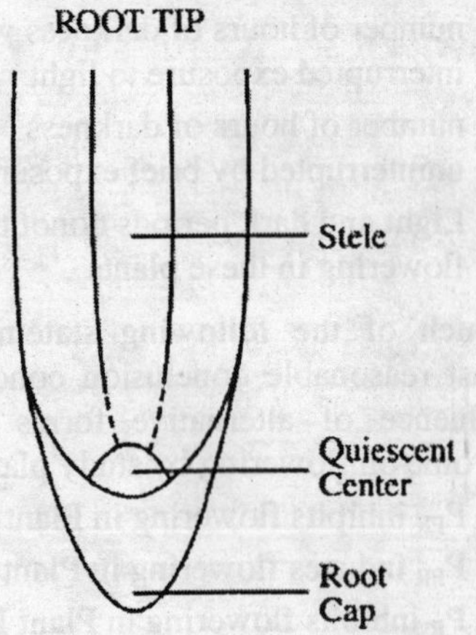

AVERAGE DURATIONS OF THE CELL CYCLE (IN HOURS) AFTER ACUTE X-IRRADIATION

Dose (rad)	Time After Irradiation (days)	Quiescent Center	Dividing Cells of Root Cap	Stele (vascular-tissue)
0	–	174	12	28
900	3	89	54	44
"	6	124	52	60
1,800	3	62	134	137
"	6	64	149	86
"	9	181	56	56

39. The first change in the quiescent center in irradiated roots, as shown in table, was

(1) an increase in the number of polyploidy cells

(2) an increase in the length of cell cycle

(3) a decrease in the number of cell cycles

(4) a decrease in the length of the cell cycle

40. Increase dosages of irradiation had which of the following effects?

(1) Shorter cell cycle in quiescent center cells at high dosage than at low dosage

(2) Longer duration of cell cycle in quiescent center cells at high dosage than at low doasage

(3) Shorter duration of cell cycle in meristematic cells of the caps and stele at high than at low dosage

(4) Cessation of division in all cells in the root tip

41. Six to nine days after irradiation, which of the following statement is correct about the cells of quiescent center?

(1) They retained a high level of mitotic activity

(2) They returned to a relatively low lovel of mitotic activity

(3) The level of mitotic activity continued to fluctuate

(4) They retained a permanent low level of mitotic activity

42. The quiescent center in roots of maize is significant because

(1) root meristems are not responsible for root growth

(2) irradiation damages all root meristem cell equally

(3) quiescent center cells can continue meristem growth if the other meristem cells are damaged by irradiation

(4) the quiescent center cells will have the most radiation damage of all the meristem cells

43. In plants, the energy of sunlight is first absorbed by the pigments present in their leaf cells followed by the fixation of carbon through photosynthesis. Consider the following statements

A. Chlorophylls a and b are abundant in green plants

B. Chlorophylls c and d are found in some protists and cyanobacteria

C. Out of different types of bacteriochlorophyll, type a is the most widely distributed

D. Out of different types of bacteriochlorophyll, type b is the most widely distributed

Which one of the following combination of above statements is correct?

(1) A, B and C

(2) A, C and D

(3) B, C and D

(4) A, B and D

44. Nitrate reductase is an important enzyme for nitrate assimilation. Given below are some statements on nitrate reductase enzyme

A. Nitrate reductase of higher plants is composed of two identical subunits

B. One subunit of nitrate reductase contains three prosthetic groups

C. One of the prosthetic groups attached to both subunits is heme

D. One of the prosthetic groups complexed with pterin is magnesium

Which one of the following combination of statements on nitrate reductase mentioned above is correct?

(1) A, B and C

(2) A, C and D

(3) B, C and D

(4) A, B and D

45. A farmer growing a particular variety of grape plants in vineyard, observes the following

A. Fruit size normally remained small

B. Natural seed abortion

C. Development of fungal infection as the pedicels are small in size due to which moisture is retained in the bunches of grapes

Experts suggested spraying gibberellic acid during the fruit development. This treatment would help in getting rid of

(1) A, B and C

(2) only A and B

(3) only A and C

(4) only B and C

46. Light is perceived by various photoreceptors in plants. The photoreceptors predominantly work at specific wavelengths of light. Some of the following statements are related to the functions of plant photoreceptors

A. Phytochrome A predominantly perceives the red and far-red light

B. Phytochrome B predominantly perceives red light

C. Cryptochromes regulate plant development

D. Phototropins are involved in blue light perception and chloroplast movements

Which one of the following combinations based on above statements is correct?

(1) A, B and C

(2) B, C and D

(3) C, D and A

(4) A, B and D

47. From the following statements

A. Triose phosphate is utilized for the synthesis of both starch and sucrose

B. Triose phosphate is translocated to cytosol from chloroplast

C. Triose phosphate is confined to chloroplast and is utilized for synthesis of starch only

D. Triose phosphate is translocated from cytosol to chloroplast

Which one of the following combinations is correct regarding starch and sucrose synthesis during day time?

(1) A and B

(2) B and C

(3) C and D

(4) D and A

48. Shown below, is a graph representing the growth of different plant species subjected to salinity relative to that of unsalinized control. Which of the following statements is not true?

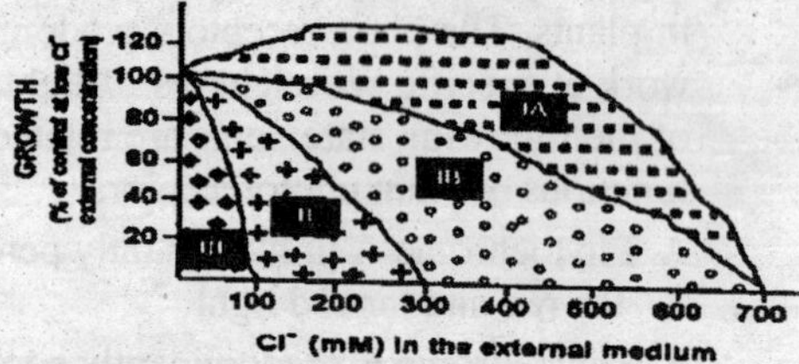

(1) Plants In group IA are extreme halophytes while very salt sensitive species will be part of group III

(2) Plants in group IA are very salt sensitive and extreme halophytes will be part of group III

(3) Halophytes, which can tolerate salt but their growth is retarded will be of part of group IB

(4) Non-halophytes, which are salt tolerant but lacks salt glands will be a part of group II

49. Experiments were carried out on plants living in different environments to measure the size of leaf stomata. Previous investigations have shown that plants transpire more water when the size of stomata is larger. Which of the graphs *best* represents a plant living in a dry environment?

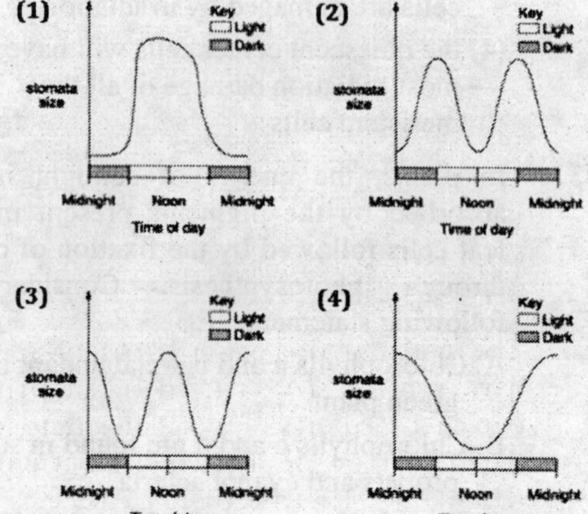

50. Bacteria often use two-component signaling systems to sense and respond to environmental signals such as nutrients or light. The C termini of phytochromes have some sequence similarity to the transmitter domains of bacterial sensor proteins

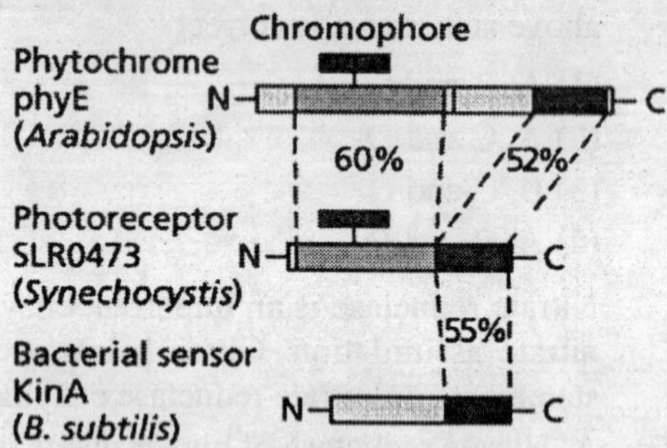

Consider the following statements

A. This sharing of sequences shows the possibility that phytochrome evolved from the sensor protein of a bacterial two-component system

B. The kinase activity associated with these phytochromes is of the histidine type similar to bacterial two component system

C. Red-light stimulation leads to auto phosphorylation of phytochrome, supporting the hypothesis that Pfr is the active form

D. Phytochromes are also located on plasma membrane similar to receptors in bacterial two component system

Which of the following statement are true?

(1) A and B

(2) A and C

(3) B and D

(4) B and C

51. The inhibition of stem or hypocotyl growth, have usually been studied in dark-grown, etiolated seedlings. The action spectrum for the inihibition of hypocotyl elongation in dark-grown lettuce seedlings is shown in Figure

Note that the main peak of activity is in the far-red rather than the red region of the spectrum, and there are peaks in the blue and UV-A regions as well. The absence of a peak in the red is unusual for a phytochrome-mediated response

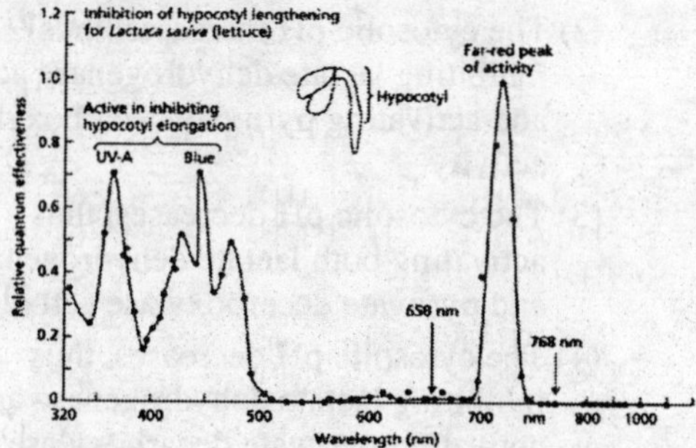

To probe such an interaction, Hartmann irradiated plants with not one but two wavelengths of light simultaneously, using two wavelengths that were inactive in the HIR for lettuce hypocotyl inhibition: 658 nm and 768 nm. Although physiologically inactive, the two wavelengths are still photochemically active in driving phytochrome conversions, each in the opposite direction. By carefully varying the irradiances of each wavelength, Hartmann was able to induce a range of ratios of Pfr to total phytochrome in the plant from near zero to higher values. He was able to show that the optimum ratio of Pfr to total phytochrome for the inhibition of lettuce hypocotyl growth is approximately 0.03. Remarkably, this is precisely the ratio that results when the plants are irradiated with a single wavelength at 720 nm, the far-red peak of the 3

(1) Very Low Fluence Response (VLFR)

(2) Low Fluence Response (LFR)

(3) High Irradiance Response (HIR)

(4) Cryptochrome Mediated Response (CMR)

52. The importance of membrane lipids for tolerance to low temperatures has been demonstrated by work with mutant and transgenic plants in which the activity of particular enzymes led to a specific change in membrane lipid composition independent of acclimation to low temperature. In one study, the genes used to modify the lipid composition of thylakoid membranes encoded glycerol-3-phosphate acyltransferases (GPATs) These enzymes select fatty acids for incorporation into PG; GPATs from different plant species have different selectivity for saturated and unsaturated fatty acids, leading to the differences in composition illustrated in Figure

(A) Tobacco

	Wild type and plants transformed with a control gene	Plants transformed with squash gene	Plants transformed with *Arabidopsis* gene
Molecular species of PG	64 36	24 76	72 28
Percent inhibition of photosynthesis after 4 hours at 1°C	25	88	7

cis-Unsaturated molecular species of phosphatidylglycerol (PG)

Saturated and *trans*-monounsaturated molecular species of PG

The degree of fatty acid saturation of phosphatidylglycerol (PG) is particularly important to the chilling resistance of chloroplasts. In one experiment, tobacco was transformed with an acyltransferase gene from either squash or Arabidopsis to modify the fatty acid saturation of PG in the thylakoid membranes of the chloroplasts

Following conclusions may be drawn

A. *Arabidopsis* is chill sensitive and Squash is chill resistance plant

B. In transgenic tobacco transformed with the squash gene, which increased the proportion of lower-melting-point (trans-monounsaturated and saturated) molecular species of PG, chilling tolerance increased (88% inhibition of photosynthesis following 4 hours at 1°C)

C. When tobacco was transformed with the *Arabidopsis* chill-tolerant gene, the proportion of high-melting-point molecular species of PG decreased, and tolerance of chilling was greater than in wild-type plants

Which of the following conclusions are true?

(1) C only (2) A and B
(3) B and C (4) A and C

53. Plants produce compounds that protect cells against intracellular ice formation. Consider the following available mechanisms

A. Accumulation of compounds such as sugars, amino acids, and other solutes, induce supercooling in the plant's tissues

B. Decrease in compounds such as sugars, amino acids, and other solutes, due to efflux, induce supercooling in the plant's tissue

C. Overwintering plants produce antifreeze proteins which provide freeze tolerance by inhibiting ice crystal growth and the nucleation of ice crystals

D. Overwintering plants accumulates more saturated fatty acids and glycerol in their membranes which provide freeze tolerance

Which of the following are correct regarding mechanisms that protect cells against intracellular ice formation?

(1) A and B (2) B and C
(3) A and C (4) B and D

54. If an *Arabidopsis* plant, mutated in lycopene biosynthetic pathway is grown in sunny tropical climate in the presence of oxygen

(1) it would accumulate higher biomass due to higher rate of photosynthesis
(2) there will not be any influence of this mutation on the rate of photosynthesis and plant growth
(3) it would show reduced biomass due to photo oxidative damage
(4) the leaves would be bluish purple in color because of higher accumulation of xanthophylls

55. During episodes of anoxia in plants, pyruvate produced in glycolysis is initially fermented to lactate. During later stage, there is an increase in the fermentation to ethanol and decrease in the fermentation to lactate, a phenomena which helps plants survive anoxia. Which of the following statements is correct about this change of fermentation flux from lactate towards ethanol?

(1) The cytosolic pH increases, thus activating both lactate dehydrogenase and pyruvate decarboxylase activity
(2) The cytosolic pH increases, thus inhibiting lactate dehydrogenase activity and activating pyruvate decarboxylase activity
(3) The cytosolic pH decreases, thus activating both lactate dehydrogenase and pyruvate decarboxylase activity
(4) The cytosolic pH decreases, thus inhibiting lactate dehydrogenase and activating pyruvate decarboxylase activity

56. Atmospheric CO_2 contains the naturally occurring stable carbon isotopes ^{12}C and ^{13}C in the proportion of 98.9% and 1.1%, respectively

Following are some of the statements regarding CO_2 assimilation

A. Both C_3 and C_4 plants assimilate less $^{13}CO_2$ than $^{12}CO_2$

B. Both C_3 and C_4 plants assimilate less $^{12}CO_2$ than $^{13}CO_2$

C. C_3 plants assimilate lesser $^{13}CO_2$ than $^{12}CO_2$ as compared to C_4 plants

D. C4 plants assimilate lesser $^{13}CO_2$ than $^{12}CO_2$ as compared to C_3 plants

Which one of the following combinations of above statements is true?

(1) A and B
(2) A and C
(3) C and D
(4) A and D

57. Which of the following statement is correct regarding activation of phytochromes?

(1) Red light converts Pr to active form Pfr
(2) Red light represses activity of Pr
(3) When red light and far red light are in equal fluence rate Pr is activated
(4) When red light and far red light are in different fluence rate Pr is activated

58. Consider the following statement regarding position of components of photosystem in chloroplast of plant

A. PSII is preferentially localized in granal thylakoids

B. PSI and PSII are co-localized in stromal thylakoid

C. PSI is preferentially localized in stromal thylakoid

D. Cytchrome b_6f complex is not a membrane resident complex

Which of the following statements are correct?

(1) A and C
(2) B and D
(3) A and D
(4) B and C

59. Consider the following statements regarding phenyl ammonia lyase (PAL) and chalcone synthase (CS) enzyme involved in synthesis of flavanoids

A. Substrate for PAL is phenyl alanine and for CS is chalcone

B. PAL catalyze conversion of phenyl alanine to cinnamic acid

C. PAL catalyze conversion of phenyl alanine to p-coumaric acid

D. Coumaryl CoA is converted to chalcone by chalcone synthase

Which of the following statement are correct?

(1) A and B
(2) B and D
(3) A and C
(4) C and D

60. Those proteins required especially for the formation of nodules, and which are synthesized by the host plant in the course of nodule formation, are called nodulins. Which of the following is not a nodulin?

(1) Leghaemoglobin
(2) Sucrose synthase
(3) Uriede synthase
(4) Lipo-chito-oligosaccharide

61. You are conducting an experiment under controlled conditions in a greenhouse in which paddy and maize are being grown. Keeping all other factors constant you increase the concentration of carbon dioxide in the greenhouse by 25% above normal atmospheric levels of this gas. The most likely outcome of this experiment would be

(1) Increase in growth rates of paddy
(2) Increase in growth rates of maize
(3) Increase in growth rates of paddy and maize
(4) Decrease in growth rates of paddy

62. Consider the following statements for different photosynthetic strategies. Which of the following is incorrect statement?

(1) The ratio of water loss to CO_2 uptake is lower in CAM plants than it is in either C_3 and C_4 plants

(2) The rate of photosynthesis attains maximum rate at lower intracellular CO_2 partial pressure in C_4 plants than in C_3 plants

(3) The compensation point in C_3 plants are always lower than C_4 plants

(4) Plants with C_4 metabolism need less rubisco than C_3 plants to achieve a given rate of photosynthesis

63. Following are certain statements regarding secondary metabolites found in plants. Among the following which is false?

(1) All terpenes are derived from a five carbon element

(2) Alkaloids are nitrogen containing compounds

(3) Pyrethroids, a monoterpene ester found in the leaves and flower of *Chrysanthemum* species, show insecticidal activity

(4) Limonoids are groups of alkaloids and have anti-herbivoral activity

64. Light is crucial for plant growth and development. Following are certain statements related to photoreceptors in model plant *Arabidopsis thaliana*. Identify the incorrect statement

(1) Among the five phytochrome genes, representing a gene family, PhyB plays a predominant role in red-light perception

(2) Cryptochromes are involved in the regulation of flowering time and circadian rythm

(3) phyA photoreceptor is predominantly involved in far-red light perception

(4) The LOV domain of phytochrome C (pHYC) is an important domain for signal transmission

65. One of the important functions of Program Cell Death (PCD) in plants is protection against pathogens. PCD also appears to occur during the differentiation of xylem tracheary elements that leads to nuclei and chromatin degradation. These changes result from the activation of certain genes. Which one of the following combinations of the above is involved in differentiation of xylem tracheary elements?

(1) Topoisomerase and nuclease

(2) RNA polymerse and DNA polymerase

(3) Topoisomerase and protease

(4) RNA polymerase and topoisomerase

66. Following are some statements to osmotic stress in plants. Which of the following is not correct?

(1) The accumulation of ions during osmotic adjustment is predominantly restricted to the vacuoles

(2) In order to maintain the water potential equilibrium within the cell, other solutes or compatible osmolytes accumulate in the cytoplasm

(3) Glucose is one of the most compatible osmolytes involved in osmotic stress in plants

(4) There are mainly four groups of molecules that frequently serve as compatible solutes

67. A transgenic lettuce plant was generated by over-expressing isopentenyl transferase (IPT) gene under the control of the promoter of senescence activator gene (SAG12) Following are some statements regarding this transgenic plant

The transgenic plants

A. exhibit delayed senescence

B. exhibit fast senescence

C. have higher amount of cytokinin during senescence

D. have higher amount of gibberellins during senescence

Which one of the following combinations of above statements is correct?

(1) A and B

(2) A and C

(3) B and D

(4) C and D

68. Which one of the following statements on plant hormones is incorrect?

(1) Salicyclic acid is intimately involved in shoot and root elongation

(2) Ethylene is involved in senescence and fruit ripening

(3) Abscisic acid plays a key role in seed germination and drought responses

(4) Jasmonic acid inhibits root growth and is involved in pathogenesis

69. Nitrate reductase is a key enzyme of nitrate assimilation process. Which of the following statements is incorrect about the nitrate reductase?

(1) Nitrate reductase-mediated reaction involves the transfer of one electron

(2) Nitrate reductase is influenced by nitrate and carbohydrate

(3) Nitrate reductase converts nitrate to nitrate in the cytosol

(4) Nitrate reductase is influenced by light

70. Following are some enzyme-substrate combinations involved in the conversion of photosynthetically formed triose-phosphate to sucrose

A. Fructose-1, 6-bisphosphate aldolase: Dihydroxyacetone phosphate

B. UDP-glucose pyrophosporylase: UDP-Glucose

C. Sucrose 6-phosphate synthase: Fructose-6-phosphate

D. Phosphofructo kinase: Fructose 6 phosphate

Which one of the following combinations of above enzyme-substrate combinations is correct?

(1) A, Band D (2) A, C and D
(3) B, C and D (4) A, B and C

Question 71-75

Two types of preparation from leaves were used to determine whether photosynthesis can provide energy for synthesis of a particular chloroplast protein (CP) Identical amounts of ^{14}C amino acids were introduced into preparations of whole-cell homogenates and of isolated chloroplast, each containing equal quantities of chlorophyll. As indicated in the table below, the reactions were carried out in light and dark. After each reaction was terminated, labeled CP was isolated and counted. Additives (A) used in treatments (T) 5, 6 and 7 were supplied before the introduction of ^{14}C amino acids

Synthesis of chloroplast proteins

T	Preparations	Condition	A	Amino acids incorporation (Counts/min/mg/ chlorophyll/hr)
1	Chloroplasts	Dark	-	56
2	Chloroplasts	Light	-	525,315
3	Whole cell homogenate	Dark	-	25,650
4	Whole cell homogenate	Light	-	150,418
5	Chloroplasts	Dark	ATP	446,519
6	Chloroplasts	Light	ATP	525,313
7	Chloroplasts	Light	DCMU	41,740

71. Which of the following is the best conclusion from results of treatments 1 and 2?

(1) The synthesis of CP by chloroplast requires light

(2) The synthesis of CP by chloroplast requires photosynthesis

(3) The synthesis of CP by chloroplast requires dark

(4) Chloroplast photosynthesize in the light

72. Which of the following is the best conclusion from the results of treatment 1, 2, 3 and 4?

(1) Whole cell homogenate contain no chloroplasts

(2) In the dark, whole cell homogenate make no CP

(3) In the dark, whole cell homogenate are less effective than chloroplasts are in synthesizing CP

(4) In the dark, a component of whole cell homogenate is required for the synthesis of CP

73. Which of the following is best conclusion about whole cell homogenate in the light?

(1) They have mitochondria and nuclei that code for the synthesis of CP

(2) They photosynthesize more than isolated chloroplasts do

(3) They make some components needed for the synthesis of CP in larger amount than dark

(4) They make none of the intermediates needed for the synthesis of CP

74. If the treatments 1 and 2 are used as controls, which of the following is best conclusion for the results of treatments 5 and 6?

(1) Exogenous ATP has no effect on the synthesis of CP in light or in dark

(2) ATP inhibits the synthesis of CP

(3) Exogenous ATP stimulates CP synthesis in whole cell homogenate

(4) ATP added to dark incubated preparations stimulates CP synthesis to a level 85% of that in light

75. Which of the following can be best concluded about the added DCMU, a known inhibitor of photosystem II?

(1) It blocked all photosynthetic ATP formation and since, some CP synthesis remained, ATP is not required for the process

(2) It blocked ATP formation associated with non cyclic electron flow, but probably not that associated with cyclic electron flow

(3) IT is required for synthesis of CP

(4) It would probably increase the synthesis of CP by chloroplast in dark

76. The Filamentous alga *Cladophora* is illuminated with light dispersed by a prism. As shown in the diagram, aerobic bacteria *Psuedomonas* included in the medium congregate where the alga is illuminated with light 656 nm and at 486 nm. *Pseudomonas* do not congregate in this manner if the *Cladophora* is removed from the medium. Which of the following is the most likely explanantion for the bacterial movement?

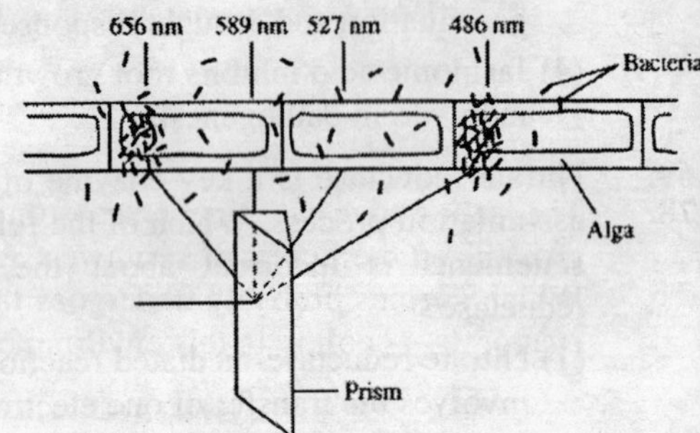

(1) Bacteria have a phototropic response stimulated by light at 656 nm and 486 nm

(2) Starch made by photosynthesis is secreted from the alga in regions illuminated by ligt at 656 nm and at 486 nm

(3) O_2 evolved by photosynthesis in the regions illuminated by light at 656 nm and 486 nm attracts bacteria

(4) Alga and bacteria have a symbiotic relationship

77. A study concerning the opening of stomata produces result that are shown in graph above. From the graph, which of the following can correctly be concluded?

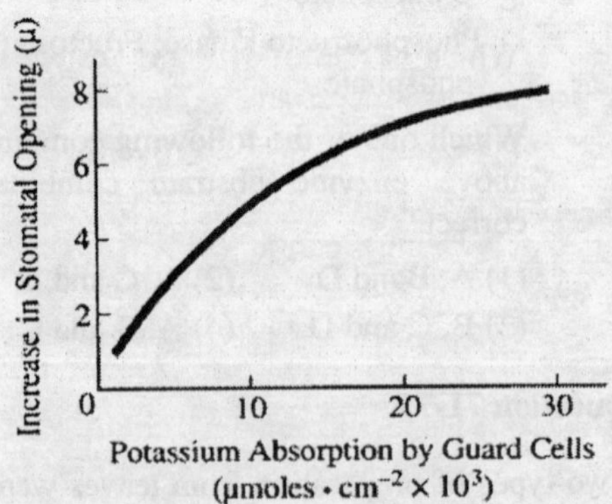

(1) Potassium absorption, rather than osmotic regulation of glucose is responsible for stomatal opening

(2) Potassium is actively transported into guard cells during the day and leak out to reach equilibrium at night

(3) An increase in K uptake by guard cell is accompanied by stomatal opening

(4) An increase in K absorption to more than 30 micromoles would lead to rupture of the membrane and closing of the stomata

78. The data above were collected from plants growing in the desert of southwest Arizona. Which species probably undergoes the most evaporative cooling in this environment?

Species	Stomata·mm^{-2} on Leaves	
	Upper Epidermis	Lower Epidermis
Encelia farinosa	248	355
Haworthia attenuata	15	25
Bursera lutiflora	0	272
Scilla violaceae	3	35

(1) *Encelia farinose*

(2) *Haworthia attenuata*

(3) *Bursera latiflora*

(4) *Scilla violaceae*

79. In the global nitrogen cycle, the following microbial organisms are involved in three important process- denitrification, nitrification and nitrogen fixation

A. *Rhizobium* B. *Nitrosomonas*

C. *Nitrobacter* D. *Pseudonomas*

E. *Azotobacter*

Which of the following is correctly matched pair of process and its causative species?

(1) Denitrification – (B); nitrogen fixation – (C) and (E); nitrification – (D)

(2) Denitrification – (D); nitrogen fixation – (A) and (E); nitrification – (C)

(3) Denitrification – (E); nitrogen fixation – (A) and (D); nitrification – (D)

(4) Denitrification – (B); nitrogen fixation – (A) and (D); nitrification – (C)

80. Following are some of the statements regarding the effect of CO_2 concentration on photosynthesis in plants

A. With elevated CO_2 levels, C_3 plants are much more responsive than C_4 plants under well watered conditions

B. In C_3 plants, increasing intracellular CO_2 partial pressure can stimulate photosynthesis only over a narrow range

C. In C_4 plants, CO_2 compensation point is nearly zero

Which one of the following combination of above statements is correct?

(1) A and B (2) B and C

(3) A and C (4) Only C

81. The quantum yield of photosynthetic carbon fixation in a C_3 plant and C_4 plant is studied as a function of leaf temperature. Following are some statements based on this study

A. At lower temperature the quantum yield of C_3 plant is lower than C_4 plant

B. In C_4 plant quantum yield does not show a temperature dependence

C. Since the photorespiration is low in C_4 plants because of CO_2 concentrating mechanism, quantum yield is not affected

D. At higher temperature the quantum yield of C_3 plant is lower than C_4 plant

Which one of the following combination of above statements is correct?

(1) A, B and D (2) B, C and D

(3) A, B and C (4) A, C and D

82. Following are some statements regarding plant growth hormones

A. Ethylene regulates abscission

B. Gibberlins do not play any role in flowering

C. Auxin and cytokinin promote cell division

D. Over expression of cytokinin oxidase would promote root growth

E. ABA inhibits root growth and promotes shoot growth at low water potential

F. ABA promotes leaf senescence independent of ethylene

Which one of the following combination of above statements is correct?

(1) A, C and F (2) B, C and D
(3) D, E and F (4) B, D and E

83. Following are some statements for synthesis of jasmonic acid in plants

A. 12-oxo-phytodienoic acid is produced in chloroplast and transported to peroxisome

B. Action of lipoxygenase, allene oxide synthase and allene oxide cyclase takes place in peroxisome

C. 12-oxo-phytodicnoic acid is first reduced and then converted to jasmonic acid by β-oxidation

D. Final production of jasmonic acid takes place in chloroplast

E. Action of allene oxide synthase and allene oxide cyclase takes place in chloroplast

Which one of the following combination of above statements is correct?

(1) A, B and C (2) B, D and E
(3) C, D and E (4) A, C and E

84. Following are some statements about low temperature stress in plants

A. Fatty acid composition of mitochondria isolated from chilling resistant and chilling sensitive plants differs significantly

B. Ratio of unsaturated fatty acids to saturated fatty acids is lower in chilling resistant species

C. The cellular water does not freeze even at -40°C, because of the presence of solutes and other antifreeze proteins

D. Heat shock proteins do not play any role during low temperature stress

Which one of the following combination of above statement is correct?

(1) A and B (2) A and C
(3) B and C (4) B and D

85. Upon absorption of a photon, a chlorophyll molecule get converted to its excited state when the energy of photon is

(1) more than that of ground state of pigment molecule
(2) equal to that of pigment molecule's excited state
(3) more than that of ground state but lesser then excited state of the pigment molecule
(4) equal to energy gap between ground state and excited state energy

86. Following are certain facts about the effect of abscisic acid (ABA) on the development and physiological effect of plants

A. ABA promotes leaf senescence independent of ethylene

B. ABA promotes shoot growth and inhibits root growth at low water potential

C. ABA inhibits gibberellin induced enzyme production

D. Seed dormancy is controlled by ratio of ABA and gibberllin

Which one of the following combination of above statements is true?

(1) A, B and C (2) B, C and D
(3) A, B and D (4) A, C and D

87. Following are some facts regarding localization of photosynthetic supramolecular complexes on plastid lamellae

A. PSII is preferentially localized on granal lamellae

B. ATP synthase and PSI is preferentially localized in stromal lamellae

C. PSI and PSII are located adjacent to each other in stromal lamellae

D. Cytchrome b_6/f complex is not a membrane bound complex

Which of the following combinations of the above statements is true?

(1) A and B (2) C and D
(3) B and D (4) B and C

88. The water content and the rate of water movement in soils depend to a large extent on soil type and soil structure

Consider the following statement

A. Because the water-filled pore spaces in the soil are interconnected, water moves to the root surface by bulk flow through these channels against the pressure gradient

B. As the water content (and hence the water potential) of a soil decreases, the hydraulic conductivity decreases drastically

C. In very dry soils, the water potential (ψ_w) may fall below what is called the permanent wilting point. At this point the water potential of the soil is so low that plants can regain turgor pressure even if all water loss through transpiration ceases

D. Water enters the root most readily in the apical part of the root that includes the root hair zone

Which of the following statement is not correct?

(1) A and B (2) A and C
(3) C and D (4) B and D

89. In the soil, water is transported predominantly by bulk flow. However, when water comes in contact with the root surface, the nature of water transport becomes more complex

Consider the following statement

A. From the epidermis to the endodermis of the root, there are three pathways through which water can flow- the apoplast, transmembrane, and symplast pathways

B. At the endodermis, water movement through the symplast pathway is obstructed by suberin in the Casparian strip and thus occurs through the apoplast

C. Water uptake decreases when roots are subjected to low temperature or anaerobic conditions, or treated with respiratory inhibitors (such as cyanide)

D. Roots generate positive hydrostatic pressure (Root Pressure) by absorbing ions from the dilute soil solution and transporting them into the xylem and it is a major force responsible for ascent of sap

Which of the following statement is correct?

(1) A and B (2) A and C
(3) C and D (4) B and D

90. Consider the following statement regarding water potential and their relation to plants

A. Water moves toward regions of more negative water potential to lesser negative water potential

B. Water potential increases continuously from the soil to the leaves

C. Inside the leaf cells, the water potential is approximately the same as in the neighboring xylem, yet the components of ψw are quite different. The dominant component of ψw in the xylem is the negative pressure (ψp), whereas in the leaf cell ψp is generally positive

D. Within the leaf cells, water potential is reduced by a high concentration of dissolved solutes (low ψs)

Which of the following statement is correct?

(1) A and B (2) A and C
(3) C and D (4) B and D

91. Hydrogen ion concentration (pH) is an important property of soils because it affects the growth of plant roots & soil microorganisms

Consider the following statements

A. Root growth is generally favored in slightly acidic soils, at pH values between 5.5 and 6.5. Soil pH determines the availability of soil nutrients. Fungi generally predominate in acidic soils; bacteria become more prevalent in alkaline soils

B. Acidity promotes the weathering of rocks that releases K^+, Mg^{2+}, Ca^{2+}, and Mn^{2+} and increases the solubility of carbonates, sulfates, and phosphates. Increasing the solubility of nutrients facilitates their availability to roots

C. Major factors that lower the soil pH are the decomposition of organic matter which release CO_2 and the amount of rainfall

D. Hydrogen ions also displace K^+, Mg^{2+}, Ca^{2+}, and Mn^{2+} from the cation exchange complex in a soil. Leaching then may remove these ions from the upper soil layers, leaving a more acid soil. By contrast, the weathering of rock in arid regions releases K^+, Mg^{2+}, Ca_{2+}, and Mn^{2+} to the soil, but because of the low rainfall, these ions do not leach from the upper soil layers, and the soil remains alkaline

Which of the following statement is correct?

(1) A, B and C (2) A, C and D
(3) B, C and D (4) A, B, C and D

92. The plasma membrane potential of a pea cell collapses when cyanide (CN^-) is added to the bathing solution as shown in figure. Washing the cyanide out of the tissue results in a slow recovery and restoration of the membrane potential. Which of the following is most likely conclusion drawn on above information?

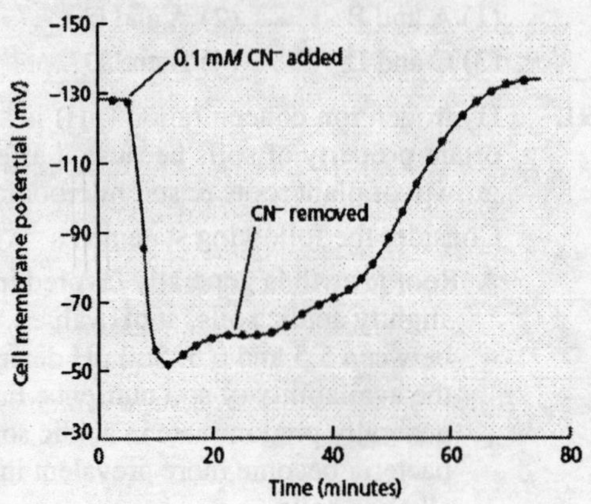

(1) Cyanide blocks ATP production in the cells by poisoning the mitochondrial electron transport chain
(2) The collapse of the membrane potential upon addition of cyanide indicates that an ATP supply is necessary for maintenance of the potential and it is a active process
(3) Membrane potential is mainly maintained by ion channels and transporters
(4) The fall in potential suggest that ATP inhibits ion channels and do not allow voltage gated ion channels to open

93. Which statement is correct regarding role and structure of thioredoxins?

(1) Thioredoxins form a family of small proteins, consisting of about 100 amino acids, which contain as a reactive group the sequence Cys-Gly-Pro-Cys, located at the periphery of the protein
(2) Thioredoxins are found in all living organisms from the archaebacteria to plants and animals. They function as protein disulfide oxido-reductases, in reducing disulfide bonds in target proteins to the SH form and reoxidizing them again to the S-S form
(3) Reduced thioredoxins transmit the signal for "illumination" to enzyme proteins
(4) All of the above

94. Following are some statements for synthesis of secondary metabolites in plants

A. Terpenes are synthesized by shikimic acid pathway and mevalonic acid pathway
B. Alkaloids are nitrogen containing compounds and are synthesized by shikimic acid pathway
C. Phenolic compounds are synthesized by shikimic acid pathway and mevalonic acid pathway
D. Both alkaloids and terpenes are synthesized by mevalonic acid pathway and MEP pathway

Which one of the following combinations of the above statement is true?

(1) A and D (2) A and C

(3) B and C (4) B and D

95. Match list-I with list-II and select the correct answer using the codes given below the lists:

List-I	List-II
Essential elements	**Deficiency symptoms**
P. Boron	1. Die-back disease
Q. Copper	2. Interveinal chlorosis
R. Iron	3. Little leaf disease
S. Zinc	4. Terminal leaf necrosis

(1) P-l, Q-4, R-3, S-2

(2) P-l, Q-4, R-2, S-3

(3) P-4, Q-l, R-3, S-2

(4) P-4, Q-l, R-2, S-3

96. Which of the following is the correct sequence of events in cytokinin signaling pathway

(1) receptor binding → activation of His kinase → phosphorylation of an aspartate residue → phosphorylation of AHP protein

(2) receptor binding → phosphorylation of an aspartate residue → activation of His kinase →phosphorylation of AHP protein

(3) receptor binding → phosphorylation of AHP protein → activation of His kinase → phosphorylation of an aspartate residue

(4) receptor binding → phosphorylation of AHP protein → phosphorylation of an aspartate residue →activation of His kinase

97. In the following biosynthetic pathway X and Y represent

X → S-adenosyl methionine → 1-amino-cyclo propane 1-carboxylic acid → Y

(1) X→ tryptophan Y→IAA

(2) X→ methionine Y→ cthylcnc

(3) X → geranylgeranyl phosphate Y→gibberrellin

(4) X → tryptophan Y → ethylene

98. The nitrate taken up into the root cells can be stored there temporarily in the vacuole. Nitrate is reduced to NH_4^+ in the epidermal and cortical cells of the root. This NH_4^+ is mainly used for the synthesis of amides which can be transported to the leaves via the xylem vessels

The main transport forms of amides in xylem are

(1) glutamine and asparagines

(2) glutamine and aspartate

(3) Citrulline and glutamate

(4) Glutamate and aspartate

99. The transport properties of a solute can change at different solute concentrations

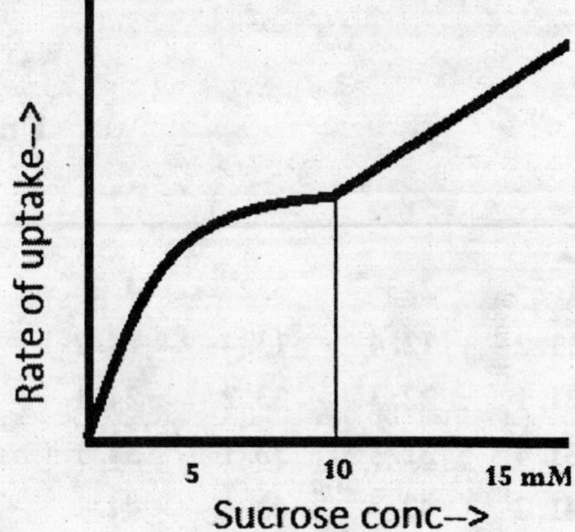

Figure above shows sucrose uptake by soybean cotyledon protoplasts as a function of the external sucrose concentration. Uptake increases sharply with concentration and begins to saturate at about 10 mM. At concentrations above 10 mM, uptake becomes linear and nonsaturable. Inhibition of ATP synthesis with metabolic poisons blocks the saturable component but not the linear one. The interpretation which is least likely to be true is that

(1) Sucrose uptake at low concentrations is an active carrier-mediated process (sucrose–H^+ symport) At higher concentrations, sucrose enters the cells by diffusion down its concentration gradient and is therefore insensitive to metabolic poisons
(2) At low concentrations (1 to 10 m*M*), the rate of uptake of sucrose by soybean cells shows saturation kinetics typical of carriers. A curve fit-ted to these data is predicted to approach a maximal rate (V_{max}) of 57 nmol per 10^6 cells per hour
(3) At higher sucrose concentrations the uptake rate continues to increase linearly over a broad range of concentrations suggesting the existence of other sucrose transporters, which might be carriers with very low affinity for the substrate
(4) At low concentration the sucrose uptake is mainly passive transport mediated by GLUT1 transporter but at high concentration sucrose start moving through space between membrane lipids

100. When ferredoxin is very highly reduced, it is possible that electrons are transferred from PS I to oxygen to form superoxide radicals ($\bullet O_2^-$) This process is called the Mehler reaction (pseudocyclic electron transport) Further conversion of reactive oxygen species $\bullet O_2^-$ into H_2O_2 and O_2 is carried out by enzyme
(1) Ascorbate peroxidase
(2) Superoxide dismutase
(3) Glutathione
(4) Glutathione reductase

ANSWER SHEET

Unit 6: Plant Physiology

1. 3	**2.** 3	**3.** 2	**4.** 2	**5.** 1	**6.** 2	**7.** 2	**8.** 4	**9.** 2	**10.** 2
11. 2	**12.** 4	**13.** 4	**14.** 2	**15.** 1	**16.** 2	**17.** 3	**18.** 4	**19.** 1	**20.** 2
21. 1	**22.** 3	**23.** 2	**24.** 1	**25.** 1	**26.** 1	**27.** 2	**28.** 3	**29.** 1	**30.** 1
31. 4	**32.** 3	**33.** 1	**34.** 1	**35.** 2	**36.** 4	**37.** 1	**38.** 3	**39.** 4	**40.** 1
41. 2	**42.** 3	**43.** 1	**44.** 1	**45.** 1	**46.** 2	**47.** 1	**48.** 2	**49.** 4	**50.** 2
51. 3	**52.** 1	**53.** 3	**54.** 3	**55.** 4	**56.** 2	**57.** 1	**58.** 1	**59.** 2	**60.** 4
61. 1	**62.** 3	**63.** 4	**64.** 4	**65.** 3	**66.** 3	**67.** 2	**68.** 1	**69.** 1	**70.** 3
71. 1	**72.** 4	**73.** 3	**74.** 4	**75.** 2	**76.** 4	**77.** 3	**78.** 1	**79.** 2	**80.** 3
81. 2	**82.** 1	**83.** 4	**84.** 2	**85.** 4	**86.** 4	**87.** 1	**88.** 2	**89.** 2	**90.** 3
91. 4	**92.** 2	**93.** 4	**94.** 3	**95.** 4	**96.** 1	**97.** 2	**98.** 1	**99.** 1	**100.** 2

UNIT-7

ANIMAL PHYSIOLOGY

1. In a neuronal culture experiment the response gradually reduced when the neurons were exposed to increasing concentration of a chemical. However, if the cells were thoroughly washed and left for sometime in normal medium and then the experiment was repeated, the cells started responding similarly as before. Which of the following could be the most probable explanation?

(1) Increased apoptosis of the cells

(2) The cells were gradually necrosed in an exponential manner

(3) The pH of the medium was changed

(4) The receptors were desensitized/down-regulated

2. Under a condition (1) a neuron showed transmembrane potential -50mV while after some treatment (2) it was -70mV. Given such a condition, which of the following statements would be correct ?

(1) Condition (1) is hyperpolarized state than condition (2)

(2) Condition (1) needs higher intensity stimulation than condition (2) for inducing a response

(3) The treatment caused depolarization of the neuron

(4) The treatment induced hyperpolarization

3. KCN is considered to be one of the potent poisons. You are doing an animal (mouse) experiment to test a new KCN sample synthesized by a chemist. The experiment is intravenous injection of KCN and checking the death of mouse. The sample was tested by many researchers and found to be pure and as potent as it should be. Although your experimental procedures were correct you failed to obtain desired results due to which one of the following

(1) KCN resistant mice were used

(2) Dose of KCN is much below the LD50 dose needed to kill the mice

(3) KCN is being degraded in mice

(4) KCN is immediately filtered by kidney

4. In a gastrocnemius-sciatic (nerve-muscle) preparation electrical stimulation of the nerve caused twitching of the muscle. Direct stimulation of the muscle also caused twitching. When curare (a cholinergic antagonist) was applied in the bath where the preparation was maintained and the nerve was stimulated, the muscle twitch was not seen. However, under these conditions, if the muscle was stimulated directly, it twitched. These observations suggest that

(1) stimulation of the nerve was directly communicated to the muscle for contraction

(2) stimulation of the nerve was communicated to the muscle through the mediation of acetylcholine

(3) after application of curare the stimulation of the nerve did not evoke muscle twitch because the nerve was fatigued

(4) after application of curare the stimulation of the nerve did not evoke muscle twitch because muscle was fatigued

5. One of the key experiments to show that gap junctions allowed the passage of small molecules was carried out by allowing cardiac muscle cells (which respond to norepinephrine by contraction) to form gap junctions with ovarian granulosa cells (which respond to FSH by undergoing various metabolic changes) The researchers then added FSH to the mixed cell culture. Which of the following observation is possible?

(1) Relaxation of muscle cells
(2) Contraction of muscle cells
(3) Contraction of ovarian granulosa cells
(4) Relaxation of granulosa cells

6. The body fluids are enclosed in compartments. Which of the following statement is not correct?

(1) The extracellular fluid is composed of the interstitial fluid (the fluid between cells) and the blood plasma. Of the extracellular fluid, 80 percent is interstitial fluid, and 20 percent is plasma
(2) Interstitial fluid and plasma have essentially the same composition except that plasma contains a much higher concentration of protein
(3) Extracellular fluid have essentially the same composition from the fluid inside cells-the intracellular fluid
(4) Approximately one-third of body water is in the extracellular compartment, and two-thirds is intracellular

7. Consider the following statement for digestion in higher animals

A. Starch is digested by amylases secreted by the salivary glands and pancreas, and the resulting products, as well as ingested disaccharides, are digested to monosaccharides by enzymes in the luminal membranes of epithelial cells in the small intestine. Most monosaccharides are then absorbed by secondary active transport

B. Proteins are broken down into small peptides and amino acids, which are absorbed by secondary active transport in the small intestine. The breakdown of proteins to peptides is catalyzed by pepsin in the stomach and by the pancreatic enzymes trypsin and chymotrypsin in the small intestine

C. Peptides are broken down into amino acids by pancreatic carboxypeptidase and intestinal aminopeptidase. Small peptides consisting of two to three amino acids can be actively absorbed into epithelial cells and then broken down to amino acids, which are released into the blood

D. Large fat globules leaving the stomach are emulsified in the small intestine by bile salts and phospholipids secreted by the liver. Lipase from the pancreas digests fat at the surface of the emulsion droplets, forming fatty acids and monoglycerides. These water-insoluble products of lipase action, when combined with bile salts, form micelles. Free fatty acids and monoglycerides diffuse across the luminal membranes of epithelial cells, within which they are enzymatically recombined to form triacylglycerol, which is released as chylomicrons from the blood side of the cell by exocytosis. The released chylomicrons enter lacteals in the intestinal villi and pass, by way of the lymphatic system, to the venous blood returning to the heart

Which of the following statements are correct?

(1) A and B (2) B and C
(3) A and D (4) All of the above

8. A person's plasma potassium concentration (a homeostatically regulated variable) is 4 mmol/L when she is eating 150 mmol of potassium per day. One day she doubles her potassium intake and continues to eat that amount indefinitely. At the new steady state, do you think her plasma potassium concentration is more likely to be?

(1) 4.4 mmol/L (2) 8 mmol/L
(3) 4 mmol/L (4) 2 mmol/L

9. Patient A is given a drug that blocks the synthesis of all eicosanoids, whereas patient B is given a drug that blocks the synthesis of leukotrienes but none of the other eicosanoids. What are the enzymes most likely blocked by these drugs?

(1) Drug A: Phospholipase A_2; Drug B: Cyclooxygenase
(2) Drug A: Cyclooxygenase; Drug B: Lipoxygenase
(3) Drug A: Phospholipase A_2; Drug B: Lipoxygenase
(4) Drug A: Lipoxygenase; Drug B: Cyclooxygenase

10. Neurons can be divided into three functional classes: afferent neurons, efferent neurons, and interneurons

Consider the following statements

A. *Afferent neurons*: They transmit information out of the central nervous system to effect or cells, particularly muscles, glands, or other neurons. In such neurons cell body, dendrites, and a small segment of the axon are in the central nervous system; most of the axon is in the peripheral nervous system

B. *Efferent neurons*: They transmit information into the central nervous system from receptors at their peripheral endings. Their cell body and the long peripheral process of the axon are in the peripheral nervous system; only the short central process of the axon enters the central nervous system. They have no dendrites

C. *Interneurons*: They function as integrators and signal changers. They integrate groups of afferent and efferent neurons into reflex circuits. They lie entirely within the central nervous system and account for 99 percent of all neurons

Which of the following statement is correct?

(1) A and C (2) B and C
(3) Only C (4) A, B and C

11. Development of the nervous system in the embryo begins with a series of divisions of precursor cells that can develop into neurons or glia

Consider the following statement regarding neural growth and regeneration

A. Neurons develop from precursor cells, migrate to their final location, and send out processes to their target cells

B. Cell division to form new neurons is markedly slowed after birth

C. After degeneration of a severed axon, damaged peripheral neurons do not regenerate or restore significant function

D. Damaged neurons of the CNS may regrow the axon to their target organ

Which of the following statements are correct?

(1) A and B (2) C and D
(3) A, C and D (4) B, C and D

12. Membrane potentials are generated mainly by diffusion of ions and are determined by (a) the ionic concentration differences across the membrane, and (b) the membrane's relative permeabilities to different ions

Consider the following statements

A. Plasma-membrane Na^+, K^+-ATPase pumps maintain intracellular sodium concentration high and potassium low

B. In almost all resting cells, the plasma membrane is much more permeable to potassium than to sodium, so the membrane potential is close to the potassium equilibrium potential-that is, the inside is negative relative to the outside

C. The Na^+- K^+-ATPase pumps also contribute directly a small component of the potential because they are electrogenic

D. The propagation of action potential though axon is largely contributed due to voltage gated Na^+ channels

Which of the following statements are correct?

(1) A and B (2) B and C
(3) C and D (4) A and C

13. An Action Potential (AP) is a rapid change in the membrane potential during which the membrane rapidly depolarizes and repolarizes. In neurons, the potential reverses and the membrane becomes positive inside. APs provide long-distance transmission of information through the nervous system

Consider the following statements

A. APs occur in excitable membranes because these membranes contain voltage-gated sodium channels, which open as the membrane depolarizes, causing a positive feedback toward the sodium equilibrium potential

B. The AP is ended as the sodium channels close and additional potassium channels open, which restores the resting conditions

C. Depolarization of excitable membranes triggers APs only when the membrane potential is below a threshold potential

D. Depending on the size of the stimulus, if the membrane is below threshold, the APs generated vary in size and are propagated with change in size from one site to another along a membrane

Which of the following statements are correct?

(1) A and B (2) B and C
(3) C and D (4) A and C

14. A monkey undergoes cerebellectomy. After the post-operative recovery, the monkey was given a task to press a bar. The possible observations are

A. Its hand would overshoot the target while reaching the bar

B. It would be unable to move forelimbs

C. It would show intention tremor while trying to press the bar

D. It would press the bar with mouth instead of hand

Which one of the following is correct?

(1) A and C (2) B only
(3) D only (4) B and D

15. Figure below represents interaction of EPSPs at the postsynaptic neuron. Arrows indicate time of stimulation

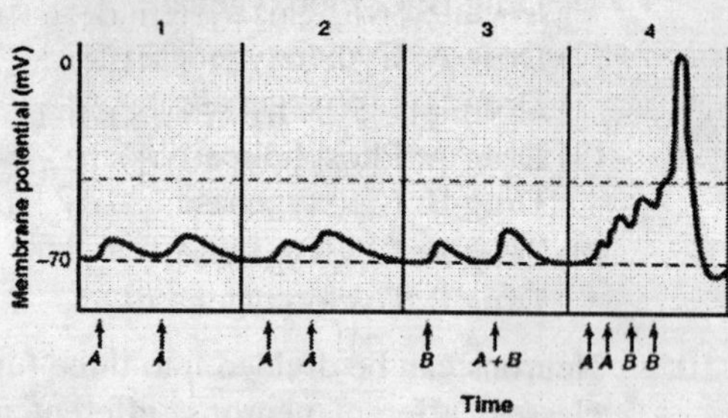

Which of the following is correct for panel 1, 2, 3 and 4 shown in diagram? 3

(1) Panel 1 and 2: Temporal summation; Panel 3: Temporal Summation; Panel 4: Spatial Summation

(2) Panel 1 and 2: Spatial Summation; Panel 3: Temporal & Spatial Summation Panel 4: Temporal Summation

(3) Panel 1 and 2: Temporal Summation; Panel 3: Spatial Summation; Panel 4: Temporal & Spatial Summation

(4) Panel 1 and 2: Spatial Summation; Panel 3: Temporal & spatial Summation; Panel 4: Spatial Summation

16. Neurotransmitters are involved in rapid communication among pre-synaptic and post-synaptic cells which may be excitatory or inhibitory in nature. Neurons are often referred to as "-ergic," where the missing prefix is the type of neurotransmitter released by the neuron. For example, "dopaminergic" applies to neurons that release the neurotransmitter dopamine

Consider the following statement regarding neurotransmitters

A. Acetylcholine is a major neurotransmitter in the peripheral nervous system, and it is also present in the brain. Fibers that release ACh are called cholinergic fibers

B. Some ACh receptors also respond to drug nicotine in brain and are known as nicotinic receptors. They are important in cognitive functions such as attention, learning, and memory. Neurons associated with this system degenerate in people with *Alzheimer's disease*

C. Serotonin (5-hydroxytryptamine, or 5-HT) is produced from tyrosine. In general, serotonin has an excitatory effect on pathways that are involved in the control of muscles, and an inhibitory effect on pathways that mediate sensations. The activity of serotonergic neurons is highest during sleep and lowest during states of alert wakefulness

D. Glutamate and aspartate, serve as neurotransmitters at the vast majority of inhibitory synapses in the central nervous system. These neurotransmitters function in learning, memory, and neural development. They are also implicated in epilepsy, Alzheimer's and Parkinson's diseases, and the neural damage that follows strokes, brain trauma, and other conditions of low oxygen availability

Which of the following statement is not correct?

(1) A and B (2) B and C
(3) C and D (4) A and D

17. Following may be altered response due to loss of some part of brain?

A. Regulation anterior pituitary gland, water balance, eating and drinking behavior

B. Coordination of movements, including those for posture and balance

C. Participation in regulation of autonomic nervous system and reinforces certain behaviors

D. Regulation reproductive system, generation and regulation of circadian rhythms, regulation of body temperature and participation in generation of emotional behavior

E. Affect on reticular formation and its various integrating centers, including those for cardiovascular and respiratory activity

F. Participation in generation of emotions and emotional behavior and essentially affect on most kinds of learning

If the hypothalamus is severely damaged or non-functional in Chimapanzee named "Sulemaan". Which of following functioning would be affected?

(1) A, C and D (2) A, E and F
(3) A, B and C (4) D, E and F

18. Person has received a severe blow to the head but appears to be all right. Over the next weeks, however, he develops loss of appetite, thirst, and sexual capacity, but no loss in sensory or motor function. What part of the brain do you think may have been damaged?

(1) Cerebellum
(2) Hypothalamus
(3) Medulla Oblongata
(4) Cerebrum

19. A person is taking a drug that causes, among other things, dryness of the mouth and speeding of the heart rate but no impairment of the ability to use the skeletal muscles. What type of receptor does this drug probably block?

(1) Cholinergic muscarinic receptors
(2) Cholinergic nicotinic receptors
(3) Dopaminergic receptors
(4) Glutamate receptors

20. Figure below represents action potentials from an afferent carotid sinus fiber leading from the pressure receptors (carotid sinus) of a single sensory unit as the receptors are subjected to pressures of different magnitudes

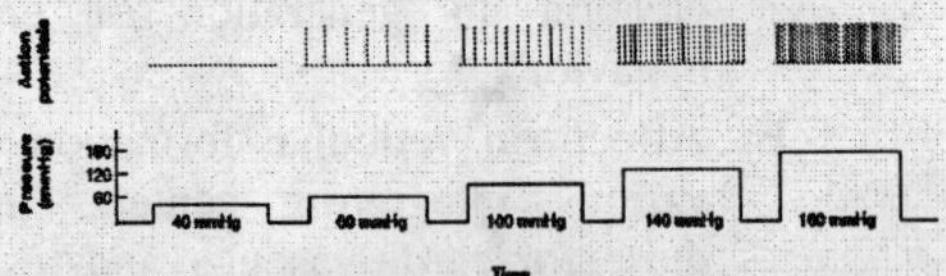

The following are the possible observations

A. No discharge when carotid sinus perfusion pressure was below 40 mm Hg

B. Linear increase in discharge frequency when carotid sinus perfusion pressure was gradually increased from 70 to 180 mm Hg

C. Increase in discharge frequency was lesser prominent in greater pulsatile changes of carotid sinus pressure keeping the mean pressure identical in all cases

Which one of the following is correct?

(1) A and B (2) A and C
(3) B and C (4) A only

21. In the figure below top line in each graph indicates the action-potential firing of the afferent nerve fiber from the receptor, and the bottom line, application of the stimulus

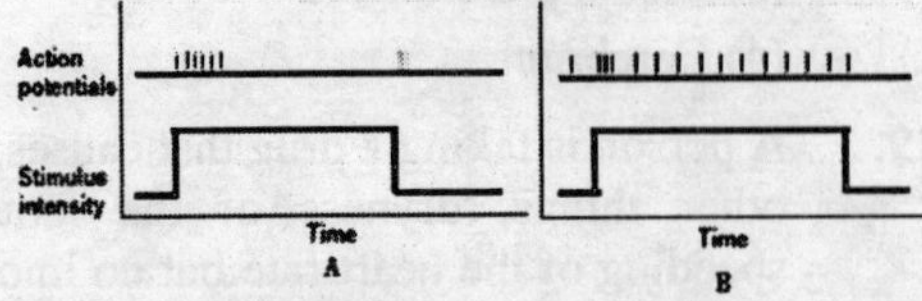

The following are the possible observations

A. A is slowly and B is rapidly adapting receptors

B A is Tactile (Merkle's) corpuscles for sensation of touch and B is Tactile (Meissner's) corpuscle for light touch

C. A is Lamellated (Pacinian) corpuscle for deep pressure and sensation of vibrations of high frequency and B is Ruffini corpuscle for warmth and sensation of low frequency vibration

Which of the following is correct?

(1) A and B (2) B and C
(3) A, B and C (4) Only C

22. The oxygen dissociation curve for hemoglobin reflects allosteric effects that result from the interaction of hemoglobin with O_2, CO_2, H^+, and BPG. Which of the following structural changes DO not occur in the hemoglobin molecule when O_2, CO_2, H^+, or BPG bind?

(1) The binding of O_2 pulls the iron into the plane of the heme and causes a change in the interaction of all four globin subunits, mediated through His F8

(2) BPG binds at a single site between the four globin subunits in deoxyhemoglobin and stabilizes the deoxyhemoglobin form by cross-linking the β subunits

(3) The deoxy form of hemoglobin has a greater affinity for H^+ because the molecular environment of His and the α-NH_2 groups of the α chains changes, rendering these groups less acidic when O_2 is released

(4) The binding of CO_2 stabilizes the oxy form of hemoglobin

23. The graph below represents the time relations between a muscle fiber action potential and the resulting shortening and relaxation of the muscle fiber

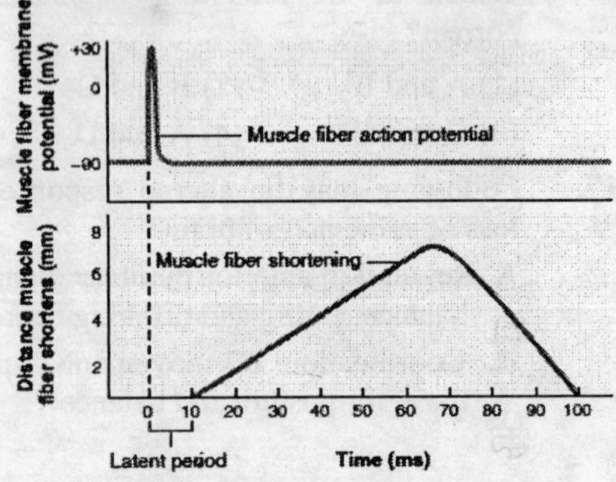

The graph represents which type of muscle(s)?

(1) Cardiac muscle
(2) Skeletal muscle
(3) Cardiac and Skeletal muscle
(4) Cardiac and Smooth muscle

24. Figure below velocity of skeletal-muscle fiber shortening and lengthening as a function of load

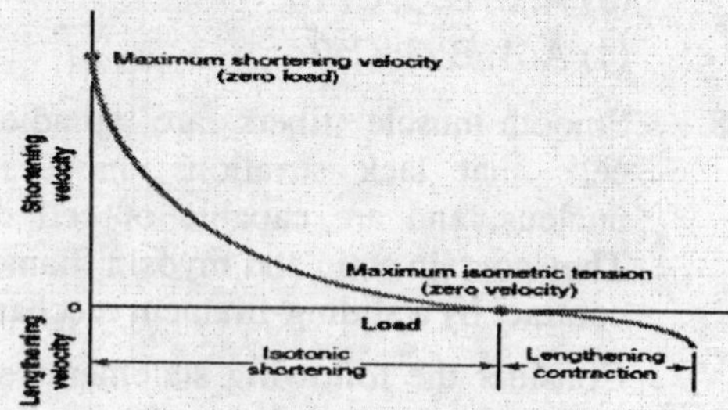

Following conclusion can be drawn

A. Force on the cross bridges during a lengthening contraction is lesser than the maximum isometric tension

B. The shortening velocity is maximal when there is no load and is zero when the load is equal to the maximal isometric tension

C. The shortening velocity is determined by the rate at which individual cross bridges undergo their cyclical activity. Because one ATP is split during each cross bridge cycle, the rate of ATP splitting determines the shortening velocity. Increasing the load on a bridge, for complex reasons, decreases the rate of ATP hydrolysis and thus the velocity of shortening

D. At loads greater than the maximal isometric tension, the fiber will shorten at a velocity that increases with load, and at very high loads the fiber will break

Which of the following statements is correct?

(1) A and B (2) B and C
(3) C and D (4) A and D

25. Consider the following statement regarding role of ATP in skeletal-muscle contraction

A. Hydrolysis of ATP by myosin energizes the cross bridges, providing the energy for force generation

B. Hydrolysis of ATP to ADP by ATPase activity of myosin dissociates cross bridges bound to actin, allowing the bridges to repeat their cycle of activity

C. Hydrolysis of ATP by the Ca-ATPase in the sarcoplasmic reticulum provides the energy for the active transport of calcium ions into the lateral sacs of the reticulum, lowering cytosolic calcium to pre-release levels, ending the contraction, and allowing the muscle fiber to relax

Which of the following statement are correct?

(1) A and B (2) B and C
(3) A and C (4) A, B and C

26. Consider the following events between a motor neuron action potential and skeletal-muscle fiber contraction

A. Action potential in transverse tubules triggers release of Ca_2^+ from lateral sacs of sarcoplasmic reticulum, Ca_2^+ binds to troponin on the thin filaments, causing tropomyosin to move away from its blocking position, thereby uncovering cross-bridge binding sites on actin

B. Local currents depolarize the adjacent plasma membrane to its threshold potential, generating an action potential that propagates over the muscle fiber surface and into the fiber along the transverse tubules

C. ACh diffuses from axon terminals to motor end plate in muscle fiber and binds to receptors on motor end plate, opening Na^+, K^+ ion channels

D. Action potential is initiated and propagates in motor neuron axon and triggers release of ACh from axon terminals at neuromuscular junction

E. Energized myosin cross bridges on the thick filaments bind to actin, cross-bridge binding triggers release of the strained conformational state of myosin, producing an angular movement of each cross bridge

F. ATP binds to myosin, breaking linkage between actin and myosin and thereby allowing cross bridges to dissociate from actin. Cross bridges repeat steps, producing movement of thin filaments past thick filaments. Cycles of cross-bridge movement continue as long as Ca_2^+ remains bound to troponin

G. Cytosolic Ca_2^+ concentration decreases as Ca_2^+ is actively transported into sarcoplasmic reticulum by Ca-ATPase, removal of Ca_2^+ from troponin restores blocking action of tropomyosin, the cross-bridge cycle ceases, and the muscle fiber relaxes

Arrange the following event according to sequence of events during skeletal muscle contraction

(1) A→B→C→D→E→G→F

(2) D→C→A→B→F→E→G

(3) D→C→A→B→E→F→G

(4) D→C→B→A→E→F→G

27. All skeletal-muscle fibers do not have the same mechanical and metabolic characteristics. Different types of fibers can be identified on the basis of (1) their maximal velocities of shortening—fast and slow fibers—and (2) the major pathway used to form ATP—oxidative and glycolytic fibers

Consider the following type of muscle fibers

A. Slow-oxidative fibers (type I)

B. Fast-oxidative fibers (type IIa)

C. Fast-glycolytic fibers (type IIb)

Here are some of characteristic of different skeletal muscle fibres

1. Combine high myosin-ATPase activity with high glycolytic capacity, fatigue fast
2. Combine high myosin-ATPase activity with high oxidative capacity, fatigue intermediate
3. Combine low myosin-ATPase activity with high oxidative capacity, Fatigue slow

Which of the following is correct match?

(1) A: 3; B: 2; C: 1

(2) A: 1; B: 2; C: 3

(3) A: 2; B: 3; C: 1

(4) A: 1; B: 3; C: 2

28. Smooth-muscle fibers are spindle-shaped cells that lack striations, have a single nucleus, and are capable of cell division. They contain actin and myosin filaments and contract by a sliding-filament mechanism

Consider the following statement related to contraction in smooth muscles

A. A decrease in cytosolic calcium leads to the release of calcium from calmodulin. Thus myosin light-chain phosphatase, inactivates the myosin kinase. Only dephosphorylated myosin is able to bind to actin and undergo cross-bridge cycling

B. Smooth-muscle myosin has a high rate of ATP splitting, resulting in a much higher shortening velocity than is found in striated muscle. However, the tension produced per unit cross-sectional area is equivalent to that of skeletal muscle

C. Two sources of the cytosolic calcium ions initiate smooth-muscle contraction: the sarcoplasmic reticulum and extracellular calcium. The opening of calcium channels in the smooth-muscle plasma membrane and sarcoplasmic reticulum, mediated by a variety of factors, allows calcium ions to enter the cytosol

D. The increase in cytosolic calcium resulting from most stimuli does not activate all cross bridges. Therefore smooth-muscle tension can be increased by agents that increase the concentration of cytosolic calcium ions

Which of the following statements are correct?

(1) A and B (2) B and C
(3) A and C (4) C and D

29. If the blood flow to a skeletal muscle were markedly decreased, which types of motor units would most rapidly lost their ability to produce ATP for muscle contraction?

(1) Slow-oxidative fibers (type I)
(2) Fast-oxidative fibers (type IIa)
(3) Fast-glycolytic fibers (type IIb)
(4) All of the above

30. As a result of an automobile accident, 50 percent of the muscle fibers in the biceps muscle of a patient were destroyed. Ten months later, the biceps muscle was able to generate 80 percent of its original force. Following are the changes that may take place in the damaged muscle that enables it to recover

A. Some new fibers can be formed by the fusion and development of undifferentiated satellite cells. This will replace some, but not all, of the fibers that were damaged
B. Some of the restored force results from hypertrophy of the surviving fibers. Because of the loss of fibers in the accident, the remaining fibers must produce more force to move a given load. The remaining fibers undergo increased synthesis of actin and myosin, resulting in increases in fiber diameter and thus their force of contraction
C. Bone marrow stem cells differentiate into myoblast and start dividing to replace the missing cells

One or a combination of the following changes most likely taking place in this condition

(1) A and B (2) B and C
(3) C only (4) A, B and C

31. The following experiments were performed on a single-unit smooth muscle in the gastrointestinal tract

A. Stimulating the parasympathetic nerves to the muscle produced a contraction
B. Applying a drug that blocks the voltage-sensitive sodium channels in most plasma membranes led to a failure to contract upon stimulating the parasympathetic nerves
C. Applying a drug that binds to muscarinic receptors, and hence blocks the action of ACh at these receptors, did not prevent the muscle from contracting when the parasympathetic nerve was stimulated

From these observations, what might one conclude about the mechanism by which parasympathetic nerve stimulation produces a contraction of the smooth muscle?

(1) Upon parasympathetic nerve stimulation, a neurotransmitter is released that binds to receptors on the membranes of smooth-muscle cells and triggers contraction. The substance released, however, is not acetylcholine (ACh)
(2) Upon parasympathetic nerve stimulation, a neurotransmitter is released that binds to receptors on the membranes of smooth-muscle cells and triggers contraction. The substance released is acetylcholine (ACh)
(3) When the muscarinic receptors for ACh were blocked, however, stimulation of the parasympathetic nerves still produced a contraction, providing evidence that ACh is being released by the neurons and producing contraction
(4) Action potentials in the parasympathetic nerves are not essential for initiating nerve-induced contraction. When the nerves were prevented from generating action potentials by blockage of their voltage-gated sodium channels, there was response to nerve stimulation

32. The electroencephalogram provides one means of defining the states of consciousness and to diagnose brain disease and damage. Electric currents in the cerebral cortex due predominantly to summed postsynaptic potentials are recorded as the EEG

Consider the following statements

A. Fast EEG waves correlate with less responsive behaviors

B. Alpha rhythms and, during EEG arousal, beta rhythms characterize the EEG of an awake person

C. NREM sleep progresses from stage 1 (slower, higher amplitude waves) through stage 4 (faster, lower amplitude waves) and then back again, followed by an episode of REM sleep. There are generally four or five of these cycles per night

D. Aminergic and cholinergic brainstem centers, via their projections forward to the cerebrum as components of the reticular activating system, interact with the thalamus to regulate the sleep-wake cycles. Hypothalamic nuclei also play a role

Which of the following statement are correct?

(1) A and C (2) B and D
(3) C and D (4) B and C

33. Match the these Major Hematopoietic Growth Factors with their target progenitor cells which are stimulated in response to them

A. Erythropoietin
B. GM-Colony-stimulating factors (GM-CSFs)
C. Interleukin 3
D. Thrombopoietin

1. Platelets (from megakaryocytes)
2. Erythrocytes
3. Granulocytes and monocytes
4. Various leukocytes

Which of the following is correct match?

(1) A: 1; B: 2; C: 3; D: 4
(2) A: 4; B: 1; C: 2; D: 3
(3) A: 4; B: 2; C: 3; D: 1
(4) A: 2; B: 3; C: 4; D: 1

34. Erythrocytes, which make up more than 99 percent of blood cells, contain hemoglobin, an oxygen binding protein. Oxygen binds to the iron in hemoglobin

Consider the following statements

A. Erythrocytes are produced in the spleen and liver and destroyed in the bone marrow

B. Iron, folic acid, and vitamin B12 are essential for erythrocyte formation

C. The hormone erythropoietin, which is produced by the kidneys in response to low oxygen supply, stimulates erythrocyte differentiation and production by the bone marrow

The correct answer is

(1) A and B (2) A and C
(3) B and C (4) A, B and C

35. The cardiovascular system consists of two circuits: the pulmonary circulation, from the right ventricle to the lungs and then to the left atrium, and the systemic circulation, from the left ventricle to all peripheral organs and tissues and then to the right atrium. Arteries carry blood away from the heart, and veins carry blood toward the heart

Consider the following statements

A. In the systemic circuit, the large artery leaving the right heart is the aorta, and the large veins emptying into the left heart are the superior vena cava and inferior vena cava. The analogous vessels in the pulmonary circulation are the pulmonary trunk and the four pulmonary veins

B. The microcirculation consists of the vessels between arteries and veins: the arterioles, capillaries, and venules

C. Flow between two points in the cardiovascular system is directly proportional to the pressure difference between the points and inversely proportional to the resistance: $F = \Delta P / R$

D. Resistance is inversely proportional to the viscosity of a fluid and to the length of the tube. It is directly proportional to the fourth power of the tube's radius, which are the major variable controlling changes in resistance

Which of the following statement are correct?

(1) A and B (2) C and D
(3) B and D (4) B and C

36. Below is ECG of a person

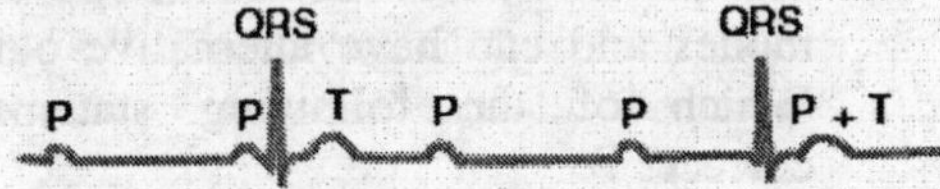

Which is correct?

(1) A normal ECG
(2) Partial block to AV node
(3) Complete block to AV node
(4) Non-functional ventricles

37. Figure below shows effects of sympathetic and parasympathetic nerve stimulation on the slope of the pacemaker potential of an SA-nodal cell

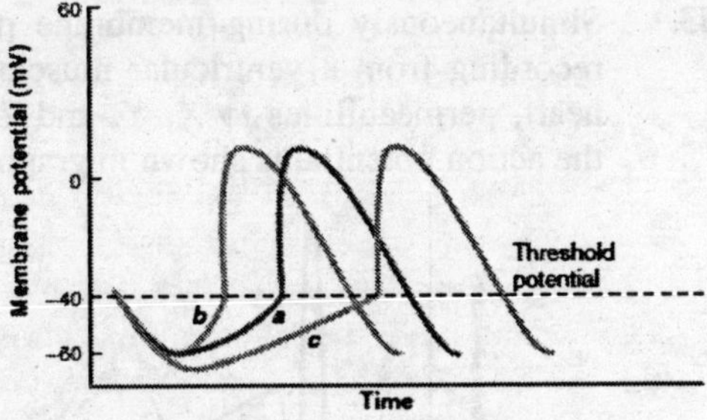

Here a, b and c represents?

(1) A: Control; B: sympathetic stimulation; C: parasympathetic stimulation
(2) A: Control; B: parasympathetic stimulation; C: sympathetic stimulation
(3) A: sympathetic stimulation; B: parasympathetic stimulation; C: Control
(4) A: sympathetic stimulation; B: Control; C: parasympathetic stimulation

38. The figure shows the effect of changing mean arterial pressure (MAP) on the firing of action potentials by afferent neurons from the carotid sinus. This experiment is done by pumping blood in a non-pulsatile manner through an isolated carotid sinus so as to be able to set the pressure inside it at any value desired

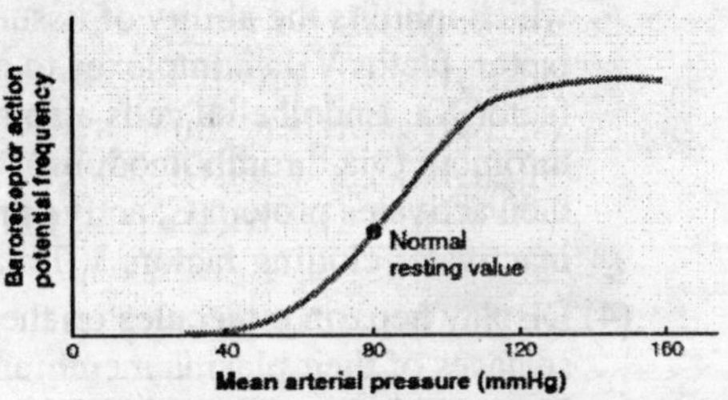

Consider the following statements?

A. the rate of discharge of the carotid sinus is directly proportional to the mean arterial pressure ranging from 0 to 160 mmHg

B. If the experiment is repeated using the same mean pressures as before but allowing pressure pulsations, it is found that at any given mean pressure, the larger the pulse pressure, the faster the rate of firing by the carotid sinus

C. At higher and lower arterial pressure there is all or none response

Which of the following statement is correct?

(1) A and B (2) B and C
(3) A and C (4) All of the above

39. Which of the following is not an anti-clotting role of endothelial cells?

(1) Normally provide an intact barrier between the blood and subendothelial connective tissue thus platelet aggregation and the formation of tissue factor--factor VIIa complexes are not triggered and synthesize and release

PGI2 and nitric oxide, these inhibit platelet activation and aggregation

(2) Injury to a vessel disrupts the endothelium and exposes the underlying connective tissue collagen molecules. Platelets adhere to collagen, largely via an intermediary called von Willebrand factor (vWF), a plasma protein secreted by endothelial cells and platelets which leads to platelet aggregation that rapidly creates a platelet plug inside the vessel

(3) Secrete tissue factor pathway inhibitor which inhibits the ability of tissue factor–factor VIIa complexes to generate factor Xa. Endothelial cells also bind thrombin (via thrombomodulin), which then activates protein C, active protein C inactivates clotting factors VIIIa and Va

(4) Display heparin molecules on the surfaces of their plasma membranes, Heparin binds antithrombin III, and this molecule then inactivates thrombin and several other clotting factors

40. A person with a heart rate of 40 has no P waves but normal QRS complexes on the ECG. What is the probable explanation?

(1) SA Node is not working

(2) Problem in working of AV node

(3) Tricuspid and mitral valve are not functioning properly

(4) There is problem in ventricle contraction and relaxation

41. The surface of the alveolar cells is moist. At an air-water interface surface tension, make the water lining like a stretched balloon that constantly tries to shrink and resists further stretching. It is extremely important, therefore, that the alveolar cells secrete a detergent-like substance known as pulmonary surfactant, which markedly reduces the cohesive forces between water molecules on the alveolar surface

Consider the following statements about surfactants

A. Pulmonary surfactant is a mixture of phospholipids and protein

B. It is secreted by type II alveolar cells

C. It increases surface tension of the water layer at the alveolar surface, which decreases lung compliance

D. A deep breath increases its secretion (by stretching the type II cells) Its concentration decreases when breaths are small

Which of the following is correct?

(1) A, B and C (2) B, C and D

(3) A, B and D (4) A, C and D

42. A patient undergoes liver transplantation and during the course of post-operative treatment, becomes susceptible to infection. The patient can be treated in two different modes and can have alternative outcomes. Which of the following statements is correct?

(1) Treatment with immunostimulatory drugs reducing the infection but rejecting the transplant

(2) Treatment with immunostimulatory drugs reducing the infection and retaining the transplant

(3) Treatment with antibiotics reducing the infection but retaining the transplant

(4) Treatment with antibiotics reducing the infection but rejecting the transplant

43. Simultaneously during membrane potential recording from a ventricular muscle cell of heart, permeabilities to X, Y, and Z during the action potential is shown in graph below

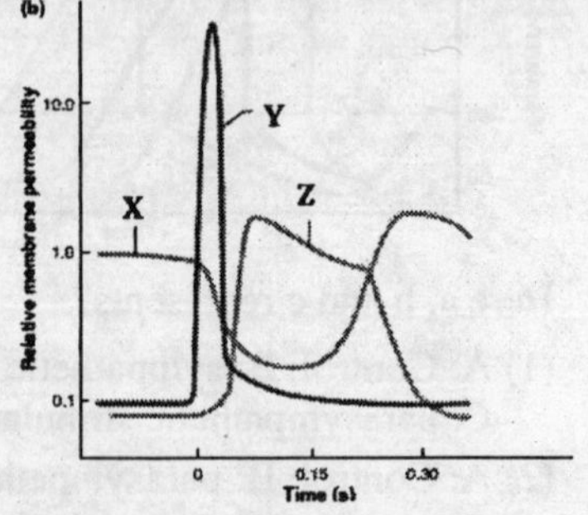

Identify X, Y and Z?

(1) X: Na^+; Y: K^+; Z: Ca_2^+
(2) X: Na^+; Y: Ca_2^+; Z: K^+
(3) X: Ca_2^+; Y: K^+; Z: Na^+
(4) X: K^+; Y: Na^+; Z: Ca_2^+

44. Which of the following is not acclimatization to the hypoxia of high altitude?

(1) Erythropoietin, secreted by the kidneys, stimulates erythrocyte synthesis, resulting in increased erythrocyte and hemoglobin concentration in blood
(2) DPG (BPG) decreases and shifts the hemoglobin dissociation curve to the left, facilitating oxygen unloading in the tissues
(3) Increases in capillary density (due to hypoxia-induced expression of the genes that code for angiogenic factors), mitochondria, and muscle myoglobin occur, all of which increase oxygen transfer
(4) The peripheral chemoreceptors stimulate an increased loss of sodium and water in the urine. This reduces plasma volume, resulting in a concentration of the erythrocytes and hemoglobin in the blood

45. Which of the following is not a function of human Kidney?

(1) Regulation of water and inorganic-ion balance and removal of metabolic waste products from the blood and their excretion in the urine
(2) Gluconeogenesis
(3) Secretion of hormones such as erythropoietin, which controls erythrocyte production and Renin, which controls formation of angiotensin, which influences blood pressure and sodium balance
(4) Production of 1,25-dihydroxyvitamin D3 from cholesterol in presence of UV-A light which influences calcium balance

46. Forces involved in glomerular filtration are shown in table. The symbol π denotes the osmotic force due to the presence of protein in glomerular capillary plasma

Forces	mmHg
Fluid pressure in Bowman's space (P_{BS})	15
Glomerular capillary blood pressure (P_{GC})	60
Osmotic force due to protein in plasma (π_{GC})	29

What would be net glomerular filtration pressure?

(1) 16 (2) 46
(3) 74 (4) 31

47. Inulin (not insulin) is an important research tool in renal physiology because, its clearance is equal to the glomerular filtration rate. It is not found normally in the body, but we can administer it intravenously to a person at a rate sufficient to maintain plasma concentration constant at 4 mg/L. Suppose a urine collected over a 1-h period has a volume of 0.1 L and an inulin concentration of 300 mg/L, thus, inulin excretion equals 0.1 L/h = 300 mg/L, or 30 mg/h. How much plasma (Volume of fluid filtered [GFR]) had to be completely cleared of its inulin to supply this 30 mg/h?

(1) 0.75 L/h (2) 0.1 L/h
(3) 1 L/h (4) 7.5 /h

48. The ability of the kidneys to produce hyperosmotic urine is a major determinant of one's ability to survive with limited amounts of water. The human kidney can produce a maximal urinary concentration of 1400 mOsmol/L, almost five times the osmolarity of plasma, which is 300 mOsmol/L. The typical daily excretion of urea, sulfate, phosphate, other waste products, and ions amounts to approximately 600 mOsmol. Therefore, the minimal volume (obligatory water loss) of urine water in which this mass of solute can be dissolved equals?

(1) 0.44 lt (2) 0.50 lt
(3) 4.66 lt (4) 3. 5 lt

49. Which of the following descriptions of mammalian nephron components is incorrect?

(1) Cells of the proximal convoluted tubules have many microvilli and mitochondria
(2) The osmolarity of the urine is controlled by changing the rate of NaCl transport in the thick ascending limb of the loop of Henle
(3) Cells of the thick segment of the ascending limb of the loop of Henle resorb NaCl out of the tubular fluid
(4) Urine is less concentrated when it leaves the loop of Henle than when it enters

50. In addition to regulating water and salt balance, kidneys also play a role in regulating the pH of blood. Which of the following statements about pH balance in the blood is false?

(1) The kidneys control the level of CO_2 in the blood
(2) HCO_3^- is the major buffer present in the blood
(3) Buffers serve to supply or absorb excess H^+, depending upon conditions
(4) Maintaining a constant pH is important since pH can alter protein structure

51. Consider the following statements is relation to potassium concentration regulation by kidney

A. A person remains in potassium balance by excreting an amount of potassium in the urine equal to the amount ingested minus the amounts lost in the feces and sweat

B. Potassium is freely filterable at the renal corpuscle and undergoes both reabsorption and secretion, the latter occurring in the cortical collecting ducts and being the major controlled variable determining potassium excretion

III. When body potassium is increased, extracellular potassium concentration increases. This increase acts directly on the cortical collecting ducts to increase potassium secretion and also stimulates aldosterone secretion, the increased plasma aldosterone then also stimulating potassium secretion

Which of the following statement is correct?

(1) A and B (2) A and C
(3) B and C (4) A, B and C

52. The figure to right shows a hemoglobin's oxygen-binding curve. The portion (a-c) of this curve representing the percent saturation of hemoglobin with oxygen in the venous blood returning from the tissues is ______, and the sigmoid shape of the curve is due to ______

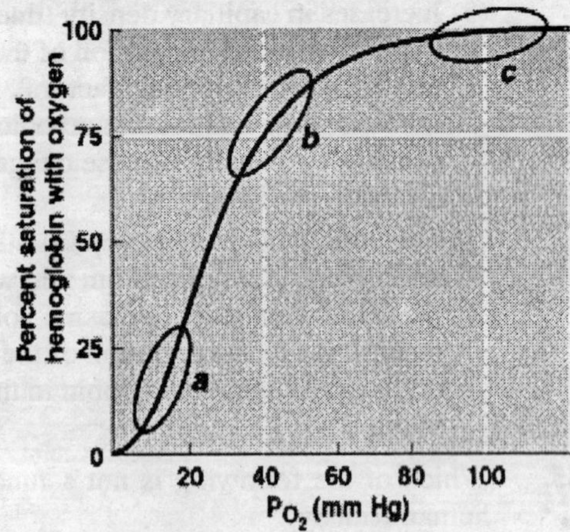

(1) a, Bohr effect
(2) a, positive cooperativity
(3) b, Bohr effect
(4) b, positive cooperativity

53. GFR, and hence the filtered load of sodium, is controlled by baroreceptor reflexes. Decreased vascular pressures cause decreased baroreceptor firing and hence increased sympathetic outflow to the renal arterioles, resulting in vasoconstriction and decreased GFR. These changes are generally relatively small under most physiological conditions

Consider the following statements regarding sodium reabsorption and GFR

A. The major control of tubular sodium reabsorption aldosterone, which stimulates sodium reabsorption in the cortical collecting ducts

B. The renin-angiotensin system is one of the two major controllers of aldosterone secretion. When extracellular volume decreases, renin secretion is stimulated by three inputs: (1) stimulation of the renal sympathetic nerves to the juxtaglomerular cells by extrarenal baroreceptor reflexes; (2) pressure decreases sensed by the juxtaglomerular cells, themselves acting as intrarenal baroreceptors; and (3) a signal generated by low sodium or chloride concentration in the lumen of the macula densa

C. Other factors influence sodium reabsorption such as atrial natriuretic factor, is secreted by cells in the atria in response to atrial distension; it inhibits sodium reabsorption and it also increases GFR

D. Arterial pressure acts locally on the renal tubules to influence sodium reabsorption, an increased pressure causing decreased reabsorption and hence increased excretion

Which of the following statement is correct?

(1) A and B (2) A and D

(3) B and C (4) A, B, C and D

54. The muscle cells listed above will most likely fatigue in what order, from first to last?

Type I: slow, oxidative, red
Type II A: fast, oxidative, red
Type II B: fast, glycolytic, white

First ⟶ Last

A. Type I→ Type II A →Type II B

B. Type I →Type II B →Type II A

C. Type II A →Type I →Type II B

D. Type II B →Type II A →Type I

55. Several oxygen dissociation curves are shown in Figure. Assuming that curve 3 corresponds to isolated hemoglobin placed in a solution containing physiologic concentrations of CO_2 and BPG at a pH of 7.0; indicate which of the curves reflects the following changes in conditions: (a) decreased CO_2 concentration (b) increased BPG concentration (c) increased pH (d) dissociation of hemoglobin into subunits

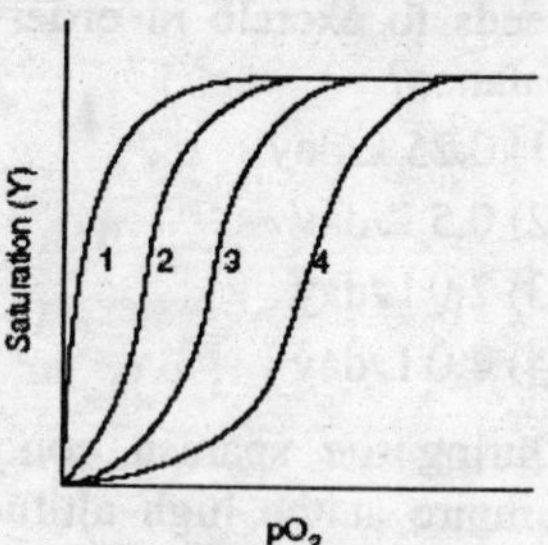

FIGURE Oxygen dissociation curves.

(1) (a) 1 (b) 2 (c) 3 (d) 4

(2) (a) 2 (b) 4 (c) 2 (d) 1

(3) (a) 2 (b) 3 (c) 4 (d) 1

(4) (a) 4 (b) 1 (c) 1 (d) 3

56. A pituitary tumor is removed from a 40-year old man with acromegaly resulting from excessive secretion of growth hormone. It is known that G proteins and adenylyl cyclase normally mediate the stimulation of growth hormone secretion produced by growth hormone-releasing hormone (GHRH) Which of the following problems is most likely to be present in the patient's tumor cells?

(1) Adenylyl cyclase activity is abnormally low

(2) The $G_{\alpha s}$ subunit is unable to hydrolyze GTP

(3) The $G_{\alpha s}$ subunit is inactivated

(4) The $G_{\alpha i}$ subunit is activated

57. A 72-year-old man with an atrial rate of 80 beats/min develops third-degree (complete) AV block. A pacemaker site located in the AV node below the region of the block triggers ventricular activity, but at a rate of only 40 beats/min. What would be observed?

(1) One P wave for each QRS complex

(2) An inverted T wave

(3) A shortened PR interval

(4) A normal QRS complex

58. A man needs to excrete 570 mOsm of solute per day in his urine and his maximum urine osmolality is 1,140 mOsm/kg H_2O. What is the minimum urine volume per day that he needs to excrete in order to stay in solute balance?

(1) 0.25 L/day

(2) 0.5 L/day

(3) 2.0 L/day

(4) 4.0 L/day

59. During the spanish conquest of the Inca Empire at the high altitude in Peru, many soldiers fell sick. It was found that the sickness was due to low partial pressure of O_2 in the atmosphere at that altitude. To determine the reason, blood was collected from those patients. The circulating erythropoietin (EPO) levels were estimated and the O_2-dissociation curve of haemoglobin were drawn and compared with the same in native people as depicted below

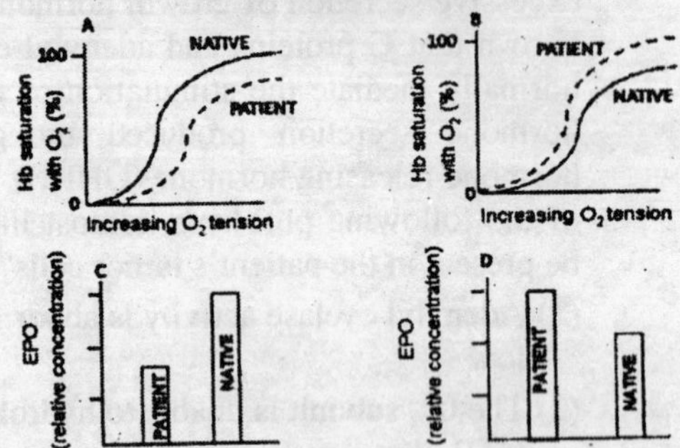

Which one of the following combinations is logically correct?

(1) A and C

(2) A and D

(3) B and C

(4) B and D

60. Batrachotoxin (BTX) is a steroidal alkaloid from the skin of *Phyllobates terribilis*, a poisonous Colombian frog (source of the poison used on blowgun darts) In the presence of BTX, sodium channels in an excised patch stay persistently open when the membrane is depolarized. They close when the membrane is repolarized. Which transition is blocked by BTX?

(1) Tansition from closed to open state

(2) Transition from open to closed state

(3) Transition is not blocked

(4) Trasition of closed to open state is slowed

61. γ-Aminobutyric acid (GABA) opens channels that are specific for chloride ions. The $GABA_A$ receptor channel is pharmacologically important because it is the target of Valium, which is used to diminish anxiety. The extracellular concentration of Cl^- is 123 mM and the intracellular concentration is 4 mM. In which direction does Cl- flow through an open channel when the membrane potential is in the -60 mV to +30 mV range?

(1) Chloride ion flow into the cell

(2) Chloride ion flow outside the cell

(3) Cloride ions are at equilibrium

(4) Chloride ions randomly moves out and in of cell

62. Acetylcholine is rapidly destroyed by the enzyme acetylcholinesterase. This enzyme, which has a turnover number of 25,000 per second, has attained catalytic perfection with a k_{cat}/K_M of 2×10^8 $M^{-1}s^{-1}$. Why is it physiologically crucial that this enzyme be so efficient?

(1) It ensures that the duration of nerve impulse will be short

(2) It helps in rapid transmission of nerve impulse

(3) It helps in long time availability of acetyl choline

(4) It increases rate of synthesis of acetyl choline

63. Acid sensing is associated with pain, tasting, and other biological activities. Acid sensing is carried out by a ligand-gated channel that permits sodium influx in response to H^+. This family of acid sensitive ion channels (ASICs) comprises a number of members. Psalmotoxin 1 (PcTX1), a venom from the tarantula, inhibits some members of this family. Below are electrophysiological recordings of cells containing one of several members of the ASIC family made in the presence of the toxin at a concentration of 10 nM. The channels were opened by changing the pH from 7.4 to the indicated values. The PcTX1 was present for a short time (indicated by the black bar above the recordings) after which time it was rapidly washed from the system

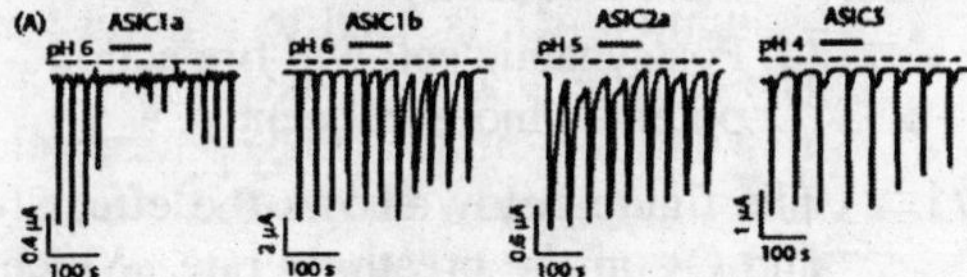

Which of the ASIC family members is most sensitive to the toxin?

(1) ASIC1a (2) ASIC1b
(3) ASIC2a (4) ASIC3

Questions 64-67 refer to the following experiment

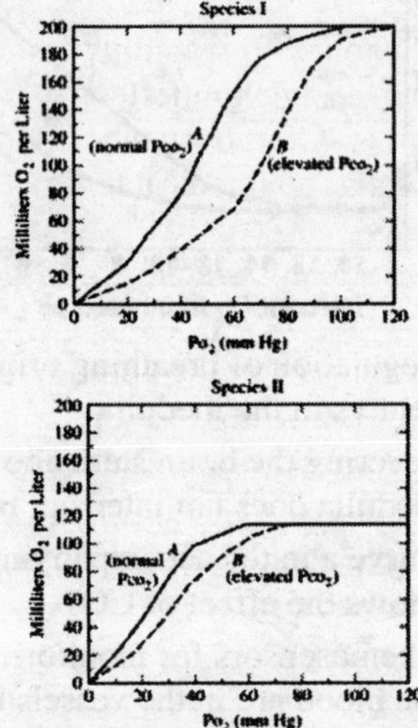

The two graphs below indicate the relationship between oxygen content and oxygen partial pressure (P_{o2}) in samples of blood taken from two species of vertebrates (species I and species II) Each sample was subjected to two levels of carbon dioxide pressure (Pco_2): curve *A* represents the values measured at normal Pco_2 and curve *B* represents the values measured at elevated Pco_2. The blood passing through the lungs of these species normally has a Po_2 of 100 millimeters of mercury and the venous blood leaving the tissues has a Po_2 of 40 millimeters of mercury

64. When curve *A* for species I is compared to curve *A* for species II, it can be predicted that the O_2 concentration of the blood will be

(1) higher in the lung for species I than for species II
(2) lower in the lung for species I than for species II
(3) higher in species I at all pressures
(4) lower in species I at all pressures

65. If deoxygenated blood samples were exposed to increasing P_{o2}, the first sample to become saturated with O_2 would be from which of the following?

(1) Species I at normal Pco_2
(2) Species I at elevated Pco_2
(3) Species II at normal Pco_2
(4) Species II at elevated Pco_2

66. Which of the following statements best describes the effect of increased carbon dioxide pressure in the blood of the two species?

(1) It affects oxygen affinity and oxygen carrying capacity in one species only
(2) It increases the oxygen affinity but has no effect on oxygen-carrying capacity in either species
(2) It reduces the oxygen affinity and the oxygen-carrying capacity in both species
(4) It reduces oxygen affinity but has no effect on oxygen-carrying capacity in either species

67. If the Pco_2 of arterial blood is equivalent to that used to obtain curve *A* and if venous

Pco_2 corresponds to that used to obtain curve *B*, how much oxygen will be released from a liter of blood as it passes through the tissues in species I ?

(1) 20 mL (2) 40 mL
(3) 120 mL (4) 160 mL

68. An isolated carotid sinus was prepared so that the pressure may be regulated by a pump and the resulting discharge in single carotid sinus nerve fibre could be recorded. The following are the possible observations

A. No discharge when carotid sinus perfusion pressure was below 30 mm Hg
B. Linear increase in discharge frequency when carotid sinus perfusion pressure was gradually increased from 70 to 110 mm Hg
C. Increase in discharge frequency was more prominent in greater pulsatile changes of carotid sinus pressure keeping the mean pressure identical in all cases
D. Increase in discharge was more prominent in the falling phase of pulsatile change of carotid sinus pressure than in the rising phase

Which one of the following is correct?

(1) A, B and C (2) A and C
(3) B and D (4) D only

69. For a normal heart, the time taken for atrial systole and diastole are A_s and A_d seconds, respectively, while the same for ventricular systole and diastole are V_s and V_d. Which one of the following equations is correct?

(1) $A_s + V_d = V_s + A_d$
(2) $A_s + A_d < V_s + V_d$
(3) $A_s + V_s = V_s + V_d$
(4) $V_d + A_d > A_s + V_s$

70. In response to membrane depolarization, the voltage-gated Ca^{2+}channels in axon terminals allow an influx of $Ca^{2+}$from the extracellular space into the presynaptic cell. The Ca^{2+} ions bind to proteins in the membrane of synaptic vesicles that interact with plasma membrane proteins, inducing membrane fusion and exocytosis of the vesicle contents into the synaptic cleft. Thus these Ca^{2+} channels transduce an electrical signal, which then induces neurotransmitter release into the synaptic cleft. What would happen to synaptic transmission if the pre- and postsynaptic cells were incubated in Ca^{2+}-free medium?

(1) It would generate impulse from depolarized presynaptic cells to postsynaptic cells
(2) It would prevent impulse transmission from depolarized presynaptic cells to postsynaptic cells
(3) Presynaptic cell will fire action potential more frequently
(4) Postsynaptic cell will fire action potential more frequently

71. The figure below shows the effects of CO_2 and O_2 on the breathing rate. Which of the following statements about the control of breathing is false?

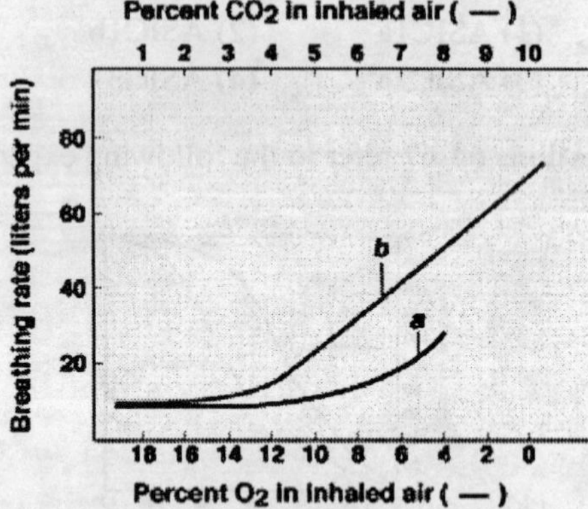

(1) Regulation of breathing originates from centers in the medulla
(2) Severing the brain stem above the medulla does not interrupt breathing
(3) Curve a in the accompanying figure shows the effect of CO_2
(4) Chemosensors for monitoring oxygen in the blood are in the vessels leaving the heart

72. To what group of vertebrates does the circulatory system diagram below apply?

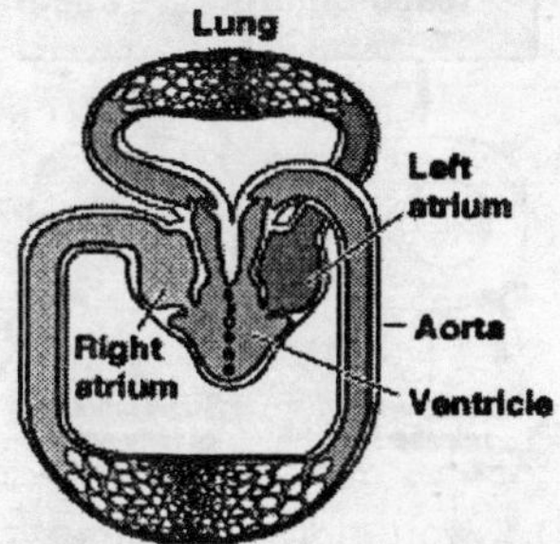

(1) Lungfish
(2) Adult frogs and toads
(3) Reptiles
(4) Crocodilians

73. The figure below shows the relationship between glomerular filtration rate and arterial pressure. Which of the following events would not be expected to occur when arterial pressure is less than 85 mm Hg?

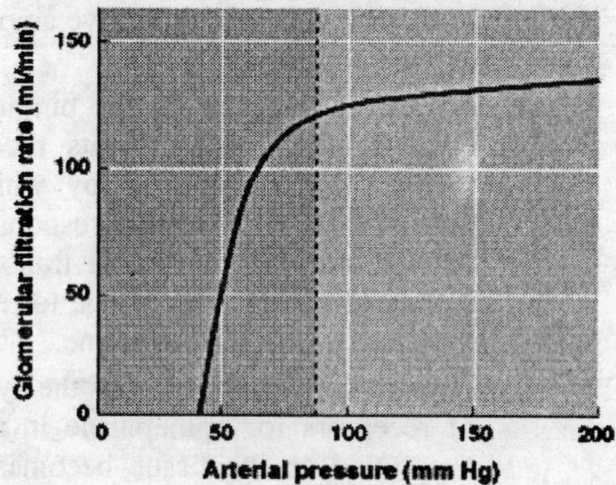

(1) Dilation of afferent renal arterioles
(2) The kidney releases the enzyme renin
(3) The efferent renal arterioles dilate
(4) Enhanced secretion of antidiuretic hormone (ADH) by the hypothalamus

74. A 12 cm sciatic nerve bundle was dissected from a frog. The bundle was electrically stimulated at position 'X' and recording done 10 cm away at *'Y'*. The record showed four peaks *A, B, C* and *D* with increased latency of response, respectively. What would be observed if the same experiment was carried out by gradually decreasing the temperature of the medium in which the nerve bundle was immersed?

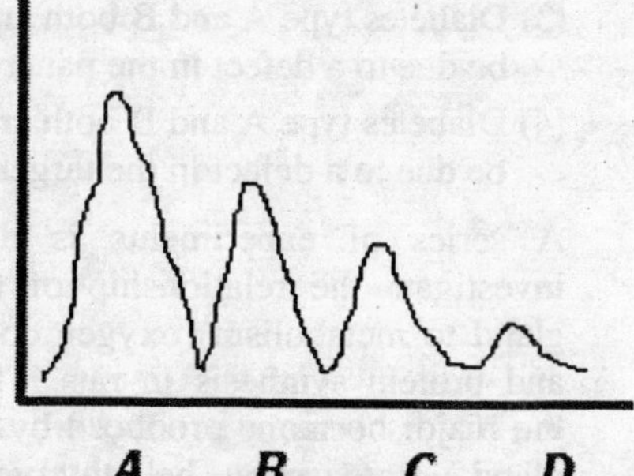

(1) All the peaks would increase
(2) Latency of all the peaks would decrease
(3) The peaks would decrease and disappear in the temporal order D → C → B → A
(4) The peaks would decrease and disappear in the temporal order A → B → C → D

75. The time taken for atrial systole and diastole in a normal heart are t_{as} and t_{ad} seconds, respectively. If ventricular systole takes t_{vs} seconds, calculate the ventricular diastolic time (seconds)

(1) $(t_{as} + t_{ad}) - t_{vs}$ (2) $(t_{as} - t_{ad}) + t_{vs}$
(3) $(t_{ad} - t_{as}) - t_{vs}$ (4) $(t_{as} + t_{ad}) \times t_{vs}$

76. Insulin is a protein that is secreted by the pancreas (an endocrine organ) when the level of blood sugar is high. Insulin acts on many distant tissues by binding and activating a receptor tyrosine kinase, leading to a reduction in blood sugar by appropriate storage of the products of sugar metabolism. Diabetes is a disease in which the level of blood sugar remains high because some part of the insulin pathway is defective. One kind of diabetes (call it type A) can be treated by giving the patient insulin. Another kind of diabetes (call it type B) does not respond to insulin treatment. Which of the following statement is correct?

(1) Diabetes type A is likely to be due to a defect in the pancreas, and type B is likely to be due to a defect in the target cells

(2) Diabetes type A is likely to be due to a defect in the target cell, and type B is likely to be due to a defect in the pancreas

(3) Diabetes type A and B both are likely to be due to a defect in the pancreas

(4) Diabetes type A and B both are likely to be due to a defect in the target cells

77. A series of experiments is designed to investigate the relationship of the thyroid gland to metabolism (oxygen consumption) and protein synthesis in rats. Thyroxine is the major hormone produced by the thyroid gland. Rats may be thyroidecotimized (THX) either surgically or chemically. Experimental and control animals are tested by measuring oxygen consumption in a respirometer for 1 hour and by measuring incorporation of injected tritiated leucine (^{3}H-leu) into the liver. At various times after injection, the rats are sacrificed and the radioactivity in the liver is measured. The data are recorded in the table below

	Oxygen Consumption ($mL \cdot gram^{-1} \cdot hr^{-1}$)	Incorporation of Label ($\mu moles \cdot gram^{-1} \cdot hr^{-1}$)
THX rats	23.4	154
Rats with thyroid gland	32.6	214

Observation suggests that thyroidectomy decreases protein synthesis and oxygen consumption. To confirm further the relationship of thyroid gland to respiration and protein synthesis, one should do which of the following?

(1) Inject thyroxine into the control rats and take measurement again

(2) Inject thyroxine into the THX rats and take measurement again

(3) Inject Ringer's solution into the THX rats and take measurement again

(4) Feed Iodine to the THX rats and take measurement again

78. Figure below represents effect of thyroid hormone and epinephirine alone or incombination on release of fatty acids from adipose tissues

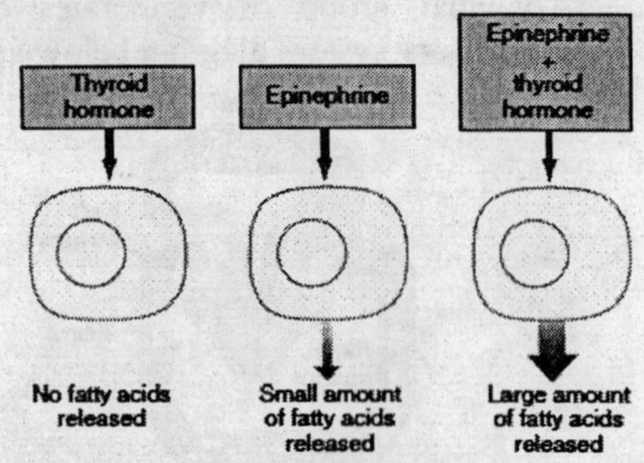

Following are possible conclusions

A. Thyroid hormone "inhibits" epinephrine-induced release of fatty acids from adipose-tissue cells by closing the transporter for fatty acid release and when epinephrine is there it inactivates thyroid hormone thus there is release of fatty acids from adipose tissues

B. Epinephrine causes a large release of fatty acids from adipose tissue, but only in the presence of permissive amounts of thyroid hormone

C. The events initiated by the binding of a epinephrine hormone to its receptor - that is, the mechanisms by which the hormone elicits a cellular response - are one or more of the signal transduction pathways that is antagonistic to response mediated by thyroid hormone

D. Thyroid hormone facilitates the synthesis of receptors for epinephrine in adipose tissue and so the tissue becomes much more sensitive to epinephrine

Which one of the following is most appropriate?

(1) A and B (2) B and C
(3) B and D (4) A and D

79. Figure below represent targets and major functions of the six classical anterior pituitary hormones

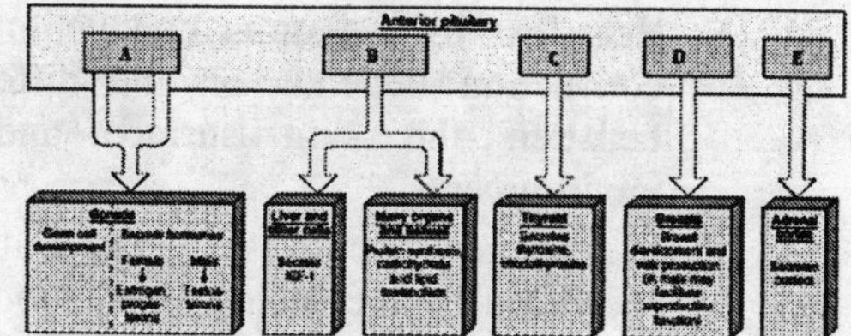

Choose the correct option?

(1) A: ACTH; B: GH; C: TSH; D: LH and FSH; E: Prolactin

(2) A: FSH and LH; B: TSH; C: GH; D: Prolactin; E: ACTH

(3) A: FSH and LH; B: GH; C: TSH; D: Prolactin; E: ACTH

(4) A: LH; B: GH; C: TSH; D: Prolactin and FSH; E: ACTH

80. SNARE complexes are important for guiding synaptic transmission and Ca^{2+} binding is essential for activation of neurotransmitter release from the pre-synaptic terminal. In lieu of the above, identify the components of the SNARE complex that are activated in a sequential manner in order to facilitate neurotransmitter uptake, docking and release

(1) Syntaxin→synaptotagmin→ synaptobrevin

(2) Synaptotagmin → synaptobrevin → SNAP-25

(3) Synaptobrevin → syntaxin → SNAP-25

(4) SNAP-25→synaptotagmin → syntaxin

81. A person with symptoms of hypothyroidism (for example, sluggishness and intolerance to cold) is found to have abnormally low plasma concentrations of T4, T3, and TSH. After an injection of TRH, the plasma concentrations of all three hormones increase. Where is the site of the defect leading to the hypothyroidism?

(1) Follicle cells of thyroid

(2) Adenohypophysis (Anterior Pituatory)

(3) Neurohypophysis (Posterior Pituatory)

(4) Hyptothalamus

82. Consider the following water conservation mechanisms

P. absence of sweat glands

Q. dependence on metabolic water

R. greater ADH level in the blood

S. lying more on protein-rich diet

The kangaroo rate (Dipodiomys) living in deserts can survive without drinking water because of

(1) P, Q and R (2) P, Q and S

(3) Q, R and S (4) P, R and S

83. *Hypertension* is defined as a chronically increased systemic arterial pressure. The dividing line between normal pressure & hypertension is set at approximately 140/90 mmHg Hypertension could result from an increase in cardiac output or in total peripheral resistance, or both

Which of the following is not a cause of hypertension?

(1) Diseases that damage a kidney or decrease its blood supply are often associated with *renal hypertension.* The cause of the hypertension is decreased release of renin from the kidneys, with subsequent decreased generation of the potent vasoconstrictor angiotensin II

(2) Excessive sodium retention is a contributing factor in genetically predisposed ("salt-sensitive") persons. Many persons with hypertension show a drop in blood pressure after being on low-sodium diets or receiving drugs, termed diuretics, that cause increased sodium loss via the urine

(3) Low dietary intake of calcium has also been implicated as a possible contributor to primary hypertension

(4) Obesity and a sedentary lifestyle are definite risk factors for primary hypertension, and weight reduction and exercise are frequently effective in causing some reduction of blood pressure in persons with hypertension. Cigarette smoking, too, is a definite risk factor

84. Match list-I with list-II and select the correct answer using the codes given below the lists

List - I	List - II
P. Dehydration	1. Aldosterone
Q. Antidiuretic hormone	2. Osmoreceptor
R. Reabsorption of Na^+ in the kidney	3. Blood hyperosmotic urine
S. Water diuresis	4. Vasopressin
	5. Blood hyposmotic urine

(1) P-4, Q-1, R-2, 5-3
(2) P-3, Q-4, R-5, 5-1
(3) P-4, Q-1, R-2, 5-5
(4) P-3, Q-4, R-1, 5-5

85. GnRH is secreted during infancy (0-6 month) and puberty onwards (4 years and above) in monkeys. However, *i. v.* injection of GnRH during pre-pubertal period (about 2 years of age) led to elevated LH and FSH in blood compared to untreated 2 years old monkey. This suggest that

A. Hypothalamus is active during pre-pubertal period
B. GnRH action on pituitary is age dependent
C. Pituitary matures during adulthood
D. Pituitary is active in all stages of development in monkeys

Which one of the following is true?

(1) A and B (2) B and C
(3) C only (4) D only

86. The hypothalamus and other brain areas contain the integrating centers for temperature-regulating reflexes, and both peripheral and central thermoreceptors participate in these reflexes. Body temperature is regulated by altering heat production and/or heat loss so as to change total body heat content

Following are some possible physiological mechanisms

A. Heat production is altered by increasing muscle tone, shivering, and voluntary activity
B. Heat loss by radiation, conduction, and convection depends on the difference between the skin surface and the environment
C. In response to cold, skin temperature is increased by increasing skin blood flow through reflex stimulation of the sympathetic nerves to the skin. In response to heat, skin temperature is decreased by inhibiting these nerves
D. Evaporation of water occurs all the time as insensible loss from the skin and respiratory lining. Additional water for evaporation is supplied by sweat, stimulated by the sympathetic nerves to the sweat glands
E. Decreased heat production is essential for temperature regulation at environmental temperatures below the thermoneutral zone and sweating is essential at temperatures below this zone

Which of the following statement are not correct?

(1) A and C (2) B, D and E
(3) C and E (4) A, B and D

87. Consider the following physiological functions in respect to sertoli cells

A. Provide Sertoli-cell barrier to chemicals in the plasma and nourish developing sperm as well as secrete luminal fluid, including androgen-binding protein
B. Receive stimulation by testosterone and FSH to secrete paracrine agents that stimulate sperm proliferation & differentiation
C. Secrete the protein hormone inhibin, which inhibits FSH Secretion and secrete paracrine agents that influence the function of Leydig cells
D. Phagocytize defective sperm

Select the functions carried out by sertoli cells

(1) A, B and C (2) C and D
(3) A and D (4) All of the above

88. The statement below represent the major feedback effects of Estrogen, Progesterone, and Inhibin

A. Estrogen in low plasma concentrations, Negative-feedback inhibition of FSH and LH secretion during the early and middle follicular phase

B. Inhibin: Negative-feedback inhibition of FSH secretion throughout the cycle

C. Estrogen in high plasma concentrations: Positive-feedback stimulation of the LH surge, which triggers ovulation

D. High plasma concentrations of progesterone: responsible for Negative-feedback inhibition of FSH and LH secretion and prevention of LH surges during the luteal phase and pregnancy

Which of the following is correct?

(1) A and C (2) B and D
(3) A, C and D (4) A, B, C and D

89. Graphs showing the metabolic response of five different species of animals at various ambient temperature exposures are shown below. The average body mass of each of the species is 25 grams. All measurements were made using respirometers in the lab with the animals at rest and postabsorptive

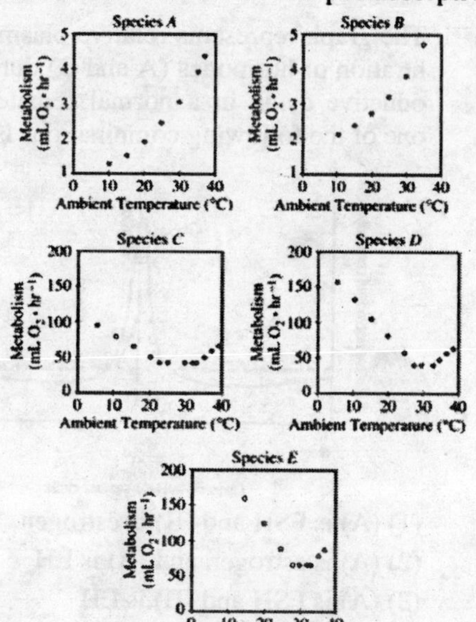

Of the species studied, which is best insulated, highest basal metabolic rate and ectotherm respectively?

(1) Best Insulated: Species *C*, Highest BMR: Species *B*, Ectotherm: Species *C* and *D*

(2) Best Insulated: Species *C*, Highest BMR: Species *C*, Ectotherm: Species *A* and *B*

(3) Best Insulated: Species *D*, Highest BMR: Species *D*, Ectotherm: Species *C* and *D*

(4) Best Insulated: Species *D*, Highest BMR: Species *D*, Ectotherm: Species *A* and *B*

90. Which of the following can be concluded for species *C*, *D*, and *E*?

(1) Metabolic costs increase with increased body mass

(2) Metabolic expenditures decrease below the lower critical temperature

(3) The lowest metabolic costs are in the thermoneutral zone

(4) The lowest metabolic costs are attained at an ambient temperature of 20°C

91. In an experiments, sperm removed from epididymis of a male mouse was added in a dish containing appropriate media and oocyte. No fertilization was seen. However, when sperm from epididymis were directly placed in uterus of an ovulated female, she became pregnant. These observation suggest that

(1) The sperm need to travel some distance to attain fertilizing ability

(2) The oocyte secrets some biochemicals or factors which help sperm to fertilize

(3) The hormones in body help sperm to attain fertilizing ability

(4) The contents of female reproductive tract interact with sperm and activate it for fertilization

92. A person suffering from thyrotoxocosis has extremely high level of thyroid hormone in blood. There is failure of feedback regulation in hypothalamic-pituitary-thyroid axis.

The detailed blood investigation exhibited high level of the following

A. Thyroid stimulating hormone (TSH)

B. Thyroid stimulating Immunoglobulin (TSI)

C. Thyroid releasing hormone

D. Parathyroid hormone

In your opinion, which one of the following is the reason for such thyrotoxicosis

(1) A only (2) B only

(3) A and C (4) C and D

93. A person has been suffering from night blindness. On consultation, the doctor advised person to eat carrots and/or cod fish oil. After some time having seen no improvement, doctor gave person vitamin A injection. Still no marked improvement was seen. The doctor mooted several suggestions indicating lack of the following enzymes for failure of treatment

A. Retinol dehydrogenase

B. Retinal reducatase

C. Retinal isomerase

D. Retinal synthase

According to your opinion which is correct reason for night blindness in above case?

(1) A only (2) B only

(3) B and C both (4) C and D both

94. Initiation of mitogenesis by epidermal growth factor and depolarization of the membrane of a skeletal muscle cell by acetylcholine are similar in that each

(1) involves, as an essential early step, an ion flux across the plasma-membrane receptor of the responding cell

(2) requires a ligand-mediated conformational change in a plasma-membrane receptor of the responding cell

(3) requires activation of a G protein on the cytoplasmic face of the plasma membrane in the responding cell

(4) is mediated by phosphorylation of the ligand receptor in the responding cell

95. The oxygen consumption of marine crabs acclimated to either 10°C or 20°C was measured at environmental temperatures of 5°C to 25°C to produce the following graph

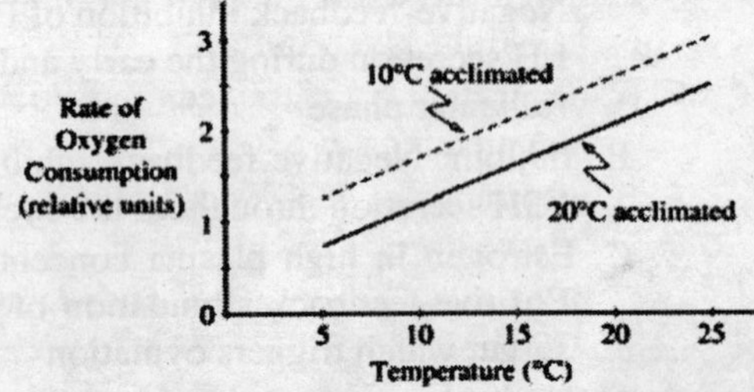

Which of the following can be correctly concluded from the information in the graph?

(1) Crabs have higher rates of oxygen consumption when measured at 10°C than when measured at 20°C

(2) Oxygen consumption is higher in the 10°C-acclimated crabs than in the 20°C-acclimated crabs at each test temperature

(3) Acclimation to a high temperature induces a high rate of oxygen consumption at all test temperatures

(4) This pattern of acclimation produces complete metabolic compensation

96. The graph represents relative plasma concentration of hormones (A and B) during reproductive cycle in a normal female. Which one of the following combinations is correct

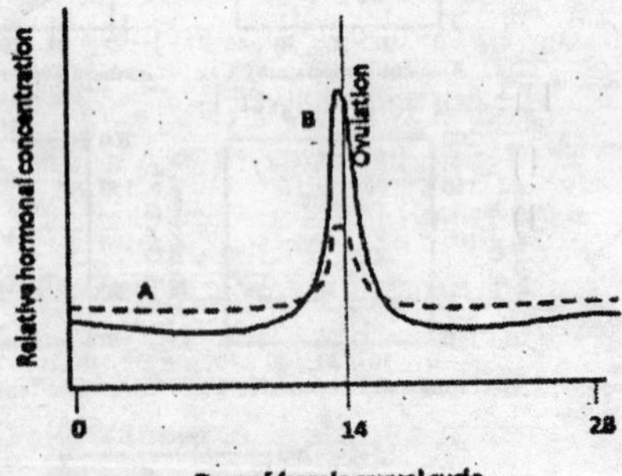

(1) (A) is FSH and (B) is estrogen

(2) (A) is estrogen and (B) is LH

(3) (A) is FSH and (B) is LH

(4) (A) is LH and (B) is FSH

97. Spinal cord of an animal was transected at C1/C2 level. The respiration of the animal stopped and it needed artificial respiration. However, the heart continued to beat although at a slower rate. Some of the explanations given were

A. respiration regulatory centre is located in the medulla

B. respiration regulatory centre is located above the C1/C2 cut

C. heart regulatory centre is above the C1/C2 cut

D. heart has autoregulation

Which one of the following is most appropriate?

(1) A only
(2) B and C only
(3) A, B and D only
(4) B, C and D only

98. In a stressful condition, ACTH secretion was increased and as a result glucocorticoid concentration was elevated in blood. One or a combination of the following changes most likely taking place in this condition

A. Decreased circulating eosinophils and basophils

B. Reduced IL2 release

C. Potentiated inflammatory response to tissue injury

D. Increased mitotic activity of lymphocytes in lymph nodes

The correct answer is

(1) B and C. (2) A and B
(3) B and D. (4) C and D

99. Experiments by Wigglesworth with the bug Rhodnius have led to our current understanding of the hormonal control of insect growth and development. Select the choice below that does not pair an experimental treatment with the expected outcome

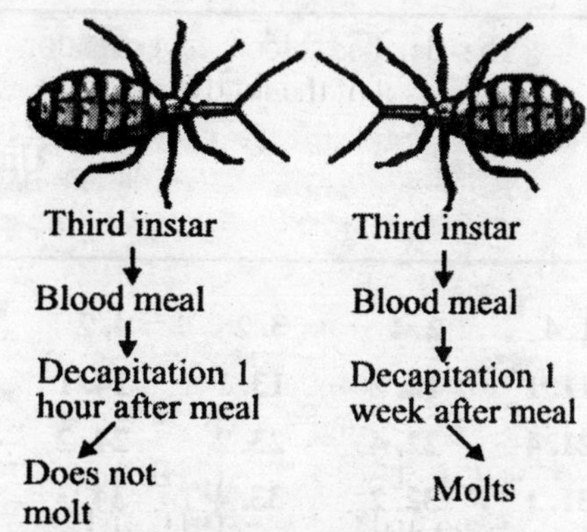

(1) A third instar bug is decapitated one week after a blood meal; molts into a miniature adult
(2) A third instar bug is decapitated one hour after a blood meal; does not molt
(3) A third instar bug is decapitated one hour after a blood meal and injected with ecdysone; molts into a miniature adult
(4) An intact fourth instar bug is given a blood meal and injected with the ground corpora allata of a first instar bug; molts into an adult

100. A steroid hormone receptor contains a functional nuclear localization signal in the ligand-binding domain. A specific hormone-based drug for that receptor, which was seen to enter the cytosol, could not efficiently carry out its function in tumors. This could be due to the following reasons

A. Mutations in the ubiquitous bipartite signal clusters

B. Lack of communication between importin-α and importin-β

C. Pairing of the NLS with proline-tyrosine

D. Presence of Ran-GTP prior to entry of the complex through the nuclear pore

Which one of the following combinations stated above is correct?

(1) A, C and D (2) A and C
(3) B and C (4) A, B and D.

ANSWER SHEET

Unit 7: Animal Physiology

1. 4	**2.** 4	**3.** 2	**4.** 2	**5.** 2	**6.** 3	**7.** 4	**8.** 4	**9.** 3	**10.** 3
11. 1	**12.** 2	**13.** 1	**14.** 1	**15.** 3	**16.** 3	**17.** 1	**18.** 2	**19.** 1	**20.** 1
21. 4	**22.** 4	**23.** 2	**24.** 2	**25.** 2	**26.** 4	**27.** 1	**28.** 4	**29.** 3	**30.** 1
31. 1	**32.** 2	**33.** 4	**34.** 3	**35.** 4	**36.** 3	**37.** 1	**38.** 2	**39.** 2	**40.** 1
41. 3	**42.** 3	**43.** 4	**44.** 2	**45.** 4	**46.** 1	**47.** 2	**48.** 1	**49.** 2	**50.** 1
51. 4	**52.** 4	**53.** 4	**54.** 4	**55.** 2	**56.** 2	**57.** 4	**58.** 2	**59.** 1	**60.** 2
61. 1	**62.** 1	**63.** 1	**64.** 1	**65.** 3	**66.** 4	**67.** 4	**68.** 1	**69.** 4	**70.** 2
71. 3	**72.** 2	**73.** 3	**74.** 3	**75.** 1	**76.** 1	**77.** 2	**78.** 3	**79.** 3	**80.** 2
81. 4	**82.** 1	**83.** 1	**84.** 4	**85.** 4	**86.** 3	**87.** 4	**88.** 4	**89.** 2	**90.** 3
91. 4	**92.** 2	**93.** 1	**94.** 2	**95.** 2	**96.** 3	**97.** 2	**98.** 2	**99.** 4	**100.** 4

UNIT - 8

GENETICS

1. **Cystic Fibrosis is the most common serious single-gene inherited disease in the Western world. Many organs are affected. Sticky mucus builds up in the reproductive tract and in the lungs. The pancreas is always affected and usually fails completely. Consider the following observations**

A. Electrical measurements show that the basic problem is in chloride transport

B. CF is X linked recessive disorder. The gene is on chromosome 7

C. Affected transporter CFTR is P-type of pump

D. CFTR is expressed as co-dominant manner on affected cells

Which of the following are correct for cystic fibrosis?

(1) A and B **(2) B and C**

(3) C and D **(4) A and D**

2. **In some sheep, the presence of horns is produced by an autosomal allele that is dominant in males and recessive in females. A horned female is crossed with a hornless male. One of the resulting F1 females is crossed with a hornless male. What proportion of the male and female progeny from this cross will have horns?**

(1) ¼ of the male progeny will be horned and ¼ female progeny will be horned

(2) ½ of the male progeny will be horned and ½ the female progeny will be horned

(3) ½ of the male progeny will be horned but none of the female progeny will be horned

(4) none of the male and female progeny will be horned

3. **When F1 female Drosophila of the genotype $a^+a\ b^+b\ c^+c$ is test crossed, the following progenies were obtained**

Progeny Classes	No. of progenies
$a^+b^+c^+$	200
a^+b^+c	200
abc^+	200
abc	200
a^+bc^+	50
a^+bc	50
ab^+c^+	50
ab^+c	50
Total	1000

The progeny has been shown as classes derived from the female gamete. Statements A to F as given below are conclusions derived from the above result

A. Genes a and b are linked in cis

B. Genes a and b are linked in trans

C. Genes a and b are linked in cis while b and c are linked in trans

D. The genotype of the parents are a^+a^+bb and aab^+b^+

E. The genotype of the parents are $a^+a^+b^+b^+$ and aabb

F. Genes a and b are 20 cM apart

Which of the above statements are correct?

(1) B, D and F **(2) A, C and F**

(3) A and E only **(4) A, E and F**

4. All of the following statements about double crossovers are true except

(1) Double-crossover frequencies can only be estimated by analyzing three or more traits simultaneously

(2) Double-crossover analysis enables the mapping of the relative distances between three or more genes

(3) For any pair of linked traits, double crossover increases the variety of gametes produced

(4) Double-crossover frequencies are generally underestimated because of interference

5. Mendel crossed tall pea plants with dwarf ones. The F1 plants were all tall. When these F1 plants were selfed to produce F2 generation, he got a 3: 1 tall to dwarf ratio in the offspring. What is the probability that out of three plants (of F2 generation) picked up at random two would be tall and one would be dwarf?

(1) 3/4 (2) 3/8

(3) 9/64 (4) 27/64

6. In tomato the following genes are located on chromosome 3

+ tall plant	d dwarf plant
+ normal leaves	m mottled leaves
+ smooth fruit	p pubescent fruit

Results of the cross: +++/ dmp X dmp/ dmp were

+++	430
dmp	452
+mp	45
d++	38
++p	16
dm+	17
+m+	1
d+p	1

Which one of the following progeny groups represents double crossovers?

(1) +mp / d++ (2) ++p / dm+

(3) +m+ / d+p (4) +++ / dmp

7.

Group I (Type of interaction)	Group II (F2 phenotypic ratio)
P. Recessive epistasis	A. 12 : 3 : 1
Q. Dominant epistasis	B. 13: 3
R. Duplicate recessive epistasis	C. 9: 6: 1
S. Dominant and recessive epistasis	D. 9: 3: 4
	E. 9: 7
	F. 15: 1

(1) P - B, Q - A, R - B, S - E

(2) P - F, Q - C, R - B, S - A

(3) P - D, Q - A, R - E, S – B

(4) P - A, Q - E, R - C, S - D

8. In a particular species of mammal black hair (B) is dominant to green hair (b) and red eyes (R) are dominant to white eyes (r) If a BbRr individual is mated with a bbrr individual the expected phenotypic ratio of the offspring is 1 black-red: 1 black-white: 1 green-red: 1 green-white. However, when you mate these individuals you find that the phenotypic ratio of the offspring is 6 black-red: 1 black-white : 1 green-red : 6 green-white. What could account for this difference?

(1) Green-haired individuals have a higher prenatal mortality than black-haired individuals

(2) The genes for hair color and the genes for eye color are carried on different chromosomes

(3) The expected results did not take genetic recombination into account

(4) The genes for hair color and eye color are linked

9. When two independent pure lines of pea with white flowers are crossed, the F_1 progeny has purple flowers. The F_2 progeny obtained on selfing shows both pruple and white flower in a ratio of 9:7. The following conclusions were made

A. Two different genes are involved, mutation in which lead to formation of white flower

B. These two genes show independent assortment

C. This is an example of complementary gene action

D. The is an example of duplicate genes

Which of the following conclusions are correct?

(1) A and C (2) A and D

(3) A, B and D (4) A, B and C

10. Four different mutant lines showing similar phenotype were identified from a genetic screen. When genetic crosses among these mutants were carried out, the first mutant was found to complement the second, third and fourth mutant lines. However, no other complementation groups do the four mutant lines belongs to?

(1) 1 (2) 2

(3) 3 (4) 4

11. Why lysogenic cycle is more beneficial to a virus than lytic cycle under certain circumstances?

(1) The Lysogenic cycle prevent local extinction of host while still retaining infectious potential

(2) By integrating with the bacterial chromosomes, the genetic instructions for the virus become refreshed after one or more replication events during binary fission

(3) Lysogenic infection cycles donot harm their host cells, so they can produce virus particles indefinitely

(4) Lysogeny causes more mutations to occur in the virus, creating more variants upon which natural selection can operate

12. The following figure depicts the relationship between a genetic map for four genes (A, B, C and D) and their corresponding physical map

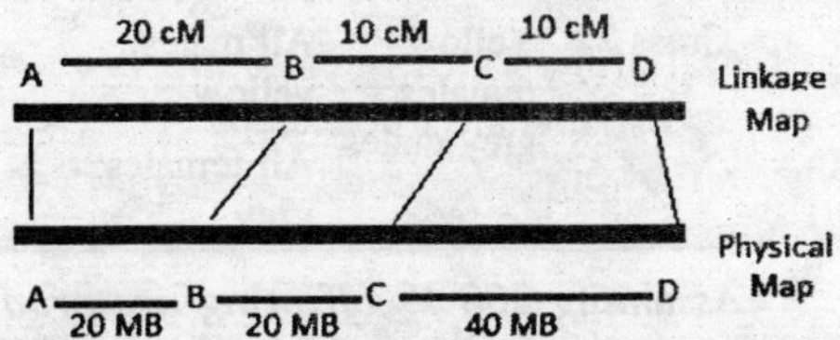

The following statements are made to explain this relationship

A. More number of recombination events occur between A and B as compared to B and C or C and D

B. Lesser number of recombination events occur between C and D as compared to B and C

C. Although the physical distance between A and B is less than that between C and D, the region between A and B is more recombinogenic

D. The physical distance between A and B is less than that between C and D, and thus the region between A and B is less recombinogenic

E. Although the physical distance between C and D is more than that between B and C, the region between C and D is less recombinogenic

F. Although the physical distance between C and D is more than B and C, the region between C and D is more recombinogenic

Which statements are correct?

(1) A and B (2) C and E

(3) D and F (4) A, C and E

13. Consider the following crosses involving grey (wild-type) and yellow body colour true breeding *Drosophila*

	Cross	F_1 progeny	F_2 progeny
Cross 1	Grey female x yellow male	All males : grey All females : grey	Grey females : 98 Yellow males : 45 Grey males : 49

Cross 2	Yellow females x grey males	All males : yellow All females: grey	?

Assuming 200 F_2 offspring are produced in cross 2, which one of the following outcome is expected?

(1) 97 grey males, 54 yellow females, 49 grey males

(2) 102 yellow males, 46 yellow females, 52 grey females

(3) 52 grey males, 49 yellow males, 48 yellow females, 51 grey females

(4) 98 grey males, 94 yellow females, 2 yellow males, 6 grey females

14. The ABO blood type in human is under the control of autosomal multiple alleles. Colour blindness is recessive X-linked trait. A male with a blood type A and normal vision marries a female who also has blood type A and normal vision. The couple's first child is a male who is colour blind and has a blood group O. What is the probability that their next male child has normal vision and a blood group O?

(1) 1/4 (2) 3/4

(3) 1/8 (4) 1

15. The following is a schematic representation of region (showing six bands) of the polytene chromosome of *Drosophila,* along with the extent of five deletions (Del1 to Del5)

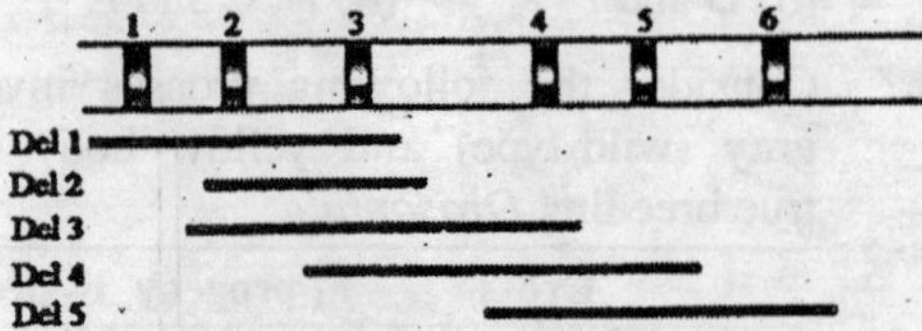

Recessive alleles a, b, c, d, e and f are known to correspond to each of the bands (1 to 6), but their order is not known. When the recessive alleles are placed against each of these deletions, the following results are obtained. The plus (+) in the table indicates wild type phenotype of the corresponding allele, while a minus (-) indicates the phenotype governed by the corresponding mutant allele

	a	*b*	*c*	*d*	*e*	*f*
Del 1	+	–	–	–	+	+
Del 2	+	+	–	–	+	+
Del 3	–	+	–	–	+	+
Del 4	–	+	+	–	–	+
Del 5	–	+	+	+	–	–

Which one of the following indicates the correct location of the recessive alleles on the bands of the polytene chromosomes?

(1) a-3; b-1; c-2; d-4; e-5; f-6

(2) a-2; b-1; c-3; d-4; e-5; f-6

(3) a-4; b-1; c-2; d-3; e-5; f-6

(4) a-6; b-2; c-3; d-4; e-1; f-5

16. Assuming a 1 : 1 sex ratio, what is the probability that three children from the same parents will consist of two daughters and one son?

(1) 0.375 (2) 0.125

(3) 0.675 (4) 0.75

17. The coat colors of mice are known to be governed by a multiple allelic series in which the allele A^y when homozygous, is lethal early in embryonic development but produces yellow color when in heterozygous condition with other alleles. Agouti (mousy color) is governed by the A allele, and black by the recessive a. The dominance hierarchy is as follows: A^y > A> a. What phenotypic and genotypic ratios are expected in the viable F, from the cross A^yA x A^ya?

(1) Phenotypic ratio: 3:1
Genotypic ratio: 1:2:1

(2) Phenotypic ratio: 2:1
Genotypic ratio: 2:1

(3) Phenotypic ratio: 2:1
Genotypic ratio: 1:1:1

(4) Phenotypic ratio: 1:2:1
Genotypic ratio: 2:1

18. Coat colors of dogs depend upon the action of at least 2 genes. At one locus a dominant epistatic inhibitor of coat color pigment (I_) prevents the expression of color alleles at another independently assorting locus, producing white coat color. When the recessive condition exists at the inhibitor locus (ii), the alleles of the hypostatic locus may be expressed, (B_) producing black and (bb) producing brown. When dihybrid white dogs are mated together, determine the phenotypic proportions expected in the progeny,

(1) 12:3:1 (2) 13:3
(3) 9:3:4 (4) 15:1

19. Red color in wheat kernels is produced by the genotype R-B-, white by the double-recessive genotype (rrbb) The genotypes R_bb and rrB_ produce brown kernels. A homozygous red variety is crossed to a white variety. What phenotypic results are expected in the F_2?

(1) 9:6:1 (2) 12:3:1
(3) 9:3:4 (4) 1:2:1

20. At least three loci are known to govern coat colors in mice. The genotype C- will allow pigment to be produced at the other two loci. The recessive genotype cc does not allow pigment production, resulting in "albino." The "agouti" pattern depends upon the genotype A-, and non-agouti upon the recessive aa. The color of the pigment may be black (B-) or chocolate (bb) Five coat colors may be produced by the action of alleles at these three loci

Wild type (agouti, black)	A-B-C-
Black (nonagouti)	aaB-C
Chocolate (nonagouti)	aabbC
Cinnamon (agouti, chocolate)	A-bbC
Albino	---cc

What phenotypic frequencies are expected in the F2 from crosses of pure black (aaBBCC) with albinos (AAbbcc)

(1) 27:16:9:9:3
(2) 27:9:9:9:3:3:3:1
(3) 9:3:4
(4) 9:7

21. Consider three yellow, round peas, labeled A, B, and C. Each was grown into a plant and crossed to a plant grown from a green, wrinkled pea. Exactly 100 peas issuing from each cross were sorted into phenotypic classes as follows

A:	51 yellow, round
	49 green, round
B:	100 yellow, round
C:	24 yellow, round
	26 yellow, wrinkled
	25 green, round
	25 green, wrinkled

What were the genotypes of A, B, and C?

(1) A: Y/y ; R/r B: Y/Y ; R/R C: Y/y ; R /R
(2) A; Y/y ; R /R B: Y/Y ; R/R C: Y/y ; R/r
(3) A; Y/y ; R /r B: Y/Y ; R/R C: Y/Y ; R/r
(4) A; Y/y ; R /R B: Y/Y; R/r C: Y/y ; R/r

22. In human beings, the absence of molars is inherited as a dominant trait. If two heterozygotes have four children, what is the probability that three will have no molars and one will have molars?

(1) 81/256 (2) 108/256
(3) 9/256 (4) 32/256

23. Following is the diagram of a pericentric inversion heterozygote ABCDEFG/ ABFEDCG involved in recombination during meiosis I

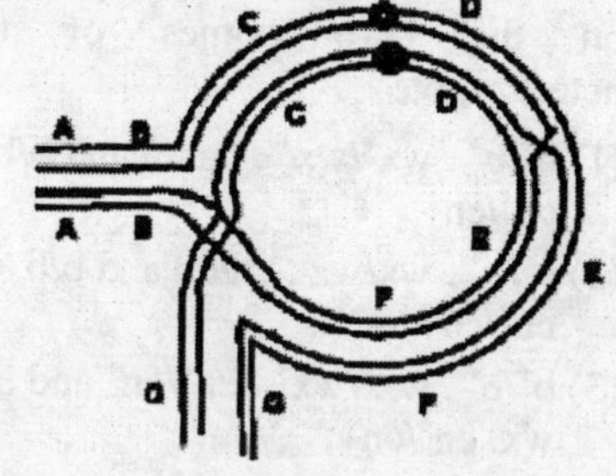

The consequence of this recombination will be the formation of

A. A dicentric and an acentric chromosome in meiosis I as the chiasmata gets terminated

B. No dicentric or acentric chromosome but appearance of deletion and duplication in both the chromosomes

C. All non-viable gametes

D. Non-viable gametes from crossover products only

Which of the above statements are correct?

(1) A and B (2) A and C
(3) B and D (4) B and C

24. The allele b gives Drosophila flies a black body, and b^+ gives brown, the wild-type phenotype. The allele wx of a separate gene gives waxy wings, and wx^+ gives nonwaxy, the wild-type phenotype. The allele cn of a third gene gives cinnabar eyes, and cn^+ gives red, the wild-type phenotype. A female heterozygous for these three genes is testcrossed, and 1000 progeny are classified as follows

- 5 wild type
- 6 black, waxy, cinnabar
- 69 waxy, cinnabar
- 67 black; 382 cinnabar
- 379 black, waxy
- 48 waxy; and
- 44 black, cinnabar

Note that a progeny group may be specified by listing only the mutant phenotypes

Predict the alleles in their proper positions on the chromosomes of the triple heterozygote

(1) b^+/b^+ . wx^+/wx^+ . cn/cn and b/b . wx / wx. cn^+/cn+

(2) b^+/b^+ . wx/wx . cn/cn and b/b. wx^+ / wx^+. cn^+/cn+

(3) b^+/b^+ . wx^+/wx^+ . cn^+/cn^+ and b/b wx / wx. cn^+/cn+

(4) b^+/b^+ . wx/wx . cn^+/cn^+ and b/b wx^+/ wx^+ . cn/cn

25. DNA from a strain of bacteria with genotype $a^+ b^+ c^+ d^+ e^+$ was isolated and used to transform a strain of bacteria that was $a^- b^- c^- d^-$ e. The transformed cells were tested for the presence of donated genes. The following genes were co-transformed

a^+ and d^+

b^+ and e^+

c^+ and d^+

c^+ and e^+

What is the order of genes a, b, c, d, and e on the bacterial chromosome?

(1) d, c, e, b, a (2) c, e, b, a, d
(3) e, b, a, d, c (4) a, d, c, e, b

26. Crosses of three different Hfr strains with separate samples of an F^- strain are carried out, and the following mapping data are provided from studies of interrupted conjugation

		Appearance of Genes in F^- cells				
Hfr1:	Genes	b^+	d^+	c^+	f^+	g^+
	Time	3	5	16	27	59
Hfr1:	Genes	e^+	f^+	c^+	d^+	b^+
	Time	6	24	35	46	48
Hfr1:	Genes	d^+	c^+	f^+	e^+	g^+
	Time	4	15	26	44	58

Construct a genetic map for these genes, indicating their order on the bacterial chromosome

(1) b d c f g e (2) g e f d c b
(3) b d c f e g (4) e f c d b g

27. Two mutations that affect plaque morphology in phages (a ⁻and b⁻) have been isolated. Phages carrying both mutations (a^- b^-) are mixed with wild-type phages (a^+ b^-) and added to a culture of bacterial cells. Subsequent to infection and lysis, samples of the phage lysate are collected and cultured on bacterial cells. The following numbers of plaques are observed

Plaque phenotype	Number
$a^+ b^+$	2043
$a^+ b^-$	320
$a^- b^+$	357
$a^- b^-$	2134

What is the frequency of recombination between the a and b genes?

(1) 15.72 (2) 13.94
(3) 6.57 (4) 6.88

28. A geneticist isolates two *r* mutants (r_{13} and r_2) that cause rapid lysis. He carries out the following crosses and counts the number of plaques listed here

Genotype of parental phage	Progeny	Number of plaques
$h^+ r_{13}^- \times h^- r_{13}^+$	$h^+ r_{13}^+$	1
	$h^- r_{13}^+$	104
	$h^+ r_{13}^-$	110
	$h^- r_{13}^-$	2
	total	216
$h^+ r_2^- \times h^- r_2^+$	$h^+ r_2^+$	6
	$h^- r_2^+$	86
	$h^+ r_2^-$	81
	$h^- r_2^-$	7
	total	180

Calculate the recombination frequencies between r_2 and h and between r_{13} and h

(1) 0.01388 and 0.0722
(2) 1.388 and 7.220
(3) 13.88 and 72.22
(4) 1.432 and 14.232

29. Phenotypic variation is analyzed for milk production in a herd of dairy cattle and the following variance components are obtained

Additive genetic variance (V_A) = 4

Dominance genetic variance (V_D) = 1

Genic interaction variance (V_I) = 2

Environmental variance (V_E) = 5

Genetic-environmental interaction variance (V_{GE}) = 0

What is the narrow-sense heritability of milk production?

(1) 0.33 (2) 0.58
(3) 0.67 (4) 0.42

30. A geneticist is working with a new bacteriophage called phage Y3 that infects *E. coli.* He has isolated eight mutant phages that fail to produce plaques when grown on *E. coli* strain K. To determine whether these mutations occur at the same functional gene, he simultaneously infects *E. coli* K cells with paired combinations of the mutants and looks to see whether plaques are formed. He obtains the following results. (A plus sign means that plaques were formed on *E. coli* K; a minus sign means that no plaques were formed on *E. coli* K)

Mutant	1	2	3	4	5	6	7	8
1								
2	+							
3	+	+						
4	+	–	+					
5	–	+	+	+				
6	–	+	+	+	–			
7	+	–	+	–	+	+		
8	–	+	+	+	–	–	+	

To how many functional genes (cistrons) do these mutations belong?

(1) 3 (2) 4
(3) 5 (4) 6

31. A farmer has two homozygous varieties of tomatoes. One variety, called *Chotte tamatum*, has fruits that average only 2 cm in diameter. The other variety, *Badde Banarasi*, has fruits that average a whopping 14 cm in diameter. The farmer crosses *Chotte tamatum* and *Badde Banarasi*; he then intercrosses the F_1 to produce F_2 progeny. He grows 2000 F_2 tomato plants and he finds eight F_2 offspring that produce fruits as small as *Chotte tamatum* or as large as *Badde Banarasi*. If we assume that the differences in fruit size of these varieties are produced by genes with equal and additive effects, what conclusion can we make about the minimum number of loci with pairs of alleles determining the differences in fruit size of the two varieties?

(1) 2
(2) 3
(3) 4
(4) Cannot be determined

32. An experiment was conducted in E. *coli* to map the following genes (pro, his, bio, met, phe and trp) on a circular map'using three different Hfr strains

- Strain 1: order of transfer (early to late): trp met his pro
- Strain 2: order of transfer (early to late): his met trp bio
- Strain 3: order of transfer (early to late): pro phe bio trp

Based on the results, what is the most likely map?

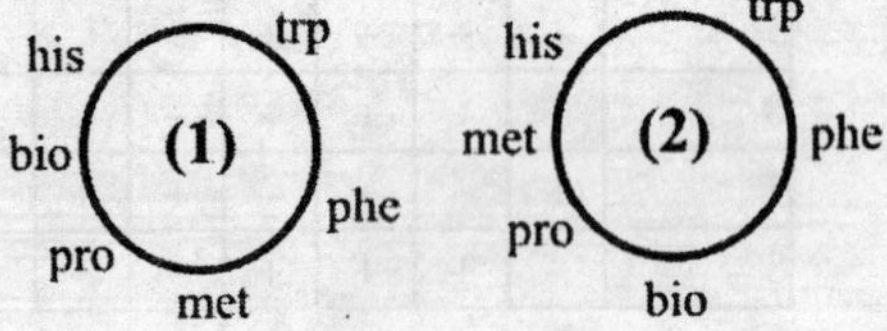

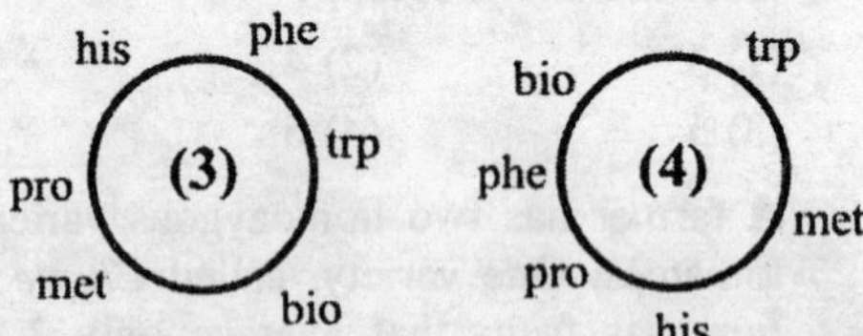

33. Following three-point test cross progeny was obtained in a gene mapping experiment

ABC - 365; abc - 367; aBc - 110;
AbC - 105; aBC - 3; Abc - 4; ABc - 25;
abC - 21; Total- 1000

From these data we can estimate that the distance between A and B is

(1) 22.2 cM (2) 5.3 cM
(3) 21.5 cM (4) 4.7 cM

34. In roses, the synthesis of a red pigment is carried out in two steps as shown below, conversions taking place by the products of genes A and B which are not linked

A colourless substrate → Magenta coloured intermediate → Red pigment
(↑ Product of gene A; ↑ Product of gene B)

A plant homozygous for a null mutation of gene A is crossed to a plant homozygous for a null mutation for gene B. In what ratio is the F_2 progeny expected from this cross?

(1) 9 red: 4 colourless: 3 magenta
(2) 9 red: 3 colourless: 4 magenta
(3) 9 red: 6 magenta: 1 colourless
(4) 15 red: 1 colourless

35. With unordered tetrads, a cross involving two genes potentially produces three types of tetrads: parental ditype tetrads (PD), which have two non-recombinant combinations of genes; tetratype tetrads (T), which have two nonr-ecombinant and two recombinant combinations of genes; and nonparental ditype tetrads (NPD), which have two recombinant combinations of genes. Consider the following statement

A. If two genes are assorting independently, the number of PD tetrads and NPD tetrads should be approximately equal
B. If two genes are assorting independently, the numbers of PD tetrads are more and NPD tetrads should be less
C. if there are more PD tetrads than NPD tetrads, then the genes are unlinked
D. if there are more PD tetrads than NPD tetrads, then the genes are linked

Which of the following statement is correct?

(1) A and B
(2) A and D
(3) B and C
(4) C and D

36. In yeast, the following cross is made: ab X AB. The resulting zygote undergoes meiosis and produces the following unordered tetrads

A B	A b	A B
A B	A b	a b
a b	a B	A b
a b	a B	a B
514	4	216

If a and b genes linked, what is the genetic distance between them?

(1) 29.42 (2) 29.99
(3) 15.25 (4) 14.98

37. A cross is made between a strain of *Neurospora crassa* with genotype nic^+ ad^+ and another strain with genotype *nic ad*. The following linear octads are observed

1	2	3	4	5	6	7
nic^+ ad^+	nic^+ ad^+	nic^+ ad	nic^+ ad	nic^+ ad	nic^+ ad^+	nic^+ ad
nic^+ ad^+	nic^+ ad^+	nic^+ ad	nic^+ ad	nic^+ ad	nic^+ ad^+	nic^+ ad
nic^+ ad^+	nic ad	nic^+ ad^+	nic^+ ad	nic ad^+	nic ad^+	nic ad^+
nic^+ ad^+	nic ad	nic^+ ad^+	nic^+ ad	nic ad^+	nic ad^+	nic ad^+
nic ad	nic ad	nic ad	nic ad^+	nic ad^+	nic^+ ad	nic^+ ad^+
nic ad	nic ad	nic ad	nic ad^+	nic ad^+	nic^+ ad	nic^+ ad^+
nic ad	nic^+ ad^+	nic ad^+	nic ad^+	nic^+ ad	nic ad	nic ad
nic ad	nic^+ ad^+	nic ad^+	nic ad^+	nic^+ ad	nic ad	nic ad
423	105	101	4	4	15	15

Determine the distance between *nic* and *ad*

(1) 11.3 (2) 36.4
(3) 47.4 (4) 1.52

38. In *Neurospora*, a fungus with mating type *A* is crossed with a fungus that has mating type *a*, and the following linear octads are produced

A A A A a a a a	212
a a a a A A A A	221
A A a a A A a a	15
a a A A a a A A	17
A A a a a a A A	19
a a A A A A a a	16
Total	500

What is the genetic distance of the mating type locus to the centromere?

(1) 6.7 (2) 13.4
(3) 3.3 (4) 1. 3

39. In sweet peas, the purple color of the petals is controlled by two genes, *B* and *D*. The pathway is

colorless precursor $\xrightarrow{\text{gene B}}$ red anthocyanin pigment
colorless precursor $\xrightarrow{\text{gene D}}$ blue anthocyanin pigment
} purple

What color petals would you expect in a plant that carries two copies of a null mutation for B?

(1) All plants will be have blue colored petals
(2) All plants will be have red colored petals
(3) All plants will be have white colored petals
(4) Plants will be have blue and Red colored petals in approximate ratio of 1:1

40. Phenylketonuria (PKU) is a human hereditary disease resulting from the inability of the body to process the chemical phenylalanine, which is contained in the protein that we eat. PKU is manifested in early infancy and, if it remains untreated, generally leads to mental retardation. PKU is caused by a recessive allele with simple Mendelian inheritance. A couple intends to have children but consults a genetic counselor because the man has a sister with PKU and the woman has a brother with PKU. There are no other known cases in their families. They ask the genetic counselor to determine the probability that their first child will have PKU. What is this probability?

(1) 0 (2) ¼
(3) 1/16 (4) 1/9

41. The *R* and *S* loci are 50 m.u. apart. If a plant of genotype $\frac{R\ S}{r\ s}$ is selfed, In what proportions genotype $\frac{R\ S}{r\ s}$ would be observed?

(1)1/4 (2) 1/8
(3) 1/16 (4) 9/16

42. In this experiment, P1 phage grown on the leu^+ thr^+ azi^r donor strain infect the leu^- thr^- azi^s recipient strain. The strategy is to select one or more donor markers in the recipient and then test these transductants for the presence of the unselected markers. Results are outlined in table

Experiment	Selected marker	Unselected markers
1	leu^+	50% are azi^r; 2% are thr^+
2	thr^+	3% are leu^+; 0% are azi^r
3	leu^+ and thr^+	0% are azi^r

What is order of genes?

(1) leu^+ thr^+ azi^r (2) leu^+ azi^r thr^+

(3) thr^+ leu^+ azi^r (4) thr^+ leu^+ azi^s

43. In *Drosophila*, the mutant alleles are *v* (vermilion eyes), *cv* (cross vein less, or absence of a cross vein on the wing), and *ct* (cut, or snipped, wing edges) The analysis is carried out by performing the following crosses

P $v^+/v^+ \cdot cv/cv \cdot ct/ct \times v/v \cdot cv^+/cv^+ \cdot ct^+/ct^+$

Gametes $v^+ \cdot cv \cdot ct$ $v \cdot cv^+ \cdot ct^+$

F_1 trihybrid $v/v^+ \cdot cv/cv^+ \cdot ct/ct^+$

Trihybrid females are testcrossed to triple recessive males

$v/v^+ \cdot cv/cv^+ \cdot ct/ct^+$ ♀ × $v/v \cdot cv/cv \cdot ct/ct$ ♂

F_1 trihybrid female Tester male

From any trihybrid, only 2 x 2 X 2 = 8 gamete genotypes are possible. These are the genotypes seen in the testcross progeny. The chart shows the number of each of the eight gametic genotypes counted out of a sample of 1448 progeny flies

Gametes	
$v \cdot cv^+ \cdot ct^+$	580
$v^+ \cdot cv \cdot ct$	592
$v \cdot cv \cdot ct^+$	45
$v^+ \cdot cv^+ \cdot ct$	40
$v \cdot cv \cdot ct$	89
$v^+ \cdot cv^+ \cdot ct^+$	94
$v \cdot cv^+ \cdot ct$	3
$v^+ \cdot cv \cdot ct^+$	5
	1448

What is order of genes?

(1) v^+. cv^+. ct^+ (2) v. cv^+. ct^+

(3) v^+. ct^+. cv^+ (4) v. ct^+. cv^+

44. In *E. coli,* four Hfr strains donate the following genetic markers, shown in the order donated

Strain 1: Q W D M T

Strain 2: A X P T M

Strain 3: B N C A X

Strain 4: B Q W D M

All these Hfr strains are derived from the same F^+ strain. What is the order of these markers on the circular chromosome of the original F^+?

(1) M, T, Q, W, D, P, X, A, C, N, B, M

(2) Q, X, A, C, W, D, M, T, P, N, B, Q

(3) Q, W, D, M, T, P, X, A, C, N, B, Q

(4) B, Q, W, C, N, D, M, T, P, X, A, B

45. In an Hfr X F^- cross, leu^+ enters as the first marker, but the order of the other markers is unknown. If the Hfr is wild-type and the F^- is auxotrophic for each marker in question, what is the order of the markers in a cross where leu^+ recombinants are selected if 27 percent are ile^+, 13 percent are mal^+, 82 percent are thr^+, and 1 percent are trp^+?

(1) *trp, mal, ile, thr, leu*

(2) *leu, thr, ile, mal, trp*

(3) *leu, mal, thr, ile, trp*

(4) *trp, mal, ile, trp, thr*

Question 46 and 47

In Neurospora, a cross is made between ab^+ and a^+ b individuals. The following one hundred ordered tetrads are obtained

Spores	I	II	III	IV	V	VI	VII	VIII
1, 2	a^+b	a^+b	a^+b	a^+b^+	a^+b^+	a^+b	a^+b	ab^+
3, 4	a^+b	a^+b^+	a^+b^+	a^+b	a^+b	ab^+	ab^+	a^+b
5, 6	ab^+	ab	ab^+	ab	ab^+	a^+b	ab^+	a^+b
7, 8	ab^+	ab^+	ab	ab^+	ab	ab^+	a^+b	ab^+
	85	2	3	2	3	3	1	1

46. Are genes a and b linked? How do you know?

(1) Yes; PD >> NPD

(2) No; PD >> NPD

(3) Yes; NPD>>PD

(4) No; Tetratype>PD

47. Calculate the gene-to-centromere distances for a and b

(1) 2.5 and 7.5 m.u

(2) 7.5 and 2.5 m.u

(3) 5.0 and 2.5 m.u

(4) 5.0 and 15 m.u

Question 48 and 49

The following three recessive markers are known in lab mice: h, hotfoot; o, obese; and wa, waved. A trihybrid of unknown origin is test crossed, producing the following offspring

hotfoot, obese, waved	357
hotfoot, obese	74
waved	66
obese	79
wild-type	343
hotfoot, waved	61
obese, waved	11
hotfoot	9
Total	**1,000**

48. If the genes are linked, determine the relative order & the map distances between them

(1) obese-waved-hotfoot

(2) waved-hotfoot –obese

(3) waved-obese-hotfoot

(4) Genes are not linked

49. Caculate crossover interference?

(1) 20 (2) 25.6

(3) 0.22 (4) 0.88

50. A variety of squash has fruits that weigh about 5 pounds each. In a second variety, the average weight is 2 pounds. When the two varieties are crossed, the F1 produce fruit with an average weight of 3.5 pounds. When two of these are crossed, their offspring produce a range of fruit weights, from 2 to 5 pounds. Of two hundred offspring, three produce fruits weighing about 5 pounds and three produce fruits about 2 pounds in weight. Approximately how many allelic pairs are involved in the weight difference between the varieties, and approximately how much does each effective gene contribute to the weight?

(1) 1 (2) 2

(3) 3 (4)4

Question 51-52

Gaucher disease (type I) is due to a defect in a gene that encodes a protein called acid β -D-glucosidase. This enzyme plays a role in carbohydrate metabolism within the lysosome. The gene is located on the long arm of chromosome 1. Persons who inherit two defective copies of this gene exhibit Gaucher disease. The symptoms include an enlarged spleen, bone lesions, and changes in skin pigmentation. Let's suppose a phenotypically normal woman, whose father had Gaucher disease, has a child with a phenotypically normal man, whose mother had Gaucher disease

51. What is the probability that this child will have the disease?

(1) 25 % (2) 50 %

(3) 75 % (4) 100 %

52. What is the probability that this child will have two normal copies of this gene?

(1) 25 % (2) 50 %

(3) 75 % (4) 100 %

53. The following mouse-human hybrid cell lines have the human chromosomes as shown in the first table and the human enzymes as shown in the second table

Table 1. Human chromosome

Clone	3	7	9	11	15	18	20
X	-	-	-	+	+	+	+
Y	+	+	-	+	-	+	-
Z	-	+	+	-	-	+	+

Table 2

	Clone		
	X	Y	Z
Glutathione reductase	+	+	-
Malate dehydrogenase	+	-	-
Galactokinase	-	+	+

Which chromosome is the carrier of galactokinase?

(1) 3 (2) 7
(3) 11 (4) 18

Question 54-55

You have a Drosophila line that is homozygous for autosomal recessive alleles a, band c, linked in that order. You cross females of this line with males homozygous for the corresponding wild-type alleles. You then cross the Fl heterozygous males with their heterozygous sisters. You obtain the following F_2 phenotypes

364	**$a^+b^+c^+$**
365	**a b c**
87	**b c^+**
84	**$^+b^+c$**
47	**a b^+c^+**
44	**a^+b c**
5	**a b^+c**
4	**a^+b c^+**

54. What is the recombinant frequency between a and b and between b and c?

(1) 9.1 m.u. and 17.1 m.u
(2) 17.1 m.u and 9.1 m.u
(3) 10 m.u. and 18 m.u
(4) 5 m.u. and 9 m.u

55. What is the coefficient coincidence in the above question?

(1) 0.18 (2) 0.5
(3) 0.36 (4) 0.68

56. Hemophilia is a recessive X-linked trait in humans. If a heterozygous woman has children with a normal man, what are the odds of the four unaffected offspring in a row?

(1) 1/4 (2) ¾
(3) 81/256 (4) ½

57. In 1941, Chaplin met a young actress named Joan Barry, with whom he had an affair. The affair ended in February 1942 but, 20 months later, Barry gave birth to a baby girl and claimed that Chaplin was the father. Barry then sued for child support. At this time, blood typing had just come into widespread use, and Chaplin's attorneys had Chaplin, Barry, and the child blood typed. Barry had blood type A, her child had blood type B, and Chaplin had blood type O. Could Chaplin have been the father of Barry's child?

(1) Yes (2) No
(3) May be (4) Data not sufficient

58. The presence of a beard on some goats is determined by an autosomal gene (B) that is dominant in males and recessive in females. In males, a single allele is required for the expression of this trait: both the homozygote (BB) and the heterozygote (Bb) have beards, whereas the bb male is beardless. In contrast, females require two alleles in order for this trait to be expressed: the homoygote BB has a beard, whereas the heterozygote (Bb) and the other homozygote (bb) are beardless. The trait for beard is

(1) Sex linked
(2) Sex influenced
(3) Sex limited
(4) All of the above

59. You set up an experiment in which you breed two populations of true-breeding pea plants. The first true-breeding population has yellow round seeds and the second has green wrinkled seeds. All of the F1 plants yield yellow round seeds. When you self fertilize the F1 the F2 generation yields a mixture of yellow round, yellow wrinkled, green round and green wrinkled seeds. What does this tell you about the alleles for seed color and shape?

(1) The recessive alleles are always expressed

(2) The alleles are on different chromosomes

(3) The two alleles for each character segregate during gamete production

(4) Both genes are on the same chromosome

60. John and Bipasha are planning a family, but since each has a brother who has sickle cell anemia, they are concerned that their childen may develop sickle-cell disease. Neither John, Bipasha nor their respective parents have the disease. They consult a genetic counselor who tells them

(1) there is very little chance that any of their children will have sickle-cell disease

(2) that one out of four of their children could be expected to have sickle cell-disease

(3) that its possible that none of their children will have the disease but blood tests on them both will be required to make sure

(4) that it is impossible to tell whether or not their children could have the disease

61. While on a field trip in the jungle you find a new species of mouse. You catch a pair and take them back to the lab. In mice, black coat color, B, is dominant to brown b, yet the female mouse gives rise to a large litter in which 9 of the offspring were black, 3 were brown and 4 were white. You conclude that

(1) a new mutation has occurred in the mice

(2) this is an example of polygenic inheritance

(3) there must be an epistatic interaction influencing coat color

(4) the coat color alleles are codominant

62. A genetic defect in humans results in the absence of sweat glands in the skin. Some men have this defect all over their bodies, but in women it is usually expressed in a peculiar way. A woman with this defect typically has small patches of skin with sweat glands and other patches where sweat glands are lacking. This pattern suggests the phenotypic effect of

(1) X-linked defect

(2) Random chromosome inactivation

(3) RNA splicing

(4) Genomic imprinting

63. In cucumbers, smooth fruit (t) is recessive to warty fruit (T) and glossy fruit (d) is recessive to dull fruit (D) Geneticists have determined that these two genes are 16 cM apart. Suppose we cross a plant homozygous for warty and dull fruit with a plant homozygous for smooth and glossy fruit and then carry out a testcross by using the F1

$$\frac{T\quad D}{t\quad d} \times \frac{t\quad d}{t\quad d}$$

What proportions of progeny will result $\frac{T\quad D}{t\quad d}$ from this testcross?

(1) 84 % (2) 16 %

(3) 8 % (4) 42 %

Questions 64-65

In Drosophila melanogaster, cinnabar eye (cn) and vestigial wing (vg) are simple recessive traits. A female, heterozygous for both genes, was crossed with a male with cinnabar eyes and vestigial wings. The offspring resulting from this cross are listed in the table below

Class	Phenotype	Number
I	Wild type	445
II	Cinnabar eye, wild-type wing	51
III	Wild-type eye, vestigial wing	49
IV	Cinnabar eye, vestigial wing	455

64. Recombinant phenotypes include which of the following classes?

(1) IV only (2) I and IV only

(3) II and III only (4) I, II, and III only

65. From the data shown, the distance between the cn and vg genes is approximately
(1) 500 map units
(2) 100 map units
(3) 50 map units
(4) 10 map units

66. If crossing-over were to occur in the male parent, the anticipated effect on phenotypic numbers would be
(1) a decrease in class I
(2) a decrease in class II
(3) a decrease in class III
(4) no significant difference

67. If cn and vg were unlinked, what percent of the offspring would be expected to have either cinnabar eyes or vestigial wings or both?
(1) 100% (2) 75%
(3) 50% (4) 25%

Question 68-70

Some individuals suffer from an autosomal recessive disorder known as "alpha$_1$-antitrypsin deficiency". The recessive homozygote for this disorder lacks the enzyme that ordinarily degrades trypsin. The genotype of the recessive disorder may be designates as aa. Normal individual are either homozygous dominant (AA) or heterozygous carriers (Aa)

Enymatic test ans colorimetric analysis reveals that a given individual possesses one of the three distinct level of alpha$_1$-antitrypsin enzyme activity. The data below show the level of trypsin inhibited per milliliter of serum (which contain the antitrypsin enzyme) in three different groups

GROUP I: The general population

GROUP II: Several families in which some member have alpha$_1$-antitrypsin deficiency

GROUP III: Patients with alpha$_1$-antitrypsin deficiency

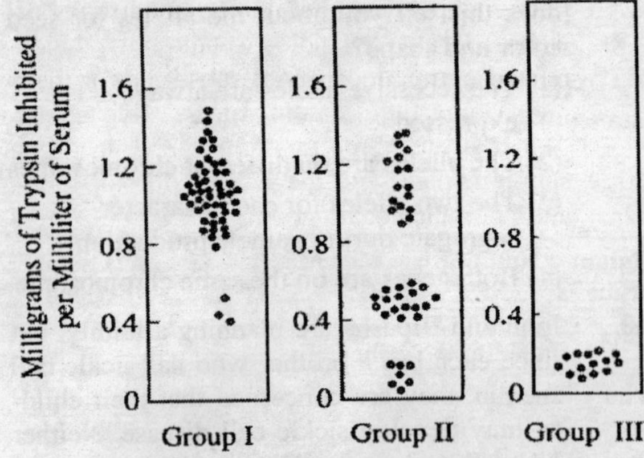

68. When the serum of an adult woman was tested colorometrically, al level of 0.5 mg of trypsin inhibited per ml of serum was found. This can be interpreted as indicating which of the following?
(1) She is homozygous dominant
(2) She is an asymptomatic heterozygous carrier
(3) She will exhibit alpha$_1$-antitrypsin deficient
(4) She will always transmit the abnormal recessive gene to her offsprings

69. If an offspring displays alpha$_1$-antitrypsin deficiency, then it can be reasonably predicted that the phenotypically normal father would have a value of trypsin inhibition of
(1) 0.24 $mg.mL^{-1}$
(2) 0.50 $mg.mL^{-1}$
(3) 1.20 $mg.mL^{-1}$
(4) 1.60 $mg.mL^{-1}$

70. The possible genotypes of a person possessing a 0.2 mg level of trypsin inhibited per ml of serum include
(1) AA only
(2) Aa only
(3) aa only
(4) AA and Aa

71. Four independent methionine auxotrophs of Neurospora were studied to determine which related compounds might substitute for their methionine requirement. In the table below "+" indicates growth on minimal medium supplemented with the indicated compound, and "-" indicates no growth

Mutant strain	Nothing added (minimal medium)	O-acetyl homo-erine	Methio-ine	Homo-ysteine	Cystat-ione
Wild type	+	+	+	+	+
Mutant A	–	+	+	+	+
Mutant B	–	–	+	+	+
Mutant C	–	–	+	+	–
Mutant D	–	–	+	–	–

Based on the results, give the order of these four compounds in the biosynthetic pathway of methionine?

(1) Homocysteine→O-Acetyl homoserine→ Cystathione → Methionine

(2) Cystathione→Homocysteine→ O-Acetyl homoserine → Methionine

(3) Homocysteine→ Cystathione→ O-Acetyl homoserine → Methionine

(4) O-Acetyl homoserine→Cystathione→Homocysteine → Methionine

72. The pedigree chart below for the ability to taste the chemical PTC (phenylthio-arbamide)

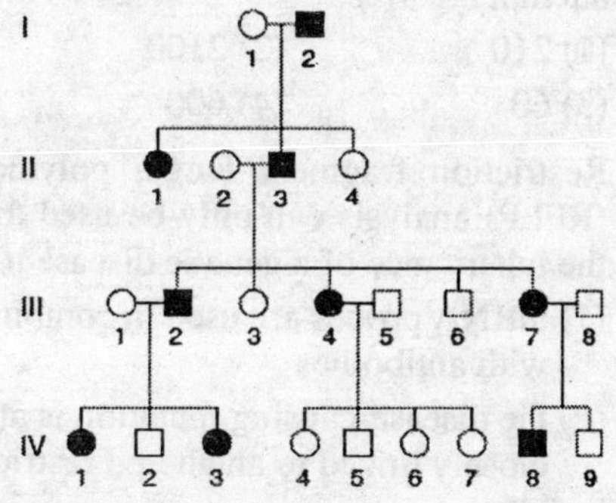

If trait is completely penetrate than what is probable mode of inheritance?

(1) Autosomal dominant trait

(2) Autosomal recessive trait

(3) X-linked dominant trait

(4) X-linked recessive trait

73. Consider the following crosses involving Red (wild-type) and white eye colour true breeding *Drosophila*

	Cross	F1 progeny	F2 progeny
Cross 1	red female X white male	All red	Red : white:: 3:1
Cross 2	White female X red male	All males: white All females: red	?

If F_2 offspring are produced in cross 2, which one of the following outcome is expected?

(1) 1 white males, 1 red females, 1 white males

(2) 2 red males, 1 red females, 1 white females

(3) 1 white males, 1 red males, 1 red females, 1 white females

(4) 1 red males, 1 red females, 1 red males, 1 white females

74. A rare human disease afflicted a family as shown in the accompanying pedigree

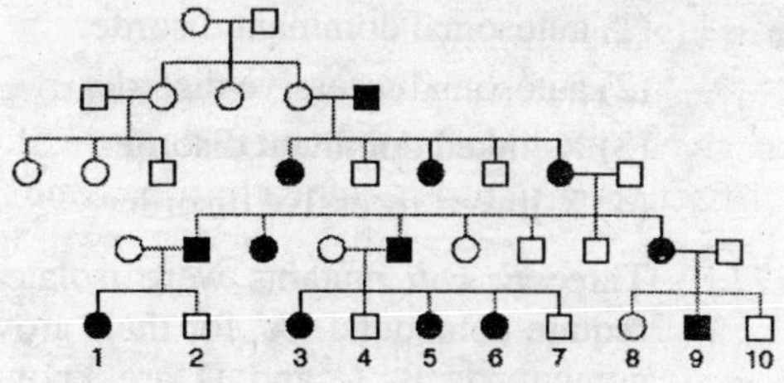

Deduce the most likely mode of inheritance

(1) Autosomal dominant trait

(2) Autosomal recessive trait

(3) X-linked dominant trait

(4) X-linked recessive trait

75. Affected individuals from the pedigree given below are suffering from albinism, an autosomal recessive disease. Identify the confirmed carrier individuals in this pedigree assuming that the members coming from outside the family are homozygous for the dominant allele

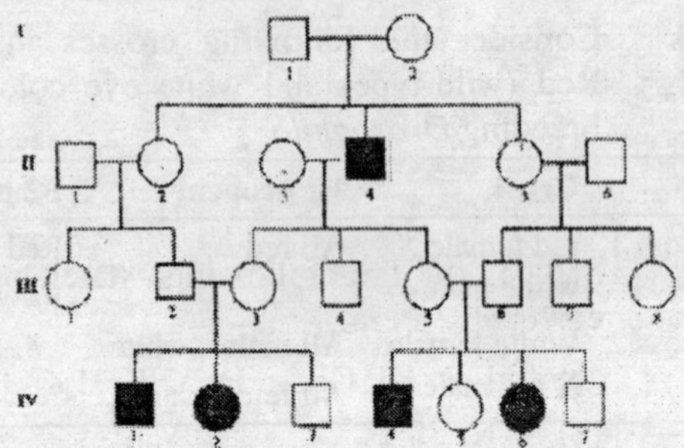

(1) III-2, III-3, III-5, III-6, II-1, II-3 and II-6

(2) III-2, III-3, III-5, III-6, II-2, II-4, I-1 and I-2

(3) III-2, 1II-3, III-5, III-6, II-2, II-4 and II-5

(4) III-I, III-4, III-7, II-2, II-4 and II-5

76. In a family consisting of 4 sons and 3 daughters, 3 sons are colour blind. All the daughters and both the parents are phenotypically normal. One of the daughters marries a normal man and has 2 sons and 1 daughter. Of them, 1 son and 1 daughter are normal. From this data, you conclude that colour blindness is

(1) autosomal dominant disorder

(2) autosomal recessive disorder

(3) X-linked dominant disorder

(4) X-linked recessive disorder

77. Three *E. coli* mutants were isolated which require compound 'A' for their growth. The compounds B, C and D are known to be involved in biosynthetic pathway to A. In order to determine pathway, the mutants were grown in a minimal medium supplemented with one of the compounds, A to D. The results obtained are summarized below

Mutant	Medium supplemented with compound			
	A	B	C	D
1	0	0	+	0
2	+	0	+	0
3	+	+	+	0

'+' = Growth on medium

'0' = No growth

Which of the following equation represents the biosynthetic pathway of A?

(1) D→B→A→C

(2) C→A→B→D

(3) B→A→C→D

(4) A→C→D→B

78. The ABO blood type in human is under the control of autosomal multiple alleles. Colour blindness is recessive X-linked trait. A male with a blood type A and normal vision marries a female who also has blood type A and normal vision. The couple's first child is a male who is colour blind and has a blood group O. What is the probability that their next female child has normal vision and a blood group O?

(1) ¼ (2) 3/4

(3) 1/8 (4) 1

79. The probability of passing an exam is 0.5, 0.6 and 0.7 in first, second and third attempt respectively. How many students out of 10000 are expected fail even after third attempt?

(1) 210 (2) 2100

(3) 60 (4) 600

80. Restriction fragment length polymorphism (RFLP) analysis can only be used to follow the inheritance of a genetic disease if

(1) mRNA probes are used in combination with antibodies

(2) the disease-causing mutation is at or closely linked to an altered restriction site

(3) proteins of mutated and normal genes migrate differently upon gel electrophoresis

(4) mutations are outside of restriction sites so that cleaving still occurs

81. Genes *a, b,* and *c* are widely spaced in the bacterial genome. Transducing phage from an *a+ b+ c+* bacterium were used to infect a culture of *a– b– c–* cells, and *b+* transductants were selected. Which of the following best describes the predicted genotypes of these transductants?

(1) Mostly *a– b+ c–*

(2) Mostly *a– b+ c+*

(3) Mostly *a+ b+ c+*

(4) Mostly *a+ b+ c–*

82. The uppermost figure below shows the locations of four genes on the genetic map of an organism; the lower figure shows the locations of the same four genes on a physical map derived from the nucleotide sequence of the DNA of that organism. The maps are not identical because

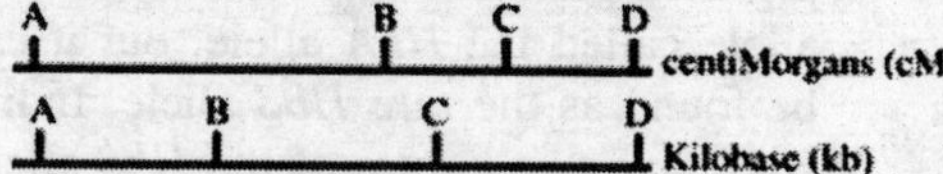

(1) there is no relationship between the position of genes in a genetic map and their positions on the DNA

(2) recombination frequencies per kb of DNA are not uniform throughout a chromosome

(3) the farther apart two genes are, the more likely they are to recombine

(4) the closer two genes are, the more likely they are to recombine

83. Consider the following steps in homologous recombination in *E. coli*

I. Resolution of Holliday junctions

II. Pairing homologous DNAs and strand invasion

III. Processing DNA breaks to generate single strands for invasion

IV. Assembly of strand exchange proteins

V. Holliday junction recognition and branch migration

Match the above steps with following proteins

A. RecA protein

B. RecBCD

C. RecBCD and RecFOR

D. RuvAB complex

E. RuvC

Which of the following statements are correct?

(1) A: I, B: III, C: IV, D: V, E: II

(2) A: II, B: III, C: IV, D: V, E: I

(3) A: II, B: I, C: IV, D: V, E: III

(4) A: II, B: IV, C: III, D: V, E: I

84. You are provided with an *E. coli* strain auxotrophic for the amino acids Trp, Leu and Val. To select for double revertants in Trp and Leu, which amino acid(s) would you include in the growth medium?

(1) Trp (2) Leu

(3) Val (4) Trp, Leu, Val

85. *E. coli* cells were simultaneously infected by two rII bacteriophage mutants. From the progeny obtained after lysis of the *E. coli* cells it was observed that some of the bacteiophages showed a wild type phenotype. These were obtained at extremely low frequency. This is due to

(1) complementation of the two mutations

(2) recombination between the two mutant chromosomes

(3) transposition of the mutation

(4) incomplete penetrance

86. loxP sites are specific 34-bp sequences consists of an 8-bp core sequence where recombination takes place and two flanking 13-bp inverted repeats. The *Cro* recom-inase is an enzyme that catalyzes recombination between two loxP sites. But what will happen to gene in question in between loxP sites depends upon orientation of two loxP sites. Study following diagram and state what will happen with gene respectively

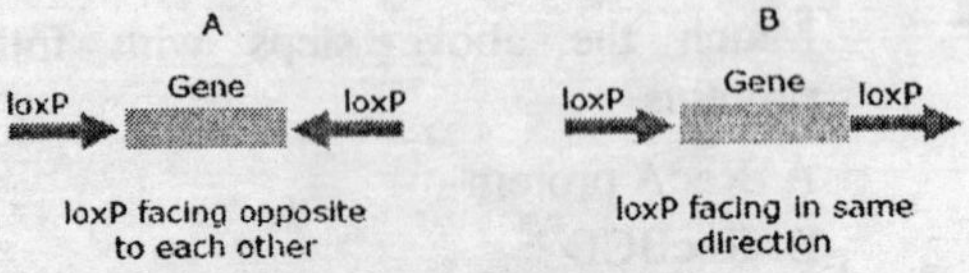

(1) A = inversion, B = deletion
(2) A = deletion, B = inversion
(3) A = translocation, B = inversion
(4) A = inversion, B = translocation

87. Molecular analysis is performed on the three copies of chromosome 21 in a child with Down's syndrome using markers of DNA polymorphism for which both parents are heterozygous for different alleles. Two of the chromosomes (#21) have the same mother's alleles. Based on this information, when did the nondisjunction event most likely occur?

(1) Maternal meiosis I
(2) Maternal meiosis II
(3) Paternal meiosis I
(4) Paternal meosis II

88. Base analogs are compounds that resemble the natural bases found in DNA and RNA but are not normally found in those macromolecules. Base analogs can replace their normal counterparts in DNA during in vitro DNA synthesis. Four base analogs were studied for their effects on in vitro DNA synthesis using *E. coli* DNA polymerase. The results were as follows, with the amounts of DNA Synthesized expressed as percentage of the DNA synthesized from normal base only,

Analog	Normal Base Substituted by the Analog			
	A	T	C	G
A	0	0	0	25
B	0	54	0	0
C	0	0	100	0
D	0	97	0	0

Consider the following inferences

A. A, B, C and D are analog for Adenine; A and C are analogs for Thymine; A, B and D are analog for cytosine and B, C, D are analog for Guanine
B. None is analog for Adenine; B and D are analogs for Thymine; C is analog for cytocine and A is analog for Guanine
C. A, B are analog for Adenine; A and C are analogs for Thymine; A, B are analog for cytocine & B, C are analog for Guanine
D. A, B, C & D are analog for Adenine; B & D are analogs for Thymine; C is analog for cytosine & A is analog for Guanine

Chose correct result(s) from above

(1) A only (2) B only
(3) A and C (4) D only

89. A human gene called the β-globin gene encodes a polypeptide that functions as a subunit of the protein known as hemoglobin. Hemoglobin is found within red blood cells; it carries oxygen. In human populations, the β-globin gene can be found as the common allele called the *HbA* allele, but it can also be found as the rare *HbS* allele. Individuals who have two copies of the *HbS* allele have the disease called sickle-cell disease. Consider the following statements

A. The *HbS* allele encodes a polypeptide that functions slightly differently from the polypeptide encoded by the *HbA* allele
B. If an individual has single copy of the *HbS* allele, that person's red blood cells form a sickle shape
C. Individuals who have two copies of the *HbA* allele do not have sickle-cell disease, but they are resistant to malaria. People who have one *HbA* allele and one *HbS* allele have sickle cell disease, and they are not resistant to malaria. People who have one copy of the *HbS* allele have sickle-cell anemia, and this disease may significantly shorten their lives

D. Individuals with sickle-cell disease have anemia because their red blood cells are easily destroyed by the body

Which of the following statement is correct?

(1) A and B (2) B and C
(3) C and D (4) A and D

90. A heterozygous pea plant that is tall with yellow seeds, *TtYy,* is allowed to self-fertilize. What is the probability that an offspring will be either tall with yellow seeds, tall with green seeds, or dwarf with yellow seeds?

(1) 25 % (2) 50 %
(3) 75 % (4) 94 %

91. For the following pedigrees, what you think is the most likely inheritance pattern. Filled (black) symbols indicate affected individuals

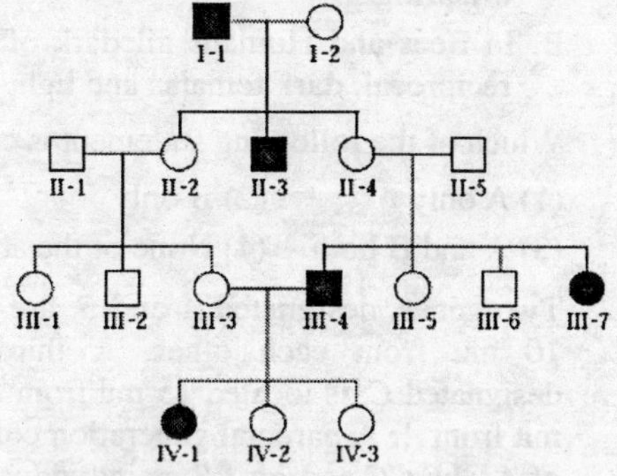

(1) X-linked dominant
(2) Autosomal dominant
(3) X-linked recessive
(4) Autosomal recessive

92. You are cloning Gene X from *S. cerevisiae* into a plasmid. You want to include the sequence 1 kb upstream and downstream from the open reading frame to try to get the gene to express under its own promoter. For this particular region of the genome, there are no other expected AUG codons upstream of the open reading frame for Gene X. To ensure that the protein is expressed, you try to complement the deletion strain for Gene X with your plasmid version of GeneX, but it is not complementing (not returning to the wild-type phenotype) You are not worried about splicing issues for *S. cerevisiae*. You suspect that you have a mutation in your cloned Gene X. The sequencing facility is down this week, but you have access to an in vitro translation system in your lab. You translate the wild-type Gene X and the cloned gene X using the in vitro translation system. You separate your products using SDS-PAGE and stain with Coomassie Brilliant Blue (data below) You expect a small protein, so you use a low molecular weight maker

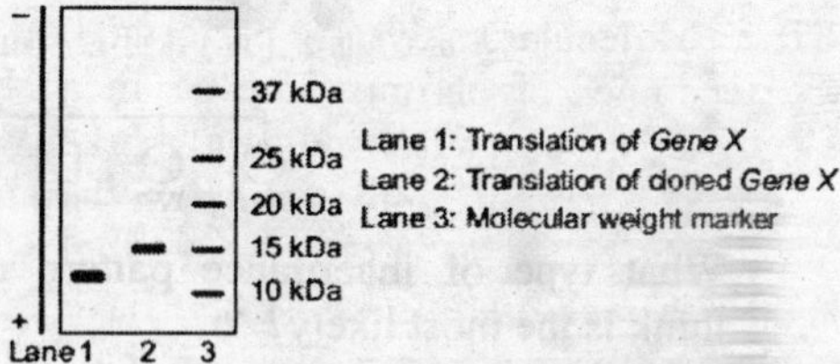

On basis of above information our cloned gene X has which type mutation?

(1) missense (2) nonsense
(3) frameshift (4) No mutation

93. Identical twins are produced from the same sperm and egg (which splits after the first mitotic division), whereas fraternal twins are produced from separate sperm and separate egg cells. If two parents with brown eyes (a dominant trait) produce one twin boy with blue eyes, what are the following probabilities, if the other twin is fraternal, he/she will have blue eyes

(1) 25 % (2) 50 %
(3) 75 % (4) 100 %

94. In cocker spaniels, solid coat color is dominant over spotted coat color. If two heterozygous dogs were crossed to each other, what would be the probability of the following combinations of offspring - In a litter of five pups, four with solid fur and one with spotted fur?

(1) 25 % (2) 39.6 %
(3) 50 % (4) 76.3 %

95. Marfan syndrome is a rare inherited human disorder characterized by unusually long limbs and digits plus defects in the heart (especially the aorta) and the eyes. Following is a pedigree for this disorder. Affected individuals are shown with filled (black) symbols

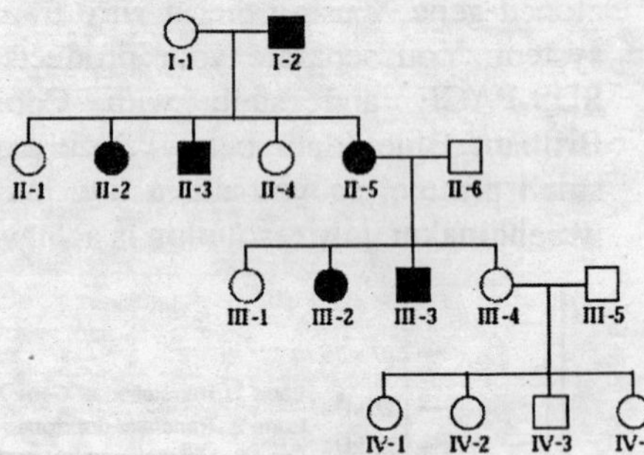

What type of inheritance pattern do you think is the most likely?

(1) Autosomal recessive
(2) Autosomal dominant
(3) X-linked recessive
(4) X- linked dominant

96. Honeybees are unusual in that male bees (drones) have only one copy of each gene, while female bees have two copies of their genes. That is because drones develop from eggs that have not been fertilized by sperm cells. In bees, the trait of long wings is dominant over short wings, and the trait of black eyes is dominant over white eyes. If a drone with short wings and black eyes was mated to a queen bee that is heterozygous for both genes, what are the predicted genotypes and phenotypes of male and female offspring?

(1) Male offspring will be *SB and sb,* and female offspring will be *SsBB, SsBb, ssBB,* and *ssBb*
(2) Male offspring will be *SB, Sb, sB,* and *sb,* and female offspring will be *SsBB, SsBb, ssBB,* and *ssBb*
(3) Male offspring will be *Sb* and *sB,* and female offspring will be *SsBB, SsBb, ssBB,* and *ssBb*
(4) Male offspring will be *SB, Sb, sB,* and *sb,* and female offspring will be *SsBb* and *ssBb*

97. Let's suppose that a gene affecting pigmentation is found on the X chromosome (in mammals or insects) or the Z chromosome (in birds) but not on the Y or W chromosome. It is found on an autosome in bees. This gene is found in two alleles, *D* (dark), which is dominant to *d* (light) What would be the phenotypic results of crosses between a true-breeding dark female and true-breeding light male, and the reciprocal crosses involving a true-breeding light female and true-breeding dark male, in the following species?

A. In birds and *Drosophila* Dark males and light females; reciprocal: all dark offspring
B. In Bees and Humans all dark offspring; reciprocal: dark females and light males

Which of the following statement is correct?

(1) A only
(2) B only
(3) A and B both
(4) None of the above

98. Two genes, designated *A* and *B*, are located 10 mu from each other. A third gene, designated *C*, is located 15 mu from *B* and 5 mu from *A*. A parental generation consisting of *AA bb CC* and *aa BB cc* individuals were crossed to each other. The F1 heterozygotes were then testcrossed to *aa bb cc* individuals. If we assume no double crossovers occur in this region, what percentage of offspring would you expect with the following genotypes?

A. *Aa Bb Cc*
B. *aa Bb Cc*
C. *Aa bb cc*

(1) A: 5 %, B: 5 %, C: 10 %
(2) A: 10 %, B: 5 %, C: 10 %
(3) A: 5 %, B: 10 %, C: 10 %
(4) A: 5 %, B: 2.5 %, C: 2.5 %

99. The pedigree shown here concerns a human disease known as familial hypercholesterolemia. This disorder is characterized by an elevation of serum cholesterol in the blood. Though relatively rare, this genetic abnormality can be a contributing factor to heart attacks. At the molecular level, this disease is caused by a defective gene that encodes a protein called the low-density lipoprotein receptor (LDLR) In the bloodstream, serum cholesterol is bound to a carrier protein known as low-density lipoprotein (LDL) LDL binds to LDLR so that cells can absorb cholesterol. When LDLR is defective, it becomes more difficult for the cells to absorb cholesterol. This explains why the levels of LDL blood cholesterol remain high. Based on the pedigree, what is the most likely pattern of inheritance of this disorder?

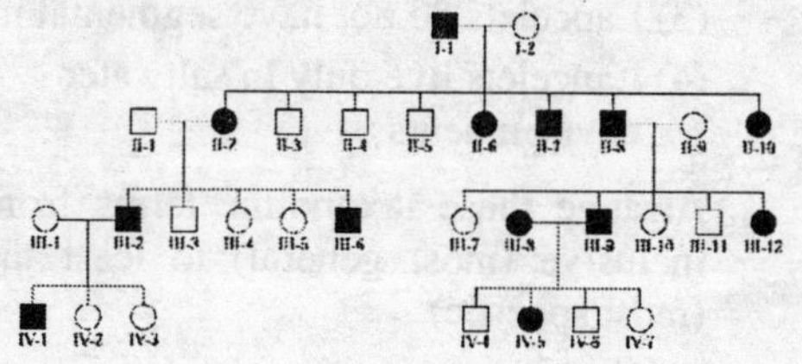

(1) Autosomal Dominant

(2) X-linked Dominant

(3) X- linked recessive

(4) Autosomal recessive

100. How the Holliday junctions present in a recombination intermediate are cleaved has a huge impact on the structure of the product DNA molecules. Products will either have the DNA flanking the site of recombination reassorted (in the splice/crossover products) or not (in the patch/non-crossover products) depending on how resolution is achieved

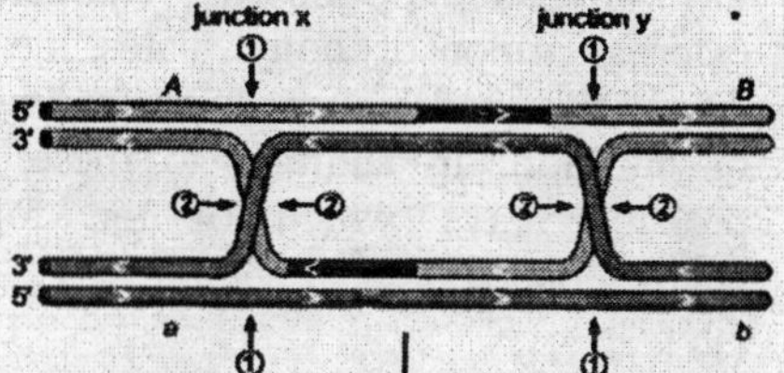

Which of the following combination can lead to formation of crossover products?

(1) 1 and 2 or 2 and 1

(2) 1 and 1 or 2 and 2

(3) 1 and 2 only

(4) 1 and 1 only

ANSWER SHEET
Unit 8: Genetics

1. 4	**2.** 3	**3.** 4	**4.** 4	**5.** 4	**6.** 3	**7.** 3	**8.** 4	**9.** 4	**10.** 2
11. 1	**12.** 4	**13.** 3	**14.** 3	**15.** 3	**16.** 1	**17.** 3	**18.** 1	**19.** 1	**20.** 1
21. 2	**22.** 2	**23.** 3	**24.** 1	**25.** 4	**26.** 3	**27.** 2	**28.** 2	**29.** 1	**30.** 1
31. 3	**32.** 4	**33.** 1	**34.** 1	**35.** 2	**36.** 3	**37.** 1	**38.** 1	**39.** 1	**40.** 4
41. 1	**42.** 3	**43.** 4	**44.** 3	**45.** 2	**46.** 1	**47.** 1	**48.** 2	**49.** 3	**50.** 3
51. 1	**52.** 1	**53.** 2	**54.** 3	**55.** 1	**56.** 3	**57.** 2	**58.** 2	**59.** 3	**60.** 3
61. 3	**62.** 2	**63.** 4	**64.** 3	**65.** 4	**66.** 4	**67.** 2	**68.** 2	**69.** 2	**70.** 3
71. 4	**72.** 1	**73.** 3	**74.** 3	**75.** 2	**76.** 4	**77.** 1	**78.** 1	**79.** 4	**80.** 2
81. 1	**82.** 2	**83.** 2	**84.** 3	**85.** 2	**86.** 1	**87.** 2	**88.** 2	**89.** 4	**90.** 4
91. 4	**92.** 3	**93.** 1	**94.** 2	**95.** 2	**96.** 2	**97.** 2	**98.** 4	**99.** 1	**100.** 1

UNIT-9

DIVERSITY AMONG LIFE FORMS

1. A new species of aquatic chordate is discovered that closely resembles an ancient form. It has the following characteristics: external armor of bony plates, no paired lateral fins, and a suspension-feeding mode of nutrition. In addition to these, it will probably have which of the following characteristics?

(1) legs
(2) no jaws
(3) an amniotic egg
(4) endothermy

2. How many of the following statements about craniates is (are) correct?

1. Craniates are more highly cephalized than are noncraniates
2. Craniates' genomic evolution includes duplication of clusters of genes that code for transcription factors
3. The craniate clade is synonymous with the vertebrate clade
4. Pharyngeal slits that play important roles in gas exchange originated in craniates
5. The two-chambered heart originated with the early craniates

(1) one of these (2) two of these
(3) three of these (4) four of these

3. A team of researchers has developed a poison that has proven effective against lamprey larvae in freshwater cultures. The poison is ingested and causes paralysis by detaching segmental muscles from the skeletal elements. The team wants to test the poison's effectiveness in streams feeding Lake Michigan, but one critic worries about potential effects on lancelets, which are similar to lampreys in many ways. Why is this concern misplaced?

(1) A chemical poisonous to lampreys could not also be toxic to organisms as ancestral as lancelets
(2) Lamprey larvae and lancelets have very different feeding mechanisms
(3) Lancelets do not have segmental muscles
(4) Lancelets live only in saltwater environments

4. Arrange these taxonomic terms from most inclusive (most general) to least inclusive (most specific)

1. lobe-fins
2. amphibians
3. gnathostomes
4. osteichthyans
5. tetrapods

(1) 4, 3, 1, 5, 2 (2) 4, 3, 2, 5, 1
(3) 4, 2, 3, 5, 1 (4) 3, 4, 1, 5, 2

5. During chordate evolution, what is the sequence (from earliest to most recent) in which the following structures arose?

1. amniotic egg
2. paired fins
3. jaws
4. swim bladder
5. four-chambered heart

(1) 2, 3, 4, 1, 5 (2) 3, 2, 4, 1, 5
(3) 3, 2, 1, 4, 5 (4) 2, 1, 4, 3, 5

6. Arrange the following taxonomic terms from most inclusive (most general) to least inclusive (most specific)

1. apes
2. hominins
3. *Homo*
4 anthropoids
5. primates

(1) 5, 1, 4, 2, 3 (2) 5, 4, 1, 2, 3
(3) 5, 4, 2, 1, 3 (4) 5, 2, 1, 4, 3

7. Fishes that have swim bladders can regulate their density and, thus, their buoyancy. There are two types of swim bladder: physostomus and physoclistus. The ancestral version is the physostomus version, in which the swim bladder is connected to the esophagus via a short tube. The fish fills this version by swimming to the surface, taking gulps of air, and directing them into the swim bladder. Air is removed from this version by "belching." The physoclistus version is more derived, and has lost its connection to the esophagus. Instead, gas enters and leaves the swim bladder via special circulatory mechanisms within the wall of the swim bladder. The presence of a swim bladder allows the typical ray-finned fish to stop swimming and still

(1) effectively circulate its blood
(2) be highly maneuverable
(3) not sink
(4) use its swim bladder as a respiratory organ

8. Rank the following fish, from most to least, in terms of the amount of energy it must use to maintain its position (depth) in the water column over the long term

1. physoclistus fish
2. physostomus fish
3. chondrichthyan fish

(1) 1, 2, 3 (2) 2, 3, 1
(3) 2, 1, 3 (4) 3, 2, 1

9. Regarding its position in the water column, the same thing that happens to a shark when it stops swimming also happens to a

(1) physostomus fish when it gulps air
(2) physoclistus fish when it moves gas from the blood into the water
(3) physoclistus fish when it moves gas from the swim bladder into the blood
(4) physoclistus fish when it moves gas from the blood into the swim bladder

10. In coelacanths, a swim bladder is present, but the swim bladder is full of adipose tissue (fat), which is there on a fairly permanent basis. If such a swim bladder is used by coelacanths to affect buoyancy, then it does so in much the same way as does the

(1) liver of a shark
(2) physoclistus swim bladder
(3) physostomus swim bladder
(4) lung of a lungfish

11. Which graph properly depicts the relationship between the amount of gas in the swim bladder and the density of the fish?

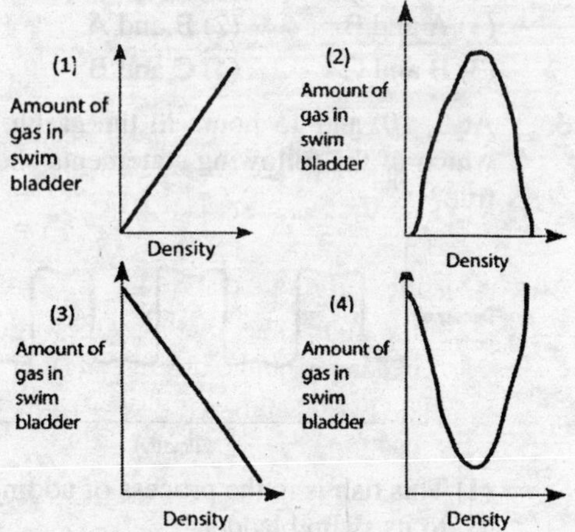

12. Which graph below best represents the way that density changes over time in a physoclistus fish and in a physostomus fish, respectively?

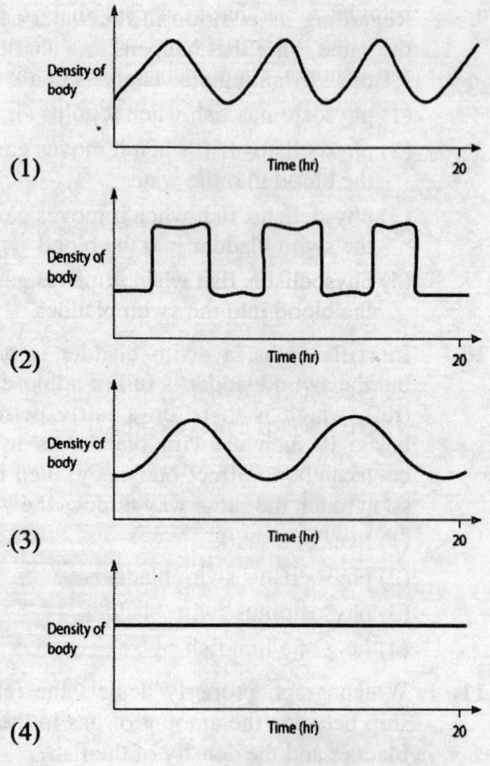

(1) A and B (2) B and A
(3) B and C (4) C and B

13. At 5, 10, and 15 hours in the graph below, which of the following statements should be true?

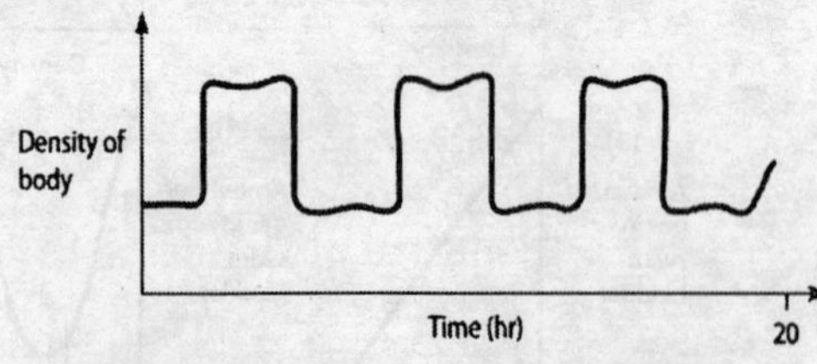

(1) This fish is in the process of adding gas to its swim bladder
(2) This fish is in the process of removing gas from its swim bladder
(3) This fish has a swim bladder that is relatively full of gas
(4) This fish has a swim bladder that contains relatively little gas

Q14-Q19. Refer to the phylogenetic tree shown in figure

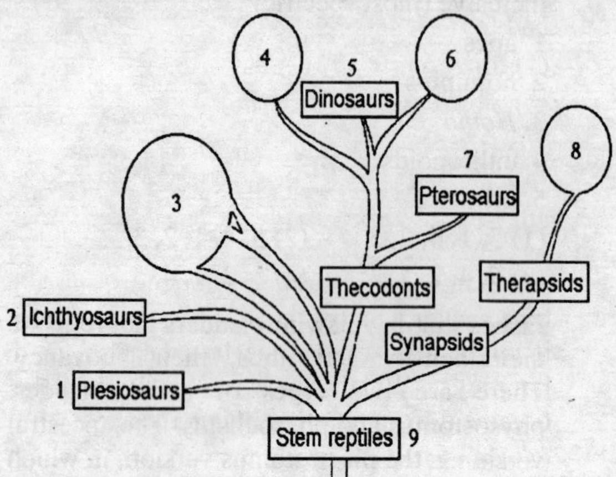

14. Which number represents the birds?
(1) 3
(2) 4
(3) 6
(4) 10, if it were at the end of a branch emerging from the dinosaurs

15. Which pair of numbers represents extinct reptiles that had returned to an aquatic life?
(1) 1 and 2 (2) 3 and 4
(3) 5 and 7 (4) 6 and 8

16. If circle 3 includes all of the extant reptiles that are ectotherms, then what do circles 4 and 6 represent?
(1) crocodiles and birds
(2) turtles and birds
(3) ratites and carinates
(4) There is not enough information to say

17. Which of these numbers represents ectotherms that were able to fly?
(1) 1 (2) 2
(3) 7 (4) 8

18. Whose DNA would have had the most sequence homologies with amphibian DNA?
(1) 5 (2) 6
(3) 7 (4) 9

19. In order for the four-chambered hearts of birds and mammals to be homologous, which other organisms would have to have had four-chambered hearts?

(1) dinosaurs (2) thecodonts
(3) plesiosaurs (4) stem reptiles

20. Amit catches a ray-finned fish from the ocean and notices that attached to its flank is an equally long, snakelike organism. The attached organism has no external segmentation, no scales, a round mouth surrounded by a sucker, and two small eyes. Amit thinks it might be a marine leech, a hagfish, or a lamprey. Which feature excludes the organism from possibly being a leech?

(1) Elongated shape
(2) Lack of scales
(3) Lack of external segmentation
(4) Round mouth

21. Amit detaches the snakelike organism from the fish and uses a knife to cut off its head. In doing so, its brain slides out onto the deck of the boat. Amit peers into the cut end of the head and notices that the brain had lain in a sort of panlike structure that only partially surrounded the brain. What is the structure Amit is observing, and what is it made of?

(1) Skull, made of bone
(2) Cranium, made of bone
(3) Cranium, made of cartilage
(4) Vertebral column, made of bone

22. While on an intersession course in tropical ecology, Kailash pulls a large, snakelike organism from a burrow (the class was granted a collecting permit) The 1-m-long organism has smooth skin, which appears to be segmented. It has two tiny eyes that are hard to see because they seem to be covered by skin. Kailash brings it back to the lab at the field station, where it is a source of puzzlement to the class. Kailash says that it is a giant oligochaete worm; Ram suggests it is a legless amphibian; Navish proposes it belongs to a snake species that is purely fossorial (lives in a burrow) The class decided to humanely euthanize the organism and subsequently dissect it. Having decided that it was probably not a reptile, two of their original hypotheses regarding its identity remained. Which of the following, if observed, should help them arrive at a conclusive answer?

(1) Presence of a closed circulatory system
(2) Presence of moist, highly vascularized skin
(3) Presence of lungs
(4) Presence of a nerve cord

23. The organism was found to have two lungs, but the left lung was much smaller than the right lung. Ram added that the herpetology instructor had said that in most snakes, the same condition exists. If the size difference between the lungs in this organism is *not* a shared ancestral characteristic with its occurrence in snakes, then its existence in this organism is explained as which of the following?

1. A result of convergent evolution
2. An example of homologous structures
3. A similar adaptation to a shared lifestyle or body plan
4. A result of having identical *Hox* genes
5. A homoplasy

(1) 3 only (2) 1 and 5
(3) 1, 3 and 5 (4) 2, 3 and 5

24. The adaptation of the body shape of snakes has resulted in one of their lungs becoming vestigial. Another adaptation (to a fossorial lifestyle) is snakes' absence of limbs. If the "mystery organism" has also become adapted to a fossorial lifestyle, though its ancestors moved about on the surface, then which structures should one expect to find upon dissecting the organism?

1. Reduced or absent pelvic and/or pectoral girdles
2. Metanephridia

3. Hydrostatic skeleton

(1) 1 only (2) 1 and 2

(3) 1 and 3 (4) 2 and 3

25. Which one of these, if found, should clear up any remaining doubt as to the identity of the organism?

(1) Vestigial pelvic girdle

(2) Blood vessels carrying oxygenated blood from both the skin and the functional lung to the heart

(3) Closed circulatory system

(4) Ability to produce toxins from glands located on the skin, or that empty into the mouth

26. Due to its system of nine air sacs connected to the lungs, the respiratory system of birds is arguably the most effective respiratory system of all air-breathers. Upon inhalation, air first flows into posterior air sacs, then into the lungs, and then into anterior air sacs on the way to being exhaled. Thus, there is one-way flow of air through the lungs, along thousands of tubules called parabronchi. If the inner lining of the air sacs is neither thin nor highly vascularized, then what can be inferred about the air sacs?

(1) They must not belong to the respiratory system

(2) They cannot be derived from endoderm

(3) They cannot be sites of gas exchange between air and blood

(4) They must obtain nutrition from some source other than the bloodstream

27. Birds generate a lot of heat, especially during flight. Yet the adipose tissue under their skin and the feathers atop their skin make it difficult to eliminate excess heat across the skin. Which of the following alternatives can absorb body heat and eliminate it from the bird most effectively?

(1) Air in the air sacs

(2) Lymph in the lymphatic vessels

(3) Blood in the vessels

(4) Blood in the heart

28. If *Archaeopteryx* had air sacs, then which of its features would have had the opposite effect on enabling *Archaeopteryx* to fly long distances from that provided by air sacs?

1. Teeth
2. Contour feathers
3. Wing claws
4. Long tail with many vertebrae

(1) 1 and 2 (2) 1 and 4

(3) 2 and 3 (4) 1, 3, and 4

(29-32) In the United States and Canada, bats use one of two strategies to survive winter: They either migrate south, or they hibernate. Recently, those that hibernate seem to have come under attack by a fungus, *Geomyces destructans* (G(4), an attack that is occurring from Missouri to southern Canada. Many infected bats have a delicate, white filamentous mat on their muzzles, which is referred to as white-nose syndrome (WNS) The fungus invades the bat tissues, causes discomfort, and awakens the bat from its hibernation. The bat fidgets and wastes calories, using up its stored fat. The bat then behaves abnormally, leaving its cave during daytime in winter to search for food. Their food, primarily insects, is scarce during the winter, and the bats ultimately starve to death. Since 2007, it is estimated that up to 1 million bats have perished from WNS

29. The Gd mat on the fur of the bats should be expected to consist of

(1) hyphae (2) haustoria

(3) arbuscules (4) yeasts

30. Gd is a cold-loving fungus. Thus, which characteristics of normal bat behavior can be most expected to favor the growth of this fungus?

(1) Southward migration during the winter

(2) Communal roosting in tightly packed clusters during hibernation

(3) Reliance on fat reserves for calories during hibernation

(4) Much-reduced metabolic rate during hibernation

31. Gd is a cold-loving fungus. Thus, which characteristics of abnormal bat behavior can be most expected to favor the growth of this fungus?

A. Shifting roosting location to the mouth of the cave during winter

B. Searching for food during winter

C. Searching for food during the day

(1) A and B (2) B and C

(3) A and C (4) A, B and C

32. Gd is a cold-loving fungus. Thus, which characteristics of normal bat behavior can be most expected to favor the spread of this fungus to uninfected bats?

(1) Southward migration during the winter

(2) Communal roosting in tightly packed clusters during hibernation

(3) Reliance on fat reserves for calories during hibernation

(4) Much-reduced metabolic rate during hibernation

33. If organisms A, B, and C belong to the same class but to different orders and if organisms D, E, and F belong to the same order but to different families, which of the following pairs of organisms would be expected to show the greatest degree of structural homology?

(1) A and B (2) A and C

(3) B and D (4) D and F

34. The common ancestors of birds and mammals were very early (stem) reptiles, which almost certainly possessed three-chambered hearts (two atria, one ventricle) Birds and mammals, however, are alike in having four-chambered hearts (two atria, two ventricles) The four-chambered hearts of birds and mammals are best described as

(1) structural homologies

(2) vestiges

(3) homoplasies

(4) the result of shared ancestry

35. The lakes of northern Minnesota are home to many similar species of damselflies of the genus *Enallagma* that have apparently undergone speciation from ancestral stock since the last glacial retreat about 10,000 years ago. Sequencing which of the following would probably be most useful in sorting out evolutionary relationships among these closely related species?

(1) Nuclear DNA

(2) Mitochondrial DNA

(3) Small nuclear RNA

(4) Ribosomal RNA

36. The reason that paralogous genes can diverge from each other within the same gene pool, whereas orthologous genes diverge only after gene pools are isolated from each other, is that

(1) having multiple copies of genes is essential for the occurrence of sympatric speciation in the wild

(2) paralogous genes can occur only in diploid species; thus, they are absent from most prokaryotes

(3) polyploidy is a necessary precondition for the occurrence of sympatric speciation in the wild

(4) having an extra copy of a gene permits modifications to the copy without loss of the original gene product

37. A large proportion of archaeans are extremophiles, so called because they inhabit extreme environments with high acidity, salinity, and/or temperature. Such environments are thought to have been much more common on the primitive Earth. Thus, modern extremophiles survive only in places that their ancestors became adapted to long ago. Which of the following is, consequently, a valid statement about modern extremophiles, assuming that their habitats have remained relatively unchanged?

(1) Among themselves, they should share relatively few ancestral traits, especially

those that enabled ancestral forms to adapt to extreme conditions

(2) On a phylogenetic tree whose branch lengths are proportional to the amount of genetic change, the branches of the extremophiles should be shorter than the non-extremophilic archaeans

(3) They should contain genes that originated in eukaryotes that are the hosts for numerous species of bacteria

(4) They should currently be undergoing a high level of horizontal gene transfer with non-extremophilic archaeans

Use Figure to answer the following questions **(38-40)**

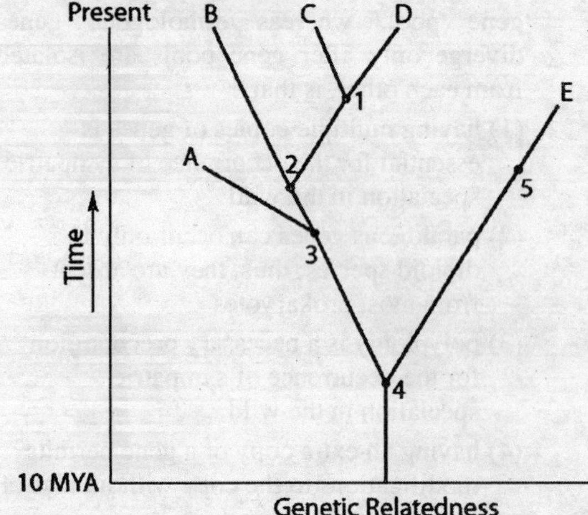

38. Which extinct species should be the best candidate to serve as the outgroup for the clade whose common ancestor occurs at position 2 in Figure?

(1) A (2) B
(3) C (4) D

39. If Figure is an accurate depiction of relatedness, then which of the following should be correct?

1. The entire tree is based on maximum parsimony
2. If all species depicted here make up a taxon, this taxon is monophyletic
3. The last common ancestor of species B and C occurred more recently than the last common ancestor of species D and E
4. Species A is the direct ancestor of both species B and species C

(1) 1 and 3
(2) 1, 3 and 4
(3) 2, 3, and 4
(4) 1, 2, and 3

40. Match the following animal with their attribute?

Animal	Attributes
A-Rotifer	(i) Naupliuas larva stage
B-Sea anemone	(ii) Radial symetry
C-Bamacle	(iii) Pseudocoelomate body cavity
D-Sea urchin	(iv) Water vascular system

(1) A- (iii), B-(ii), C- (i), D- (iv)
(2) A-(ii), B- (iii), C- (i), D- (iv)
(3) A- (iii), B- (iv), C -(i), D- (ii)
(4) A- (iv), B- (Hi), C- (ii), D- (i)

(41-47) Traditionally, whales and hippopotamuses have been classified in different orders, the Cetacea and the Artiodactyla, respectively. Recent molecular evidence, however, indicates that the whales' closest living relatives are the hippos. This has caused some zoologists to lump the two orders together into a single clade, the Cetartiodactyla. There is no consensus on whether the Cetartiodactyla should be accorded order status or superorder status. This is because it remains unclear whether the whale lineage diverged from the lineage leading to the hippos before or after the other members of the order Artiodactyla (pigs, camels, etc.) diverged (see figure)

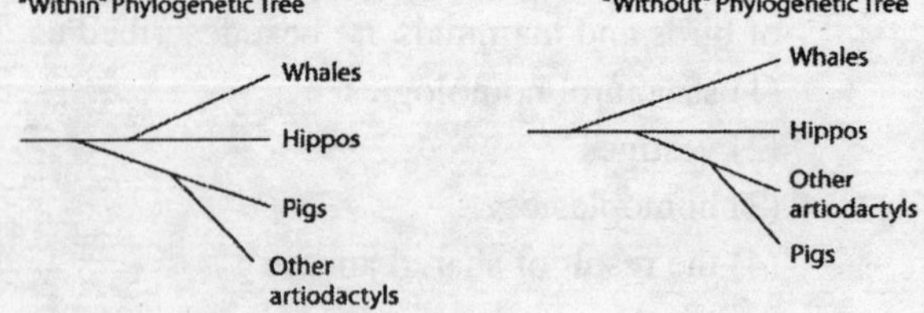

Figure contrasts the "Within the artiodactyls" origin of the whale lineage with the "Without the artiodactyls" origin of the whale lineage

41. Placing whales and hippos in the same clade means
(1) that these organisms are phenotypically more similar to each other than to any others shown on the trees in figure
(2) that their morphological similarities are probably homoplasies
(3) that they had a common ancestor
(4) that all three of the responses are correct

42. If it turns out that the whale lineage diverged from the lineage leading to hippos after the divergence of the lineage leading to the pigs and other artiodactyls, and if the whales continue to be classified in the order Cetacea, then what becomes true of the order Artiodactyla?
(1) It becomes monophyletic
(2) It becomes paraphyletic
(3) It becomes polyphyletic
(4) It is incorporated into the order Cetacea

43. If it turns out that the whale lineage diverged from the lineage leading to hippos after the divergence of the lineage leading to the pigs and other artiodactyls, and if the whales continue to be classified in the order Cetacea, then what becomes true of the taxon Cetartiodactyla?
(1) It should be considered as one monophyletic order
(2) It should be considered a superorder that consists of two monophyletic orders
(3) It should be established as a paraphyletic order
(4) It should be thrown out or modified by taxonomists if classification is to reflect evolutionary history

44. One morphological feature of modern cetaceans is a vestigial pelvic girdle. If it is determined that cetacean lineage diverged from the artiodactyls' lineage after the divergence of pigs and other artiodactyla, then what should be true of the vestigial pelvic girdle of cetaceans?
(1) It should be considered a shared ancestral character of the cetartiodactyls
(2) It should be considered a shared derived character of the cetartiodactyls
(3) It should be considered a shared ancestral character of the cetaceans
(4) It should be considered a shared derived character of the cetaceans

45. If cetaceans are determined to have diverged from the lineage leading to the artiodactyls *before* the divergence of lineages leading to the modern artiodactyls (including hippos), then the cetaceans can be considered
1. a sister order to the order Artiodactyla
2. an ingroup of the order Artiodactyla
3. the common ancestor of the order Artiodactyla
(1) 1 only (2) 3 only
(3) 1 and 2 (4) 1 and 3

46. It was once thought that cetaceans had evolved from an extinct group of mammals called the mesonychids. If, in the future, it is determined that some organisms currently classified as cetaceans did actually evolve from mesonychids, whereas other cetaceans evolved from artiodactyl stock, then what will be true of the order Cetacea?
(1) It will be paraphyletic
(2) It will be polyphyletic
(3) It will need to be thrown out or modified if classification is to reflect evolutionary history
(4) Two of the responses are correct

47. What can be properly inferred from figure above?
(1) In the "Without" tree, pigs are more distantly related to hippos than is depicted in the "Within" tree
(2) In the "Without" tree, pigs are more closely related to hippos than are whales

(3) In the "Within" tree, pigs are more closely related to whales than they are to hippos

(4) The "Without" tree is more consistent with molecular evidence than is the "Within" tree

	position 1									position 10										position 20
Species A	T	T	T	A	C	A	C	A	G	T	C	C	C	G	A	T	T	G	T	A
Species B	T	T	A	C	A	C	A	G	T	C	C	C	G	A	T	T	G	T	A	A
Species C	T	T	A	C	A	C	A	G	T	C	C	C	G	G	T	T	A	T	A	A
Species D	T	T	A	C	A	C	A	G	T	C	C	C	G	A	T	T	A	T	A	A
Species E	T	T	A	C	A	C	A	G	C	C	C	C	G	G	T	T	A	T	A	A

Figure for question 48. Morphologically, species A is very similar to four other species, B-E. Yet the nucleotide sequence deep within an intron in a gene shared by all five of these eukaryotic species is quite different in species A compared to that of the other four species when we study the nucleotides present at each position

48. If the sequence of species A in Figure differs from that of the other four species due to simple misalignment, then what should the computer software find when it compares the sequence of species A to those of the other four species?

(1) The nucleotide at position 1 should be different in species A, but the same in species B-E

(2) The nucleotide sequence of species A should have long sequences that are nearly identical to those of the other species, but offset in terms of position number

(3) The sequences of species B-E, though different from that of species A, should be identical to each other, without exception

(4) If the software compares the amino acid sequence of the actual protein product rather than the nucleotide sequence, then the amino acid sequences of species B-E should be similar to each other, but very different from that of species A

49. Which of the following items does not necessarily exist in a simple linear relationship with the number of gene-duplication events when placed as the label on the vertical axis of the following graph?

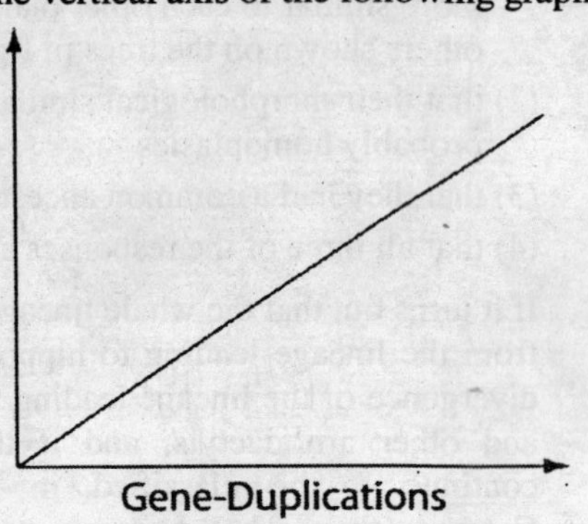

(1) Number of genes

(2) Number of DNA base pairs

(3) Genome size

(4) Phenotypic complexity

50. Following is a diagram showing phylogenetic relationship among plant taxa

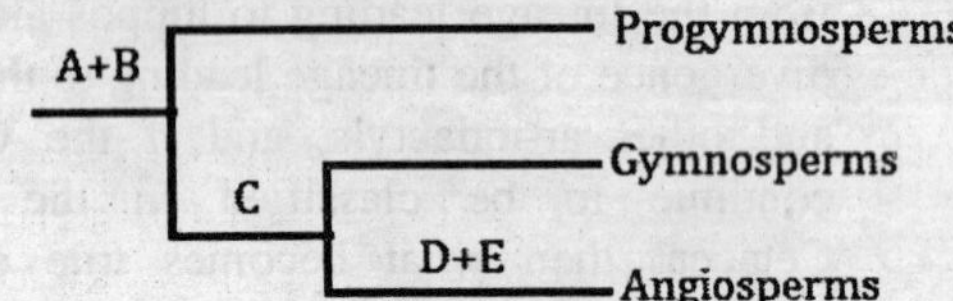

In the above diagram, symbols A, B, C, D and E represent

(1) A=Cambium, B= Carpel, C = Euphyll D= Seed (ovule), E= Double Fertilization

(2) A=Euphyll, B= Cambium, C = Seed (ovule) D= Carpel, E= Double Fertilization

(3) A= Carpel, B= Euphyll, C = Cambium D= Double Fertilization E= Seed (ovule)

(4) A= Carpel, B= Double Fertilization, C = Euphyll D=Seed (ovule), E= Cambium

Question 51-53. The following questions refer to this phylogenetic tree, depicting the origins of life and of the three domains. Horizontal lines indicate instances of gene or genome transfer

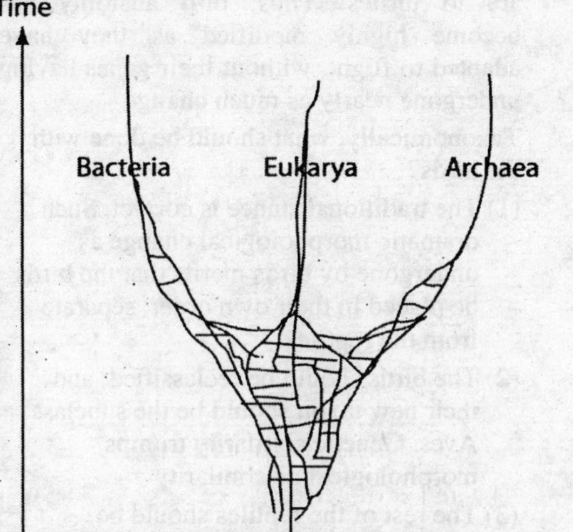

Figure A possible phylogenetic tree for the three domains of life

51. If the early history of life on Earth is accurately depicted by Figure, then which statement is *least* in agreement with the hypothesis proposed by this tree?
(1) The last universal common ancestor of all extant species is better described as a community of organisms, rather than an individual species
(2) The origin of the three domains appears as a polytomy
(3) Archaean genomes should contain genes that originated in bacteria, and vice versa
(4) Eukaryotes are more closely related to archaeans than to bacteria

52. Which of these processes can be included among those responsible for the horizontal components of Figure?
(1) endosymbiosis
(2) mitosis
(3) binary fission
(4) point mutations

53. Which portion of Figure may ultimately be better depicted as a "ring"?
(1) the bacterial lineage
(2) the archaean lineage
(3) the eukaryotic lineage
(4) the trunk of the tree

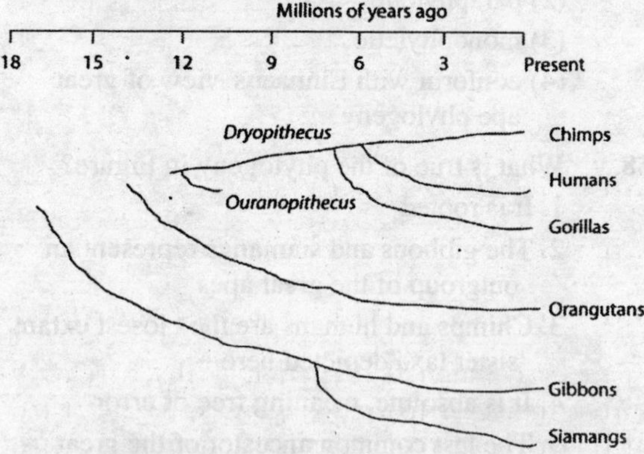

Figure 54-59 Humans, chimpanzees, gorillas, and orangutans are members of a clade called the great apes, which shared a common ancestor about 18 million years ago (Figure) Gibbons and siamangs comprise a clade called the lesser apes. Tree-branch lengths indicate elapsed time

54. The great apes comprise the family Hominidae, whereas the lesser apes comprise the family Hylobatidae. If the extant organisms on the far right side of comprise the next-most exclusive (i.e., specific taxon, then they comprise different
(1) subspecies (2) species
(3) genuses (4) genera

55. Together, the lesser apes and great apes shared a common ancestor most recently with other members of their
(1) order (2) class
(3) subclass (4) subfamily

56. From Figure, to which of the extant apes are orangutans most closely related?
(1) Gibbons and siamangs
(2) *Dryopithecus* and *Ouranopithecus*
(3) Gorillas
(4) Chimps, gorillas, and humans

57. Assuming chimps and gorillas are humans' closest relatives, removing humans from the great ape clade and placing them in a different clade has the effect of making the phylogenetic tree of the great apes

(1) polyphyletic
(2) paraphyletic
(3) monophyletic
(4) conform with Linnaeus' view of great ape phylogeny

58. What is true of the phylogeny in Figure?

1. It is rooted
2. The gibbons and siamangs represent an outgroup of the great apes
3. Chimps and humans are the closest extant sister taxa depicted here
4. It is absolute, meaning free of error
5. The last common ancestor of the great apes lived about 14 million years ago

(1) 1, 2, and 3
(2) 1, 2, and 5
(3) 1, 2, 3, and 4
(4) 1, 2, 3, and 5

59. From figure, what is true of *Dryopithecus* and *Ouranopithecus*?

1. They were great apes
2. They shared a common ancestor more recently with the orangutans than with the other great apes
3. They appear to be part of a polytomy
4. Their closest common ancestor with all of the extant great apes is the one they share with the orangutans
5. They were about as different from each other genetically as humans are different from chimps

(1) 1, 2, and 4
(2) 1, 3, and 4
(3) 2, 4, and 5
(4) 1, 2, 3, and 4

60. Traditionally, zoologists have placed birds in their own class, Aves. More recently, molecular evidence has shown that birds are more closely related to reptiles than their anatomy reveals. Genetically, birds are more closely related to crocodiles than crocodiles are to turtles. Thus, bird anatomy has become highly modified as they have adapted to flight, without their genes having undergone nearly as much change

Taxonomically, what should be done with the birds?

(1) The traditional stance is correct. Such dramatic morphological change as undergone by birds merits that the birds be placed in their own order, separate from the reptiles
(2) The birds should be reclassified, and their new taxon should be the subclass Aves. Genetic similarity trumps morphological dissimilarity
(3) The rest of the reptiles should be reclassified as a subclass within the class Aves
(4) Science is consensual. Taxonomy is a science. Variant classification schemes involving the birds should be tolerated until consensus is reached

Q. 61-64 refer to the following table, which compares the % sequence homology of four different parts (two introns and two exons) of a gene that is found in five different eukaryotic species. Each part is numbered to indicate its distance from the promoter (e.g., Intron I is the one closest to the promoter) The data reported for species A were obtained by comparing DNA from one member of species A to another member of species A

% Sequence Homology

Species	Intron I	Exon I	Intron VI	Exon V
A	100%	100%	100%	100%
B	98%	99%	82%	96%
C	98%	99%	89%	96%
D	99%	99%	92%	97%
E	98%	99%	80%	94%

61. Based on the tabular data, and assuming that time advances vertically, which cladogram (a type of phylogenetic tree) is the most likely depiction of the evolutionary relationships among these five species?

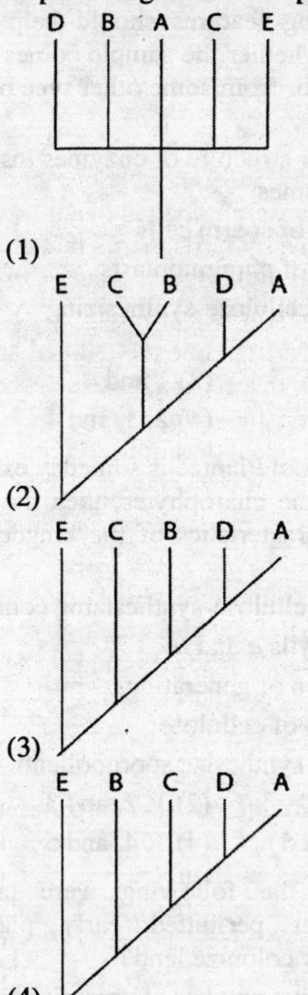

62. Which of the following is the best explanation for the high degree of sequence homology observed in Exon I among these five species?

(1) It is the most-upstream exon of this gene
(2) Due to alternative gene splicing, this exon is often treated as an intron
(3) It codes for a polypeptide domain that has a crucial function
(4) These five species must actually constitute a single species

63. Regarding these sequence homology data, the principle of maximum parsimony would be applicable in

(1) distinguishing introns from exons
(2) determining degree of sequence homology
(3) selecting appropriate genes for comparison among species
(4) inferring evolutionary relatedness from the number of sequence differences

64. Which of these four gene parts should allow the construction of the most accurate phylogenetic tree, assuming that this is the only part of the gene that has acted as a reliable molecular clock?

(1) Intron I (2) Exon I
(3) Intron VI. (4) Exon V

65. Three living species X, Y, and Z share a common ancestor T, as do extinct species U and V. A grouping that consists of species T, X, Y, and Z (but not U or V) makes up

(1) a valid taxon
(2) a monophyletic clade
(3) an ingroup, with species U as the outgroup
(4) a paraphyletic group

66. Based on this tree, which statement is *not* correct?

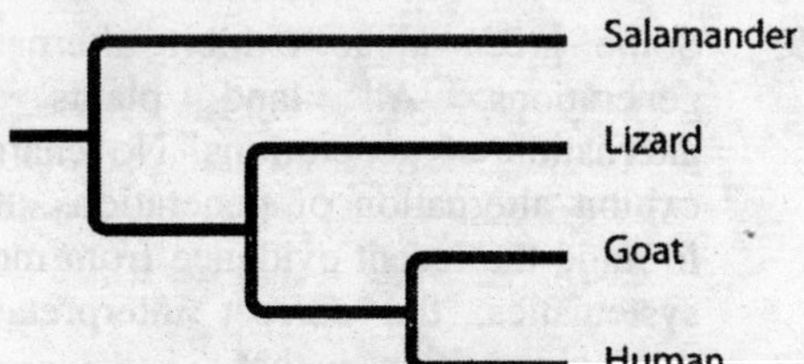

(1) The salamander lineage is a basal taxon
(2) Salamanders are a sister group to the group containing lizards, goats, and humans
(3) Salamanders are as closely related to goats as to humans
(4) Lizards are more closely related to salamanders than to humans

67. In animal cells and in the meristem cells of land plants, the nuclear envelope disintegrates during mitosis. This disintegration does not occur in the cells of most protists and fungi. According to our current knowledge of plant evolution, which group of organisms should feature mitosis most similar to that of land plants?

(1) Unicellular green algae
(2) Cyanobacteria
(3) Charophytes
(4) Red algae

68. On a field trip, a student in a marine biology class collects an organism that has differentiated organs, cell walls of cellulose, and chloroplasts with chlorophyll *a*. Based on this description, the organism could be a brown alga, a red alga, a green alga, a charophyte recently washed into the ocean from a freshwater or brackish water source, or a land plant washed into the ocean. The presence of which of the following features would definitively identify this organism as a land plant?

(1) Alternation of generations
(2) Sporopollenin
(3) Rings of cellulose-synthesizing complexes
(4) Embryos

69. Some green algae exhibit alternation of generations. All land plants exhibit alternation of generations. No charophytes exhibit alternation of generations. Keeping in mind the recent evidence from molecular systematics, the correct interpretation of these observations is that

(1) Charophytes are not related to either green algae or land plants
(2) Plants evolved alternation of generations independently of green algae
(3) Alternation of generations cannot be beneficial to charophytes
(4) Land plants evolved directly from the green algae that perform alternation of generations

70. A student encounters a pondweed which, judging from its appearance, seems to be a charophyte. She brings a sample back to her biology lab. Using only a compound light microscope to study the sample, which of the following features should help her to determine whether the sample comes from a charophyte or from some other type of green alga?

1. Molecular structure of enzymes inside peroxisomes
2. Structure of sperm cells
3. Presence of phragmoplasts
4. Rings of cellulose-synthesizing complexes

(1) 1 and 3 (2) 1 and 4
(3) 2 and 3 (4) 2, 3, and 4

71. If the kingdom Plantae is someday expanded to include the charophytes, then the shared derived characteristics of the kingdom will include

1. rings of cellulose-synthesizing complexes
2. chlorophylls *a* and *b*
3. alternation of generations
4. cell walls of cellulose
5. ability to synthesize sporopollenin

(1) 1 and 5 (2) 1, 2, and 3
(3) 1, 3, and 5 (4) 1, 4, and 5

72. Which of the following were probably factors that permitted early plants to successfully colonize land?

1. The relative number of potential predators (herbivores)
2. The relative number of competitors
3. The relative availability of symbiotic partners
4. Air's relative lack of support, compared to water's support

(1) 1 and 2
(2) 2 and 3
(3) 3 and 4
(4) 1, 2, and 3

73. Assuming that they all belong to the same plant, arrange the following structures from largest to smallest

1. Antheridia
2. Gametes
3. Gametophytes
4. Gametangia

(1) 1, 4, 3, 2 (2) 3, 1, 2, 4
(3) 3, 4, 2, 1 (4) 3, 4, 1, 2

74. As is true of the gametophytes of all land plants, the gametophytes of true mosses lack stomata. Yet, the feather moss *Pleurozium* harbors nitrogen-fixing cyanobacteria. Which of the following is a feature of moss gametophytes that is most important for the continued survival of these cyanobacteria in the tissues of the feather moss gametophyte?

(1) Lack of multiple cell layers in "leaves" of "buds"
(2) Lack of vascular tissues
(3) Lack of true leaves or roots
(4) Lack of an independent sporophyte

75. Beginning with the germination of a moss spore, what is the sequence of structures that develop after germination?

1. Embryo
2. Gametes
3. Sporophyte
4. Protonema
5. Gametophore

(1) 4 → 1 → 3 → 5 → 2
(2) 4 → 3 → 5 → 2 → 1
(3) 4 → 5 → 2 → 1 → 3
(4) 3 → 4 → 5 → 2 → 1

76. Two small, poorly drained lakes lie close to each other in a northern forest. The basins of both lakes are composed of the same geologic substratum. One lake is surrounded by a dense *Sphagnum* mat; the other is not. Compared to the pond with *Sphagnum*, the pond lacking the moss mat should have

(1) lower numbers of bacteria
(2) reduced rates of decomposition
(3) reduced oxygen content
(4) less-acidic water

77. Arrange the following terms from most inclusive to least inclusive

1. Embryophytes
2. Green plants
3. Seedless vascular plants
4. Ferns
5. Tracheophytes

(1) 1, 2, 5, 3, 4 (2) 2, 1, 5, 3, 4
(3) 2, 5, 1, 3, 4 (4) 1, 4, 2, 5, 3

78. Evidence indicates that plants increase the number of stomata in their leaves as atmospheric CO_2 levels decline. Increasing the number of stomata per unit surface area should have the effect of doing which of the following?

1. Increasing dehydration of leaf tissues
2. Decreasing dehydration of leaf tissues
3. Countering the effect of declining CO_2 on photosynthesis
4. Reinforcing the effect of declining CO_2 on photosynthesis
5. Decreasing the O_2 content of air next to the leaves lower than it would otherwise be
6. Increasing the O_2 content of air next to the leaves higher than it would otherwise be

(1) 1, 3, and 5 (2) 1, 3, and 6
(3) 1, 4, and 5 (4) 2, 3, and 6

79. Of the following list, flagellated (swimming) sperm are generally present in which groups?

1. Lycophyta
2. Bryophyta
3. Angiosperms
4. Chlorophyta
5. Pterophyta

(1) 1, 2, and 3 (2) 1, 2, 4, and 5
(3) 1, 3, 4, and 5 (4) 2, 3, and 5

80. If intelligent extraterrestrials visited Earth 475 million years ago, and then again 300 million years ago (at the close of the Carboniferous period), what trends would they have noticed in Earth's terrestrial vegetation over this period?

1. A trend from dominant gametophytes to dominant sporophytes
2. A trend from sporangia borne on modified leaves (sporophylls) to sporangia borne on stalks (set 1)
3. A trend from no true leaves, to microphylls, to megaphylls
4. A trend from soil-surface-hugging plants to "overtopping" plants
5. A trend toward increased lignification of conducting systems

(1) 1 and 3 (2) 3, 4, and 5
(3) 1, 2, 4, and 5 (4) 1, 3, 4, and 5

81. What is thought to be the correct sequence of the following events during the Carboniferous period?

1. Vascular plants become more prevalent
2. Megaphylls with large surface areas become more prevalent
3. Atmospheric CO_2 levels decline by a factor of five
4. Global cooling occurs, leading to widespread glaciations

(1) 1, 2 ,3, 4 (2) 2, 1, 3, 4
(3) 2, 1, 4, 3 (4) 1, 2, 4, 3

82. Researchers tested nitrogen loss from soil where the moss *Polytrichum* was growing, and compared it to soil from which *Polytrichum* had been removed. The data are presented below

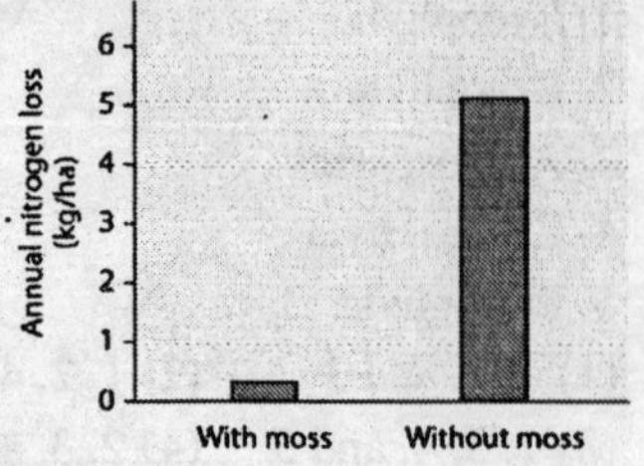

Loss of soil nitrogen via "gaseous emission" was found to be negligible. Rather, most loss of soil nitrogen was due to water erosion of the soil. Which of these hypotheses is least likely to account for the observed results?

(1) If rhizoids had helped stabilize the soil, then less erosion and less loss of nitrogen would occur
(2) If protonemata had absorbed, and stored, nitrogen from the soil, then they would have reduced loss of nitrogen by erosion
(3) If the overlying mat of gametophores had slowed the entry of water into the soil, then it would have reduced water's ability to erode the soil and carry away its nitrogen
(4) If sporophyte stomata had absorbed nitrogen from the soil, then they would have reduced loss of nitrogen by erosion

83. Researchers decided to test the hypothesis that if the 2-m tall *Polytrichum* gametophyte-sporophyte plants had acted as a physical buffer, then they would have reduced water's ability to erode the soil and carry away its nitrogen. They began with four equal-sized areas where *Polytrichum* mosses grew to a height of 2 m above the soil surface. One of the four areas was not modified. In the second area, the mosses were trimmed to a height of 1 m above the soil surface. In the third area, the mosses were trimmed to a height of 0.5 m above the soil surface. In the fourth area, the mosses were trimmed all the way to the ground, leaving only the rhizoids. Water, simulating rainfall, was then added in a controlled fashion to all plots over the course of one year. Figure presents four graphs that depict potential results of this experiment

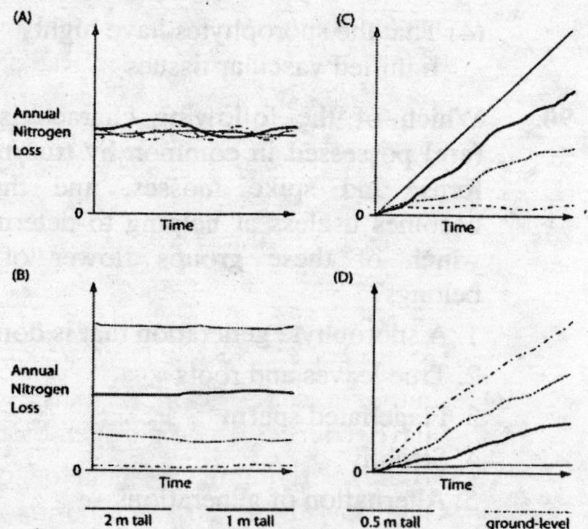

Which graph of soil nitrogen loss over time in Figure most strongly supports the hypothesis that if the 2-m tall Polytrichum gametophyte-sporophyte plants had acted as a physical buffer, then they would have reduced water's ability to erode the soil and carry away its nitrogen?

(1) A (2) B
(3) C (4) D

84. If the actual results most closely resembled those in Figure above, then a further question arising from these data is: "Do the *Polytrichum* rhizoids have to be alive in order to reduce soil nitrogen loss, or do dead rhizoids have the same effect?" Arrange the following steps in the correct sequence to test this hypothesis

1. Add metabolic poison to the soil of the experimental plot of mosses
2. Apply water equally to the experimental and control plots
3. Measure initial soil nitrogen contents of control and experimental plots
4. Determine nitrogen loss from soil of control and experimental plots
5. Establish two identical plots of *Polytrichum* mosses; one as a control, the other as the experimental treatment

(1) 5 → 1 → 3 → 2 → 4
(2) 5 → 2 → 3 → 1 → 4
(3) 5 → 3 → 1 → 2 → 4
(4) 4 → 5 → 1 → 3 → 2

85. Which of these potential results of applying a metabolic poison to the rhizoids of *Polytrichum* should interfere the least with the ability to draw valid conclusions from this experiment?

(1) If, upon dying, the rhizoids leak nitrogenous compounds into the soil before final nitrogen content is measured
(2) If, upon dying, decomposition of the rhizoids introduces nitrogenous compounds to the soil before final nitrogen content is measured
(3) If the metabolic poison is hydrogen cyanide (HCN) or sodium azide (NaN_3), and much of the poison remains in the soil
(4) If the metabolic poison acts against the mitochondria of the rhizoid cells

86. What is true of the phylogenetic tree in figure?

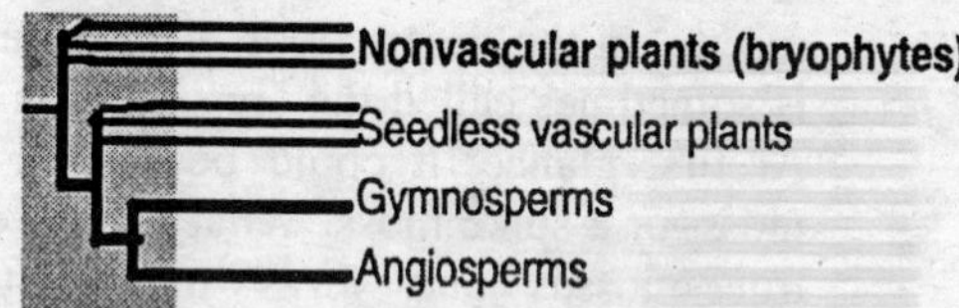

1. It depicts uncertainty about whether the bryophytes or the vascular plants evolved first
2. It is hypothetical
3. It includes polytomies
4. It shows that ferns (phylum Pterophyta) are the closest living relatives to the seed plants
5. It indicates that seeds are a shared ancestral character of all vascular plants

(1) 1 and 2
(2) 2 and 3
(3) 1, 2, and 3
(4) 1, 2, and 4

87. A biology student hiking in a forest happens upon an erect, 15-cm-tall plant that bears microphylls and a strobilus at its tallest point. When disturbed, the cone emits a dense cloud of brownish dust. A pocket magnifying glass reveals the dust to be composed of tiny spheres with a high oil content. This student has probably found a(n)

(1) immature pine tree
(2) bryophyte sporophyte
(3) fern sporophyte
(4) lycophyte sporophyte

88. Big Bend National Park in Texas is mostly Chihuahuan desert, where rainfall averages about 10 inches per year. Yet, it is not uncommon when hiking in this bone-dry desert to encounter mosses and ferns. One such plant is called "flower of stone." It is not a flowering plant, nor does it produce seeds. Under arid conditions, its leaflike structures curl up. However, when it rains, it unfurls its leaves, which form a bright green rosette on the desert floor. Consequently, it is sometimes called the "resurrection plant." At first glance, it could be a fern, a true moss, or a spike moss. What feature of both true mosses and ferns makes it most surprising that they can survive for many generations in dry deserts?

(1) Flagellated sperm
(2) Lack of vascular tissues
(3) Lack of true roots
(4) Lack of cuticle

89. Which of the following features is most important in order for true mosses and ferns to survive and reproduce in the desert?

(1) That the sporophytes occupy only permanently shady, north-facing habitats
(2) That the sporophytes hug the ground, growing no taller than a couple of inches
(3) Either that their gametophytes grow close together, or that they be hermaphroditic
(4) That the sporophytes have highly lignified vascular tissues

90. Which of the following characteristics is (are) possessed in common by true mosses, ferns, and spike mosses, and therefore becomes useless at helping to determine to which of these groups flower of stone belongs?

1. A sporophyte generation that is dominant
2. True leaves and roots
3. Flagellated sperm
4. Strobili
5. Alternation of generations

(1) 5 only (2) 1 and 5
(3) 2 and 3 (4) 3 and 5

91. Upon closer inspection of the leaves of flower of stone, one can observe tiny, cone-like structures. Each cone-like structure emits spores of two different sizes. Based on this information, which of the following can be properly inferred about flower of stone?

1. It is heterosporous
2. It is a fern
3. The cone-like structures are sori
4. It is a lycophyte
5. It has separate male and female gametophytes

(1) 1 and 5 (2) 2 and 3
(3) 1, 2, and 3 (4) 1, 4, and 5

92. Upon closer inspection of the leaves of flower of stone, one can observe tiny, cone-like structures. Each cone-like structure emits spores of two different sizes. Consequently, flower of stone should be expected to possess which other characteristics?

1. A gametophyte generation that is dominant
2. Lignified vascular tissues
3. Microphylls
4. Filamentous rhizoids, but not true roots
5. Spores that are diploid when mature

(1) 1 and 2 (2) 1 and 5
(3) 2 and 3. (4) 2, 3, and 4

93. In which combination of locations would one who is searching for the gametophytes of flower of stone have the best chance of finding them?

1. Moist soil
2. Underground, nourished there by symbiotic fungi
3. South- or west-facing slopes
4. Permanently shady places
5. Far from any flower of stone sporophytes

(1) 1 only (2) 1 and 2
(3) 1, 2, and 4 (4) 1, 2, and 5

94. Suppose that the cells of seed plants, like the skin cells of humans, produce a pigment upon increased exposure to UV radiation. Rank the following cells, from greatest to least, in terms of the likelihood of producing this pigment

1. Cells of sporangium
2. Cells in the interior of a subterranean root
3. Epidermal cells of sporophyte megaphylls
4. Cells of a gametophyte

(1) 3, 4, 1, 2 (2) 3, 4, 2, 1
(3) 3, 1, 4, 2 (4) 3, 2, 1, 4

95. The fruit of the mistletoe, a parasitic angiosperm, is a one-seeded berry. In members of the genus *Viscum*, the outside of the seed is viscous (sticky), which permits the seed to adhere to surfaces, such as the branches of host plants or the beaks of birds. What should be expected of the fruit if the viscosity of *Viscum* seeds is primarily an adaptation for dispersal rather than an adaptation for infecting host plant tissues?

(1) It should be drab in color
(2) It should be colored so as to provide it with camouflage
(3) It should be nutritious
(4) It should secrete enzymes that can digest bark

96. The cycads, a mostly tropical phylum of gymnosperms, evolved about 300 million years ago and were dominant forms during the Age of the Dinosaurs. Though their sperm are flagellated, their ovules are pollinated by beetles. These beetles get nutrition (they eat pollen) and shelter from the microsporophylls. Upon visiting megasporophylls, the beetles transfer pollen to the exposed ovules. In cycads, pollen cones and seed cones are borne on different plants. Cycads synthesize neurotoxins, especially in the seeds, that are effective against most animals, including humans

Which feature of cycads distinguishes them from most other gymnosperms?

1. They have exposed ovules
2. They have flagellated sperm
3. They are pollinated by animals

(1) 1 only (2) 2 only
(3) 3 only (4) 2 and 3

97. Oviparous (egg-laying) animals have internal fertilization (sperm cells encounter eggs within the female's body) Yolk and/or albumen is (are) provided to the embryo, and a shell is then deposited around the embryo and its food source. Eggs are subsequently deposited in an environment that promotes their further development, or are incubated by one or both parents. The yolk and/or albumen of an animal egg is what type of angiosperm analog?

(1) Endosperm
(2) Pollen tube and sperm nuclei
(3) Carpels
(4) Fruit

98. The Brazil nut tree, *Bertholletia excels* (n = 17), is native to tropical rain forests of South America. It is a hardwood tree that can grow to over 50 meters, is a source of high-quality lumber, and is a favorite nesting site for harpy eagles. As the rainy season ends, tough-walled fruits, each containing 8-25 seeds (Brazil nuts), fall to the forest floor. About $50 million worth of nuts are harvested each year. Scientists have discovered that the pale yellow, self-incompatible flowers of Brazil nut trees admit only female orchid bees as pollinators. Animals that consume Brazil nuts are deriving nutrition mostly from tissue whose nuclei have how many chromosomes?

(1) 17 (2) 34
(3) 51 (4) 68

99. The agouti (*Dasyprocta* spp.), a cat-sized rodent, is the only animal with teeth strong enough to crack the hard wall of Brazil nut fruits. It typically eats some of the seeds, buries others, and leaves still others behind inside the fruit, which moisture can now enter. The uneaten seeds may subsequently germinate. Consequently, which terms describe the relationship between the Brazil nut tree and the agouti?

1. Parasitic
2. Commensalistic
3. Symbiotic
4. Endosymbiotic
5. Mutualistic

(1) 1 and 3 (2) 2 and 4
(3) 2 and 5 (4) 3 and 5

100. For several decades now, amphibian species worldwide have been in decline. A significant proportion of the decline seems to be due to the spread of the chytrid fungus, *Batrachochytrium dendrobatidis* (BD) Chytrid sporangia reside within the epidermal cells of infected animals, animals that consequently show areas of sloughed skin. They can also be lethargic, which is expressed through failure to hide and failure to flee. The infection cycle typically takes four to five days, at the end of which zoospores are released from sporangia into the environment. In some amphibian species, mortality rates approach 100%; other species seem able to survive the infection. Apart from direct amphibian-to-amphibian contact, what is the most likely means by which the zoospores spread from one free-living amphibian to another?

(1) By wind-blown spores
(2) By flagella
(3) By cilia
(4) By pseudopods

ANSWER SHEET

Unit 9: Diversity Among Life Forms

1. 4	**2.** 4	**3.** 4	**4.** 4	**5.** 1	**6.** 2	**7.** 1	**8.** 4	**9.** 4	**10.** 1
11. 3	**12.** 4	**13.** 4	**14.** 3	**15.** 1	**16.** 4	**17.** 3	**18.** 4	**19.** 4	**20.** 3
21. 3	**22.** 3	**23.** 3	**24.** 1	**25.** 2	**26.** 3	**27.** 1	**28.** 4	**29.** 1	**30.** 4
31. 4	**32.** 2	**33.** 4	**34.** 3	**35.** 2	**36.** 4	**37.** 2	**38.** 1	**39.** 4	**40.** 1
41. 3	**42.** 2	**43.** 1	**44.** 4	**45.** 1	**46.** 4	**47.** 2	**48.** 2	**49.** 4	**50.** 2
51. 4	**52.** 1	**53.** 4	**54.** 4	**55.** 1	**56.** 4	**57.** 2	**58.** 4	**59.** 2	**60.** 4
61. 4	**62.** 3	**63.** 4	**64.** 3	**65.** 4	**66.** 4	**67.** 3	**68.** 4	**69.** 2	**70.** 4
71. 1	**72.** 4	**73.** 4	**74.** 1	**75.** 3	**76.** 4	**77.** 2	**78.** 2	**79.** 2	**80.** 4
81. 1	**82.** 4	**83.** 4	**84.** 3	**85.** 4	**86.** 2	**87.** 4	**88.** 1	**89.** 3	**90.** 4
91. 4	**92.** 2	**93.** 3	**94.** 3	**95.** 3	**96.** 4	**97.** 1	**98.** 3	**99.** 4	**100.** 2

UNIT-10

ECOLOGY

1. Ecological succession may be defined in terms of the following parameters

A. It is an orderly process of community development that is non-directional and, therefore, unpredictable

B. It results from modification of the physical environment by the community; that is, succession is community controlled even though the physical environment determines the pattern, the rate of change, and often sets limits as to how far development can go

C. It culminates in a stabilized ecosystem in which maximum biomass (or high information content) and symbiotic function between organisms are maintained per unit of avalaible energy flow

Which of the following statement is/are correct?

(1) A and B (2) B and C
(3) A and C (4) Only B

2. An endangered frog species is found in only three isolated, closed populations. Suppose the probability of local extinction in a time period of 10 years of each of these three population is the same and equals to 0.7, what is probability that at the end of 10 year, the species does not become extinct?

(1) 0.7 (2) 0.343
(3) 0.657 (4) 0.133

3. Two species of aquatic predators A and B were introduced into a lake along with three prey species C, D and E. Individuals of the three prey species were introduced in the following proportion: C=0.6, D=0.3 and E=0.1. Predators and prey encountered each other randomly. Shown below are the proportion of the three prey species found in diet of the two predators

	Proportion of prey species in the diet		
Predator	C	D	E
A	0	0.75	0.25
B	0.5	0.1	0.4

Which of the following statements is true?

(1) Predator A does not discriminate between prey species D and E
(2) Predator A prefers prey species D over E
(3) Predator B prefers species C over E
(4) Predator B does not discriminate between prey species D and E

4. The Andaman and Nicobar Island have many endemic species, including geckos of the genus *Phelsuma*. Genus *Phelsuma* is found no where else except Madagascar and adjoining islands. Recent molecular studies suggest that *Phelsuma* in Andaman and Nicobar Islands diverged from their counterparts in Madagascar around 2 million year ago. It is well known that the center of origin of genus *Phelsuma* is Madagascar

Given the information the most likely explanation for the presence of Phelsuma in Andaman and Nicobar Islands is

(1) Recent human mediated transport
(2) Continental drift
(3) Long distance transoceanic dispersal
(4) Dispersal by land via Africa and Asia

5. Consider a foraging individuals feeding in a habitat where food occurs as patches. Suppose that travel costs between patches are minimal and the individual has perfect information. Then, an optimally foraging individuals seeking to maximize food acquisition rate should stop feeding in one patch and move to next when

(1) no prey items are left in Patch 1

(2) when very few prey items are left in Patch 1

(3) when there are fewer prey items in Patch 1 than the average for the habitat

(4) when the habitat average for prey items is high

6. The webs of individuals of a social spider species are arranged in concentric circles radiating outward from a central point. Given below are the probabilities of predation (being eaten) and prey capture for large and small spiders in peripheral and central locations in the colony

	Large spiders	Small spiders
Predation risk at periphery	0.8	0.3
Predation risk at centre	0.2	0.2
Prey capture probability at periphery	0.7	0.8
Prey capture probability at centre	0.3	0.2

The results predicts that

(1) Large spiders will try to occupy position at the centre and small one at the periphery of the colony

(2) Large spiders will try to occupy positions at the periphery and small ones at the centre of the colony

(3) Both large and small spider will try to occupy central position in the colony

(4) The distribution of large and small spiders in the colony will be random

7. A species of gecko is found on two group of islands. In one group of islands, the islands are distributed as a network at a varying distance from each other. In other group, the islands are an archipelago, distributed in a linear chain. There is no difference in the average pair wise distance between nearest neighbors. Occasionally, a species goes extinct on one of the islands and is recolonised from neighboring island. If all islands are occupied it begin with

(1) the archipelago is likely to have more empty islands

(2) the network is likely to have more empty islands

(3) both will have the same number of empty islands

(4) the archipelago is likely to have higher immigration

8. In bottom sediments of lakes and streams, organic matter is broken down by heterotrophic microorganisms and ammonia is released. Under aerobic conditions, specialized bacteria convert ammonia to nitrate (nitrification) This nitrate, together with nitrate from other sources, diffuses into deeper sediments, where it may undergo anaerobic conversion to nitrogen gas (denitrification) These sediments typically contain oligochaete worms that live with their heads buried and their tails waving back and forth in the overlaying water. Plastic columns were packed with freshly collected stream sediment and then covered with layers of nitrate enriched water. A similar set of columns was packed with sediments that had been sterilized and then covered with either nitrate enriched water or distilled water. Oligochaete worms were collected and acclimated at 20°C. Following acclimation, worms were rinsed in distilled water and then added to three sediments columns. The columns were incubated in the dark at 20°C and monitored every three days for changes in the concentration of nitrate in the overlying water. Nitrate concentrations in each of the experimental treatments were plotted against time, as shown in graph below

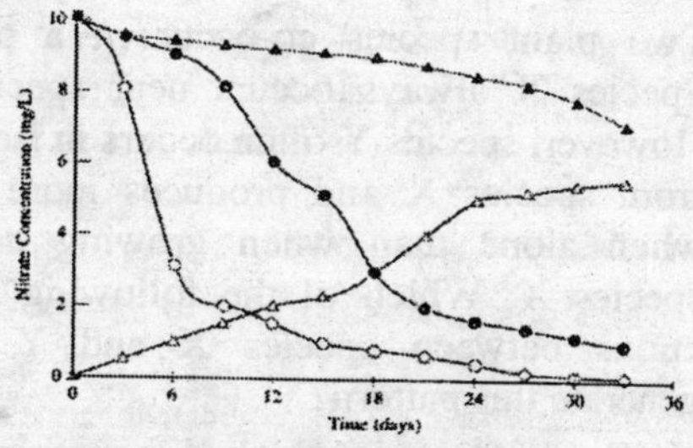

●——● Freshly Collected Sediments + Nitrate
○——○ Freshly Collected Sediments + Nitrate + Worms
▲——▲ Sterile Sediments + Nitrate + Worms
△——△ Sterile Sediments + Distilled Water + Worms

The best explanation for the increase in nitrate concentration in one treatment is that

(1) ologichaete worms may harbor microorganisms that are capable of carrying out nitrification
(2) oligochaete worms are capable of carrying out a limited amount of denitrification in the absence of sediment microorganisms
(3) some nitrifying bacteria must have survived the autoclaving in form of spores
(4) some denitrifying bacteria may have been added to the columns along with the oligochaete worms

9. The figure above depicts the intensity versus frequency curve of an acoustical receptors of the larva of a noctuid moth. The larva is preyed on by a wasp whose vibrating wings produces a vuzz. The acoustic receptors are destroyed on a population of larvae and rates of wasp predation are measured on the "deaf" larvae and on a population of normal larvae

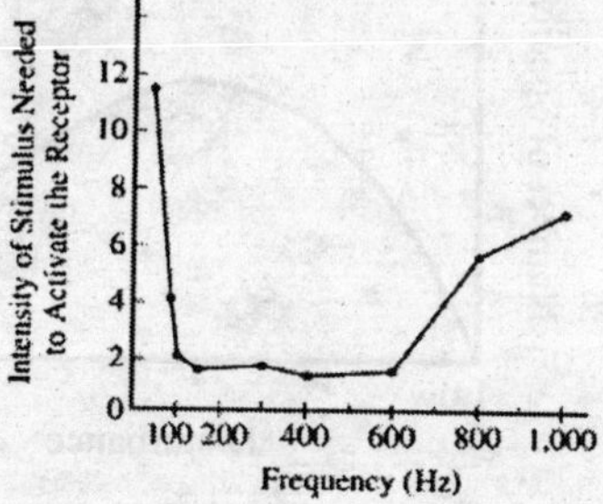

Based on the curve in the figure and on the concept that animals posses sensory mechanisms that enhance survival, one could predict which of the following?

(1) Predation intensity on normal and "deaf" larvae would not differ because wasps hunt visually and both larval population would be equally visible
(2) Predation intensity would be greater on "deaf" larvae because the wasps could approach more closely before being detected and there would be less time for escape
(3) Predation intensity would be greater on the normal larvae because they are attracted to wasp buzz
(4) Predation intensity on normal and on "deaf" larvae would not differ because the larvae are too slow to escape the approaching wasp

10. Ten pairs of laboratory mice were kept in research station located at high altitude (3000 m) and ten pairs of the same strain at a low altitude station located at sea level. Both were allowed to breed and their progeny were tested at 4 weeks of age on a treadmill for aerobic performance. The mice born and raised at high altitude out performed the low altitude reared ones when subjected to experimentally-induced low oxygen levels. Mice born and raised at low altitudes out performed their high altitude cohorts when oxygen levels were experimentally raised. This experiment demonstrates that

(1) low oxygen levels experienced during development are always adaptive
(2) high oxygen level experienced during development are always adaptive
(3) mice are adapted to the oxygen levels experienced during their development
(4) mice are not adapted to the oxygen levels experienced during their development

11. Ant colonies often have colony-specific hydrocarbon cocktails on their bodies. When queen from one colony was removed, then re-introduced a few hour later after being coated with the hydrocarbon cocktail of another colony, she was aggressively attacked. Presentation of a paper soaked in the cuticular cocktail of the other colony elicited strong aggressive responses. In the control situation, where the queen was removed and re-introduced after the same time without any treatment, she was immediately reaccepted. This experiment shows that the cuticular hydrocarbon coktail

(1) is necessary to detect strangers

(2) is sufficient to detect stranger

(3) is both necessary and sufficient to detect strangers

(4) is neither necessary nor sufficient to detect strangers

12. Match list-I with list-II and select the correct answer using the codes given below the lists

List-I

P. The amount of energy accumulation in green plants through the process of photosynthesis

Q. The total organic matter synthesized by the producers in the process of photosynthesis per unit time and area

R. Rate of storage of organic matter in plant tissues in excess of the respiratory utilization during the measurement period

S. Rate of energy storage at consumer level

List-II

1. Gross primary productivity
2. Primary productivity
3. Secondary productivity
4. Net primary productivity

(1) P-1, Q-2, R-3, S-4

(2) P-2, Q-1, R-3, S-4

(3) P-1, Q-3, R-4, S-2

(4) P-2, Q-1, R-4, S-3

13. Two plant species co-occur in a prairie. Species X always occurs near species Y. However, species Y often occurs in isolation from species X and produces more seeds when alone than when growing next to species X. Which of the following interactions between species X and Y could generate this pattern?

(1) Competition, in which Y is superior to X in accumulating resources

(2) Mutualism, in which both X and Y benefit by exchanging resources

(3) Parasitism, in which X benefits from resources produced by Y and reduces the growth of Y by doing so

(4) Commensalism, in which X benefits from resources produced by Y but does not affect the growth of Y by doing so

14. Barring the Himalayan region India can be divided into following four major vegetational zones

P. tropical deciduous forests

Q. tropical rain forests

R. thorn and scrub forests

S. tidal forests

The correct sequence of the descending order in terms of the areas occupied by these zones is

(1) P, Q, R, S (2) P, Q, S, R

(3) Q, P, R, S (4) Q, P, S, R

15. The graph below illustrates a simplified description of the effect of disturbance on species diversity. Which of the following best explains the graph?

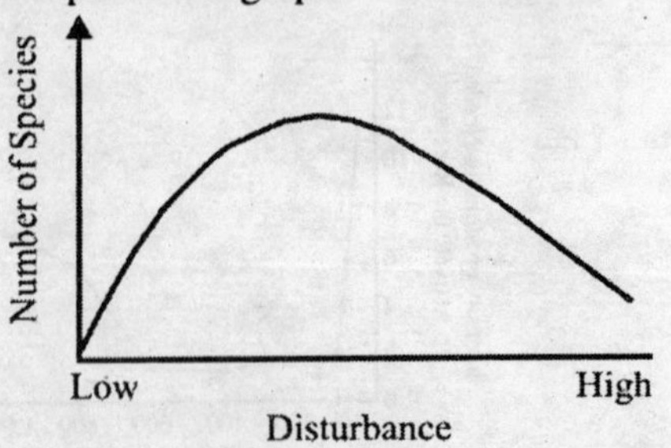

(1) Productivity is enhanced at intermediate levels of disturbance

(2) Stabilizing selection occurs at intermediate levels of disturbance

(3) The interaction of physical factors and biological competition increases diversity at intermediate levels of disturbance

(4) Directional selection is enhanced by increased disturbance

16. The graphs below show that the relationship between atmospheric CO_2 and mean global temperature is

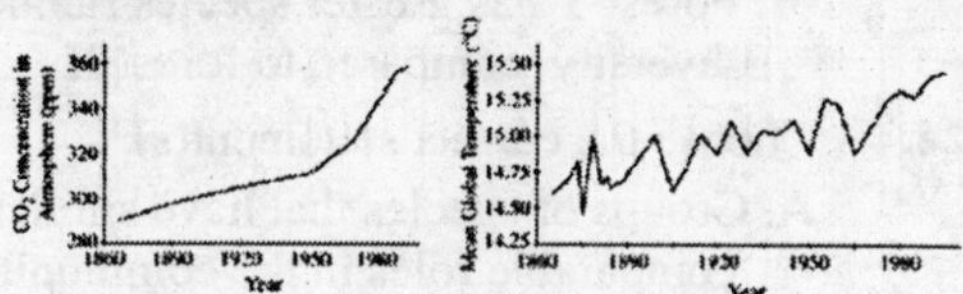

(1) a positive correlation

(2) an inverse relationship

(3) a cause-and-effect relationship

(4) evidence of an impending ice age

17. The graphs below show the survivorship and fecundity curves for two congeneric species, X and Y, that are found in different habitats. Of the two species, Y is more likely to

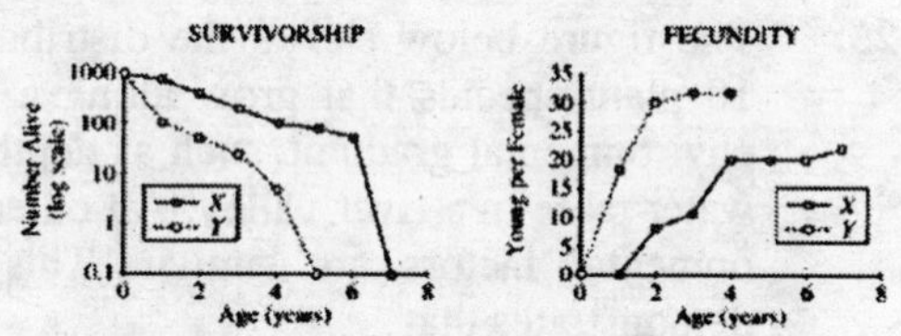

(1) come from a disturbed habitat in early succession

(2) have a longer generation time

(3) have a lower fecundity

(4) have a longer life span

18. For green house effect, which of the following statements are correct?

P. Intensive plantation can produce large amount of CO_2 which traps heat and produces the green house effect

Q. Extensive use of fossil fuel and intensive plantation result in the production of large amount of CO_2 and H_2O forming a thick layer that prevents the heat from being reradiated out

R. Excessive use of fossil fuel produces heat, which is absorbed by the CO_2 and water vapour in the atmosphere and causes green house effect

S. Increasing level of CO_2 tends to warm air in the lower layer of the atmosphere on global scale resulting in global warming

Codes

(1) P and Q (2) Q and R

(3) R and S (4) P and D

19. In the population growth curve given below, in which segment the growth rate dN/dt will be maximum?

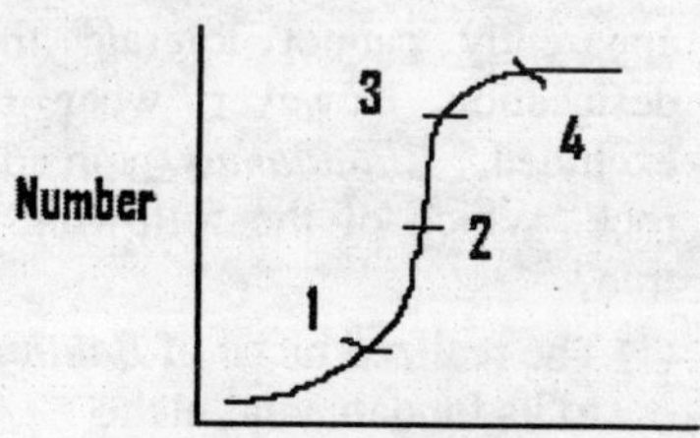

(1) 1 (2) 3

(3) 2 (4) 4

20. The Shannon's index of diversity is calculated as $H' = -\Sigma\ p_i \ln p_i$, where p_i is the proportion of the ith species. From the table below, we may conclude that

Species identity	Community A	Community B
A	3	1
B	4	2
C	8	1
D	2	3
E	5	2
F	6	1
G	4	15
H	5	13
I	3	3
J	5	4
Total	45	45

(1) Community A has a higher index of diversity
(2) Community B has a higher index of diversity
(3) The index of diversity is equal for the two communities
(4) It is not possible to calculate this

21. There is a well known study on the distribution of barnacles in intertidal zone in Scotland. Two species, *Banaus balanoides* and *Chthamalus stellatus* have stratified distribution on intertidal rocks. *Chthamalus* occurs higher than *Balanus* in the intertidal zones. In the experiment, each species was excluded to see the effect on the other. When *Chthamalus* was excluded, *Balanus* did not spread higher up the rocks because it apparently cannot tolerate the stress of desiccation. However, when *Balanus* was excluded, *Chthalamus* spread down the rock. Which of the following statement is true?

(1) The realized niche of *Balanus* is similar to its fundamental niche
(2) The realized niche of *Chthamalus* in the presence of *Balanus* is larger than its fundamental niche
(3) The fundamental niche of *Balanus* is smaller than its realized niche
(4) The fundamental niche of *Chthamalus* is same as its realized niche

22. While conducting studies of amphibians in a forest patch, a researcher lays quadrats in which he searches for amphibian species. In order to get a good estimate of species richness, the number of quadrats he lays should be determined by

(1) the amount of time spent in each quadrat
(2) the size of each quadrat
(3) the total species richness in the forest patch
(4) the number of new species encountered in each quadrat

23. Forest X has 5 species of trees with 20 individuals in each species and forest Y has 6 species of tree with 75, 10, 5, 5, 3 and 2 individuals respectively. Which of the following statements is true?

(1) Forest X has greater species richness and less species diversity, compared to forest Y
(2) Forest Y has greater species richness and less species diversity, compared to forest X
(3) Forest X has greater species richness and diversity, compared to forest Y
(4) Forest Y has greater species richness and diversity, compared to forest X

24. Choose the correct statement(s)

A. Groups of species that have similar or comparable roles in the community is termed guild
B. No two species in the same general territory can occupy for long the same ecological niche
C. When one population produces a substance harmful to a competiting population then it is called allelopathy

(1) A and B (2) B and C
(3) A, B and C (4) None

25. The figure below shows the distribution of 10 plant species that grow along a regular environmental gradient, such as depth of the water table in a river valley. All other environmental factors are similar. This figure demonstrates that

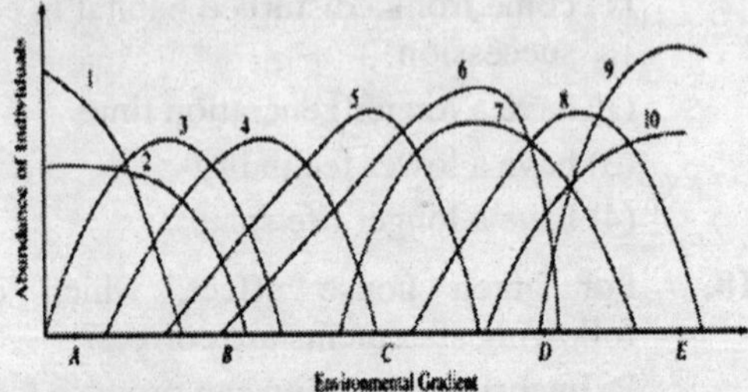

(1) species richness increases along the gradient from A to E
(2) species richness decreases along the gradient from A to E

(3) discrete communities are formed at sites A, C, and D
(4) no discrete communities are formed along the gradient

26. When numbers of organisms and amounts of living material in successively higher trophic levels are compared, the values usually take the form of a pyramid, with the largest numbers and greatest biomass in the producer trophic level. However, in some marine ecosystems, the consumer trophic levels contain significantly greater amounts of living material than does the primary-producer trophic level. The best explanation for this is which of the following?
(1) The main primary producers in marine ecosystems are microscopic algae with extremely high rates of population turnover
(2) Most consumers in marine ecosystems are filter feeders that must maintain large basketlike structures for extracting food from the water
(3) The increased availability of solar radiation in marine ecosystems means that fewer primary producers are required to support marine food chains
(4) Marine zooplankton often produce large extensions of their bodies in order to increase buoyancy

27. The location of a sound source in horizontal space (azimuth) is determined by the human auditory system using the difference in arrival time of sound between the two ears, which ranges from zero when the source is at the midline to 180 μs when the source is located 90 degrees laterally. For the system to have an angular resolution of one degree in azimuth, what should be the time resolution of the auditory system?
(1) 1 μs
(2) 2 μs
(3) 4 μs
(4) 0.5 μs

28. The total number of species of birds in two large forest areas is the same. Within each of these forests are two smaller patches. Individual patches in forest 2 have more species of birds than individual patches in forest 1. Which of the following is true?
(1) In Forest 1, species are more evenly distributed than in Forest 2
(2) The number of species shared by individual patches within a forest is the same for both forests
(3) Species are distributed uniformly in both forests
(4) In Forest 2, the two patches share more species than do the patches in forest 1

29. When examining whether individuals (of a forest tree species, for example) are distributed randomly in space or not, a common approach used is to lay many spatial plots and estimate the density of the focal species in each plot. The ratio of the variance in local density to mean local density is used as a measure of whether data points are distributed randomly in space. A ratio of 1 suggests that individuals are randomly distributed. A ratio much smaller that 1 would indicate that individuals are most likely to be
(1) Uniformly distributed
(2) Clumped
(3) Low in density
(4) Scattered at random

30. An ecologist studies seed rain in a tropical forests. The method includes the placement of a sticky board that is 1 m by 1 m in size at different distances from the tree. The board has a circle painted on it, which touches four sides of the square. If seed fall randomly at a given distance, what percentage of seeds that falls on ball falls outside the circle?
(1) About 10
(2) About 20
(3) About 30
(4) About 50

31. Which of the graphs below describes the per capita growth rate of population showing exponential growth ($N_t=N_0e^{rt}$)

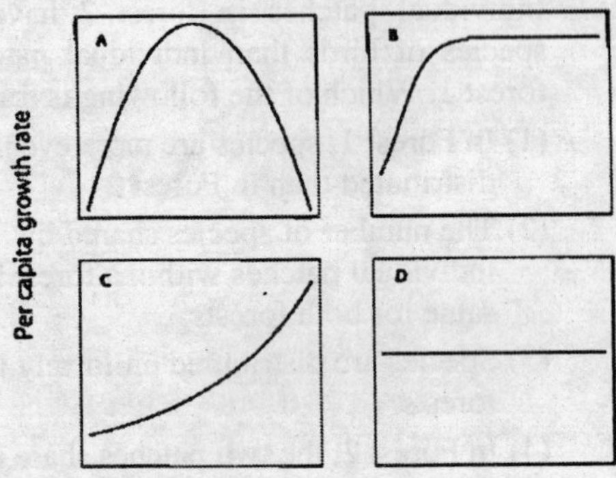

(1) A
(2) B
(3) C
(4) D

32. In replicated experiments, plots are planted with flax seeds in densities of 50, 1440 and 3660 per square meter, respectively. The factors of soil, water and light are similar for each plot. Dry weight of mature plants are obtained

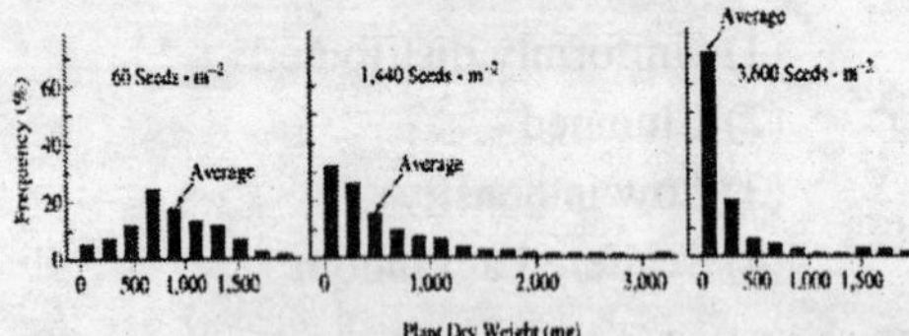

The experiments are most likely performed to test for

(1) interspecific competition
(2) intraspecific competition
(3) competitive exclusion
(4) founder effect

33. Biologist randomly sampled about 300 insects from a newly found island. The distribution of their abundance in the sample was un the figure given below

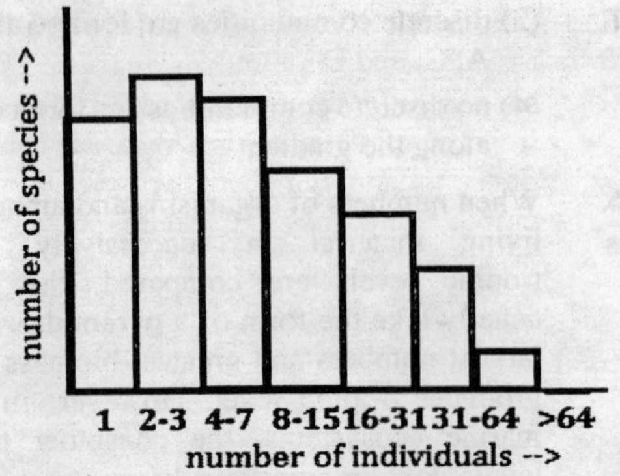

Which of the following can be correctly inferred from the graph?

(1) Many species have only one individual each on the island
(2)The bar on the extreme right represents a large number of species with very few individuals
(3) Summation of the height of all columns will be exactly equal to the total number of species present on the island
(4) All species from the island may not be represented in the sample

34. The order of decomposition of various organic compounds is as from fastest decomposition to last decomposition

(1) Sugar→Starch→Hemicellulose→ Pectin→ Cellulose→Lignin
(2) Suberin →Cutin → Sugar → Starch → Pectin →Cellulose → Lignin
(3) Sugar→Cellulose→Pectin→Lignin→ Suberin→Cutin
(4) Cellulose→Hemicellulose→Lignin→ Pectin→Suberin→Cutin

35. In a sample from an African population of 10,000, number of person suffering from sickle cell anemia is 100. What are expected frequency of individual who are sensitive to malaria

(1) 0.81
(2) 0.18
(3) 0.34
(4) 0.01

36. As a result of competition, the overall fitness of interacting individuals may be reduced, so generally competitors avoid face to face interaction. Competition that in absence of direct interaction, involves removal or pre-emption of a resource, leaving less for others is said to be

(1) interference competition
(2) exploitative competition
(3) interspecific competition
(4) intra specific competition

37. The average daily temperature change on an exposed south-facing slope and a north-facing slope for a late summer day in the Northern Hemisphere are shown below

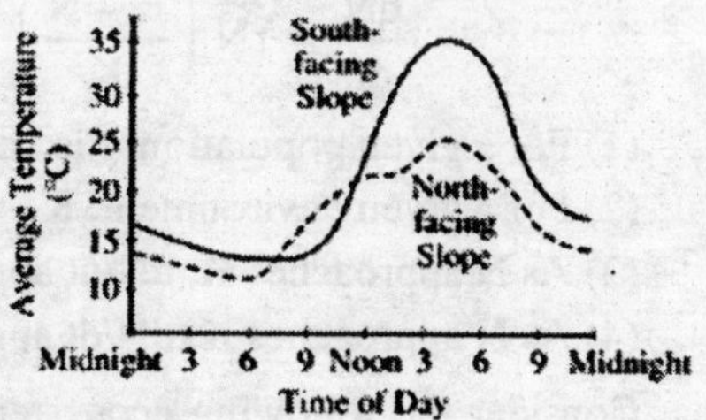

Compared with a south-facing slope, the climatic conditions on a north-facing slope favor the formation of a vegetation community best adapted to

(1) colder temperatures and wetter environments
(2) colder temperatures and drier environments
(3) hotter temperatures and similar moisture environments
(4) hotter temperatures and wetter environments

38. In an experiment, a microbiologist put equal numbers of each of the following organisms into a flask of sterile broth consisting mostly of sugar and a few amino acids. She then placed the flask in the dark. Which of the organisms would be best able to survive and reproduce in this environment?

(1) Chemoheterotrophic bacteria
(2) Cyanobacteria
(3) Diatoms
(4) Thermo-acidophile bacteria

39. Based on the theory of island biogeography, you expect the equilibrium number of species to be

(1) Higher on a large island close to mainland than on a large island farther from the mainland
(2) Smaller on a large island close to mainland than on a large island farther from the mainland
(3) Unrelated to size and distance from the mainland because it only applies to islands
(4) Higher on a small island close to a large island than on the larger island

40. Predator-prey model proposed by Lotka and Volterra in 1927; in words, states that

A. Each prey gives rise to a variable number of offspring per year; In other words, there are many factors limiting prey population growth apart from predation
B. Each predator eats a constant proportion of the prey population per year; In other words, doubling the prey population will double the number eaten per predator, regardless of how big the prey population is
C. Predator reproduction is directly proportional to prey consumed; another way of expressing this is that a certain number of prey consumed results in one new predator; or that one prey consumed produces some fraction of a new predator
D. A constant proportion of the predator population dies per year. In other words, the predator death rate is independent of the amount of food available

Which of the following statements are correct?

(1) A, B and C (2) B, C and D
(3) A, B and D (4) A, C and D

41. Consider the following ecosystem attributes concerning community structure from pioneering community to climax stage

A. Total organic matter for pioneering community is high and for climax community it is low

B. Inorganic nutrients for pioneering community are largely extrabiotic and for climax community they are intrabiotic

C. Species diversity: variety and equitability component for pioneering community high and for climax community is low

D. Stratification and spatial heterogeneity (pattern diversity) for pioneering community is poorly organized for climax community are well-organized

Which of the following statement is/are not correct?

(1) D and B (2) B and C
(3) A and C (4) A and D

42. Consider the following ecosystem attributes concerning Community Life history and selection pressure from pioneering community to climax stage

A. Mineral cycles for pioneering community are open for climax community are closed

B. Nutrient exchange rate, between organisms and environment for pioneering community are slow and for climax community are fast

C. Role of detritus in nutrient regeneration for pioneering community is not important but for climax community it is very important

D. Growth form for pioneering community is for feedback control (K-selection) and for rapid growth (r-selection) during climax stage

Which of the following statement is/are correct?

(1) A and D (2) B and C
(3) A and C (4) B and D

43. Life-history characteristics associated with K-selected organisms include

(1) rapid reproductive rates, short generation times, and large body size
(2) repeated reproduction, few progeny, and large body size
(3) inhabiting early successional state communities, rapid maturation rates, and numerous large offspring
(4) inhabiting climax communities, many small offspring, and short life span

44. The logistic equation below is used to describe the rate of change of a population, N, with time, t, where r is the intrinsic rate of increase and K is the carrying capacity. Which of the following statements is true for this equation?

$$\frac{dN}{dt} = rN\left(\frac{K-N}{K}\right)$$

(1) For a given population, r is variable
(2) For a given environment, K is variable
(3) As N approaches K, dN/dt approaches 0
(4) As N approaches K, dN/dt approaches K

45. Consider the following ecosystem attributes concerning Community energetics from pioneering community to climax stage

A. Gross production/community respiration (P/R ratio) for pioneering community is often greater than 1 and for climax community it approaches 1

B. Gross production/standing crop biomass (P/B ratio) for pioneering community is low for climax community is low

C. Biomass supported/unit energy flow (B/E ratio) for pioneering community is low for climax community it is high

D. Net community production (yield) for pioneering community high for climax community it is low

E. Food chains for pioneering community community is weblike, predominantly detritus and for climax it is linear and pre-dominanting grazing

Which of the following statement is/are correct?

(1) A, B and E (2) B, C and E

(3) A, B and D (4) A, C and D

46. Consider the following ecosystem attributes concerning overall homeostasis from pioneering community to climax stage

A. Internal symbiosis for pioneering community is undeveloped and for climax community are developed

B. Nutrient conservation for pioneering community is good and for climax community it is not good

C. Entropy for pioneering community low and for climax community it is high

D. Information for pioneering community low and for climax community it is high

Which of the following statement is/are correct?

(1) A and C (2) B and C

(3) B and D (4) A and D

47. Suppose you discovered a new species about which you know only two facts: it is large sized (>100 cm) and long lived (>10 years) Which of the following strategies is most likely to be true for this species?

(1) Breed early and more than once in life and produces large number of small sized offspring

(2) Breed late and only once in life and produces large number of small sized offspring

(3) Breed early and only once in life and produces large number of small sized offspring

(4) Breed late and more than once in life and produces small number of large sized offspring

48. Autotrophs in the aquatic ecosystem, unlike their counterparts in the terrestrial ecosystem are mostly microscopic and very low in indigestible (to the herbivores) matter. This explains the fact that compared to the terrestrial ecosystem, in the aquatic ecosystem

(1) Productivity/Biomass ratios are higher and energy transfer rates to higher trophic levels are faster

(2) Productivity/Biomass ratios are lower and the energy transfer rates to higher trophic levels are slower

(3) Productivity Biomass ratios are lower and the energy transfer rate to higher trophic levels are faster

(4) Productivity/Biomass ratios are higher and the energy transfer rate to higher trophic levels are slower

49. A common method of measuring community productivity in a pond is by measuring oxygen production or consumption in light (transparent) and dark (opaque) bottles filled in pond water. The oxygen concentration in each bottle is measured, the bottles are sealed, and then suspended from floats near the surface of the pond. After 24 hours, the oxygen concentration is again measured in each bottle. Representative data are presented below

Initial oxygen concentration (Initial) = 10 mg/L

Oxygen concentration in light bottle after 24 hours (light) = 12 mg/L

Oxygen concentration in dark bottle after 24 hours (Dark) = 7 mg/L

After 24 hours, the light bottle had a higher concentration of oxygen (12 mg/l) than did initially (10 mg/l) This difference represents

(1) Gross primary production

(2) Net primary production

(3) Secondary production

(4) Net community respiration

50. Five similar populations of herbaceous annual plant are monitored for three years; then, the number of individuals in population I-IV are increased by an experimenter. Subsequent changes in the sizes of all the populations are shown in the graph below

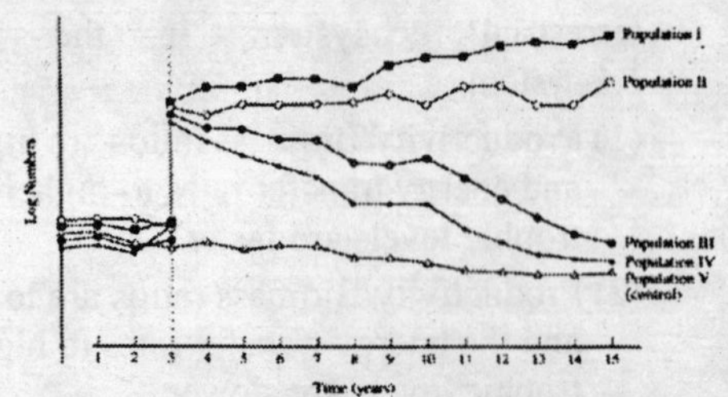

If size distribution of individual do not vary among the populations, which of the following relationships best illustrates the occurrence of population regulation?

(1) Population I and II compared with population V
(2) Population I and II compared with population III and IV
(3) Population III and IV compared with population V
(4) Population I compared with population II

51. The diagram below depicts three general types of survivorship curves. Only the form of the curve is important, not its absolute scale. Which of the following statements is LEAST likely to be true?

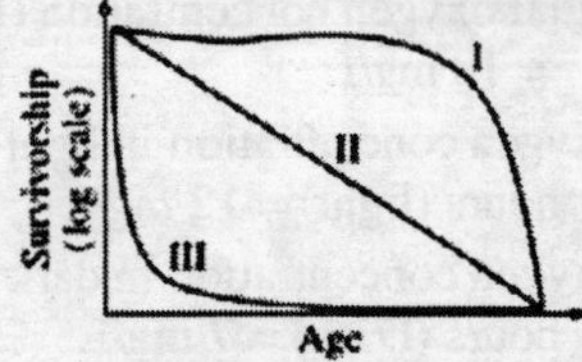

(1) Nearly constant mortality results in a Type II curve
(2) A fish that lays vast numbers of eggs exhibits a Type III curve
(3) Mortality that is initially low, then increases rapidly during old age, results in a Type I curve
(4) A species with high juvenile mortality and high adult survivorship results in a Type I curve

Question 52-55 refer to the following experiment, which is designated to test the co-evolutionary relationships among an unpalatable butterfly (the monarch), a palatable butterfly (the viceroy), and a butterfly predator (the jay) Monarch butterflies are reared on three diets: milkweed (their natural food), cabbage, and cabbage treated with an extract from milkweed leaves. Viceroy butterflies, mimics of monarchs, also reared on three diets: willows (their natural food), cabbage, and cabbage treated with an extract from milkweed leaves

In trial 1 of the first experiment, adult butterflies reared on a particular diet are presented one at a time at 1-hour intervals to jays and the jays are allowed to feed. Each jay is fed until it refuses to eat the butterfly presented, but no more than 12 butterflies are presented to a jay during a particular test. Five birds are used for each test; therefore upto 60 butterflies can be consumed in each diet test. The observer records the actual number of butterflies eaten. In trial 2, the experiment is repeated 2 weeks later

In the second experiment, the butterflies are reared on same diets as in experiment 1. However, when they are offered to jays, some jays receive a monarch reared on milkweed before being offered the butterflies reared on the experimental diets; the other group of jays is first given a viceroy reared on willow before being offered the butterflies reared on the experimental diets. The initial butterfly offered is included in the total number eaten, but no more than 12 butterflies are presented to each jay

TOTAL NUMBER OF BUTTERFLIES INGESTED BY JAYS

Butterfly and Diet	Experiment 1		Experiment 2	
	Trial 1	Trial 2	Fed Monarch First	Fed Viceroy First
Monarch				
Milkweed	5	5	5	10
Cabbage	56	58	5	57
Cabbage with extract	11	10	5	10
Viceroy				
Willow	58	55	5	56
Cabbage	53	57	5	52
Cabbage with extract	54	56	5	59

52. The data in the table indicate which of the following?

(1) The diet of viceroy butterfly protects them from being eaten by jays
(2) Cabbage contain defensive compounds

(3) Monarch butterflies synthesize a compound that protects them for being eaten by jays

(4) Monarch butterflies obtain a compound from their natural diet that protects them from being eaten by jays

53. Which of the following supports the hypothesis that viceroys receive protection from predation because they resemble monarchs?

(1) Fewer monarchs than viceroys are eaten when butterflies are reared on their natural diets

(2) Fewer viceroys are eaten when jays are fed monarch first

(3) More monarch are eaten when jays are fed viceroy first

(4) More viceroys than monarch are eaten when reared on cabbage with extract

54. The experimental design and the data indicate which of the following about the jays?

(1) Jays learn to avoid eating monarchs by experience

(2) Jays have a built-in instinct than enables them to recognize and avoid eating monarchs

(3) Eating a single monarch causes a jay to avoid eating another monarch for indefinite period

(4) A jay needs to eat many monarchs before it decides that monarchs are unpalatable

55. In the experiment, which data in the table most strongly indicate the viceroys are not obtaining compounds that will make them unpalatable?

(1) Equal numbers of monarch and viceroys are eaten when each species is reared on cabbage

(2) Fewer monarch than viceroy are eaten when each species is reared on its natural diet

(3) Jays eat as many viceroys reared on cabbage with extract as on cabbage alone

(4) Jays eat as many viceroys that are reared on willow, their natural food, as those reared on cabbage

56. Life-history characteristics not associated with r-selected organisms include

(1) rapid reproductive rates, short generation times, and large body size

(2) repeated reproduction, few progeny, and large body size

(3) inhabiting early successional state communities, rapid maturation rates, and numerous large offspring

(4) inhabiting climax communities, few small offspring, and long life span

57. Consider a population that shows exponential growth over time: $N_t = N_o e^{rt}$ where N_o is initial population size at time = 0, N_t is population size at time t, r is the intrinsic rate of increase, and e is the constant 2.7 1 8. . . In such a population

(1) Population growth rate is highest when the population is small and declines over time

(2) Population growth rate is smallest when the population is small, increases with time and reaches an asymptote

(3) Population growth rate is smallest when the population is small and increases with time

(4) Population growth rate is constant

58. A forest patch has 100 deer on Sunday, May 1, 2007. Hunting season opens on May 8,2007. Hunters kill 20 deer every Sunday. Each deer eats 1 kg of grass per day. If none of these populations are reproducing, how much grass would there need to be for the hunters to wipe out the deer population before the grass is wiped out?

(1) 700 kg

(2) 1.4 tonnes

(3) 2.1 tones

(4) 1.8 tonnes

59. Bird species A and B both nest in marshes. Species A is dominant over B when they are found together. Species B will occupy the entire marsh if Species A is absent, but if Species A is present, it will exclude Species B from the marsh center, and restrict Species B to the edge of the marsh. For Species B, the entire marsh would represent its

(1) realized niche (2) habitat
(3) range (4) fundamental niche

60. In population A, each woman bears exactly two female children, born at maternal ages 20 and 25. Population B is the same, except that the children are born when the mother is 25 and 30. Assume that these are the only differences between A and B. When the sizes of the two populations are monitored over time, we will find that

(1) both are stationary
(2) both grow at the same rate
(3) b grows faster than A
(4) a grows faster than B

61. The number of individuals or biomass at each trophic level is usually depicted as an ecological pyramid. Sometimes, inverted pyramids such as produced by a host parasite system, may be observed. These can be seen in

(1) pyramids of numbers and biomass, but never energy
(2) pyramids of biomass and energy, but never numbers
(3) pyramids of energy and numbers, but never biomass
(4) all kinds of pyramids

62. A bacterial culture undergoing balanced growth requires a first-order reaction. In other words, the rate of increase in bacteria at any particular time is proportional to the number or mass of bacteria present at that time. If '*N*' is the number of cells at any given time '*t*' and 'μ' is 'growth rate constant' this process can be expressed mathematically as

(1) $dN/dt = \mu N$
(2) $dt/dN = \mu N$
(3) $d\mu/dt = N$
(4) $dt/dN \times N = \mu$

63. Severe direct competition between species causes extinction of one of them in a situation known as competitive exclusion. In competitive exclusion, one species' population out-competes another species' population vying for the same space and food, leading to a local extinction of the loser

Consider the following statements

A. Usually the more adaptive species wins, and the loser is then considered the less fit organism for the ecosystem. The defeated species must then migrate to a new habitat or become extinct

B. Resource partitioning allows two different animal species to share the same resource in a slightly different manner and thus both species can co-occur in same area without competition

C. In niche specialization, two species develop specialties that allow them to occupy same niches in the same ecosystem. As they evolve toward these distinct specialties, they reach a point in which neither species need compete directly with the other

D. species to avoid competition occupies a large fundamental niche even though it can occupy a smaller, more specific niche, usually to avoid competition or predators

Which of the following statements are true?

(1) A and B
(2) C and D
(3) A, B and C
(4) A, C and D

64. A generalized illustration of the performance of a species in respect of an environmental condition is shown below. What does range P, Q and R most likely represent?

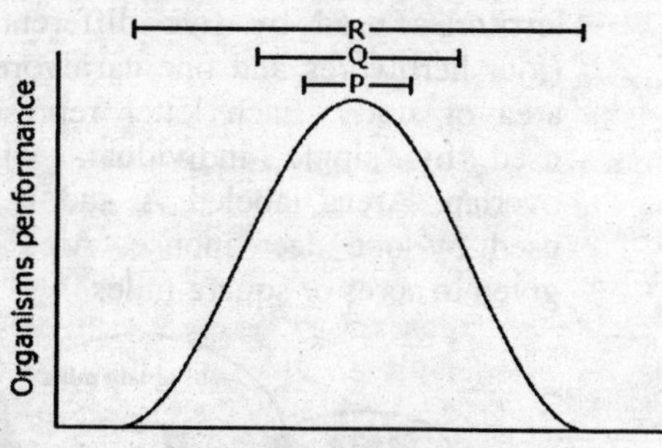

(1) P survival, Q growth, R reproduction
(2) P growth, Q survival, R reproduction
(3) P reproduction, Q survival, R growth
(4) P reproduction, Q growth, R survival

65. Two plant species co-occur in a prairie. Species X always occurs near species Y. However, species Y often occurs in isolation from species X and produces more seeds when alone than when growing next to species X. Which of the following interactions between species X and Y could generate this pattern?

(1) Competition, in which Y is superior to X in accumulating resources
(2) Mutualism, in which both X and Y benefit by exchanging resources
(3) Parasitism, in which X benefits from resources produced by Y and reduces the growth of Y by doing so
(4) Commensalism, in which X benefits from resources produced by Y but does not affect the growth of Y by doing so

66. In an experiment, honey bees is first exposed to its predator kept in box of blue colour and then kept in cage for different interval of time; t = 1 day and t = 5 day. During subsequent exposure of honey bees it was found that it choosed the blue background to enter 2/3 times in total of 99 such attempts, when exposed after t = 5 days however it choosed the blue background 40 times when same number of attempts were made after t = 1 day. Following conclusion can be drawn

A. Honey failed to associate background colour and predator presence in box
B. Memory to relate colour background with predator lasts only for 1 day in honey bees
C. Honey bees have ability to relate colour background with predator's presence for upto and above 5 days

Which of the following conclusion is/are correct?

(1) only A (2) only B
(3) only C (4) A and B

67. Figure below represents an outline life history for a unitary organism. Reproductive output is plotted on the vertical axis. The figures below (A–D) are variations on this basic theme

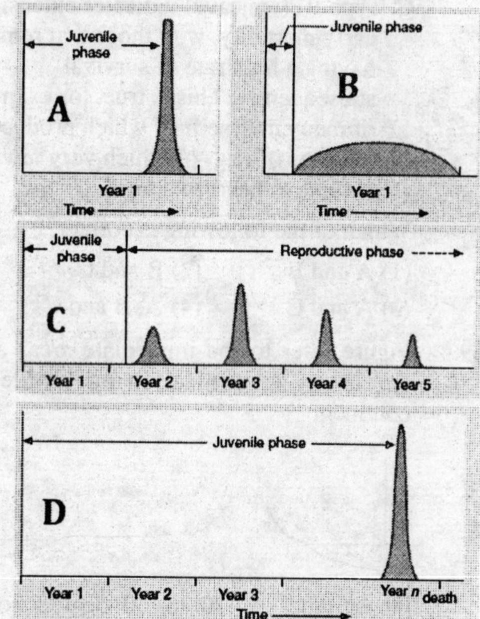

(1) A semelparous annual species
(2) An iteroparous annual species
(3) A long-lived iteroparous species with seasonal breeding
(4) A semelparous species living longer than a year

Which of the following is correct match?

(1) A-4, B-3, C-2, D-1
(2) A-3, B-4, C-2, D-1
(3) A-1, B-3, C-2, D-4
(4) A-1, B-2, C-3, D-4

68. Consider the classification of survivorship curves

A. Type I (convex) – epitomized perhaps by humans in rich countries, cosseted animals in a zoo or leaves on a plant – describes the situation in which mortality is concentrated at the end of the maximum lifespan

B. Type II (straight) indicates that the probability of death remains constant with age, and may well apply to the buried seed banks of many plant populations

C. Type III (concave) indicates extensive early mortality, with those that remain having a high rate of survival subsequently. This is true, for example, of many marine fish, which produce millions of eggs of which very few survive to become adults

Which of the following is correct?

(1) A and B (2) B and C
(3) A and C (4) A, B and C

69. Figure refer to the immediate areas around the letters in the following profile of a continent

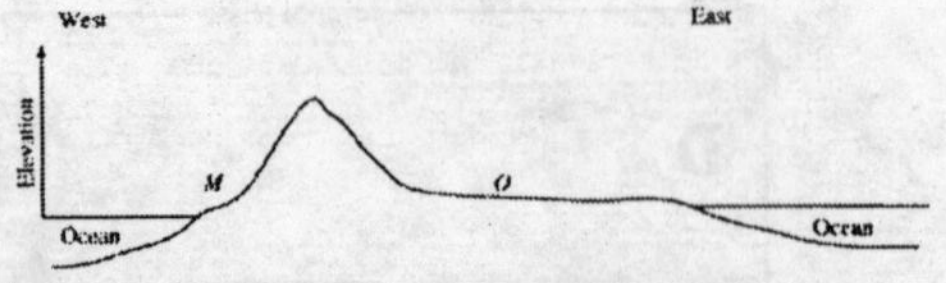

At latitudes of 20 to 30 degree north or south and with prevailing winds from east to west, the most likely biome to develop at M would be a

(1) monsoon forest (2) tundra
(3) desert (4) rainforest

70. Figure refer to the diagram below which represents the respective home ranges and territories used by five different animals (four herbivores and one carnivores) in the area of study. Each letter represents area used by single individual, with some overlap. Areas labeled A and B are each used by one deer mouse. Area sizes are given in acres or square miles

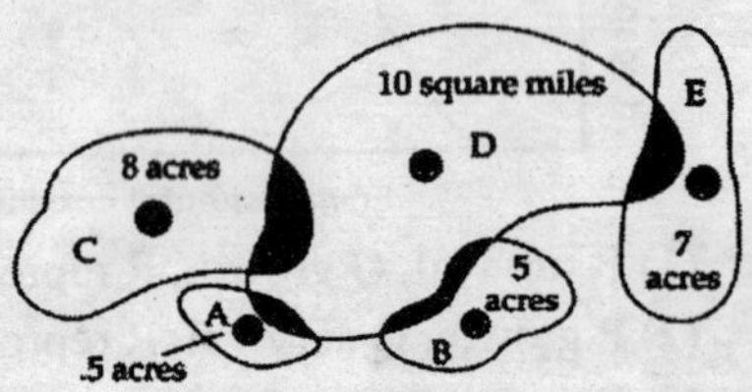

(Note: figure not drawn to scale.)

Which of the following area is most likely associated with a poor habitat?

(1) Area A (2) Area B
(3) Area C (4) Area D

71. An animal is most likely to defend an area of land against other members of its species if the

(1) defended area is extensive and has far-ranging borders
(2) defended area contains an evenly dispersed food source that is also available elsewhere
(3) defended area contains a concentrated food source that is difficult to locate
(4) benefit derived from exclusive use of the area is equal to the çost in energy of defending the area

72. A graph of population growth is shown below for two different animal species, X and Y, each of which produces a characteristics growth curve

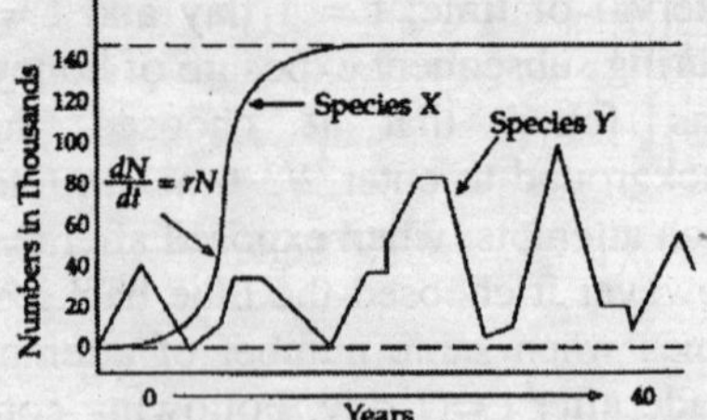

Which of the following statements is best supported by the data?

(1) Species X is characterized by a negative feedback control mechanism

(2) Species X is representative of an r-selected species

(3) Species Y is representative of an K-selected species

(4) The number of individuals in the population of species X has surpassed the carrying capacity of the environment once it reached a density of 140,000 individuals

73. Interactions between species may have a neutral effect upon the interacting populations, or they may act to increase or decrease the population sizes. The effects of different interspecies interactions possible between a two populations (population A and B) on population size, relative to the respective population sizes in the absence of the interactions, are presented in chart

Effect on Population Growth and Survival of Two Populations, *A* and *B*

Interaction	When Not Interacting A	When Not Interacting B	When Interacting A	When Interacting B	Outcome of Interaction
I	0	0	0	0	Neither *A* nor *B* Affects The Other
II	0	0	–	–	Population Most Affected is Driven from Niche
III	–	–	+	+	Interaction is Obligatory for *A* and *B*
IV	0	0	+	+	Interaction Favorable to *A* and *B*, but Not Obligatory
V	–	0	+	0	Obligatory for *A*; *B* Unaffected
VI	0	0	–	0	*A* Inhibited; *B* Unaffected
VII	–	0	+	–	Obligatory for *A*; *B* is Inhibited

+ = population growth increases
– = population growth decreases
0 = population growth remains the same

Interactions II characterizes interspecies competition between population A and B. What results are expected according Gause's competitive exclusion theory?

(1) If population A and B compete for the same limited resources, one population will succeed in gaining access to resources and the other population will be excluded

(2) If population A and B compete for the same limited resources, the population that succeed in gaining access to resources establishes new niche for itself

(3) If population A and B compete for the same limited resources, neither will excluded, so that both eventually will co-occupy the same niche

(4) If population A and B compete for the slightly different resources, the more fit population will cause the exclusion of the less fit population from the same area

74. A food web is a more accurate representation of the inter-relationship of predator and prey species than is a food chain. In California chaparral, a select group of resident producers and consumers were observed in the field and a food web was compiled based on their interactions. The diagram above indicates the inter-relationship between organism studies. Golpher snakes were noted to occasionally scavenge in addition to hunting their prey

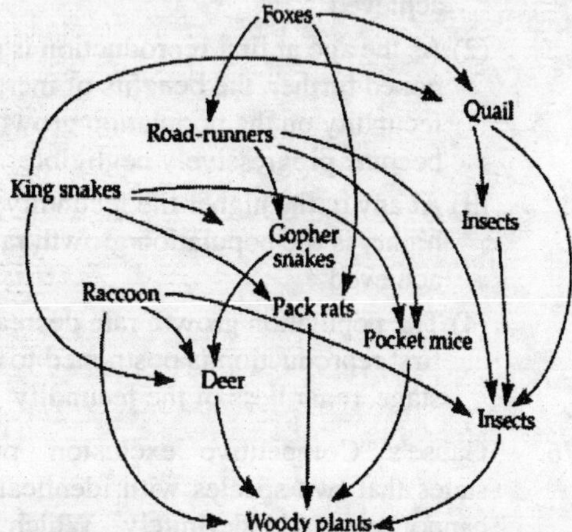

As the trophic levels represented become successively higher, which of the following takes place?

(1) Total biomass increases
(2) Amount of available energy increases
(3) Non-biodegradable pesticides residues become less concentrated
(4) Home range requirement of predators increase

75. The graph below shows the relationships of per capita population growth rate *(r)*, fecundity *(b)* and age at first reproduction *(α)* in an animal species

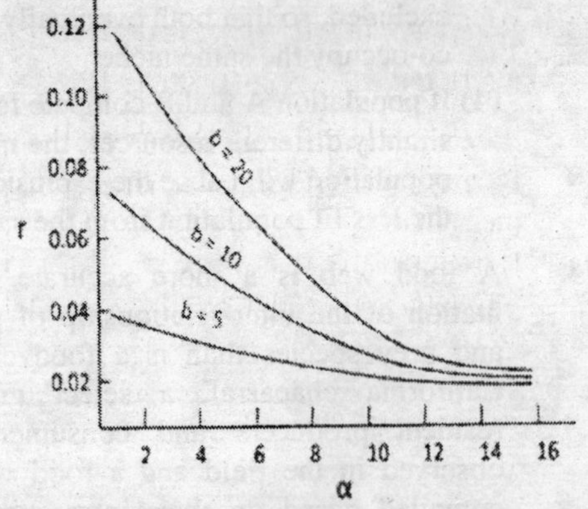

What is the most important conclusion to be drawn from the graph?

(1) The later the age of first reproduction, the lower is the population growth rate achieved
(2) As the age at first reproduction is postponed further, the benefits of increasing fecundity on the population growth rate become progressively negligible
(3) At any α, the higher the fecundity, the higher is the population growth rate achieved
(4) The population growth rate decreases as first reproduction is postponed to a later stage, regardless of the fecundity

76. Gause's 'Competitive exclusion' principle states that two species with identical niches cannot coexist indefinitely. Which of the following statements is the most appropriate regarding the validity of the principle?

(1) It depends on how one defines niche
(2) There are in nature many instances of continued coexistence of closely related species
(3) The principle is universally true
(4) It does not predict the outcome where, both the species are equally strong competitors

77. In a lake ecosystem, bottom-up effects (B) refers to control of a lower trophic level by the higher trophic levels and top down effects (T) refer to the opposite. In a lake with three trophic levels - Phytoplankton (P), Zooplankton (Z) and Carnivore (C),

(1) P is controlled by T and C are controlled by B, and Z is controlled by T
(2) P, Z and C are all controlled by T
(3) P is controlled by B, Z is controlled by T and C is controlled by B
(4) P is controlled by T, Z is controlled by B and C is controlled by T

78. Which one of the following trait set characterizes best a *k*-selected species?

(1) Usually a type III survivorship curve, short life span and density dependent mortality
(2) Usually a type I survivorship curve, long life span and density dependent mortality
(3) Usually a type I survivorship curve, long life span and density independent mortality
(4) Usually a type III survivorship curve, short life span and density independent mortality

79. The following graphs show the population growth of two species P and Q, each growing either alone (a) or in the presence of other species (b)

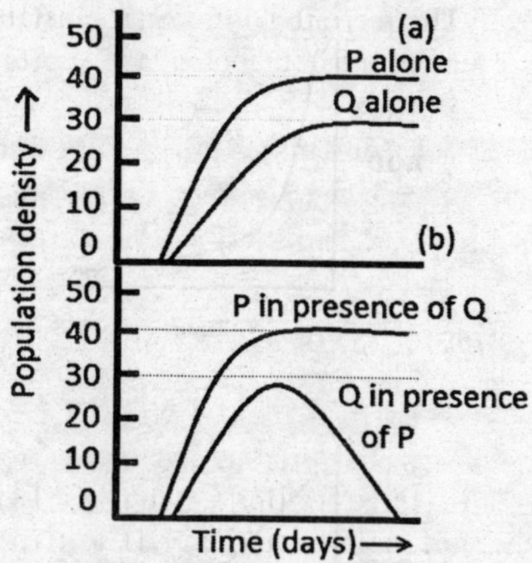

The most important conclusion to be drawn from the graph is

(1) P and Q are equally competitive

(2) In competition, the growth of both species is adversely affected

(3) There is no evidence of competitive exclusion

(4) In competition, species P remains unaffected while Q suffers

80. Three important biological parameters – generation time, population growth rate (r) and metabolic rate per gram body weight are a function of the organism's body size

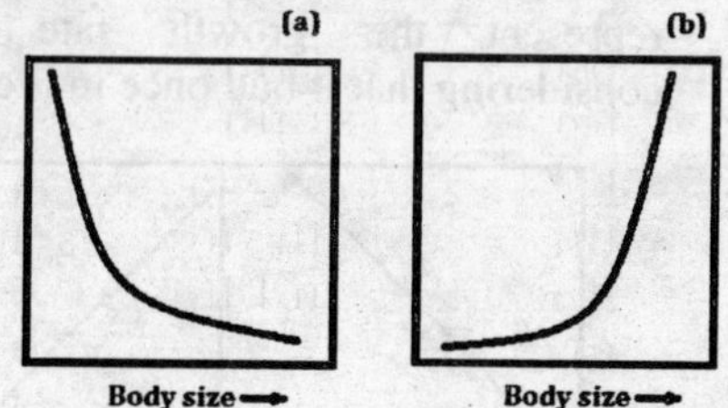

Which of the curves (a) or (b) represents the correct relation of each of the parameters to body size?

(1) Generation time – (b); population growth rate – (a); metabolic rate/g bw – (a)

(2) Generation time – (a); population growth rate – (b); metabolic rate/g bw – (a)

(3) Generation time – (a); population growth rate – (b); metabolic rate/g bw – (b)

(4) Generation time – (b); population growth rate – (b), metabolic rate rate/g bw – (a)

81. Consider the following statement regarding *r*-selected species

A. survivorship curve convex

B. low fecundity

C. density fairly constant at or near carrying capacity

D. good dispersal process

E. low level of social organization

Which of the following is correct?

(1) B and E (2) D and E

(3) A, B and C (4) C, D and E

82. Primary production in aquatic ecosystem is measured using Light-and-Dark-Bottle technique. In this method, as an indirect measure of photosynthetic production, dissolved oxygen concentration of the pond water enclosed in a BOD bottle is measured initially (I) and after a fixed duration of incubation in a light bottle (L) and a dark bottle (D) Then, the gross and net primary productions are estimated is

(1) (L-D) and (L-I), respectively

(2) (L-I) and (L--D), respectively

(3) (L-I) and (I-D), respectively

(4) (L-D) and (I-D), respectively

83. In column I are given equations and in column II what the equations represent. Column I and II are incorrectly matched

Column I	*Column II*
P. $\frac{dN_1}{dt} = r_1N_1\left(\frac{K_1 - N_1}{K_1}\right)$	1. population growth of species 1 in the presence of competing species 2
Q. $\frac{dN_1}{dt} = r_1N_1$	2. logistic population growth of species 1
R. $\frac{dN_1}{dt} = r_1N_1\left(\frac{K_1 - N_1 - \alpha N_2}{K_1}\right)$	3. population growth of prey species 1 in the presence of predator species 2
S. $\frac{dN_1}{dt} = N_1b_1 - d_1N_1N_2$	4. exponential population growth of species 1.

The correct match in sequence of the four equations from column I is

(1) 2, 4, 1 and 3 (2) 4, 2, 1 and 3

(3) 2, 4, 3 and 1 (4) 4, 2, 3 and 1

84. Which of the following statements are *incorrect* on ecological point of view?

P. primary succession involving xerosere is initiated in a wet habitat

Q. halones commonly found in electronic equipment are one of the active force destroying the protective ozone layer in the stratosphere

R. sympatric speciation occurs when the new species evolves in geographic isolation from the parent species

S. α-diversity is the diversity of species within a habitat or community

(1) P, Q (2) P, R
(3) Q, R (4) Q, S

85. In a lake low nutrients occur in upper strata due to consumption by algae while high nutrients are present at bottom due to activity of decomposer. Which is correct graphical representation of nutrients with depth for such lakes under steady state?

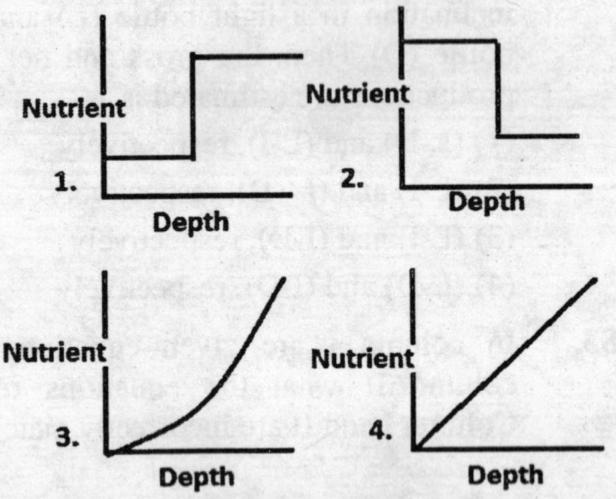

86. If water is oxygenated and inoculated by bacteria, then the graph of BOD with time for duration of five days will be

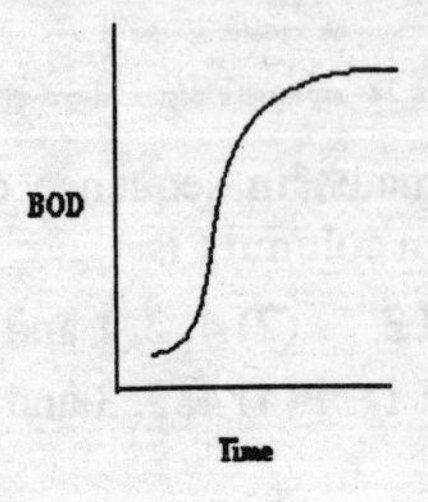

(1)

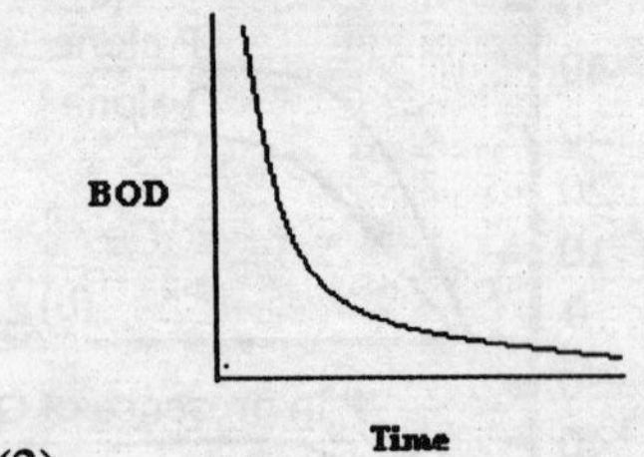

(2)

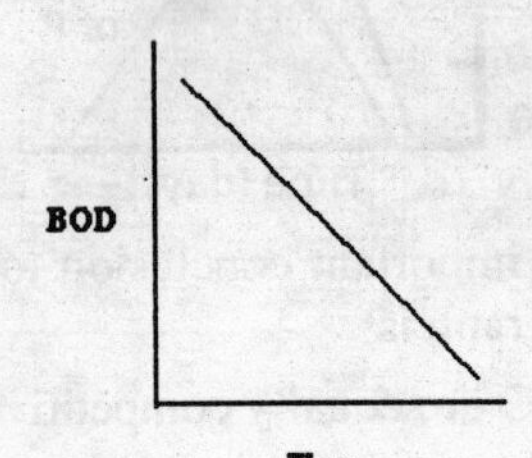

(3)

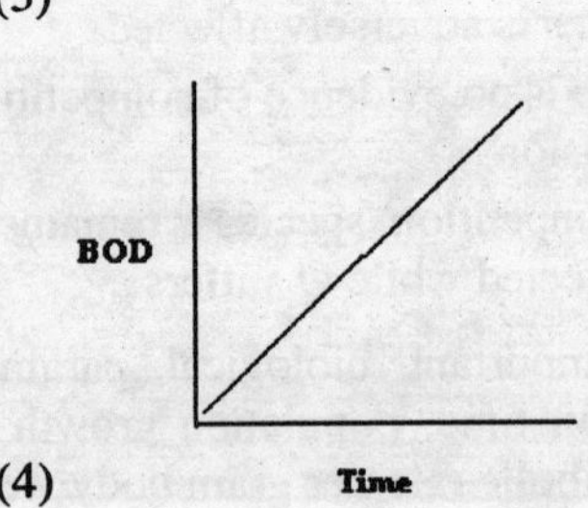

(4)

87. Among the following which graph correctly represent the growth rate in yeast considering that it bud once in life

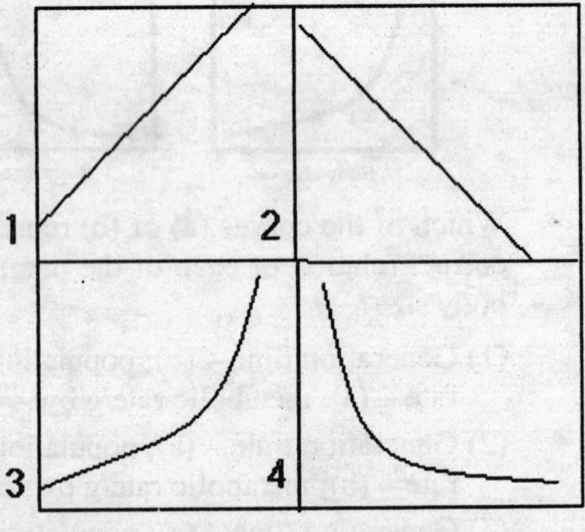

88. Survivorship Curve representing constant mortality at every stage of life would be

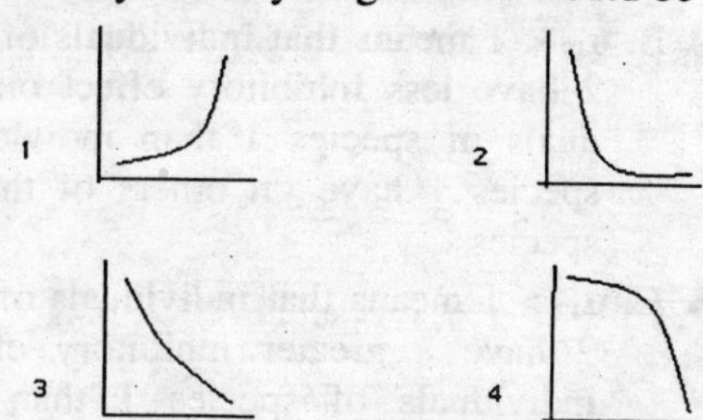

89. Curve which best explains the a population growing logistically is

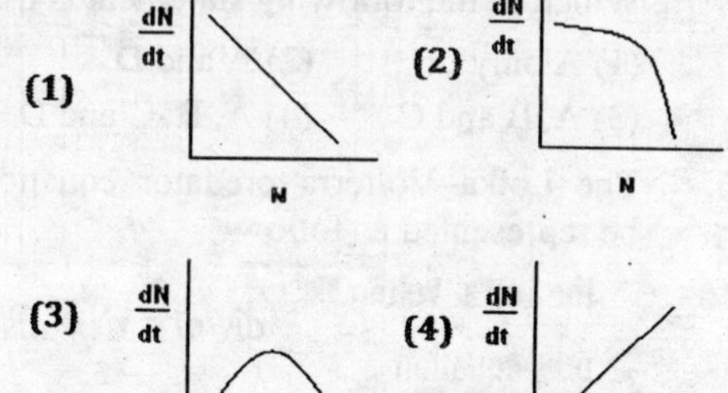

90. Consider the following statement regarding niche

A. Tolerances and requirements interact to define the conditions and resources needed by an individual or a species in order to practice its way of life is modern concept of niche given by Hutchinson

B. Niche to be an *n-dimensional hypervolume*, where *n* is the number of dimensions that make up the niche

C. Realized niche is overall potentialities of a species; fundamental niche describes the more limited spectrum of conditions and resources that allow it to persist, even in the presence of competitors and predators

Which of the following statement is correct?

(1) A and B (2) B and C
(3) A and C (4) B only

91. The reproductive value of an individual of age *x* (RV*x*) is the currency by which the worth of a life history in the hands of natural selection may be judged. Which is correct representation of reproductive value?

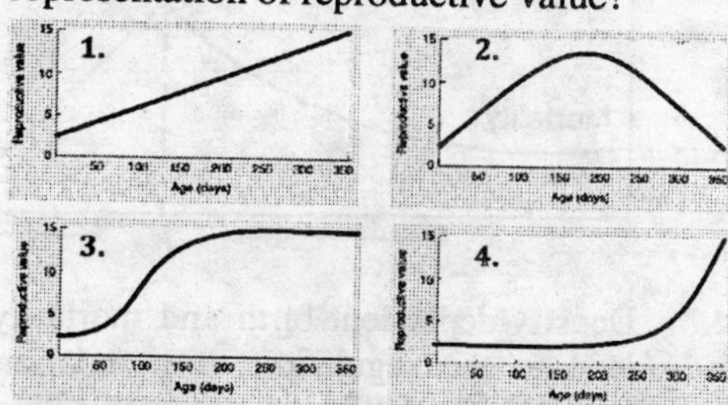

92. If the fitness of each individual offspring decreases as total clutch size increases, then the total fitness of a clutch (the product of number and individual fitness) must be maximized at some intermediate. However, if there is also a cost of reproduction, then the 'optimum' clutch size is that where the net fitness is greatest. In the graph below which point represents optimum clutch size?

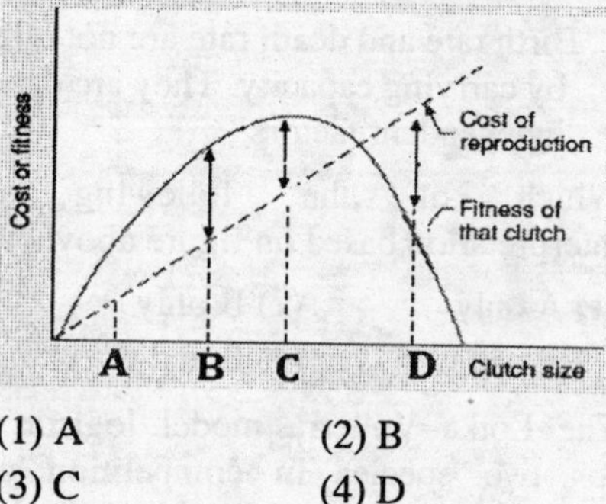

(1) A (2) B
(3) C (4) D

93. A competing grass plant is adversely affected by the presence of close neighbors, because the zone from which it extracts resources (light, water, nutrients) has been overlapped by the 'resource depletion zones' of these neighbors, making it more difficult to extract those resources. In such cases, competition may be described as

(1) interference competition
(2) exploitation competition
(3) apparent competition
(4) inter specific competition

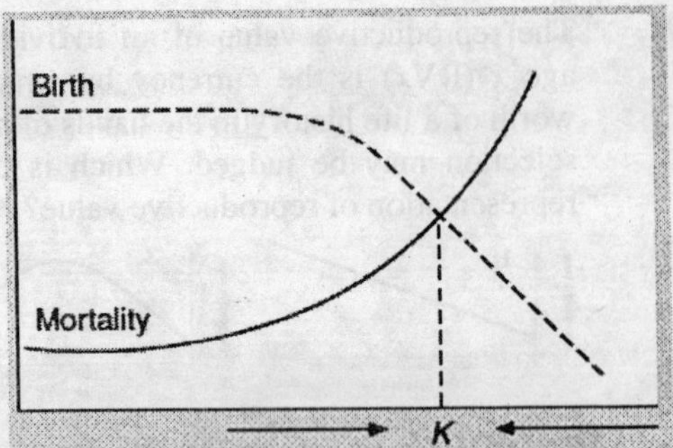

94. Density-dependent birth and mortality rates lead to the regulation of population size. Consider the following statements?

A. The density at which density dependent factor balance each other is called the carrying capacity (*K*) Below this the population increases, above it the population decreases: *K* is a stable equilibrium

B. Where mortality rate broadly increases, and birth rate also broadly increases, with density

C. Birth rate and death rate are not effected by carrying capacity. They are density independent factors

Which of the following correct interpretation based on figure above?

(1) A only (2) B only
(3) A and C (4) B and C

95. The Lotka–Volterra model logistic model for two species in competition can be represented as

$$\frac{dN_1}{dt} = r_1 N_1 \frac{(K_1 - (N_1 + \alpha_{12} N_2))}{K_1}$$

Where, the population size of one species can be denoted by N_1, and that of a second species by N_2. Their carrying capacities and intrinsic rates of increase are K_1, K_2, r_1 and r_2, respectively

Consider the following statement

A. α_{12} ('alpha-one two') is a measures the per capita competitive effect on species 1 of species 2. Thus, multiplying N_2 by α_{12} converts it to a number of '*N*1-equivalents

B. $\alpha_{12} < 1$ means that individuals of species 2 have less inhibitory effect on individuals of species 1 than individuals of species 1 have on others of their own species

C. $\alpha_{12} > 1$ means that individuals of species 2 have a greater inhibitory effect on individuals of species 1 than do the species 1 individuals themselves

D. This equation cannot be applied for intraspecific competition

Which of the following statement is true?

(1) A only (2) A and D
(3) A, B and C (4) A, B, C and D

96. The Lotka–Volterra predator equation can be represented as follow

the Lotka–Volterra prey equation $dN/dt = rN - aPN.$

the Lotka–Volterra predator equation $dP/dt = faPN - qP.$

The model has two components: P, the numbers present in a predator (or consumer) population, and N, the numbers or biomass present in a prey or plant population. Where 'r' in intrinsic rate of growth for prey and 'a' is a, called the 'attack rate' by predator, 'q' is the predator mortality rate and 'f' is the predator's efficiency

Consider the following statements

A. prey individuals are removed by predators at a rate that depends on the frequency of predator–prey encounters. Encounters will increase with the num-bers of predators (*P*) and the numbers of prey (*N*) However, the exact number encountered and successfully consumed will depend on the searching and attacking efficiency of the predator

B. Predator birth rate is *faPN* which is the rate at which food is consumed, *aPN*, and the predator's efficiency, *f*, at turning this food into predator offspring

C. Since *r* and *a* are constants, the prey zero isocline is a line for which *P* itself is a constant. Below it, predator abundance is low and the prey increase; above it, predator abundance is high and the prey decrease

Which of the following statement is correct?

(1) A and B (2) B and C
(3) A and C (4) A, B and C

97. Decomposer communities are, in their composition and activities, as diverse as or more diverse than any of the communities more commonly studied by ecologists. Consider the following statements

A. Decomposers and detritivores tend to have low levels of activity when temperatures are low, aeration is poor, soil water is scarce and conditions are acid

B. The activity of decomposers unlocks the mineral resources such as phosphorus and nitrogen that are fixed in rocks of oceans

C. The activities of the decomposers and detritivores are intimately interlocked, and are mostly antagonistic

D. Many of the decomposers and detritivores are specialists and the decay of dead organic matter results from the combined activities of organisms with widely different structures, forms and feeding habits

Which of the following statements are true?

(1) A and B (2) B and C
(3) C and D (4) A and D

98. A Population is growing logistically with a growth rate (r) of 0.15/week, in an environment with a carrying capacity of 800. What is the maximum growth rate (N_o) of individuals/week that this population can achieve?

(1) 15 (2) 30
(3) 22.5 (4) 60

99. Which of the following has maximum potential to exert green house potential per molecule

(1) CO_2 (2) CFC
(3) NOx (4) CH_4

100. Which of the following is not character of extinction prone species

(1) High trophic level
(2) Specialized niche
(3) Low intrinsic rate of growth
(4) Opportunitist

ANSWER SHEET
Unit 10: Ecology

1. 2	2. 3	3. 1	4. 2	5. 3	6. 1	7. 1	8. 1	9. 2	10. 3
11. 3	12. 4	13. 3	14. 2	15. 3	16. 1	17. 1	18. 3	19. 2	20. 1
21. 1	22. 2	23. 2	24. 3	25. 4	26. 1	27. 2	28. 4	29. 2	30. 2
31. 3	32. 2	33. 4	34. 1	35. 1	36. 2	37. 1	38. 1	39. 1	40. 2
41. 3	42. 3	43. 2	44. 3	45. 4	46. 4	47. 4	48. 1	49. 2	50. 3
51. 4	52. 4	53. 2	54. 1	55. 3	56. 4	57. 3	58. 3	59. 4	60. 4
61. 1	62. 1	63. 1	64. 4	65. 4	66. 3	67. 4	68. 4	69. 3	70. 3
71. 3	72. 3	73. 1	74. 1	75. 4	76. 2	77. 3	78. 2	79. 4	80. 4
81. 1	82. 1	83. 1	84. 1	85. 2	86. 1	87. 1	88. 1	89. 3	90. 1
91. 4	92. 2	93. 2	94. 1	95. 3	96. 4	97. 4	98. 2	99. 2	100. 4

UNIT-11

EVOLUTION

1. Which of the following statement is not true?
   (1) Natural selection is gradual process which brings change in allele frequency toward adaptive peaks
   (2) Genetic drift leads to homozygosity and random fixation of alleles over long period of time
   (3) Hardy weinberg can be applied only to non-evolving population
   (4) Sexual reproduction alone is sufficient for a population to evolve

2. This diagram represents two sedimentary rock layers with one intrusion/dike and a set of boulders. According to the principles used by 19th-century geologists to establish the relative ages of the major rock formations of Europe, what is the correct sequence of the illustrated structures (oldest to youngest)?

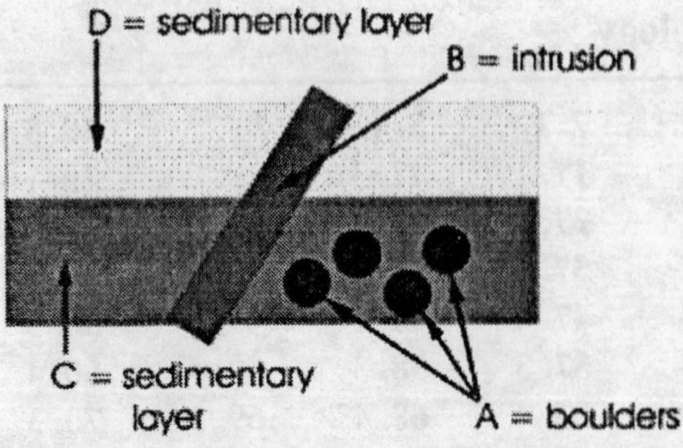

   (1) C, D, A, B
   (2) A, C, B, D
   (3) A, B, C, D
   (4) A, C, D, B

3. Match the following with correct term
   a. A group that contains an ancestor and all of its descendants
   b. A group that contains an ancestor and some, but not all, of its descendants
   c. The occurrence of shared traits that were inherited from a common ancestor
   d. The occurrence of shared traits that were not inherited from a common ancestor
   e. Modified from an ancestral state to a new state
   f. A shared trait that was modified (changed from an older ancestral state) in the most recent common ancestor
   g. A close relative of the groups under study, but one that is known to have branched off earlier than all the other groups
   h. A change of a modified trait back to its ancestral state

   Terms

   1. paraphyletic group 2. homoplasy
   3. outgroup 4. homology
   5. derived 6. synapomorphy
   7. reversal 8. monophlyetic group

   (1) a-4; b-1; c-8; d-5; e-7; f-6; g-3; h-2
   (2) a-8; b-2; c-1; d-3; e-4; f-5; g-6; h-7
   (3) a-7; b-1; c-2; d-5; e-4; f-6; g-3; h-8
   (4) a-8; b-1; c-4; d-2; e-5; f-6; g-3; h-7

4. Evolution as such is the simple proposition that the characteristics of lineages of organi-sms change over rune. Darwin was the first to argue that species had diverged from

common ancestors and that all of life could be portrayed as one great family tree

Consider the following points in light of Darwinian view

A. The differences between even radically different organisms have evolved incrementally, by small steps through intermediate forms

B. Large differences among organisms evolve by leaps, or saltations, without intermediates

C. Evolution occurs by changes in the proportions of individuals within a population that have different inherited characteristics

D. The sudden origin of new species by saltation and or due evolutionary change by transformation of individuals

E. Changes in the proportions of different types of individuals are caused by differences in their ability to survive and reproduce-and that such changes result in the evolution of adaptations, features that appear "designed" to fit organisms to their environment

Which of the following statement is correct?

(1) A, B and C (2) A, C and E
(3) B, C and D (4) A, D and E

5. "Absolute" ages of geological events can often be determined by radiometric dating, which measures the decay of certain radioactive elements in minerals that form in igneous rock. The decay of radioactive parent atoms into stable daughter atoms (e.g., Uranium 235 to Lead-207) occurs at a constant rate, such that each element has a specific half-life. The half-life of U-235, for example, is about 0.7 billion years

Consider the following statements

A. In each 0.7 billion year period, half the U-235 atoms present at the beginning of the period will decay into Pb-207

B. The ratio of parent to daughter atoms (Uranium 235 to Pb-207) in a rock sample thus provides an estimate of the rock's age

C. Only igneous rocks can be dated radiometrically, so a fossil-bearing sedimentary rock must be dated by bracketing it between younger and older igneous formations

D. Igneous rocks has majority of fossils in petrified forms

Which of the following statements are correct?

(1) A, B and C (2) A, B and D
(3) B, C and D (4) A, C and D

6. Although evolution is described by paleontological examples of gradual transitions through intermediate states, these kinds of transitions are by no means universally found in the fossil record. Intermediate stages in the evolution of many higher taxa are not known, and many closely related species in the fossil record are separated by smaller but nonetheless distinct gaps. Most paleontologists have followed Darwin in ascribing these gaps to the great incompleteness of the fossil record. Long periods in which species exhibit little or no detectable phenotypic change, interrupted by rapid shifts from one such state to another; can be explained by

(1) Punctuated gradualism
(2) Punctuated equilibrium
(3) Phyletic gradualism
(4) Adaptive radiation

7. Consider a situation in which a Uranium deposit is laid down in the strata and quickly isolated geologically. Uranium decays to Lead with a half-life of 100,000 years. (Note the actual half-life is different, but use this fictional value for this test question) If the deposit remains isolated for 400,000 years and is then uncovered and analyzed, what is the expected Uranium: Lead ratio?

(1) 1:3 (2) 1:7
(3) 1:15 (4) 3:15

8. Which of the following is a prediction of neutral theory of molecular evolution that is supported by data?

(1) Human and chimps differ more in DNA sequences of psuedogenes than in coding regions of functional genes

(2) Human and chimps differ more in DNA sequences of coding regions of functional genes than of pseduogenes

(3) Human have a faster evolution of most of their DNA sequences than chimps

(4) The more advanced species have more number of functional genes

9. Consider the following statements regarding evolution

A. Genes mutate, individuals are selected and populations evolve

B. Evolution is equilibrium in the gene pool of a population over time. The gene pool is the set of all genes in a species or population

C. Evolution cannot occur without morphological change; and morphological change cannot occur without evolution

D. Evolution is not progress. Populations simply adapt to their current surroundings. They do not necessarily become better in any absolute sense over time. A trait or strategy that is successful at one time may be unsuccessful at another

Which of the following statement is correct?

(1) A and D (2) B and C

(3) A and B (4) B and D

10. Linkage disequilibrium is a measure of association between alleles of two different genes. If there two loci in an organism (A and B) and two alleles at each of these loci (A_1, A_2, B_1 and B_2) linkage disequilibrium (D) is calculated as $D = f(A_1B_1) * f(A_2B_2) - f(A_1B_2) * f(A_2B_1)$ (where f(X) is the frequency of X in the population)

Consider the following statements

A. If two alleles were found together in organisms more often than would be expected, the alleles are in linkage equilibrium

B. Location of a gene on a chromosome, D varies between -1/4 and 1/4; the greater the deviation from zero, the greater the linkage. The sign is simply a consequence of how the alleles are numbered

C. Linkage disequilibrium can be the result of physical proximity of the genes. Or, it can be maintained by natural selection if some combinations of alleles work better as a team

Which of the following statement is correct?

(1) A and B (2) B and C

(3) A and C (4) All of the above

11. Sexual selection, the selection process acting on traits involved in the competition for reproductive opportunities, is thought to be stronger in male mammals than in female mammals. Which of the following is explanation of this phenomenon?

(1) Males experience greater energy expenditure through maintenance of weapons, such as horns and antlers, then do females

(2) Male reproductive success is limited by access to forage more than female reproductive success is by access to forage

(3) Isogamy

(4) Male reproductive success is more limited by access to females than female reproductive success is access to males

12. *Geospiza fortis* lives on the Galapagos islands along with fourteen other finch species. It feeds on the seeds of the plant *Tribulus cistoides*, specializing on the smaller seeds. Another species, *G. magnirostris*, has a larger beak and specializes on the larger seeds. The health of these bird populations depends on seed production. Seed production, in turn, depends on the arrival of wet season. In 1977, there was a drought. Rainfall was well below normal and fewer seeds were produced. As the season progressed, the *G. fortis* population depleted the supply of small seeds. Eventually, only larger seeds remained. Most of the finches starved; the population plummeted from about twelve hundred birds to less than two hundred

Under the given scenario which would be the probable outcome?

(1) A decrease in the proportion of large beaked birds in the population the next generation
(2) An increase in the proportion of large beaked birds in the population the next generation
(3) An increase in the proportion of small beaked birds in the population the next generation
(4) Both species will remain approximately in equal frequency

13. What will be the approximate effective population size in a panmictic population of 100 with 50 females and 50 polygamous males?

(1) 400 (2) 100
(3) 50 (4) 67

14. Some types of organisms within a population leave more offspring than others. Over time, the frequency of the more prolific type will increase. The difference in reproductive capability is called natural selection. Natural selection is the only mechanism of adaptive evolution; it is defined as differential reproductive success of prexisting classes of genetic variants in the gene pool

Consider the following statements

A. Natural selection can maintain or increase genetic variation depending on how it acts
B. Natural selection favors selfish behavior because any truly altruistic act increases the recipient's reproductive success while lowering the donors which would be difficult to evolve
C. Natural selection favors traits or behaviors that decrease a genotype's inclusive fitness
D. Natural selection does not have any foresight. It only allows organisms to adapt to their current environment

Which of the following statement about natural selection is not correct?

(1) A and D (2) B and C
(3) A and C (4) B and D

15. The theory of punctuated equilibrium is an inference about the process of macroevolution from the pattern of species documented in the fossil record. Consider the following statemetents

A. In the fossil record, transition from one species to another is usually abrupt in most geographic locales - no transitional forms are found. It appears that species remain unchanged for long stretches of time and then are quickly replaced by new species
B. Most major morphological change occurs (relatively) quickly in small peripheral population at the time of speciation. New forms will then invade the range of their ancestral species. Thus, at most locations that fossils are found, transition from one species to another will be abrupt
C. For a long time that rates of macroevolution vary over time, that is not controversial

Which of the following statement is correct as per punctuated equilibrium hypothesis?

(1) A and B (2) B and C
(3) A and C (4) A, B and C

16. Biologists studying evolution do a variety of things: population geneticists study the process as it is occurring; systematists seek to determine relationships between species and paleontologists seek to uncover details of the unfolding of life in the past. Life evolved in the sea. It stayed there for the majority of the history of earth

Consider the following statements regarding origin of life on earth?

A. The first replicating molecules were most likely DNA

B. These are: the prokaryotes ("ordinary" bacteria), archaebacteria (thermophilic, methanogenic and halophilic bacteria) and eukaryotes. Eukaryotes include protists (single celled organisms like amoebas and diatoms and a few multi-cellular forms such as kelp), fungi (including mushrooms and yeast), plants and animals. Eukaryotes and archaebacteria are the two most closely related of the three

C. The first cells must have been anaerobic because there was no oxygen in the atmosphere. In addition, they were probably thermophilic ("heat-loving") and fermentative

D. Photosynthesis evolved around 1.4 billion years ago. Photosynthesis is a process that allows organisms to harness sunlight to manufacture sugar from simpler precursors. The first photosystem to evolve, PSI, uses light to convert carbon dioxide (CO_2) and hydrogen sulfide (H_2S) to glucose. This process releases sulfur as a waste product

Which of the following statement is not correct?

(1) A and B (2) B and C
(3) C and D (4) A and D

17. Soft selection could be envisioned as selection that might act on aggressive behavioral genotypes in some lemming and field mouse species. When population density and frequency of the genotypes are low, these animals survive and reproduce. As population density increases, there can be a selection for more aggressive genotypes because they may be more successful in obtaining resources. As density increases further and the frequencies of the aggressive genotypes increase, they may be selected against because of the preoccupation of these aggressive individuals with territory defense under crowded conditions. This mode of selection is termed as

(1) Directional selection
(2) Stabilizing selection
(3) Disruptive selection
(4) Frequency dependent selection

18. Imagine that you go to the mountaintop this year, sample these beetles, and determine that 80% of the genes in the population are for green coloration and 20% of them are for brown coloration. You go back the next year, repeat the procedure, and find a new ratio: 60% green genes to 40% brown genes

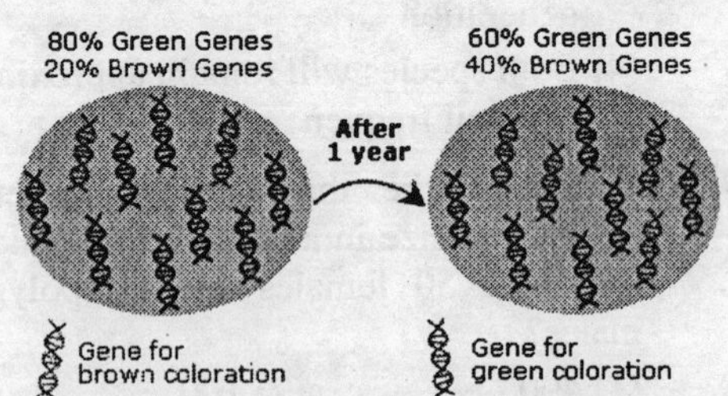

You have detected a microevolutionary pattern: a change in gene frequency. A change in gene frequency over time means that the population has evolved. The big question is, how did it happen?

(1) Natural Selection
(2) Genetic Drift
(3) Mutation
(4) Information is not complete to predict cause

19. In most of the Rocky Mountains, red squirrels are an important predator of lodgepole pine seeds. They harvest pinecones from the trees and store them through the winter. However, the pine trees are not defenseless: squirrels have a difficult time with wide pinecones that weigh a lot but have fewer seeds. Crossbill birds live in these places and also eat pine seeds, but the squirrels get to the seeds first, so those birds don't get as many seeds. However, in a few isolated places, there are no red squirrels, and crossbills are the most important seed predator for lodge poles. Again, the trees are not defenseless: crossbills have more difficulty getting seeds from cones with large, thick scales. But the birds have a mode of counter attack: crossbills with deeper, shorter, less curved bills are better able to extract seeds from tough cones. This plot gives an idea of

(1) Competition
(2) Symbiosis
(3) Co-evolution
(4) Divergent evolution

20. The term coevolution is used to describe cases where two (or more) species reciprocally affect each other's evolution. So for example, an evolutionary change in the morphology of a plant, might affect the morphology of an herbivore that eats the plant, which in turn might affect the evolution of the plant, which might affect the evolution of the herbivore and so on. Coevolution is likely to happen when different species have close ecological interactions with one another

These ecological relationships include:

A. Predation	B. Competitive species
C. Mutualistic species	D. Proto-cooperation
E. Amensalism	F. Parasitism

Which of the of the following relationship will result into co-evolution

(1) A, C and D
(2) D, E and F
(3) A, B, C and F
(4) B, C, D and E

21. Adult male frogs produce loud calls from a stationary calling position, which are detected by distant females of their species, who then approach them. Given that a male frog calls with a sound pressure level of 80 decibels (measured at a distance of 1 m), that sound attenuates by 6 dB for every doubling of distance and that females need a minimum sound pressure level of 50 dB to hear the male, what is maximum distance at which a female will hear the male?

(1) 16 m (2) 32 m
(3) 64 m (4) 128 m

22. Selection makes many organisms go to extreme lengths for sex: peacocks (top left) maintain elaborate tails, elephant seals (top right) fight over territories, fruit flies perform dances, and some species deliver persuasive gifts. After all, what female Mormon cricket (bottom right) could resist the gift of a juicy sperm-packet? Going to even more extreme lengths, the male red-back spider (bottom left) literally flings itself into the jaws of death in order to mate successfully. Such selection is often powerful enough to produce features that are harmful to the individual's survival. For example, extravagant and colorful tail feathers or fins are likely to attract predators as well as interested members of the opposite sex. Such a evolutionary process is termed as

(1) Natural selection
(2) Artificial selection
(3) Co-evolution
(4) Sexual selection

23. Because natural selection can produce amazing adaptations, it's tempting to think of it as an all-powerful force, urging organisms on, constantly pushing them in the direction of progress. Consider the following statement about natural selection and which is least likely to be true?

(1) Natural selection is not all-powerful; it does not produce perfection. If your genes are "good enough," you'll get some offspring into the next generation - you don't have to be perfect. This should be pretty clear just by looking at the populations around us: people may have

genes for genetic diseases, plants may not have the genes to survive a drought, a predator may not be quite fast enough to catch her prey every time she is hungry. No population or organism is perfectly adapted

(2) Natural selection is a process rather than as a guiding hand. Natural selection is the simple result of variation, differential reproduction, and heredity — it is mindless and mechanistic. It has no goals; it's not striving to produce "progress" or a balanced ecosystem

(3) The population or individual does not "want" or "try" to evolve, and natural selection cannot try to supply what an organism "needs." Natural selection just selects among whatever variations exist in the population. The result is evolution

(4) Natural selection is interpreted as a random process. The genetic variation that occurs in a population because of mutation is random and selection acts on that variation in a very random way

24. *Biston betularia*, the "peppered" moths living near English industrial cities. These insects have varieties that vary in wing and body coloration from light to dark. During the 19th century, sooty smoke from coal burning furnaces killed the lichen on trees and darkened the bark. When moths landed on these trees and other blackened surfaces, the dark colored ones were harder to spot by birds who ate them and, subsequently, they more often lived long enough to reproduce. Over generations, the environment continued to favor darker moths. As a result, they progressively became more common. By 1895, 98% of the moths in the vicinity of English cities like Manchester were mostly black. Since the 1950's, air pollution controls have significantly reduced the amount of heavy particulate air pollutants reaching the trees, buildings, and other objects in the environment. As a result, lichen has grown back, making trees lighter in color. In addition, once blackened buildings were cleaned making them lighter in color. Now, selection favors lighter moth varieties so they have become the most common. This is example is in support of

(1) Cryptic coloration
(2) Adaptive coloration
(3) Natural selection
(4) Loss of biodiversity

25. On returning to England, Darwin and an ornithologist associate identified 13 species of finches that he had collected on the Galápagos Islands. This was puzzling since he knew of only one species of this bird on the mainland of South America, nearly 600 miles to the east, where they had all presumably originated. He observed that the Galápagos species differed from each other in beak size and shape. He also noted that the beak varieties were associated with diets based on different foods. He concluded that when the original South American finches reached the islands, they dispersed to different environments where they had to adapt to different conditions. Over many generations, they changed anatomically in ways that allowed them to get enough food and survive to reproduce. This observation in beak shape and size is termed as

(1) Inheritance of acquired character
(2) Natural selection
(3) Genetic Drift
(4) Character displacement

26. Caterpillar of an insect species are specialized to feed on the leaves of single host plant species, which is toxic and contain a chemical which releases cyanide when degraded. Concentration of this chemical were measured in the leaves as well as the blood, gut and faeces of the caterpillars. The chemical was found to be higher in concentration in the gut and faeces of the caterpillar than in the leaves but absent in blood. The caterpillars were not seen to suffer any ill effects from feeding on the leaves but were instantly killed on artificial exposure to cyanide. This suggests that the caterpillars have evolved a mechanism to

(1) absorb and detoxify the cyanide precursor by metabolizing it
(2) selectively forage on leaves that do not contain the cyanide precursor
(3) resist the toxic affects of cyanide
(4) inhibit the cyanide precursor from breaking down to release cyanide

27. Of three lizard species A, B and C that form a monophyletic group, A and B possess the ability to glide in the air, while C does not. Molecular genetic studies indicate that B and C are more closely related to each other than either is to A. Which of the following statements is more likely to be true, given the principle of parsimony?

(1) Gliding in the air is an ancestral trait for this group
(2) Gliding in the air is a derived trait for this group
(3) Gliding in the air is an ancestral trait for all lizards
(4) The ability to glide has been lost only in this group of lizards

28. Individually marked bees were trained to forage from a feeder kept 100 m away from their hive. In test trails, the feeders were removed and the distance at which bees started searching for the feeders were noted in each trial. Small weights of 2, 5 and 10 mg were then glued on to each bee's thorax and the trials repeated with each of these weights

The results are shown below

Weight added to the bees (mg)	Distance from hive at which searching behaviour was initiated (m)
0	100
2	85
5	75
10	50

This experiment suggests that

(1) The estimation of distance by flying bees is positively correlated with the energy expended during flight
(2) The estimation of distance by flying bees is negatively correlated with the energy expended during flight
(3) The estimation of distance by flying bees is uncorrelated with the energy expended during flight
(4) Bees are unable to perform distance estimation when weights are added to their bodies

29. Newborn chicks will react by fleeing for cover when a model of a hawk is drawn above them in the air. Experiments reveal that this response is completely non-specific and can be elicited by moving objects of any shape in the air. Ehen the chicks are three week old, they respond selectively to objects that have the shapes of birds of prey and ignore others. In an experiment, newly hatched chicks were divided into two groups: One was exposed twice a day to models of different shape and sizes and the other was reared with no exposure to such models at any points. After three weeks, both groups were tested for their responses to different shaped models. Chicks of both group showed selective avoidance responses only to models shaped like birds of prey. This experiments shows that

(1) The avoidance behavior by fleeing for cover is learned in first three week post hatching
(2) The selective avoidance of birds of prey is learned in the first three weeks post hatching
(3) The selective avoidance of birds of prey develops in the first three week post hatching and is not learned
(4) The selective avoidance of birds of prey is learned before hatching

30. A species of bird prey on two insects species, A and B that occur in the same habitat and breed in the same season. At the height of breeding season, the relative abundance of insect species A and B in the habitat is 0.2 and 0.8 respectively. When the proportion of insect species A and B in the diet of birds was examined, however, the ratio (A:B) was found to be 0.5:0.5. Which of the following statement is correct?

(1) The birds preys randomly on the two insect species

(2) The bird prefers species A over species B

(3) The bird prefers species A over species A

(4) The bird preys in accordance with their relative abundance

31. In an experiment with red-winged black birds, which are territorial songbird, females were presented a choice between normal control males and males whose wings were artificially made redder. Females were found to mate more often with males with artificial redder wings than normal males, who attacked them. The above experiment suggest that redness of male wings evolves due to

(1) Intersexual selection rather than intrasexual selection

(2) Intrasexual selection rather than intersexual selection

(3) Both intra- and inter-sexual selection

(4) Neither intra- nor inter-sexual selection

32. A songbird species produces specific alarm calls in the presence of aerial predators such as eagles. On hearing an alarm call, individual songbirds drop what they are doing and fly into trees. In an observational field study, the number of eagles alarm calls produced in different contexts was recorded and the results are sho'vn below

Behavioural context	Number of eagle alarm calls observed/day
Presence of eagles	35
Absence of eagles, songbirds foraging	10
Absence of eagles, songbirds not foraging	0

The above results may be construed as evidence for

(1) Mimicry (2) Aposematism

(3) Deception (4) Camouflage

33. Males of a grasshopper species attracts mates using songs. Females reply to each verse of male song with a reply song, enabling males to find them. Song rates are species-specific and are used by moth males and females to identify their species. Being poikilotherms (no physiological control of body temperature) their song rates changes with temperature, rising linearly with rising temperature in both males and females. Individual singing in sunny patches have a body temperature 35°C, those singing in the shade have a temperature of 20°C. Which of the following is most likely to be true?

(1) A female in the sun will respond most to a male singing in the shade

(2) A female in the sun will be found most easily by a male singing in the shade

(3) A female in the sun will respond most to a male singing in the sun

(4) A female in the sun will not respond to a male singing in the sun

34. Several species of moths belonging to a variety of lepidopteran families have evolved chemical communication are use chemical signals (pheromones) to attract mates. In some species, females call (release pheromone) and males approach whereas in other species, males produce pheromones and females approach them. Individual of species with female signaling produce pheromones in tiny quantities while those with male signaling produce pheromone in large quantities. Which of the following is not a likely explanation for this difference between male and female signaling strategies across species?

(1) Resources are more limiting for females than for males

(2) The ability to locate odour sources evolve faster in males than in females

(3) Females usually employ low cost, low risk strategies while males often employ high cost, high risk strategies for reproduction

(4) Males that can locate females using minute quantities of odour may have been favoured by sexual selection as this ability may be an honest signal of male quality

35. In order for continuing evolution there must be mechanisms to increase or create genetic variation and mechanisms to decrease it

The mechanisms of evolution are

A. Mutation

B. Natural selection

C. Genetic drift

D. Recombination and gene flow

Which of the following mechanisms decrease genetic variation?

(1) A and B (2) B and C

(3) C and D (4) A and D

36. R.A. Fisher proposed that females may have an innate preference for some male trait before it appears in a population. Females would then mate with male carriers when the trait appears. The offspring of these matings have the genes for both the trait and the preference for the trait. As a result, the process snowballs until natural selection brings it into check

Suppose that female birds prefer males with longer than average tail feathers. Mutant males with longer than average feathers will produce more offspring than the short feathered males. In the next generation, average tail length will increase

As the generations progress, feather length will increase because females do not prefer a specific length tail, but a longer than average tail. Eventually tail length will increase to the point were the liability to survival is matched by the sexual attractiveness of the trait and an equilibrium will be established

This concept indicates about?

(1) Handicap principle

(2) Run away selection model

(3) The good genes model

(4) Intra sexual selection

37. The genetic relatedness (r) of an individual to his nephew is 0.25. The alleles that cause uncles to care for nephews will spread, according to Hamilton's Rule, only if the fitness benefit is

(1) equal to cost of care

(2) more than the cost of care by 25%

(3) double the cost of care

(4) four times the cost of care

38. The frequencies of two alleles p and q for a gene locus in a population at Hardy-Weinberg equilibrium are 0.3 and 0.7, respectively. After a few generations of inbreeding, the heterozygote frequency was found to be 0.28, The inbreeding coefficient in this case is

(1) 0.42 (2) 0.28

(3) 0.33 (4) 0.67

39. Number of trials required for rates to learn a task when they were exposed to various conditions were as follows

Experimental conditions	Observation
A. Light : Dark cycle-12 h: 12h	N- trials
B. Bright light- 24 h	Significantly more trails than 'N'
C. Bright light-24 h + continuous physical disturbances	Significantly more trails than 'N'
D. Dark light-24 h + continuous physical disturbances	Significantly more trails than 'N'

Which of the following inferences is most appropriate?

(1) Continuous light enhanced learning

(2) Continuous darkness inhibited learning

(3) Physical activity inhibited learning

(4) Learning was reduced by sleep loss

40. Shown in the graph below are the fitness costs and benefits of four alternatives behavioural phenotypes (P,Q,R,S) Given sufficient evolutionary time, which phenotype(s) is least likely to evolve as an adaptation?

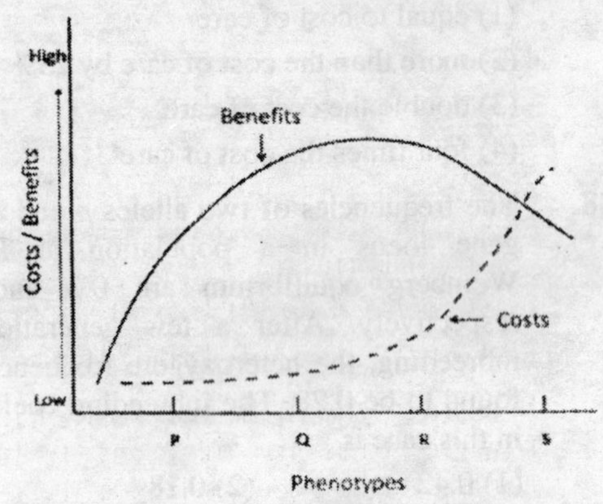

(1) Q and R (2) Q only
(3) P, Q and R (4) S only

41. The following geological eras mark the advent of important events in the history of earth - origin of terrestrial plants, origin of mammals and break up of the land mass Pangaea

A. Early Cambrian
B. Late Devonian
C. Late Cretaceous
D. Late Jurassic

Identify the correct match of the events with the geological era

(1) Origin of terrestrial plants - (c);
Origin of mammals (d);
Break-up of Pangaea - (a)
(2) Origin of terrestrial plants - (b);
Origin of mammals (d);
Break up of Pangaea –(c)
(3) Origin of terrestrial plants - (d);
Origin of mammals (c);
Break up of Pangaea –(b)
(4) Origin of terrestrial plants - (b);
Origin of mammals (c);
Break up of Pangaea –(d)

42. A species of moth that pollinates a night-flowering plant species is subjected to the following experiment. In a Flowering patch of the plant species outdoors, four treatments are applied in equal numbers

A. Black cloth bags enclosing individual natural flowers
B. Artificial unscented model flowers
C. Artificial unscented model flowers covered with black cloth bags and D Control unmanipulated natural flowers

The number of moths approaching each flower between 6 p.m. and 10 p.m. is noted. The results are shown below

Treatment	Number of moths approaching
Control unmanipulated natural flowers	35
Natural flowers covered by cloth bags	1
Artificial unscented model flowers	32
Artificial unscented model flowers covered with cloth bags	0

The results suggest that

(1) olfactory cues are necessary for the moths to recognize the flowers
(2) olfactory cues are sufficient for the moths to recognize the flowers
(3) visual cues are sufficient for the moths to recognize the flowers
(4) visual cues are necessary but not sufficient for the moths to recognize the flowers

43. In several birds and mammals, offspring from one year stay back and help parents bring up offspring in the next year. If the main benefit they get from such helping is inclusive fitness benefits, then which of the following processes is most likely to favor helping behaviours?

(1) Extensive intraspecific brood parasitism
(2) Long reproductive tenure and genetic monogamy
(3) Frequent turn-over in reproducing adults
(4) Extensive extra-pair fertilizations

44. Nearly 25% of all insect species are known to be herbivores. Yet, in spite of such heavy herbivore pressure, globally green plants tend to persist, contributing to a green earth". Which of the following account for the relative success of green plants?

A. Herbivore insects are inefficient feeders

B. Herbivore insect densities are kept low by predators

C. Plants secrete herbivore-deterrent chemicals

(1) B and C (2) A only

(3) B and C (4) A, B and C

45. In a species of frog, large males are susceptible to predation by bats but make up for its being preferred by the females. Small males are less preferred by the females but make up for it by also not being preferred by bats. Intermediate sized males seems to have the worst of both-they are often eaten by bats and are not preferred by females. In a such a population, male body size will evolve by

(1) sexual selection

(2) directional selection

(3) stabilizing selection

(4) disruptive selection

46. The flightless birds ostrich, rhea and emu are distributed on different continents. What is the most plausible explanation that is given by an evolutionary biologist for this discontinuous observation?

(1) The birds were able to fly earlier, but lost their flight ability later

(2) Prehistoric humans transported these birds to different continents

(3) The birds, although flightless, may have used drifting logs to cross the ocean and reach other continents

(4) All the continents used to be one single supercontinent earlier and the flightless birds were isolated after the break up of the landmass

47. Inspite of its two-fold cost, sexual reproduction is the most dominant mode of reproduction among the living organisms. Which of the following reasons might account for this?

A. Sexual reproduction generates genetic heterogeneity through recombination

B. Sexual reproduction helps in purging deleterious mutations

C. Sexual reproduction evolved to stay evolutionarily ahead of fast evolving internal parasites

(1) A only (2) A and B

(3) C only (4) A, B and C

48. In a bird species A, the male alone builds the nest, incubates the eggs and feeds the nestling. In bird species B, it is the female that does all that. In bird species C, both sexes contribute equally to the above activities. In species A and B, the uninvolved partner may fly away and mate again. Which sex among A, B and C is most likely to develop colourful plumage during breeding season?

(1) Male in species A and B, both sexes in species C

(2) Female in species A, male in species B and C

(3) Female in species A and B, neither in species C

(4) Female in species A, male in species B, neither in species C

49. A species of bird specializes in capturing live snakes for food. When young fledglings of this species that have never encountered live snakes before are presented with long, thin rods, they readily approach and peck at them. However, if these rods are painted with alternate red and yellow rings to resemble the deadly coral snake, the same fledglings avoid these rods. This experiment shows that

(1) both the attraction to rod-like objects and avoidance of red and. yellow ones are behaviors that are not learned

(2) the attraction towards rod-like objects is not a learned behavior but avoidance of red and yellow objects is learned
(3) the attraction towards rod-like objects is a learned behavior whereas the avoidance of red & yellow objects is not learned
(4) both the attraction towards rod-like objects and the avoidance of red and yellow ones are learned behaviors

50. Chickadees were given an opportunity to store seeds in 72 possible storage sites consisting of holes drilled in small trees that had been placed in an aviary. Typically, the seeds were placed in only 4 or 5 storage sites. The chickadees were then removed from the aviary, the seeds removed from the storage sites, and each hole covered. A day later, the chickadees were returned to the aviary and they spent nearly 5 times as long pulling at the covers on storage sites than at the covers on sites where they had not stored food. This experiment indicates that chickadees
(1) use auditory cues to find stored food
(2) have a spatial memory that enables them to locate storage sites
(3) have a propensity to inspect and pull at the covers wherever they are located
(4) cannot find a storage site if they cannot see the food in it

51. 1 Tyrannosaurus (dinosaur)
2. Lobe-finned fishes
3. Great apes
4. Eohippus (horse)
5. Trilobites

Which of the following represents the proper sequence, from the earliest in the fossil record to the latest, of the animals listed above?
(1) 1, 2, 5, 4, 3
(2) 2, 5, 1, 3, 4
(3) 5, 1, 2, 3, 4
(4) 5, 2, 1, 4, 3

52. The mussel *Mytilus edulis* thrives in saline habitats, in both the highly salty seawater of tidal zones and the less salty estuaries. This result in two kinds of populations: one adapted to the higher salt concentrations of the tidal zone, and one adapted to the lower salt concentrations of the estuary. It has been found that the more salt-tolerant populations have high frequencies of an allele that produces an enzyme involved in maintaining osmotic equilibrium

Conversely, estuarine mussels having the same enzyme seem to be disfavored and have a much higher death rate than mussels without the allele. Adult estuarine populations do have lower frequencies of this allele. Each spring, large numbers of larvae from the salty habitats pour into the estuaries

More salt-tolerant populations have high frequencies of an allele that produces an enzyme involved in maintaining osmotic equilibrium. Such a type of selection in which one extreme is preferred is termed as
(1) directional selection
(2) stabalizing selection
(3) cyclic selection
(4) disruptive selection

Question 53-54

Birds were placed in a small room maintained at a constant temperature of 20ºC and were provided with food. Lights were turned on at 0800 hours and were turned off at 2000 hours. Three experiments were conducted as follows

In experiment 1, after 3 days of the light/dark cycle; the birds were placed in constant darkness and their activity cycles were recorded

In experiment 2, after 3 days of the light/dark cycle; the birds were kept in dark for another 12 hours. The lights were turned on at 2000 hours and turned off at 0800 hours for a number of days and bird's activity cycle were recorded

In experiment 3, after three days of light/dark cycle, the bird's pineal glands were removed. The birds were returned to constant darkness and their activity cycles were recorded. These experiments are illustrated in the following figures

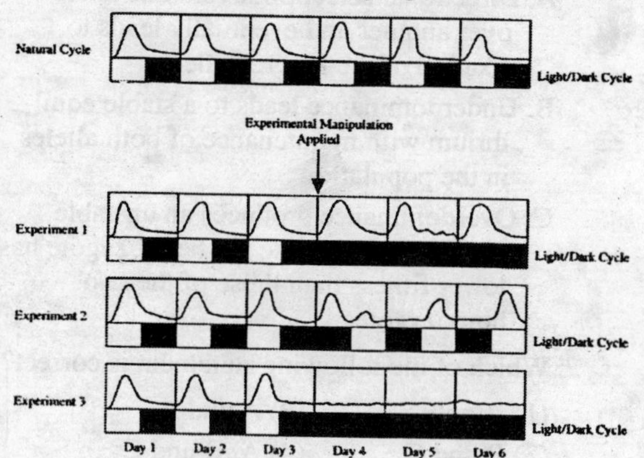

53. All of the following characteristics of an endogenous activity cycle are indicated by these experiments except

(1) The cycles persist under constant conditions

(2) The onset of activity shifts as a result of changing day length

(3) The timing of activity can be reset by environmental stimuli

(4) The period of cycle approximates 24 hours

54. If all of the following experiments were performed, which would best hypothesis that the pineal gland controls the endogenous activity cycle?

(1) Transplanting intact pineal gland into pinealectomized birds to determine whether the activity cycle is reestablished

(2) Repeating experiment 3 with different species of bird

(3) Repeating experiment 2, but using pinealectomized birds to determine whether they can shift their activity to the changed light/dark cycles

(4) Repeating experiment 3 several more times to establish the same results always occur in pinealectomized birds

55. A small, isolated population of moths are nearly decimated by a tornado. The few surviving moths by chance happened to be located in an area tangential to the direct path of the tornedo when it touched down. Thirty percent of the tangential path moth survived in the tornado's wake. Those moths that happened to be in the direct path of the tornado weren't as fortunate: only 1% of their original number survived. Which of the following is true of the subpopulations of tangential–path moths which survived in the greatest numbers?

(1) They were more fit than the subpopulations of direct path moths that suffered the greater loss

(2) They were acted upon by natural selection while the direct path moths were not

(3) They demonstrated Darwinian competition with regard to the direct path members of their species

(4) The higher rate of survival compared to the direct path moths was due to chance

56. A Florida dentist tested positive for HIV. At a later date, a female patient also tested positive for HIV. Genetic analysis determined that the dentist had infected some of his patients. The principle behind this analysis is

(1) that all strains of HIV are identical.

(2) that the virus isolated from an infected individual will be most similar to the infecting virus than to any other

(3) all cases of HIV infection yield a unique strain of the virus

(4) that a viral strain passed from one individual to another will be identical in each person

57. The human MN blood type antigens are determined by two codominant alleles, L^M and L^N. The MN blood types and corresponding genotypes of 398 villagers from Piparlia are tabulated here

Phenotype	Genotype	Number
MM	$L^{M}L^{M}$	182
MN	$L^{M}L^{N}$	172
NN	$L^{N}L^{N}$	44

Calculate the allele frequencies at the MN locus for the Piparlia population

(1) p =0.327; q = 0.673
(2) p =0.673; q = 0.327
(3) p =0.457; q = 0.111
(4) p =0.432; q = 0.568

58. Genetic drift is change in allelic frequency due to chance factors. The amount of change in allelic frequency due to genetic drift is inversely related to the effective population size (the equivalent number of breeding adults in a population) Effective population size decreases when there are unequal numbers of breeding males and females. Consider the following effects of genetic drift

A. It produces change in allelic frequencies within a population. Because drift is random, allelic frequency is just as likely to increase as it is to decrease and will wander with the passage of time
B. It increases genetic variation within populations. Through random change, an allele may eventually reach a frequency of either 1 or 0, at which point all individuals in the population are homozygous for one allele
C. Different populations diverge genetica-lly with time and finally lead to forma-tion of new species
D. It may occur in large panmicitic population if part of population crash due to catastrophs

Which of the following statement is correct?

(1) A and D (2) A and B
(3) B and C (4) C and D

59. Natural selection changes allelic frequencies; the direction and magnitude of change depends on the intensity of selection, the dominance relations of the alleles, and the allelic frequencies. Consider the following statements

A. Directional selection favors one allele over another and eventually leads to fixation of the favored allele
B. Underdominance leads to a stable equilibrium with maintenance of both alleles in the population
C. Overdominance produces an unstable equilibrium because the heterozygote has lower fitness than those of the two homozygotes

Which of the following statement is correct?

(1) A only (2) A and B
(3) B and C (4) A, B and C

60. Following is the phylogeny showing relationships between six bird species. Species A and F are nectar feeding birds and possess long, thin, curvy beaks, while the remaining species are seed eaters and possess thick beaks. The evolution of beak character in species A and F can be explained by

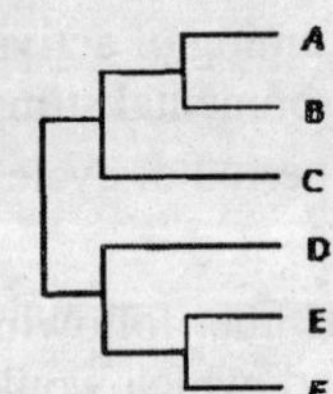

(1) convergent evolution
(2) directional selection
(3) disruptive selection
(4) reversal

61. The Weber-Fechner law states that the magnitude of a perceived sensation increases as the logarithm of the stimulus intensity. Assume a background stimulus level of 1 unit. Suppose this is increased to 10 units (situation A) or to 100 units (situation B). Then the perceived sensation in situation 0 is stronger than the sensation perceived in situation A by a factor of

(1) 1 (2) 2
(3) 10 (4) 100

62. The diagram below represents a section of undistributed layers of sedimentary rocks in the Western Ghats and shows the location of fossils of several closely related species. According to currently accepted evolutionary theory, which is the most probable assumption about species A, B and C?

Species A
Species B and A
Species B
Species C

→ Surface of ground

(1) Species B is more abundant than species C

(2) Species C existed before species B

(3) Species A and B are genetically identical

(4) Species B descended from species A

63. Stalk-eyed flies have their eyes at the end of pair of stalks on the head. In a mate choice experiment, females were offered a choice between pair of males. Each pair was represented by one male with shorter than average eye stalks and one with longer average eye stalks. It was found that female with short eye stalks preferred makes with short eye stalks, and those with long eye stalks preferred with males with long eye stalks. This suggests that male eye stalk are under

(1) Directional selection

(2) Stabilizing selection

(3) Frequency dependent selection

(4) Disruptive selection

64. Researchers studying red-vented bulbuls catch birds and place colored bands around the legs of individual birds in order to be able to identify them individually. What is the maximum number of unique color codes that are possible if 3 bands are placed on the right leg alone, 5 different colors are used (red, blue, green, yellow, black), repeated colors are not used on an individual bird, and a given set of three colors can be used as different arrangement (i.e. one individuals can get red blue and green another green red blue, or any other arrangements of these three colors) but each arrangement can be used once?

(1) 60 (2) 120

(3) 3125 (4) 125

65. A moth species occurs as two distinct morphs based on wing colour – pale and dark. In the forest there are trees with dark coloured trunks as well as those with light coloured trunks and the moths can rest on either tree. Birds capture the resting moths and eat. In a field experiment, the proportion (%) of dark and pale morphs captured from dark and light trunk trees was recorded

Tree Trunk Colour	Moth morph Dark	Pale
Dark	45	55
Light	48	52

The most plausible conclusion to be drawn from the results is

(1) Natural selection favours dark morphs in forests where trees with dark trunks are dominant

(2) Birds can detect dark morphs better than light morphs

(3) Pale morphs prefer to rest on light coloured trunks

(4) Birds detect the moths by cues other than their wing colour

66. It is found that people with the genetic disease called sickle cell anaemia are resistant to malaria. Which of the following best describes the underlying mechanism?

(1) Frequency-dependent selection

(2) Superiority of heterozygotes

(3) Transient polymorphism

(4) Balanced polymorphism

67. Following are the characteristics of species that make them more or less prone to extinction

Rare – a

Common – b

Good dispersal rate – c

Poor dispersal rate – d
Low specialization – e
High specialization – f
High variability – g
Low variability – h
Low trophic status – i
High trophic status – j
Long life span – k
Short life span – l
High reproductive output – m
Low reproductive output – n

Which of the following is the correct combination of characteristics that makes the species more prone to extinction?

(1) a d f g j l n (2) a e f h i k m
(3) b d e g i l n (4) b c f h j k m

68. The phosphoglucomutase (Pgm) locus in California populations of Nuttall's white-crowned sparrow has allelic frequencies of 0.8 and 0.2, respectively, for the alleles Pgm-A and Pgm-B. What is the probability that two heterozygous individuals will mate, given that Pgm is in Hardy-Weinberg equilibrium?

(1) 0.01 (2) 0.10
(3) 0.16 (4) 0.32

69. In a certain species of lizard, there are two color morphs associated with a single locus (Y) having two alleles (Y, y): yellow (yy) and green (YY, Yy) The fitness of the morphs is frequency-dependent, as indicated in the graph above. If the frequency of the yellow morph in a population is 0.70 and individuals are randomly mating, then the expected equilibrium population due to natural selection is

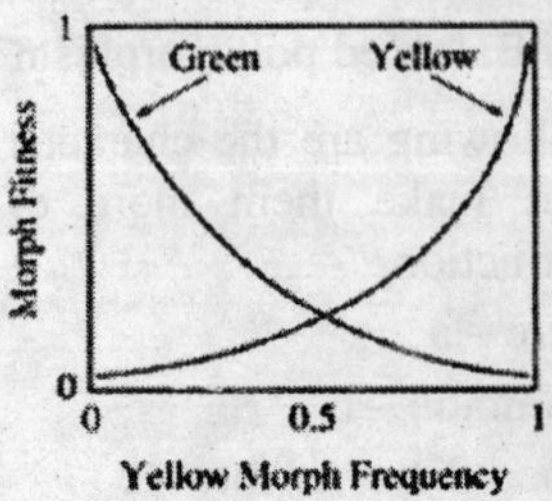

(1) monomorphic for yellow morphs
(2) monomorphic for green morphs
(3) polymorphic with a stable ratio of about 1:1 of yellow to green morphs
(4) polymorphic with oscillating frequencies of yellow and green morphs

70. Two populations of sunflowers, A and B, differ in the heritability of the oil content in their seeds; H^2 equals 0.40 in population A, while H^2 equals 0.70 in population B. Which of the following statements about these sunflower populations is correct?

(1) Population A's seeds have a lower oil content than those of B
(2) The variability in seed oil content is higher in population B than in A
(3) Population A grows in a more diverse and unstable environment than does B
(4) Proportionately more phenotypic variation is environmental variation in population A than in B

71. Sewall Wright's metaphor of evolution on an adaptive landscape describes a population as occupying a field of genotypes graded by contours delimiting gene combinations of equal fitness. The evolutionary mechanism most likely to move a population consistently "uphill" toward adaptive peaks on this landscape is

(1) chance
(2) directed mutation
(3) natural selection
(4) gene flow between populations

72. Several distinct time periods and different routes might explain the entrance of marsupials into Australia

(i) Late Jurassic - early therians arrived in Antarctica - Australia where the marsupials subsequently evolved
(ii) Early to middle Cretaceous - early marsupials arrived in Australia from northern regions and then radiated in isolation
(iii) Paleocene - marsupials entered Australia from South-East Asia

(iv) Eocene - chance dispersal of marsupials into Australia

Which of the following is the correct combination?

(1) (i) (ii) (iii) (2) (i) (iii) (iv)
(3) (ii) (iii) (iv) (4) (i) (ii) (iv)

73. According to fossil record, the earliest fossils of liverworts are found in late Devonian, of mosses in early cretaceous, and vascular plants in the later Silurian/early Devonian. *Anthoceros* (hornworts) fossils have not been recovered. Reading fossil records we would say that vascular plants appeared first and then liverworts

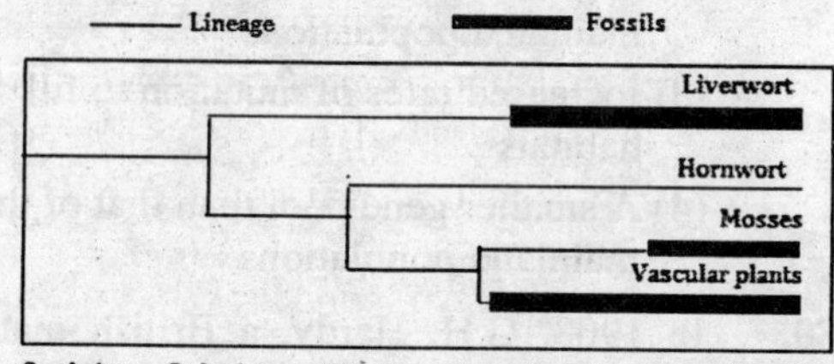

However, phylogenetic relationship (shown in figure) suggest otherwise. It may be that

A. evolutionary history can be read directly from fossil record
B. The moss lineage goes back to at least early Silurian/early Devonian
C. Fossil can only set a maximum age for a lineage
D. Fossil can only sets a minimum age for a lineage
E. the divergence between liverworts and rest of land plants goes back to at least early Ordovician

Which of the following statements is correct?

(1) A, B, C and E (2) B, D and E
(3) A, B, D and E (4) B, C and E

74. A species of moth feeds on nectar from the white flowers of a particular plant species. In an experiment, individual moths were presented with white model flowers mimicking real ones and scent extracts of the flowers presented separately on paper. The two cues were separated by different distances in different treatments. The cues were placed at the end of a flight tunnel in the laboratory and moths were released one at a time and their behaviour observed. Moths were scored as showing a positive response if they attempted to feed from either the visual or the scent cue. The results are shown below

Distance between visual and scent cues (cm)	Percentage of tested moths attempting to feed on either models or scented paper
0	95
2	95
10	50
50	0

This experiment shows that

(1) Visual and olfactory cues from flowers are equally effective in attracting moths
(2) Visual cues are more effective than olfactory ones for flower recognition
(3) Olfactory cues are more effective than visual ones for flower recognition
(4) Moths integrate visual and olfactory cues in the process of flower recognition

75. Suppose evolution on each had occurred in such a way that there are 96 amino acids instead of 20. DNA has 12 different types of bases and DNA synthesis occurs in the same way as today. The minimum number of bases per codon would be

(1) 12 (2) 8
(3) 2 (4) 3

76. You visit an island and drop off 2,000 individuals of a sexually reproducing species. 5,000 years later you revisit the island and discover that there are now only 500 individuals on the island, and that there is a range of body sizes from 2 cm to 100 cm tall, some of which look nothing like what you dropped off 5,000 years ago. Your best guess as to what happened is

(1) a population bottleneck led to homozygous conditions and then mutations

(2) adaptive radiation led to the speciation process occurring
(3) mosaic evolution with some individuals being small and others larger
(4) polytomy and inbreeding depression leading to mutations

77. An animal was first maintained in a constant environmental condition for several days until a consistent biological rhythm (B) was established. The animal was then exposed to an experimental physical rhythm (E)which modulates the phase and period of B. However, upon withdrawal of E, the B gradually regained its pattern of pre-exposure condition. From these observations which one or more of the following should be the most logical inference?

A. E is a Zeitgeber
B. E is a masking agent
C. E causes entrainment of B
D. B is a conditioned to E

The correct answer is

(1) A and C (2) B and D
(3) B only (4) D only

78. Which of the following statements are correct?

A. Organisms produce many more offspring than can survive, which results in a struggle for existence
B. Natural selection will operate among any entities that reproduce, show inheritance of their characteristics from one generation to the next
C. Selection may be directional, stabilizing, or disruptive
D. In the absence of natural selection, and with random mating in a large population in which inheritance is Mendelian, the genotype frequencies at a locus move in one generation to the Hardy–Weinberg ratio; the genotype frequencies are then stable

(1) A and B (2) A, B and C
(3) B and D (4) A, B, C and D

79. One of the classical examples of evolution occurs on the Galápagos Islands with Darwin's finches. All the organisms living on these islands are descendants of species that have emigrated there, primarily from South America. In studying the finch populations, researchers have identified fourteen species, none of which are found on the mainland

The initial colonizing population of finches most likely exhibited which of the following?

(1) Hybridization with bird species already existing on the islands
(2) High rates of interbreeding with mainland populations
(3) Increased rates of mutation to fill habitats
(4) A smaller gene pool than that of the mainland populations

80. In 1908, G.H. Hardy, a British mathematician, and W. Weinberg, a German physician, independently discovered a rule that relates allelic and genotypic frequencies in a population of diploid, sexually reproducing individuals if that population has random mating, large size, no mutation or migration, and no selection

Consider the following aspects

A. The allelic frequencies at an autosomal locus in a population will not change from one generation to the next (allelic-frequency equilibrium)
B. The genotypic frequencies of the population are determined in a predictable way by the allelic frequencies (genotypic-frequency equilibrium)
C. The equilibrium is neutral. That is, if it is perturbed, it will be reestablished within one generation of random mating at the new allelic frequencies (if all the other requirements are maintained)
D. If population is allele and genotype frequency equilibrium there would be phyletic evolution which gradual and slow

Which of the following statement are correct in relations to Hardy-Weinberg equilibrium?

(1) A and D only (2) B and C only
(3) A, B and C (4) A, B, C and D

81. In the sample of one hundred individuals having allele frequency p=0.7 and q=0.3 and segregating the A^1 and A^2 alleles at the A locus: A^1A^1, fifty-four; A^1A^2, thirty-two; and A^2A^2, fourteen. Calculate the inbreeding coefficient of this population

(1) 0.42 (2) 0.24
(3) 0.76 (4) 0.50

82. At a particular locus, there are two alleles, *B* and *b*. The mutation rate of *B* to *b* is 3.5 x 10^{-4}, whereas the mutation rate of *b* to *B* is 6 x 10^{-8}. What is the equilibrium frequency of the *b* allele, assuming no other factor is operating in this population to disturb the Hardy-Weinberg equilibrium?

(1) 0.9998 (2) 0.8828
(3) 0.1742 (4) 0.5833

83. Digger wasps paralyze cockroach prey by injecting venom into their head and thorax via their stings. In an experiment, "hot" wasps were produced, whose venom component were radioactively labeled. Cockroaches were divided into two groups: one group was stung by "hot" wasps, while the other was artificially injected in the head and thorax capsules with the same amount of radioactive venom. Examination of the cockroaches revealed high radioactivity in the head and thoracic ganglia compared with the rest of the head and thorax in stung cockroaches but only low levels in the ganglia as compared to other tissue in injected cockroaches. This experiment demonstrate that

(1) wasps exhibit extreme precision in their stinging behaviour
(2) radiolabeling enables wasps to produce large quantities of venom
(3) radiolabeling increase the rate of diffusion of venom into the nervous system
(4) radiolabeling increases the precision of was stinging behaviour

84. Consider the following periods of geological time scale

(i) Paleoarchean (ii) Cambrian
(iii) Ordovician (iv) Silurian

Link above periods with the following events in flora and fauna

A. First green plants and fungi on land
B. First Vascular plants (the rhyniophytes); First jawed fishes
C. First known oxygen-producing bacteria. Oldest definitive microfossils
D. Major diversification of life, numerous fossils; First chordates appear

(1) (i)- C; (ii)-D; (iii)-B; (iv)-A
(2) (i)- D; (ii)-C; (iii)-A; (iv)-B
(3) (i)- A; (ii)-D; (iii)-B; (iv)-C
(4) (i)- C; (ii)-D; (iii)-A; (iv)-B

85. Consider the following periods of geological time scale

(i) Permian (ii) Triassic
(iii) Jurassic (iv) Createceous
(v) Devonian

Link with the following events in flora and fauna

A. Flowering plants proliferate. Monotremes, marsupials and placental mammals appear. Break up of Gondwana
B. Gymnosperms (especially conifers and cycads) and ferns common. First birds and lizards. Atmospheric CO_2 levels 4–5 times the present day levels
C. First mammals and crocodilia appear
D. Landmasses unite into supercontinent Pangaea, 95% of life on Earth becomes
E. First clubmosses, horsetails and ferns appear, as do the first seed-bearing plants (progymnosperms) First amphibians still aquatic

(1) (i)- E; (ii)-D; (iii)-C; (iv)-B; (v)-A
(2) (i)- C; (ii)-D; (iii)-A; (iv)-B; (v)-E
(3) (i)- C; (ii)-D; (iii)-B; (iv)-A; (v)-E
(4) (i)- E; (ii)-C; (iii)-D; (iv)-B; (v)-A

86. The Galapagos finches were an important clue to Darwin's thinking about the origin of species. These finches are believed to have descended from a single ancestral species that colonized the Galapagos archipelago, America, over a short period of time. The Galapagos finches differ in their beak shape and size. Different species feed on seeds that vary in size and hardness

Which of the following is the most likely explanation for these patterns?

(1) The finches represent an example of directional trend in beak size from small to big

(2) Beak shapes changed in response to different seed types and these changes were inherited by subsequent generations

(3) The ancestral finch already had all the beak variations and different lineages formed that were specialized to eat different seed

(4) The finches represent an example of adaptive radiation in which beak variation was generated by mutation followed by selection by different seed types

87. In order to demonstrate that the long tails of males attracted females in a bird species, experimenters captured and cut the tails of 'n' number of males and monitored the number of females mated by each male

They had two types of controls in the experiment

(i) 'n' males that were not captured

(ii)'n' males that were captured, had their tails cut and then stitched back to attain the original size

The males with cut tails mated with a significantly smaller number of females than both the controls. Which of the following alternative explanations is not ruled out by the, experiment?

(1) The stress of cutting tails affected the performance of males

(2) The time wasted in the capture reduced mating opportunities of males

(3) Females avoided any deviation from normal

(4) Females chose males randomly

88. In an experiment that has continued for more than 50 years, corn has been propagated by breeding only from plants with the height amount of oil in the seeds. The average oil content is now much greater than any of the plants in the original population

The following hypotheses were proposed as explanations for this observation

A. Mutations occurred that increased the oil content in seeds

B. Plants with high oil content were stimulated to produce offspring with more oil in their seeds

C. The breeding led to increased frequency of alleles at multiple loci, so that new combinations of genes for even higher oil content were formed

Which of the following represents a combination of correct statements?

(1) A and B only

(2) A and C only

(3) B and C only

(4) A, B and C

89. The following questions refer to Figure, which shows an outcrop of sedimentary rock whose strata are labeled A-D

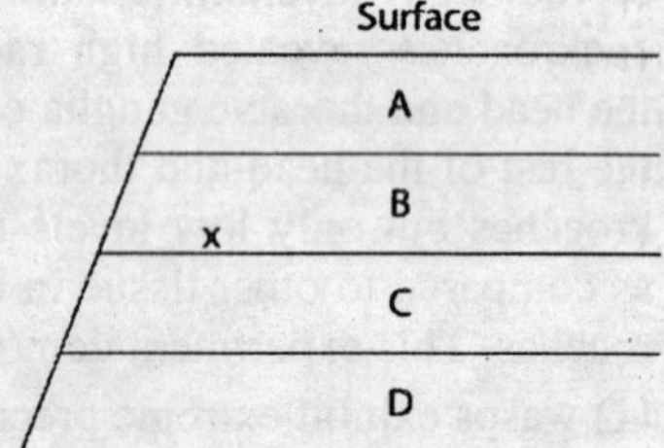

If x indicates the location of fossils of two closely related species, then fossils of their most-recent common ancestor are most likely to occur in which stratum?

(1) A (2) B

(3) C (4) D

90. **Currently, two extant elephant species (X and Y) are placed in the genus *Loxodonta* and a third species (Z) is placed in the genus *Elephas*. Assuming this classification reflects evolutionary relatedness, which of the following is the most accurate phylogenetic tree?**

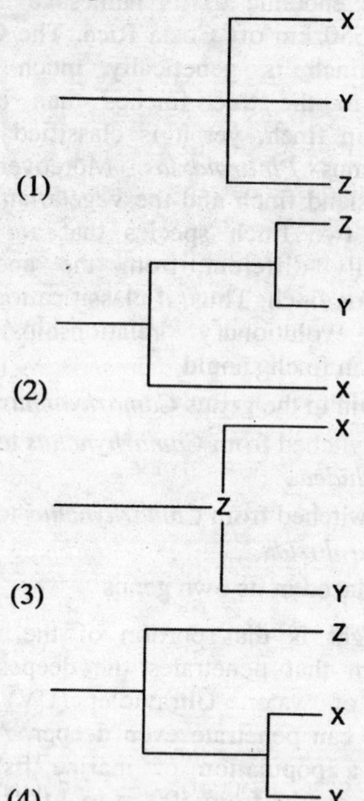

91. **In life history evolution there is generally a trade off between the size and number of offsprings produced. Some conditions are listed below**

A. Scarcity of food during the early stages of life

B. Provision of parental care

C. High mortality during early stages of life

D. Predator's preference for large sized prey

What are the above two conditions that would favour the production of a small number of large-sized offspring?

(1) B and C **(2) B and D**

(3) A and B **(4) A and C**

92. **The horizontal axis of the cladogram depicted below is a timeline that extends from 100,000 years ago to the present; the vertical axis represents nothing in particular. The labeled branch points on the tree (V—Z) represent various common ancestors. Let's say that only since 50,000 years ago has there been enough variation between the lineages depicted here to separate them into distinct species, and only the tips of the lineages on this tree represent distinct species**

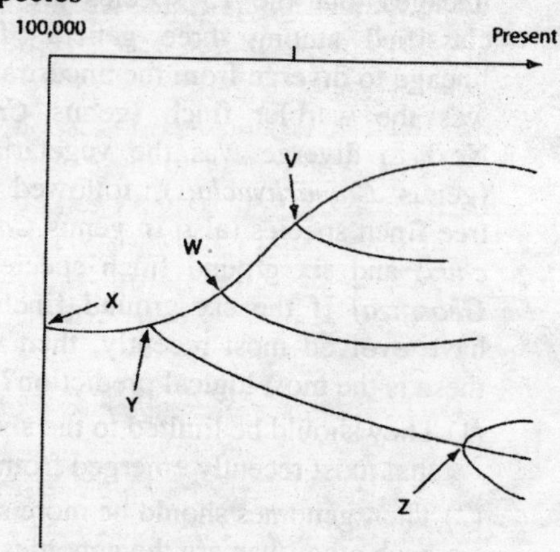

How many separate species, both extant and extinct, are depicted in this tree?

(1) two **(2) three**

(3) four **(4) six**

93. **Two kinds of natural selection (A and B) acting on a trait are shown in the figure below. In each, the top graph shows the trait frequency before and the bottom graph frequency after the action of natural selection**

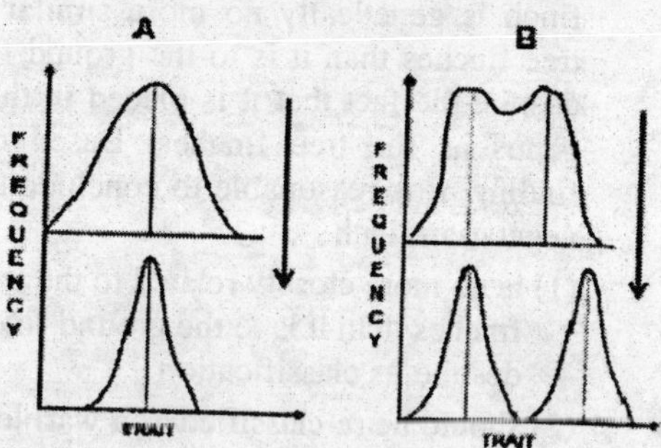

The kind of natural selection in A and B are

(1) A- Directional, B-Disruptive

(2) A- Neutral, B- Disruptive

(3) A- Stabilizing, B- Disruptive

(4) A-Disruptive, B- Stabilizing

94. About 13 different species of finches inhabit the Galápagos Islands today, all descendants of a common ancestor from the South American mainland that arrived a few million years ago. Genetically, there are four distinct lineages, but the 13 species are currently classified among three genera. The first lineage to diverge from the ancestral lineage was the warbler finch (genus *Certhidea)* Next to diverge was the vegetarian finch (genus *Camarhynchus*), followed by five tree finch species (also in genus *Camarhynchus*) and six ground finch species (genus *Geospiza)* If the six ground finch species have evolved most recently, then which of these is the most logical prediction?

(1) They should be limited to the six islands that most recently emerged from the sea

(2) Their genomes should be more similar to each other than are the genomes of the five tree finch species

(3) They should share fewer anatomical homologies with each other than they share with the tree finches

(4) The chances of hybridization between two ground finch species should be less than the chances of hybridization between two tree finch species

95. According to a 1999 study, the vegetarian finch is genetically no more similar to the tree finches than it is to the ground finches, despite the fact that it is placed in the same genus as the tree finches. Based on this finding, it is reasonable to conclude that the vegetarian finch

(1) is no more closely related to the tree finches than it is to the ground finches, despite its classification

(2) should be re-classified as a warbler finch

(3) is not truly a descendent of the original ancestral finch

(4) is a hybrid species, resulting from a cross between a ground finch and a tree finch

96. A 14th species that descended from the original ancestral finch, the Cocos Island finch, is endemic to its namesake island, located 550 km off Costa Rica. The Cocos Island finch is genetically much more similar to the tree finches than is the vegetarian finch, yet it is classified in its own genus *Pinarolaxias*. Moreover, the Cocos Island finch and the vegetarian finch are the two finch species that are most genetically different from the ancestral Galápagos finch. Thus, if classification is to reflect evolutionary relationships, the vegetarian finch should

(1) remain in the genus *Camarhynchus*

(2) be switched from *Camarhynchus* to *Certhidea*

(3) be switched from *Camarhynchus* to *Pinarolaxias*

(4) be placed in its own genus

97. Blue light is that portion of the visible spectrum that penetrates the deepest into bodies of water. Ultraviolet (UV) light, though, can penetrate even deeper. A gene within a population of marine fish that inhabits depths from 500 m to 1,000 m has an allele for a photopigment that is sensitive to UV light, and another allele for a photopigment that is sensitive to blue light. Which of the following graphs best depicts the predicted distribution of these alleles within a population if the fish that carry these alleles prefer to locate themselves where they can see best?

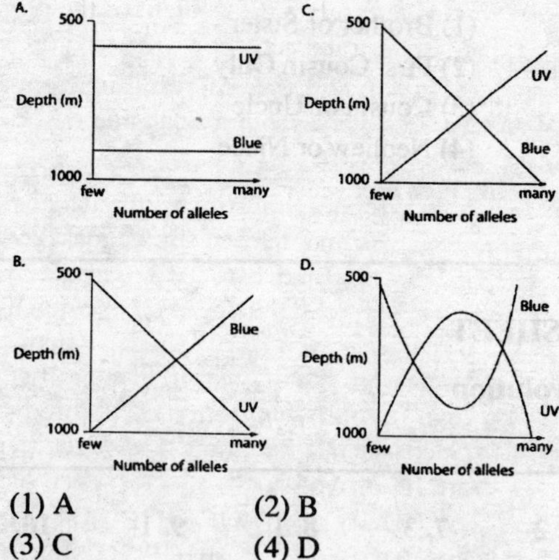

(1) A (2) B
(3) C (4) D

98. Anopheles mosquitoes, which carry the malaria parasite, cannot live above elevations of 5,900 feet. In addition, oxygen availability decreases with higher altitude. Consider a hypothetical human population that is adapted to life on the slopes of Mt. Kilimanjaro in Tanzania, a country in equatorial Africa. Mt. Kilimanjaro's base is about 2,600 feet above sea level and its peak is 19,341 feet above sea level. If the incidence of the sickle-cell allele in the population is plotted against altitude (feet above sea level), which of the following distributions is most likely, assuming little migration of people up or down the mountain?

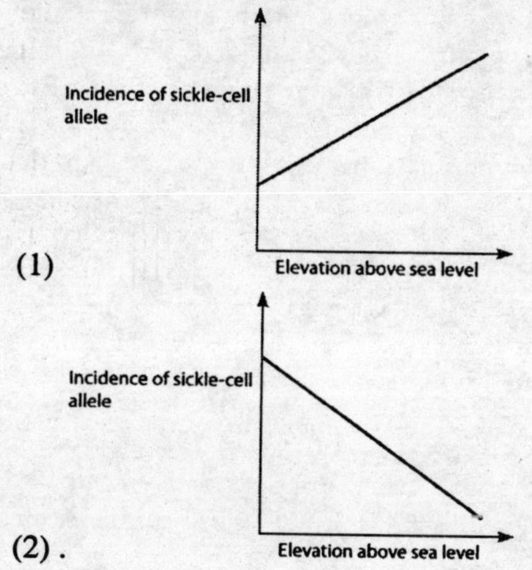

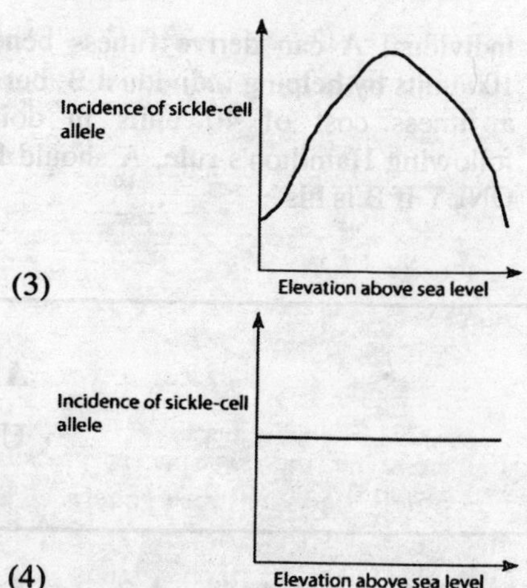

99. There are three species of frogs - A, B and C. Species A does not provide parental care for its eggs and larvae. Species B is subjected to predation by a predator that selectively feeds only on small-sized larvae. Species C faces progressively decreasing opportunities for breeding with increasing age. Assuming that resources available for reproduction are similar for A, B and C. Which of the following strategies would have been favored?

(1) A should produce large number of small-sized offspring; B should produce a small number of large-sized offsprings; C should breed earlier in life
(2) Species A and B should produce a small number, of large-sized offspring and C should breed earlier in life
(3) Both species A and B should produce a large number of small-sized offspring and C should breed later in life but increase its clutch size
(4) Species A should produce a small number of large-sized offspring; B should 'produce a large number of small sized offspring and C should breed earlier in life with a small clutch size

100. Individual A can derive fitness benefit of 100 units by helping individual B, but incurs a fitness cost of 40 units in doing so following Hamilton's rule, A should help B ONLY if B is his

(1) Brother or Sister
(2) First Cousin Only
(3) Cousin or Uncle
(4) Nephew or Niece

ANSWER SHEET

Unit 11: Evolution

1. 4	**2.** 4	**3.** 4	**4.** 2	**5.** 1	**6.** 2	**7.** 3	**8.** 1	**9.** 1	**10.** 2
11. 4	**12.** 2	**13.** 2	**14.** 3	**15.** 4	**16.** 4	**17.** 4	**18.** 4	**19.** 3	**20.** 3
21. 2	**22.** 4	**23.** 4	**24.** 3	**25.** 4	**26.** 4	**27.** 1	**28.** 1	**29.** 2	**30.** 2
31. 1	**32.** 3	**33.** 3	**34.** 2	**35.** 2	**36.** 2	**37.** 4	**38.** 3	**39.** 4	**40.** 4
41. 4	**42.** 3	**43.** 2	**44.** 4	**45.** 4	**46.** 4	**47.** 4	**48.** 4	**49.** 1	**50.** 2
51. 4	**52.** 1	**53.** 1	**54.** 1	**55.** 4	**56.** 2	**57.** 2	**58.** 2	**59.** 1	**60.** 1
61. 1	**62.** 2	**63.** 4	**64.** 2	**65.** 4	**66.** 2	**67.** 1	**68.** 4	**69.** 4	**70.** 4
71. 3	**72.** 3	**73.** 2	**74.** 4	**75.** 3	**76.** 2	**77.** 1	**78.** 4	**79.** 4	**80.** 3
81. 2	**82.** 1	**83.** 1	**84.** 4	**85.** 3	**86.** 4	**87.** 3	**88.** 2	**89.** 3	**90.** 4
91. 3	**92.** 4	**93.** 3	**94.** 2	**95.** 1	**96.** 4	**97.** 2	**98.** 2	**99.** 1	**100.** 1

UNIT-12

BIOTECHNOLOGY

1. A transgenic organism is one that carries transferred genetic material (i.e. the transgene) that has been inserted into its genome at a random site. Gene targeting – the replacement or mutation of a particular gene provides the means for creating strains of rats with mutations in virtually any gene are termed as
   (1) Knock out mouse
   (2) Knock in mouse
   (3) Nude Mouse
   (4) SCID Mouse

2. The pedigree below shows an autosomal dominant disease in a family. All affected individuals are shown. Below the pedigree are Southern blot results for a closely linked polymorphic Short Tandem Repeat (STR) with four alleles (W through Z) Each blot result is directly below the tested individual. What is the predicted status of the unborn child II-3? Assume no recombination has taken place

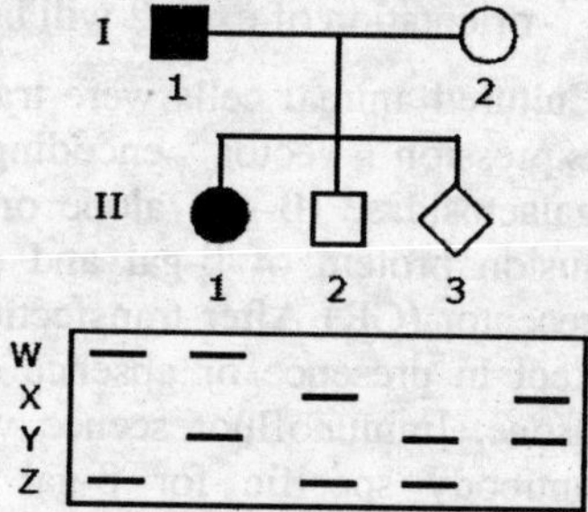

   (1) almost certainly affected, and heterozygous for the disease gene
   (2) almost certainly affected, and homozygous for the disease gene
   (3) almost certainly unaffected
   (4) the marker is uninformative for predicting the status of II-3

3. Which of the following are the properties of transformed cells
   A. anchorage independent
   B. contact inhibition; monolayer culture
   C. dependent on external growth factor
   D. immortal or continuous cell lines
   (1) A and B (2) B and C
   (3) A and D (4) A, C and D

4. Potential advantages of adenovirus based vector for gene delivery in human
   A. integrates into host cell genome providing stable gene expression
   B. can contain >30 kb of nonviral DNA
   C. infects both non-dividing and dividing cells
   D. does not provide long-term gene expression due to lack of integration
   (1) A and B (2) B and D
   (3) B and C (4) A, C and D

5. Which of the following is/are correct about phytoextraction?
   A. Process used by the plants to accumulate contaminants into the roots and above ground shoots
   B. Uptake of organic contaminants from soil or water and, subsequently, their transformation to more stable, less toxic form

C. Process in which plants reduce the mobility and migration of contaminants and contaminated soil

D. Process of the breakdown of contaminants due to the presence of proteins and enzymes produced by the plants

(1) only A (2) only B
(3) A and C (4) C and D

6. The following table lists some of the enzymes of fungi and bacteria having wide variety of industrial applications, including alcoholic beverages, food, detergents and pharmaceuticals, along with their microbial original

Enzyme	**Microorganism**
A. Amylase	E. *Azotobacter vinelandii*
B. Asparginase	F. *Serratia marcescens*
C. Lipase	G. *Aspergillus aureus*
D. Pectinase	H. *Aspergillus oryzae*

The correct combinations are

(1) A and H (2) B and G
(3) C and E (4) D and F

7. Animal cell cultures are frequently used for production of therapeutic proteins. NIH3T3 (a fibroblast cell line) and CHO (Chinese hamster ovarian cell line) are some of the popular cell lines used. Choose the best combination of cell line (for transfection) and starting material for purification of human growth hormone, a secretary protein

(1) NIH3T3 cell pellet
(2) CHO and cell pellet
(3) NIH3T3 and culture medium
(4) CHO and culture medium

8. Marker-Assisted selection (MAS) defined as selection based on molecular markers should have some important criterion for plant breeding activities. Some statements about these criteria are mentioned below

A. Marker should co-segregate with the desired trait of interest

B. Marker should not co-segregate with the desired trait of interest

C. Marker should be un-linked with the desired trait of interest

D. Marker is used for indirect selection of a genetic determinant or determinants of a trait of interest

Which one of above combinations is correct?

(1) A and B (2) B and C
(3) C and D (4) A and D

9. In a transgenic mice line, lox P sites are introduced in the target gene A in the following manner

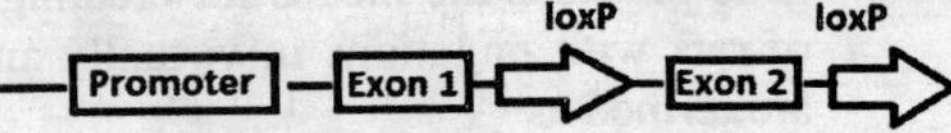

This transgenic mice line was mated with another transgenic mice line where Cre recombinase is expressed only in B cells. What will be the expression profile of gene A in Cre/lox recombinant mice?

(1) Gene will not be expressed in B cells, as orientation of exon I will be inverted by Cre

(2) Gene will not be expressed in B cells, as exon 2 will be deleted by Cre

(3) Gene will only be expressed in B cells of the recombinant mice where Cre removes the two lox P sites

(4) Gene will not be expressed in B cells as orientation of exon 2 will be inverted

10. Cultured animal cells were transfected with expression vector encoding either β-galactosidase (β-gal) alone or expressing a fusion protein of β-gal and glucocorticoid receptor (GR) After transfection, cells were kept in presence or absence of Dexamethasone. Immunofluorescence with a labeled antibody specific for β-gal was used to detect the expressed protein in cytoplasm or nucleus of transfected cells. Possible results of the experiments are

A. Expression of β-gal alone in the cytoplasm in both absence or presence of Dexamethasone

B. Expression of β-gal-GR in the cytoplasm in the absence of Dexamethasone

C. Expression of β-gal alone in the nucleus both in the presence or absence of Dexamethasone

D. Expression of β-gal-GR in the nucleus in presence of Dexamethasone

E. Expression of β-gal alone in both cytoplasm and nucleus in presence or absence of Dexamethasone

F. Expression of β -gal-GR in both cytoplasm and nucleus in presence of Dexamethasone

Choose the correct combination of results from the following options

(1) B, C and D (2) A, B and D

(3) B, D and E (4) A, B and F

11. Following are certain statements regarding somatic hybridization, a technique used for plant improvement

A. Protoplasts of only sexually compatible plant species can be fused

B. Hybrids are produced with variable and asymmetric amounts of genetic material of parental species

C. Protoplast fusion permits transfer of gene block or chromosomes

D. Genes to be transferred need to be identified and isolated

Which one of the following combinations of the above statements is correct?

(1) A and C (2) B and C

(3) A and D (4) B and D

12. The average size of human genome is about 3.2×10^9 base pairs. Considering 60% of it is made up of A-T base pairs, what will be the frequency of the following restriction sites to be present in human genome in high to low order?

X= 5'-GAATTC-3'

Y= 5'-ACCGGT-3'

Z= 5'-CGCG-3'

(1) X>Y>Z (2) Z> X> Y

(3) Y>Z>X (4) Z>Y>X

13. Microorganisms having nitrilase activity can be enriched from soil samples by supplying nitriles as the sole nitrogen or carbon source in the growth medium. A problem with this approach is that false-positive strains may be isolated, which use nitrites to produce amide with the help of a different enzyme activity such as

(1) phosphorylases

(2) phosphatases

(3) hydratases

(4) phosphoestarases

14. In order to create an animal whose prostate gland is devoid of the p53 gene, engineered animals were created by cloning specially designed knock-out cassettes. Identify the most appropriate cassette combination for creating the knock out animal by breeding

	MICE 1		MICE 2
1.	MMTV — CRE	X	GFP ◄ GOI — GFP ◄
2.	MMTV — CRE	X	GFP — GOI — GFP ◄►
3.	MMTV — CRE ◄►	X	GOI — GFP — GSP
4.	MMTV — CRE	X	GFP ◄ GOI — GFP

MMTV= Mouse Mammary Tumor Virus
CRE = Cre Recombinase
GSP=Gene Specific Promoter
GOI=Gene of Interest

15. Following are some statements related to raising transgenic plants

A. T-DNA of *Agrobacterium* contains oncogenes and opine synthesis genes

B. All 8 operons present in vir region of *Agrobacterium* is essential for virulence

C. The CaMV 35S RNA promoter is a very strong constitutive promoter used in plant transformation

D. Conjugative transfer (oriT or tra) region functions in conjugative transfer of the Ti-plasmids

Which one of the following combinations of above statements is correct?

(1) B, C and D (2) A, C and D
(3) A, B and D (4) A, B and C

16. Electrons are transferred from reduced c enzymes like NADH, NADPH to pyruvic acid its derivatives during fermentation. Those fin electron acceptors are reduced to the end-product for example latic acid, propionic acid, etc. end products depend on the particular microorganism and the substrate

Organism	End product (s)
A. *Streptococcus*	a. Butanediol, formic acid
B. *Clostridium*	b. Lactic acid
C. *Salmonella*	c. Butyric acid and iso proanol
D. *Enterobacter*	d. Succinic acid, acetic acid

The correct match is

(1) A - a, B - b, C - c, D – d
(2) A - d, B - c, C - a, D - c
(3) A - b, B - c, C - d, D - a
(4) A - c, B - d, C - a, D – b

17. In resting cells, proteins X and Y are localized in the cytosol. Upon stimulation with lipopolysaccharide (LPS), both of them are phosphorylated and translocate to the nucleus. You have used antibodies against phosphorylated forms of proteins X and Y which are conjugated to either red, or green or blue dye. Keeping optical aberration of light in mind, which one of the following will be the best for visualizing X and Y In the nucleus by fluorescence microscopy?

(1) Anti green X and anti red Y
(2) Anti red X and anti green Y
(3) Anti red X and anti blue Y
(4) Anti blue X and anti green Y

18. A transgenic plant is developed with the following T-DNA construct

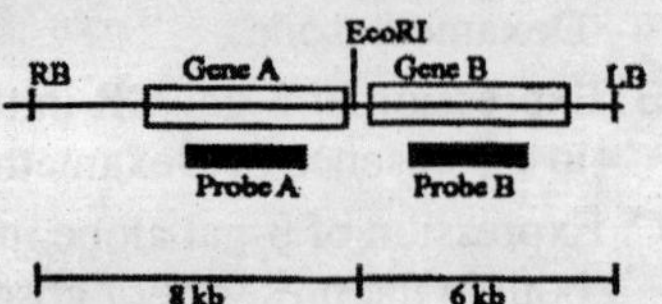

In order to analyze the nature of integration, genomic DNA digested with EcoRI was used for Southern hybridization using either probe A or B. The result obtained is as

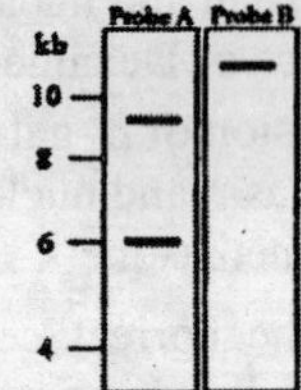

The following conclusions were made

A. There are two copies of the T-DNA cassette integrated at one loci and a third copy at another loci
B. There are two copies of the T-DNA cassette integrated at one loci
C. Complete T-DNA cassette has been integrated in all cases
D. In one T-DNA cassette there is a truncation towards the RB
E. In one T-DNA cassette there is a towards the LB
F. The arrangement of the T-DNA cassettes integrated at the same loci is

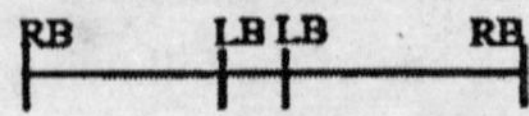

G. The arrangement of the T-DNA cassette integrated at the same loci is

Which of the above are correct?

(1) B, E and G
(2) B, D and F
(3) A, D and F
(4) B, C and F

19. While designing an experiment for *Agrobacterium* mediated plant transformation, a student noted down the following points

A. Ti and Ri plasmids induce crown gall and hairy root disease, respectively

B. Enzymes octopine synthase and nopaline synthase involved in the synthesis of octopine and nopaline, respectively are encoded by T-DNA

C. All the six vir genes, vir A, vir B, vir C, vir D, vir E and vir G are absolutely required for virulence

D. Almost perfect 25 bp direct repeat sequences flanking all Ti and Ri plasmids in the T-DNA region is essential for T-DNA transfer

Which one of the following combinations of the above statement is correct?

(1) A, B and C (2) B, C and D
(3) A, C and D (4) A, B and D

20. The following are certain facts regarding bioremediation

A. Biodegradable plastics are made using polyhydroxy alkanoates (PHAs) such as polyhydroxybutyrate (PHB)

B. *Pseudomonas putida* F1 bacterial strain is involved in degradation of aromatic hydrocarbon

C. The bacterium *Deinococcus radiodurans* consume and digest toluene and ionic mercury from highly radioactive nuclear waste

D. Bioaugmentation is a process of improving the microorganisms already existing in the system for degradation of xenobiotic compound

Which one of the following combination of above statements is correct?

(1) A, B and C
(2) A, B and D
(3) A, C and D
(4) B, C and D

21. The autoradiogram below (after the electrophoresis and southern blotting) shows human DNA digested with a specific restriction enzyme and probed with labelled rRNA. In the autoradiogram on the left, the probe was 28S rRNA; at the right, the probe was 18S rRNA. If the arrows in the following maps locate the recognition site of the restriction enzyme, which map best explains the results shown below

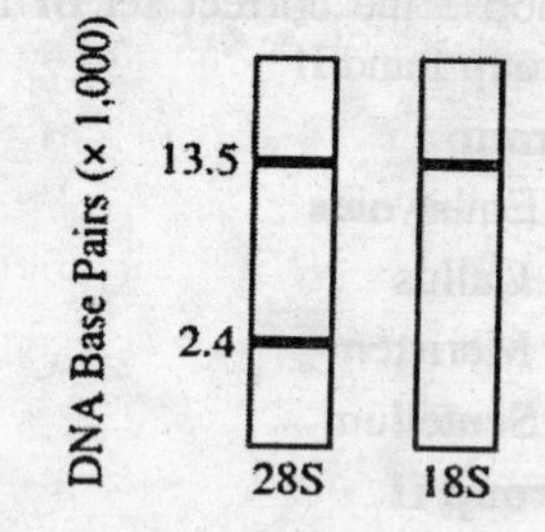

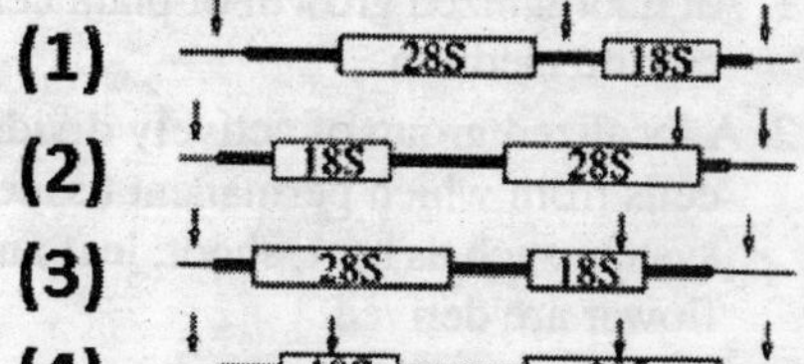

22. Which of the following statements are correct? Very long DNA fragments (>100 kb) from eukaryotic genomes

I. can be efficiently packaged in a λ phage vector

II. can be propagated in yeast artificial chromosomes

III. must have cohesive ends for cloning

IV. can be separated by pulsed field gel electrophoresis techniques

(1) I and II (2) I and IV
(3) II and III (4) II and IV

23. All cloning vectors have certain basic properties in common

A. selectable marker

B. replication origin

C. cloning sites (restriction sites)

D. capacity for screening for inserts (e.g., insertional activation)

E. capacity for selection for inserts (e.g., selection against red-gram stuffers in lambda vectors)

Which of the following is not a basic feature of all cloning vectors?

(1) A, B and C (2) C, D and E
(3) A, C and D (4) D and E

24. Choose the correct set of matches between Group I and II

Group I

P. Embryoids

Q. Callus

R. Meristem

S. Scutellum

Group II

1. An unorganized growth of plant cells in a culture medium
2. A localized group of actively dividing cells from which permanent tissue system such as root, shoot, leaf and flower are derived
3. A process whereby specialized, non-dividing cells begin to proliferate by mitotic division, presumed to involve regression to an undifferentiated state
4. Mass of cells which has external morphology resembling proembryo
5. The embryonic cotyledons of monocots

(1) P-4, Q-2, R-3, S-5
(2) P-4, Q-1, R-2, S-5
(3) P-3, Q-4, R-2, S-1
(4) P-2, Q-1, R-3, S-4

25. Immobilization of enzymes using entrapment method requires

P. photosensitive polyethylene glycol dimethacrylate

Q. CNBr activation of sepharose

R. polyfunctional reagent like hexamethylene diisocyanate

S. radiation of polyvinyl alcohol

(1) P, Q (2) R, S
(3) P, S (4) Q, S

26. Which of the following would result in somaclonal variation in micropropagated plants?

P. Propagation by axillary branching in the absence of plant growth regulators

Q. Cell suspension maintained for five years before induction of somatic embryogenesis

R. Callus induction using 20μM 2,4-Dichlorophenoxyacetic acid followed by shoot organogenesis

S. Shoot organogenesis from an explant in the absence of an intermediate callus phase

(1) P, Q (2) Q, R
(3) P, S (4) Q, S

27. Lignocellulosic biomass was subjected to microbial composting. The microbial consortium produced an extracellular enzyme xylanase, which was a glycoprotein having a molecular weight of 68 kDa and a positive charge. An aqueous extract of the enzyme could be easily prepared from the compost. What techniques would you recommend for confirming the molecular weight of the purified enzyme?

(P) Isoelectric focusing

(Q) SDS-PAGE

(R) Native PAGE

(S) Gel filtration

(1) P, Q (2) Q, S
(3) R, S (4) P,S

28. If Concavalin A sepharose column was used for the purification of enzyme, the separation would be based on

(1) molecular exclusion
(2) ion exchange
(3) affinity binding
(4) hydrophobic interaction

29. During cultivation of microorganisms in a fermenter, various parameters are controlled by appropriate sensor (probe) Match each probe in group 1 with the appropriate response mechanism in group 2

Group 1 (Probe)

P. Thermistor

Q. Oxygen electrode

R. Metal rod

S. pH electrode

Group 2 (Response)

1. Activation of acid / alkali pump
2. Activation of vegetable oil pump
3. Activation of hot / cold water pump
4. Increase / decrease in stirrer motor speed

(1) P-2, Q-3, R-1, S-4

(2) P-3, Q-2, R-4, S-1

(2) P-1, Q-2, R-4, S-3

(4) P-3, Q-4, R-2, S-1

30. Match items in group 1 with correct options from those given in group 2

Group 1

P. VNTR sequence

Q. Leader sequence

R. SD sequence

S. cis-acting sequence

Group 2

1. Gene regulation on the same chromosome
2. Ribosome binding site
3. DNA finger printing
4. Functions in attenuation

(1) P-3, Q-1, R-4, S-2

(2) P-3, Q-4, R-2, S-1

(3) P-2, Q-3, R-1, S-4

(4) P-3, Q-1, R-2, S-4

31. Majority of the cereals are highly recalcitrant to *Agrobacterium*-mediated transformation, and so direct transformation methods have been developed to transform such plants. Which of the following direct transformation methods is applicable to intact plant tissues?

(1) Calcium chloride and PEG-mediated transformation

(2) Liposome-mediated transformation

(3) Electroporation

(4) Transformation using microprojectiles

32. Match items in group 1 with correct options from those given in group 2

Group 1	Group 2
P. Tissue plasminogen activator	1. Immunomodulator
Q. Gamma interferon	2. Biodegradable plastic
R. Podophyllotoxin	3. Clot dissolution
S. Polyhydroxyalkanoate	4. Anti-tumor agent

(1) P-4, Q-3, R-l, S-2

(2) P-3, Q-l, R-2, S-4

(3) P-l, Q-3, R-4, S-2

(4) P-3, Q-l, R-4, S-2

33. Match items in group 1 with correct options from those given in group 2

Group 1

P. Amperometric biosensor

Q. Evanescent wave biosensor

R. Calorimetric biosensor

S. Potentiometric biosensor

Group 2

1. Light beam
2. Flux of redox electrons
3. Field effect transistors
4. Exothermic reaction

(1) P-3, Q-4, R-2, S-1

(2) P-3, Q-2, R-4, S-1

(3) P-2, Q-1, R-4, S-3

(4) P-2, Q-4, R-3, S-1

34. In a transgenic mouse experiment a founder male produces 100 pups and only 20 of these are transgenic. This result leads to the conclusion that the

(1) founder transgenic animal is chimeric

(2) founder animal is mosaic

(3) transgene is integrated on Y–chromosome

(4) transgene is integrated on X–chromosome

35. A nutritionist wanted to study the effect of storage time on the amount of vitamin C present in freeze-dried amla stored for 6, 12 and 18 months. Six fruit packs were randomly assigned to each of the three storage times and vitamin C was measured in milligrams per 100 milligrams of dried amla. The treatment, experimental unit and response were respectively

(1) a specific storage time, amount of vitamin C, a fruit pack

(2) a fruit pack, amount of vitamin C, a specific storage time

(3) random assignment, a fruit pack, amount of vitamin C

(4) a specific storage time, a fruit pack, amount of vitamin C

36. A mixture of lipids is applied to a silica gel column, and the column is then washed with increasingly polar solvents. The mixture consists of phosphatidylserine, phosphatidylethanolamine, phosphatidylcholine, cholesteryl palmitate (a sterol ester), sphingomyelin, palmitate, *n*-tetradecanol, triacylglycerol, and cholesterol. In what order will the lipids elute from the column?

(1) First cholesteryl palmitate and triacylglycerol → Cholesterol and *n*-tetradecanol→ phosphatidylcholine and phosphatidylethanolamine → Sphingomyelin→ phosphatidylserine and palmitate elute last

(2) First cholesteryl palmitate and triacylglycerol → phosphatidylcholine and phosphatidylethanolamine → Cholesterol and *n*-tetradecanol → Sphingomyelin→ phosphatidylserine and palmitate elute last

(3) First phosphatidylserine and palmitate → phosphatidylcholine and phosphatidylethanolamine → Cholesterol and *n*-tetradecanol→ Sphingomyelin→ cholesteryl palmitate and triacylglycerol elute last

(4) First phosphatidylserine and palmitate → Cholesterol and *n*-tetradecanol→ phosphatidylcholine and phosphatidylethanolamine → Sphingomyelin→ cholesteryl palmitate and triacylglycerol elute last

37. The MALDI-TOF mass spectrometer determines the MW of the protein. Because it also analyzes protein mixtures (separates protein mixtures substantially better than SDS gel electrophoresis), it can perform many functions of SDS gel electrophoresis. For example, the MALDI-TOF provides convincing data about the purity of proteins. As far as the sensitivity is concerned, you need 10 to 100 fM less protein for the MALDI-TOF spectrum than for the silver staining of an SDS gel. However, the MALDI-TOF mass spectrometer offers more than the SDS gel electrophoresis

Consider the following statements

A. If you know the sequence (e.g., from cDNA cloning) and have purified protein. Then you can determine the MW of the protein via MALDI-TOF and compare to the MW calculated from the cDNA sequence

B. If the MW are identical, the primary structure of your protein is correct. Deviations are evidence of impure protein

C. You can trace the effect of glycosidases through the decrease in MW of a glycosylated protein. At the same time, you measure the number of separated sugar groups. Because of the heterogeneity of the glycoproteins

D. The MALDI-technique is limited to proteins/peptides

Which of the following statements are correct?

(1) A and B (2) A and C
(3) A, B and C (4) B and D

38. A patient with the autosomal dominant Gardner syndrome (tooth changes, bony tumors, and high rates of colonic polyps/colon cancer) was found to have a nucleotide substitution within the causative adenomatous polyposis coli (APC) gene. This mutation changed the DNA sequence from CAGAGGT to CAGGGGT and ablated the splice junction at the border of exon 7 and its adjacent intron. A partial sequencing result using the Sanger method is shown in the figure below, and supports which of the following conclusions about the child's APC genotype and expression?

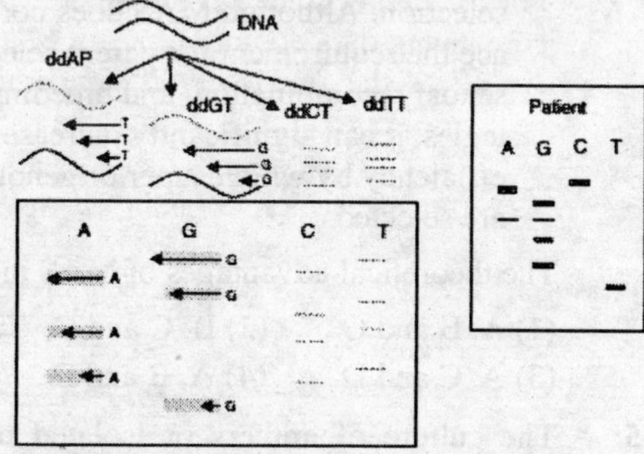

(1) Homozygous for the normal Gardner allele, some smaller mRNA molecules
(2) Heterozygous for normal and mutant Gardner allele, some larger mRNA molecules
(3) Heterozygous for normal and mutant Gardner allele, some smaller mRNA molecules
(4) Homozygous for mutant Gardner allele, no change in mRNA molecules

39. A recombinant viral DNA containing a 300 bp human gene sequence is replicated in vitro using media containing bromo-deoxy-uridine triphosphate, completely saturating the DNA with bromo-deoxyuridine residues. The bromo-substituted viral DNA is then switched to media with regular deoxyuridine triphosphate, and allowed to undergo two rounds of DNA replication. The human DNA fragment is then cleaved by restriction enzyme digestion and subjected to ultracentrifugation in cesium chloride under conditions that will separate bromo-substituted DNA duplexes, hybrid bromo-substituted/normal duplexes, and normal duplexes

Which of the results depicted in the figure below (with their interpretations) should be obtained?

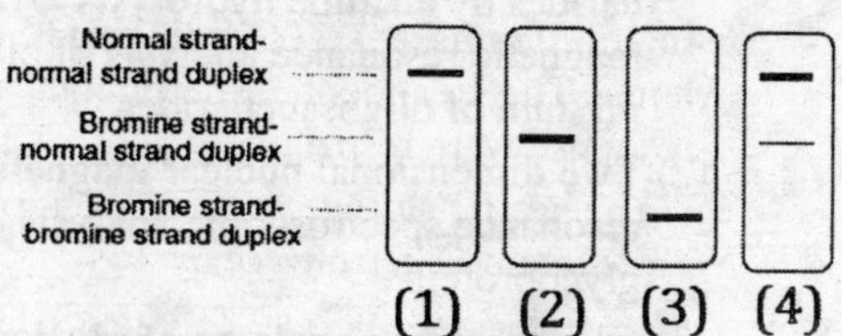

40. Which of the following statements are correct for somatic cell hybridization?

P. For fusion of protoplast, dimethylsulfoxide (DMSO) is used as a fusogen

Q. The enzyme 'Cellulase Onozuka' used for protoplast isolation is sourced from *Trichoderma viride*

R. The first report of somatic hybrid plants resulted from the fusion of protoplasts of *Nicotiana glauca* and *N. tabacum*

S. Viability of isolated protoplasts can be determined by Evan's blue staining

(1) P, Q (2) Q, S
(3) R, S (4) Q, R

41. When a small amount of sodium dodecyl sulphate ($Na^+CH_3(CH_2)_{11}OSO^{-3}$) is dissolved in water, the detergent ions enter the solution as a monomeric species. As more detergent is added, a concentration is reached (the critical micelle concentration) at which the monomers associated to form micelles. The critical micelle concentration of SDS is 8.2mM. The micelle has an average particle weight (the sum of molecular weight of the constituent monomer) of 18,000. Calculate the number of detergent molecules in the averaged micelle

(1) 23 SDS molecules/micelle
(2) 43 SDS molecules/micelle
(3) 63 SDS molecules/micelle
(4) 83 SDS molecules/micelle

42. The sequence of monosaccharides including position and configuration of glycosidic bonds in a glycoprotein is to be determined. Which one of the following methods can be employed?

(1) Glycoprotein → removal of oligosaccharides by alkaline hydrolysis → nuclear magnetic resonance analysis of cleaved mixture of oligosaccharides
(2) Two dimensional nuclear magnetic resonance spectroscopic analysis of the glycoprotein
(3) Glycoprotein → release of oligosaccharides with endoglycosidases followed by purification to separate oligosaccharides → enzymatic hydrolysis of purified oligosaccharides with specific glycosidases → mass spectroscopic analysis of smaller oligosaccharides
(4) Glycoprotein →treat with trypsin followed by MALDI analyses of tryptic peptides

43. A disease free tomato plant was planted in soil contaminated with *Agrobacterium tumefaciens* harboring Ti plasmid that lacks VirA gene. Provided all other conditions are optimum for the bacterial infection, identify the appropriate consequence

P. octapine synthesis by the bacterium will enhance
Q. acetylsyringone receptor will not be synthesized
R. the bacteria will fail to transfer the T-DNA to the plant
S. a fragmented T-DNA will be transferred to the tomato plant

(1) P, S
(2) Q, S
(3) P, R
(4) Q, R

44. Rather than selecting traits, which are the outcome of many genes, Marker assisted selection (MAS) is based on selecting specific alleles at marker loci that are known to be linked to the genes that cause the desired trait

Consider the potential advantages of MAS

A. avoids errors caused by environmental variance
B. can be applied only at adult stage once a trait is expressed
C. can be applied on a single plant, whereas phenotypic selection of some traits might require seedor tissue from many plants to be effective
D. may be less expensive than phenotypic selection. Although MAS does not replace the requirement for parent selection, sexual recombination, and breeding strategies, it can significantly increase the efficiency by which superior genotypes are selected

The theoretical advantages of MAS includes

(1) A, B and C (2) B, C and D
(3) A, C and D (4) A, B and D

45. The culture of anthers or isolated microspores to produce haploid plants is known as anther culture or microspore culture. Microspore culture has developed into a powerful tool in plant breeding. Embryos can be produced via a callus phase or be a direct recapitulation of the developmental stages characteristic of zygotic embryos

Consider the following statements

A. Late uninucleate to early binucleate microspores are the best explants for embryogenesis. In this case, the somatic embryos develop into haploid plants
B. Doubled haploids can then be produced by chromosome-doubling techniques. Thus microspore culture enables the production of homozygous (at every locus) plants in a relatively short period as compared to conventional breeding techniques

C. These haploid plants are useful tools in plant breeding as there is increase in productivity and resistance properties in them as compare to diploid plants

D. Haploid embryos are used inmutant isolation, gene transfer, studies of storage product biochemistry, and physiological aspects of embryo maturation

Which of the following is correct statement?

(1) A, B and C (2) B, C and D
(3) A, C and D (4) A, B and D

46. Protoplast fusion and somatic hybridization techniques provide the opportunity for bypassing reproductive isolation barriers, thus facilitating gene flow between species

Consider the following statements

A. Fusion of protoplasts is accomplished by the use of PEG [poly(ethylene glycol)]

B. Protoplast fusion has helped in the development of somatic hybrids or cybrids (cytoplasmic hybrids)

C. DNA uptake has been found to be difficult in protoplasts than into intact plant cells

D. Protoplasts have to be cultured on a medium with a high osmoticum such as sucrose or mannitol; otherwise the protoplasts will burst open, and plant regeneration from protoplasts has proven to be difficult

Which of the following are true for protoplast culture?

(1) A, B and C (2) B, C and D
(3) A, C and D (4) A, B and D

47. A range of processes mediated by plants or algae are useful in treating environmental problems

Consider the following statements

A. "Phytosequestration"- Phytochemical complexation in the root zone, reduce the fraction of the contaminant that is bioavailable.Transport protein inhibition on the root membrane-preventing contaminants from entering the plant.Vacuolar storage in the root cells - Contaminants can be sequestered into the vacuoles of root cells

B. Phytostimulation — uptake and concentration of substances from the environment into the plant biomass

C. Phytostabilization — reducing the mobility of substances in the environment, for example, by limiting the leaching of substances from the soil

D. Phytotransformation — chemical modification of environmental substances as a direct result of plant metabolism, often resulting in their inactivation, degradation (phytodegradation), or immobilization (phytostabilization)

Which of the following statements are correct?

(1) A, B and C (2) B, C and D
(3) A, C and D (4) A, B and D

48. Phytoremediation describes the treatment of environmental problems (bioremediation) through the use of plants that mitigate the environmental problem without the need to excavate the contaminant material and dispose of it elsewhere

Consider the following statements

A. Phytoextraction — enhancement of soil microbial activity for the degradation of contaminants, typically by organisms that associate with roots

B. Phytovolatilization — removal of substances from soil or water with release into the air, sometimes as a result of phytotransformation to more volatile and/or less polluting substances

C. Rhizofiltration — filtering water through a mass of roots to remove toxic substances or excess nutrients. The pollutants remain absorbed in or adsorbed to the roots

Which of the following statements are true for phytoremediations?

(1) A and C (2) B and C
(3) A and B (4) C only

49. A biosensor is an analytical device, used for the detection of an analyte, that combines a biological component with a physico-chemical detector

A. In a biosensor, the bioreceptor is designed to interact with the specific analyte of interest to produce an effect measurable by the transducer

B. An immunosensor utilizes the very specific binding affinity of antibodies for a specific compound or antigen

C. Commercially available glucose monitors rely on amperometric sensing of glucose by means of glucose oxidase, which oxidises glucose producing hydrogen peroxide which is detected by the electrode

D. Microbial biosensors for example the arsenic biosensos detects arsenic by using the Ars operon

Which of the following statements are correct?

(1) A, B and C (2) B, C and D
(3) A, C and D (4) A, B, C and D

50. A widely used method for detecting either antigen or antibody in body fluids is known as ELISA (Enzyme-Linked Immunosorbent Assay) The figure below demonstrates an ELISA for detection of antigen (Double antibody sandwich method of EIA for assay of antigen. A, antibody is adsorbed to surface. B, test solution containing antigen is added. C, enzyme-labeled specific antibody is added. D, enzyme substrate is added). One ofthe problems with ELISA is nonspecific reactivity due to nonspecific antibody present in the reaction

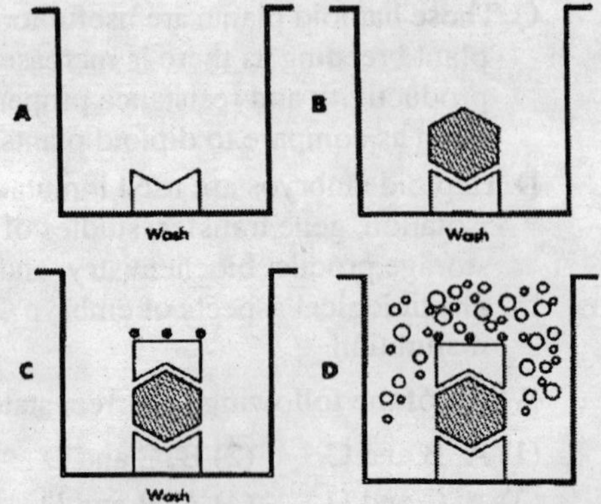

Of the four steps depicted, A, B, C, and D, which one may be the major cause of nonspecificity?

(1) A (2) B
(3) C (4) D

51. HeLa cell extract was used to study transcription of a gene X having six introns. RNA Pol II complex containing all associated proteins was isolated from actively transcribing system and subjected to proteome analysis. Results showed the presence of both splicing and capping enzymes in the complex. When transcription elongation was inhibited by flavopiridol, polymerase complex contained only capping enzymes. When phosphorylation of CTD domain of Pol II was inhibited by a kinase inhibitor, the complex contained neither splicing nor capping enzymes

From these results, following conclusions were made

A. Transcription of gene X is coupled to mRNA capping. B. Transcription elongation is coupled to splicing

C. Phosphorylation of CTD is required for the recruitment of capping and splicing enzymes

D. Both capping and splicing of mRNAs occurs simultaneously

Which of the following conclusions are correct?

(1) A, B and C (2) B, C and D
(3) A, C and D (4) A, C and D

52. The organization of an eukaryotic gene expressed at high levels in liver is diagrammatically represented below

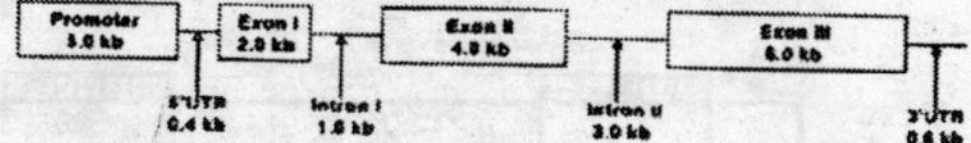

The size of mature mRNA generated by the transcription followed by normal splicing of this gene will be (assume that this mRNA is not polyadenylated; 5'-UTR and 3'-UTR refers to the 5' and 3' untranslated regions, respectively)

(1) 12.4 kb (2) 13.0 kb
(3) 12.0 kb (4) 12.6 kb

53. Spores of the soil bacterium *Bacillus thuringiensis* (Bt) contain a crystal (cry) protein (δ-endotoxin) Inside insect gut, the crystals break apart and release a toxin that binds to and creates pores in the intestinal lining. Instead of whole gene truncated cry gene is used in Bt crops

A. B*t* gene is used in the production of insect resistant crops (genetically modified) and biological insecticides as well

B. *Bacillus thuringiensis*, during sporulation produces a toxic protein which possesses insecticidal activity against Lepidoptera, Coleoptera, Hymenoptera, Diptera and Nematode

C. This crystal protein is called as Cry protein, encoded by cry gene present on bacterial chromosome

D. As soon as the Cry protein crystals reach the digestive tract of the insect, the prevailing acidic condition there causes denaturation of toxin

Which of the following are true statements?

(1) A and B
(2) B and C
(3) C and D
(4) A and D

54. Consider the few examples of transgenic rice developed for therapeutic purposes such as

A. A rice variety was developed to overcome the iron deficiency by introducing the soybean ferritin protein gene in rice genome

B. Golden Rice is capable of producing β–carotene (provitamin-A) in the endosperm (main grain) which is a pre cursor for vitamin A. The golden rice was developed by introducing one gene from daffodil and one gene from the bacterium *Erwinia uredovora* were into rice genome

C. Rice has been as an expression system to express human milk proteins (lactoferrin, lysozyme and 1-antitrypsin)

D. A gene from Amaranth has been used in developing a variety with high protein content rice

E. Genetically modified rice has been developed which carries the gene for the hepatitis B surface antigen which can be used as edible vaccine

Which of the following transgenic development are true for rice?

(1) A and B only (2) B, C and D
(3) A and D only (4) A, B and C

55. Embryonic stem (ES) cells are pluripotent stem cells isolated from inner cell mass (ICM) of blastocyst stage of embryo. Use of ES cell mediated transgenic animal production is quite effective specially to generate targeted gene modification at precise location

Consider the following steps

A. Transformed cells are injected into blastocyst followed by implantation in a surrogate female

B. Transgene introduction to ES cells using methods like DNA micro-injection, electroporation, precipitation reaction, transfection etc. DNA transfection to ES cells can be carried out using different vectors likeliposome, retroviral vectors etc

C. Isolation of pluripotent ES cells from ICM of blastocyst and culturing of the cells *in vitro*

D. Selection of transformed ES cells for either knockin or knockout constructs

E. After birth, the offsprings are screened for chimerism. Inbreeding of the genetic chimeras is performed to obtain homozygous transgenic animal carrying both the mutated alleles for that character

The correct steps to generate transgenic animals using modified ES cell technology are

(1) A→B→D→C→E
(2) C→B→D→A→E
(3) C→D→B→A→E
(4) C→A→D→B→E

56. Consider the following statements for gene transfer methods using bacteria as biological means

A. Bactofection is direct gene transfer using bacteria into the target cell, tissue, organ or organism

B. The genes located on the bacterial chromosome of the transformed bacterial strains are delivered and expressed into the cells

C. The efficiency of bactofection mediated gene transfer can be increased using integrin receptors

D. Lipofectamine-mediated bactofection has also been employed for enhancing the gene transfer efficiency in *E. coli* strains, particularly in the transfer of large intact DNA for gene expression

Which of the following statements are correct?

(1) A, B and C (2) B, C and D
(3) A, C and D (4) A, B and D

57. Retroviruses are RNA viruses that replicate via a ds-DNA intermediate and have been widely modified as vector for gene delivery in animals. Which of the following is correct representation of prototypical retroviral vector?

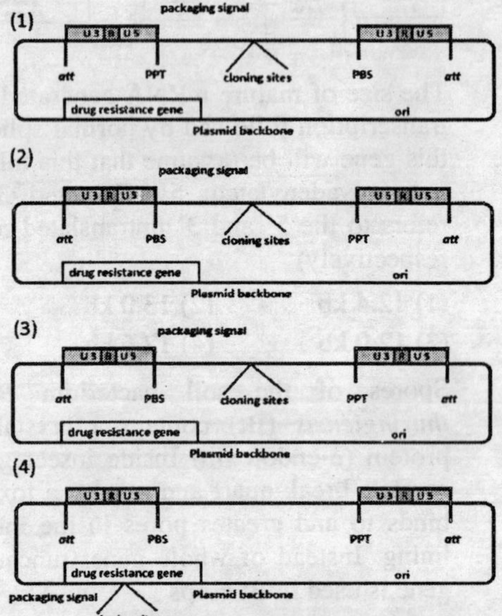

58. In a transgenic mice line, lox P sites are introduced in the target gene in the following manner and cre is under control of tissue specific promoter. Among the following construct where we can expect inversion of gene 1?

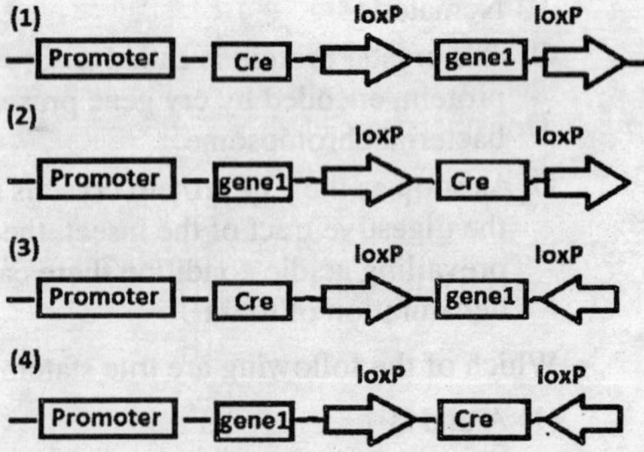

59. Genomic DNA of transgenic plants (P1, P2 and P3) obtained by transforming with binary vector A whose map is depicted below, was digested with BamH I and Sal I and hybridized with genomic fragment X

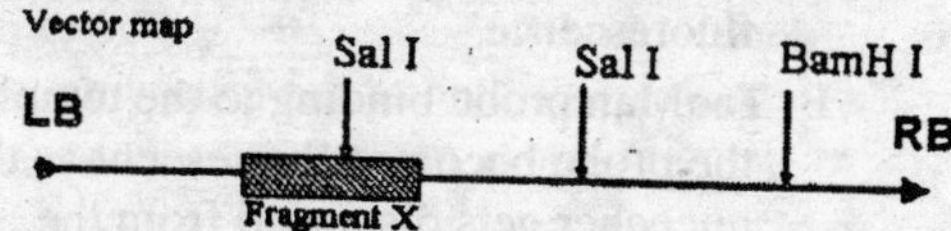

Pattern obtained in Southern hybridization is shown below

P1		P2		P3	
BamH I	Sal I	BamH I	Sal I	BamH I	Sal I

Based on the above, which of the following interpretation is correct

(1) All the plants (P_1, P_2 and P_3) contains two copies of the transgene

(2) P_1, P_3 contains one and P_2 contains two copies of the transgene

(3) P_1 contains two, whereas P_2 and P_3 contains one copy of transgene each

(4) P_1 and P_2 contains two and P_3 contains one copy of the trangene

60. Cre/loxP system is used by phage P1 to remove terminally redundant sequences that arise during packaging of the phage DNA. Cre-lox system can be used to create targeted deletions, insertion and inversion in genomes of transgenic animals and plants. Consider a series of genetic markers A to K. How should the Lox P sites be positioned in order that Cre recombinase can create a deletion in the DEFG segment relative to ABC and HIJK?

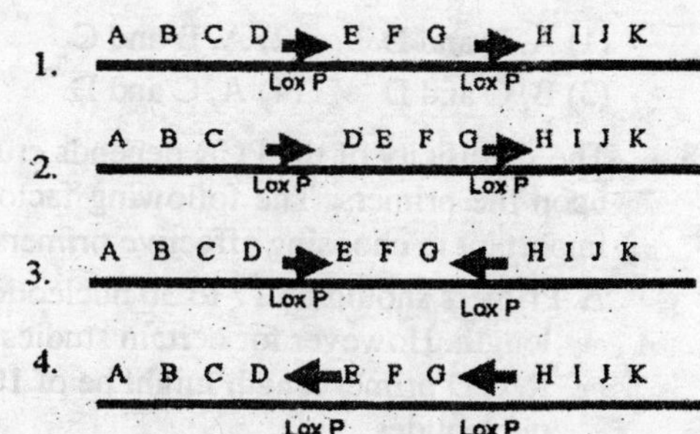

61. Gene expression involves the synthesis of mRNA through transcription followed by synthesis of protein through translation. If the purpose of cloning is to express a foreign gene, it is necessary to check the expression both at the mRNA and protein level in terms of quality and quantity. Following techniques could be used for assessing gene experession

A. Western blotting

B. Southern blotting

C. Reverese Transcriptase-PCR

D. Activity/functional testing assay or staining

Screening for foreign gene mRNA transcript can be done by

(1) A and B (2) B and C

(3) C and D (4) B and D

62. DNA molecule used for carrying an exogenous DNA into a host organism and facilitates stable integration and replication inside the host system is termed as Vector

Consider the following properties

A. Capable of replicating inside the host and have compatible restriction site for insertion of DNA molecule (insert)

B. Capable of autonomous replication inside the host (*ori* site)

C. Larger in size and able to incorporate small specific insert size

D Have a selectable marker for screening of recombinant organism

In general, vectors should have following characteristics

(1) A, B and D (2) A, B and C
(3) B, C and D (4) A, C and D

63. The specificity of the PCR depends crucially upon the primers. The following factors are important in choosing effective primers

A. Primers should be 17 to 30 nucleotides in length. However for certain studies the RAPD primer length might be of 10-12 nucleotides

B. Ideally GC content of the primers is 10% .For primers with a high GC content, it is desirable to choose a long primer so as to avoid a low melting temperature

C. The basic formula to calculate melting temperature is Tm = 4°C x (number of G's and C's in the primer) + 2°C x (number of A's and T's in the primer) Two primers must have a similar Tm value

Which of the following are true?

(1) A and C (2) A and B
(3) C only (4) B and C

64. Consider the following statements for primers used in PCR

A. Sequences with long runs (i.e. more than three or four) of a single nucleotide should be avoided

B. Primers with significant secondary structures (self –hair pin, loop formation) are undesirable

C. There should be high base complementarities between the two primers

Which of the following are true?

(1) A and B (2) B and C
(3) A only (4) A and C

65. Real-time or Quantitative PCR (qPCR) uses the linearity of DNA amplification to quantify absolute or relative amounts of target sequence in a sample. With the help of a fluorescent reporter, the amount of generated DNA can be measured

Consider the following statements

A. In TaqMan probe, when Reporter is adjacent to Quencher –there is no fluorescence

B. TaqMan probe binding to the template, the probe becomes fluorescent as the quencher gets distanced from the fluorescent reporter. The amount of PCR product amplified is directly proportional to the amount of fluorescence

C. Molecular Beacons probes are labeled with a fluorescent dye or quencher and do not exhibit any significant fluorescence in the free, unhybridized condition

D. During extension polymerase hydrolyzes the Molecular beacon probe leading to separation of Reporter from the Quencher and thus fluorescence is exhibited

E. Scorpion primers are similar to Molecular Beacons, but they contain a PCR primer sequence. When the target DNA is amplified during PCR, the beacon fragment binds to the newly synthesized DNA, and thus, separates the fluorophore from the quencher

Which of the following statements are correct?

(1) A, C and E (2) A, B and E
(3) C, D and E (4) A, B, C and D

66. In multiplex PCR assay, more than one target sequence can be amplified by using multiple primer pairs in a single reaction mixture. Since multiple genes are the target in a single reaction, various information comes out from a single test run otherwise it would require several reactions. However, the primer sets must have annealing temperatures within a narrow range and are optimized to work properly within a single reaction

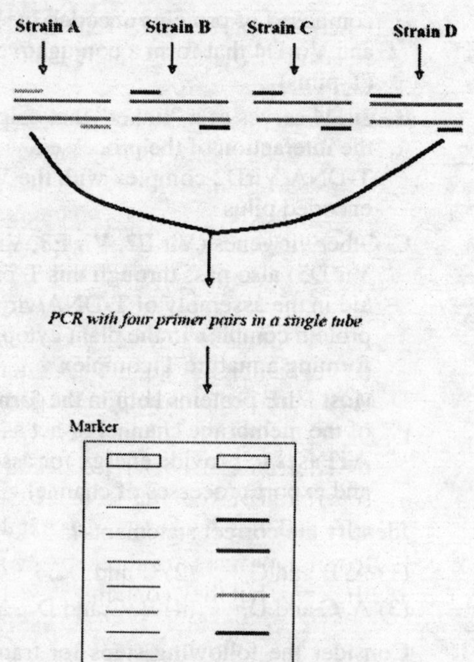

Diagram above represents the multiplex PCR. Consider the following statements

A. In multiplex assays, false positive results are often obtained because each amplified fragment provides an internal control for the other amplified fragments

B. It is ideal for conserving costly polymerase and templates in short supply

C. Multiplex PCR can be utilized to determine the amount of a particular template in a sample by exponential amplification and internal standards

D. It is widely used in Pathogen identification, High throughput SNP genotyping, Mutation analysis, Gene deletion Analysis, Template quantitation, Linkage analysis, RNA detection and Forensic studies etc

E. Multiplex PCR is a modified form of PCR designed to reduce the contamination in products occurs due to non-specific amplification or non-specific primer binding sites

(1) A. B and E only
(2) B, C and E only
(3) A, B and D only
(4) A, B, C and D

67. The virulent strains of *A. tumefaciens* harbor large plasmids (140–235 kbp) known as tumor-inducing (Ti) plasmid involving elements like T-DNA, vir region, origin of replication, region enabling conjugative transfer and o-cat region (required for catabolism of opines)

Consider the following statemetns

A. T-DNA is a small, specific segment of the plasmid, about 24kb in size and found integrated in the plant nuclear DNA at random site. This DNA segment is flanked by right and left borders

B. The T-DNA contains two groups of genes oncogenes for synthesis of auxins and cytokinins (phytohormones) The overproduction of phytohormones leads to proliferation of callus or tumour formation and Opine synthesizing genes for the synthesis of opines (a product from amino acids and sugars secreted by the crown gall infected cells and utilized by *A. tumefaciens* as carbon and nitrogen sources) Thus opines act as source of nutrient for bacterial growth, e.g. Octopine, Nopaline

C. Virulence genes aid in the transfer of T-DNA into the host plant cell. Ti plasmid contains 35 *vir* genes arranged in 8 operons

Which of the following are true?

(1) A only (2) A and C
(3) B and C (4) A, B and C

68. Consider the following statements for T-DNA border sequences

A. T-regions are defined by direct repeats known as T-DNA border sequences (Right and Left Border i.e. RB and LB of 25 bp each)

B. These are not transferred intact to the plant genome, but are involved in the transfer process. TheRB is rather precise, but the LB can vary by about 100 nucleotides

C. Deletion of the LB repeat abolishes T-DNA transfer, but the RB seems to be non-essential. The LB repeat has little transfer activity alone

Identify the correct statements

(1) A and B only (2) B and C only
(3) A and C only (4) A, B and C

69. *Agrobacterium* perceives signaling molecules (phenolic compounds and sugars) released by the wounded plant cells. These signaling molecules act as chemotactic attractants for *Agrobacterium*. This is followed by recognition by the bacterial VirA/VirG2-component signal transduction system

Consider the following statements

A. Vir G, a kinase present in the inner bacterial membrane senses these phenolics (e.g. acetosyringone), gets autophosphorylated and then activates Vir A

B. Vir A protein is a transcriptional activator of other vir genes and is inactive in non-phosphorylated form. The activation of Vir A thus induces the expression of the other vir genes

C. Vir D1/D2 border-specific endonucleases recognize the left and right borders of T-DNA. Vir D2 induces single stranded nicks in Ti plasmid causing the release of the ss-T DNA

D. Vir D2 attaches to the 5'-end of the displaced ss-T DNA forming an immature T-complex

Identify the correct statements?

(1) A and B (2) C and D
(3) A, C and D (4) B, C and D

70. Consider the following steps for transfer of T-DNA and integration into the plant cell

A. The transfer of T-DNA to the plant cell is mediated by Type-IV secretion system composed of proteins encoded by *Vir B* and Vir D4 that form a conjugative pilus (T-pilus)

B. VirD4 serves as a "linker" that helps in the interaction of the processed T-DNA/VirD2 complex with the VirB-encoded pilus

C. Other vir genes (Vir E2, Vir E3, VirF, Vir D5) also pass through this T-pilus to aid in the assembly of T-DNA/vir protein complex in the plant cytoplasm forming a mature T-complex

D. Most VirE proteins help in the formation of the membrane channel or act as ATPases to provide energy for assembly and export processes of channel

Identify the correct statements?

(1) A, B and C (2) C and D
(3) A, C and D (4) B, C and D

71. Consider the following steps for transfer of T-DNA and integration into the plant cell

A. Ss-TDNA is coated with VirE2, a non-sequence specific single stranded DNA binding protein

B. Vir D2 and Vir E2 protect the ss-T strand from nucleases inside the plant cytoplasm by attaching to the 5'end

C. Both VirD2 and VirE2 proteins have Nuclear Localization Signals (NLS) which serves as pilot proteins to guide the mature T-complex to the plant nucleus

D. The efficiency of transfer is enhanced by VirC2 proteins, which recognize and bind to the overdrive enhancer element

Identify the correct statements?

(1) A, B and C (2) B, C and D
(3) A, C and D (4) All of the above

72. Consider the statements for *Agrobacterium tumefaciens* T-DNA transfer into plant system

A. VIP1 and VirF importins may interact with the T-strand, either directly or indirectly, to form larger T-complexes in the plant cell

B. Vir F directs the proteins coating T-complex (VIP1 and Vir E2) for destruction in proteasome

C. Inside the nucleus, ss-TDNA is converted into ds-TDNA which gets integrated into the plant genome via process called homologous recombination

Identify the correct statements

(1) A and B (2) B and C

(3) A and C (4) A, B and C

73. Consider the following statements regarding binary vector developed for *Agrobacterium tumefaciens* Ti plasmid

A. It utilizes the trans- acting functions of the vir genes of the Ti-plasmid and can act on any T-DNA sequence present in the same cell

B. Binary vector contains transfer apparatus (the vir genes) and the disarmed T-DNA containing the transgene on separate plasmids

C. The advantage of binary vectors is its small size due to the absence of border sequences needed to define T-DNA region and vir region

D. Binary vectors are the deletion derivatives of Ti-plasmids. The DNA to be introduced into the plant transformation vector is sub cloned in a conventional *Escherichia coli* plasmid vector for easy manipulation, producing a so-called intermediate vector

Which of the following are true in reference to *Agrobacterium* based binary vectors?

(1) A, B and C (2) B, C and D

(3) A, C and D (4) A, B and D

74. Chloroplast transformation offer several advantages compared with nuclear transformation

Consider some of potential advantages

A. It decreases risk of transgene escape

B. It exhibits higher level of transgene expression and thus higher level of protein production due to the presence of multiple copies of chloroplast transgenes per cell

C. There is high level of pre and post transcriptional gene silencing due to multiple copy gene insertion

D. Chloroplast transformation involves homologous recombination and is therefore precise and predictable. This minimizes the insertion of unnecessary DNA that accompanies in nuclear genome transformation. This also avoids the deletions and rearrangements of transgene DNA, and host genome DNA at the site of insertion

Which of the following is correct statement in reference to chloroplast transformation?

(1) A and B (2) B and C

(3) B, C and D (4) A, B and D

75. Diagram below shows a vector containing a selectable marker under the control of plastid expression signals (promoter, 5´ UTR and 3´ UTR) flanked by chloroplast sequences. Homologous recombination takes place between the flanking targeting arms and recipient plastid genome (plastome)

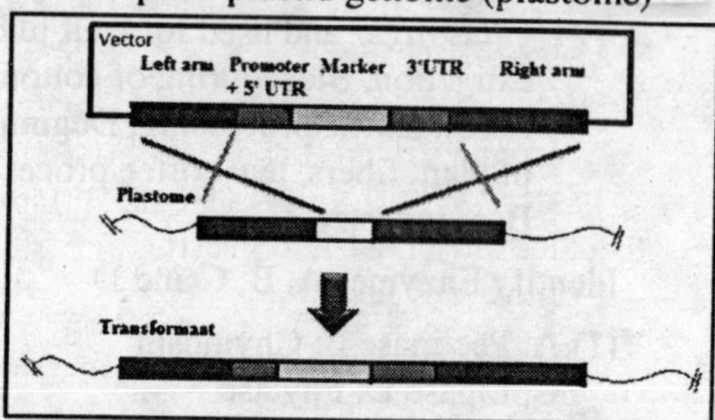

Consider the following statements

A. Gene silencing or RNA interference does not occur in genetically engineered chloroplasts

B. Absence of position effect due to lack of a compact chromatin structure and efficient transgene integration by homologous recombination in chloroplasts. It avoids inadvertent inactivation of host gene by transgene integration

C. Chloroplast has ability to form disulfide bonds and folding human proteins results in high-level production of biopharmaceuticals in plants

D. Multiple transgene expression is possible due to polycistronic mRNA transcription

Identify the correct statements?

(1) A and B (2) C and D

(3) A, C and D (4) All of the above

76. Enzymes are nowadays used extensively in bio-processing industries

- Enzymes A is obtained from *Bacillus sp, Aspergillus niger, E. coli* and is Widely used in animal feeding to improve phosphorus nutrition and to reduce phosphorus pollution of animal waste
- Enzyme B is obtained from *Aspergillus niger, E. coli, K. lactis* and used by cheese producers
- Enzyme C is obtained from *E. coli* and widely used in detergent, paper and food industry
- Enzyme D is obtained from *Aspergillus niger* and used for fruit juice extraction, Bioscouring of cotton fibres, textile processing, Degumming of plant fibers, tea/ coffee processing, Beer industry

Identify Enzymes A, B, C and D

(1) A: Pectinase B: Chymosin
C: Lipase D: Phytase

(2) A: Phytase B: Lipase
C: Chymosin D: Pectinase

(3) A: Phytase B: Chymosin
C: Lipase D: Pectinase

(4) A: Chymosin B: Lipase
C: Pectinase D: Phytase

77. Enzymes are now a days widely used for analytical purposes

- Enzyme A is obtained from *Aspergillus niger* and has been exploited to develop biosensor in nano-diagnostics
- Enzyme B is obtained from *Lactobacillus brevis* and used in fuel cells, it can be used to catalyze the breakdown of fuel for an ethanol fuel cell
- Enzyme C is obtained from *E. coli* and used to control cholesterol level inside the cell
- Enzyme D is obtained from *E. coli* and often used as a conjugate to detect the target antigen or other bio-molecules
- Enzyme E is obtained from *E. coli* and used to hydrolase enzyme responsible for removing phosphate groups from many types of molecules, including nucleotides, proteins or alkaloids

Identify the enzymes A-E

(1) A: Alkaline phosphatases
B: Alcohol dehydrogenase (ADH)
C: Glucose oxidases
D: cholesterol oxidases
E: horseradish peroxidase (HRP)

(2) A: Glucose oxidases
B: ADH
C: choles-terol oxidases
D: HRP
E: Alkaline phosphatases

(3) A: cholesterol oxidases
B: ADH
C: Glucose oxidases
D: HRP
E: Alkaline phosphatases

(4) A: Glucose oxidases
B: ADH
C: Alkaline phosphatases
D: cholesterol oxidases
E: HRP

78. Among the following which graph best depicts the antibiotic production by *Penicillum*?

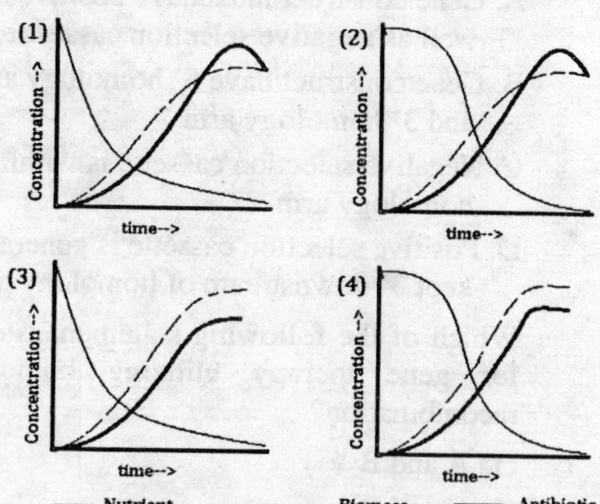

79. Stem cell therapy is the use of stem cells to treat or prevent a disease or condition. Bone marrow transplant is the most widely used stem cell therapy, but some therapies derived from umbilical cord blood are also in use. Which of the following is safest way for stem cell therapy in a patient with damaged liver?

(1) Transplanting liver cells from healthy donor to patient

(2) Transforming stem cells from patient into iPS cells and allowing them to re-differentiate into liver cells under in vitro conditions and then graft in patient

(3) Grafting ES cells from donor directly to patient

(4) Grafting stem cells from umbical cord directly to liver of patient

80. Graph below represents relation between growth rate and growth yield in microbial cultures for randomly variable conditions

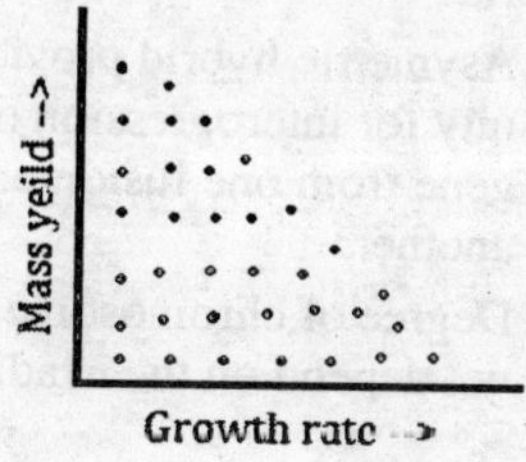

Which of the following conclusion is most appropriate?

(1) Growth rate and growth yield are not related

(2) Growth rate and growth yield are inversely proportional

(3) Growth yield is negatively correlated with growth rate

(4) High growth yield cannot be obtained with high growth rate in above microbial cultures

81. During investigation of gluco-corticoid receptor by Radio Immuno Assay (RIA), tritiated glucocorticoid is added to RIA cocktail. But no radioactive counts were observed. Some of the possible reasons are

A. Amount of radioactive hormone was two low

B. During storage of radioactive hormone radioactivity may have been lost

C. Antibody may not have been added to cocktail

D. The specific activity of radio isotope was too low

The correct reason may be

(1) A only (2) C only

(3) A and D (4) B and C

82. Four different types of receptors namely A, B, C and D bind to same ligand 'X'. In order to determine binding affinity of a receptor for ligand, receptors A, B, C and D were expressed as fusion proteins with GFP on different cells and then incubated with red fluorescent protein tagged ligand 'X' and FACS analysis was carried out to sort different types of cell. The result can be shown as below

	A		B		C		D	
↑	25 %	5 %	50%	20%	20%	50%	5 %	25%
RFP	50 %	20%	25%	5%	5 %	25%	20%	52%

GFP→

Based on above experiment which receptor has highest binding affinity for ligand 'X'?

(1) A (2) B
(3) C (4) D

83. In microarray which sequence of gene must be preferred for preparing target sequence on microarray chip to investigate genome wide expression?

(1) 5' region from coding DNA sequence and 5' untranslated region
(2) 3' region from coding DNA sequence and 3' untranslated region
(3) Any part from coding DNA sequence
(4) Sequence specifically from intron 1 from gene

84. Transgenic plant was prepared by introduction of gene 'X'. It was observed that out of 100 transgenic plant only 70 of them are showing high level expression of gene 'X' while rest of 30 display very low level of expression. The probable reason is

(1) Suppression effect of gene X
(2) Knock down effect of gene X
(3) Gene silencing effect
(4) Co-suppression effect of gene X

85. Four different types of receptors namely P, Q, R and S bind to same ligand 'A'. In order to determine binding affinity of a receptor for ligand, receptors 1, 2, 3 and 4 were expressed as fusion proteins with GFP on different cells and then incubated with red fluorescent protein tagged ligand 'A' and FACS analysis was carried out to sort different types of cell. The result can be shown as below

	1		2		3		4	
↑	25 %	65%	50%	10%	20%	15%	65%	15%
RFP	10 %	20%	25%	65%	5 %	25%	20%	52%

GFP→

Based on above experiment which receptor has highest binding affinity for ligand 'A'?

(1) 1 (2) 2
(3) 3 (4) 4

86. Consider the following statements for homologous recombination based gene transfer

A. Gene construct must have positive as well as negative selection cassestte
B. Gene construct have 5' homology arm and 3' homology arm
C. Negative selection cassette is within homology arm
D. Positive selection cassette is generally kept 3' downstream of homology arm

Which of the following statement is correct for gene therapy utilizing homologous recombination?

(1) A and B
(2) C and D
(3) A, B and D
(4) All of the above

87. Consider the following statements for plant tissue culture

A. Somatic embryogenesis protocol is often pre-requisite for gene transfer in plants
B. Somatic embryo is single cell which give rise to complete plantlet
C. Somatic emryos are formed on MS medium supplemented with auxins and cytokinins
D. Somatic embryo and zygotic embryo are similar in genetic composition

Which of the following are true?

(1) A and B
(2) A and C
(3) B and D
(4) B and C

88. Which is not correct about asymmetric hybrid?

(1) Asymetric hybrid provide and opportunity for interogression of few desirable gene from one fusion partner into another
(2) Degree of chromosome elimination does not depend on the irradiance dose, it

seem to depend mainly on the species combination used in fusion

(3) Asymetric hybrid can be produced by fusion of cytoplast of a species with irradiated protoplast of another species

(4) Several asymmetric hybrid have been produced in *Nicotiana, Brassica* etc

89. Consider the following statements for somatic cell hybridization

A. Most commonly used chemical method for protoplast fusion is use of PEG

B. Cybrids has nuclear information from both species used in protoplast fusion

C. Somatic hybrids could be produced only fusion of related species

D. Somatic hybrids help in interogression of traits from unrelated species for crop improvement

Which of the following are true?

(1) A and B (2) A and C
(3) A and D (4) B and C

90. Organ culture, the cultivation of whole organs or parts thereof, is particularly suitable for studies of development, of inductive interactions, and of the effects of chemical and physical agents upon the physiological functions of specific organs

Consider the following statements

A. The explants remain comparable to the in vivo organs both in structure and function, which makes them more suitable than cell cultures for physiological studies

B. The development of foetal organs in vitro is comparable to that in vivo. Hormone dependent organs remain so, while endocrine organs secrete the specific hormones

C. Organ cultures provide information on the patterns of growth, differentiation and development, and on the influences of various factors on these features

D. Results from organ cultures are often comparable to those from whole animal studies

Which of the following are true statements?

(1) A, B and C (2) B, C and D
(3) A, C and D (4) A, B and D

91. Gene therapy is the therapeutic delivery of nucleic acid polymers into a patient's cells as a drug to treat disease

Consider the following statements

A. The most common form uses DNA that encodes a functional, therapeutic gene to replace a mutated gene

B. A number of viruses have been used for human gene therapy, including retrovirus, adenovirus, lentivirus, herpes simplex, vaccinia and adeno-associated virus

C. Increased understanding of nuclease function has led to more direct DNA editing, using techniques such as zinc finger nucleases and CRISPR

D. In Germ cell gene therapy (GCGT), the therapeutic genes are transferred into any of any cell other than a gamete, germ cell, gametocyte or undifferentiated stem cell

Which of the following is correct?

(1) A, B and C (2) B, C and D
(3) A, C and D (4) A, B and D

92. The primary explant technique was the original method developed by Harrison [1907], Carrel [1912], and others for initiating a tissue culture

A. Mechanical and enzymatic disaggregation of the tissue avoids problems of selection by migration and yields a higher number of cells that are more representative of the whole tissue in a shorter time

B. Crude trypsin is by far the most common enzyme used in tissue disaggregation, as it is tolerated quite well by many cells and is effective for many tissues

C. The cold trypsin method usually gives a lower yield of viable cells, with lesser survival after 24 h culture as compare to warm trypsin

D. In general, trypsin is more severe than collagenase, but is sometimes more effective in creating a single-cell suspension

(1) A, B and C (2) B, C and D
(3) A, C and D (4) A, B and D

93. Friend (murine ery- throleukemia) cells were fixed in methanol as a single-cell suspension and stained with chromomycin A. The suspension was then mixed with chicken erythrocytes (RBC), fixed and stained in the same way, and run through a FACS II. The printout plots cell number on the vertical axis and channel number (fluorescence) on the hor izontal axis. As fluorescence is directly proportional to the amount of DNA per cell, the trace gives a distribution analysis of the cell population by DNA content. The lowest DNA content is found in the chicken erythrocytes, included as a standard. The major peak around channel 30 represents those cells

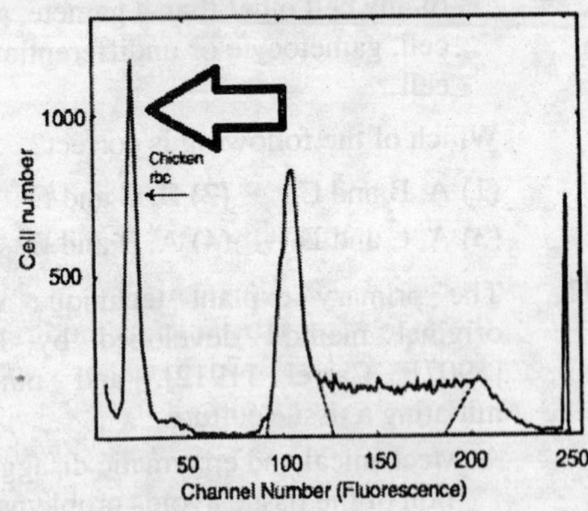

(1) G1 stage of cell cycle
(2) S stage of cell cycle
(3) G2 stage of cell cycle
(4) M stage of cell cycle

94. Consider the following cell separation method and basis of selection

A. FACS: Cell surface area, fluorescent markers, fluorogenic enzyme substrates, multiparameter

B. Immunopanning: Cell surface antigens, cell surface carbohydrate

C. Affinity choromatography: Surface antibodies

D. Countercurrent distribution: Affinity of cell surface constituents for solvent phase

Which of the following is correct match?

(1) A and D (2) B and C
(3) A, B and C (4) B, C and D

95. Transformation is associated with genetic instability major classes of phenotypic change, one or all of which may be expressed in one cell strain

A. Immortalization, the acquisition of an infinite life span

B. Aberrant growth control, the loss of contact inhibition of cell motility, density limitation of cell proliferation, and anchorage dependence

C. Malignancy, as evidenced by the growth of invasive tumors in vivo

D. Genetic stability and controlled growth

Which of the following properties are associated with transformation of animal cells?

(1) A, B and C (2) B, C and D
(3) A, C and D (4) A, B and D

96. Telomeres play an essential role in chromosome stability and determining cellular life span

A. Telomerase or terminal transferase is composed of two main subunits, RNA component (hTR) and a protein catalytic subunit (hTERT) The RNA subunit is ubiquitously expressed in malignant tissues, whereas hTERT is expressed in all cells and tissues

B. The primary cause of senescence appears to be telomeric shortening, followed by telomeric fusion and the formation of dicentric chromosomes and subsequent apoptosis

C. Transfecting cells with the telomerase gene htrt extends the life span of the cell line, and a proportion of these cells become immortal but not malignantly transformed. As a high proportion of the $htrt^+$ clones become immortal, this appears to be a promising technique for immortalization

D. Although the functionality of some of these lines has yet to be demonstrated, there are some encouraging reports of uncompromised differentiation, e.g., in keratinocytes and in myocytes

Which of the following are true statements?

(1) A and B (2) B and C
(3) B and D (4) A and D

97. The graph below represents semilog plot of the increase in cell concentration (left axis) and cell density (right axis) after subculture of animal cells

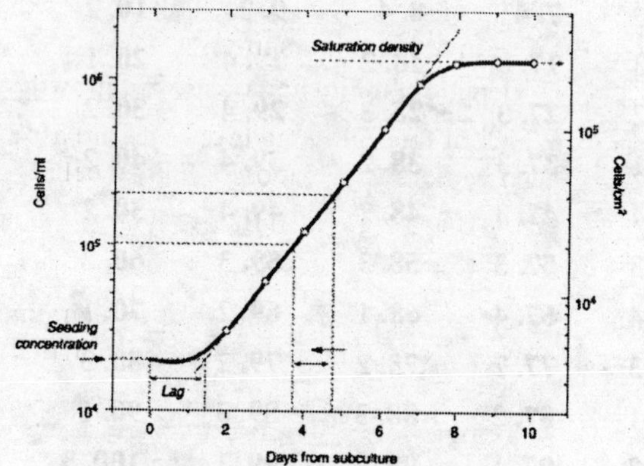

What is estimated doubling time?

(1) Less then 2 days
(2) Exactly 2 days
(3) Exactly 4 days
(4) In between 4-6 days

98. Diagraom below represents semilog plot of cell survival against the concentration of cytotoxin. The slope increases with increasing sensitivity and decreases with reduced sensitivity until it becomes totally flat for complete resistance. Partial resistance can be expressed as the resistant fraction shown by the curve flattening out at the lower end

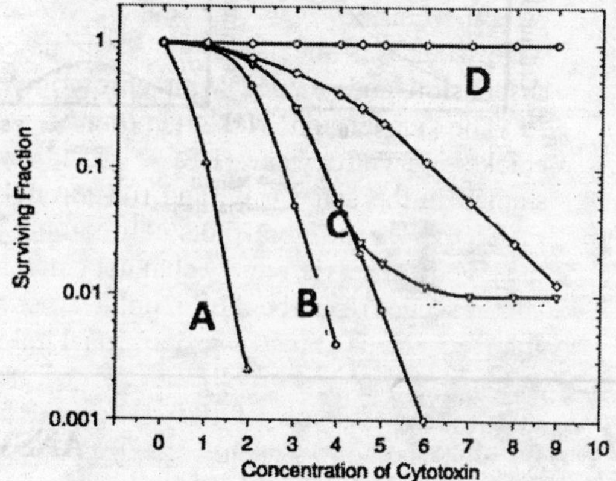

Based on graph which interpretation is correct?

(1) A is most resitant and D is most sensitive cell line to toxin
(2) D is most resitant and A is most sensitive cell line to toxin
(3) B and C shows no response to toxin
(4) No interpretation can be made

99. During cell cycle regulation in eukaryotes, there are post-translational modifications of protein factors, which act as switches for different phases of cell cycle. A cell population of yeast was transfected with gene for wee 1 kinase (modifies cdc2 protein) Assuming that the transfection efficiency was 50% only, which of the following graphical representation of the results is most appropriate?

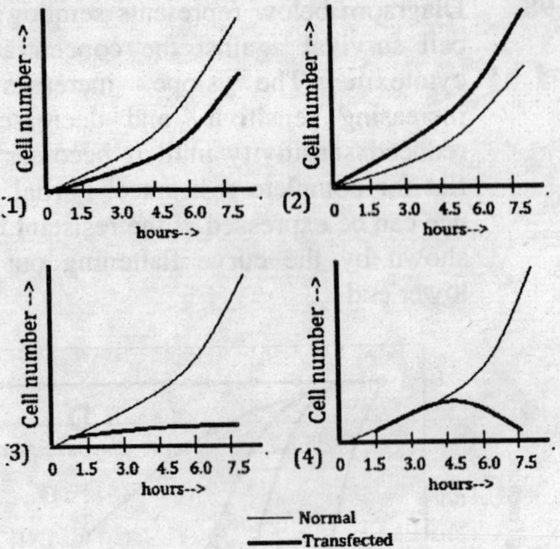

100. Level of follicle stimulating hormone (FSH) during infancy and adulthood is the same but spermatogenesis is seen only during adulthood, mRNA levels coding for FSH receptor are also found to be the same in testis of both age groups

Which of the following investigations will clarify this paradox a little more?

(1) Culture testicular cells and add LH to see testosterone production

(2) Culture testicular cells and add testosterone to see comparative rise in FSH mRNA from both age groups

(3) Culture testicular cells and add FSH to see comparative rise in cAMP production by both age groups

(4) Add both LH and FSH to testicular cells and evaluate cAMp production

ANSWER SHEET

Unit 12: Biotechnology

1. 1	**2.** 3	**3.** 3	**4.** 3	**5.** 1	**6.** 1	**7.** 4	**8.** 4	**9.** 2	**10.** 2
11. 2	**12.** 2	**13.** 3	**14.** 1	**15.** 2	**16.** 3	**17.** 3	**18.** 2	**19.** 4	**20.** 1
21. 2	**22.** 4	**23.** 4	**24.** 2	**25.** 1	**26.** 2	**27.** 3	**28.** 3	**29.** 4	**30.** 2
31. 4	**32.** 4	**33.** 3	**34.** 1	**35.** 4	**36.** 1	**37.** 3	**38.** 2	**39.** 4	**40.** 2
41. 3	**42.** 1	**43.** 4	**44.** 3	**45.** 4	**46.** 4	**47.** 3	**48.** 2	**49.** 4	**50.** 2
51. 1	**52.** 2	**53.** 1	**54.** 4	**55.** 2	**56.** 3	**57.** 3	**58.** 3	**59.** 3	**60.** 2
61. 2	**62.** 1	**63.** 1	**64.** 1	**65.** 1	**66.** 4	**67.** 4	**68.** 1	**69.** 2	**70.** 1
71. 1	**72.** 3	**73.** 1	**74.** 4	**75.** 4	**76.** 3	**77.** 2	**78.** 2	**79.** 2	**80.** 3
81. 4	**82.** 3	**83.** 1	**84.** 3	**85.** 1	**86.** 1	**87.** 2	**88.** 3	**89.** 3	**90.** 1
91. 1	**92.** 4	**93.** 1	**94.** 1	**95.** 1	**96.** 3	**97.** 1	**98.** 2	**99.** 1	**100.** 3

UNIT-13

APPLIED BIOTECHNOLOGY & BIOTECHNIQUES

1. In an *in vitro* experiment using radio-labeled nucleotides, a researcher is trying to analyze the possible products or DNA replication by resolving the products using urea-polyacrylamide gel electrophoresis. In one experimental set up RNase H was added (Set 1), while in another set no RNase H was added (Set 2)

The possible observations of this experiment could be

A. There is no difference in the mobility of labelled DNA fragments between the Set I and Set 2

B. There is distinct difference in the mobility of the newly synthesized labeled DNA fragments between Set I and Set 2

C. The mobility of the newly synthesized labelled DNA fragments in case of Set 1 is faster as compared to the Set 2

D. The mobility of the newly synthesized labelled DNA fragments in case of Set 1 is slower as compared to the Set 2

Which of the following combinations represent correct observations?

(1) A and B (2) B and C

(3) A and D (4) B and D

2. You have a polypeptide that has been degraded with cyanogen bromide (cleaves on the C terminal side of methionines) and trypsin (cleaves on the C-terminal side of lysines and arginines) The sequences of the fragments are listed below (using the one-letter codes)

Cyanogen bromide treatment:	*Trypsin treatment*
(1) K	(5) FMK
(2) RFM	(6) GMNIK
(3) MLYCRGM	(7) GLMR
(4) NIKGLM	(8) MLYCR

What is the sequence of original polypeptide?

(1) NIKGLMMLYCRGMRFMK

(2) KMLYCRGMRFMNIKGLM

(3) NIKGLMRFMKMLYCRGM

(4) MLYCRGMNIKGLMRFMK

3. Double stranded nucleic acids will tend to fold up into complex topological structures. The migration of the molecule will be greatly influenced by the way it folds. Following are some of DNA molecules differing in topological conformations

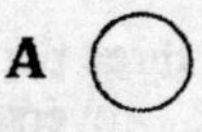

Although all three forms A, B and C are the same size (in terms of the number of base pairs) What would be distance migrated by A, B and C during agarose gel electrophoresis?

(1) A> B> C

(2) C> B > A

(3) C>A> B
(4) All will migrate to same distance

4. In equilibrium or density gradient centrifugation, a density gradient is produced by centrifuging a solution of caesium chloride (CsCl) at a very high speed. Macromolecules present in the CsCl solution when it is centrifuged form bands at distinct points in the gradient. Exactly where a particular molecule bands depends on its buoyant density

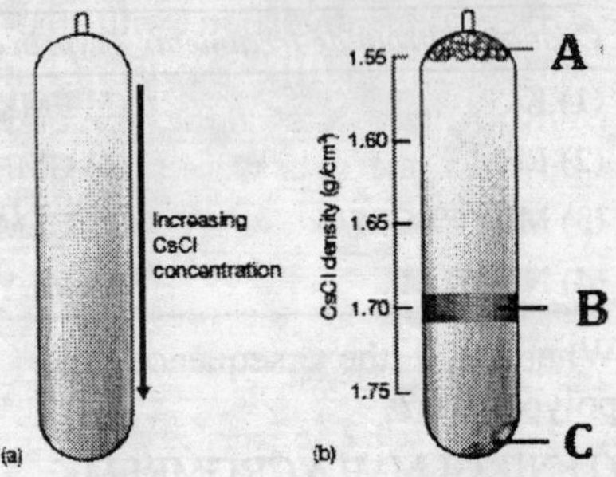

Suppose your friend have separated DNA, RNA and protein through this technique and three separate bands A, B and C were obtained as shown in diagram above. Which of the following inference is most correct?

(1) A is DNA, B is RNA and C is protein
(2) A is protein, B is DNA and C is RNA
(3) A is RNA, B is protein and C is DNA
(4) A is Protein, B is RNA and C is DNA

5. A mixture of the following three peptides A, B and C are loaded on C_{18} reverse phase, chromatography column

Peptide A = L L W G Y S G
Peptide B = L S G G S G K
Peptide C = K A K A K L G

A mixture of water and acetonitrile is then used as a solvent with increasing concentration of the latter in a gradient mode to facilitate separation of the peptides

The peptides A, B and C will elute in the following order

(1) A followed by B followed by C
(2) B followed by C followed by A
(3) C followed by B followed by A
(4) A followed by C followed by B

6. A biochemist discovers and purifies a new enzyme, generating the purification table below. Which of the purification procedures is least effective?

Procedure	*Total protein (mg)*	*Activity (units)*
1 Crude extract	20,000	4,000,000
2 Precipitation (salt)	5,000	3,000,000
3 Precipitation (pH)	4,000	1,000,000
4 Ion-exchange chromatography	200	800,000
5 Affinity chromatography	50	750,000
6 Size-exclusion chromatography	45	675,000

(1) 1 (2) 3
(3) 4 (4) 6

7. Partial hydrolysis of the hexapeptide followed by chromatographic separation and sequence analysis yielded the following di- and tripeptides (the amino-terminal amino acid is always at the left)

Leu-Phe Phe-Pro Orn-Leu Val-Orn
Van-Orn-Leu Phe-Pro-Val Pro-Val-Orn

Given the above information, deduce the amino acid sequence of the peptide antibiotic

(1) Val-Orn-Leu-Phe-Pro-Phe
(2) Phe-Pro-Val-Orn-Leu-Phe
(3) Phe-Pro-Val-Orn-Leu-Val
(4) Val-Orn-Leu-Phe-Pro-Val

8. In X-ray studies of crystalline peptides, Linus Pauling and Robert Corey found that the C-N bond in the peptide link is intermediate in length (1.32 Å) between a typical C-N single bond (1.49 Å) and a C-N double bond (1.27 Å). They also found that the peptide bond is planar (all four atoms attached to the C-N group are located in the same

plane) and that the two α- carbon atoms attached to the C-N are always trans to each other (on opposite sides of the peptide bond): What does the length of the C-N bond in the peptide linkage indicate about its strength and its bond order

(1) It has single bond character

(2) It has double bond character

(3) It has partial double bond character

(4) It has triple bond character

9. Glucose labeled with ^{14}C at C-1 is incubated with the glycolytic enzymes and necessary cofactors. If the specific activity of the glucose substrate is 10 mCi mM^{-1}, what is the specific activity of the pyruvate that is formed?

(1) 10 mCi mM^{-1} (2) 5 mCi mM^{-1}

(3) 55 Ci μM^{-1} (4) 27.5 µCi mM^{-1}

10. Representations of sequencing gels obtained from Sanger's dideoxy chain termination method for variants A and B of the α chain of human hemoglobin are shown here. What is the nature of the amino acid?

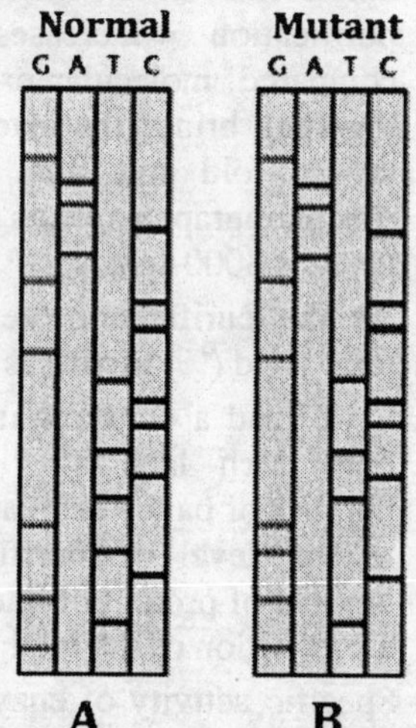

(1) Residue 2, L→R, CTG→CGG

(2) Residue 5, A→P, GCC→CCC

(3) Residue 6, D→G, GAC→GGC

(4) Residue 6, Q→P, CAG→CCG

11. A series of people are found to have difficulty eliminating certain types of drugs from their bloodstreams. The problem has been linked to a gene *X*, which encodes an enzyme Y. Six people were tested with the use of various techniques of molecular biology. Person A is a normal control, person B is asymptomatic but some of his children have the metabolic problem, and persons C through F display the trait. Tissue samples from each person were obtained. Southern analysis was performed on the DNA after digestion with the restriction enzyme *Hin*dIII. Northern analysis of mRNA also was done. In both types of analysis, the gels were probed with labeled *X* cDNA. Finally, a Western blot with an enzyme-linked monoclonal antibody was used to test for the presence of protein Y. The results are shown here

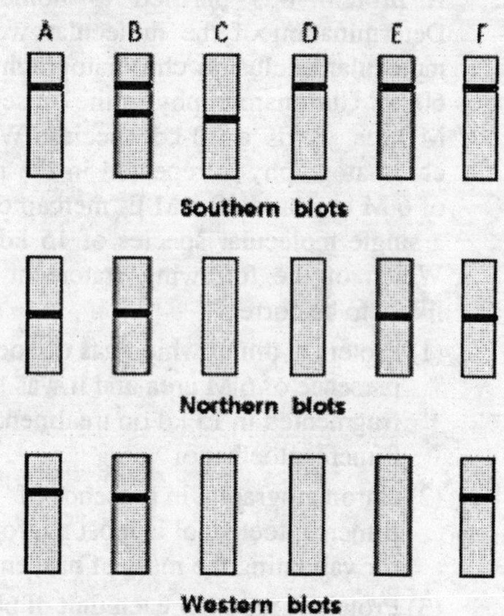

Following conclusion can be drawn

A. A mutation in person B has altered one if the of allele of gene X, leaving the other intact. The one functional allele is transcribed and translated enough protein to render person asymptomatic

B. Person C has only the smaller version (having deletion) of gene X. This gene is transcribed but not translated

C. Person D has a normal copy of m-RNA but no RNA or protein. There may be mutation is promoter region of the gene which prevents transcription

D. Person E has normal copy of gene that is transcribed but no protein is made, which suggest mutation prevents translation, which may be due to incorporation of premature stop codon in m-RNA

E. Person F has a normal copy of gene that is transcribed and translated but still possess metabolic problem. It is unlikely and probably error has been made in experimentation

Which of the following conclusion are not valid?

(1) A and C (2) B and E
(3) C and D (4) A and D

12. A protein was purified to homogeneity. Determination of the molecular weight by molecular exclusion chromatography yields 60 kd. Chromatography in the presence of 6 M urea yields a 30-kd species. When the chromatography is repeated in the presence of 6 M urea and 10 mM β- mercaptoethanol, a single molecular species of 15 kd results. Which of the following statement is most likely to be correct?

(1) Protein is dimer whic.i has dissociated in presence of 6 M urea and it was further fragmented in 15 kd on treatment with β-mercaptoethanol
(2) Chromatography in presence of β-mercaptoethanol is most appropriate for validating the mass of nascent protein
(3) Protein is tetramer each unit of 60 kd which fragmented into different mass peptides on vigorous treatment with urea and β-mercaptoethanol
(4) The original 60 kd protein is made up of two 30 kd subunits associated with non-covalent interaction, when these each subunit were subsequently treated with urea and β-mercaptoethanol, a single band of 15 kd suggesting that disulphide bonds links these 30 kd subunits

13. To determine the success of a protein purification scheme, we monitor the procedure at each step by determining specific activity and by performing an SDS-PAGE analysis. Consider the results for the purification of a fictitious protein, summarized in table below

Step	Total protein (mg)	Total activity (units)
Homogenization	15,000	150,000
Salt fractionation	4,600	138,000
Ion-exchange chromatography	1,278	111,500
Molecular exclusion chromatography	68.8	75,000
Affinity chromatography	1.75	52,500

Consider the following statements

A. The first purification step, salt fractionation, leads to an increase in purity of only 3-fold, an ion-exchange column, the purification increases to 9-fold compared, molecular exclusion chromatography brings the level of purification to 110-fold and final step is affinity chromatography results in a purification level of 3000-fold

B. As the purification steps are followed total yield (%) increases

C. If we load a constant amount of protein onto each lane after each step, the number of bands decreases in proportion to the level of purification, and the amount of protein of interest increases as a proportion of the total protein present

D. Specific activity of enzyme decreases as we carry out different purification step and finally becomes constant

Which of the following statement are correct?

(1) A and B (2) C and D
(3) A and C (4) B and D

14. To construct a restriction map, a series of restriction digests must be performed. First, the number and sizes of the fragments produced by each restriction endonuclease must be determined by gel electrophoresis followed by comparison with size markers (Figure) This information must then be supplemented by a series of double digestions, in which the DNA is cut by two restriction endonucleases at once. It might be possible to perform a double digestion in one step if both enzymes have similar requirements for pH, Mg^{2+} concentration, etc. Alternatively, the two digestions may have to be carried out one after the other, adjusting the reaction mixture after the first digestion to provide a different set of conditions for the second enzyme. Comparing the results of single and double digests will allow many, if not all, of the restriction sites to be mapped. Results obtained during single and double digestion of lambda phage with restriction enzymes are summarized in table below

Enzymes	Number of fragments	Sizes (kb)
*Xba*I	2	24.0, 24.5
*Xho*I	2	15.0, 33.3
*Kpn*I	3	1.5, 17.0, 30.0
*Xba*I + *Xho*I	3	9.0, 15.0, 24.5
*Xba*I + *Kpn*I	4	1.5, 6.0, 17.0, 24.0

Which of the following inference in not correct?

(1) As λ DNA is linear, the number of restriction sites for each enzyme is *Xba*I 2, *Xho*I 2, *Kpn*I 3

(2) The *Xba*I and *Xho*I sites can be mapped

*Xba*I fragments: 24.0 | 24.5
*Xba*I – *Xho*I fragments: 9.0 | 15.0 | 24.5
*Xho*I fragments: 15.0 | 33.5

The only possibility is: 15.0 (*Xho*I) 9.0 (*Xba*I) 24.5

(3) All the *Kpn*I sites fall in the 24.5 kb *Xba*I fragment, as the 24.0 kb fragment is intact after *Xba*I–*Kpn*I double digestion The order of the *Kpn*I fragments can be determined only by partial digestion

(4) Total genome size of λ phage is 48.5 kb

15. Bacterial cell number can be estimated by measurement of optical density (OD) A sample of the culture is placed in a glass cuvette and light with a wavelength of 600 nm shone through. The amount of light that passes through the culture is measured and the OD (also called the absorbance) calculated as

$$10\text{ D unit} = \log_{10}\frac{\text{intensity of transmitted light}}{\text{intensity of incident light}}$$

The operation is performed with a spectrophotometer. The cell number corresponding to the OD reading is calculated from a calibration curve. This curve is plotted from the OD values of a series of cultures of known cell density as shown below

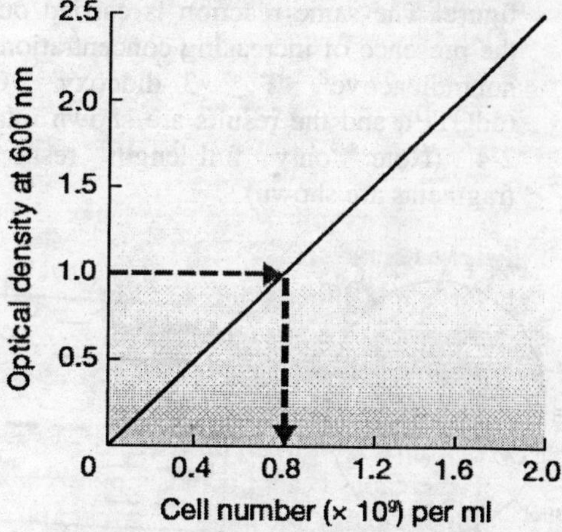

E. coli stock was diluted 100 times and 1ml of sample was used to take OD. It shows OD of one. Estimate the number of *E. coli* in original stock?

(1) 0.8×10^9 cells/ml

(2) 8.0×10^9 cells/ml

(3) 80×10^9 cells/ml

(4) 0.8×10^7 cells/ml

16. A quantitative amino acid analysis reveals that bovine serum albumin (BSA) contains 0.58% tryptophan (*M*r 204) by weight. assuming there is only one tryptophan residue per protein molecule

(1) 35.7 kd (2) 64.5 kd
(3) 58.0 kd (4) 118.2 kd

17. An *in vitro* system is used to study replication of a viral DNA genome that is a double-stranded, covalently closed circle. The location of the sites for the restriction endonuclease MboI on this DNA are shown in figure. Replication reactions are carried out using viral DNA as a template, extracts of infected cells as a source of enzymes, and other exogenous nucleotides (dGTP, dCTP, dATP and ATP), all of which are labeled with 32P. The reaction products are digested with MboI, analyzed by agarose gel electrophoresis, and visualized by autoradiography, producing the results shown in lane 1 of figure. The same reaction is carried out in the presence of increasing concentrations of nonradioactive 2`, 3`-dideoxy GTP (ddGTP), and the results are shown inlanes 2-4 (Note: only full-length restriction fragments are shown)

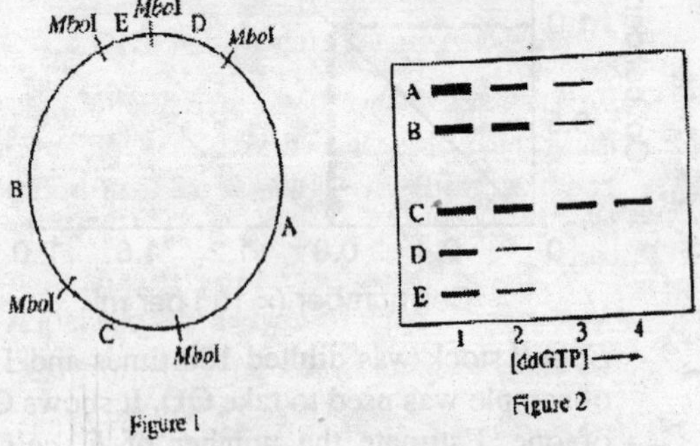

Figure 1 Figure 2

The data in Figure indicate that the origin of replication of this DNA is in fragment

(1) A (2) B
(3) C (4) D

18. Match the following microscopic techniques with their respective functions

1. Phase contrast microscopy	a. view gram stained microbial cells
2. Fluorescent microscopy	b. view nano-sized details of section through interior of bacteria
3. Transmission electron microscopy	c. view internal structure of unstained living cells
4. Bright field microscopy	d. view antibody tagged cells
5. Dark field microscopy	e. view translucent microbes
6. Scanning electron microscopy	f. view nano-sized details of exterior of bacteria

(1) 1-c; 2-d; 3-b; 4-a; 5-e; 6-f
(2) 1-d; 2-c; 3-a; 4-b; 5-e; 6-f
(3) 1-c; 2-d; 3-a; 4-b; 5-f; 6-f
(4) None of these

19. A solution containing NAD^+ and NADH had an optical density in a 1 cm cuvette of 0.311 at 340 nm and 1.2 at 260 nm. Calculate the concentrations of the oxidized and reduced forms of the coenzyme in the solution. Both NAD^+ and NADH absorb at 260 nm, but only NADH absorbs at 340 nm. The extinction coefficients are given below

Compound	ε ($M^{-1} \times cm^{-1}$) 260 nm	340 nm
NAD^+	18,000	~0
NADH	15,000	6220

(1) [NADH] = 5×10^{-5} M and [NAD^+] = 2.50×10^{-5} M
(2) [NADH] = 2.5×10^{-5} M and [NAD^+] = 5×10^{-5} M
(3) [NADH] = 1×10^{-5} M and [NAD^+] = 2×10^{-5} M
(4) [NADH] = 6.5×10^{-5} M and [NAD^+] = 2×10^{-5} M

20. In a two-colour microarray hybridization experiment, yeast cells were treated with diclonine and the cDNA was labelled with cyanine-3 and the control sample was labelled with cyanine-5. Upon scanning, merging the images and data extraction, ten spots were found to exhibit Cy3/Cy5 log2 ratio $\geq$ 3 whereas two spots had log2 ratio $\leq$ -2. What would you infer about the expression of these ORFs?

(1) The ten ORFs are down-regulated and two ORFs are up-regulated by diclonine treatment

(2) The ten ORFs are induced while the two ORFs are repressed two-fold or more

(3) Can not assess gene expression from this data

(4) The behaviour of the ORFs does not fit the criteria of two-fold regulation

21. Which of the following peptide fragments will be formed if a polypeptide X with the sequence Leu -Met - His- Tyr -Lys-Arg - Ser- Val-Cys- Ala - Lys- Asp-Gly- Ile- Phe-Ile is cleaved with trypsin?

(1) Leu-Met-His-Tyr, Lys-Arg-Ser-Val-Cys-Ala and Lys-Asp-Gly-Ile-Phe-Ile

(2) Leu- Met and His- Tyr- Lys-Arg-Ser-Val-Lys-Ala -Lys-Asp-Gly-Ile- Phe- lIe

(3) Leu-Met-His-Tyr-Lys, Arg, Ser-Val-Cys-Ala,-Lys,Asp-Gly-Ile-Phe-Ile

(4) Leu-Met-His-Tyr, Lys-Arg-Ser-Val-Cys-Ala-Lys-Asp-Gly-Ile-Phe, lIe

22. Mark the correct sequ.ence in which myoglobin (16 kDa), catalase (221 kDa), cytochrome c (13 kDa), myosin (524 kDa), and serum albumin (68 kDa) proteins would be eluted from a molecular-exclusion column with a protein-fractionation range of 5000 to 400000

(1) Myoglobin, catalase, cytochrome c, myosin and serum albumin

(2) Serum albumin, myoglobin, cytochrqme c, catalase and myosin

(3) Myosin, catalase, serum albumin, myoglobin and cytochrome c

(4) Cytochrome c, myosin, myoglobin, catalase and serum albumin

23. To label a nicked DNA a researcher mixes DNA with ($^{32P}\alpha$) dATP. Which of the following statements is true?

(1) Addition of T4 polynucleotide kinase to the reaction mix will label the DNA at 5' end

(2) Addition of T4 polynucleotide kinase to the reaction mix will label the DNA at 3' end

(3) Addition of Klenow fragment to the reaction mix will label the DNA via nick translation reaction

(4) Addition of DNA pol I to the reaction mix will label the DNA via nick translation reaction

24. A researcher wants to set up a ligation reaction. The reaction mix should contain

(1) one double-strand linear DNA with 3' PO_4, one double-strand DNA with 5' OH, T4 DNA ligase and ATP at pH 7.5

(2) one double-strand linear DNA with 3 OH, one double-strand DNA with 5' PO_4, T4 DNA ligase, and ATP at pH 7.5

(3) one double-strand linear DNA with 3' PO_4, one double-strand DNA with 5' OH, T4 DNA ligase and dATP at pH 7.5

(4) one double-strand linear DNA with 3' PO_4' one double-strand DNA with 5' OH, T4 DNA ligase and GTP at pH 7.5

25. Prokaryotes do not have the splicing machinery and thus fail to express eukaryotic genes with introns. However, such limitation is overcome by

(1) splicing the transcript in *vitro* and then putting it back to *E. coli* for translation

(2) generating an intron less gene in *vitro* by polymerase chain reaction using genomic DNA as template

(3) genetically modifying *E. coli* so that it can splice mRNA

(4) generating an intron-free version of the gene by manipulating the mRNA isolated from the test organism

26. The genome-wide comparative mRNA expression profile between normal and cancer cells can be studied using the microarray technology. The basic approach of this technology is

(1) hybridization of fluorescently labelled genomic DNA from human cells to normal and cancer cell cDNAs immobilized onto two glass slides
(2) hybridization of fluorescently labelled cDNAs from normal and cancer cells to a single glass slide containing immobilized clones of human genomic DNA
(3) hybridization of fluorescently labelled cDNAs from normal and cancer cells to a single glass slide containing immobilized clones of human cDNA
(4) hybridization of fluorescently labelled genomic DNA from normal and cancer cells to immobilized human cDNAs on a single glass slide

27. GA induced starch degradation in endosperm essentially requires gene expression, protein synthesis and protein transport. In an experiment to understand this, GFP tagged protein is seen to be moving from cytoplasm into the nucleus before alpha amylase could be detected. What do you think this GFP is tagged with?

(1) It is in fusion with a gene regulator
(2) It is in fusion with a signaling kinase
(3) It is in fusion with alpha-amylase
(4) It is tagged with a cell wall protein

28. The average molecular mass of nucleotide residue in DNA of random base composition is 330 daltons. You have a protocol of transfection that yields 10^8 colonies per micrograms of a 7.5 kbp vector. Which of the following would be the approximate efficiency of transfection?

(1) 5% (2) 10%
(3) 0.8% (4) 8%

29. While measuring the UV-visible absorption spectrum of a protein using a 1 mm path length cell, you find that the signal is weak because the concentration of the protein is low. You would overcome this problem by

(1) increasing viscosity by adding glycerol
(2) changing the wavelength range
(3) changing the temperature of sample
(4) increasing the path length of the cell

30. The MALDI spectrum of a peptide shows a peak at *m/z* corresponding to 3600. When the ESI spectrum is recorded, peaks at *m/z* corresponding to 721,904 and 1801 were obtained. When the MALDI MS/MS spectrum was recorded, large number of peaks with *m/z* less than 3600 were observed. The spectral data indicate that the peptide is

(1) highly impure
(2) pure with molecular mass of 3600 and partial sequence of the peptide can be determined
(3) highly unstable and degrades rapidly
(4) degraded under condition employed for recording ESI spectrum

31. The resonance condition that must be satisfied for a transition between energy states in an ESR experiment (for which *g* is the spectroscopic splitting factor, β the Bohr magneton and H is the strength of the field applied) is given by

(1) hν = gβH (2) *hν* = gβ/H
(3) hν = g/βH (4) hν = (gβH)-1

32. T cell proliferation *in vivo* is to be analyzed. The cells are labeled with CFSE (a fluorescent probe) and injected in CD86-deficient mice and BALB/c mice along with the required antigens. Three days later, the cells are recovered and analyzed by flow cytometry. Which one of the following is logically correct?

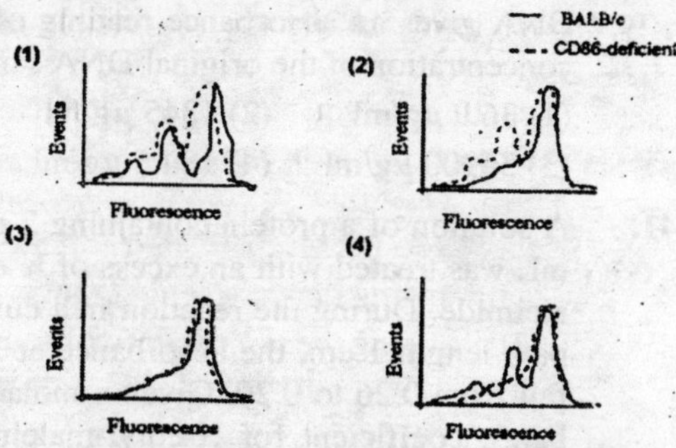

33. The most important property of any microscope is its power of resolution, which is numerically equivalent to D; the minimum distance between two distinguishable objects. D depends on three parameters namely, the angular aperture, α, the refractive index, N, and wavelength, λ, of the incident light. Below are given few options to increase the resolution of the microscope

A. Decrease the value of λ or increase either N or α to improve resolution

B. Moving the objective lens closer to the specimen will decrease sin α and improve the resolution

C. Using a medium with high refraction index between specimen and the objective lens to improve the resolution

D. Increase the wavelength of the incident light to improve resolution

Which of the following combinations of above statement are correct?

(1) A and C (2) B and C

(3) A and D (4) C and D

34. In the following statement taken from a research paper, what does p in the parenthesis stands for?

"The mean temperature of this region now is significantly higher than the one 50 years ago (p<0.05, t-test)"

(1) Ratio of the mean temperature of the two times periods tested

(2) Probability of the error of rejecting a true null hypothesis

(3) Probability of the error of accepting a false null hypothesis

(4) Probability of the t-test being effective in detecting significant difference in the mean annual temperatures of the two periods

35. 4A 30-residue peptide was treated with trypsin and the tryptic peptides were separated by HPLC. Four peaks A, B, C and D were obtained. Peptides corresponding to A, B, C and D were reduced and alkylated selectively at cysteine residues. The sequences obtained from A, B, C and D after reduction and alkylation were: A, AEK; B, C(S-alkyl)EPGYR and WC(S-alkyl)SPPK; C,C(Salkyl) EHFR and C(S-alkyl)GGK; D, C(S-alkyl)EAFC(S-alkyl)L. The sequence of the 30-residue peptide is

(1) AEKCEPGYRCEHFRWCSPPKCGGKCEAFCL

(2) AEKCEPGYRCEHFRWCSPPKCGGKCEAFCL

(3) AEKCEPGYRCEHFRWCSPPKCGGKCEAFCL

(4) AEKCEPGYRCEHFRWCSPPKCGGKCEAFCL

36. A student is asked to isolate and purify various subcellular organelles from a liver cell sample. Following sucrose density gradient centrifugation technique, and taking utmost care mitochondrial fraction is found to be contaminated with nuclear fraction. Which technique out of the choices given below should be appropriate to obtain a clean mitochondrial fraction?

(1) Gel filtration

(2) Percoll density gradient

(3) Pulse field electrophoresis

(4) Immuno-magnetic separation

37. The values of molar absorption coefficient (ε) of Trp and Tyr at 240 nm and 280 nm are the following

Waveleught (nm)	E_{Tyr} (M^{-1} cm^{-1})	ε_{Trp} (M^{-1} cm^{-1})
240	11,300	1,960
280	1,500	5,380

A 10-mg sample of a protein is hydrolyzed to its constituent amino acids and diluted to 100 ml. The absorption of this solution in a 1-cm path length, is 0.717 at 240 nm and 0.239 at 280nm. The estimated content of Trp and Tyr in μmol / g protein respectively are

(1) 586 and 28.1 (2) 58.6 and 281
(3) 586 and 281 (4) 58.6 and 28.1

38. The structure of a protein is known from X-ray diffraction studies which gave 30% α-helix, 50% β-sheet and 20% random coil. Circular dichroism (CD) measurements gave 50% α-helix, 40% β-sheet and 10% random coil. What could not be a possible explanation for these observations

(1) Protein structure in the crystal is different from that in the solution
(2) CD analysis for structural components is not appropriate for this protein
(3) Contributions from other chromophores also contribute to the CD spectrum of the protein
(4) Protein contains high content of disulphide bonds

39. A homogeneous protein of native molecular weight 100,000 gave a single band of molecular weight 50,000 on SDS-PAGE in presence of β-mercaptoethanol. N-terminal analysis gave two amino acids alanine and leucine in equal proportions. Hence the

(1) protein is a homodimer
(2) protein is contaminated with another protein
(3) protein has two polypeptides linked by disulfide bridges
(4) none of the above

40. A DNA sample when it is diluted by a factor of 100 gives an optical density reading at 260 nm of 0.369. Assuming 50 μg/ml of DNA gives an absorbance reading of 1, the concentration of the original DNA sample is

(1) 36.9 μg/ml (2) 1845 μg/ml
(3) 36900 μg/ml (4) 184.5 μg/ml

41. A solution of a protein containing 2 mg in 1 mL was treated with an excess of N-ethylm-aleimide. During the reaction in a cuvette of path length 1 em, the absorbance at 305 nm fell from 0.26 to 0.20. Given a molar absorbance coefficient for N-ethyl-maleimide of 620 L $mol^{-1}cm^{-1}$. Calculate the concentration of sulfhydryl groups in the solution

(1) $2.8\times10^{-5}\, molL^{-1}$
(2) $4.8\times10^{-5}\, molL^{-1}$
(3) $6.8\times10^{-5}\, molL^{-1}$
(4) $9.8\times10^{-5}\, molL^{-1}$

42. The absorbance *A* of a solution is defined as

$$A = \log_{10}\ (I_o / I)$$

in which I_0 is the incident light intensity and I is the transmitted light intensity. The absorbance is related to the molar absorption coefficient (extinction coefficient) ε (in M^{-1} cm^{-1}), concentration c (in M), and path length l (in cm) by

$$A = \varepsilon lc$$

The absorption coefficient of myoglobin at 580 nm is 15,000 M^{-1} cm^{-1}. What is the absorbance of a 1 mg ml^{-1} solution across a 1-cm path?

(1) 0.14 (2) 0.44
(3) 0.84 (4) 1.14

43. A bottle contains 1 mCi of L-glutamic acid-C^{14} (uniformly labeled) in 2.0 ml of solution. The specific activity of the labeled amino acid is given as 150 mCi/m mole. Calculate the concentration of L-glutamic acid in the solution

(1) 1.435 X 10^{-3} mmol/ml
(2) 3.335 X 10^{-3} mmol/ml
(3) 4.430 X 10^{-3} mmol/ml
(4) 6.660 X 10^{-3} mmol/ml

44. Determine the sequence of hexapeptide based on the following data. Note: When the sequence is not known, a comma separates the amino acids

- Amino acid composition: (2R,A,S,V,Y)
- N-terminal analysis of the hexapeptide: A
- Trypsin digestion: (R,A,V) and (R,S,Y)
- Carboxypeptidase digestion: No digestion
- Chymotrypsin digestion: (A,R,V,Y) and (R,S)

(1) AVRSYR (2) AVRYSR
(3) ARVYSR (4) AVYSRR

45. A solution of a protein whose sequence includes three tryptophan residues, no tyrosine residues, and no phenylalanine residues has an absorbance of 0.1 at 280 nm in a cell with a path length of 1 cm. Estimate the concentration of the protein in units of molarity. If the protein has a molecular mass of 100 kd, estimate the concentration in units of milligrams of protein per milliliter of solution ($\varepsilon = 3400\ M^{-1}\ cm^{-1}$)

(1) 1 mg/ml (2) 5 mg/ml
(3) 10 mg/ml (4) 100 mg/ml

46. When sequencing proteins, one tries to generate overlapping peptides by using cleavages at specific sites. Which of the following statements about the cleavages caused by particular chemicals or enzymes is not true?

(1) Cyanogen bromide cleaves at the carboxyl side of threonine
(2) Trypsin cleaves at the carboxyl side of Lys and Arg
(3) Chymotrypsin cleaves at the carboxyl side of aromatic and bulky amino acids
(4) 2-Nitro-5-thiocyanobenzoate cleaves on the amino side of cysteine residues

47. A mixture of three peptides A, B and C is loaded in a C_{18} reverse phase chromatography column. A mixture of water and acetonitrile is used as the solvent with increasing concentration of the latter in a gradient mode. If the hydrophobicities of the peptides are in the order A < B < C, the order in which they will elute from the column would be

(1) A followed by B followed by C
(2) B followed by C followed by A
(3) C followed by B followed by A
(4) A followed by C followed by B

48. Chymotrypsinogen in the native state and reduced unfolded states is loaded in well A and B respectively. They are electrophoreses and stain, one of the following results is likely to be observed

(1) protein in well A would have moved longer distance than that in well B
(2) protein in well A would have moved shorter distance than that in well B
(3) protein in well A and B would have moved the same distance
(4) protein in well A and B would not move at all

49. loxP sites are specific 34-bp sequences consists of an 8-bp core sequence where recombination takes place and two flanking 13-bp inverted repeats. The *Cro* recombinase is an enzyme that catalyzes recombination between two loxP sites. But what will happen to gene in question in between loxP sites depends upon orientation of two loxP sites. Study following diagram and state what will happen with gene respectively

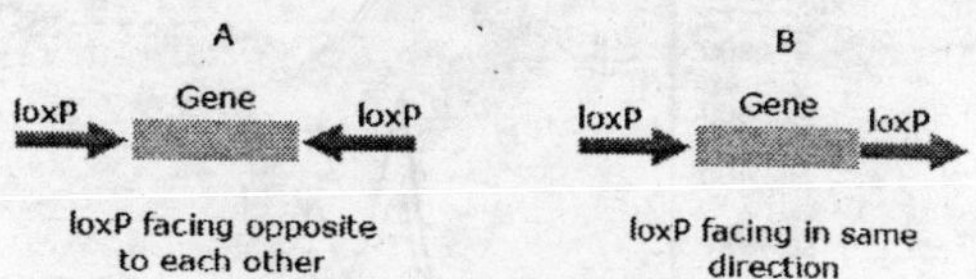

(1) A = inversion, B = deletion
(2) A = deletion, B = inversion
(3) A = translocation, B = inversion
(4) A = inversion, B = translocation

50. A quantitative amino acid analysis reveals that Bovine Serum Albumin (BSA) contains 0.58% tryptophan (*Mr* 204) by weight.

assuming there is only one tryptophan residue per protein molecule

(1) 35.1 kd (2) 64.5 kd
(3) 58.0 kd (4) 118.2 kd

51. For the study of biosynthesis of any secretory glycoprotein, the first step is to fractionate a crude RNA extract using an oligo-dT column. The RNA bound to the column are then eluted and translated in vitro in the presence of [^{3}H] leucine and [^{14}C] mannose. An antibody specific for the secretory glycoprotein are added and the resulting immunoprecipitate is analyzed by Sodium Dodecyl Sulfate (SDS) polyacrylamide gel electrophoresis. Distribution of radioactivity as a function of position in the gel is then analyzed as shown in figure 1

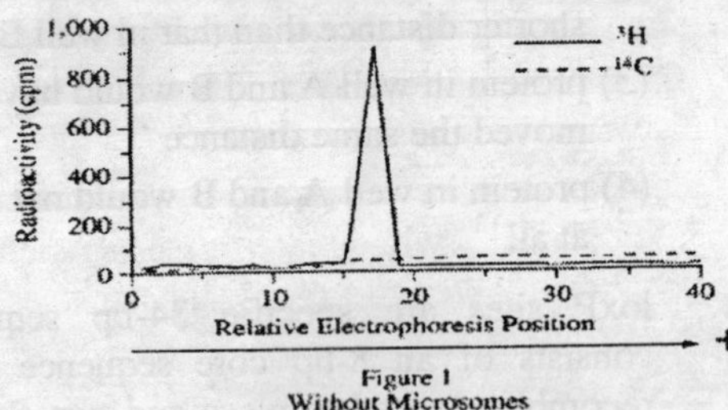

Figure 1
Without Microsomes

The in vitro translation experiment was repeated in the presence of rough microsomes, which were then solubilized. Immunoprescipitation and electrophoresis were performed as in the previous experiment. The results are shown in figure 2

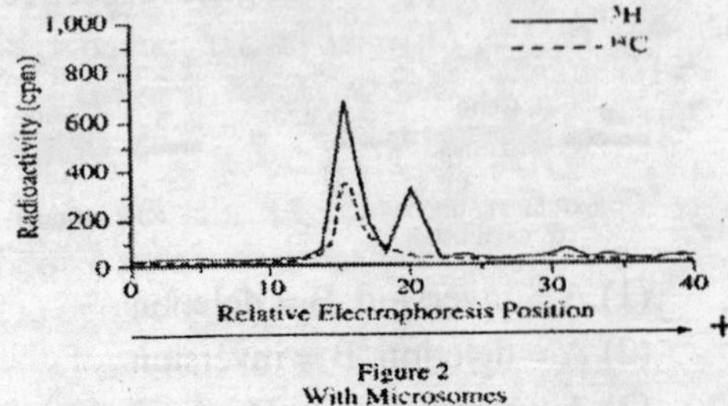

Figure 2
With Microsomes

The protein in the peak of ^{3}H at position 20 in Figure 2 is of lower monomeric molecular weight than the protein in figure 1. This suggests that microsomes cause proteolysis but only one of the product of proteolysis appears on the gel in figure 2. All of the following can account for the absence of a second product of proteolysis except

(1) It was not recognized by the antiserum
(2) It contains no leucine or mannose residues
(3) It was bound by, but not released from, the oligo-dT column
(4) It was rapidly degraded

52. Green Fluorescent Protein (GFP)

A. encoded by gene of bioluminescent jellyfish *Aequorea victoria*
B. discovered by Shimomura
C. amino acid residues 65-67 (Ser-Tyr-Gly) form the fluorescent chromophore
D. it fluoresces with a peak wavelength of 508 nm (green light) when irradiated by UV or blue light (optimally 400 nm)

(1) A and B (2) B and C
(3) A and D (4) A, B, C and D

53. The yeast Gal 4 gene activator protein comprises the activation and DNA binding domains. Whereas the DNA binding domain binds to specific DNA sequence element, the activation domain interacts with components of the transcriptional machinery. In the absence of Gal 4, the *GAL* gene promoters normally bound by Gal 4 are turned off, and when Gal 4 is normally expressed in medium with galactose as the carbon source, the same promoters are highly activated, and when Gal 4 is massively overexpressed the galactose genes are switched off. Which of the following mechanisms is the most likely cause for the over expression-mediated *GAL* gene repression?

(1) Gal 4 over-expression reflects a transcriptional repressor to *GAL* gene promoters
(2) Enhanced Gal 4 protein occupancy at the *GAL* promoter cause formation of repressive higher order chromatin fibres

(3) Over-expressed Gal 4 levels cause titration of coactivators critical for Gal 4 activation function

(4) Gal 4 binds to CG-rich regions of *GAL* genes and causes gene repression

54. In a physiological buffer, Sephadex molecular sieve column chromatography (= size or gel exclusion) demonstrates that a protein has a molecular weight of 180 kD. When this protein is analyzed by SDS-PAGE gel electrophoresis, the result shown in lane A is obtained without the reducing agent beta-mercaptoethanol (2- mercaptoethanol), the result in lane B with the reducing agent. Lane C has standards of known MW

A -BME

B +BME

C standards

200kD

100kD

50kD

<---40 kD

Which of the following is valid conclusion?

(1) Protein is composed of two subunits held by weak interaction

(2) Protein is composed of two subunits held by disulphide bonds

(3) Protein is composed of three subunits

(4) Single sub unit protein has been degraded into two peptide fragments

55. A mixture of Cytochrome-C (MW 11.7 KD) and Myoglobin (MW 17.2 KD) are to be separated by polyacrylamide gel electrophoresis. Their isoelectric pH (pI) values are 9.6 and 7.2 respectively. In which direction will each protein migrate at pH 8.5?

(1) Myoglobin will migrate to anode and Cytochrome-C will migrate to cathode

(2) Myoglobin will migrate to cathode and Cytochrome-C will migrate to anode

(3) Both will migrate to anode

(4) Both will migrate to cathode

56. Centrifugation is a method used to separate cellular components. Breaking up the cells either through mechanical or chemical means results in a cell extract. This extract is then placed in a tube and spun at varying rates. Different components of the cell sediment, or settle towards the bottom, at different speeds. In one experiment, a cell extract was divided between four tubes and centrifuged at different speeds, shown below. When the pellets (sedimented material at the bottom of the tube) from all the tubes were examined, they all contained nuclei, but only the pellet from the 100,000 x g tube contained microsomes. This is because

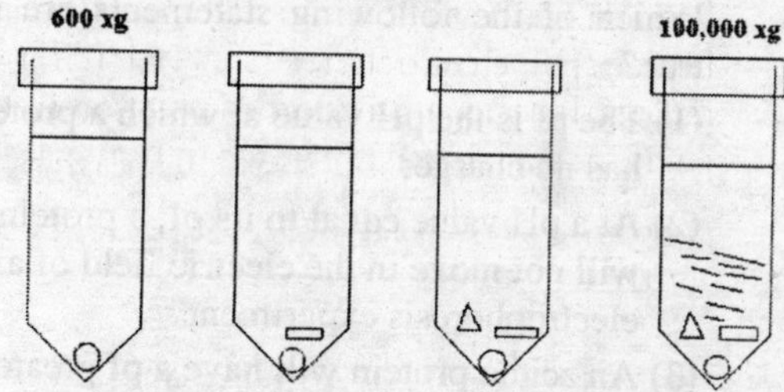

(1) Particles with greater mass sediment with lower g forces, while particles with very little mass require a great deal of gravitational pull to settle to the bottom of a tube

(2) The tiny size of the nuclei allows them to more easily settle to the bottom of the centrifuge tubes

(3) The large size of microsomes means they need high g forces to drive them to the bottom of the tube

(4) The microsomes just happened to be in the tube that was centrifuged at 100,000 xg

57. If the five proteins in question were separated in an isoelectric-focusing experiment by loading toward low pH [positive (+) end] to high pH [negative (-)] ends of the gel? Indicate which will migrate to longest distance?

	Molecular weight (daltons)	pI
(a) α-antitrypsin	45,000	5.4
(b) cytochrome c	13,400	10.6
(c) myoglobin	17,000	7.0
(d) serum albumin	69,000	4.8
(e) transferrin	90,000	5.9
Top		Bottom

(1) cytochrome C
(2) serum albumin
(3) transferring
(4) myoglobin

58. Which of the following statements are not true?
(1) The pI is the pH value at which a protein has no charges
(2) At a pH value equal to its pI, a protein will not move in the electric field of an electrophoresis experiment
(3) An acidic protein will have a pI greater than 7
(4) A basic protein will have a pI greater than 7

59. A peptide was found to have a molecular mass of about 650 and upon hydrolysis produced Ala, Cys, Lys, Phe, and Val in a 1:1:1:1:1 ratio. The peptide upon treatment with Sanger's reagent (FDNB) produced DNP-Cys and exposure to carboxypeptidase produced valine. Chymotrypsin treatment of the peptide produced a dipeptide that contained sulfur and has a UV absorbance, and a tripeptide. Exposure of the peptide to trypsin produced a dipeptide and a tripeptide. Deduce the sequence of the peptide
(1) val-ala-lys-phe-cys
(2) cys-lys-phe-ala-val
(3) cys-ala-lys-phe-val
(4) Cys-phe-lys-ala-val

60. Which enzyme was used to produce the molecule in figure?

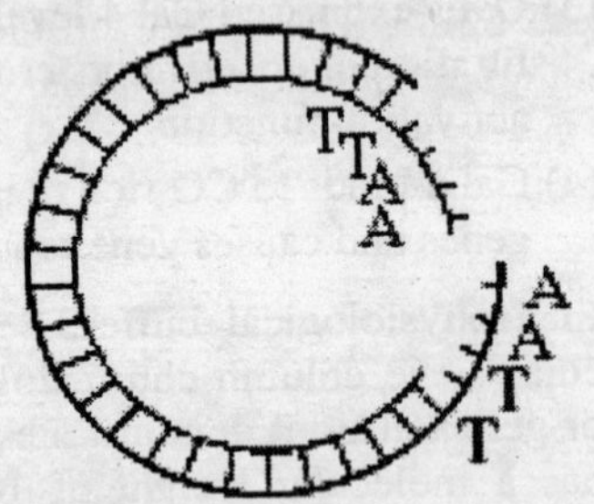

(1) Ligase
(2) Transcriptase
(3) A restriction enzyme
(4) RNA polymerase

61. The segment of DNA shown in Figure has restriction sites I and II, which create restriction fragments A, B, and C. Which of the gels produced by electrophoresis shown below best represents the separation and identity of these fragments?

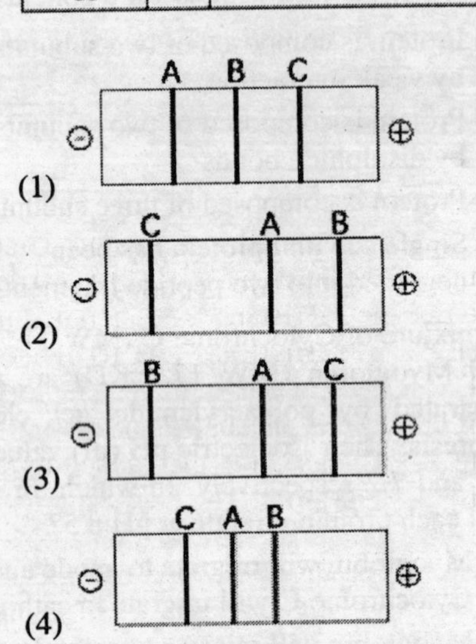

Question 62-63 The DNA profiles that follow represent four different individuals

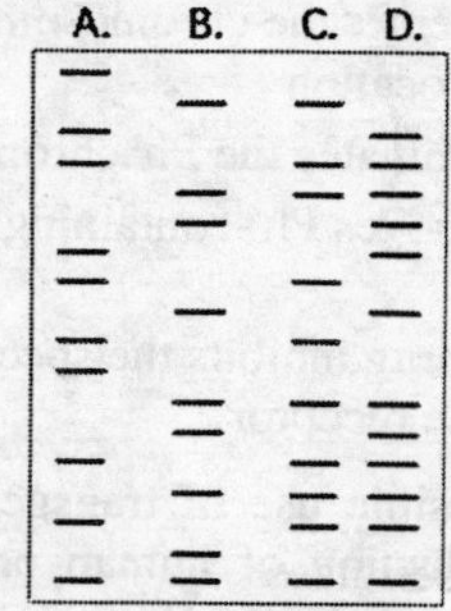

62. Which of the following statements is consistent with the results?
(1) B is the child of A and C
(2) C is the child of A and B
(3) D is the child of B and C
(4) A is the child of B and C

63. Which of the following statements is most likely true?
(1) D is the child of A and C
(2) D is the child of A and B
(3) D is the child of B and C
(4) A is the child of C and D

Question 64-66 A eukaryotic gene has "sticky ends" produced by the restriction endonuclease *Eco*RI. The gene is added to a mixture containing *Eco*RI and a bacterial plasmid that carries two genes conferring resistance to ampicillin and tetracycline. The plasmid has one recognition site for *Eco*RI located in the tetracycline resistance gene. This mixture is incubated for several hours, exposed to DNA ligase, and then added to bacteria growing in nutrient broth. The bacteria are allowed to grow overnight and are streaked on a plate using a technique that produces isolated colonies that are clones of the original. Samples of these colonies are then grown in four different media: nutrient broth plus ampicillin, nutrient broth plus tetracycline, nutrient broth plus ampicillin and tetracycline, and nutrient broth without antibiotics

64. Bacteria that contain the plasmid, but not the eukaryotic gene, would grow
(1) in the nutrient broth plus ampicillin, but not in the broth containing tetracycline
(2) only in the broth containing both antibiotics
(3) in the broth containing tetracycline, but not in the broth containing ampicillin
(4) in all four types of broth

65. Bacteria containing a plasmid into which the eukaryotic gene has integrated would grow in
(1) the nutrient broth only
(2) the nutrient broth and the tetracycline broth only
(3) the nutrient broth, the ampicillin broth, and the tetracycline broth
(4) the ampicillin broth and the nutrient broth

66. Bacteria that do not take up any plasmids would grow on which media?
(1) the nutrient broth only
(2) the nutrient broth and the tetracycline broth
(3) the nutrient broth and the ampicillin broth
(4) the tetracycline broth and the ampicillin broth

Question 67-69 A group of six students has taken samples of their own cheek cells, purified the DNA, and used a restriction enzyme known to cut at zero, one, or two sites in a particular gene of interest

67. Why might they be conducting such an experiment?
(1) To find the location of this gene in the human genome
(2) To prepare to isolate the chromosome on which the gene of interest is found
(3) To find which of the students has which alleles
(4) To collect population data that can be used to assess natural selection

68. Their next two steps, in order, should be
(1) use of a fluorescent probe for the gene sequence, then electrophoresis
(2) electrophoresis of the fragments followed by autoradiography
(3) electrophoresis of the fragments, followed by the use of a probe
(4) use of a ligase that will anneal the pieces, followed by Southern blotting

69. Analysis of the data obtained shows that two students each have two fragments, two students each have three fragments, and two students each have one only. What does this demonstrate?
(1) Each pair of students has a different gene for this function
(2) The two students who have two fragments have one restriction site in this region
(3) The two students who have two fragments have two restriction sites within this gene
(4) The students with three fragments are said to have "fragile sites"

Question 70-72 CML (Chronic Myelogenous Leukemia) results from a translocation between human chromosomes 9 and 22. The resulting chromosome 22 is significantly shorter than usual, and it is known as a Philadelphia (Ph') chromosome. The junction at the site of the translocation causes overexpression of a thymine kinase receptor. A new drug (Gleevec or imatinib) has been found to inhibit the disease if the patient is treated early

70. Which of the following would be a reasonably efficient technique for confirming the diagnosis of CML?
(1) searching for the number of telomeric sequences on chromosome 22
(2) looking for a Ph' chromosome in a peripheral blood smear
(3) enzyme assay for thymine kinase activity
(4) FISH study to determine the chromosomal location of all chromosome 22 fragments

71. Why would Gleevec most probably cause remission of the disease?
(1) It reverses the chromosomal translocation
(2) It eliminates the Ph' chromosome
(3) It removes Ph'-containing progenitor cells
(4) The drug inhibits the specific thymine kinase receptor

72. One possible use of transgenic plants is in the production of human proteins, such as vaccines. Which of the following is a possible hindrance that must be overcome?
(1) prevention of transmission of plant allergens to the vaccine recipients
(2) prevention of vaccine-containing plants being consumed by insects
(3) use of plant cells to translate non-plant-derived mRNA
(4) inability of the human digestive system to accept plant-derived protein

Question 73-74 Pharmacogenetics is an increasingly important discipline that uses genetic information to tailor the prescription of drug treatments to individuals. In the case of chemotherapy for breast cancer, for example, different patients need and/or respond to different treatments

73. Patients whose tumors are HER-2 positive respond to herceptin whereas other patients do not. Patients whose tumors have estrogen receptors will be best served if
(1) their estrogen receptors are blocked by using RNAi
(2) their estrogen release is activated and/or elevated
(3) the estrogen receptors are blocked by other molecules that can use the same receptors
(4) they are given herceptin as well as estrogen

74. Breast tumor biopsy specimens can be typed for a number of gene expression patterns. Together, these can provide risk analysis for the likely aggressive growth and metastasis of the tumor. How can this most help the physician and patient?

(1) Some patients want to know as much as possible

(2) This can help them to decide whether & what kind of chemotherapy is warranted

(3) This can help them decide whether the tumor should be removed

(4) Some physicians may use the information to decide what to do, but not tell the patient

75. Consider the following properties of an ideal vector

A. Autonomously replication absent i.e. should not have ori (origin of replication) region

B. Contain at least one selectable marker *e. g*. gene for antibiotic resistance

C. May contain a scorable marker (β-galactosidase, green fluorescent protein etc.)

D. Presence of unique restriction enzyme site

A good vector should have the following characteristics

(1) A, B and C (2) A, C and D
(3) B, C and D (4) A, B and D

76. Consider the following properties of an ideal vector

A. Have multiple cloning sites

B. Preferably small in size & easy to handle

C. Stringent control of replication

D. Presence of appropriate regulatory elements for expression of foreign gene

E. low copy number

A good vector should have the following characteristics

(1) A, B and E (2) A, C and E
(3) B, C and D (4) A, B and D

77. Molecular cloning serves as a tool for developing of various recombinant cells and organisms which has application both in basic and applied biological science as discussed below

A. Production of recombinant proteins: Genes encoding proteins with diagnostic, therapeutic or commercial value can be cloned, expressed and purified to obtain recombinant proteins in bulk with limited space and lesser time

B. Gene therapy is used for correcting a disorder or deficiency

C. To study the structure and function of a particular gene using a model host organism

D. To study the regulation of gene expression during developmental stages

E. Complete genome sequence of an organism having large genome size can be facilitated by BAC or YAC library formation using rDNA technology

Which of the following are true in concenrn to molecular cloning?

(1) A, B and C (2) B, C and D
(3) C, D and E (4) All of the above

78. In an attempt to detect protein expression profile in a cell, Western blot technique is employed. Expression of two new proteins is to be followed by probing with respective high affinity antibodies (raised in rabbit) Unfortunately, the two proteins were found to co-migrate in SDS-PAGE profile. Under this situation, using one dimensional SDSPAGE and by Western blot, which one of the following is the best way to demonstrate the presence of both the proteins?

(1) Develop Western blots with their antibodies in the same gel

(2) Prior to doing SDS-PAGE/Western blot, one protein could be removed by immuno precipitating in the cell extracts

(3) Silencing the expression of one protein at a time by siRNA and performing Western blotting

(4) Subjecting the technique of stripping/reprobing of the gel after transferring to nitrocellulose membrane while doing Western blotting

79. Nichrome coated stainless steel electrodes were implanted in a rat brain for chronically recording the electrical activity of deep brain structures. During a study of 3 months the intensity of electrical signals gradually decreased

The following statements may explain the cause of this observation

A. The deposition of metallic iron from the electrode tips caused degeneration of some neurons

B. The gradual accumulation of microglia at the electrode tips increased the resistance of electrodes

C. The neurons at the electrode tips were hyperpolarized gradually

D. The threshold for firing action potential in the neurons at the electrode tips was increased due to prolonged presence of electrodes

Which one of the following is correct?

(1) A only (2) A and B
(3) C only (4) C and D

80. The number of seeds in the fruit of a plant species, Ho : μ=30. A random sample of 9 fruits gives the mean number of seeds as 24 with a standard deviation of 6.12

A. What are the confidence limits for the sample mean?

B. Would your reject or accept the null hypothesis at 95% confidence level?

(1) (a) 18 and 30, (b) reject the hypothesis
(2) (a) 20 and 28, (b) reject the hypothesis
(3) (a) 20 and 28, (b) accept the hypothesis
(4) (a) 18 and 30, (b) accept the hypothesis

81. A Chromatin Iminuno Precipitation (ChIP) assay was performed to determine specific transcription factor binding sites on the promoter of a gene. Pull down was done using either IgG or anti-bodies against c-myc. A DNA containing c-myc binding regions was used as a control for PCR amplification (input) Which one of the following PCR representations of DNA is correct?

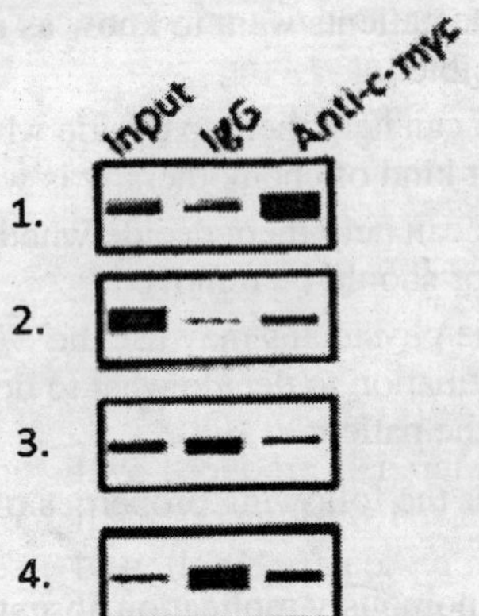

82. A researcher was repeating a FACS experiment but somehow got confused with the labeling of the tubes. There are four tubes, one control, C (with no fluorescent label), one standard 1, S1 (with FITC label), one standard 2, S2 (with PE label) and the last one test, T (which should be FITC positive) Given below is the result of the FACS experiment

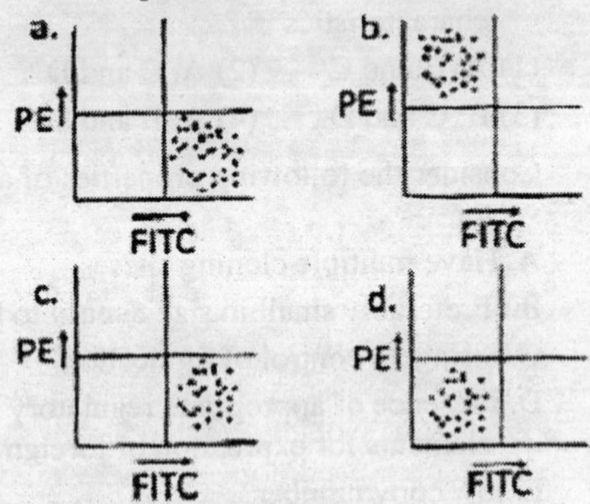

What should be the correct labeling?

(1) a,S2; b,S1; c,T; d,C
(2) a,S1; b,T; c,C; d, S2
(3) a,S2; b,S1/T; c, C; d,S1/T
(4) a,S1/T; b,S2; c, s1/T; d,C

83. A new protein "X" has been discovered in erythrocyte plasma membrane. The following observation are made about "X"

A. It can be extracted from disrupted erythrocyte membranes into a concentrated salt solution (NaCl), and it can be cleaved into fragments by proteinase K

B. Treatment of erythrocytes with proteinase K followed by disruption and extraction of membrane components yields intact X

C. However, treatment of erythrocyte "ghosts" (which consist of just plasma membranes, produced by disrupting the cells and washing out the hemoglobin) with proteinase K followed by disruption and extraction yields extensively fragmented "X"

What do these findings indicate about the location of "X" in the plasma membrane? Identify the correct answer

(1) Integral membrane protein with both N- and C- terminal exposed outside the cell

(2) Peripheral outer membrane

(3) Integral membrane protein

(4) peripheral inner membrane

84. Ribonuclease A (RNase A) consists of 124 amino acid residues. RNase A is cleaved by subtilisin (a protease) in a quantitative fashion at Ala 20 - Ser 21 peptide bond generating two fragments of 20 and 104 residues respectively. However, gel-filtration chromatography of the subtilisin-treated reaction mixture yields a single peak that elutes at the same position of the full-length RNase A. Notably, under identical chromatographic conditions as above, similar size fragments (20, 104, 124 residues) are seen to separate well. Which of the following is the most likely explanation for this observation?

(1) The smaller fragment is formed in vanishingly small amounts and cannot be detected

(2) The smaller 20-residue fragment is unstable and degraded

(3) The two fragments are non-covalently bound despite the cleavage of 20-21 peptide bond

(4) The two fragments are linked by disulfide bonds after subtilisin cleavage

85. Tryptic digestion of a heptapeptide produced a dipeptide and a pentapeptide. Chymotryptic cleavage of the same heptapeptide yielded a tripeptide and a tetrapeptide. The tripeptide was found to be composed of two Thr (T) and one Lys (K) residue. The tetrapeptide contained one Ala (A), one Met (M), one Arg (R) and one Tyr (Y) residue. Which of the following sequence fits the heptapeptide?

(1) AKMYTTR (2) ARMYTTK

(3) AYMKTTR (4) MYARTKT

86. An unknown peptide was isolated from the leaf of a medicinal plant and purified to homogeneity. The peptide did not yield any sequence when subjected to Edman degradation. However, tryptic digest of the peptide produced a unique sequence. The mass of the intact peptide was found to be 18 Da less than that obtained from the trypsin treated sample

The possible interpretation of the above experimental results could be that the

A. the N-terminus of the peptide was blocked by acetylation or methylation

B. the peptide was cyclic and contained at least one internal arginine or lysine residue

C. the peptide was cyclic and contained a lysine or arginine residue at the C terminus

D. the peptide was cyclized by peptide bond formation between α-amino group and α-carboxyl group

Which of the above statements is true?

(1) A and D

(2) A and B

(3) B and C

(4) B and D

87. A gene is regulated by a novel transcription factor. The following techniques may be used to identify the cis- regulatory element in the 1 kb promoter sequence of the gene where the novel transcription factor binds

A. Bioinformatics analysis
B. Cell based reporter assay
C. S 1 nuclease assay
D. Electrophoretic mobility shift assay
E. DNAse-I foot-printing analysis

Which one of the following can help to identify the cis element?

(1) A and B (2) C and E
(3) D only (4) E only

88. A protein D is encoded by a gene, which is 5 Kb long and has three Hind III restriction enzyme sites. The first one is 0.5 Kb from the transcription start site, the second one is 2.5 Kb from the first site and the third one is 0.5 Kb internal to the stop codon. The second site is polymorphic. In order to find out whether fetal cells contain the normal or the mutated gene, total genomic DNA from fetal cells was isolated, completely digested with Hind III, separated in an agarose gel, transferred to membrane and detected by a probe against the region between the second and third restriction site. Which one of the following band patterns will be obtained if the fetal cell is heterozygous?

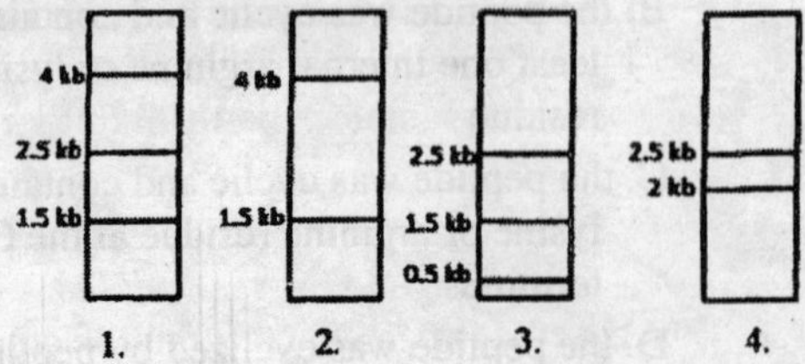

89. Given below are the experimental protocols to find out the exact location of repetitive DNA sequence in mitotic chromosome by FISH (Fluorescence In situ Hybridization) Which one of the protocols will give the correct result?

(1) Mitotic chromosomes were fixed on glass slide → incubated with biotinylated telomeric DNA → denatured → incubated with fluorescently labeled avidin → localization observed under fluorescence microscope

(2) Mitotic chromosomes were fixed on glass slide →denatured →incubated with FITC labeled unrelated non-repetitive DNA sequence → counterstained with propidium iodide → localization observed under fluorescence microscope

(3) Mitotic chromosomes were fixed on glass slide → denatured → incubated with biotinylated satellite DNA→ incubated with fluorescently labeled avidin → localization → observed under fluorescence microscope

(4) mitotic chromosomes were fixed on glass slide → incubated with repetitive DNA sequence binding protein → denatured → FITC labeled antibody against the protein → localization observed under fluorescence microscope

90. At 25°C values of $[\theta]_{222}$, the mean residue ellipticity at 222 nm, are - 33,000 and -3,000 deg cm^2 $dmol^{-1}$ for a polypeptide existing in α-helical (α) and β-structure (β), respectively. If this polypeptide undergoes a two-state heat-induced α → β transition, and a value of $[\theta]_{222}$ = -18,000 deg cm^2 $dmol^{-1}$ is observed at 60°C, then this observation leads to the conclusion that the α helix conversion to β-structure

(1) 40% (2) 50%
(3) 55% (4) 60%

91. An EEG was recorded and its power spectrum analyses were done in rats with implant electrode for a long time. The power of the EEG waves decreased two months after electrode implantation

This observation may be due to the following

A. Glial cells accumulate surrounding exposed tips of electrodes

B. Degeneration of neurons occur surrounding the electrode tips due metal ion deposition

C. Coating of electrodes are destroyed with time

D. The microsocket becomes loose time

Which one of the following is true?

(1) Only A (2) A and B
(3) Only C (4) C and D

92. A biologist studied the environmental impact of a herbicide on squirrel reproduction. He selected two small island with similar vegetation and similar-sized squirrel populations (about 200 squirrel) with equal reproductive rates. On island, selected at random, was sprayed with herbicide, and the other was used as a control. Squirrel reproduction rates were measured on both islands before and after treatment. What is the greatest flaw (mistake) in this investigation?

(1) Lack of control
(2) Lack of replication
(3) Lack of randomization
(4) Lack of varaibles

93. A DNA molecule needs to display several features to be able to act as a vector for gene cloning. Most importantly it must be able to replicate within the host cell, so that numerous copies of the recombinant DNA molecule can be produced and passed to the daughter cells. Plasmids are circular molecules of DNA that lead an independent existence in the bacterial cell and shows various features. Some of them are as follows

A. Conjugation and plasmid transfer are controlled by a set of transfer or *tra* genes, which are present on conjugative plasmids but absent from the non-conjugative type

B. To be able to coexist in the same cell, different plasmids must be compatible. If two plasmids are incompatible then one or the other will be rapidly lost from the cell

C. Some plasmids, especially the larger ones, are stringent and have a low copy number of perhaps just one or two per cell; others, called relaxed plasmids, are present in multiple copies of 50 or more per cell

D. plasmids are widespread in bacteria and are not found in any of the eukaryotic organisms

Which of the following statements are corrects for plasmid?

(1) A and B (2) A, B and C
(3) B, C and D (4) A, B, C and D

94. Consider the following steps in a gene cloning experiment

A. A fragment of DNA, containing the gene to be cloned, is inserted into a circular DNA molecule called a vector, to produce a recombinant DNA molecule

B. Within the host cell the vector multiplies, producing numerous identical copies, not only of itself but also of the gene that it carries

C. After a large number of cell divisions, a colony, or clone, of identical host cells is produced. Each cell in the clone contains one or more copies of the recombinant DNA molecule

D. The vector transports the gene into a host cell, which is usually a bacterium, although other types of living cell can be used

E. When the host cell divides, copies of the recombinant DNA molecule are passed to the progeny and further vector replication takes place

What of the following is correct sequence of steps in gene cloning experiment?

(1) A→B→C→D→E
(2) D→E→A→B→C
(3) D→B→C→A→E
(4) A→D→B→E→C

95. The polymerase chain reaction is very different from gene cloning. Rather than a series of manipulations involving living cells, PCR is carried out in a single test tube simply by mixing DNA with a set of reagents and placing the tube in a thermal cycler, a piece of equipment that enables the mixture to be incubated at a series of temperatures that are varied in a preprogrammed manner

A. In order for the primers to anneal to the correct positions, either side of the gene of interest, the sequences of these annealing sites must be known. It is easy to synthesize a primer with a predetermined sequence

B. If the sequence of a gene is not known, it is not possible to determine the appropriate sequences for a pair of primers, even what is known about the sequence of the equivalent gene in a different organism

C. There is a limit to the length of DNA sequence that can be copied by PCR. Five kilobases (kb) can be copied fairly easily, and segments up to forty kb can be dealt with by using specialized techniques

D. PCR technique can detect viruses at the earliest stages of an infection, increasing the chances of treatment being successful

Which of the following statement is incorrect?

(1) A only (2) B only
(3) B and C only (4) B and D only

96. Digestion of 4kb DNA molecule with *Eco*RI yields two fragments of 1 kb and 3 kb each. Digestion of same molecule with *Hind*III yields fragments of 1.5kb and 2.5kb. Finally digestion with *Eco*RI and *Hind*III in combination yields fragments of 0.5 kb. 1 kb, and 2.5 kb. Select from the following the correct restriction map indicating the position of the *Eco*RI and *Hind*III cleavage sites

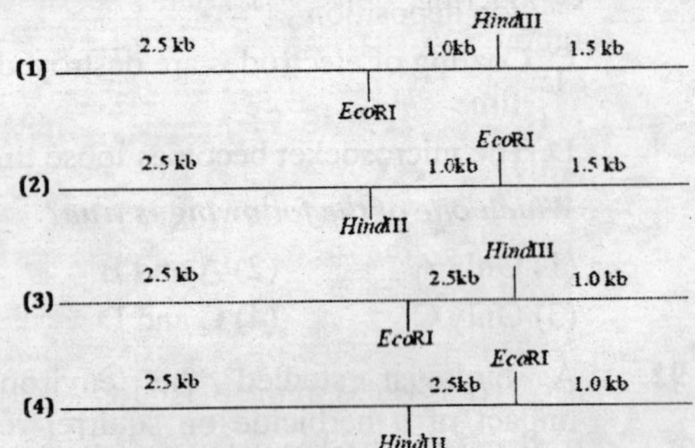

97. Viral infections are frequently detected using tests that assess the presence or absence of patient antibodies against viral proteins. If antibodies against the viral proteins are present, then this is taken as evidence that the patient is infected with the virus. However, a flaw in this method of detection is the existence of a "latent period", i.e., a time period after infection with the virus has occurred, but before the patient's body begins to make antibodies. It is frequently possible to detect viral nucleic acids in blood during the latent period using which of the following techniques?

(1) Enzyme-Linked Immunosorbent Assay (ELISA)
(2) Western blot (immunoblot analysis)
(3) Polymerase Chain Reaction (PCR)
(4) Transmission Electron Microscopy (TEM)

98. Which of the following are the characteristics of type II restriction endonucleases?

P. Nature of enzyme : Bifunctional enzyme with both endonuclease and methylase activity

Q. Protein structure : Two identical subunits

R. Restriction requirement : ATP, Mg^{2+}

S. Cleavage sites : At or near restriction site

(1) P and Q (2) Q and S
(3) P and S (4) R and S

99. I want to know the function of unknown gene 'K', which I have recently cloned, by conducting gene-knockout experiments in rabbit. Do study this initially, I prepared a plasmid containing the cloned gene K and two drug sensitivity genes, *neo* and *tk,* which confer resistance, to neomycin and sensitivity to gancylovir, respectively. I then introduce the recombinant plasmid into rabbit embryonic stem (ES) cells

What might be the purpose of including the neo and tk genes in the plasmid?

A. *tk* gene is used to make cells with non-homologous recombination gancylovir sensitive

B. *neo* gene is used to selected transfected cell

C. *neo* gene is used to make cells with non-homologous recombination neomycin sensitive

D. *tk* gene is used to selected transfected cell

(1) B and D (2) A and B

(3) A and C (4) A, C and D

100. Affinity chromatography is a method that can be used to purify cell surface receptors, while they retain their hormone binding ability. A ligand (hormone) for a receptor of interest is chemically linked to polystyrene beads. A solubilized preparation of membrane proteins is passed over a column containing these beads. Only the receptor binds to the beads

When an excess of the ligand (hormone) is poured through the column after the receptor binding step, what do you expect will occur?

(1) the ligand will attach to those beads that have the receptor and remain on the column

(2) the ligand will cause the receptor to be displaced from the beads and eluted out

(3) the ligand will attach to the bead instead of the receptor

(4) the ligand will cause the bead to lose its affinity by changing shape

ANSWER SHEET

Unit 13: Applied Biotechnology Biotechniques

1. 2	**2.** 4	**3.** 2	**4.** 2	**5.** 1	**6.** 4	**7.** 2	**8.** 3	**9.** 2	**10.** 4
11. 2	**12.** 4	**13.** 3	**14.** 1	**15.** 3	**16.** 1	**17.** 3	**18.** 1	**19.** 1	**20.** 2
21. 3	**22.** 3	**23.** 4	**24.** 2	**25.** 4	**26.** 3	**27.** 2	**28.** 3	**29.** 4	**30.** 2
31. 1	**32.** 4	**33.** 1	**34.** 2	**35.** 3	**36.** 1	**37.** 2	**38.** 1	**39.** 3	**40.** 2
41. 4	**42.** 3	**43.** 2	**44.** 2	**45.** 1	**46.** 1	**47.** 1	**48.** 1	**49.** 1	**50.** 1
51. 3	**52.** 4	**53.** 2	**54.** 3	**55.** 1	**56.** 1	**57.** 1	**58.** 3	**59.** 4	**60.** 3
61. 2	**62.** 2	**63.** 1	**64.** 4	**65.** 4	**66.** 1	**67.** 3	**68.** 2	**69.** 2	**70.** 4
71. 4	**72.** 1	**73.** 3	**74.** 2	**75.** 3	**76.** 4	**77.** 4	**78.** 4	**79.** 2	**80.** 2
81. 1	**82.** 4	**83.** 4	**84.** 4	**85.** 2	**86.** 4	**87.** 2	**88.** 2	**89.** 3	**90.** 2
91. 2	**92.** 2	**93.** 2	**94.** 4	**95.** 2	**96.** 4	**97.** 3	**98.** 2	**99.** 2	**100.** 2